D1300010

About Island Press

Island Press is the only nonprofit organization in the United States whose principal purpose is the publication of books on environmental issues and natural resource management. We provide solutions-oriented information to professionals, public officials, business and community leaders, and concerned citizens who are shaping responses to environmental problems.

In 1994, Island Press celebrated its tenth anniversary as the leading provider of timely and practical books that take a multidisciplinary approach to critical environmental concerns. Our growing list of titles reflects our commitment to bringing the best of an expanding body of literature to the environmental community throughout North America and the world.

Support for Island Press is provided by Apple Computer, Inc., The Bullitt Foundation, The Geraldine R. Dodge Foundation, The Energy Foundation, The Ford Foundation, The W. Alton Jones Foundation, The Lyndhurst Foundation, The John D. and Catherine T. MacArthur Foundation, The Andrew W. Mellon Foundation, The Joyce Mertz-Gilmore Foundation, The National Fish and Wildlife Foundation, The Pew Charitable Trusts, The Pew Global Stewardship Initiative, The Rockefeller Philanthropic Collaborative, Inc., and individual donors.

Creating a Forestry for the 21st Century

Creating a Forestry for the 21st Century

The Science of Ecosystem Management

Edited by Kathryn A. Kohm and Jerry F. Franklin

Foreword by Jack Ward Thomas

ISLAND PRESS

Washington, D.C. • Covelo, California

Grateful acknowledgment is expressed for permission to publish the following copyrighted material: Figure 3.1, on page 112, from Peet, R.K. 1992. "Community Structure and Ecosystem Function." In *Plant Succession: Theory and Prediction,* edited by D.C. Glenn-Lewin, R.K. Peet, and T.T. Veblen. New York: Chapman and Hall.

Library of Congress Cataloging-in-Publication Data

Creating a forestry for the 21st century: the science of ecosystem management / edited by Kathryn A. Kohm & Jerry F. Franklin.
p. cm.
Includes bibliographical references and index.
ISBN 1-55963-398-0 (cloth). — ISBN 1-55963-399-9 (pbk.)
1. Forest management—Northwest, Pacific. 2. Forest ecology—Northwest, Pacific. 3. Forests and forestry—Northwest, Pacific. 4. Ecosystem management—Northwest, Pacific. 5. Forest management. 6. Forest ecology. 7. Forests and forestry. 8. Ecosystem management. I. Kohm, Kathryn A. II. Franklin, Jerry F.
SD144.A13C74 1997
634.9—dc20 96-32771
CIP

Printed on recycled, acid-free paper ♻

Manufactured in the United States of America

10 9 8 7 6 5 4 3

Contents

Foreword by Jack Ward Thomas ix
Acknowledgments xiii
Contributors xv

1. Introduction 1
   Kathryn A. Kohm and Jerry F. Franklin

Section I. Ecological Processes and Principles 7

2. Forest Stand Structure, Composition, and Function 11
   Thomas Spies
3. Disturbance, Recovery, and Stability 31
   David A. Perry and Michael P. Amaranthus
4. The Biological Landscape 57
   Malcolm L. Hunter Jr.
5. Riparian Management in the 21st Century 69
   Stanley V. Gregory
6. Biodiversity of Old Forests of the West: A Lesson from Our Elders 87
   Bruce G. Marcot

Section II. Silvicultural Systems and Management Concerns 107

7. Alternative Silvicultural Approaches to Timber Harvesting: Variable Retention Harvest Systems 111
   Jerry F. Franklin, Dean Rae Berg, Dale A. Thornburgh, and John C. Tappeiner
8. Shaping Stand Development Through Silvicultural Practices 141
   Dean S. DeBell, Robert O. Curtis, Constance A. Harrington, and John C. Tappeiner
9. Silvicultural Systems and Regeneration Methods: Current Practices and New Alternatives 151
   John C. Tappeiner, Denis Lavender, Jack Walstad, Robert O. Curtis, and Dean S. DeBell
10. The Role of Extended Rotations 165
    Robert O. Curtis
11. Integrating the Ecological Roles of Phytophagous Insects, Plant Pathogens, and Mycorrhizae in Managed Forests 171
    Timothy Schowalter, Everett Hansen, Randy Molina, and Yanli Zhang
12. Fire Management for the 21st Century 191
    James K. Agee

13. Forest Genetics for Ecosystem Management 203
Sharon Friedman

Section III. Approaches to Management at Larger Spatial Scales 213

14. Ecosystem Management: Managing Natural Resources in Time and Space 215
Thomas R. Crow and Eric J. Gustafson

15. The Physical Environment as a Basis for Managing Ecosystems 229
Frederick J. Swanson, Julia A. Jones, and Gordon E. Grant

16. Approaches to Management at the Watershed Scale 239
Robert J. Naiman, Peter A. Bisson, Robert G. Lee, and Monica G. Turner

17. Landscape Analysis and Design 255
Nancy M. Diaz and Simon Bell

18. Implementing Spatial Planning in Watersheds 271
John Sessions, Gordon Reeves, K. Norman Johnson, and Kelly Burnett

Section IV. Forest Economics: Products and Policies 281

19. The Context for Forest Economics in the 21st Century 285
Richard W. Haynes and James F. Weigand

20. Changes in Wood Products Manufacturing 303
Clive G. Whittenbury

21. Special Forest Products: Integrating Social, Economic, and Biological Considerations into Ecosystem Management 315
Randy Molina, Nan Vance, James F. Weigand, David Pilz, and Michael P. Amaranthus

22. The Public Interest in Private Forests: Developing Regulations and Incentives 337
Frederick W. Cubbage

Section V. Institutions in Transition 357

23. Organizational and Legal Challenges for Ecosystem Management 361
Errol E. Meidinger

24. Building Bridges Across Agency Boundaries 381
Steven L. Yaffee and Julia M. Wondolleck

25. Science-Based Assessments of the Forests of the Pacific Northwest 397
K. Norman Johnson

26. Scarcity, Simplicity, Separatism, Science—and Systems 411
R. W. Behan

27. Making Decisions in a Complex and Dynamic World 419
Gordon R. Smith

28. Open Institutions: Uncertainty and Ambiguity in 21st-Century Forestry 437
Margaret A. Shannon and Alexios R. Antypas

29. The Emerging Role of Science and Scientists in Ecosystem Management 447
John C. Gordon and James Lyons

Index 455

Foreword

Two separate but connected factors have combined in recent decades to dramatically alter the practice of forestry in the United States. The first has been a rising environmental consciousness among a significant and politically effective segment of the population. The second was a spate of environmental legislation enacted in the 1960s and 1970s. Among laws that changed the practice of forestry are the National Environmental Policy Act of 1969, which requires the federal government to perform a detailed assessment of costs and benefits of all federally financed activities; the Endangered Species Act of 1973, which established a government policy that species should be preserved; and the National Forest Management Act of 1976, which sets high standards for management of national forests.

Of particular note is the statement of purpose included in the Endangered Species Act. Twenty years after passage of the act, that statement has emerged full-blown, with far-reaching consequences for federal land management. That statement is: "The purposes of this act are to provide a means whereby the ecosystems upon which endangered species and threatened species may be conserved" and not merely the welfare of the single species identified as 'endangered' or 'threatened.' In addition, there were regulations promulgated by the federal government pursuant to the National Forest Management Act that called for the retention of viable populations of vertebrate species well-distributed on national forests with particular emphasis on habitat.

In combination, these laws and regulations have had a profound effect on forest management on federal lands. These changes came about over a period of 25 years as governmental agencies (most notably the Forest Service and Bureau of Land Management) struggled to maintain or increase historic timber sale levels, satisfy the needs of traditional constituencies (such as grazers, hunters, fishers, and recreationists), and simultaneously remain in compliance with environmental laws. Under the U.S. legal system, citizens may challenge the government's compliance with law. Over the past 25 years, there have been numerous such challenges to federal land management activities—many of which have been successful. These federal court decisions have forced federal land management agencies to change some of their traditional approaches to forest management.

The most noted of these legal challenges is the case of the northern spotted owl (Thomas et al. 1990). This subspecies of the genus *Strix* was declared by the U.S. Fish and Wildlife Service to be "threatened" in early 1990. It is considered closely associated with the habitat conditions most commonly found in the late-successional/old-growth forests of the Pacific Northwest. These forests have been diminished significantly in amount and quality through timber harvesting (most commonly clearcutting) and losses to fire, blowdown, and other natural events since the late 1800s (Thomas et al. 1990). The late-successional/old-growth forests of the Pacific Northwest are extremely valuable as a source of large volumes of high quality timber and as a significant source of employment (FEMAT 1993). The reservation of significant amounts of old-growth from timber harvesting to maintain a range-wide viable population of a subspecies of owl (Thomas et al. 1990, FEMAT 1993) has been fraught with social, economic, and ecological consequences that have translated into prolonged legal and political battles which continue today. The details surrounding this continuing controversy have been described by Thomas et al. (1993).

As the political and legal drama over old-growth forests and the northern spotted owl evolved, it became more and more obvious that the issue, as clearly foreseen and prescribed in the Endangered Species Act, was not one of saving or maintaining viable populations of an individual subspecies. Rather, it was centered on public and scientific concerns with the maintenance of ecosystem functions. These evolving concerns of scientists (and in turn the public) also began to surface under other names and allied concepts such as "sustainable forestry," "biodiversity retention," "new perspectives in forestry," and "new forestry." But most recently and predominantly, these concepts have come to be known as "ecosystem management" (Thomas 1993).

Ecosystem Management

By mid-1993, both the Forest Service and the Bureau of Land Management announced that they were embarking on a course of ecosystem management. That pronouncement was made without a detailed assessment of what such a management approach might entail or how it might be accomplished. However, the chief of the Forest Service did say that the agency would move away from clearcutting (except in certain circumstances) as the primary silvicultural prescription for stand regeneration.

By 1993, repeated successful lawsuits by organized environmental groups essentially brought timber sales on federal lands in the Pacific Northwest to a halt. Federal court judges ordered federal land management agencies to cease selling timber on lands designated by the U.S. Fish and Wildlife Service as critical habitat for the northern spotted owl. This impasse prompted vocal public concern, and received attention from all three major candidates during the presidential election of 1992. In the course of that campaign, candidate Governor Bill Clinton of Arkansas promised that, if elected, he would convene a conference to devise a means of ending the court-ordered injunction—that is, he would break the "gridlock."

Shortly after his inauguration, President Clinton convened a forest conference in Portland, Oregon, on April 2, 1993. At the close of that conference, the president promised a solution to the impasse over forest management in the Pacific Northwest within 60 days. He instructed the secretaries of Agriculture, Interior, Commerce, and Labor to carry out that promise. Three teams were organized to formulate management options for the president's consideration. The instructions given to one of those teams, the Forest Ecosystem Management Assessment Team (FEMAT), stated that an "ecosystem management approach" was to be included in their report, and that late-successional/old-growth ecosystems and species associated with those ecosystems that were listed by the U.S. Fish and Wildlife Service as "threatened" (northern spotted owls and marbled murrelets) were to receive specific consideration.

Approximately 90 days after the conference, the president selected an option from among 10 presented to him (FEMAT 1993). The consequences of that selection have been ecologically, economically, and socially profound. Of the land in federal ownership within the assessment area (the range of the northern spotted owl), 7.05 million acres (2.85 hectares) of reserves were established where late-successional/old-growth forest conditions are to be preserved and enhanced over time. An additional 2.23 million acres (.90 million hectares) were designated as riparian reserves to meet water quality standards and protect and enhance habitat for native fishes—particularly anadromous fishes considered to be "at risk" of being listed as "threatened" or "endangered."

These late-successional/old-growth and riparian reserves were established in addition to 6.98 million acres (2.83 hectares) already designated as wilderness or national parks or otherwise withdrawn from timber management activities for reasons such as soil stability, scenic corridors, or recreation needs. Approximately 7.34 million acres (2.97 hectares) out of 24.26 million acres (9.22 million hectares) in the analysis area remained available for timber harvest (about 30 percent of the total area). However, it should be noted that significant portions of the total area support no trees, offer little potential for growing trees, or have fragile soils, steep slopes, or other circumstances that preclude timber harvesting.

The acreage available for harvesting yields a probable sale quantity of approximately 1.2 billion board feet, with an additional 100–150 million board feet potentially available from thinning stands younger than 80 years for silvicultural purposes. This probable annual timber sale level compares to 4.6 billion board feet cut annually from 1980–1989, and 2.4 billion board feet cut annually from 1990–1992. However, a significant portion of the decline can be attributed to the accumulated experience of managers with conditions that precluded maintaining the sale quantities projected in the initial modeling efforts for forest plans.

Scientists doing the ecosystem assessments for President Clinton noted that despite the political and economic advantages of stable timber yields over time, experience has shown that this is unlikely over the long term. The world of forest management in the Pacific Northwest is, clearly and simply, inherently unstable—ecologically, economically, legally, and politically. Forest management plans are frequently changed and often unpredictable. They are subject to the vicissitudes of droughts, fires, insect and disease outbreaks, and volcanic eruptions, as well as funding shortfalls, frequent changes in laws and their interpretation, legal actions, court orders, public acceptance, and changes in policy. The only certainty seems to be the certainty of changing conditions—biological, social, economic, and legal.

The case of the spotted owl and old-growth/late-successional forests in the Pacific Northwest is but one example of the dramatic changes in forest management that are occurring in the United States. State after state has, or is in the process of, tightening up regulations defining appropriate forestry practices for private and state lands. Much of this revision seems to be a response to public demands for forestry practices that are more aesthetically acceptable and more sensitive (realistically or perceptually) to actual multiple-use values—primarily those associated with fish and wildlife habitat—than past practices almost solely directed toward profit and job maximization from timber production, harvesting, processing, and utilization.

Of particular interest to ecologists is the emphatic shift in public interest toward concern for all species of wildlife along with an increasingly holistic sense of ecosystems. This broadened perspective replaces the historic, rather single-minded emphasis on habitat for game species such as white-tailed deer, mule deer, and black-tailed deer. There is every reason to believe that this trend will continue. As a result, biologists will have to broaden their interests, increase their expertise, and work with foresters to produce habitat conditions for a myriad of life forms and ecosystems.

There are, however, countervailing pressures at play. The changes in forestry currently underway come about at significant costs. Those costs are measured in higher prices for wood, jobs lost or foregone, loss of revenue to federal and county treasuries, and disproportionately negative impacts on rural communities dependent on the timber industry and timber harvest levels that existed from 1980–1992 on federal lands (FEMAT 1993).

Yet the trend toward ecosystem management and forestry that is more benign in environmental and aesthetic effects seems likely to continue for the foreseeable future. These changes reflect evolving public demand and current law as interpreted by the courts. There seems to be a distinct and growing distrust of natural resource managers—particularly government and corporate managers—by at least a vocal portion of the public. That distrust must be allayed if land managers are to retain any semblance of their historic management prerogatives. One lesson to be learned is that, in a democracy, forests are managed at the sufferance of the citizenry or at least by the majority of the minority of that citizenry that cares about the issue. The greatest challenge that foresters and other natural resource management professionals face in the practice of their professions may not be the technical aspects of forest management, but public acceptance of those practices.

These are among the many issues raised and discussed in *Creating a Forestry for the 21st Century.* In this volume, well-qualified experts have combined to produce a comprehensive view of ecosystem management.

Ecosystem management is a concept whose time has come. But ecosystem management is only a concept for dealing with larger spatial scales, longer time frames, and many more variables (ecological, economic, and social) than have commonly been considered in past management approaches. To be useful, a concept must be rendered operational. That requires placing the concept in context and in operational terms.

This book is an attempt to take that critical next step—to move the concept of forest ecosystem management into an operational context. Other such efforts are underway. But this, in my opinion, is the best of such efforts to date.

This book can be likened to a river that is fed by many streams. It flows more strongly with the addition and mixing into the current of each stream. This volume has identified many of the contributing factors that must be considered and integrated to make up the first efforts of forest ecosystem management. It is an exciting prospect.

Jack Ward Thomas
Chief, USDA Forest Service

Literature Cited

FEMAT. 1993. *Forest ecosystem management: An ecological, economic, and social assessment.* Report of the Forest Ecosystem Management Assessment Team (FEMAT). 1993-793-071. Washington, DC: GPO.

Johnson, K.N., J.F. Franklin, J.W. Thomas, and J.C. Gordon. 1991. Alternatives for management of late-successional forests of the Pacific Northwest: A report to the U.S. House of Representatives, Committee on Agriculture, Subcommittee on Forests, Family Farms, and Energy, Committee on Merchant Marine and Fisheries, Subcommittee on Fisheries and Wildlife, Conservation, and the Environment. Unpublished report.

Thomas, J.W. 1993. Addressing the retention of biodiversity in the national forests in the United States. In *Our living legacy: Proceedings of a symposium on biodiversity,* ed. M.A. Fenger, E.H. Miller, J.A. Johnson, and E.J.R. Williams. Victoria, British Columbia: Royal British Columbia Museum.

Thomas, J.W., E.D. Forsman, J.B. Lint, E.C. Meslow, B.R. Noon, and J. Verner. 1990. *A conservation for the northern spotted owl.* Portland, Oregon: U.S. Dept. of Agriculture, Forest Service, U.S. Fish and Wildlife Service, National Park Service, Bureau of Land Management.

Thomas, J.W., M.G. Raphael, R.G. Anthony, E.D. Forsman, A.G. Gunderson, R. S. Holthousen, B.G. Marcot, G.H. Reeves, J.R. Sedell, and D.M. Solis. 1993. *Viability assessments and management considerations for species associated with late-successional and old-growth forests of the Pacific Northwest: A report of the Scientific Assessment Team.* Portland, Oregon: USDA Forest Service.

Acknowledgments

Like most projects of this size and scope, this book is the result of the work, support, and cooperation of many people and organizations. The rallying point for this project was a conference entitled "Creating a Forestry for the 21st Century" held in Portland, Oregon, in September 1993. Sponsors of that event included the Olympic Natural Resources Center at the University of Washington College of Forest Resources, the Cascade Center for Ecosystem Management, the USDA Forest Service Pacific Northwest New Perspectives Program, the Consortium for the Social Values of Natural Resources, the Coastal Oregon Productivity Enhancement Program at Oregon State University's College of Forestry, the Washington State Department of Natural Resources, the British Columbia Forest Service, and the USDI Bureau of Land Management.

Substantial support for the editing of the book as a whole and preparation of some chapters came from The Pew Charitable Trusts. The grant was administered by The Nature Conservancy.

Carrie Bayless, Chuck Ayers, and Gordon Smith deserve special thanks not only for their careful reading of the manuscripts, but also for their support and encouragement throughout the project. In addition, Virginia Lore and Cynthia Cimperman provided valuable editorial assistance.

Finally, we thank all of the authors who stuck with us through many drafts. Without their ideas, insights, and expertise, the book truly would not have been possible.

Contributors

James K. Agee, Professor of Forest Ecology, University of Washington College of Forest Resources, Seattle, Washington

Michael P. Amaranthus, Research Biologist, U.S. Forest Service Pacific Northwest Research tation, Grants Pass, Oregon

Alexios R. Antypas, Doctoral Candidate, University of Washington College of Forest Resources, Seattle, Washington

R. W. Behan, Dean Emeritus, Northern Arizona University, Flagstaff, Arizona

Simon Bell, Chief Landscape Architect, Forest Authority, British Forestry Commission, Edinburgh, Scotland

Dean Rae Berg, Forest Contractor, Seattle, Washington

Peter A. Bisson, Research Fisheries Biologist, U.S. Forest Service Pacific Northwest Research Station, Olympia, Washington

Kelly Burnett, Fish Biologist, U.S. Forest Service Pacific Northwest Research Station, Corvallis, Oregon

Thomas R. Crow, Project Leader, U.S. Forest Service North Central Forest Experiment Station, Rhinelander, Wisconsin

Frederick W. Cubbage, Department Head and Professor, North Carolina University Department of Forestry, Raleigh, North Carolina

Robert O. Curtis, Principal Mensurationist, U.S. Forest Service Pacific Northwest Research Station, Olympia, Washington

Dean S. DeBell, Chief Silviculturist, U.S. Forest Service Pacific Northwest Research Station, Olympia, Washington

Nancy M. Diaz, Area Ecologist, Mt. Hood National Forest, Gresham, Oregon

Jerry F. Franklin, Professor of Ecosystem Science, University of Washington College of Forest Resources, Seattle, Washington

Sharon Friedman, Research Coordinator, U.S. Forest Service Resources Assessment Program, Washington, D.C.

John C. Gordon, Pinchot Professor, Yale School of Forestry and Environmental Studies, New Haven, Connecticut

Gordon E. Grant, Research Hydrologist, U.S. Forest Service Pacific Northwest Research Station, Corvallis, Oregon

Stanley V. Gregory, Professor, Oregon State University, Department of Fisheries &and Wildlife, Oregon State University Corvallis, Oregon

Eric J. Gustafson, Research Ecologist, U.S. Forest Service North Central Forest Experiment Station, Rhinelander, Wisconsin

Everett Hansen, Professor of Botany and Plant Pathology, Oregon State University, Corvallis, Oregon

Constance A. Harrington, Research Forester, U.S. Forest Service Pacific Northwest Research Station, Olympia, Washington

Richard W. Haynes, Research Forester, U.S. Forest Service Pacific Northwest Research Station, Portland, Oregon

Malcolm L. Hunter Jr., Libra Professor of Conservation Biology, University of Maine Department of Wildlife Ecology, Orono, Maine

K. Norman Johnson, Professor of Forest Resources, Oregon State University Department of Forest Resources, Corvallis, Oregon

Julia A. Jones, Associate Professor, Oregon State University Department of Geosciences, Corvallis, Oregon

Kathryn A. Kohm, Editor, University of Washington College of Forest Resources, Seattle, Washington

Denis Lavender, Professor Emeritus, Oregon State University, Corvallis, Oregon

Robert G. Lee, Associate Dean for Academic Affairs, University of Washington College of Forest Resources, Seattle, Washington

James Lyons, Undersecretary of Agriculture for Natural Resources and Environment, U.S. Department of Agriculture, Washington, D.C.

Bruce G. Marcot, Wildlife Biologist, U.S. Forest Service Pacific Northwest Research Station, Portland, Oregon

Errol E. Meidinger, Professor of Law, State University of New York, Buffalo, New York

Randy Molina, Team Leader for Mycology and Research Botanist, U.S. Forest Service Pacific Northwest Research Station, Corvallis, Oregon

Robert J. Naiman, Professor and Director of the Center for Streamside Studies, University of Washington College of Forest Resources, Seattle, Washington

David A. Perry, Professor of Ecology, Oregon State University Department of Forest Sciences, Corvallis, Oregon

David Pilz, Botanist, U.S. Forest Service Pacific Northwest Research Station, Corvallis, Oregon

Gordon Reeves, Research Fish Biologist, U.S. Forest Service Pacific Northwest Research Station, Corvallis, Oregon

Timothy Schowalter, Professor of Entomology, Oregon State University, Corvallis, Oregon

John Sessions, Professor of Forest Engineering, Oregon State University, Corvallis, Oregon

Margaret A. Shannon, Corkery Professor of Forest Resources, University of Washington College of Forest Resources, Seattle, Washington

Gordon R. Smith, Research Manager, University of Washington College of Forest Resources, Seattle, Washington

Thomas Spies, Research Forester, U.S. Forest Service Pacific Northwest Research Station, Corvallis, Oregon

Frederick J. Swanson, Ecosystem Team Leader, U.S. Forest Service Pacific Northwest Research Station, Corvallis, Oregon

John C. Tappeiner, Professor of Forest Ecology and Silviculture, Oregon State University, Corvallis, Oregon

Dale A. Thornburgh, Professor of Forestry, Humboldt State University, Arcata, California

Monica G. Turner, Associate Professor of Zoology, University of Wisconsin, Madison, Wisconsin

Nan Vance, Plant Physiologist, U.S. Forest Service Pacific Northwest Research Station, Corvallis, Oregon

James F. Weigand, Forester, U.S. Forest Service Pacific Northwest Research Station, Portland, Oregon

Clive G. Whittenbury, Chairmen of the Aerial Forest Management Foundation, Member of Board of Directors, The Erickson Group, Ltd., Yuba City, California

Julia M. Wondolleck, Assistant Professor of Resource Policy, University of Michigan School of Natural Resources and Environment, Ann Arbor, Michigan

Jack Walstad, Department Head and Professor of Silviculture, Oregon State University Department of Forest Resources, Corvallis, Oregon

Steven L. Yaffee, Professor of Resource Policy, University of Michigan School of Natural Resources and Environment, Ann Arbor, Michigan

Yanli Zhang, English Language Editor of the Journal of the Northeastern Forestry University, Harbin, People's Republic of China, Visiting Scholar in Entomology, Oregon State University, Corvallis, Oregon

Creating a Forestry for the 21st Century

1 Introduction

Kathryn A. Kohm and Jerry F. Franklin

Appreciating the Complexity of Ecological and Organizational Systems 2
Developing Site-Specific Knowledge 3
Linking Social, Biological, and Technical Information 4
Uncertainty, Humility, and Adaptive Management 5

A student of forestry who picked up a textbook in the 1950s or 1960s would have found information on converting old-growth stands into even-aged regulated forests, preventing and suppressing fire, creating habitat for game species, or calculating optimum rotations. Little mention was made of institutional or social issues. The forester of the 20th century could go to his post in the woods, plan for a sustained flow of timber, mitigate the negative effects of harvesting, provide for other values where possible, and feel secure in the knowledge that he had carried out his professional duties.

Of course, the 21st century will not be such a time. And this is not such a book. As a whole, or in its component parts, this book is designed as a conversation about the future of forestry. If 20th-century forestry was about simplifying systems, producing wood, and managing at the stand level, 21st-century forestry will be defined by understanding and managing complexity, providing a wide range of ecological goods and services, and managing across broad landscapes. These are the themes that reappear throughout the book.

Those looking for a how-to book with patent answers are likely to be frustrated. There are still a great number of unanswered questions about the future of forestry, and many of the answers that we do have start with "It depends. . . ." So, how do we prepare for a future that we cannot know? The answer in part lies in drawing the distinction between looking for what to do and looking for how to be—that is, looking for new ways of thinking about our problems. The following chapters lead readers through a thinking process about what forestry might become and what considerations should be taken into account. In many cases, the authors do not tell us what to do—such as the exact number of trees to retain in a harvest operation or the best forum for resolving disputes. Rather, they give us an array of tools and ideas. That is the beginning of the conversation. Where it leads is largely up to those who read and debate the following pages.

The book has been divided into five sections. The chapters in the first section focus on ecological processes and principles. They cover a broad range of topics from stand structure and function and disturbance processes to the movement of organisms across landscapes. Together they portray the rich complexity of forest ecosystems. This knowledge provides much of the impetus for major changes in the practice of forest resources management.

Section two addresses silvicultural systems. The science of silviculture has focused on a narrow set of prescriptions throughout much of its history. The chapters in this section look beyond those bounds. The authors challenge such long-held assumptions as the rationale for clearcutting, the wisdom of short rotations, and exclusion of fire. In addition, some traditional silvicultural tools are examined in light of new and expanded goals for forest landscapes.

The chapters in section three explore different aspects of managing at larger spatial scales. To protect and provide for a wide array of ecosystem goods and services, we must manage in the context of large landscapes over long time periods. This has become one of the tenets of ecosystem management. Developing agreement over the concept is relatively easy; garnering the knowledge and resources to implement landscape-level management is far more difficult. Authors in this section offer some practical information and ideas to that end.

Finally, sections four and five take up economic, organizational, and political issues critical to ecosystem management. In the past, forestry professionals have relegated social issues in forest management to a secondary status—an approach that has erupted in controversies that have shaken the very foundation of the U.S. Forest Service and the forestry profession as a whole. As several authors point out, ecosystem management cannot be built on biological and technical science alone. Ecosystem management will succeed to the degree that we can integrate biological, technical, and social solutions and develop institutions to transform knowledge into action.

The geographic focus of many of the chapters is the Pacific Northwest of the United States. This is not because ecosystem management concepts and practices are confined to this region—quite the contrary. Examples and ideas can be drawn from other regions within the United States, as well as from around the world. However, Pacific Northwest forest resources (particularly old-growth and the spotted owl) have been at the center of recent controversies, and these have propelled us toward a future based on ecosystem management. The Pacific Northwest is also an area in which there have been tremendous advances in ecological research—particularly in such fields as ecosystem structure and function, disturbance and recovery, and landscape ecology. Most important, many of the underlying concepts and thinking processes presented in the following chapters are widely applicable.

As with any attempt to define such a broad and rapidly changing field, there are missing pieces and style differences. We have done what we could to ameliorate these and to stress common themes. Of these, four themes stand out as fundamental to the practice of ecosystem management.

Appreciating the Complexity of Ecological and Organizational Systems

The single most salient theme of this volume is complexity. The adoption of ecosystem management as a guiding philosophy for 21st-century forestry represents a move from simplified to complex conceptions of ecological and organizational systems.

Discovery and recognition of complexity have been particularly dramatic in the study of ecological processes and principles. For example, forest invertebrates and fungi traditionally have been treated as pests and pathogens. Yet these small organisms, which comprise a surprisingly large portion of the biomass of a forest (particularly belowground), perform critical functions such as nutrient capture and recycling, thinning, and biological control. One of the tasks faced by 21st-century foresters is to understand the interdependencies of these species and move from simply trying to eradicate them to managing a delicate balance between their positive and negative effects (Schowalter et al., Chapter 11).

Similarly, disturbance and recovery processes are far more intricate than we had ever imagined. One of the icons of 20th-century forestry was that clearcutting mimics fire. Yet we have learned that fire and other natural disturbances do not destroy everything in their wake. Rather they leave an array of biological legacies that typically provide strong linkages between old and new ecosystems (see Perry and Amaranthus, Chapter 3).

But the complexity issue is not limited to ecology. As Meidinger points out in Chapter 23, simple top-down organizational structures often are poorly suited to ecosystem management. Achieving new ecological and social objectives will require exploration of varied models of organization and politics.

Appreciating the complexity of systems and managing for wholeness rather than for the efficiency of individual components place forestry in the context of a much broader movement toward systems thinking. Systems theory has permeated such diverse fields as business, medicine, education, and physics. In these disciplines, as well in forestry, discreet disciplinary research has served us well throughout much of the 20th century. No doubt tremendous strides in scientific understanding have been achieved. But we are reaching the limits of such thinking. In the future we will need to take much more information into account when managing forested landscapes. This will require new, innovative computer models such as the SNAP II model described by Sessions et al. in Chapter 18, new analytical methods such as those identified by Haynes and Weigand in Chapter 19, and new organizational structures of the type discussed by Meidinger in Chapter 23.

Finally, appreciation of the vast complexity of forest ecosystems means confronting the limitations of our knowledge. This will require both learning to cope with uncertainties so that we are not paralyzed to act, and adopting a sense of humility in forestry endeavors.

Developing Site-Specific Knowledge

A corollary of the importance of recognizing complexity is improving site-specific knowledge and management prescriptions. This is a key lesson of 20th-century forestry: Beware of simple formulas applied over broad areas.

Americans have had a pragmatic tendency to find a strategy that works in one place and to apply it extensively over large areas. In forestry, the most visible example of this is clearcutting. By the 1950s, all new Bureau of Land Management and Forest Service timber sales in the Douglas-fir region called for clearcutting. The result is a fragmented landscape in which species have been pushed to the edge of extinction, the productive capacity of many sites has been depleted, and public outcries have originated over landscape aesthetics.

Perhaps under the singular goal of wood fiber production such a broad brush strategy made sense. But management goals have dramatically increased in number and complexity—ranging from providing for specific levels of ecosystem processes to restoring old-growth habitats to helping local economies diversify. This, coupled with our growing understanding of the heterogeneous character of natural forests (see Spies, Chapter 2), means that managers as well as scientists increasingly will be called upon to tailor their research and prescriptions to local ecological and social situations.

For example, the theoretical concept of forest succession has moved away from simple generalizations toward more site-specific constructs (Spies, Chapter 2). Understanding the ecological dynamics of a site and tailoring appropriate management strategies will require more local information on environmental conditions, site history, disturbance regimes, community dynamics, and species habitat requirements.

From a silvicultural perspective, variable retention harvesting strategies recognize the immense flexibility that has always existed but has been stunted by the traditions and terminology of silviculture. For example, the shape of small patches of forest (or aggregates) retained as part of a harvest operation can be customized to modify microclimatic conditions or reduce the risk of wind throw. Structural features can be retained in varying configurations and amounts to provide habitat for specific organisms or lifeboat ecological processes from one stand to the next (see Franklin et al., Chapter 7). In addition, individual sites need to be assessed and managed in the context of much larger landscapes (Diaz and Bell, Chapter 17).

All of this will require resource managers to be schooled in a growing body of ecological and social research and to be familiar with the idiosyncrasies of a site. Indeed, carrying out customized harvest prescriptions means that woods workers of the 21st century often will be called upon to understand as much about forest ecology as professional resource managers have in the 20th century.

Such requirements raise the issue of institutional decentralization. The more we expect managers to be better educated and to fit prescriptions to local conditions, the more flexibility they will need and demand. One of the central institutional questions for forestry in the next century will be building organizational structures that give managers the flexibility to be innovative and accommodate local conditions while maintaining national goals and standards of quality. Striking a balance between centralized and decentralized planning is certainly not unique to forestry—it is a key issue defining the move from an industrial age to an information age.

Linking Social, Biological, and Technical Information

Our increasing inability to isolate biological, social, and technical issues is reflected in years of bitter controversy over the management of national forest lands. Despite technical expertise and volumes of data, the $3 billion forest planning effort of the 1980s has become mired in court battles. If ecosystem management simply replaces one technical fix with another, we are bound to end up in the same quagmire.

We have learned that "good science" is necessary, but insufficient. Forestry in the 21st century involves "creating and using information networks, facilitating effective multiparty decision making, building broad coalitions of political support, and participating in cross-jurisdictional management arrangements" (Yaffee and Wondolleck, Chapter 24).

To the degree that the debate has shifted from whether integration of biological and social concerns is necessary to how we should accomplish such integration, we have already made some progress. But linking social and biological issues runs counter to deep-rooted disciplinary, professional, and organizational divisions (Behan, Chapter 26). Bridging those gaps will require experimentation and time.

One route toward improved integration will be the development and use of new decision models such as the SNAP program described by Sessions et al. in Chapter 18. Unlike past models that focused primarily (or exclusively) on timber harvesting, newer models allow users to manipulate a wide range of ecological, economic, and social parameters to come up with alternative management strategies. Similarly, the landscape analysis and design process described by Diaz and Bell in Chapter 17 can be used by managers to synthesize information about landscape character and resources with issues and policies that direct land management.

Technological advances in data management and manipulation also have made information sharing possible on a much broader scale than ever before. As part of a move toward ecosystem management, we need to develop ways to make forest resource information, modeling programs, and education on how to use them broadly available. By sharing data with stakeholders, public agencies not only can tap into a much larger knowledge base, but also can get out of the bind of having to single-handedly know and integrate all social and biological factors (see Yaffee and Wondolleck, Chapter 24).

Finally, science-based assessments have the potential to be vehicles for integrating biological, social, and technical concerns. The Gang-of-Four and FEMAT reports have pointed the way toward developing comprehensive measures for ecological, economic, and social effects of alternative policies (Johnson, Chapter 25). Future assessments no doubt will build upon and improve these efforts.

Uncertainty, Humility, and Adaptive Management

If nothing else, the most important result of ecological research on forest landscapes and ecosystems has been an appreciation of their complexity and the limitations of our knowledge. Surprises and basic new insights into forest composition, structure, and function have been the hallmark of recent decades. And there is no reason to believe that the number and importance of new discoveries is likely to change.

From this we are reminded of the very tentative state of our current knowledge and the iterative nature of learning. We begin, finally, to appreciate that each management prescription is a working hypothesis whose outcome is not entirely predictable. And, hopefully, we adopt humility as a basic attitude in all approaches to forests—whether as scientists, advocates, managers, or policy makers.

Adaptive management is the only logical approach under the circumstances of uncertainty and the continued accumulation of knowledge. Management must be designed to enhance the learning process and provide for systematic feedback from monitoring and research to practice. Building the institutional structures and securing the necessary resources to accomplish this will be major challenges for forestry in the next century.

Finally, scientists, managers, and stakeholders must appreciate the idiosyncratic nature of each forest stand and landscape, as well as the limitations of general theory. All participants must understand and accept that new knowledge will almost certainly require continuing (and sometimes major) adjustments in our perceptions and treatment of forest ecosystems.

We hope that the following chapters will allow students as well as professionals to deepen and broaden their interests in ecosystem management—to become better informed and more effective. For skeptics, we hope this book will provide a compelling case for thinking creatively beyond the bounds of traditional forest resource management.

Section

I Ecological Processes and Principles

Advances in the natural sciences, particularly in ecology, have both stimulated and informed the paradigm shift that is occurring in the management of forest resources. The body of new knowledge generated by ecological scientists during the last 30 years has become, ultimately, so overwhelming in quantity and so fundamental in quality, that it can no longer be discounted or denied. We can never look at forests and forest landscapes in the same simplified way, for the lesson, ultimately, has been complexity and its importance.

The information harvest of the last three decades has been truly enormous. We have studied trees and forests for nearly a century, but from the limited perspectives of autecology and silvicultural manipulations. However important this knowledge, it did not provide insight into the forest as an ecosystem. Investigations during recent decades have taken us into such rich arenas as the structure and function of ecosystems, landscape ecology, and the varied nature of disturbances.

Structure and Function of Forest Ecosystems

Ecosystem science has deep roots in American forestry, beginning with the watershed research efforts pioneered by the U.S. Forest Service and going all of the way back to the Wagon Wheel Gap, Colorado, experiments of the early 20th century. The Hubbard Brook project marks the beginning of the modern era, however, in which there were major expansions in ecosystem research. Much of this expansion was fueled by the U.S. International Biological Program and succeeding activities, including the Long-Term Ecological Research Program sponsored by the National Science Foundation and other federal agencies.

From this science we began to learn how forest ecosystems work. The functional aspects of ecosystems are where much of the science began—cycles and flows of carbon or energy, nutrients and other mineral materials, and water. Again, the watershed experiments of the Hubbard Brook, Coweeta, and H. J. Andrew Experimental Forests provided many insights: the linkages between revegetation following timber harvest and nutrient losses, for example, or the role of old-growth forests in reducing the potential for rain-on-snow flood events.

Although compositional diversity is rarely a focus of ecosystem science, it has been a continuing and important part of the learning process (see Marcot, Chapter 6). As our knowledge of the dependence of wildlife (defined here as vertebrates) on forests has grown, concerns with biological diversity have spilled into the much larger world of invertebrates, fungi, and other small but functionally important organisms (see Perry and Amaranthus, Chapter 3). Sustaining these elements of diversity—a necessity for sustainability—has become a major challenge to silviculturists (see Franklin et al., Chapter 7).

An important related and unresolved concern involving biodiversity has to do with organisms traditionally considered to be pathogens and insect pests (see Schowalter et al., Chapter 11). Facets of this issue include understanding the balance between the positive and negative effects of these organisms, recognizing that so-called pests and pathogens are only a tiny percentage of the invertebrate, fungal, and microbial organisms that perform such important roles in forest ecosystems, and understanding that silvicultural treatments designed to eliminate pests and pathogens can have large negative impacts on other desirable organisms. The linkage between drastic forest site treatments to control diseases and other pests and the potential loss of nontarget ecologically critical organisms from the same phyla seems not to have occurred to many resource managers.

Perhaps the greatest revelations in forest ecosystem science have been the recognition of structural complexity and its importance. With rare exception, natural forests have proven very heterogeneous in terms of individual structures and their spatial arrangements within stands (see Spies, Chapter 2). The fact that dead trees and logs are as important as living trees in ecosystem function has been a particular challenge to foresters who could no longer view such material as waste, fire hazards, and mechanical impediments; this information has created an immense set of new questions about how many snags and how much coarse woody debris is needed to fulfill given levels of ecological function (see Franklin et al., Chapter 7).

Spatial heterogeneity in forest stands—patchiness—has also proven to be a pervasive and important characteristic of natural forest ecosystems (see Spies, chapter 2). Included are areas of open canopies (gaps) and heavy shade (anti-gaps) and the multiple canopy layers that increase biological diversity by dramatically expanding the number of niches that are available.

Some particularly important subsystems of forests have finally begun to yield their secrets—riparian areas, canopies, and below ground (see Perry and Amaranthus, Chapter 3, and Gregory, Chapter 5). All are extremely important for their richness in organisms and for their role in critical ecosystem processes. Understanding is most advanced for the riparian zone, with its strong mutual interactions between the terrestrial and aquatic subsystems. Only recently have foresters begun to acknowledge the overwhelming influences of forest conditions on processes, including productivity, in associated streams and rivers. Recognition of this dependence and the associated responsibilities is certainly incomplete, however, as evidenced by an entire recent issue of the *Journal of Forestry* on sustainability that contained not a single reference to aquatic resources.

Disturbances, Recovery, and Biological Legacies

Other claims notwithstanding, ecologists and foresters have known that forests are dynamic for a very long time. This recognition is not new in any sense. The last several decades, however, have seen an increased understanding of the importance and diversity of disturbance and recovery processes. Disturbances not only are not all alike, they are, in fact, dramatically different in their impacts. And, most particularly, clearcuts are quite unlike almost all other natural disturbances.

Comparative studies of various types of natural and human disturbances have been particularly revealing (see Perry and Amaranthus, Chapter 3). Disturbances vary dramatically in type, intensity, spatial pattern, frequency, and, most important, the legacy of organisms and structures that are left behind to be incorporated into the recovering ecosystem. These residual organisms and organically derived structures are sometimes referred to as biological legacies, and their type and level are very important influences on the composition, structure, and function of the post-disturbance ecosystem.

Disturbances are not only not all alike, they are, in fact, dramatically different in their impacts. And, most particularly, clearcuts are quite unlike almost all other disturbances in the intensity of their impacts, including their very low levels of biological legacies and uniformity. The increased scientific understanding of disturbances, biological legacies, and recovery processes is a major contribution to efforts to create silvicultural systems that incorporate more natural elements (see Perry and Amaranthus, Chapter 3, and Franklin et al., Chapter 7).

Forest Landscapes

Forestry traditionally has concerned itself with individual stands and has been most reluctant to deal with issues at larger spatial scales, even though some of these—such as cumulative effects and fragmentation—are of overwhelming importance (see Hunter, Chapter 4). Perhaps this reluctance has been due partially to the lack of critical information. In the last 20 years, however, landscape ecology has begun to provide critical concepts and empirical data, and technology has provided essential tools, such as geographic information and global positioning systems.

The basic lesson has been that context does matter! Spatial patterns are important, and foresters and resources are at risk when they ignore this principle (see Hunter, Chapter 4). The fact that landscapes, like ecosystems, have compositional, functional, and structural features is increasingly understood. Important concepts include the patch (including its size and shape), edges and edge effects (especially where there are high levels of contrast between adjacent patches), the matrix or dominant patch type in the landscape, and connectivity, which turns out to be much more than just a concern with corridors.

Of course, not all parts of a landscape are of equal importance, and areas associated with streams, rivers, lakes, ponds, and wetlands have emerged as some of the most important ecologically. Forests along streams and rivers are powerful influences on the aquatic ecosystems, affecting physical conditions, controlling energy and nutrient bases, and providing critical structure (see Gregory, Chapter 5). At the landscape level, roads and harvest treatments control rates and patterns of critical geomorphic processes, including production and transport of sediments (see Swanson et al., Chapter 15).

Foresters have been slow to respond to their responsibilities within the stream and river influence zones. However, the importance of these areas to social and ecological goals has been very clear in every major forest policy or planning effort carried out in the last decade: Protective buffers for aquatic ecosystems are always a central element in proposed management scenarios. Yet, much remains unknown about such protective zones, including the degree to which active management and commodity removal are consistent with ecological objectives.

2 Forest Stand Structure, Composition, and Function

Thomas Spies

Forest Stands and Ecosystems 12
Modern Concepts of Forest Succession 13
General Models of Forest Development 14
Establishment Phase 14
Thinning 15
Transition 17
Shifting Mosaic 19
Variability of Developmental Stages in Space and Time 19
Environmental and Ecosystem Characteristics of Stands 21
Resources and Microclimate 21
Composition 22
Ecosystem Functions 22
Time as a Factor in Forest Compositional Development 24
Perspectives on Old Growth 24
Implications for Forest Management 25
Literature Cited 27

Over the last 100 years, steep increases in human populations and the development of global market economies have rapidly changed the values of forests and increased societal conflicts over their use. The cutting and widespread exploitation and management of forests for fiber production, agriculture, and development have changed the face of many forest landscapes across the globe (Nelson et al. 1987, Ellenberg 1978, Perlin 1989). Forest lands have been converted to agricultural uses and human developments in large areas of the world (Perlin 1989). In other areas, unsound forestry practices have resulted in poor or delayed tree regeneration. Where forests have been exploited for fiber production, economically driven forestry practices have shifted the composition and structure of forests toward younger stands dominated by commercially valuable tree species, or shifted the composition toward other species that may have lower economic value than the previous dominant species (Graham et al. 1963, Seymour and Hunter 1992).

In some landscapes, such as in parts of the northeastern United States, the amount and age of the forest is increasing following declines resulting from logging in the 19th century (Foster 1993). The net effect globally, however, appears to be a decline in the amount of older forests or forests that have not been strongly influenced by modern agricultural and industrial human cultures. For example, worldwide the amount of primary forest is estimated to be 16 percent of its preindustrial amount (Postel and Ryan 1991). In the Pacific Northwest, where forest-use conflicts are especially severe, the amount of old-growth forest has declined over 50 percent in the last 60 years (Bolsinger and Wadell 1993). What remains has become highly fragmented in the last 20 years (Spies et al. 1994). The decline in older forest, or loss of primary forest, such as in the tropics and parts of the temperate zone, has lead to conflicts between segments of societies that value forest land for forest products and other land uses and those that value forests for biological diversity, ecosystem productivity and quality, and aesthetic and spiritual values (Aplet et al. 1993). These concerns have arisen not just from changes in the amount of old forest. For example, the quality of many forests has changed because of fire control, grazing, introduced forest pathogens and species, and effects of roads and agriculture on water quality. However, the loss of old forests has become symbolic of changes and losses in forest biodiversity resulting from human activities.

The kind and rate of human-induced changes in forests are superimposed on and interact with changes that result from natural processes, including disturbance, dispersal, establishment, competition, herbivory, and environmental change. The ability of society to understand, manage, and sustain the diversity of values associated with forests depends in part on our ability to understand ecological changes that forests undergo over time in natural, seminatural, or unmanaged stands and landscapes and understand the ecology of changes associated with direct and indirect human activity.

The primary objective of this chapter is to provide an ecological basis for this understanding by reviewing ecological changes associated with natural processes of disturbance and succession from young to old forests. Secondarily, the chapter will examine some of the variability in late-successional forests and contrast plantation forests with natural or wild stands. I assume that society's goals for forest ecosystem diversity include ecosystems dominated by old trees and associated structures. The way to achieve this goal is a subject of considerable debate. Options range from reserve-based approaches in which land managers seek to maintain at stand or landscape levels old forest ecosystems without direct human manipulations, to objectives that call for direct intervention to "restore" older forests at stand or landscape scales, to strategies that call for providing both wood fiber and at least some components of old-forest ecosystems. The ability of managers to meet any of these objectives will depend in part on how well they understand the ecology and dynamics of forest change in stands and landscapes.

Forest Stands and Ecosystems

Stands and ecosystems are the fundamental units of forest management and ecosystem science, respectively. With the advent of forest ecosystem management, attention needs to be paid to reconciling these two disciplinary building blocks. The definitions and spatial boundaries of stands and ecosystems are typically determined for specific purposes of management and science. A stand typically has been defined as a unit of trees that is relatively homogeneous in age, structure, composition, and physical environment (Smith 1962, Oliver and Larson 1990). The characteristics used to delineate stands often refer to the tree layer since this traditionally has been the focus of forest management and is relatively easily mapped using aerial photographs. Soil and topographic features also frequently are used to delineate stands, especially if they have a strong effect on stand productivity or harvesting operations. Specific stand definitions, sizes, and shapes will vary depending on management intensity and objectives and the spatial heterogeneity of the vegetation, soil, and topography.

Ecosystems, in contrast to forest stands, typically have been more conceptual than real physical entities (Kimmins 1987). Whittaker (1962) defined an ecosystem as "the functional system comprising a community of interacting organisms—plants, animals, and saprobes—and the environment that affects them and is affected by them." Since ecosystems are open systems, defined by functions, interactions, and flows, the spatial boundaries of ecosystems are not easily defined (Kimmins 1987). Typically they are delineated based on specific purposes and limitations of a study.

Relatively rapid changes in ecosystem structure and function can facilitate the definition of spatial boundaries. Many of these boundaries are defined by topography (e.g., watersheds and wetlands), by distinct changes in forest structure, or by vegetation patchiness (e.g., forest woodlots in an agricultural landscape or old-growth forest stands in intensively managed forests). Thus, although forest stands are components of forest ecosystems, forest ecosystems are not necessarily forest stands—unless those stands are defined in a way that includes organisms other than trees as well as environmental factors and functions. The boundaries of stands and ecosystems will not necessarily coincide unless the criteria used to identify the two units are very similar. This chapter focuses on the ecological attributes of forest stands and assumes that the stands are defined and delineated based on ecosystem criteria such as soil, topography, vegetation structure, and landscape pattern and position in addition to management considerations.

Modern Concepts of Forest Succession

Popular and scientific views of forest succession have changed over the last century. Early concepts emphasized relatively predictable cycles of plant communities, behaving almost like organisms, that developed toward climax communities whose characteristics were controlled by the regional climate (Clements 1916). This view was modified as it became clear that the species comprising communities frequently behave individualistically and that local site factors give rise to a diversity of climaxes in a region. The concept of climax, a relatively stable community condition toward which succession proceeds, lost much of its value as ecologists began to describe climax vegetation as "varying continuously across a continuously varying landscape" (Whittaker 1953). Ecosystem-level concepts of succession that arose in the 1960s and 1970s emphasized theoretical generalities of ecosystem change in terms of information theory (e.g., species diversity measures) and flows of energy and matter. However, these ecosystem perspectives were in many ways another expression of climax, equilibrium concepts (Glenn-Lewin and van der Maarel 1992).

More recently, ecologists conceptualize succession as a nonequilibrium spatial process (dynamic mosaic of Watt 1947) that is the outcome of disturbance and population processes such as birth, death, dispersal, and growth under changing environmental conditions (Peet and Christensen 1980, Glenn-Lewin and van der Maarel 1992). Thus the theoretical concept of succession has moved away from simple generalizations toward more complex constructs. Site-specific predictions about vegetation dynamics cannot be made from broad general theory; rather, they require a more complex scientific framework and specific information about disturbance, environment, propagule availability, and species biology (Pickett et al. 1994).

The maturation of ecological thinking about succession and the increased involvement of ecologists in conservation and management issues has moved ecology toward site-specific, mechanistic approaches. Such approaches are similar to those used by forest scientists and silviculturists who typically have not been concerned with generating broad general theories, but rather with developing the ability to make site-specific predictions for growing stands for forestry applications.

The empirical studies and approaches of applied forest ecologists provide a rich source of information about growing commercially valuable species under relatively short rotations; this will be useful to ecologists seeking to develop and test new theoretical frameworks. However, while applied forest scientists have been focused on developing information relevant to the dominant forestry objectives (primarily timber production), they have not kept up with advances in ecology (Oliver and Larson 1990). The emergence of a broader set of objectives related to ecosystems and biodiversity, such as the maintenance of old-growth ecosystems, and new fields, such as landscape ecology and conservation biology, have caught most applied forest ecology professionals by surprise. Consequently, information is lacking about how to implement forest management practices that will meet new broader ecological objectives while providing for the more traditional objectives of wood fiber production.

General Models of Forest Development

While stand development is complex and diverse—a function of initial disturbance, environmental patterns, species mix, and intermediate disturbances—some general patterns of development appear across a wide range of forest types and locations (Oliver and Larson 1990). Several authors have proposed general models of natural stand development following major disturbances. Four major phases of forest development are typically identified: (1) establishment, (2) thinning, (3) transition, and (4) shifting mosaic (Bormann and Likens 1979, Oliver 1981, Oliver and Larson 1990, Peet 1981, Peet and Christensen 1987, Spies and Franklin 1995). These phases are idealized states of relatively homogeneous areas of forests called stands. They range from a fraction of a hectare to thousands of hectares in size, depending on forest structure, composition, and the objectives of stand classification and mapping.

Establishment Phase

This first phase following a major disturbance, which is also termed "stand initiation" (Oliver and Larson 1990), is characterized by establishment of new individuals, release of surviving seedlings and saplings, and vegetative reproduction of injured plants from belowground structures (Figure 2.1). It is marked by relatively rapid changes in species dominance, environment, structure, and levels of competition and high mortality among small individuals. Initially this phase is characterized by an abundance of sites free from competition of established plants. Consequently, many species can establish during this period, including nonnative species (Halpern 1989) and hybrids (Spies and Barnes 1982). At the same time, many individuals of the original understory that survived the disturbance but lost aboveground parts may resprout and begin to reoccupy their previous aboveground spaces. Consequently, a stand or patch can achieve high species diversity for a time with the mixture of new and old individuals. Establishment of trees may begin immediately following the disturbance or be delayed for short to long periods depending on the availability of seed sources and competition from shrubs and herbs. The processes of tree invasion may be highly patchy depending on microsite availability, seed and seedling mortality from competition, climatic factors, or herbivory and disease (Veblen 1992). Patches of trees may eventually initiate in places and grow together in a spatial process of nucleation and coalescence. Horizontal patchiness may be very high, composed of a mosaic of patches of trees, shrubs, herbs, dead wood, and bare ground. Vertical heterogeneity, while high at fine scales and increasing at larger scales with the growth of shrubs and trees, is still low overall because of the short stature of the vegetation. As shrub and tree canopies grow together, shade-intolerant herbaceous and shrub species are typically lost from the site—although they may persist for varying lengths of time in gaps that do not close and in the soil seed bank.

Figure 2.1 Establishment stage following fire in *Pinus sylvestris* forests in northern Sweden. (Photo: Thomas Spies)

Although this stage lacks live tree structure, standing and down dead trees can be prevalent during this phase depending on the previous stand's composition and structure and the type of disturbance (Harmon et al. 1986, Spies et al. 1988). Dead wood is prevalent following death from blowdown, insects, disease, drought, standing water, and fire (Figure 2.2). Dead wood may be completely absent in the establishment phase following avalanches, high-en-

(a) (b)

Figure 2.2 (a) Carryover of coarse woody debris following fire in old-growth mixed-conifer forest in southwestern Oregon. (b) Carryover of coarse woody debris following blowdown in old-growth forests on the Mount Hood National Forest in Oregon. (Photos: Thomas Spies)

ergy floods, and fires in stands with small trees or low biomass of live trees (Figure 2.3).

Thinning

The next phase, which is also termed the stem exclusion phase (Oliver and Larson 1990), is characterized by the closing together of tree canopies. This results in steep declines in understory establishment and growth, increases in mortality of many understory plants, and the onset of mortality in the tree layer associated with competition for light and water. Understory vegetation may become sparse or absent in many areas beneath the dense tree canopy (Figure 2.4). Species diversity typically declines relative to the establishment phase due to loss of shade-intolerant species (Schoonmaker and Mckee 1988, Halpern 1989, Elliot and Swank 1994). Populations of shade-tolerant forest species that were eliminated by the disturbance or the unsuitable environment of the establishment phase may not recolonize during this phase because of loss of seed sources, slow dispersal rates, or competitive barriers to establishment. The degree to which understories decline or disappear depends on the density and species of the overstory trees, spatial heterogeneity of the site, and tree regeneration. Where the tree layer is dominated by shade-intolerant species, enough light typically penetrates to the understory to maintain many shrub and herb species. In addition, site conditions and regen-

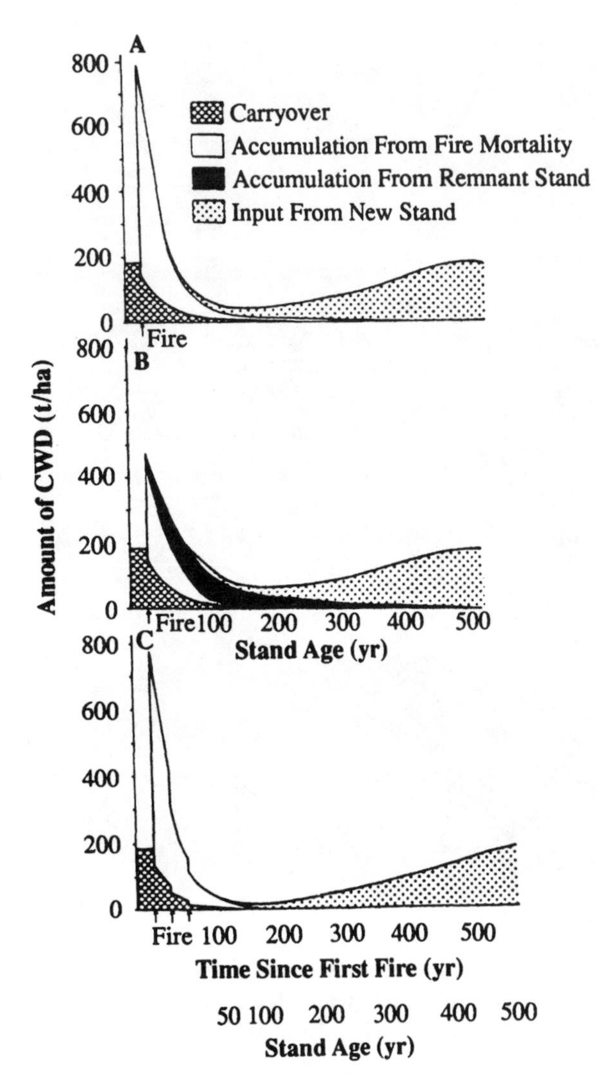

Figure 2.3 Alternative models of the dynamics of coarse woody debris in stands with different disturbance histories (from Spies et al. 1988).

Figure 2.4 Dense Douglas-fir stand in stem exclusion stage illustrating sparsely developed ground layer of vegetation. (Photo: Thomas Spies)

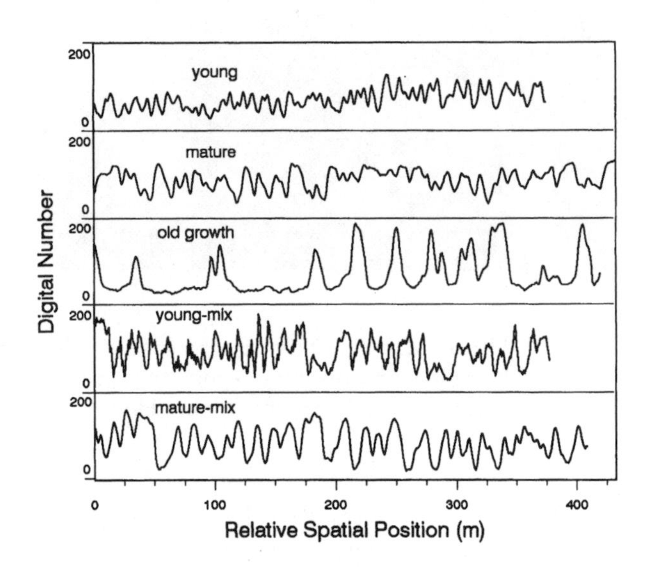

Figure 2.5 Spatial variability in canopy brightness (digital number) recorded along transects by video camera above five conifer stands of different ages in Oregon. Areas of high brightness are tree crowns and areas of low brightness are gaps in the upper canopy (from Cohen et al. 1990).

eration may provide relatively large canopy openings or gaps that allow understory vegetation to survive and grow in many places in stands (Bradshaw and Spies 1992) (Figure 2.5). Thus, the understory "exclusion" process of this phase may not occur in many stands dominated by shade-intolerants or may be very short-lived as tree canopies rapidly differentiate and open up. Yet, on some very low productivity sites, stands may stagnate in this phase, creating a dense, slowly changing stand of relatively small trees with sparse understory (Oliver and Larson 1990). Typically, however, the number of canopy trees declines during this phase as mortality occurs from competition, diseases, and other disturbances.

Vertical heterogeneity in the vegetation increases rapidly as the height of the forest increases. Canopy

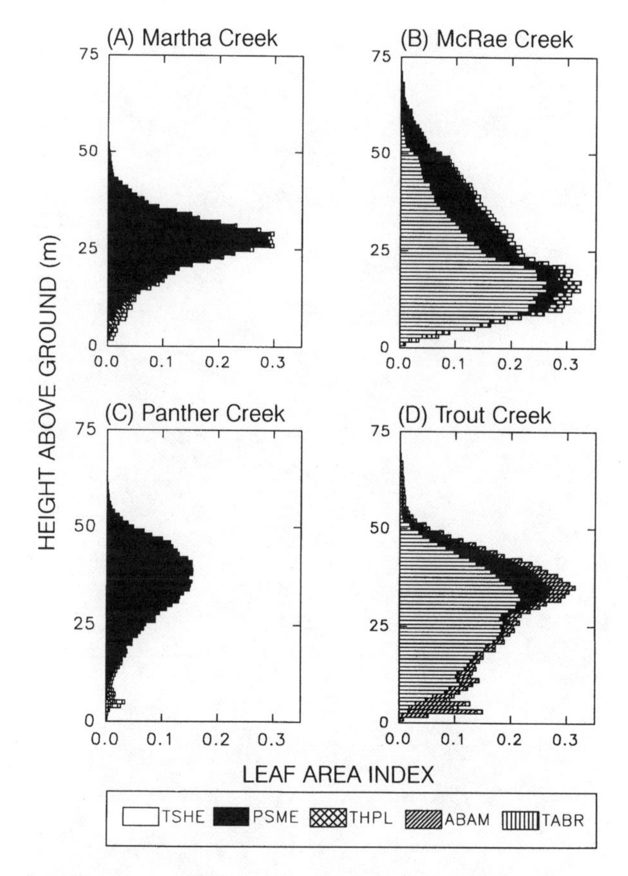

Figure 2.6 Vertical distribution of leaf area index in two mature stands (A and C) and two old-growth stands (B and D) in Oregon and Washington: TSHE, Tsuga heterophylla; PSME, Pseudotstuga Menziesii; THPL, Thuja plicata; ABAM, Abies amabilis; TABR, Taxus brevifolia (from Easter and Spies 1994).

growth often creates two relatively distinctive vegetation layers: an uppermost layer of canopy trees and a sparse layer of dying and surviving shrubs and herbs (Figure 2.6). The canopy tree layer is composed of dense, rapidly growing tree crowns with high foliage-to-branch-biomass ratios. Not much light penetrates to mid and lower portions of the crowns. In dense stands, lower-side branches in tree crowns die in the low-light environment.

Amounts of wood debris will be variable during this phase. Large well-decayed standing dead and fallen trees may still be present from the preceding stand if it contained many large decay-resistant trees that were not removed or destroyed by disturbance. Canopy trees of the new cohort that die during this thinning phase do not contribute much to the accumulated mass of dead wood because they are relatively small and decay rapidly without decay-resistant heart wood.

Transition

Stand development following the first two phases is less understood (Oliver and Larson 1990) and has been classified in different ways. Two major phases can be recognized: a transition phase and a steady-state or shifting gap phase (Peet 1992, Spies and Franklin 1995). Of these, the transition phase may be the most variable and least understood. It is marked by a variety of gradual changes in population, stand structure, and vegetation processes that can last from less than 100 to over 1,000 years depending on the forest type and disturbance history. The original cohort of trees slowly breaks up, tree establishment and release of suppressed understory trees increases, and a new cohort of trees gradually grows into the canopy gaps. Tree size, live biomass, and diversity of tree sizes and canopy layers typically peak, while dead wood biomass may decline to a low point before increasing.

Because of the diversity and gradual nature of changes during this stage, it has been subdivided in several different ways. Oliver and Larson (1990) identify an understory reinitiation phase that coincides with the beginning of the transition phase. It is characterized by the appearance and growth of new trees and other plants in the understory. This understory reinitiation phase ends when these newer cohorts of trees begin to reach the overstory. At this point a transition old-growth stage begins. It is characterized by canopies containing trees that established in the understory reinitiation phase and relics from the original establishment phase (Oliver and Larson 1990). When these relic trees finally die, the stand enters a "true old-growth stage" in which all of the live trees in the stand originated from smaller-scale gap disturbances resulting from "autogenic" factors such as pathogens, insects, and wind.

Recent studies and observations of long-lived Douglas-fir forests in the Pacific Northwest suggest that the transition phase may be divided into four subphases based on changes in structure and processes: (1) understory release, (2) maturation, (3) early transition, and (4) old-growth and late-transition old-growth phases (Spies et al. 1988, Spies and Franklin 1991, Spies and Franklin 1995).

In the understory release subphase, existing individuals of shrubs, herbs, and some trees that survived through the thinning phase, either because of their tolerance of low light levels or because of their persistence in microsites with higher light levels, grow and expand from existing roots and rhizomes. This occurs as light levels increase as a result of the numerous small gaps forming in the canopy with the death of weak and suppressed trees. Stands during this subphase may appear as single-storied stands of trees with well-developed shrub and herb layers. In some cases, such as in the salmonberry/alder/conifer stands in coastal Oregon, these well-developed understory layers may inhibit the establishment of new trees from seed.

The maturation subphase begins as overstory trees become more widely spaced, dominant individual trees approach their maximum size, gaps that form from tree death grow larger and more persistent, trees that have established from seed in the previous stage move above the shrub layers, and more tree seedling establishment occurs (Figure 2.7). Dead woody biomass in the stand may be at a low point during this phase because dead wood from the previous stand has lost most of its mass and new inputs of large new trees are just beginning (Spies et al. 1988).

Figure 2.7 Mature Douglas-fir forest about 130 years old on the Willamette National Forest in Oregon. (Photo: Jerry Franklin)

The early transition old-growth phase begins as trees that established during the understory release and maturation stages dominate the upper canopy layers (Figure 2.8). Trees that established during the first establishment phase may still dominate the uppermost layer—such as in old-growth Douglas-fir–western hemlock stands where large Douglas-firs occupy a scattered layer of emergent trees above western hemlock and other shade-tolerants. The presence of remnant trees and new cohorts of trees beneath them and in gaps creates a multilayered foliage canopy that is distinctive of many old-growth types. This phase is also characterized by the buildup of relatively large amounts of dead woody debris as trees from the original cohort die and dead biomass accumulates more rapidly than it is lost from decay processes (Spies et al. 1988). This phase typically requires 100 to over 500 years to develop (Oliver and Larson 1990) for forest types in North America.

A late transition old-growth subphase, which may occur following the early transition old-growth phase, is distinguished by the absence of live remnant trees from the establishment phase and the continued presence of dead wood from trees that originated during this phase. This phase may be marked by a lower live biomass relative to the early transition old-growth phase, but with a relatively high mass of dead wood. A late transition old-growth phase is probably most distinctive in forest types with long-lived trees and decay-resistant wood or in climates that result in slow decay. It may last centuries depending on the decay rates of large wood and the frequency of disturbance.

Figure 2.8 Transition-phase eastern white pine and eastern hemlock old-growth forest in the Upper Peninsula of Michigan. (Photo: Thomas Spies)

Shifting Mosaic

The last phase that is recognized in most classifications of stand development has been termed the "steady state phase" (Peet and Christensen 1987), "true old growth" (Oliver and Larson 1990), "shifting mosaic" (Borman and Likens 1979) or "shifting gap" (Spies and Franklin 1995). Despite the variety of terms, all authors agree that this last stage is characterized by a shifting pattern of relatively small patchy disturbances (death of individual canopy trees or groups of trees forming gaps of various sizes and shapes) which provide resources for new establishment of trees in the understory and increased height growth of individuals in lower and mid-canopy positions. By aggregating these small disturbances and vegetative responses to a larger spatial area such as a stand, the net changes in forest composition and structure typically appear very small or nonexistent.

This aggregate behavior of small patch dynamics has been frequently described as a steady state or equilibrium. The term "climax" has also been used to characterize a condition similar to this in which vegetation composition and structure are considered stable. However, the term "climax" has fallen out of favor among forest ecologists. It has been used in many different ways (Oliver and Larson 1990), often to mean a simplistic and unrealistic end point of succession that lacks disturbance and is stable. The current view of the last stage of forest development recognizes the occurrence and role of small disturbances and the presence of instability resulting from larger disturbances and climate change. This last phase typically requires many hundreds to over a 1,000 years to develop in a stand that originated following a large, severe disturbance—due to longevity of the trees that form the original cohort and the decay resistance of their wood. Consequently, this phase is uncommon in many current landscapes where logging and natural disturbances occur more frequently than the typical life span of the major tree species.

Structural and compositional differences between this phase and the transition phase are typically small (Spies and Franklin 1991). Variation in live and dead tree size or diversity of size classes may be lower if large seral dominants were present in the previous phase. Composition may shift toward shade-tolerant plants that can regenerate in relatively small canopy gaps. However, where early phases of stand development are already dominated by shade intolerants, composition may not change (Oliver and Larson 1990). As such, the composition of tree regeneration becomes more similar to that of the mature overstory trees (McCune and Allen 1985).

Variability of Developmental Stages in Space and Time

The four major phases of forest development described above are idealized, and although understanding these stages is often helpful, there is considerable variability in a population of real stands that blurs boundaries (Spies and Franklin 1991). Variation in disturbance history and site productivity, for example, result in stands being composites of two or more phases. In some cases, multiple disturbances may leave patches of several different previous stands, creating a variety of phases mixed within a stand horizontally or vertically or both. Consequently, actual site-specific management practices should be based on actual conditions within stands rather than on broad classes of stand development. The classes may be useful, however, in broad landscape-level planning and mapping.

The temporal pattern of stand developmental stages may also be quite complex—controlled by disturbance regimes and composition of propagules available at critical periods during stand development. Stand dynamics over time in complex landscapes can result in multiple pathways of development (Cattelino et al. 1979) for a given site or landscape (Figure 2.9). Over long periods, these trajectories might assume the patterns of chaotic systems with "strange attractors" (Gleick 1987). In the modern scientific view, with its emphasis on nonequilibrium dynamics, the strange attractors may represent something close to a climax state—a condition toward which stands may move statistically, but not attain for very long (Figure 2.10). In a landscape with a variety of environments, several strange attractors may occur, and under climate change these points may vary over time.

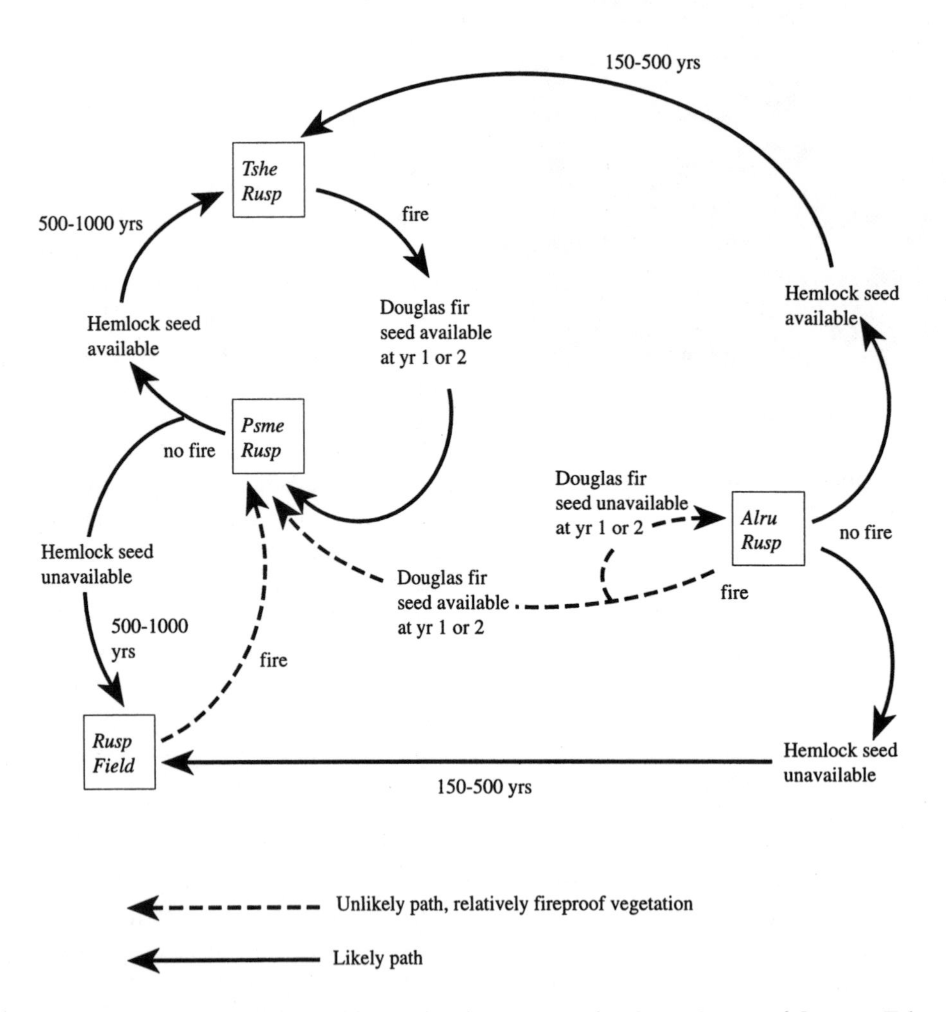

Figure 2.9 Multiple pathways of forest development in the Coast Range of Oregon. Tshe: Tsuga heterophylla; Rusp: Rubus spectabilis; Alru: Alnus rubra; Psme: Pseudotsuga menziesii (from Hemstrom and Logan 1986).

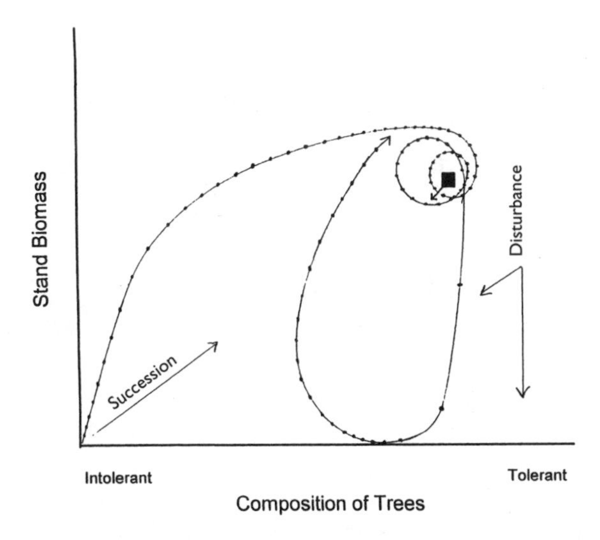

Figure 2.10 Theoretical phase-state diagram for forest succession. Intervals between points indicate equal time intervals.

Environmental and Ecosystem Characteristics of Stands

We know much less about the structure and composition of nontree biological communities and ecosystem functions than we do about the tree layer. The general models of stand development described above are based largely on the population dynamics and structure of tree layers. The degree to which this framework translates to other components of ecosystems is not well documented, although we have learned much in the last 15 to 20 years. Where organisms and ecosystem functions are strongly linked to the structure and dynamics of the live and dead tree components of the ecosystem, stand development can be a major driver of ecosystem change over space and time. On the other hand, the patterns of some ecosystem components and functions may be more strongly determined by the physical environment, competition, or accidents of migration and mortality than by stand characteristics.

Despite the relatively poor documentation of the role of stand structure and development in biological diversity and ecosystem function, we have a good conceptual model of stand structure linkages between trees and other biological and ecological characteristics (Peet 1992, Franklin and Spies 1991, Perry 1994).

Resources and Microclimate

After the death of many trees in an area, a variety of changes occur in the microclimate and resources of stands. Above- and belowground resource availability is characteristically high in the establishment phase. The loss of the canopy layer typically results in more extreme environmental conditions across the patch, such as in temperature, radiation, and soil moisture (Chen et al. 1992). The role of microsites and fine-scale heterogeneity becomes important in modifying the overall climate of the site and creating "safe sites" for the establishment of species that are more sensitive to temperature and moisture extremes and high radiation.

As tree canopies close, available light and moisture near the ground become low. Nutrient availability may be low as well. Microclimates of the understory probably have higher relative humidity and a smaller range of temperatures compared to the extremes of the open establishment phase. However, they may not be as buffered relative to later stages when canopy heights, protective phytomass, and number of foliage layers are greater.

During the transition phase, as tree mortality occurs in the overstory and height and crown growth slow, available light increases in the understory, and belowground resources such as moisture and nutrients may increase as well (Oliver and Larson 1990, Easter and Spies 1994). In later stages of the transition phase, the invasion of the understory and midstory layers by new cohorts of trees results in patchy variation in resource availability from very low levels beneath understory trees to relatively high levels in canopy openings. Canopy gaps may increase in size (Spies et al. 1990, Bradshaw and Spies 1992) between the mature phase and the early transition old-growth phase, although large gaps may also be found in earlier phases. Microclimate variation beneath gaps and canopies can be high, and variation within gaps can control the survival of establishing tree species (Gray and Spies, in press). Surface soil moisture might be lower in the understory release and mature phase compared with the later stages of the transition phase. This is probably due to moisture uptake by well-developed shrub and herb layers in these early phases.

Environmental characteristics of the shifting mosaic phase are poorly known and may differ little from the latter stages of transition. Resources such as available light and moisture are distributed in a spatially heterogeneous manner with high levels in gaps (small areas of the establishment phase) and low levels beneath canopies in small areas of thinning and transition phases (Canham et al. 1990). This mosaic of resources selects for plants and animals that are adapted to relatively small variations in resources and habitat over time, such as organisms with low mobility and greater competitive ability (Bazzaz 1991, Tilman et al. 1994).

Composition

The high resource availability during the establishment phase allows many understory shrub and herbaceous species that were formerly suppressed by the shade of the tree layer to grow rapidly and increase in biomass and productivity. Flower and fruit production typically increases for many shrubs and herbs. High vegetative productivity, phytomass, and flower and fruit production near the ground increase herbivory and foraging by invertebrates and vertebrates. Many large ungulate species that are adapted to living in forest landscapes find important sources of energy in this phase—particularly where they can venture into it from the protective cover of nearby closed or semiclosed canopy forests (Thomas 1979). Herbivory from invertebrates may also be high during this time (Schowalter 1989). Numerous bird and mammal species forage in stands of this phase of development (Brown 1985).

Vertebrate use of stands during the thinning stage is generally thought to be low relative to earlier and later stages. If understories are sparse, not much cover or food will be available. Dense tree canopies may be used by those species, such as Kirkland's warbler, that find cover and food there (Probst 1987). Canopy epiphytes are less common in younger stands (Spies 1991, McCune 1993).

Vertebrate and invertebrate species may become relatively abundant during the transition phase of forest development in some forest types. For example in Douglas-fir– western hemlock stands in the Pacific Northwest, several species of vertebrates are more abundant in the mature and old-growth phases than in earlier forest development phases (Ruggiero et al. 1991). These species include groups that use relatively large dead trees, deep or multilayered canopies, and deep forest floor layers and organic matter–rich soils. Changes in invertebrate communities may also occur during the transition phase, including an increase in the relative abundance of arthropod predators (Schowalter 1989). There is some indication that some species of fungi, lichens, and bryophytes are also more abundant in old-growth forests compared with earlier phases, although quantitative studies are limited (FEMAT 1993).

Little is known about general differences in vertebrate and invertebrate species between the transition phase and the shifting mosaic phase. Differences in animal communities might be expected if vegetative structure and composition of the shifting mosaic phase differ from the transition phase. For example, if large live and standing-dead Douglas-firs drop out of a stand over a long period of time and are replaced by western hemlocks, which have different canopy architectures and snag decay rates, then differences in species that use canopies and standing dead wood might be expected.

Ecosystem Functions

Following a major disturbance, overall ecosystem biomass and net primary productivity (NPP) are low but increasing rapidly (Peet 1992) (Figure 2.11). During early stages when mineral soil layers are exposed, sediment losses from erosion may be relatively high on steep slopes with high rainfall or where soil particles are fine and winds are high. As vegetation covers the soil surface, these outputs decline rapidly. Soil water and nutrient outputs typically increase during this period and remain relatively high until roots recolonize the belowground environment and leaf area regenerates toward previous levels. On steep slopes that are prone to mass movements because of unstable soils and high precipitation, the loss of the tree-root network may result in mass movements (Swanson and Dyrness 1975).

During the thinning phase, NPP and the rate of carbon sequestration may be very high. However, live biomass, total live and dead biomass, and storage of

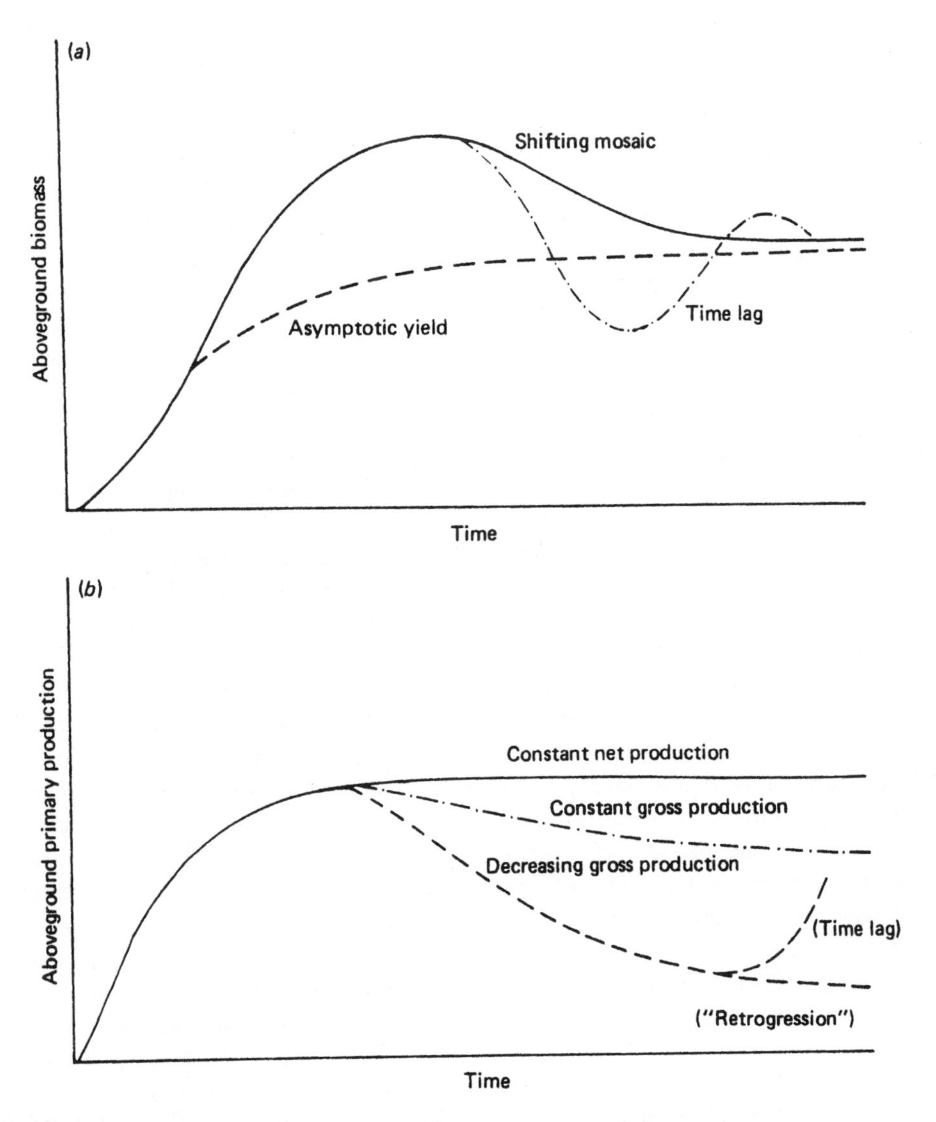

Figure 2.11 Idealized changes in biomass and net plant yield in relation to time. Curves represent different conceptual models reflecting variability in processes and environment (from Peet 1992 modified from Peet 1981).

carbon are not yet close to their maximum (Harmon et al. 1990). Nutrient losses from stands in this phase are expected to be low because of the rapidly aggrading biomass and near full occupancy of sites by trees (Vitousek and Reiners 1975, Vitousek 1977). Leaf area is thought to peak during this stage (Perry 1994), although few field studies have been made. Herbivory and mortality from pathogens are common in suppressed, weakened trees and may spread to more dominant trees through root grafts. Moreover, dense stands with relatively narrow-diameter tall trees may be more susceptible to breakage from wind and ice.

Live biomass typically peaks during the early part of the transition phase and then may level off or decline as mortality breaks up the dominant canopy of trees and slowly replaces them with tree species with smaller stature (Peet 1981). Carbon storage in live and dead components may also peak during the transition phase (Harmon et al. 1990). Nutrient retention is thought to decline as stands develop from the thinning phase into the transition and shifting mosaic phase (Vitousek 1977) where gaps are more persistent and available nutrients may be exported from the ecosystem following gap disturbances. However, nutrient retention in early transition old-growth Douglas-fir–western hemlock stands appears to be high (Sollins et al. 1980), and sediment outputs from watersheds dominated by transition phase forests appear to be low (Swanson et al. 1982). Tall, deep canopies of transition old growth may intercept more moisture from clouds than the establishment phase (Harr 1982), which lacks deep, well-developed canopies.

Live stand biomass may be lower in the shifting mosaic phase than the transition phase as the proportion of gaps or areas of the establishment phase increases in the stand (Peet 1992, Shugart 1984). Biomass may also be lower if species composition shifts toward lower-stature shade-tolerant species such as occurs in the shift from Douglas-fir to western hemlock. Theoretically, net primary productivity is close to zero because mortality and growth are roughly equal. Nutrient retention is hypothesized to be relatively low during the shifting mosaic phase because biomass and nutrient stores are near equilibrium and therefore inputs roughly equal outputs (Vitousek and Reiners 1975).

Biomass and productivity do not always increase with stand development. In boreal ecosystems, increasing accumulations of organic matter in the forest floor and soil can lead to shallower permafrost and declines in soil productivity and biomass in the latter stages of stand development (Kimmins 1987). In these ecosystems, site productivity typically declines with stand age and is enhanced through disturbances that remove or mix the surface organic matter (Spies et al. 1991).

Time as a Factor in Forest Compositional Development

The general model of stand development described above views stand development from the perspective of changes in structure, composition, and ecosystem processes associated with tree population dynamics. In this perspective, stands are a dynamic structural template or filter that allows some species to establish and grow while preventing or inhibiting others. Thus, changing structural and environmental templates can result in a changing species composition (Spies 1991, McCune 1993). For many species, however, especially plant species, the most important factor associated with their establishment and growth may simply be the amount of time since the last disturbance that removed the species from the site (Matlack 1994). Populations of these species are typically limited in landscapes and stands more by dispersal than particular stand structures. For example, some understory and epiphytic plant species with limited dispersal capabilities may move across landscapes at rates measured in meters per year (Mesler and Lu 1983). Where disturbance has removed these species from a stand, many decades or centuries may be required for enough propagules to land and successfully establish new populations (Duffy and Meier 1992, Peterkin and Game 1984). While some of these species may be limited by or favored by particular stand structures (such as nurse logs for tree seedlings or particular branch characteristics for epiphytes), many simply may be favored by particular combinations of light, moisture, and nutrients that occur through a variety of structural-environmental pathways. For example, shady moist microsites with high levels of soil organic matter, common in old-growth forests, may be found in very young stands. Similarly, open, relatively exposed areas may be found within the mosaic of vegetation of the shifting gap phase. Some species may function independently of stand development, but may be responding to the same factor as stand development—time.

Perspectives on Old Growth

Recently, considerable attention has been focused on the ecology, conservation, and management of old-growth forests. Old growth has been described variably in terms of stand structures (Franklin et al. 1981), stand development processes (Oliver and Larson 1990), and a combination of perspectives, including genetic, population, ecosystem, and landscape levels (Spies and Franklin 1995) as well as aesthetic and timber management perspectives. The variety of old-growth definitions is symptomatic of the complexity of forest stand development and the diversity of scientific and societal views of forests. In this chapter, I use the most encompassing simple definition: an ecosystem distinguished by the presence of populations of old trees that is not necessarily in a late-successional condition or free from evidence of human activity. In this definition, "a population of old trees" means the presence of at least several trees that are close to their maximum age for the particular site and climatic conditions and exhibit characteristics of old trees, such as crown and bole senescence, relatively slow growth, and relatively large size for site, species, and climate.

Using this age-based definition, old growth occurs in the last two major phases of stand development (transition and shifting gap). The onset of old growth during the transition phase depends on the species and site conditions; it can occur in as short a span as 50 years or after 500 or more years in forest types of North America. More-detailed, structure-based definitions for specific forest types have been based on tree size, tree composition, amount and size of dead trees, and vertical and horizontal heterogeneity (Old-Growth Definition Task Group). Process-based definitions (Oliver and Larson 1990) have been based on population dynamics of tree cohorts. Both perspectives have advantages and disadvantages. Structure-based definitions are particularly useful for inventory, wildlife habitat, recreation, and timber management. In these situations it is essential to have objectives described in terms of relatively easily measured objects such as the size and species of live trees and dead wood, and vegetation distribution. However, structure-based definitions do not provide information on the processes that created the structures; the same structure may be achieved through different developmental pathways. Furthermore, because of variation across forest types and landscapes, it is difficult to extrapolate structural definitions from one forest type to another without detailed sampling. In most areas of the world, detailed knowledge of old-growth conditions does not exist due to the lack of older forests. In addition, the old growth that does exist in a landscape may have developed from a particular disturbance history under a climate that no longer exists.

Process-based definitions have the appeal of focusing on the way forests develop rather than the way they appear at particular stages. This perspective is valuable in cases where old forest conditions do not currently exist and the objective is to have them in the future. It is also valuable in situations in which management is directed toward maintaining old growth across a landscape as a whole, including maintaining processes within current old-growth stands and insuring that future ones will develop. Process-based definitions also provide a stronger conceptual link with the idea that vegetation is a dynamic entity. On the other hand, it is difficult to incorporate processes into inventory or management based on habitat-suitability models. A particular process, such as the population dynamics of canopy trees or gap-phase reproduction, can lead to very different structures and ecological and management conditions across a diverse forest environment.

Implications for Forest Management

Forest management is determined by particular human objectives for stands and landscapes. Where these objectives are determined by short-term economic considerations—as they frequently are—the range of stand conditions and silvicultural practices is limited at stand and landscape scales (Davis and Johnson 1987). This has often lead to development of increasingly intensive plantation management practices. Such practices tend to limit the diversity of stand development phases and greatly shorten the time between logging disturbances relative to the potential range of natural structures and disturbance regimes. In many cases, stand development is stopped by clearcut logging somewhere near the end of the thinning phase when tree growth rates begin to level off (Oliver and Larson 1990). In addition, the establishment phase may be shortened by planting trees at high densities and controlling or removing competing vegetation to maximize resources capture by commercial tree species. Precommercial and commercial thinnings also may be applied to control species composition and tree size and obtain intermediate forest products. This type of intensive forest management will tend to produce stands that are structurally and compositionally homogeneous. However, some degree of heterogeneity often develops through patchy mortality of planted seedlings and unwanted and unexpected physical and biotic disturbances such as wind, insects, and disease. Nevertheless, stands that result from intensive timber management are not representative of the full potential range of stand development and variability of a landscape. Moreover, they may not even be close to conditions that occurred prior to forest-use practices of modern industrial economies, especially where plantations of introduced species are used. The implications of differences between intensive forest plantations and more natural stands only can be assessed in terms of a particular forest ecosystem, the management objectives for that ecosystem, and the context of the larger landscape.

Where areas are managed primarily for wood fiber production, differences between managed stands and their wilder predecessors may not be of interest. However, if basic ecosystem processes that maintain long-term site and ecosystem productivity (such as soil organic matter development and disease and insect dynamics) are altered through site-level or landscape-level changes in stand structure and composition, then intensive forest plantation practices may not lead to desirable outcomes even in terms of wood fiber production.

In situations where forest management objectives include maintenance of a broader array of ecological and human values, current intensive forest plantation practices will not meet those objectives. Species, processes, and values associated with the transition and shifting gap phases of stand development will be inhibited or absent within stands managed to maximize wood fiber production. Species and processes associated with the establishment phase may occupy sites for shorter periods of time before canopy closure, although the frequency of this phase of forest development will be greater with short rotations.

This does not necessarily mean that intensive forest management practices are incompatible with multiple forest objectives at a landscape scale. Maintenance of species and ecosystems associated with early and later forest developmental phases can only be assessed over large areas such as landscapes, subregions, and regions. All species are adapted to disturbance and environmental heterogeneity, and all have evolved life history traits that increase population survival in dynamic landscapes. However, some are better adapted than others. Hence the maintenance of many species and processes associated with later phases of forest development is probably a nonlinear function of the amount and distribution of that habitat across the landscape. This means that populations of some species require landscapes that have relatively large proportions or absolute areas of later developmental phases. For example, long-term maintenance of northern spotted owl populations in the Pacific Northwest is expected to require many large forest areas at least 20,000 ha in size that are dominated by multistoried old-growth Douglas-fir–western hemlock forests (Thomas et al. 1990). The overall landscape that spotted owls and other species live in will contain early successional stages and intensively managed stands as well as areas of human development. In landscapes where the maintenance of many forest values is the objective, the challenge to management is to determine the mix and pattern of various stand conditions and disturbance regimes that will meet the diversity of objectives.

It may be possible to modify silvicultural practices at the stand level to meet multiple objectives at landscape and larger scales. For example, thinning in young plantations may decrease the time required to develop large trees and multiple-canopy forests for late-phase species such as the spotted owl. This practice may produce suitable owl habitat in shorter time frames as well as produce some economic benefits from thinnings and mid-rotations instead of very long rotations. It may not, however, produce desired amounts of dead wood or heavy-limbed tree crowns. In addition, for species that are limited by dispersal rather than by particular habitat structures, accelerating the rate of stand development may not help if logging disturbances are too frequent and severe to allow populations of those species to reestablish between major disturbances. These species may benefit from practices such as tree and patch retention designed to reduce mortality through maintenance of refugia. Similarly, allowing for areas of low-severity disturbances within a stand at the time of logging and lengthening rotations are compatible with slow recolonization rates. For some dispersal-limited species, it may be possible to actively move propagules around the landscape. However, this may be practical for only a few species such as some of the canopy lichens.

Given our imperfect knowledge of forest stand structure and function and poor understanding of forest management effects on biodiversity and long-term ecosystem function, it is uncertain how well we can sustain ecosystem values while providing commodity resources. Prudence calls for an adaptive ecosystematic approach that spreads environmental risks across management strategies ranging from reserves to restoration to intensive plantations. The challenge to management is to find the mix of stand and landscape practices that meet biological and social objectives.

Literature Cited

Alaback, P. B. 1982. Dynamics of understory biomass in Sitka spruce–western hemlock forests of southeastern Alaska. *Ecology* 63:1932–1948.

Aplet, G. H., N. Johnson, J. T. Olson, and V. A. Sample, eds. 1993. *Defining sustainable forestry.* Washington, DC: Island Press.

Bazzaz, F. A. 1991. Habitat selection in plants. *American Naturalist* 137:116–130.

Bolsinger, C. L., and K. L. Waddell. 1993. Area of old-growth forests in California, Oregon, and Washington. USDA Forest Service. PNW Station Resource Bulletin PNW-RB-197.

Bormann, F. H., and G. E. Likens. 1979. *Pattern and process in a forested ecosystem.* New York: Springer-Verlag.

Bradshaw, G. A., and T. A. Spies. 1992. Characterizing canopy gap structure in forests using wavelet analysis. *Journal of Ecology* 80:205–215.

Brown, E. R., ed. 1985. *Management of wildlife and fish habitats in forests of western Oregon and Washington.* Portland, OR: USDA Forest Service, Pacific Northwest Region.

Canham, C. D., J. S. Denslow, J. S. Platt, W. J. Runkle, T. A. Spies, and P. S. White. 1990. Light regimes beneath closed canopies and tree-fall gaps in temperate and tropical forests. *Canadian Journal of Forest Research* 20:620–631.

Cattelino, P. J., I. R. Noble, R. O. Slayter, and S. R. Kessell. 1979. Predicting the multiple pathways of plant succession. *Environmental Management* 3:41–50.

Chen, J., J. F. Franklin, and T. A. Spies. 1992. Vegetation responses to edge environments in old-growth Douglas-fir forests. *Ecological Applications* 2:387–396.

Clements, F. E. 1916. *Plant succession: An analysis of the development of vegetation.* Publication 242. Washington, DC: Carnegie Institute of Washington.

Cohen, W. B., and T. A. Spies. 1992. Estimating structural attributes of Douglas-fir–western hemlock forest stands from Landsat and spot imagery. *Remote Sensing of Environment* 41:1–17.

Cohen, W. B., T. A. Spies, and G. A. Bradshaw. 1990. Semivariograms of digital imagery for analysis of conifer canopy structure. *Remote Sensing of Environment* 34: 167–178.

Daniel, T. W., J. A. Helms, and F. S. Baker. 1979. *Principles of silviculture.* 2nd ed. New York: McGraw-Hill.

Davis, L. S., and K. N. Johnson. 1987. *Forest management.* 3rd ed. New York: McGraw-Hill.

Duffy, D. C., and A. J. Meier. 1992. Do Appalachian herbaceous understories ever recover from clearcutting? *Conservation Biology* 6:196–201.

Easter, M. J., and T. A. Spies. 1994. Using hemispherical photography for estimating photosynthetic photon flux density under canopies and gaps in Douglas-fir forests of the Pacific Northwest. *Canadian Journal of Forest Research* 24:2050–2058.

Ellenberg, H. 1978. *Vegetation Mitteleuropas mit den Alpen.* Stuttgart, Germany: Verlag Eugen Ulmer.

Elliot, K. J., and W. T. Swank. 1994. Impacts of drought on tree mortality and growth in a mixed hardwood forest. *Journal of Vegetation Science* 5(2): 229–236.

Esseen, P. A., B. Ehnstrom, L. Ericson, and K. Sjoberg. 1992. Boreal forests—the focal habitats of Fennoscania. In *Ecological principles of nature conservation: Applications in temperate and boreal environments,* ed. L. Hansson. London, England: Elsevier.

FEMAT. 1993. *Forest ecosystem management: An ecological, economic, and social assessment.* Report of the Forest Ecosystem Management Team (FEMAT). 1993-793-071. Washington, DC: GPO.

Foster, D. 1993. Land-use history and forest transformations in central New England. In *Humans as components of ecosystems: The ecology of subtle human effects and populated areas,* ed. M. J. McDonnel and S. T. A. Pickett. New York: Springer-Verlag.

Franklin, J. F. 1989. Toward a new forestry. *American Forestry* 11:37–44.

Franklin, J. F., and M. A. Hemstrom. 1981. Aspects of succession in coniferous forests of the Pacific Northwest. In *Forest succession: Concepts and application,* ed. D. C. West, H. H. Shugart, and D. B. Botkin. New York: Springer-Verlag.

Franklin, J. F., and T. A. Spies. 1991. Composition, function, and structure of old-growth Douglas-fir forests." In *Wildlife and vegetation of unmanaged Douglas-fir forests,* tech. coords. L. F. Ruggiero, K. B. Aubry, A. B. Carey, and M. H. Huff. General technical report PNW-GTR-285. Portland, OR: USDA Forest Service.

Franklin, J. F., K. Cromack Jr., W. Denison, A. McKee, C. Maser, J. Sedell, F. Swanson, and G. Juday. 1981. *Ecological characteristics of old-growth Douglas-fir forests.* General technical report PNW-118. Portland, OR: USDA Forest Service.

Gleick, J. 1987. *Chaos.* New York: Viking.

Glenn-Lewin, D. C., and E. van der Maarel. 1992. *Patterns and processes of vegetation dynamics.* New York: Chapman and Hall.

Glenn-Lewin, D. C., R. K. Peet, and T. T. Veblen, eds. 1992. *Plant succession: Theory and prediction.* New York: Chapman and Hall.

Graham, S. A., R. P. Harrison Jr., and C. E. Westell Jr. 1963. *Aspens: Phoenix trees of the Great Lakes region.* Ann Arbor: University of Michigan Press.

Gray, A. N., and T. A. Spies. In press. Gap size, within-gap position, and canopy structure effects on conifer seedling establishment in forest canopy gaps. *Journal of Ecology.*

Grier, C. C., and R. S. Logan. 1977. Old-growth *Pseudotsuga menziesii* communities of a western Oregon watershed: Biomass distribution and production budgets. *Ecology Monographs* 47 (4):373–400.

Halpern, C. B. 1989. Early successional patterns of forest species: Interactions of life history traits and disturbance. *Ecology* 70:704–720.

Hanley, T. A. 1984. *Relationships between Sitka black-tailed deer and their habitat.* General Technical Report PNW 168. Washington, DC: USDA Forest Service.

Harmon, M. E., J. F. Franklin, F. J. Swanson, P. Sollins, S. V. Gregory, J. D. Lattin, N. H. Anderson, S. P. Cline, N. G. Aumen, J. R. Sedell, G. W. Lienkaemper, K. Cromack, Jr., and K. W. Cummins. 1986. Ecology of coarse woody debris in temperate ecosystems. In *Advances in ecological research,* ed. A. MacFadyen and E. D. Ford. Orlando, FL: Academic Press.

Harmon, M. E., W. K. Ferrell, and J. F. Franklin. 1990. Effects on carbon storage of conversion of old-growth forests to young forests. *Science* 247:699–702.

Harr, R. D. 1982. Fog drip in the Bull Run municipal watershed, Oregon. *Water Resources Research* 18:785–789.

Hemstrom, M. A., and S. E. Logan. 1986. *Plant association and management guide, Siuslaw National Forest.* R6-Ecol 220 1986a. Portland, OR: Pacific Northwest Region, USDA Forest Service.

Kimmins, J. P. 1987. *Forest ecology.* New York: Macmillan Publishing Co.

Lesica, P., B. McCune, S. V. Cooper, and W. S. Hong. 1991. Differences in lichen and bryophyte communities between old-growth and managed second-growth forests in the Swan Valley, Montana. *Canadian Journal of Botany* 69:1745–1755.

Matlack, G. R. 1994. Plant species migration in a mixed-history forest landscape in eastern North America. *Ecology* 75 (5):1491–1502.

McCune, B. 1993. Gradients in epiphyte biomass in three Pseudostuga-Tsuga forests of different ages in western Oregon and Washington. *The Bryologist* 96 (3):405–411.

McCune, B., and T. F. H. Allen. 1985. Will similar forests develop on similar sites? *Canadian Journal of Botany* 63:367–376.

Mesler, M. R., and K. L. Lu. 1983. Seed dispersal of *Trillium ovatum (Lilliaceae)* in second-growth redwood forests. *American Journal of Botany* 70(10):1460–1467.

Mooney, H.A., and M. Godron, eds. 1983. *Disturbance and ecosystems: Components of response.* Berlin, Germany: Springer-Verlag.

Morrison, P. H., and F. J. Swanson. 1990. *Fire history in two forest ecosystems of the central western Cascade Range, Oregon.* General technical report PNW-GTR-254. Portland, OR: USDA Forest Service.

Nelson, R., et al. 1987. Determining the rate of forest conversion in Mato Grosso, Brazil, using Landsat MSS and AVHRR data. *International Journal of Remote Sensing* 8:1767–1784.

Old-Growth Definition Task Group. 1986. *Interim definitions for old-growth Douglas-fir and mixed-conifer forests in the Pacific Northwest and California.* Research note PNW-447. Portland, OR: USDA Forest Service.

Oliver, C. D. 1981. Forest development in North America following major disturbances. *Forest Ecology Management* 3:153–168.

Oliver, C. D., and B. C. Larson. 1990. *Forests stand dynamics.* New York: McGraw-Hill.

Peet, R. K. 1981. Changes in biomass and production during secondary forest succession. In *Forest succession:Concepts and applications.* NewYork: Springer-Verlag.

———. 1992. Community structure and ecosystem function. In *Plant Succession: Theory and Prediction,* ed. D. C. Glenn-Lewin, R. K. Peet, and T. T. Veblen. New York: Chapman and Hall.

Peet, R. K., and N. L. Christensen. 1980. Succession: A Population process. *Vegetatio* 43:131–40.

———. 1987. Competition and tree death. *BioScience* 37:586–595.

Perlin, J. 1989. *A forest journey: The Role of wood in the development of civilization.* Cambridge, MA: Harvard University Press.

Perry, D. A. 1994. *Forest ecosystems.* Baltimore: Johns Hopkins University Press.

Peterkin, G. F. 1993. *Woodland conservation and management.* 2nd. ed. London, England: Chapman and Hall.

Peterkin, G. F., and M. Game. 1984. Historical factors affecting the number and distribution of vascular plant species in the woodlands of central Lincolnshire. *Journal of Ecology* 72:155–182.

Peterkin, G. F., and E. W. Jones. 1987. Forty years of change in Lady Park Wood: The old-growth stands. *Journal of Ecology* 75:477–512.

Pickett, S. T. A., and P. S. White, eds. 1985. *The ecology of natural disturbances and patch dynamics.* Orlando, FL: Academic Press.

Pickett, S. T. A., J. Kolasa, and C. G. Jones. 1994. *Ecological understanding.* Orlando, FL: Academic Press.

Postel, S., and J. C. Ryan. 1991. Reforming forestry. In *State of the world,* ed. L. R. Brown. NewYork: W. W. Norton.

Probst, J. R. 1987. Kirkland's warbler breeding biology and habitat management. In *Integrating forest management for wildlife and fish,* ed. W. Hoekstra and J. Capp. Paper NC-122. Rhinelander, WI: USDA Forest Service. North Central Forest Experiment Station.

Ruggiero, L. F., L. C. Jones, and K. B. Aubry. 1991. Plant and animal habitat associations in Douglas-fir forests of the Pacific Northwest: An overview. In *Wildlife and vegetation of unmanaged Douglas-fir forests,* tech. coord. L. F. Ruggiero, K. B. Aubry, A. B. Carey, and M. H. Huff. General technical report PNW-GTR-285. Portland, OR: USDA Forest Service.

Schoonmaker, P., and A. McKee. 1988. Species composition and diversity during secondary succession of coniferous forests in the western Cascade Mountains of Oregon. *Forest Science* 34 (4):960–979.

Schowalter, T. D. 1989. Canopy arthropod community structure and herbivory in old-growth and regenerating forests in western Oregon. *Canadian Journal of Forest Research* 19:318–322.

Seymour, R. S., and M. L. Hunter. 1992. *New forestry in eastern spruce-fir forests: Principles and applications to Maine.* Maine Agriculture and Forest Experiment Station, Misc. Pub. 716. Orono: University of Maine.

Shugart, H. H. 1984. *A Theory of forest dynamics.* NewYork: Springer-Verlag.

Smith, D. M. 1962. *The Practice of silviculture.* 7th ed. New York: John Wiley & Sons.

Sollins, P., C. C. Grier, F. M. McCorrison, K. Cromack Jr., R. Fogel, and R. L. Fredriksen. 1980. The internal element cycles of an old-growth Douglas-fir ecosystem in western Oregon. *Ecological Monographs* 50 (3):261–285.

Spies, T. A. 1991. Plant species diversity and occurrence in young, mature, and old-growth Douglas-fir stands in western Washington and Oregon. In *Wildlife and vegetation of unmanaged Douglas-fir forests,* tech. coord. L. F. Ruggiero, K. B. Aubry, A. B. Carey, and M. H. Huff. General technical report PNW-GTR-285. Portland, OR: USDA Forest Service.

Spies, T. A., and B. V. Barnes. 1982. Natural hybridization between *Populus alba* and the native aspens in southeastern Michigan. *Canadian Journal of Forest Research* 12: 653–660.

Spies, T. A., and J. F. Franklin. 1991. The structure of natural young, mature, and old-growth forests in Washington and Oregon. In *Wildlife and vegetation of unmanaged Douglas-fir forests,* tech. coord. L. F. Ruggiero, K. B. Aubry, A. B. Carey, and M. H. Huff. General technical report PNW-GTR-285. Portland, OR: USDA Forest Service.

———. 1995. The diversity of maintenance of old-growth forests. In *Biodiversity in Managed Landscapes,* ed. R. C. Szaro and D. W. Johnson. New York: Oxford University Press.

Spies, T. A., J. F. Franklin, and T. B. Thomas. 1988. Coarse woody debris in Douglas-fir forests of western Oregon and Washington. *Ecology* 69 (6):1689–1702.

Spies, T. A., J. F. Franklin, and M. Klopsch. 1990. Canopy gaps in Douglas-fir forests of the Cascade Mountains. *Canadian Journal of Forest Research* 20:649–658.

Spies, T. A., J. Tappeiner, J. Pojar, and D. Coates. 1991. Trends in ecosystem management at the stand level. *Transaction of North American Wildlife and Natural Resources Conference* 56:628–639.

Spies, T. A., W. J. Ripple, and G. A. Bradshaw. 1994. Dynamics and pattern of a managed coniferous forest landscape in Oregon. *Ecological Applications* 4(3):555–568.

Swanson, F. J., and C. T. Dyrness. 1975. Impact of clear-cutting and road construction on soil erosion by landslides in the western Cascade Range, Oregon. *Geology* 3:393–396.

Swanson, F. J., R. L. Fredricksen, and F. M. McCorrison. 1982. Material transfer in a western Oregon forested watershed. In *Analysis of coniferous forest ecosystems in the western United States,* ed. R. L. Edmonds. Stroudsburg, PA: Hutchinson Ross Publishing Co.

Thomas, J. W., ed. 1979. *Wildlife habitats in managed forests: The Blue Mountains of Oregon and Washington.* Agriculture handbook 553. Washington, DC: USDA Forest Service.

Thomas, J. W., E. D. Forsman, J. B. Lint, E. C. Meslow, B. R. Noon, and J. Verner. 1990. *A conservation strategy for the northern spotted owl.* Portland, OR: USDA Forest Service and USDI National Park Service.

Tiegerstedt, P. M. A., D. Rudin, T. Niemela, and J. Tammisola. 1982. Competition and neighboring effect in a naturally regenerating population of Scots pine. *Silvae Fennica* 16:122–129.

Tilman, D., R. M. May, C. L. Lehman, and M. A. Nowak. 1994. Habitat destruction and the extinction debt. *Nature* 371:65–66.

USDA Forest Service. 1989. Generic definition and description of old-growth forests. Report on file at PNW Research Station, Forestry Sciences Laboratory, Corvallis, OR.

Veblen, T. T. 1992. Regeneration dynamics. In *Plant succession: Theory and prediction,* ed. D. C. Glenn-Lewin, R. K. Peet, and T. T. Veblen. New York: Chapman and Hall.

Vitousek, P. M. 1977. The regulation of element concentrations in mountain streams in the northeastern United States. *Ecology Monographs* 47:65– 87.

Vitousek, P. M., and W. A. Reiners. 1975. Ecosystem succession and nutrient retention: A hypothesis. *BioScience* 25 (6):376–381.

Waring, R. H., and W. H. Schlesinger. 1985. *Forest ecosystems.* Orlando, FL: Academic Press.

Watt, A. S. 1947. Pattern and process in the plant community. *Journal of Ecology* 35:1–22.

Whitney, G. G., and D. R. Foster. 1988. Overstory composition and age as determinants of the understory flora of woods of central New England. *Journal of Ecology* 76:867–876.

Whittaker, R. H. 1953. A consideration of climax theory: The climax as a population and pattern. *Ecology Monographs* 23:41–78.

———. 1962. Classification of natural communities. *Botanical Review* 28(1):1–239.

3 Disturbance, Recovery, and Stability

David A. Perry and Michael P. Amaranthus

Nature to be commanded, must be obeyed.
—*Francis Bacon*

A New View of Nature: Dynamic, Interconnected, Robust Yet Fragile 32
Complexity, Self-Organization, and Thresholds: The View from the Cliff 34
New Concepts of Disturbance 35
Global, Regional, and Landscape Processes: No Ecosystem Is an Island 35
The Influence of Regional and Landscape Patterns on Disturbances 37
Mechanisms of Resistance and Recovery 39
Stability of What? 39
Two Aspects of Stability: Resistance and Resilience 40
Species Diversity and Ecosystem Function 40
Keystones 42
Threads of Continuity: Biological Legacies and Guilds 43
When Recovery Mechanisms Break Down: Threshold Transitions 46
Diversity and Stability: What Can We Conclude? 48
Management Implications 49
Literature Cited 50

Disturbance and the set of reactions that it engenders are common in nature, and the character of ecosystems results in large part from an interplay between these two forces of change. Resource managers have much to gain from understanding historic disturbance patterns, the ecological factors that act to contain and control disturbances, and the mechanisms that confer resilience (the ability to recover following disturbance). Forestry, like any renewable resource management, involves disturbing natural systems to one degree or another and has as a central objective to do so in a sustainable way. For the ecologist, by far the most reliable strategy for achieving sustainability is to understand the historic forces that have shaped ecosystems and to work within the mechanisms by which ecosystems sustain themselves.

Although most people have an intuitive sense of what constitutes a disturbance, precise definition can be elusive. Forman (1987) defines disturbances as "events that cause a significant change in the existing pattern in a system" (see also Pickett and McDonnell 1989). For foresters and plant ecologists, this often translates to stress or death of the dominant plants in numbers great enough to affect system functioning. There is good ecological rationale for this view: The dominant plants (in forests, trees) fix carbon, cycle water, build and stabilize soils, ameliorate climate, and provide the basic habitat structure for other organisms (Jones et al. 1994). However, growing knowledge of functional interdependencies in nature makes it clear that plants are not the only points of vulnerability. Ecological systems can also be disrupted by factors such as loss of large and small predators, reduced populations of mycorrhizal fungi, and spread of exotics, to name a few. Moreover, while disturbances are often thought of as dramatic events such as crown fires or hurricanes, they may also be more subtle and slower acting, as with soil compaction or the erosion of soil fertility resulting from acid rain or overly short rotations. Some changes in the existing pattern of a system may have little or no adverse impact on immediate system functioning, but increase vulnerability to future disturbances; genetic homogenization is one example of what might be termed an accident waiting to happen (Gibson et al. 1982), but it is far from the only example. Any simplification of a natural system risks eroding the system's capacity to resist and recover from disturbances, though few, if any, ecosystems are well enough understood to quantify that risk.

Recovery is another seemingly simple concept that becomes more complex with further examination. Succession is the most obvious and well-known manifestation of system recovery following disturbance. Less obvious but equally important are the many small recoveries—the controls that keep small disturbances from turning into large ones, as when particular species of mycorrhizal fungi enable their hosts to recover quickly from drought (Parke et al. 1983a) or when predators limit a growing pest population before it becomes an epidemic (e.g., Mason et al. 1983). The upshot is that, like disturbances, recoveries occur at many scales, from the small ones we never notice to the large ones that get our attention and that often come into play only when the small ones have failed. As we discuss later, these resistance and resilience mechanisms are frequently based in biological diversity.

This chapter explores these issues in more detail. We begin with the revolution in the scientific view of the structure and dynamics of complex systems (of which ecosystems are the prime example). We then move beyond these general considerations to examine more specifically what has been learned about the dynamics of disturbance and recovery in ecosystems. We close with a brief discussion of management implications.

A New View of Nature: Dynamic, Interconnected, Robust Yet Fragile

The 1980s saw the rise of what some have called the new paradigm in ecology (Pickett et al. 1992) and what we refer to as the "dynamic view." Before, the focus of community ecology was characterized by two things in particular (Pickett et al. 1992): (1) an equilibrium view, where "equilibrium" (as used here) refers to constancy in the species composition of a community (while disturbance and succession clearly alter communities, they are subordinate in the equilibrium view to the major theme of nature, which is the relatively unchanging climax community) and (2) the belief that ecosystems were self-contained, that their trajectories were determined solely by internal interactions or, put another way, that you could understand the dynamics of an ecosystem without ever looking outside its boundaries. Since the late 1970s, ecologists have increasingly focused on community and ecosystem dynamics rather than static endpoints such as the climax (DeAngelis and Waterhouse 1987). Disturbance and response to disturbance are now recognized as natural processes that lay at the core of ecosystem dynamics, a concept that plant ecologists have variously expressed as "patch dynamics" (Pickett and Thompson 1978), "the shifting landscape mosaic" (Bormann and Likens 1979), and "the mosaic cycle" (Remmert 1991). This is also known as the nonequilibrium view (DeAngelis and Waterhouse 1987), reflecting the increasing recognition that disturbances and other factors mitigate against ecological communities attaining a lasting equilibrium in species composition (though equilibria may still be attained at landscape and regional scales).

Because disturbances limit the ability of one or a few species to dominate sites, they limit the importance of competition in structuring communities (DeAngelis and Waterhouse 1987). On the other hand, the need to maintain continuity in key processes during and in the recovery stages following disturbance is hypothesized to increase the prevalence of mutually supportive interactions among species (Perry et al. 1989a, 1989b, 1992, Tilman and Downing 1994, Perry 1995a). The species diversity of a given region is strongly influenced by the interplay among disturbance (which simultaneously destroys old habitats and creates new ones), recovery (which reverses that process), and biological legacies (which act to maintain continuity of certain habitats through disturbance). Bunnell (1995) documented the relationship between disturbance history and vertebrate species composition within different forest zones in British Columbia. There, forest types characterized by small wildfires contain a higher proportion of vertebrates that breed in late-successional stages and a lower proportion that breed in early stages than forest types with a history of large fires. In British Columbia, the proportion of birds and mammals within a given forest type that use cavities drops sharply as one goes from forests typified historically by small fires to forests where large fires have been more common. However, the proportion of vertebrates that use coarse woody debris (CWD) follows the opposite trend: CWD generated by fire constitutes a biological legacy that provides habitat. In all likelihood, Bunnell's findings can be generalized beyond British Columbia.

Species diversity is generally enhanced by disturbances that occur at intermediate levels of frequency and intensity—those that do not exceed the capacity of the system to recover fully between disturbances (Petraitis et al. 1989). Diversity is reduced by disturbances to which species composing the system in question are not adapted, and also by disturbances that destroy habitats faster than they can recover (as when stands are harvested before they attain old-growth characteristics). If disturbances disrupt key processes, loss of system integrity and qualitative shifts in community type can result. The frequency and severity of disturbances on a given site are, in turn, strongly influenced by processes, both ecological and socioeconomic, that occur at the scale of landscapes, regions, and the globe.

Because natural disturbances perform critical functions that maintain ecosystem structure and processes, excluding them can lead to unforeseen and sometimes undesirable consequences. Many forest species persist only because of periodic disturbances. For example, without wind throw, which exposes mineral soil seedbeds, some northern forests convert to bogs. Some scientists believe that much of the rich diversity of moist tropical forests reflects historic patterns of shifting cultivation. Numerous forest types throughout the globe—including many pine, eucalypti, and dry tropical woodlands—are maintained by periodic fire. Fire suppression in the ponderosa pine forests of western North America provides a classic case study of the importance of periodic disturbance to stability of some ecosystems. Eliminating fire in these systems allowed late-successional trees (mostly true firs) to establish under the ponderosa pine, a dynamic that was accelerated by highgrading the older ponderosa pine. The subsequent change in forest structure from relatively widely spaced old-growth pine to densely stocked stands of younger pines and firs has increased stand susceptibility to crown fires, defoliating insects, and root pathogens (see various papers in Sampson and Adams 1995).

While the move to a more dynamic view of nature represents a significant advance in our ability to understand how natural systems function, it has yet to offer a coherent set of guiding principles to replace the equilibrium view. Consider the old idea of nature's balance. Some ecologists argue that the concept of balance as applied to nature is unrealistic and should be discarded (Botkin 1990, Pickett et al. 1992). This is certainly true if balance is interpreted as "no change." However, if we take balance to mean that some changes in the state of a system are consistent with maintaining species and processes, while other changes are not, then there clearly are balances in nature, and different kinds of disturbances can have quite different implications for the integrity of natural and managed systems. Angermeir and Karr (1994) define biological integrity as "a system's wholeness, including presence of all appropriate elements and occurrence of all processes at appropriate rates." But what is the whole? How robust are communities and ecosystems when faced with loss of species, alteration of processes, and changes in landscape and regional patterns? We must seek further understanding to guide sustainable resource management. The first step is to probe deeper into the nature of complex ecosystems. What are these unique entities that, though perhaps not organismic, are far more than the sum of their parts?

Complexity, Self-Organization, and Thresholds: The View from the Cliff

The shift among ecologists to a more dynamic view of nature is part of a larger revolution that has occurred over the past 25 years as scientists began to grapple with complexity in physical, biological, and social systems. Although a daunting subject, developing sustainable social-economic-ecological systems is unlikely to succeed without an appreciation of the nature of complexity (Levins and Lewontin 1985, Lubchenco et al. 1991, Costanza et al. 1993). It is well beyond our scope to review the subject in detail, but we summarize the basic characteristics of what scientists call complex systems:

1. Complex systems such as ecosystems are dissipative, they are maintained far from thermodynamic equilibrium by large throughputs of energy and (usually) matter (Odum 1992). Interactions among system parts include positive feedback, or what physical scientists call autocatalysis (literally, "making oneself") (Prigogine and Stengers 1984, DeAngelis et al. 1986). Autocatalysis is also referred to as "bootstrapping," after the old story of the little boy who fell in the bog and pulled himself up by his bootstraps (Davies 1983, Perry et al. 1989a, 1989b).

2. Even very simple dissipative systems (i.e., those with few parts) can exhibit highly complex and often unpredictable behavior (May and Oster 1976).

3. In a complex system, small changes in driving variables can amplify through positive feedback loops and produce large changes in system behavior. This is, by definition, chaotic behavior, often exemplified by the hypothetical (and probably spurious) example of a butterfly flapping its wings in Brazil and creating a tornado in Kansas.

4. The potential for chaotic change in complex systems is constrained by two factors: (a) the buffering effect of higher order, more stable systems (O'Neill et al. 1986, Odum 1992), such as landscape patterns (which we discuss later), and (b) internal interactions (i.e., among system components) that maintain system integrity by constraining change, a dynamic referred to as self-organizing or self-reinforcing behavior. In simplest terms, a self-organizing system is one in which the structure and the processes mediated by that structure reinforce one another: through various positive and negative feedbacks the system creates conditions that increase the probability that it will persist.

5. Self-organizing systems are stable over a range of conditions, but change rapidly when their bounds of adaptability are exceeded—a dynamic O'Neill et al. (1989) call metastability. Moreover, the change is often to a new state that is itself self-reinforcing. Note that stability as used here does not mean no change: A metastable system may be quite dynamic, but its changes are maintained within certain bounds.

While this may sound like so much esoteric gibberish, the characteristics and behaviors listed above accurately describe the world we move in every day. A system "far from thermodynamic equilibrium" is one that changes significantly when its energy source is removed: organisms, ecosystems, and processes such as the hydrologic cycle, major oceanic currents, and the earth's climate. Positive feedback is common at a variety of scales in ecological systems (DeAngelis et al. 1986, Perry et al. 1989a, 1989b), including population growth, the ubiquitous mutualistic relationships among species, nutrient cycling and other food-web dynamics, landscape and regional phenomena such as disturbance spread and the hydrologic cycle, and global phenomena such as the "Atlantic conveyor belt," a major oceanic current that influences weather globally. Moreover, complex interactions between positive and negative feedbacks create both the potential for extreme sensitivity to environmental conditions (i.e., chaos) and the evolutionary impetus to dampen that potential. Bak et al. (1988) put it this way:

> [E]cological systems are organized such that the different species "support" each other in a way which cannot be understood by studying the individual constituents in isolation. The same interdependence of species also makes the ecosystem very susceptible to small changes or "noise." However, the system cannot be too sensitive since then it cannot have evolved into its present state in the first place.

While nature is dynamic, not all changes are consistent with maintaining system integrity (Pickett et al. 1992). Numerous examples can be cited of changes in system state (e.g., species composition and processes) that are quite distinct from normal successional changes; in essence, sites convert to a new community that itself may be self-reinforcing (DeAngelis et al. 1986, Perry 1995a, 1995b, Perry et al. 1989a, 1989b, Schlesinger et al. 1990, Archer and Smeins 1991, Turner et al. 1993). Perhaps the most widely known contemporary examples are the collapse of some forests due to excessive pollution (Bormann 1985) and the worldwide desertification of arid grasslands (Schlesinger et al. 1990, Archer and Smeins 1991, Milton et al. 1994), a phenomenon underlain by loss of soil integrity due to the combined effects of overgrazing and drought, exacerbated in many cases by an intensified disturbance regime due to the spread of flammable exotic grasses (D'Antonio and Vitousek 1992). Lakes and coral reefs have also exhibited catastrophic losses of species diversity and productivity (Kaufman 1989, Knowlton 1992, Hughes 1994). Both models and real world experience show that factors other than pollution can trigger forest ecosystems into threshold transitions from one state to another (Gatto and Rinaldi 1987, Pastor and Post 1988, Perry 1988, Perry et al. 1989b, Loehle 1989, Overpeck et al. 1990).

To give an example from our own work, inability to reforest high-elevation clearcuts in the western United States has been linked to complex changes in soil structure and biology that essentially reflect a breakdown of self-organizing mechanisms (Amaranthus and Perry 1987, Perry et al. 1989b, Friedman et al. 1989, Colinas et al. 1994), the same general dynamic that underlies desertification of arid grasslands. At a much larger scale, considerable paleochemical and paleoecological data show that earth's climate is subject to rapid transitions from one state to another, at least some of which result from large-scale and relatively rapid (in less than five years, perhaps as little as one or two) reorganization of the ocean–atmosphere system (Broecker 1987, Allen and Anderson 1993, Grootes et al. 1993, Bond et al. 1993, Weaver and Hughes 1994). The geological record for the upper Mississippi basin shows that modest changes in climate can trigger large changes in the magnitude of floods (Knox 1993). Kauffman and Uhl (1990) suggest that relatively small changes in regional climate would greatly increase the probability of fire in the primary forests of Amazonia.

Threshold changes in ecological systems may be rapid, but not necessarily. In what they call the extinction debt, Tilman et al. (1994) argue that habitat losses can trigger extinctions that do not manifest for decades. The distinguishing characteristic of thresholds in complex systems is that, once the self-reinforcing mechanisms have failed, changes—whether rapid or gradual—cannot be stopped without great difficulty and expense, and in some (perhaps many) cases, not even then. Scientists working in rangelands, forests, and coral reefs have concluded that simply removing the perturbing force is unlikely to allow recovery of the original community. As Knowlton (1992) aptly puts it, once the straw has broken the camel's back, simply removing the straw does not allow the camel to rise again. A central challenge of management is to stay clear of threshold boundaries, which requires understanding and working within the natural mechanisms that buffer systems against environmental fluctuations.

New Concepts of Disturbance

Global, Regional, and Landscape Processes: No Ecosystem Is an Island

Over the past 20 years, ecologists have devoted a great deal of attention to issues of scale, or the spatial and temporal domains over which natural phenomena play out (Forman and Godron 1981, Allen and Starr 1982, Allen and Hoekstra 1991, O'Neill et al. 1986, O'Neill 1989, Franklin and Forman 1987, Urban et al. 1987, Perry 1988, Turner 1987, Wiens 1989, Levin 1992, Wiens et al. 1993, Franklin 1993). Disturbances occur at many intensities and scales, from the death of a single old dominant tree to a catastrophic forest fire to volcanic eruptions and impacts by comets, all of which are embedded within climatic cycles that may span decades to millions of years. The disturbance regime of a particular forest usually consists of a complex mixture of infrequent, large-scale events (e.g., a large fire or windstorm) and more frequent, small-scale events (e.g., small fires, the fall of a single tree). Therefore, one might draw very different conclusions about the disturbance history of a given piece of ground depending on what time period is considered. In boreal forests, for example, "few major fires that occur in extreme fire years account for the vast majority of forests burned. Sixty to 80 percent of all fires in northwest Canada and Alaska are less than 5 ha in area . . . and 85 percent of all fires in Canada between 1961–1967 were 4 ha or smaller . . . yet in severe years, individual fires can cover 50,000 to 200,000 ha" (Bonan and Shugart 1989). A similar pattern of frequent small fires and infrequent larger fires characterizes many temperate and boreal forests (Johnson 1992).

Any given disturbance may be the result of numerous, interconnected factors; moreover, the disturbance regime of a particular ecosystem cannot be understood in isolation, but rather it is tied to the systems to which it is linked and ultimately to events within surrounding landscapes, the region, and the globe (Forman 1987, Pickett et al. 1992). Climate, which strongly influences a variety of disturbances, is in large measure a global phenomenon modified by regional and local patterns of topography and vegetation. Under extreme conditions, macroclimate becomes the driving force. For example, Fryer and Johnson (1988) found that large, highly destructive fires in the subalpine forests of southern Alberta tended to occur during transitions from high-pressure systems to low-pressure systems. Hot, dry weather associated with the high-pressure system dried fuels and increased the chances of ignition, and winds associated with the incoming low-pressure system drove the fire. Johnson (1991) argue that, under such conditions, fire intensity is little affected by stand structure: "Under these conditions fires can crown in all forest types. . . ."

Disturbances at any one place may be linked through atmospheric teleconnections and global oceanic currents to events thousands of miles away (Neilson et al. 1992, Gray and Landsea 1993). Perhaps the most obvious example is climate change resulting from the buildup of greenhouse gases, which is predicted to result in more severe convective storms (thunderstorms and tornadoes) at mid and high latitudes, more severe hurricanes in the tropics (Overpeck et al. 1990), and sharp drops in available water (hence maximum site leaf area index and runoff) throughout eastern North America and interior Eurasia (Neilson et al. 1990, Neilson and Marks 1994).

However, disturbance regimes are strongly influenced by global phenomena other than the buildup of greenhouse gases. One of the more significant of these is the Atlantic conveyor belt, a massive current that flows from the North Pacific through the tropics to the North Atlantic and back (Broecker 1987). Changes in the flow of this current, which may be triggered by pulses of freshwater into the North Atlantic (Rahmstorf 1994, Weaver and Hughes 1994), are believed to influence climatic phenomena throughout the globe (Gray and Landsea 1993), including rainfall in the Sahel region of Africa, hurricanes in the Atlantic Basin, the frequency and intensity of El Niños, and multidecadal changes in global surface temperatures (Gray and Landsea 1993). Hurricanes in the Caribbean and along the U.S. East Coast have been most severe during periods of higher than average rainfall in the Sahel region of western Africa, both reflecting alterations in flow of the Atlantic conveyor belt (Gray 1990).

Excepting the ice ages, perhaps the best-known example of a link between global climatic phenomena and disturbances on land is the so-called southern oscillation, the alternate warming and cooling of the southern Pacific Ocean that produces El Niños and La Niñas and ramifies to alter climate throughout the globe. During El Niño years, spring and fall are wet in the southwestern United States, hence fire activity is low. However, the region often has severe winter and spring droughts during La Niña years, and fire activity is correspondingly high (Swetnam and Betancourt 1990). The situation is quite the opposite in the U.S. Pacific Northwest, where La Niñas tend to precede wetter than normal winters, and El Niños drier than normal winters. The latter pattern also holds in Indonesia and eastern Australia: La Niña years are accompanied by abundant rainfall, while El Niño years are drier than average. Millions of hectares burned in Australia and Indonesia during the particularly strong El Niño of 1982–1983 (Swetnam and Betancourt 1990). The strong relationship between drought and susceptibility of trees to at least some species of herbivorous insects and pathogens means that the southern oscillation (or indeed any factor that produces drought cycles) can also influence outbreaks of these disturbance agents. To complicate matters, particular sequences of wet and dry years can trigger epidemics. For example, during the past 10 years about one-half of the *Octea whitei*, a common tropical tree that grows on moist slopes and ravines, have been killed on Barro Colorado Island. Gilbert et al. (1994) traces the phenomenon to a sequence of one especially wet year, which promoted an outbreak of a native pathogen that attacks vascular tissue, followed by a severe drought (due to the 1982–83 El Niño), which killed the trees because the pathogen had reduced their ability to transport water.

Longer-term cycles of climate are known to profoundly influence fire patterns. Wildfires have been most frequent in North American forests during warm, dry periods, and less frequent but more severe during cool periods (at least in some areas). Clark (1988, 1990) used layers of charcoal buried in lake sediments to determine fire severity in northwestern Minnesota over the past 750 years. Fires were most frequent during the 15th and 16th centuries, a warm, dry period, which also saw the massive fires that established many of the old-growth Douglas-fir forests that once covered most of the Oregon and Washington Cascades. Clark found that fire frequency dropped sharply with the advent of much cooler climates beginning about A.D. 1600. Climate warmed once again during the 20th century; however, fire activity remained low because of suppression. The longevity of giant sequoia trees allowed Swetnam (1993) to study fire history in the northern Sierra Nevada as recorded in fire scars over a 2,000-year span. In that area, frequent small fires burned during a warm period from A.D. 1000 to 1300, while fires during cooler periods (A.D. 500 to 1000 and after A.D. 1300) were less frequent but more severe.

The Influence of Regional and Landscape Patterns on Disturbances

The importance of global climate in influencing disturbance regimes does not minimize the critical role of regional and local patterns of topography, soils, and vegetation. Under a given set of macroclimatic conditions, regional and local factors may be decisive in determining disturbance spread. Topography has long been known to influence stand susceptibility to fire and wind. In their study of Alberta forests, Fryer and Johnson (1988) found that "fires are oriented by wind direction, wind speed, and major topographical features. Valleys upwind of or at right angles to the fire spread are often not burned. Consequently, depending on wind direction, small valleys may escape burning largely by accident, and unless passes between large valleys are aligned with the wind and fire direction, the fire may be confined to a single large valley."

Through effects on moisture and perhaps nutrients, topography and soils also affect stand susceptibility to insects. Studies in both Quebec and Michigan found that stands growing at either end of a moisture gradient were most susceptible to spruce budworm (Hix et al. 1987, Dupont et al. 1991). Fuels generated by insect or pathogen outbreaks in turn increase susceptibility to fire (Knight 1987), and these two disturbance agents acting in concert can exert significant influence over regional forest patterns, as hypothesized by Schowalter et al. (1981) for pine forests along the U.S. Atlantic coastal plain.

More recently, attention has turned to how spatial patterns of vegetation influence a variety of ecological processes, including species interactions, regional climates, and the spread of disturbances. The growing body of work on the relation between vegetation patterns and landscape and regional processes is particularly significant for managers, who may be at the mercy of global processes and local topography, but who can directly affect vegetation patterns across landscapes.

As with global phenomena, rigorously understanding the relationships between landscape patterns and processes presents a formidable scientific challenge. These issues are not easily studied experimentally in forested landscapes (though there are a few landscape-level experiments in forests, most notably the Critical Size of Ecosystems Project in Amazonia (Bierregaard et al. 1992). Nevertheless, both experience and modeling indicate that landscape and regional patterns can profoundly influence species interactions, regional climates, and the propagation of disturbance (Turner 1987, Turner et al. 1989, 1993, Knight 1987, Perry 1988, Pielke and Avissar 1990, Franchito and Rao 1992, Mylne and Rowntree 1992, O'Neill et al. 1992, Turner et al. 1993, Tilman 1994, Castello et al. 1995, Perry 1995a).

Landscapes can have self-reinforcing patterns, that is, some landscapes tend to absorb and dampen the spread of disturbances, while other patterns magnify them, with the process reinforcing the structure in both cases. In some forest types—including Douglas-fir, ponderosa pine, and moist tropical—large intact blocks of healthy, mature forests or nondecadent, old-growth forests are less susceptible to catastrophic fires than young or fragmented forests. Landscapes dominated by these types buffer and dampen the spread of crown fires and hence preserve the forest structure (Perry 1988, Franklin et al. 1989, Kauffman and Uhl 1990). Once some threshold proportion of the landscape becomes fragmented and permeated by flammable young forests or grasses, the potential exists for a self-reinforcing cycle of catastrophic fires—an absorbing landscape crosses a threshold and becomes a magnifying one. Invasion of ecosystems by flammable exotic grasses, a growing problem throughout the world, is greatly facilitated by positive feedbacks between grass cover and landscape flammability (Hughes et al. 1991, D'Antonio and Vitousek 1992). Similarly, opening intact forests reduces their ability to buffer wind (Chen et al. 1993, 1995) and creates edges susceptible to blowdown, a self-reinforcing dynamic in which blowdown along edges creates more edge to be blown down. This happened in a portion of the Mount Hood National Forest in Oregon, where clearcut patches initiated a dynamic in which severe windstorms every several years blew down progressively more forest, resulting in an increasingly fragmented landscape (Franklin and Forman 1987).

Through albedo and transpiration, vegetation significantly influences regional climates, and even relatively slight changes in forest cover can have a pronounced effect on climate (Pielke and Avissar 1990). A particularly dramatic example of how forests can affect regional climate is provided by China's "Great Green Wall," the result of a massive tree-planting program throughout northern China beginning in the 1950s. During the 1950s, Beijing experienced 10 to 20 dust storms each spring, reducing visibility to less than 1 km for 30 to 90 hours per month; by the 1970s, Beijing had fewer than 5 dust storms each spring, with visibility below 1 km on the average only 10 hours per month (Parungo et al. 1994).

The relationship between vegetation cover and regional climate can be self-reinforcing. One of the more frequently cited examples is the Amazon rain forest, which receives roughly twice as much precipitation as can be accounted for by moisture moving in from the ocean (Salati 1987). By transpiring and keeping water in circulation, the rain forest generates the rain that supports the forest. Computer models suggest that if grasslands were to replace forests as the major vegetation type in that region, the Amazon basin would become too dry to support tree growth (Lean and Warrilow 1989, Shukla et al. 1990). This raises a question: At what point would deforestation alter the climate of the entire basin sufficiently that remaining forests could not persist? Drying related to landscape conversion would create other stresses, most notably vulnerability to fire. In moist tropical forests, wildfire is a foreign disturbance to which few trees are adapted (Kauffman 1991), and extensive wildfires would almost certainly accelerate conversion of these forests to grasses, which would then perpetuate themselves through frequent burning (D'Antonio and Vitousek 1992).

Like most things in nature, the relation between landscape pattern and the spread of disturbances is complex and may vary widely depending on the disturbance under consideration (Turner 1987, Perry 1988). However, two basic principles can be identified: (1) the homogeneity of susceptible vegetation types, which creates the potential for growth and spread of disturbance, and (2) the degree to which landscape patterns provide sources of control. Not surprisingly, the spread of herbivorous insects and pathogens is greatly influenced by the available food. Both the eastern and western spruce budworms have been more aggressive during the latter half of the 20th century than they were in the 19th century, a phenomenon that entomologists attribute to fire exclusion and high-grade logging practices that inadvertently promoted the spread of tree species susceptible to these insects (Blais 1983, Anderson et al. 1987, Swetnam and Lynch 1989, Wickman et al.

1993). *Armillaria* root rot is also spreading in these forest types for the same reason (Castello et al. 1995). The spread of ash yellows, a bacterial disease of white ash, has been linked to the abundant regeneration of its host tree on abandoned farms (Castello et al. 1995). In addition to food, populations of herbivorous insects (broadly, any insect that eats plant tissues) are controlled by the combined actions of weather, plant defenses (chemical and structural), and a complex of natural enemies including vertebrates (birds and small mammals), invertebrates (e.g., spiders, ants, parasitic wasps, and parasitic flies), microbes, and viruses. It follows that landscape patterns, as expressed both in the available habitat for natural enemies and in the available refuges from natural enemies, significantly influence herbivore dynamics (Pacala et al. 1990, Comins et al. 1992, Torgerson et al. 1990, Hochberg and Hawkins 1992, Hawkins et al. 1993), a phenomenon long recognized by European foresters, who install bird boxes and cultivate habitat for small mammals. In most cases, this is a function of habitats at a landscape and regional scale, but for migratory birds, which consume large numbers of defoliating insects, the issue becomes hemispheric: Loss of habitat in the wintering grounds (for American birds, neotropical forests) weakens part of the herbivore-control complex in northern forests (Holling 1988).

Mechanisms of Resistance and Recovery

A central question facing ecologists is how complex systems maintain integrity (DeAngelis and Waterhouse 1987). For resource managers, this question translates into what types and levels of anthropogenic disturbances are most likely to maintain desired forest structure and processes in the long run. Answers to these questions, which in the final analysis are the same question, require understanding the complexity of natural dynamics. Clearly, species evolve mechanisms to persist, and ecologists increasingly recognize that such mechanisms are not solely individualistic, but involve dynamic interactions among species, other aspects of system structure (e.g., old dead wood and soils), and critical processes such as the nutrient and hydrologic cycles.

Recall that biological integrity is defined by Angermeier and Karr (1994) as "a system's wholeness, including presence of all appropriate elements and occurrence of all processes at appropriate rates." In the following discussion, we refer to the maintenance of integrity as stability, with the understanding that stability does not mean no change, but rather maintaining change within certain bounds. The issue of stability in nature is fraught with difficulty, largely because the relationship between the structure of natural systems (within and across scales) and their resistance and resilience is poorly understood. This difficulty is compounded by the fact that change at some level is common in ecosystems, including not only successional changes, but longer-term fluctuations in the relative abundance of species. No system is likely to recover to exactly the same state as existed prior to disturbance for at least two reasons: First, the longevity of trees coupled with significant fluctuations in average climatic conditions at cycles spanning decades to centuries means that in some cases the dominant species in a given locale are out of equilibrium with their current environment. Recent studies indicate this is the case in forests of southern Ontario (Campbell and McAndrews 1993). It is unknown in how many other cases this may be true. However, should climatic warming play out as predicted, the phenomenon will become more common.

Second, variation in initial conditions (e.g., type and season of disturbance, climatic factors during early recovery) may alter the initial species composition, which in turn can influence the whole successional trajectory (Drake 1990, Lawton and Brown 1992). Note that the latter case is an example of the potential for ecosystems to behave chaotically—small differences in initial conditions creating large effects. On the other hand, as we discuss later, mechanisms exist within ecosystems to constrain the chaotic potential.

Stability of What?

It is useful to distinguish three different sets to which the concept of stability can be applied (Perry 1994):

1. Species Stability: the maintenance of viable populations or metapopulations of individual species.

2. Structural Stability: the stability of various aspects of ecosystem structure, such as food-web organization, species numbers, or soils.

3. Process Stability: the stability of processes such as primary productivity and nutrient cycling.

The stability of each of the above is generally linked to the stability of the others, although the strength of the linkages is, with some exceptions, poorly understood. One case in which the link is clearly understood is in soils, whose integrity without doubt underpins the integrity of the ecosystem as a whole. Keep in mind that in all of these, but particularly the first two, stability may include considerable change (as commonly occurs, for example, during disturbance and succession). Instability results when the system crosses some threshold from which recovery to a former state either is impossible (e.g., extinction) or, if possible, occurs only over relatively long time periods or with outside subsidies of energy and matter (e.g., loss of topsoil).

Two Aspects of Stability: Resistance and Resilience

Ecosystem stability can be divided into two separate but interrelated properties: (1) resistance, or the ability of a system to absorb small perturbations and prevent them from amplifying into large disturbances, and (2) resilience or recovery, which is the capacity to return to some given system state (defined by structure and processes) following disturbance. Succession is the classic example of resilience (some of the recent thinking regarding succession is reviewed by Huston and Smith 1987, Pickett et al. 1987, and Pickett and McDonell 1989). While the state to which the stable ecosystem recovers is unlikely to be an exact replica of what existed before, it nevertheless contains the same basic elements (species richness, habitats, and soil fertility) and supports the same key processes (e.g., photosynthetic capacity and nutrient and hydrologic cycles). In other words, system integrity is maintained.

Resistance mechanisms may be thought of as those properties—of the system and of individuals—that maintain relative constancy in processes and that prevent organisms from succumbing to some stress. A common example of the latter is regulation of herbivorous insects by the combination of tree chemical defenses and natural enemies. Resilience (or recovery) mechanisms come into play when trees are weakened or killed. While succession is the most obvious manifestation of recovery, it is far from the only one. Moreover, whereas succession manifests as change in the dominant species on a site, other aspects of recovery involve constraining change, such as retaining nutrients and stabilizing mycorrhizal fungi and belowground food webs. Resistance and resilience are closely interlinked and act jointly to stabilize ecosystems. Resistance mechanisms may be thought of as filters that reduce the potential for large disturbances. If these mechanisms are weakened, the frequency of large scale disturbances increases; if disturbance frequency becomes shorter than the time required to recover, the system enters a highly vulnerable state in which threshold transitions become probable (Turner et al. 1993).

Both resistance and resilience include mechanisms that operate at the stand and landscape scales. Tree size and vigor are important stand-level components of resistance, as are stand structure and the invertebrate and microbial food webs that cycle nutrients and help control pests and pathogens. Resistance is tied to landscape patterns through factors such as the homogeneity of susceptible vegetation types, the habitat provided for natural enemies of herbivores and pathogens, and the various climatic factors influenced by vegetation. With regard to recovery, probably the most important aspect is rapidly stabilizing the soil ecosystem, including nutrients, physical structure, and food webs. However, landscape and regional patterns also contribute significantly to recovery by providing habitat refugia which serve as sources of future colonists for the recovering site (Amaranthus et al. 1994) and, again, by affecting regional climatic factors. As discussed earlier, all of these factors are embedded within and are strongly influenced by global processes.

Species Diversity and Ecosystem Function

Species diversity and ecosystem function have been a subject of much debate among ecologists. Over the past 30 years, the weight of opinion has shifted from "diverse systems are more stable in their functioning" to "diverse systems are less stable" and, most recently, back again. By Angermier and Karr's (1994) definition, loss of a single indigenous species diminishes integrity; however, without minimizing the tragic finality of any extinction, the issue of concern here is how loss of species might affect ecological functioning, including, in particular, key processes and the ability of the system to sustain remaining species. In a 1986 review, Pimm concluded that "[m]ost of the possible questions about the relationship between diversity and stability have not been asked. Those that have yield a variety of answers." Progress has been made since Pimm's 1986 review (e.g., see reviews by Pimm 1991, Vitousek and Hooper 1993, Lawton 1994), but until recently much of that has been in modeling. While ecological modeling is important to do, its value is limited unless models are verified by observation and experiments in the real world. In the latter regard, we still have a long way to go. (Pimm has a good discussion of model limitations in Chapter 16 of his 1991 book.)

Lawton (1994) lists four hypotheses concerning the relation between species diversity and ecological functioning (see also Vitousek and Hooper 1993):

1. The redundant-species hypothesis holds that most species are redundant in their ecological roles; it is only a few keystones whose presence is critical for the rest of the system (Walker 1992, Lawton and Brown 1992, Berryman 1993).

2. The rivet hypothesis takes the position that all species contribute to proper ecological function, or to use Berryman's (1993) phrase, "everything depends on everything else" (Berryman believes not). "Rivet" comes from the analogy, first used by Aldo Leopold and later by Ehrlich and Ehrlich (1981), with rivets in an airplane, each of which is presumed to be incrementally important in holding the plane together.

3. The idiosyncratic-response hypothesis holds that diversity and function are linked, but the complexity of ecosystem organization makes it impossible to predict how a given system will respond to species deletions (except in obvious cases like removing all the green plants).

4. The null hypothesis holds that species deletions or additions do not influence ecosystem function.

Although we haven't done a survey, it is probably safe to say that few, if any, ecologists would agree with the null hypothesis; the evidence for the contrary is simply too overwhelming. The idiosyncratic hypothesis certainly has an element of truth: It would be a rare field ecologist who couldn't attest to the unpredictability of nature. It does not follow, however, that valid generalities do not exist. The current debate centers largely around hypotheses 1 and 2: What level of diversity is necessary to maintain stable functioning?

Experiments on the relation between species diversity and ecosystem function have generally involved either removing a species from an ecosystem or creating systems from scratch with different species richness. Quoting once again from Pimm's (1986) review:

> Experimental studies show that few natural communities are species-deletion stable; most removals cause further species losses. The data also show wide variation in the consequences of species removals. Communities are differentially resistant. . . .

The few experiments that have been done show that primary productivity and nutrient retention are related to diversity. In Costa Rica, soils beneath diverse natural successions had higher soil nitrogen, extractable cations, and phosphorus sorption capacity than those beneath monocultures (Ewel et al. 1991). Two recent experiments (neither in forests) support what we will term the modified rivet hypothesis—which takes the middle position that, while ecosystem function (measured as the rate at which certain processes occur) may not depend on every species, it certainly depends on more than a few. Tilman and Downing (1994) found that the ability of prairie communities to sustain primary productivity during drought increased with plant species diversity up to 12 species, after which additional species had no effect. In a series of controlled environment experiments that manipulated diversity of both plants and invertebrate animals, Naeem et al. (1994, 1995) found that plant productivity and overall system carbon dioxide flux declined with declining species richness, while litter decomposition rate and soil nutrient retention varied idiosyncratically with species richness. Wood decomposition was unaffected; however, this may have been because of the short time the experiment was run (Lawton 1994).

There is considerable evidence for the modified

rivet hypothesis in forests. One clear example is the ability of early successional plants to stabilize soil structure and biology, retain nutrients, and in the case of some species, restore nitrogen (Bormann et al. 1974, 1993; Amaranthus and Perry 1989a, 1989b; Amaranthus et al. 1990; Borchers and Perry 1990; Perry et al. 1989b). There are also many examples in which species richness dampens the spread of disturbances. Some species of hardwoods are relatively nonflammable and, when admixed with conifers, help protect the latter from fire (Perry 1988). In some cases, conifers are less affected by herbivorous insects or pathogens when admixed with hardwood trees or herbaceous plants than when growing in monocultures (Zutter et al. 1987; R. Gagnon, personal communication, Morrison et al. 1988, Schowalter and Turchin 1993, Simard and Vyse 1994). Atsatt and O'Dowd (1976) drew on a number of studies to hypothesize that plant species associate in defense guilds, in which each species within the guild benefits from reduced herbivore pressure. Researchers in this field emphasize that it is not diversity per se that stabilizes plant associations, but associations of species that fill specific functional niches.

Defenses that reside in plant species or guilds of plant species fall under the heading of what are called bottom-up controls, where bottom refers to the base of the food chain. Numerous studies have shown that predators also contribute significantly to maintaining herbivore populations below outbreak thresholds, a dynamic termed top-down control. Most research in forests has dealt with birds and invertebrate predators of tree-eating insects (Holmes et al. 1979, Torgerson et al. 1979, Kroll unpublished in Thatcher et al. 1980, Mason et al. 1983, Carlson et al. 1984, McClure 1986, Crawford and Jennings 1989, Torgerson et al. 1990, Way and Khoo 1992, Marquis and Whelan 1994), but even large predators can indirectly benefit tree growth by controlling herbivores, as recently demonstrated on Isle Royale National Park (Michigan), where tree growth correlates positively with the size of wolf populations (presumably because wolves control populations of foliage-feeding moose)(McLaren and Peterson 1994).

As discussed earlier, the importance of top-down controls means by extension that functional stability is influenced by the availability of habitat for predators at both the stand and landscape scales. Several researchers have identified old growth or old-growth components such as large dead wood as important habitat for the natural enemies of herbivores. Torgerson et al. (1990) argued that protecting habitat for foliage-gleaning birds and ants (e.g., large dead wood) increased the overall health of forests. Schowalter (1989) found that old-growth canopies in both the Pacific Northwest and the southern Appalachians supported a greater diversity of spiders and a much more favorable balance between herbivorous insects and their invertebrate predators than plantation canopies. He suggested that old growth serves as a source of predatory insects for plantations, which, if true, means that herbivore dynamics are influenced by the proximity of younger to older forests. Similarly, McCutcheon et al. (1993) found that, compared to young Douglas-fir separated from old growth, old trees and young trees adjacent to old trees had a much greater diversity of foliar endophytes, a microfungus that lives symbiotically in plant leaves and helps defend plants against pathogens and herbivores. In all likelihood, these observations can be traced to the structural complexity of old growth, which affords more niches for animals and microbes.

Keystones

Webster defines *keystone* as "that one of a number of associated parts or things that supports or holds together the others." In ecology, keystones are species, groups of species, habitats (e.g., large dead wood), or abiotic factors (e.g., fire) that play a pivotal role in ecosystem (or landscape) processes and "upon which a large part of the community depends" (Noss 1991). (Mills et al. 1993 discuss the need to consider keystones in management and policy decisions.) Some landscape features, such as riparian zones or migration corridors, may also be keystones. Loss of a keystone produces cascade effects: the loss of other species and the disruption of processes.

There undoubtedly are keystone species within ecosystems. Frequently cited examples include the so-called builder, or ecosystem engineer, species, such as beavers, gophers, tortoises, termites, and earthworms that, through their activities, physically modify the environment in ways important to other species (Noss 1991, Jones et al. 1994, Lawton 1994, Perry 1994). There are also keystone groups of species that perform some unique function: Walker (1992) calls these functional groups. These might include nitrogen-fixing plants, plants that provide unique food resources (e.g., nut trees and flowering plants in conifer-dominated forests), plants that provide food during otherwise lean periods (e.g., various species of figs), and epiphytes (which support a long food chain in tropical forests because their cup-like leaves catch water). There are also keystone structures, such as large dead wood. The list could go on, but the point is that critical functions exist that are filled by one or a few species or structures.

One problem with the keystone concept is that interrelationships are so complex and intertwined within ecosystems that numerous species and mutualisms might qualify as keystones; in fact, given our limited knowledge of how species function in ecosystems, it is impossible at this point to say with certainty which species are not keystones. If, to give just one example, whitebark pine is a keystone food resource for various animals, then surely Clarkes nutcracker, which distributes whitebark seeds, is also a keystone. And how about the mycorrhizal fungi and other microbes that the pine must have to survive and grow, or invertebrates that cycle nutrients? Without the tiny and unapparent residents of the soil, most or all natural ecosystems on land (certainly all forests) would collapse.

There are relative degrees of "keystoneness"—that is, the loss of some species creates a ripple, the loss of others a tidal wave. For example, extirpation of chestnut trees from forests of the eastern United States (due to chestnut blight) led to the extinction of perhaps seven species of moths (Lepidoptera). From an ethical standpoint, seven species may be seven too many; however, from a functional standpoint those seven were only 12 percent of the Lepidoptera species that fed on chestnut, no vertebrate extinctions occurred, and processes such as primary productivity, nutrient cycling, and hydrologic cycling apparently emerged intact (Pimm 1991). Loss of fig trees from tropical forests would create a much bigger wave because numerous frugivorous species depend on them. The degree to which that wave spread throughout the system would depend at least in part on secondary effects arising from loss of the frugivores. In the final analysis, the most serious keystones—the ones whose loss creates a tidal wave—are those that play a singular role in key system processes: photosynthesis, food-web dynamics, nutrient and water cycling, controls over herbivores and pathogens, and maintenance of biological legacies following disturbance.

Threads of Continuity: Biological Legacies and Guilds

Within three years after the eruption of Mount St. Helens, 230 plant species—90 percent of those in pre-eruption communities—had been found within the area affected by the blast deposit and mudflows (Franklin et al. 1985). Quoting from Franklin et al. (1985):

> Successional theory traditionally emphasizes invading organisms or immigrants . . . but this script for ecosystem recovery could be played out at only a few sites, as surviving organisms over most of the landscape provided a strong and widespread biological legacy from the pre-eruption ecosystem. In fact, essentially no post-eruption environment outside the crater was completely free of pre-eruption biological influences, although there were substantial differences in the amounts of living and dead organic material that persisted.

Webster defines *legacy* as "anything handed down from . . . an ancestor." In an ecological context, legacies are anything handed down from a predisturbance ecosystem, including green trees, surviving propagules and organisms (e.g., buried seeds, seeds stored in serotinous cones, surviving roots and basal buds, mycorrhizal fungi and other soil microbes, invertebrates, and mammals), dead wood, and certain aspects of soil chemistry and structure, such as soil organic matter, large soil aggregates, pH, and nutrient balances. Most, if not all legacies probably influence the successional trajectory of the recovering system to one degree or another (although much research is needed). That is clearly the case with surviving plant propagules, which directly affect composition of the early successional community. Other legacies may shape successional patterns in more subtle ways. For example, the composition of the soil biological community following disturbance is a legacy that potentially influences the relative success of different plant species during succession (Amaranthus and Perry 1989a, 1989b, Perry et al. 1989b, Perry 1994).

Patches of mature and older forests that survive a given disturbance may be thought of as ecosystem-level legacies that allow certain species to persist while the rest of the landscape recovers. Even small fragments of mature forest may stabilize food webs and provide refuge for some species. In a study in southwest Oregon, mature forest fragments no larger than 3.5 ha produced 30 times more truffles per hectare than plantations, and had nearly twice as many truffle species (Amaranthus et al. 1994). The difference was particularly striking in August, the height of the summer drought in these forests. Truffles, the belowground fruiting bodies of mycorrhizal fungi, are an important part of the food chain, composing a major portion of the diet for some small mammals. In maintaining truffle production during summer drought, mature forest patches provide animals such as northern flying squirrels and California red-backed voles with food that is not available in plantations—in all likelihood contributing significantly to maintaining populations of these animals during a period of food shortage. Maintaining small mammal populations increases, in turn, the chances of maintaining predator populations.

Large dead wood is one of the more obvious structural legacies of a natural disturbance, and a major reason why clearcuts are not the ecological equivalent of natural disturbance. Dead wood can influence system recovery in several ways. Standing dead trees mitigate environmental extremes within disturbed areas by shading and preventing excessive heat loss at night. Decaying logs are centers of biological activity, including not only decay organisms, but also roots, mycorrhizal hyphae, nitrogen-fixing bacteria, amphibians, and small mammals (Harmon et al. 1986, Franklin et al. 1985). After disturbance, logs reduce erosion by acting as physical barriers to soil movement (Franklin et al. 1985) and provide cover for small mammals that disseminate mycorrhizal spores from intact forest into the disturbed area (Maser et al. 1978). The sponge-like water-holding capacity of old decaying logs helps seedlings rooted in them survive drought (Harvey et al. 1987). In the Amaranthus et al. (1994) study previously mentioned, the highest truffle production occurred in older, decayed logs, in all likelihood because the logs retained water during drought.

Soil aggregates and soil organic matter are important legacies. Large soil aggregates, which are created and sustained by roots and the hyphae of mycorrhizal fungi, are essentially little packages of mycorrhizal propagules, other microbes, and nutrients that are passed from the old forest to the new (Borchers and Perry 1990). Soil organic matter in general, whether contained in aggregates or not (most is), provides a legacy of nutrients for the new stand. Depending on its origin and stage of decay, soil organic matter can either stimulate or inhibit plant pathogens (Linderman 1989, Schisler and Linderman 1989).

The threads of continuity provided by biological legacies significantly influence the diversity of animal communities during the recovery phase. In the Pacific Northwest, richness of vertebrate species differs little among successional stages resulting from natural disturbances (fire, wind), and the majority of species that have been studied occur throughout the sere (see the review by Hansen et al. 1991). Hansen concludes:

> A likely explanation for the similarity in species distribution is that structural differences among these natural forest stages are insufficient to strongly influence most species of plants and animals. The natural disturbance regime and structural legacy in all . . . age-classes provide the resources and habitats required by many species. The important conclusion is that the canopy structures, snag densities, and levels of fallen trees found in unmanaged young, mature, and old-growth stands appear to make all three of these seral stages suitable habitat for most species of forest plants and vertebrate animals.

Note that "most species" does not mean all species; some are still restricted to one or another seral stage, and the proper mix of seral stages must be maintained if the objective is to protect regional biodiversity.

A plant that survives disturbance in one form or another (e.g., thick bark, living roots, buried seeds) clearly promotes the continuity of its own genome on a site, and the surest way to eliminate a species from a successional trajectory is to eliminate its seed source (as has happened with eastern hemlock throughout much of northern Wisconsin (White and Mladenoff 1994). But one species may also provide legacies that either promote or inhibit the continuity of others. Different plant species can affect soils quite differently through the soil organisms they support, the particular array of nutrients they accumulate, their effect on soil acidity, or allelochemicals they release. Sprouters and other pioneering plants often become foci for the establishment of other plants, a phenomenon called the island effect (Perry 1994). An established plant might provide perches for birds that disseminate seeds (Nepstad et al. 1990), shelter establishing seedlings from climatic extremes and herbivores, provide nutrient-rich microsites, or support mycorrhizal fungi or other beneficial soil organisms (Amaranthus and Perry 1989a, 1989b, Perry et al. 1989a, 1989b). Whatever the mechanisms, pioneers that sprout from roots or buried seeds constitute legacies that influence the recovery of other species within the system.

One hypothesis holds that species within a given community form into guilds based on common interests in mycorrhizal fungi and perhaps other beneficial soil organisms (Perry et al. 1989, Read 1994). According to this view, early colonizers during secondary succession facilitate subsequent colonization by members of the same guild by providing a legacy of mycorrhizal fungi (and perhaps other beneficial soil organisms). Studies in southern Oregon and northern California clearly show that Douglas-fir establishes most successfully in the vicinity of certain species of shrubs and hardwood trees that support the same mycorrhizal fungi as the conifer (Wilson 1982, Amaranthus and Perry 1989a, 1989b; Tom Parker, personal communication). The beneficial effect of hardwoods on Douglas-fir has been associated in one or more studies with accelerated root tip formation, greater numbers of total and mycorrhizal root tips, shifts in mycorrhiza type, increased associative nitrogen fixation in seedling rhizospheres, increased ratios of iron to manganese in seedling foliage, and faster rates of nitrogen cycling in soils beneath rather than apart from hardwoods (Amaranthus and Perry 1989a, 1989b, Amaranthus et al. 1990, Borchers and Perry 1990).

In one Oregon clearcut, soils beneath hardwood saplings had only 10 percent as many *Streptomyces* colonies as grass-covered soils in between hardwoods. *Streptomyces,* a common soil actinomycete that is the source of the antibiotic streptomycin, allelopathically inhibits a variety of other microorganisms and some plants, and has been implicated in regeneration failures in Oregon (Friedman et al. 1989, Colinas et al. 1994b). Borchers and Perry (1990) hypothesize that hardwoods discourage growth of *Streptomyces* through their ability to concentrate manganese, a known inhibitor of streptomycin. The possibility that hardwoods are simply growing in nutrient-rich microsites cannot be totally excluded. However, all evidence points to the likelihood that hardwoods are imposing a biological pattern on soils that benefits Douglas-fir. It does not follow, though, that relationships among plants within a guild are strictly cooperative. All plants need nutrients, water, and sunlight to survive, and guild members may compete for resources as well as benefit one another. Such multifaceted interactions are probably common in nature, and the outcome of a given interaction will vary depending on specific conditions, such as what resources are most limiting, and how dense one plant is compared to another.

Some plants actively inhibit others during the recovery phase. For example, grass seeded onto sites burned by wildfire in southern Oregon inhibited recovery of the native shrub community (Amaranthus et al. 1993). Plants able to produce dense, uneven-aged stands can effectively exclude others; such is the case with the Pacific Coast shrub salmonberry, which produces pure stands of 30,000 or more stems per hectare following disturbance, and which then maintains itself in uneven-aged stands through sprouting from basal buds and rhizomes (Tappeiner et al. 1991). Once a pure stand attains a sufficiently high density, plants with the reproductive potential of salmonberry likely will persist until weakened by pathogens or insects or confronted with a disturbance for which they are not adapted. However, caution should be used in inferring long-term trends from short-term observations. Early successional stages are frequently dominated by fast-growing shrubs and trees that are replaced in time by species with slower initial growth (Oliver 1981). Moreover, there can be tradeoffs between growth and survival. For example, Berkowitz et al. (1995) found that shrubs, herbs, and grasses reduced growth but enhanced survival of planted maples, especially during a drought year. Long-term competitive exclusion is likely to occur only when one species can prevent another from establishing at all. The patchiness of natural disturbances (which translates into varied regeneration niches) and the threads of continuity provided by biological legacies act to preserve diversity in the plant community. Grazing animals and pathogens can also promote plant diversity by keeping one or a few species from dominating. Elk and deer, for example, have been shown to reduce the cover of salmonberry relative to that of Douglas-fir seedlings in early successional stands in Washington State (Hanley and Taber 1980).

In some cases, a change in disturbance regime has altered a balance and allowed one plant species (or group of species) to spread at the expense of others. A series of foreign disturbances contributed to regeneration failures in parts of Pennsylvania (Horsely 1977): Settlement and widespread forest clearing for farms led to the extirpation of large predators such as wolves and mountain lions. The loss of predators combined with abundant food on abandoned farms to produce an explosion in deer populations, and deer overgrazed establishing tree seedlings, which were also subjected to fires burning through clearcuts. Herbs, ferns, and grasses attained dominance in clearcuts and further inhibited tree seedlings through the production of allelochemicals.

Allelopathy has emerged as a common theme where one or more species begin to aggressively exclude others. In California, overgrazing by domestic livestock allowed unpalatable *Wyethia mollis* to dominate early seral stages on burned sites; *Wyethia,* in turn, excluded tree seedlings, in part at least through allelopathy (Parker and Yoder-Williams 1989). In Sweden, excluding wildfires contributed to the spread of the dwarf shrub *Empetrum hermaphroditum,* which then allelopathically inhibited tree regeneration (Nilsson et al. 1993). Allelopathy has also been linked to poor tree regeneration in the Alps (Pellisier and Trosset 1989). More than 30 years ago, Handley (1963) suggested that some plant species may serve as nurse plants for others by ameliorating the allelochemical environment in one fashion or another. In support of this, our own work in southern Oregon led us to hypothesize that the spread of allelopathic *Streptomyces sp.* in clearcuts was facilitated by herbiciding early successional hardwoods that inhibit the spread of Streptomyces through their ability to accumulate manganese (Borchers and Perry 1990, Perry et al. 1992).

When Recovery Mechanisms Break Down: Threshold Transitions

Determining ecological significance is not always straightforward in cases where once-dominant species appear to be excluded by one or more others—as when *Wyethia mollis* or *Empetrum hermaphroditum* spreads and inhibits tree regeneration, or exotic grasses dominate clearcuts. Are these temporary digressions that will eventually revert to the former successional sequences without outside subsidies? Or are they threshold transitions from one persistent community type to another? If the former is true, managers may lose time but not sites; but if the latter is true, sites will be lost, perhaps irretrievably.

As we discussed earlier, there is no doubt that threshold transitions can and do occur. Conversion of arid grasslands to desert shrublands is the clearest example on land (Schlesinger et al. 1990). However, there are also examples in forests, particularly in stressful environments (cold or dry). For example, shifting cultivation has converted large areas of dry miombo woodland in East Africa to semidesert scrub (Jummane Maghembe, personal communication). In parts of the western United States, numerous high-elevation clearcuts have been planted three to four times each over the past two to three decades, with most plantings failing. With sufficient expense, these areas might be eventually reforested, but there is little doubt that these sites are not following their historic recovery patterns. Rather, they have been converted from forests to persistent communities characterized primarily by exotic grasses and herbs.

What mechanisms underlie the degradation of forested sites? Experiments on one degraded site in southwest Oregon (Cedar Camp) showed that planted Douglas-fir seedlings would survive and grow if given 150 ml (1/2 cup) of soil from an established stand, whereas control seedlings not given forest soil died (Amaranthus and Perry 1987). A number of factors appear to be involved: Compared to control seedlings, those given forest soils form root tips faster, form more mycorrhizae, and are less needy of fertilizer (Amaranthus and Perry 1989a, 1989b, Colinas et al. 1994a, 1994b). Experiments have shown that much of the beneficial effect of forest soils was associated with invertebrate grazers (protozoa, microarthropods), organisms that occupy the top of the belowground food chain and that are critically important in the nutrient cycle. But evidence indicates that the transferred soils had other benefits as well, such as providing "safe" sites in which developing roots and mycorrhizae were protected from allelopathic *Streptomyces* (which had proliferated in the clearcut), and in reintroducing a source of microbially produced plant hormones, such as ethylene, that stimulate root and mycorrhiza formation. Whatever the mechanisms, forest soils collected from the rooting zone of vigorous Douglas-fir contained factors that Douglas-fir seedlings needed to survive and that had been lost from the old clearcut.

How is it that these fire-adapted forests cannot regenerate after clearcutting? We believe the answer lies in the destruction of biological legacies: Unlike on a site burned by wildfire, there were no snags to ameliorate the environment for establishing seedlings, a factor likely to be critically important on these high-elevation sites. Herbiciding early successional shrubs removed an important stabilizing agent for the belowground ecosystem, and, unlike the native shrubs, the exotic annual grass (cheatgrass) that came to dominate the site did not stabilize elements of soil biology and structure required by the establishing seedlings. The apparent reduction in predators at the top of the soil food chain likely reflects sharply reduced energy flow to belowground food chains accompanying the transition from forest to annual grassland. As soil biology and structure changed, seedlings were unable to form roots and gather resources quickly enough to become established before the summer drought arrived. A vicious cycle ensued in which loss of soil integrity inhibited seedling establishment and the lack of regeneration led to further loss of soil integrity (Perry et al. 1989a, 1989b).

What happened at Cedar Camp illustrates a general ecological principle that was first demonstrated experimentally more than 20 years ago at Hubbard Brook Experimental Forest in New Hampshire (Bormann et al. 1974), where clearcutting followed by herbiciding early successional vegetation led to large nutrient losses to streams. The principle, applicable to any forested ecosystem is this: Trees, along with most if not all perennial plants, divert large amounts of energy belowground, where it fuels processes that feed back positively to plant growth (Perry et al. 1989a, 1989b); the "bioregulation" of soils by plants, to use Bormann and Likens's (1979) term, acts to retain nutrients, support food chains that cycle nutrients, support microbial mutualists of plants (e.g., mycorrhizal fungi and certain types of rhizosphere bacteria), and structure soils to improve their aeration and water-holding capacity. In other words, plants and soils compose a self-reinforcing system in which energy flow from plants maintains the soil ecosystem, which feeds back to promote plant growth. This general dynamic—energy flow creating nonequilibrium pattern—can be extended beyond the plant–soil interaction to include other food chains within communities. The stability of such systems depends critically on the ability to dampen fluctuations and maintain processes within certain bounds (Prigogine and Stengers 1984, Perry et al. 1989a, 1989b). While details will undoubtedly vary from one site to another, when energy flow from plants to soils is disrupted, changes ensue in soil biology, chemistry, and physical structure that, if allowed to go too far, progressively diminish the ability of soils to support the original plant community. Sites may become dominated by annual weeds, which, because of their low productivity and life history, are likely to require less integrity in the soil system than perennials. Little is known about the potential for such changes in forests, however they are best guarded against by maintaining a cover of trees or ecologically equivalent shrubs at all times. Tree cover at the level of a traditional shelterwood is probably sufficient, though research is needed to verify this.

Cedar Camp illustrates a threshold transition resulting from the disruption of biological legacies and the consequent breakdown of key processes. Threshold transitions are also likely if seed sources are lost, or if the environment (especially climate or the disturbance regime) is altered to favor new species over the previous dominants. These factors tend to reinforce one another. In northern Wisconsin, for example, logging and slash fires (a change in disturbance regime) combined to sharply reduce the cover of eastern hemlock, which had dominated many old-growth forests of the region (White and Mladenoff 1994). Hemlock was not adapted to survive intense fires, which prior to settlement had occurred at intervals as great as 2,500 years or more (Frelich and Lorimer 1991). With the hemlock seed source gone, White and Mladenoff (1994) conclude that "[t]he pre-European settlement pattern of hemlock forests may be irretrievably lost due to the consequences of logging, fire disturbances, and regeneration failure." Similarly, factors that favor an intensified fire regime in forests (e.g., forest clearing and spread of flammable grasses in the tropics, and buildup of fuels in the dry forest types of western North America) will, unless checked, almost certainly result in threshold conversion to plant communities more tolerant of dry conditions and fire. Moreover, fires that burn too frequently may degrade soils and trigger a transition to annual weeds (Perry 1995a), a possibility that should be taken seriously both in dry forest types and in the moist tropics.

Diversity and Stability: What Can We Conclude?

Does everything depend on everything else? Intuitively it seems unlikely that an ecological system without some redundancy in key processes would persist long in the highly variable environments that characterize the real world (cf. Wilson and Botkin 1990). The loss of chestnut trees from forests of the eastern United States would seem to verify that, at least in that system, major processes were buffered against the loss of a dominant species. On the other hand, the more modest assertion that everything depends on something else seems irrefutable and can be extended without too much risk to say that everything is tied into a network of interdependence that encompasses from a few to many other species, some of which may be keystones and others members of functional groups (Berryman 1993). The glue that ties these networks together comprises the key processes—photosynthesis, nutrient cycling, hydrology, and population regulation (to name a few)—that run the system. It is the interaction between species and processes that not only creates interdependence, but defines the ecosystem, and which must be the basis of ecosystem management. Unfortunately, the scientific knowledge needed to understand the complex interactions that underpin ecological dynamics is in its infancy.

Considerable caution should be used when evaluating whether the loss of species has impacted the larger system or not. In many cases the stabilizing effect of diverse species is manifested during infrequent periods of stress: It was drought that revealed the importance of diversity in stabilizing prairie productivity (Tilman and Downing 1994) and wildfire that showed how some hardwoods shield conifers from flames (Perry 1988). Moreover, redundancy in key processes can actually mask a weakening of ecosystem buffering capacity associated with species loss—that is, the plane that loses a few rivets may fly just fine in good weather but break apart in a storm. Holling (1988) spoke to this in his analysis of whether the loss of insect-eating birds could, by itself, lead to increased insect outbreaks in North America. He concluded that resilience provided by redundancies in the controls over insect populations would prevent this from happening, and went on to generalize:

> This great resilience demonstrates a property common in ecological systems. First, the stability domains are large and the variables within them can fluctuate extensively. Second, the regulatory processes that are present are remarkably robust to external changes. . . . This is not to say, however, that ecological systems are infinitely resilient nor that loss of robustness of regulation short of producing a qualitative flip in behavior has no costs. Loss of resilience from one cause can make the system more vulnerable to changes in other events that otherwise could have been absorbed. . . . Ironically, the great resilience of ecological systems masks slow erosion of their capacity to renew and in those circumstances leave managers ill-prepared for surprises.

If it is unlikely that everything depends on everything else, it is equally unlikely that there is absolute redundancy in any process—that is, that any two species maintain the same process equally well at all times and under all conditions. A more accurate view would be relative redundancy (Perry 1994): Species within a functional group do the same job, but in different ways or at different times. The result can be a complex control structure that includes both redundancy and keystoneness. Controls over herbivorous insects provide a good example of what appears on the surface to be redundancy, but when viewed more closely is not. Berryman (1993) says that—

> populations of leaf-eating insects are sometimes controlled by guilds of insectivorous vertebrates when their densities are very sparse, by larval parasitoids if they escape from vertebrate imposed limitation, by pathogens if parasitoid regulation fails, and by competition for food in the absence of all the above.

The greatest functional diversity in nature is provided by the organisms that receive the least attention from both scientists and managers, the invertebrates and microbes—E. O. Wilson (1987) called invertebrates and microbes "the little things that run the world." Consider mycorrhizal fungi. Because each species of mycorrhizal fungi has its own set of physiological characteristics (Trappe 1987), none can be said to be strictly redundant to any other. Some may be active during cool or moist periods, others when it is warm and dry. Some are particularly effective at gathering nutrients from mineral soil, others at extracting nutrients bound in organic matter. Some thrive in large dead wood, others in humus, yet others in mineral soil. Healthy forests typically support a highly diverse ectomycorrhizal flora (Arnolds 1991, Fogel 1976, Hunt and Trappe 1987, Luoma et al. 1991, Menge and Grand 1978). To give one example, over 200 ectomycorrhizal types have been described from mature conifer and hardwood stands in southwest Oregon (Eberhart, Luoma, and Amaranthus 1995). It is highly probable that the array of functional niches provided by this diversity contributes significantly to the ability of long-lived trees to cope with fluctuating and often unpredictable environments (Perry 1995a).

Management Implications

The shift to a more dynamic view of nature has important implications for both forestry and conservation. In forestry, or for that matter any area of natural resource management, it means that, in theory at least, humans don't have to be a blight on the earth: With the proper understanding of nature's dynamics, management systems can be devised that sustain productivity and some, perhaps much, biological diversity. (It does not follow, however, that biological diversity can be sustained solely on managed lands, or that sustainability can be achieved without controlling human numbers and consumption.) While ecosystem dynamics are complex and for the most part poorly understood, the emerging scientific view has clear management implications.

1. Protect indigenous biological diversity. No forest ever has been or ever will be free from the threat of rapid swings in climate, outbreaks of tree-eating insects and pathogens, fire, wind, etc. Today's world carries new risks: widespread exotic pests, pollution, harmful levels of ultraviolet radiation, greenhouse-related climate change, and the risk of overutilization arising from population growth and an economic system that encourages consumption and rewards exploitation. From a practical standpoint, it makes no difference to managers whether the rivet hypothesis (everything depends on everything else) is false, ecologists are nowhere near being able to point to any one species and say with certainty that that one can be removed with no significant effects on the rest of the system, and ecologists won't be able to do that for a long time, if ever. Protecting biological diversity is the best insurance foresters and society can buy to protect the long-term integrity of the world's forests.

2. Protect soils. If soil integrity is lost, so is the ball game.

3. Plan at the landscape scale. No forest is an island, nor is any reserve or park. The critical role of landscapes and regions in buffering the spread of disturbances, providing pathways of movement for organisms, altering climate, and mediating key processes such as the hydrologic cycle means that the fate of any one piece of ground is intimately linked to its larger spatial context (Hansson and Angelstam 1991, Saunders et al. 1991, Franklin 1993, Pickett et al. 1992, Berg et al. 1994). One unavoidable implication is the necessity to view regional landscapes as wholes rather than disconnected pieces of different ownerships and land-use categories.

4. To conserve species in the long run, plan for the future. Once change is seen as inevitable, it becomes obvious that species cannot be preserved simply by reserving existing habitat (Franklin 1993). At some point that habitat will burn up, blow down, or be destroyed by

some anthropogenic insult that originates beyond its borders, such as pollution or climate change. Where will new habitat come from? For species that persist in early successional stages, this is not a problem—their habitat is created by disturbance. Species that require later successional stages are another matter; management strategies must provide for the development of their habitat, with or without human assistance.

Literature Cited

Allen, B. D., and R. Y. Anderson. 1993. Evidence from North America for rapid shifts in climate during the last glacial maximum. *Science* 260:1920–1923.

Allen, T. F., and T. W. Hoekstra. 1991. Role of heterogeneity in scaling of ecological systems under analysis. In *Ecological heterogeneity,* ed. J. Kolasa and S. T. A. Pickett. New York: Springer.

Allen, T. F,. and T. B. Starr. 1982. *Hierarchy: Perspectives for ecological complexity.* University of Chicago Press.

Amaranthus, M. P., and D. A. Perry. 1987. Effect of soil transfer on ectomycorrhiza formation and the survival and growth of conifer seedlings on old, nonreforested clear-cuts. *Canadian Journal of Forest Research* 17:944–950.

———. 1989a. Interaction effects of vegetation type and Pacific madrone soil inocula on survival, growth, and mycorrhiza formation of Douglas-fir. *Canadian Journal of Forest Research* 19:550–556.

———. 1989b. Rapid root tip and mycorrhiza formation and increased survival of Douglas-fir seedlings after soil transfer. *New Forestry* 3:259–264.

Amaranthus, M. P., C. Y. Li, and D. A. Perry. 1990. Influence of vegetation type and madrone soil inoculum on associative nitrogen fixation in Douglas-fir rhizospheres. *Canadian Journal of Forest Research* 20:368–371.

Amaranthus, M. P., J. M. Trappe, and D. A. Perry. 1993. Soil moisture, native revegetation, and *Pinus lambertiana* seedling growth, and mycorrhiza formation following wildfire and grass seeding. *Restoration Ecology* 9:188–195.

Amaranthus, M. P., J. M. Trappe, L. Bednar, and D. Arthur. 1994. Hypogeous fungal production in mature Douglas-fir forest fragments and surrounding plantation and its relation to coarse woody debris and animal mycophagy. *Canadian Journal of Forest Research* 24:2157–2165.

Anderson, L., C. E. Carlson, and R. H. Wakimoto. 1987. Forest fire frequency and western spruce budworm outbreaks in western Montana. *Forest Ecology Management* 22:251–260.

Angermeier, P. L., and J. R. Karr. 1994. Biological integrity versus biological diversity as policy directives. *BioScience* 44:690–697.

Archer, S., and F. E. Smeins. 1991. Ecosystem-level processes. In *Grazing Management: An Ecological Perspective,* ed. R. K. Heitschmidt and J. W. Stuth. Portland, OR: Timber Press.

Arnolds, E. 1991. Decline of ectomycorrhizal fungi in Europe. *Agriculture, Ecosystems and Environment* 35:209–244.

Atsatt, P. R., and D. J. O'Dowd. 1976. Plant defense guilds. *Science* 193:24–29.

Bak, P., C. Tank, and K. Wiesenfeld. 1988. Self-organized criticality. *Physics Review* 38:364–374.

Berg, A., B. Ehnstrom, L. Gustafsson, T. Hallingback, M. Jonsell, and J. Weslien. 1994. Threatened plant, animal, and fungus species in Swedish forests: Distribution and habitat associations. *Conservation Biology* 8:718–731.

Berkowitz, A. R., C. D. Canham, and V. R. Kelly. 1995. Competition vs. facilitation of tree seedling growth and survival in early successional communities. *Ecology* 76: 1156–1168.

Berryman, A. A. 1993. Food-web connectance and feedback dominance, or does everything really depend on everything else? *Oikos* 68:183–185.

Bierregaard Jr., R. O., T. E. Lovejoy, V. Kapos., et al. 1992. The biological dynamics of tropical rainforest fragments. *BioScience* 42:859–866.

Blais, J. R. 1983. Trends in the frequency, extent, and severity of spruce budworm outbreaks in eastern Canada. *Canadian Journal of Forestry Research* 13:539–547.

Bonan, G. B., and H. H. Shugart. 1989. Environmental factors and ecological processes in boreal forests. *Annual Review of Ecological Systems* 20:1–28.

Bond, G., W. Broecker, S. Johnsen, J. McManus, L. Labeyrle, J. Jouzel, and G. Bonani. 1993. Correlations between climate records from North Atlantic sediments and Greenland ice. *Nature* 365:143–147.

Borchers, S. L., and D. A. Perry. 1990. Growth and ectomycorrhiza formation of Douglas-fir seedlings grown in soils collected at different distances from pioneering hardwoods in southwest Oregon. *Canadian Journal of Forest Research* 20:712–721.

Borchers, J. G., D. A. Perry, P. Sollins, et al. 1992. The influence of soil texture and aggregation on carbon and nitrogen dynamics in southwest Oregon forests and clearcuts. *Canadian Journal of Forest Research* 22:298–305.

Bormann, F. H. 1985. Air pollution and forests: An ecosystem perspective. *BioScience* 35:434–441.

Bormann, F. H., and G. E. Likens. 1979. *Pattern and process in a forested ecosystem.* New York: Springer-Verlag.

Bormann, F. H., G. E. Likens, T. G. Sicama, et al. 1974. The export of nutrients and recovery of stable conditions following deforestation at Hubbard Brook. *Ecological Monographs* 44:255–277.

Bormann, B. T., F. H. Bormann, W. B. Bowden, R. S. Pierce, S. P. Hamburg, D. Wang, M. C. Snyder, C.Y. Li, and R. C. Ingersoll. 1993. Rapid N_2 fixation in pines, alder, and locust: Evidence from the sandbox ecosystem study. *Ecology* 74(2):583–598.

Botkin, D. B. 1990. *Discordant harmonies.* New York: Oxford University Press.

Broecker, W. S. 1987. Unpleasant surprises in the greenhouse? *Nature* 328:123–126.

Bunnell, F. L. 1995. Forest-dwelling vertebrate faunas and natural fire regimes in British Columbia: Patterns and implications for conservation. *Conservation Biology* 9: 636–644.

Campbell, I. D., and J. H. McAndrews. 1993. Forest disequilibrium caused by rapid Little Ice Age cooling. *Nature* 366:336–338.

Carlson, C. E., R. W. Campbell, L. J. Theroux, and T. H. Egan. 1984. Ants and birds reduce damage to small Douglas-fir and western larch in Montana. *Forest Ecology Management* 9:185–192.

Castello, J. D., D. J. Leopold, and P. J. Smallidge. 1995. Pathogens, patterns, and processes in forest ecosystems. *BioScience* 45(1):16–24.

Chen, J., J. F. Franklin, and T. A. Spies. 1993. Contrasting microclimates among clearcut, edge, and interior of old-growth Douglas-fir forest. *Agriculture and Forest Meteorology* 63:219–237.

———. 1995. Growing-season microclimatic gradients from clearcut edges into old-growth Douglas-fir forests. *Ecological Applications* 5(1):74–86.

Clark, J. S. 1988. Effect of climate change on fire regimes in northwestern Minnesota. *Nature* 334:233–235.

———. 1990. Fire and climate change during the last 750 years in northwestern Minnesota. *Ecological Monographs* 60:135–159.

Colinas, C., R. Molina, J. Trappe, and D. Perry. 1994a. Ectomycorrhizas and rhizosphere microorganisms of seedlings of *Pseudotsuga menziessi* (Mirb.) Franco planted on a degraded site and inoculated with forest soils pretreated with selective biocides. *New Phytology* 127:529–537.

Colinas, C., D. Perry, R. Molina, and M. Amaranthus. 1994b. Survival and growth of *Psuedotsuga menziesii* seedlings inoculated with biocide-treated soils at planting in a degraded clearcut. *Canadian Journal of Forest Research* 24:1741–1749.

Comins, H. N., M. P. Hassell, and R. M. May. 1992. The spatial dynamics of host-parasitoid systems. *Journal of Animal Ecology* 61:735–748.

Costanza, R., L. Wainger, C. Folke, and K-G. Maler. 1993. Modeling complex ecological economic systems. *BioScience* 43(8):545–555.

Crawford, H. S., and D. T. Jennings. 1989. Predation by birds on spruce budworm, *Choristoneura fumiferana:* Functional, numerical, and total responses. *Zoology* 70:152–163.

D'Antonio, C. M., and P. M. Vitousek. 1992. Biological invasions by exotic grasses, the grass/fire cycle. *Annual Review of Ecology and Systematics* 23:63–88.

Davies, P. 1983. *God and the new physics.* New York: Simon and Schuster.

DeAngelis, D. L., and J. C. Waterhouse. 1987. Equilibrium and nonequilibrium concepts in ecological models. *Ecological Monographs* 57:1–21.

DeAngelis, D. L., W. M. Post, C. C. Travis. 1986. *Positive Feedback in Natural Systems.* Berlin: Springer-Verlag.

Drake, J. A. 1990. The mechanics of community assembly and succession. *Journal of Theoretical Biology* 147:213–233.

DuPont, A., L. Belanger, and J. Bousquet. 1991. Relationships between balsam fir vulnerability to spruce budworm and ecological site conditions of fir stands in central Quebec. *Canadian Journal of Forest Research* 21:1752–1759.

Eberhart, J. L., D. L. Luoma, and M. P. Amaranthus. 1996. Response of ectomycorrhizal fungi to forest management treatments: A new method for quantifying morphotypes. In *Mycorrhizas in integrated system: From genes to plant development,* ed. C. Azcon-Aguilar and J. M. Barea. European Commission, Luxemborg: Office for Official Publications of the European Communities.

Ehrlich, P. R., and A. H. Ehrlich. 1981. *Extinction: The causes and consequences of the disappearance of species.* New York: Random House.

Ewel, J. J., M. J. Mazzarino, and C. W. Berish. 1991. Tropical soil fertility changes under monocultures and successional communities of different structure. *Ecological Applications* 1:289–302.

Fogel, R. D. 1976. Ecological studies of hypogeous fungi. II. Sporocarp phenology in a western Oregon Douglas-fir stand. *Canadian Journal Botany* 54:1152–1162.

Forman, R. T. T. 1987. The ethics of isolation, the spread of disturbance, and landscape ecology. In *Landscape heterogeneity and Disturbance,* ed. M. G. Turner. New York: Springer-Verlag.

Forman, R. T. T., and M. Godron. 1981. Patches and structural components for a landscape ecology. *BioScience* 31(10):733–740.

Franchito, S. H., and B. Rao. 1992. Climatic change due to land surface alterations. *Climate Change* 22:1–34.

Franklin, J. F. 1993. Preserving biodiversity: Species, ecosystems, or landscapes. *Ecological Applications* 3:202–205.

Franklin, J. F., and R. T. T. Forman. 1987. Creating landscape patterns by forest cutting: Ecological consequences and principles. *Landscape Ecology* 1:5–18.

Franklin, J. F., J. A. MacMahon, F. J. Swanson, and J. R. Sedell. 1985. Ecosystem responses to the eruption of Mount St. Helens. *National Geographic Research* (Spring): 198–216.

Franklin, J. F., D. A. Perry, T. D. Schowalter, et al. 1989. Importance of ecological diversity in maintaining long-term site productivity. In *Maintaining the long-term productivity of Pacific Northwest forest ecosystems,* ed. D. A. Perry, R. Meurisse, B. Thomas, et al. Portland, OR: Timber Press.

Frelich, L. E., and C. G. Lorimer. 1991. Natural disturbance regimes in hemlock-hardwood forests of the upper Great Lakes region. *Ecological Monographs* 61:145–164.

Friedman, J., A. Hutchins, C. Y. Li, and D. A. Perry. 1989. Actinomycetes inducing phytotoxic or fungistatic activity in a Douglas-fir forest and in an adjacent area of repeated regeneration failure in southwestern Oregon. *Biologia Plantarum* 31:487–495.

Fryer, G. I., and E. A. Johnson. 1988. Reconstructing fire behavior and effects in a subalpine forest. *Journal of Applied Ecology* 25:1063–1072.

Gatto, M., and S. Rinaldi. 1987. Some models of catastrophic behavior in exploited forests. *Vegetatio* 69:213–222.

Gibson, I. A. S., J. Burley, and M. R. Spreight. 1982. The adoption of heritable resistance to pests and pathogens in forest crops. In *Resistance to diseases and pests in forest trees,* ed. H. M. Heybrock, B. R . Stephan, and K. von Weissenberg. Proceedings Third International Workshop on the Genetics of Host-Parasite Interactions in Forestry, September 1980. Pudoc, Wageningen: The Netherlands.

Gilbert, G.S., S.P. Hubbell, and R.B. Foster. 1994. Density and distance-to-adult effects of a canker disease of trees in a moist tropical forest. *Oecolgia* 98:100–108.

Gray, W. M. 1990. Strong association between west African rainfall and U.S. landfall of intense hurricanes. *Science* 249:1251–1255.

Gray, W. M., and C. W. Landsea. 1993. West African rainfall and Atlantic basin intense hurricane activity as proxy signals for Atlantic conveyor belt circulation strength. In *Conference on Applied Climatology, American Meteorology Society,* January 1993. Anaheim, Calif.

Grootes, P. M., M. Stuiver, J. W. C. White, S. Johnsen, and J. Jouzel. 1993. Comparison of oxygen isotope records from the GISP2 and GRIP Greenland ice cores. *Nature* 366:552–554.

Handley, W. R. 1963. Mycorrhizal associations and Calluna heathland afforestation. *Bulletin of the Forestry Commission of London* 36.

Hanley, T. A., and R. D. Taber. 1980. Selective plant species inhibition by elk and deer in three conifer communities in western Washington. *Forest Science* 26:97–107.

Hansen, A. J., T. A. Spies, F. J. Swanson, and J. L. Ohmann. 1991. Conserving biological diversity in managed forests. *BioScience* 41:382–392.

Hansson, L., and P. Angelstam. 1991. Landscape ecology as a theoretical basis for nature conservation. *Landscape Ecology* 5:191–201.

Harmon, M. E., J. F. Franklin, F. J. Swanson, P. Sollins, S. V. Gregory, J. D. Lattin, N. H. Anderson, S. P. Cline, N. G. Aumen, J. R. Sedell, G. W. Lienkaemper, K. Cromack, Jr., and K. W. Cummins. 1986. Ecology of coarse woody debris in temperate ecosystems. *Advances in Ecological Research* 15:133–302.

Harvey, A. E., M. F. Jugensen, M. J. Larsen, and R. T. Graham. 1987. Relationships among soil microsite, ectomycorrhizae, and natural conifer regeneration of old-growth forests in western Montana. *Canadian Journal of Forest Research* 17(1):58–62.

Hawkins, B. A., M. B. Thomas, and M. E. Hochberg. 1993. Refuge theory and biological control. *Science* 262:1429–1432.

Hix, D. M., B. V. Barnes, A. M. Lynch, and J. A. Witter. 1987. Relationships between spruce budworm damage and site factors in spruce–fir-dominated ecosystems of western upper Michigan. *Michigan Forest Ecology Management* 21:129–140.

Hochberg, M. E., and B. A. Hawkins. 1992. Refuges as a predictor of parasitoid diversity. *Science* 255:973–976

Holling, C. S. 1988. Temperate forest insect outbreaks, tropical deforestation, and migratory birds. In *Memoirs of the Entomological Society of Canada.* 146:21–32.

Holmes, R. T., J. C. Schultz, and P. Nothnagle. 1979. Bird predation on forest insects: An exclosure experiment. *Science* 206:462–463.

Horsely, S. B. 1977. Allelopathic inhibition of black cherry by fern, grass, goldenrod, and aster. *Canadian Journal of Forest Research* 7:205–216.

Hughes, T. P. 1994. Catastrophes, phase shifts, and large-scale degradation of a Caribbean coral reef. *Science* 265:1547–1551.

Hughes, F., P. M. Vitousek, and T. Tunison. 1991. Alien grass invasion and fire in the seasonal submontane zone of Hawai'i. *Ecology* 73(2):743–746.

Hunt, G. A., and J. M. Trappe. 1987. Seasonal hypogeous sporocarp production in a western Oregon Douglas-fir stand. *Canadian Journal of Botany* 65:438–445.

Huston, M., and T. Smith. 1987. Plant succession: Life history and competition. *American Naturalist* 130:168–198.

Johnson, E. A. 1991. Climatically induced change in fire frequency in the southern Canadian Rockies. *Ecology* 72(1):194–201.

———. 1992. *Fire and vegetation dynamics.* Cambridge, England: Cambridge University Press.

Jones, C. G., J. H. Lawton, and M. Shachak. 1994. Organisms as ecosystem engineers. *Oikos* 69:373–386.

Kauffman, J. B. 1991. Survival by sprouting following fire in tropical forests of the eastern Amazon. *Biotropica* 22:219–224.

Kauffman, J. B., and C. Uhl. 1990. Interactions and consequences of deforestation and fire in the rainforests of the Amazon Basin. In *Fire in the tropical and subtropical biota,* ed. J. G. Goldhammer. Berlin, Germany: Springer-Verlag.

Kaufman, L. 1989. Catastrophic change in species-rich freshwater ecosystems. *BioScience* 42:846–858.

Knight, D. H. 1987. Parasites, lightning, and the vegetation mosaic in wilderness landscapes. In *Landscape heterogeneity and disturbance,* ed. M. G. Turner. New York: Springer-Verlag.

Knowlton, N. 1992. Thresholds and multiple stable states in coral reef community dynamics. *American Zoology* 32:674–682.

Knox, J. C. 1993. Large increases in flood magnitude in response to modest changes in climate. *Nature* 361:430–432.

Lawton, J. H. 1994. What do species do in ecosystems? *Oikos* 71:367–374.

Lawton, J. J., and V. K. Brown. 1992. Redundancy in ecosystems. In *Biodiversity and ecosystem function,* ed. E. D. Schulze and H. A. Mooney. New York: Springer-Verlag.

Lean, J., and L. Warrilow. 1989. Simulation of the regional climatic impact of Amazon deforestation. *Nature* 342: 411–413.

Levin, S. A. 1992. The problem of pattern and scale in ecology. *Ecology* 73(6):1943–1967.

Levins, R., and R. Lewontin. 1985. *The dialectical biologist.* Cambridge, MA: Harvard University Press.

Linderman, R. G. 1989. Organic amendments and soil-borne diseases. *Canadian Journal of Plant Pathology* 11:180–183.

Loehle, C. 1989. Forest-level analysis of stability under exploitation: Depensation responses and catastrophe theory. *Vegetatio* 79:109–115.

Lubchenco, J., A. M. Olson, L. B. Brubaker, S. R. Carpenter, M. M. Holland, S. P. Hubbell, S. A. Levin, J. A. MacMahon, P. A. Matson, J. M. Melillo, H. A. Mooney, C. H. Peterson, H. R. Pulliam, L. A. Real, P. J. Regal, and P. G. Risser. 1991. The sustainable biosphere initiative: An ecological research agenda. *Ecology* 72:371–412.

Luoma, D. L., R. E. Frenkel, and J. M. Trappe. 1991. Fruiting of hypogeous fungi in Oregon Douglas-fir forests: Seasonal and habitat variation. *Mycology* 83:335–353.

Marquis, R. J., and C. J. Whelan. 1994. Insectivorous birds increase growth of white oak through consumption of leaf-chewing insects. *Ecology* 75:2007–2014.

Maser, C., J. M. Trappe, and R. A. Nussbaum. 1978. Fungal–small mammal interrelationships with emphasis on Oregon coniferous forests. *Ecology* 59:799–809.

Mason, R. R., T. R. Torgerson, B. E. Wickman, H. G. Paul. 1983. Natural regulation of a Douglas-fir tussock moth (*Lepidoptera lymantriidae*) population in the Sierra Nevada. *Environmental Entomology* 12:587–594.

May, R. M., and G. F. Oster. 1976. Bifurcations and dynamic complexity in simple ecological models. *American Naturalist* 110:573–599.

McClure, M. S. 1986. Role of predators in regulation of endemic populations of Matsucoccus matsumurae (*Homoptera: Margarodidae*) in Japan. *Environmental Entomology* 15(4):976–983.

McCutcheon, T. L., G. C. Carroll, and S. Schwab. 1993. Genotypic diversity in populations of a fungal endophyte from Douglas fir. *Mycologia* 85(2):180–186.

McLaren, B. E., and R. O. Peterson. 1994. Wolves, moose, and tree rings on Isle Royale. *Science* 266:1555–1558.

McManus, J. F., G. C. Bond, W. S. Broecker, S. Johnsen, L. Labeyrie, and S. Higgins. 1994. High-resolution climate records from the North Atlantic during the last interglacial. *Nature* 371:326–329.

Menge, J. A., and L. F. Grand. 1978. Effects of fertilization on production of basidiocarps by mycorrhizal fungi in loblolly pine plantations. *Canadian Journal of Botany* 56:2357–2362.

Mills, L. S., M. E. Soile, and D. F. Doak. 1993. The keystone-species concept in ecology and conservation. *BioScience* 43(4):219–224

Milton, S. J., W. R. J. Dean, M. A. du Plessis, and W. R.

Siegfried. 1994. A conceptual model of arid rangeland degradation. *BioScience* 44(2):70–76.

Morrison, D. J., G. W. Wallis, and L. C. Weir. 1988. *Control of* Armillaria *and* Phellinus *root diseases: 20-year results from the Skimikin stump-removal experiment.* Victoria, BC: Canadian Forestry Service, Pacific Forestry Centre.

Mylne, M. F., and P. R. Rowntree. 1992. Modelling the effects of albedo change associated with tropical deforestation. *Climate Change* 21:317–343.

Naeem, S., L. J. Thompson, S. P. Lawler, J. H. Lawton, and R. M. Woodfin. 1994. Declining biodiversity can alter the performance of ecosystems. *Nature* 368:734–737.

———. 1995. Empirical evidence that declining biodiversity may alter the performance of terrestrial ecosystems. *Philosophical Transactions of the Royal Society of London (B)* 347:249–262.

Neilson, R. P., and D. Marks. 1994. A global perspective of regional vegetation and hydrologic sensitivities and risks from climate change. *Journal of Vegetative Science* 5:715–730.

Neilson, R. P., G. A. King, and G. Koerper. 1990. Toward a rule-based biome model. In *Proceedings of the Seventh Annual Pacific Climate (PACLIM) Workshop,* ed. J. L. Betancourt and V. L. Tharp. Sacramento: California Department of Water Resources.

Neilson, R. P., G. A. King, R. L. DeVelice, and J. M. Leniham. 1992. Regional and local vegetation patterns: The responses of vegetation diversity to subcontinental air masses. In *Landscape boundaries: Consequences for biotic diversity and ecological flows: Ecological studies 92,* ed. A. J. Hansen and F. di Castri. New York: Springer-Verlag.

Nepstad, D., C. Uhl, and E. A. Serrao. 1990. Surmounting barriers to forest regeneration in abandoned, highly degraded pastures: A Case study from Paragorninas, Para, Brazil. In *Alternatives to deforestation,* ed. A. B. Anderson. New York: Columbia University Press.

Nilsson, M. C., P. Hogberg, O. Zackrisson, and W. Fengyou. 1993. Allelopathic effects by *Empetrum hermaphroditum* Hagerup on development and nitrogen uptake by roots and mycorrhizae of *Pinus silvestris* L. *Canadian Journal of Botany* 71:620–628.

Noss, R. F. 1991. From endangered species to a biodiversity. In *Balancing on the brink of extinction: The Endangered Species Act and lessons for the future,* ed. by K. Kohm. Washington, DC: Island Press.

Odum, E. P. 1992. Great ideas in ecology for the 1990s. *BioScience* 42:542–545.

Oliver, C. D. 1981. Forest development in North America following major disturbances. *Forest Ecology and Management* 3:153–168.

O'Neill, R. V. 1989. Perspectives in hierarchy and scale. In *Perspectives in Ecological Theory,* ed. J. Roughgarden, R. M. May, and S. A. Levin. Princeton, NJ: Princeton University Press.

O'Neill, R.V., D. L. DeAngelis, J. B. Waide, and T. F. H. Allen. 1986. A hierarchical concept of ecosystems. Princeton, NJ: Princeton University Press.

O'Neill, R. V., A. R. Johnson, and A. W. King. 1989. A hierarchical framework for the analysis of scale. *Landscape Ecology* 3:193–205.

O'Neill, R. V., R. H. Gardner, M. G. Turner, and W. H. Romme. 1992. Epidemiology theory and disturbance spread on landscapes. *Landscape Ecology* 7(1):19–26.

Overpeck, J. T., D. Rind, and R. Goldberg. 1990. Climate-induced changes in forest disturbance and vegetation. *Nature* 343:51–53.

Pacala, S. W., M. P. Hassell, and R. M. May. 1990. Host-parasitoid asso ciations in patchy environments. *Nature* 344:150–151.

Parke, J. L., R. G. Linderman, and C. H. Black. 1983a. The role of ectomycorrhizas in drought tolerance of Douglas-fir seedlings. *New Phytology* 95:83–95.

Parke, J. L., R. G. Linderman, and J. M. Trappe. 1983b. Effects of forest litter on mycorrhiza development and growth of Douglas-fir and western red cedar seedlings. *Canadian Journal of Forest Research* 13:666–671.

Parker, V. T., and M. P. Yoder-Williams. 1989. Reduction of survival and growth of young *Pinus jeffreyi* by an herbaceous perennial, *Wyethia mollis. American Midland Naturalist* 121:105–111.

Parungo, F., Z. Li, X. Li, D. Yang, and J. Harris. 1994. Gobi dust storms and the Great Green Wall. *Geophysical Research Letters* 21 (11):999–1002.

Pastor, J., and W. M. Post. 1988. Response of northern forests to CO_2-induced climate change. *Nature* 334:55–58.

Pellissier, F., and L. Trosset. 1989. Effect of phytotoxic solutions on the respiration of mycorrhizal and nonmycorrhizal spruce roots (*Pices abies* L. Karst). *Annals of Scientific Forestry* (46 suppl.):731s–733s.

Perry, D. A. 1988. An overview of sustainable forestry. *Journal of Pesticide Reform* 8:8–12.

———. 1994. *Forest ecosystems.* Baltimore: The Johns Hopkins University Press.

———. 1995a. Self-organizing systems across scales. *Trends in Ecology and Evolution* 10:241–244.

———. 1995b. Landscapes, humans, and other system-level considerations: A discourse on ecstasy and laundry. In *Proceedings of a workshop on ecosystem management in western interior forests,* ed. D. Baumgartner and

R. Everett. Pullman: Washington State University Cooperative Extension Unit.

Perry, D. A., and J. Maghembe. 1989. Ecosystem concepts and current trends in forest management: Time for reappraisal. *Forest Ecology and Management* 26:123–140.

Perry, D. A., et al. eds. 1989a. *Maintaining long-term productivity of Pacific Northwest forests.* Portland, OR: Timber Press.

Perry, D. A., M. P. Amaranthus, J. G. Borchers, et al. 1989b. Bootstrapping in ecosystems. *BioScience* 39:230–237.

Perry, D. A., T. Bell, and M. A. Amaranthus. 1992. Mycorrhizal fungi in mixed-species forests and other tales of positive feedback, redundancy, and stability. In *The ecology of mixed-species stands of trees,* ed. M. G. R. Cannell, D. C. Malcolm, and P. A. Robertson. Special publication no. 11 of the British Ecological Society. London, England: Blackwell.

Petraitis, P. S., R. E. Latham, and R. A. Niesenbaum. 1989. The maintenance of species diversity by disturbance. *Quarterly Review of Biology* 64:393–418.

Pickett, S. T. A., and M. J. McDonnell. 1989. Changing perspectives in community dynamics: A theory of successional forces. *TREE* 4:241–245.

Pickett, S. T. A., and J. N. Thompson. 1978. Patch dynamics and the design of nature reserves. *Biological Conservation* 13:27–37.

Pickett, S. T. A., S. L. Collins, and J. J. Armesto. 1987. Models, mechanisms and pathways of succession. *The Botanical Review* 53(3):335–371.

Pickett, S. T. A., V. T. Parker, and P. L. Fiedler. 1992. The new paradigm in ecology: Implications for conservation biology above the species level. In *Conservation Biology,* ed. P. L. Fiedler and S. K. Jain. New York: Chapman and Hall.

Pielke, R. A., and R. Avissar. 1990. Influence of landscape structure on local and regional climate. *Landscape Ecology* 4(2/3):133–155.

Pimm, S. L. 1986. Community stability and structure. In *Conservation Biology: The science of scarcity and diversity,* ed. M. E. Soulé. Sunderland, MA: Sinauer Associates.

———. 1991. *The balance of nature? Ecological issues in the conservation of species and communities.* Chicago: University of Chicago Press.

Prigogine, I., and I. Stengers. 1984. *Out of chaos.* Toronto, Canada: Bantam.

Rahmstorf, S. 1994. Rapid climate transitions in a coupled ocean-atmosphere model. *Nature* 372:82–85.

Read, D. J. 1993. Plant-microbe mutualisms and community structure. In *Ecological studies 99 Biodiversity and ecosystem function,* ed. E. D. Schulze and H. A. Mooney. New York: Springer-Verlag.

Remmert, H. ed. 1991. *The mosaic-cycle concept of ecosystems.* Berlin: Springer-Verlag.

Salati, E. 1987. The forest and the hydrological cycle. In *The geophysiology of amazonia,* ed. R. E. Dickinson. New York: John Wiley & Sons.

Sampson, R. N., and D. L. Adams. 1995. *Assessing forest ecosystem health in the inland Northwest.* New York: Food Products Press.

Saunders, D. A., R. J. Hobbs, and C. R. Margules. 1991. Biological consequences of ecosystem fragmentation: A review. *Conservation Biology* 5:18–25.

Schisler, D. A., and R. G. Linderman. 1989. Influence of humic-rich organic amendments to coniferous nursery soils on Douglas-fir growth, damping-off and associated soil microorganisms. *Soil Biology and Biochemistry* 21(3):403–408.

Schlesinger, W. H., J. F. Reynolds, G. L. Cunningham, L. F. Huenneke, W. M. Jarrell, R. A. Virginia, and W. G. Whitford. 1990. Biological feedbacks in global desertification. *Science* 247:1043–1048.

Schowalter, T. D. 1989. Canopy arthropod community structure and herbivory in old-growth and regenerating forests in western Oregon. *Canadian Journal of Forest Research* 19:318–322.

Schowalter, T. D., and P. Turchin. 1993. Southern pine beetle infestation development: Interaction between pine and hardwood basal areas. *Forest Ecology* 39:201–210.

Schowalter, T. D., R. N. Coulson, and D. A. Crossley Jr. 1981. Role of southern pine beetle and fire in maintenance of structure and function of southeastern coniferous forest. *Environmental Entomology* 10:821–825.

Shukla, J., C. Nobre, and P. Sellers. 1990. Amazon deforestation and climate change. *Science* 247:1322–1325.

Simard, S., and A. Vyse. 1994. Paper birch: Weed or crop tree in the interior cedar-hemlock forests of South British Columbia. In *Symposium proceedings of interior cedar–hemlock–white pine forests: Ecology and management,* March 2–4, Spokane, Washington. Pullman, WA: Department of Natural Resource Sciences, Washington State University.

Swetnam, T. W. 1993. Multicentury, regional-scale patterns of western spruce budworm outbreaks. *Ecological Monographs* 63(4):3999.

Swetnam, T. W., and J. L. Betancourt. 1990. Fire–southern oscillation relations in the southwestern United States. *Science* 249:1017–1020.

Swetnam, T. W., and A. M. Lynch. 1989. A tree-ring reconstruction of western spruce budworm history in the southern Rocky Mountains. *Forestry Science* 35:962–986.

Tappeiner, J. C., and A. A. Alm. 1975. Undergrowth vegetation effects on the nutrient content of litterfall and soils in red pine and birch stands in northern Minnesota. *Ecology* 56:1193–1200.

Tappeiner, J., J. Zasada, P. Ryan, and M. Newton. 1991. Salmonberry clonal and population structure: The basis for persistent cover. *Ecology* 72:609–618.

Taylor, K. C., G. W. Lamorey, G. A. Doyle, R. B. Alley, P. M. Grootes, P. A. Mayewski, J. W. C. White, and L. K. Barlow. 1993. The "flickering switch" of late Pleistocene climate change. *Nature* 361:432–436.

Thatcher, R. C., J. L. Searcy, J. E. Coster, and G. D. Hertel, eds. 1980. *The southern pine beetle.* Technical bulletin 1631. Washington, DC: USDA Forest Service.

Thouveny, N., J. L. de Beaulieu, E. Bonifay, K. M. Creer, J. Gulot, M. Icole, S. Johnsen, J. Jouzel, M. Rellle, T. Williams, and D. Williamson. 1994. Climate variations in Europe over the past 140 kyr deduced from rock magnetism. *Nature* 371:503–506.

Tilman, D. 1994. Competition and biodiversity in spatially structured habitats. *Ecology* 75(1):2–16.

Tilman, D., and J. A. Downing. 1994. Biodiversity and stability in grasslands. *Nature* 367:363–365.

Tilman, D., R. M. May, C. L. Lehman, and M. A. Nowak. 1994. Habitat destruction and the extinction debt. *Nature* 371:65–66.

Torgerson, T. R., D. L. Dahlsten, M. H. Brookes, et al., eds. 1979. *The Douglas-fir tussock moth: A synthesis.* Technical bulletin 1585. Washington, DC: USDA Forest Service.

Torgerson, T. R., R. R. Mason, and R. W. Campbell. 1990. Predation by birds and ants on two forest insect pests in the Pacific Northwest. *Studies in Avian Biology* 13:14–19.

Trappe, J. M. 1987. Phylogenetic and ecological aspects of mycotrophy in the angiosperms from an evolutionary standpoint. In *Ecophysiology of VA mycorrhizal plants,* ed. G. R. Safir. Boca Raton, FL: CRC Press.

Trappe, J. M., and D. L. Luoma. 1992. The ties that bind: Fungi in ecosystems. In *The fungal community: Its organization and role in the ecosystem,* ed. G. C. Carroll and D. T. Wicklow. New York: Marcel Dekker, Inc.

Turner, M. G. 1987. *Landscape heterogeneity and disturbance.* New York: Springer-Verlag.

Turner, M. G., R. H. Gardner, V. H. Dale, and R. V. O'Neill. 1989. Predicting the spread of disturbance across heterogeneous landscapes. *Oikos* 55:121–129.

Turner, M. G., W. H. Romme, R. H. Gardner, R. V. O'Neill, and T. K. Kratz. 1993. A revised concept of landscape equilibrium: Disturbance and stability on scaled landscapes. *Landscape Ecology* 8:213–227.

Urban, D. L., R. V. O'Neill, and H. H. Shugart Jr. 1987. Landscape ecology. *BioScience* 37(2):119–127.

Vitousek, P. M., and D. U. Hooper. 1993. Biological diversity and terrestrial ecosystem biogeochemistry. In *Biodiversity and ecosystem function,* ed. E. D. Schulze and H. A. Mooney. New York: Springer-Verlag.

Walker, B. H. 1992. Biodiversity and ecological redundancy. *Conservation Biology* 6:18–23.

Way, M. J., and K. C. Khoo. 1992. Role of ants in pest management. *Annual Review of Entomology* 37:479–503.

Weaver, A. J., and T. M. C. Hughes. 1994. Rapid interglacial climate fluctuations driven by North Atlantic ocean circulation. *Nature* 367:447–450.

White, M. A., and D. J. Mladenoff. 1994. Old-growth forest landscape transitions from pre-European settlement to present. *Landscape Ecology* 9:191–205.

Wickman, B. E., R. R. Mason, and T. W. Swetnam. (1993). In *Proceedings: Individuals, Populations, and Patterns,* September 7–10, Norwich, England.

Wiens, J. A. 1989. Spatial scaling in ecology. *Functional Ecology* 3:385–397.

Wiens, J. A., N. C. Stenseth, B. Van Horne, and R. A. Ims. 1993. Ecological mechanisms and landscape ecology. *Oikos* 66:369–380.

Wilson, E. O. 1987. The little things that run the world. *Conservation Biology* 1:344–346.

Wilson, M. V. 1982. Microhabitat influences on species distributions and community dynamics in the conifer woodland of the Siskiyou Mountains, Oregon. Ph.D. thesis, Cornell University, Ithaca, NY.

Wilson, M. V., and D. B. Botkin. 1990. Models of simple microcosms: Emergent properties and the effect of complexity on stability. *American Naturalist* 135:414–434.

Zutter, B. R., D. H. Gjerstad, and G. R. Glover. 1987. Fusiform rust in cidence and severity in loblolly pine plantations following herbaceous weed control. *Forestry Science* 33(3):790–800.

4 The Biological Landscape

Malcolm L. Hunter Jr.

The Movements of Organisms 59
Box: Metapopulations 60
Fragmentation and Its Effects 62
Dissection 62
Perforation 63
Fragmentation and Attrition 63
Maintaining Landscape Movements 64
Dissection 64
Perforation 64
Fragmentation 64
Attrition 65
Summary 65
Acknowledgments 66
Literature Cited 66

It is unfortunate that forest ecologists and managers cannot fly. If we could travel above the landscape in small, leisurely flocks, observing large-scale patterns and discussing them with colleagues, some common understanding about these patterns and their importance would soon emerge. Such understanding is not as easily obtained by driving forest roads in a van full of people, crowding a few people into a fast noisy plane, or scanning aerial photos. One consequence of not being able to share experiences at large spatial scales is that we do not have a well-accepted common language for large-scale patterns, and thus it is necessary to begin with some definitions.

Figure 4.1 provides a starting point. Does it depict a natural community, an ecosystem, or a landscape? We can define a community as an array of interacting species, an ecosystem as a community and the physical environment it occupies, and a landscape as an array of interacting ecosystems. The key word here is "interacting." At some level every organism in this scene is interacting with every other, albeit indirectly in many cases, and we could call the whole thing an

Figure 4.1 Does this scene portray a natural community, an ecosystem, or a landscape? At the scales often used by ecologists, this would be a landscape composed of several different ecosystems. (Reprinted with permission from *Fundamentals of conservation biology,* published by Blackwell Scientific Publishers.)

ecosystem. However, by the same logic we could say that all the organisms on earth interact with one another (through global carbon and oxygen cycles, for example) and thus the whole earth is one ecosystem. Alternatively, we could shift downward in scale and say that all the salamanders, invertebrates, mosses, and fungi interacting within the confines of a single fallen log, constitute an ecosystem. We have to make some arbitrary decisions, and in practice this usually comes down to a decision about scale. Most ecologists recognize ecosystems at the scale of the patches of vegetation that one can easily see from a small plane; patches one would usually measure in acres or hectares, rather than square miles or kilometers, or square meters or feet. From this perspective, Figure 4.1 would represent multiple ecosystems, and thus it depicts a landscape.

Again, it must be emphasized that this is an arbitrary decision. Ecosystems defined at this scale are artificial constructs of convenience to people, allowing us to organize our thinking and communicate with one another. In recent years there has been a growing tendency, especially among natural resource managers, to define ecosystems at quite large scales, as in the "Greater Yellowstone Ecosystem." This tendency can probably be traced to the increasing emphasis on ecosystem management, for which one of the key principles is thinking at larger spatial scales. The emphasis on thinking at large spatial scales is laudable, but one must remember that when advocates of ecosystem management use the term "ecosystem" they are often referring to something that is so extensive that most ecologists would call it a landscape, or perhaps even an ecological region.

In many ways it was the artificiality of these decisions about scale that spawned landscape ecology as a distinct subdiscipline of ecology. As ecology progressed, ecologists became more sensitive to processes that cut across the artificial barriers they recognized between ecosystems. For example, in Figure 4.1 leaves and wood that grow in the forest, but then fall into the river, may be a primary source of energy for the riverine ecosystem. Conversely, some energy moves from the stream to the forest—a salmon in the jaws of a bear, larval amphibians metamorphosing into terrestrial adults. The study of these transecosystem processes is a cornerstone of landscape ecology (Dunning et al. 1992).

The single most important thing to know from landscape ecology is actually quite obvious. It is a tautology based on the artificiality of ecosystem definitions: boundaries between ecosystems are permeable and many important processes happen across these boundaries. In this chapter we will focus on biological transboundary processes, specifically the movements of organisms among ecosystems.

The Movements of Organisms

Organisms move at many different scales—from a plant turning to orient its leaves to the sun to a bird undertaking intercontinental migrations—but at a landscape scale there are four basic, widespread types of movement.

1. *Home range movements* occur when animals move around a defined area over a relatively short period, usually days or weeks. Virtually all plants and most animals are confined to a single ecosystem, and thus the home range movements of most animals would not constitute landscape-scale movements. On the other hand, many large birds and mammals, especially predatory species, have home ranges that extend beyond the boundaries of a single ecosystem (as commonly defined), and thus their home range movements involve traveling across landscapes. Additionally, some species with modest-sized home ranges routinely move between different ecosystems to meet different needs. For example, the American woodcock requires open fields for roosting and displaying, plus nearby young forests for foraging (Sepik et al. 1981). Similarly, many riparian species regularly cross the interface between terrestrial and aquatic ecosystems.

2. *Migration* is the seasonal movement of animals back and forth between two areas. These movements are driven by seasonal changes in the availability of resources, particularly food, or by seasonal changes in animals' needs for different resources (e.g., resources for reproduction are often needed for only one season per year). With few exceptions, migration will involve moving across different ecosystems even if the animal ends up in an ecosystem with a structure similar to the one that it left.

3. *Dispersal* is the movement of young organisms (usually juvenile animals or plant propagules) out of their natal area. Among many plants and smaller animals, especially invertebrates, dispersal may be within a single ecosystem, but for larger, more mobile organisms it will often involve landscape-scale movement.

4. *Geographic range shifts* are long-term movements of plant and animal populations in response to environmental change, particularly climate change. They are based on multiple generations of dispersal movements and thus can cover considerable distances. By definition, a shift in geographic range will be of sufficient scale to constitute a landscape-scale movement.

Conservationists have become quite concerned about how human-induced alterations of landscapes may inhibit these movements. They are asking four basic questions corresponding to the four types of movement:

1. *Can animals move freely over an area large enough to find the resources they require?* For animals that confine their home range movements to a single ecosystem, it can clearly be a problem if ecosystems are reduced in size by human alterations to a landscape. This would seem to be less of an issue for species, notably predators, that travel over such large home ranges that they regularly must move among multiple ecosystems. Nevertheless, even for a species like the wolf, which is a habitat generalist routinely traversing hundreds of square kilometers, having landscapes that are relatively free of human disturbance can be important (Fuller et al. 1992). Those species that use multiple ecosystems in a single home range may be a minority, but they include many of the species that we value most and that can have keystone roles in ecosystems.

2. *Can animals migrate freely between seasonal ranges?* Many migratory animals are highly mobile and thus are easily able to traverse ecosystems even if the ecosystems are not particularly suitable habitat. The ability of terrestrial birds to cross wide bodies of water is the best example of this. Nevertheless, barriers to migration are an issue for many nonflying migrants. For example, the altitudinal movements of large mammals such as bighorn sheep and elk can easily be blocked or impeded by landscape changes (Baker 1978). On a smaller scale, amphibians moving to breeding ponds may be hindered by landscape alterations (Langton 1989). In the aquatic realm, dams are often a barrier to the migration of fishes. Even with flying species one needs to be concerned with the availability of stopover habitat where

individuals can rest safely and forage during migration. Otherwise, whole migration routes could be fragmented by the loss of some strategic habitat.

3. *Can organisms disperse among subpopulations and habitat patches?* This question can be subdivided into some more specific questions, but to understand the underlying issues it is necessary to be familiar with the concept of metapopulations (see the box). First, is there sufficient immigration of young organisms from source populations into sink populations? Without this influx, many sink populations will disappear. Similarly, can dispersing individuals colonize unoccupied habitat after the local population has gone extinct or after the process of succession and disturbance creates new habitat? Finally, is there sufficient genetic interchange among habitat patches to avoid problems, notably inbreeding and random genetic drift, that can develop in small, isolated populations? As a rule of thumb, one successful interchange per generation is thought to be sufficient to avoid genetic isolation (Frankel and Soulé 1981, Schonewald-Cox et al. 1983).

Because fragmentation can reduce vast ecosystems to small, isolated ecosystems, it can subdivide species that once had large, regionwide populations into much smaller groups. If the species has reasonably good dispersal abilities, these groups may persist as a metapopulation; if it does not, the species may disappear from all the habitat patches one by one (Templeton et al. 1990). Even when habitat patches are naturally small and isolated (e.g., cattail marshes), fragmentation can further reduce the sizes of habitat patches and increase the distances between them, thus making subpopulations smaller, more isolated, and more vulnerable to extinction.

4. *Can organisms shift their geographic ranges in response to climate change?* Most species have a demonstrated ability to shift their range in response to long-term climate change. They would not have survived the

Metapopulations

Consider the cattail, a species found throughout much of the northern hemisphere, but which occurs only in discrete patches of habitat—freshwater wetlands—that are usually only a small portion of the overall landscape. Within their patches of habitat, cattails are often exceedingly abundant, but between these patches there are large stretches of land without any cattails. Cattails are an extreme example of an attribute that is common to many species: patchy distributions. Patchy distributions are the basis for the metapopulation model of population structure, and understanding metapopulations is key to understanding the biological structure of landscapes. In other words, you can think of the habitat patches as different examples of the same type of ecosystem separated in a landscape of other types of ecosystems.

In metapopulation terms, each patch of habitat contains a different population of the species in question, and a group of different patch populations is collectively called a metapopulation. To put it another way, a metapopulation is a "population of populations" (Hanski and Gilpin 1991). Metapopulations exist at a spatial scale where individuals can occasionally disperse among different populations (patches) but do not make frequent movements because the patches are separated by substantial expanses of unsuitable habitat. Inhabitants of different patches would usually not constitute separate populations in terms of genetic isolation, but they might. We can recognize this uncertainty and still use the metapopulation idea by calling the inhabitants of each patch a subpopulation.

Our brief examination of metapopulation dynamics will focus on two types of subpopulation—sources and sinks—and two processes—extinctions and colonizations. Some subpopulations are *sources* because they produce a substantial number of emigrants that disperse to other patches. Some subpopulations are *sinks* because they cannot maintain themselves without a net immigration of individuals from other subpopulations. In other words, some subpopulations are saved from extinction by immigration from other subpopulations; this process has been called the rescue effect (Brown and Kodric-Brown 1977, Harrison 1991). Sinks and sources are useful concepts, but in practice it is hard to distinguish between them with confidence because it is difficult to monitor the movements of individuals among subpopulations.

Despite the balancing effect of immigration and emigration, subpopulations sometimes appear and disappear in a manner often compared to small lights winking on

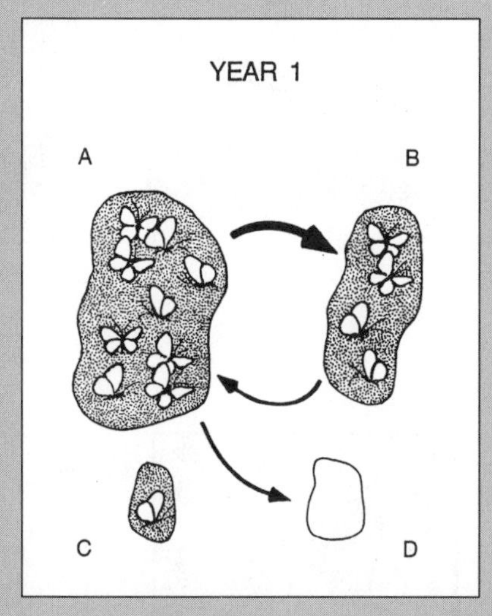

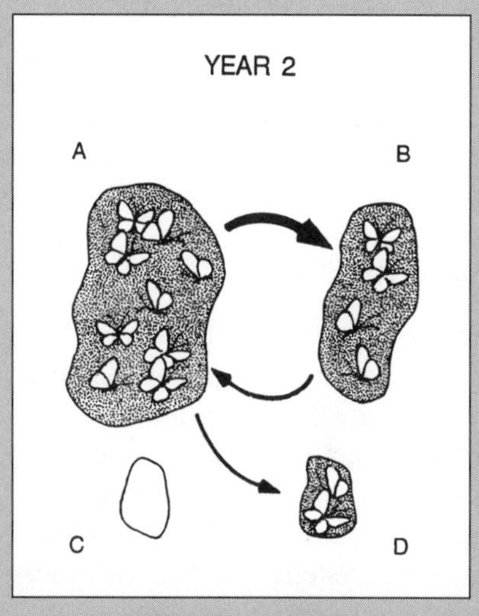

A schematic depiction of metapopulations. Occupied patches are shaded, and empty ones are unshaded. Arrows represent movement among patches, with the width of arrow corresponding to the number of dispersers. Patch A is a source of butterflies because it is a net producer of emigrants, but patch B is a sink because it is a net recipient of immigrants. The butterfly subpopulation in patch C has become extinct, while a new subpopulation in patch D has begun to develop from dispersers that have colonized the patch. Patch A is probably a core subpopulation because of its size and persistence, whereas C and D are satellites. We would need data from more years to say if B is a core or a satellite.

and off in a large darkness. More formally, these appearances and disappearances are called turnover (Hanski and Gilpin 1991). Each appearance is a colonization event, for example when a species of bark beetle colonizes a forest patch after some of its preferred host trees die. Each disappearance is a local extinction event, for example when the bullfrogs in a pond are killed by a disease. These processes occur at an ecological time scale and may be quite rapid (e.g., a windstorm drops a swarm of spiders and seeds to colonize a recently burnt forest) or quite slow (e.g., after the burn, an annual plant species restricted to recently burnt sites gradually disappears). These events may be interwoven with the whole pattern of disturbance and succession operating in a given ecosystem (e.g., the spider and plant examples of the preceding sentence), or they may affect only one or a few species (e.g., the bullfrogs and their pathogen). Subpopulations that persist for relatively long periods are often called *core* subpopulations, whereas those that are more likely to wink on and off are often called *satellite* subpopulations (Boorman and Levitt 1973). Core subpopulations are likely to be large and a net source of individuals, and satellite subpopulations are likely to be small and a net sink, but no doubt there are exceptions to these generalizations.

To the extent that these small-scale extinction and colonization events are reasonably in balance with one another, they need not worry conservation biologists. However, to state the all-too-obvious problem, the rate of subpopulation extinctions often exceeds the rate of colonizations in the lands and waters dominated by people.

In sum, the metapopulation concept offers a useful framework for understanding the dynamics of populations in patchy landscapes, and patchy landscapes are becoming more and more common because of human activity.

(Reprinted with permission from Malcolm Hunter. 1996. *Fundamentals of conservation biology,* Cambridge, Mass.: Blackwell Scientific Publishers.)

dramatic changes of the last 20,000 years if they did not. This does not, however, mean that they will survive future climate changes if human activities have limited their opportunities to disperse and generally reduced and stressed their populations (Hunter 1992).

Fragmentation and Its Effects

Conservationists worry about these questions because they realize that we have fragmented many landscapes in ways that are likely to inhibit these movements. Note that fragmentation is not exclusively an anthropogenic phenomenon: Avalanches, fires, tornadoes, and other disturbance factors can also separate a single ecosystem into two or more isolated parts. Furthermore, some ecosystems are intrinsically isolated from other ecosystems of the same type: mountaintops and desert springs, for example. This said, it is clear that people are now the primary agent of fragmentation. Before examining the effects of fragmentation, we need to describe the process, using some ideas of Richard Forman (1995) about the typical progress of the human-induced fragmentation process (Figure 4.2).

People usually initiate fragmentation by building a road or trail into a natural landscape, thereby dissecting it. Next, they perforate the landscape by converting some natural ecosystems into agricultural lands. As more and more lands are converted to agriculture, these patches coalesce and the natural ecosystems are isolated from one another; at this stage fragmentation has occurred. Finally, as more of the natural patches are converted, becoming smaller and farther apart, attrition is occurring. In sum, fragmentation involves two closely related processes: a reduction in total area of the type of ecosystem in question and increasing isolation among different examples of the same type of ecosystem.

In the following pages, we examine some of the interplay among the four types of movement and the three stages from Figure 4.2: dissection, perforation, and fragmentation and attrition.

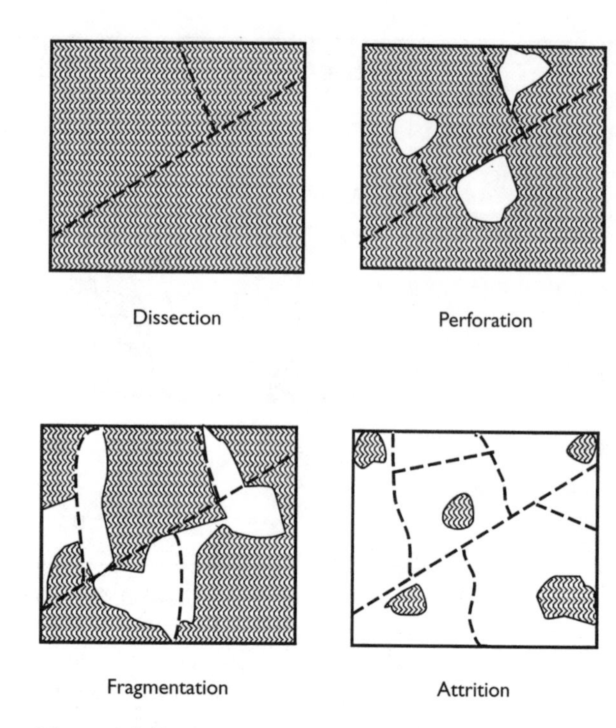

Figure 4.2 Four stages in the process of fragmentation. The shaded areas represent areas dominated by natural ecosystems while the white areas represent areas, such as agricultural lands, shaped by people. (Based on ideas of Forman 1995. Reprinted with permission from *Fundamentals of conservation biology*, published by Blackwell Scientific Publishers.)

Dissection

Here the key question is to what extent do roads and other linear features constitute barriers to movement? In extreme cases, linear features may be a nearly absolute physical barrier. A highway with a high concrete curb or lane dividers can stop the movement of many less mobile animals such as turtles and amphibians. In parts of southern Africa, livestock quarantine fences have severely impeded the migration of large mammals such as wildebeest (Williamson et al. 1988). Linear features may also constitute a psychological barrier to animals; in other words, animals that may be physically able to cross the barrier in a few seconds may be unwilling to do so. For example, a German study found that roads inhibited the movements of small mammals and ground beetles (Mader 1984). In one case, a forest road closed to traffic and only 3 m wide appeared to inhibit small mammal movement.

In most cases, linear features probably constitute a filter, not an absolute barrier. Some individuals cannot or will not cross, but others do. Motivation is likely to be critical in determining what percentage of potential crossers actually make it across. A road that would stop a turkey from extending its home range to include another forest stand, would be unlikely to stop a juvenile turkey intent on dispersing. A filtering effect can go beyond turning back some individuals; some may die while attempting to cross a linear barrier. The best-known example of this is roadkills. For example, from 1969 to 1975 road maintenance crews in Nebraska picked up about 3,500 bird and mammal carcasses per year along a 732-km highway (Case 1978)—a number that would extrapolate into millions of carcasses on all the world's highways. In the Nebraska case, the smallest items recorded were rabbits, and thus the numbers would have been much higher if small birds and mammals had been included, to say nothing of amphibians, reptiles, and invertebrates (Oxley et al. 1974). Human hunters and trappers who use roads to access remote areas can also exert a filtering effect on animals attempting to cross the road (Fuller et al. 1992).

Perforation

In the next stage of fragmentation, small human-induced patches appear, but these are not likely to be large enough to constitute a barrier to movement for most species. Home range movements by relatively immobile species might be affected, but most species can simply go around a small patch if they cannot cross it directly. The most significant effect of perforation is that it generates extensive areas of edge where natural and human-generated ecosystems come together. Much has been written about edges (see Hunter 1990) and a thorough review is beyond our scope here. It is now widely known that the old maxim, "Create more edge," is a generalization from the days when wildlife management was synonymous with game management and must be scrutinized carefully. The best-documented negative effects associated with edges involve predation on bird nests plus brood parasitism by brown-headed cowbirds (see Paton 1994 for a review). Increased vulnerability to predation near edges has also been documented for reptiles (Temple 1987) and mammals (Prins and Iason 1989). Changes in microclimate near edges are likely to be problematic for some amphibians and plants (Ranney et al. 1981, Chen et al. 1992).

There is an important caveat about research on edge effects: A large portion of it has been undertaken along edges in agricultural landscapes where the forests have been reduced to isolated woodlots (i.e., the fragmentation or attrition stage of Figure 4.2), rather than in a perforated forest landscape. Further replication of this work in forested landscapes is necessary. For example, some studies have not shown a predation/edge effect in a perforated forest landscape (Ratti and Reese 1988, Rudnicky and Hunter 1993).

Fragmentation and Attrition

In landscapes that have reached these stages of fragmentation, virtually all home range movements will be inhibited; a few raptor species might provide an exception to this generalization. Similarly, migration by most nonflying animals will be significantly restricted. At these levels of fragmentation, the most interesting questions will revolve around the effects on dispersal and ultimately range shift movements. If dispersal among the residual ecosystems is extremely limited, the occupants of these small patches may not even constitute a metapopulation. They will effectively be small, isolated populations, and therefore they may suffer from all the problems associated with being a small population.

These problems are often expressed in terms of how the probability of a population becoming extinct is affected by four kinds of uncertainties or, more technically, stochasticities (Shaffer 1981). In brief, these are (1) *demographic stochasticity,* principally based on problems that derive from a having an unbalanced sex or age structure; (2) *genetic stochasticity* associated with genetic problems such as inbreeding and random genetic drift; (3) *environmental stochasticity,* random fluctuations in habitat quality based on issues such as predators, water and food availability, and weather; and (4) *catastrophes,* events such as droughts and hurricanes that can occur at unpredictable intervals.

In a landscape that has experienced attrition, not only are the populations occupying small, isolated ecosystems likely to become extinct, but after an extinction event it will be difficult for colonists to occupy the site. Ultimately, inhibition of dispersal in a fragmented landscape will also inhibit geographic range shifts because these shifts occur largely through the dispersal of juveniles into new range along the species's frontier (Peters and Lovejoy 1992).

Note that our discussion on dissection and perforation focused on organisms that make home range and migratory movements, and these are all animals. When the discussion turns to dispersal and geographic range shifts in fragmented landscapes, we must not overlook plants. Some plant species that are wind-, water-, or animal-dispersed can be reasonably mobile, but many species are not particularly mobile, especially those that generally reproduce vegetatively. Many plants have evolved adaptations to being relatively immobile, such as a tolerance for inbreeding, but with few exceptions they do need to move in response to climate change (Woods and Davis 1989).

In sum, dispersal is a key process for maintaining population viability, and ultimately the persistence of an entire species may depend on dispersal as the mechanism for shifting geographic ranges. Although dissection and perforation are unlikely to affect dispersal significantly for most organisms, dispersal may be inhibited for many species living in landscapes that have reached a high level of fragmentation or attrition. Unfortunately, we know relatively little about dispersal because it is a difficult phenomenon to study (Chepko-Sade and Halpin 1987).

Maintaining Landscape Movements

In this section we will briefly examine some ways to maintain movements across landscapes that have reached each of the four stages in Figure 4.2. This discussion will focus on proximate solutions and will not address what most conservationists recognize as preferable: ultimate solutions such as reducing human populations and consumption.

Dissection

Conceptually, it is easy to redesign human-engineered linear barriers to allow more organisms to cross them. Underpasses beneath roads have been built for animals ranging from frogs to Florida panthers. Fences have been designed to hold livestock without stopping wild ungulates. Fishways are often constructed around dams. The effectiveness and cost of some of these engineering solutions can be challenged, but the basic idea is sound.

Perforation

Generation of edges is often associated with perforation, and one solution to this may seem counterintuitive—make the perforations larger. It is a basic rule of geometry that a single large area will have less edge than several small areas of the same total size. Recognition of this fact has led some authors to advocate having larger, clustered clearcuts rather than many, scattered small ones (Franklin and Forman 1987, Hunter 1993).

We could also argue for land-use practices that do not perforate the landscape in the first place. For example, replacing traditional clearcuts with clearcuts in which some trees and snags are retained may mean that the landscape is not perforated from the perspective of some species, or at least will recover from perforation more quickly.

Fragmentation

As with perforation, fragmentation can be mitigated in some cases by employing land-use practices that do not drastically change the original ecosystems. Under some circumstances it might also be desirable to concentrate human extraction of natural resources on a relatively small area. This idea is based on a "triad model" that recognizes three basic types of land use: (1) ecological reserves with no extra commodity extraction, (2) seminatural ecosystems managed for multiple uses, for both commodities and ecological values, and (3) agricultural-style ecosystems intensively managed for commodity production (Seymour and Hunter 1992, Hunter and Calhoun 1996). To take one example, in some landscapes timber can be grown three times as fast in plantations as in managed seminatural forests. This means that extensive timber extraction from seminatural forests may fragment three times as much area as extraction from tree plantations (for a given level of timber extraction).

In fragmented landscapes, conservationists often advocate the idea of connecting natural ecosystems with corridors—ribbons of protected land—to facilitate the movement of organisms (Noss and Harris 1986). Some commonsense approaches suggest themselves, especially "piggy-backing" corridors onto other efforts to maintain linear strips of natural ecosystems, such as hiking trail corridors and strips of riparian vegetation retained to protect water quality. The corridor idea has attracted critics (Simberloff and Cox 1987, Knopf 1992, Simberloff et al. 1992). They particularly question the cost-effectiveness of corridors. Because it will cross many ownerships, a strip of land half a kilometer wide by 10 km long is likely to be much more expensive to purchase and difficult to protect than a compact area of the same size. Furthermore, corridors are particularly vulnerable to external disturbances because of their shape, and they may even facilitate the spread of diseases and exotic species from one natural ecosystem to another. Perhaps the most telling argument in favor of corridors is that natural landscapes are far more "connected" than those heavily shaped by humans (Noss 1987).

In a perfect world, natural ecosystems would be protected by corridors of natural ecosystems and they would also be surrounded by seminatural ecosystems (Figure 4.3). Given a choice between having two natural ecosystems separated by a corridor plus highly disturbed land versus two separated by a wide swath of seminatural ecosystems, I think the latter would be preferable under most circumstances. In terms of the triad model, this means having natural ecosystems and production areas imbedded in a matrix of multiple-use ecosystems and assuring that management of this matrix is done in an ecologically sensitive way that will allow organisms to move through it freely.

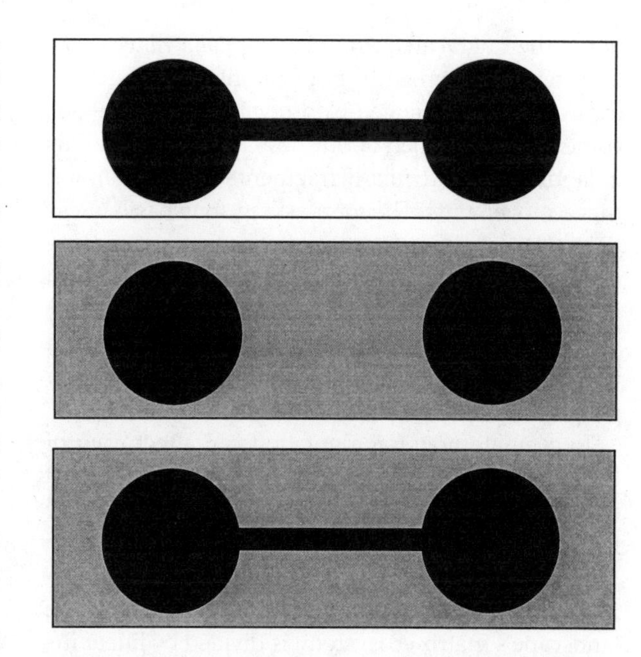

Figure 4.3 In these figures, the black circles represent natural ecosystems that may be connected with just a strip of natural vegetation (*top*), a strip of natural vegetation surrounded by semi-natural ecosystems (*bottom*), or just semi-natural ecosystems (*middle*). To allow organisms to move between the natural ecosystems, the bottom may be best, but the middle situation is probably preferable to the top.

Attrition

A landscape that has been degraded to the stage of attrition will require extraordinary means to restore its connectivity for organisms. It will mean ecosystem restoration—replacing ecosystems highly altered by people with natural, or at least seminatural, ecosystems. In recent years, restoration ecology has become a whole new discipline unto itself.

One extreme solution to "fixing" a landscape that has experienced attrition is collecting organisms and moving them from one isolated ecosystem to another. Needless to say, translocation projects are expensive and highly susceptible to failure (Griffith et al. 1989). Therefore, they can be undertaken only under unusual circumstances for a very limited set of species.

Summary

If we define a landscape as an array of interacting ecosystems, then the movement of organisms among ecosystems is a key process, for this is one of the primary ways in which ecosystems interact. Four basic types of movements can occur at a landscape scale: (1) *Home range movements,* movement within a defined area over a short period, usually days or weeks; (2) *Migration,* seasonal movement back and forth between two areas; (3) *Dispersal,* movement of young organisms (juvenile animals and plant propagules) out of their natal area; and (4) *Geographic range shifts,* long-term movement of populations in response to environmental change. Conservationists have become very concerned about how human alterations of landscapes, particularly fragmentation, may inhibit these movements. They are asking four basic questions: (1) Can animals move freely over an area large enough to find the resources they require? (2) Can animals migrate freely between seasonal ranges? (3) Can organisms disperse among subpopulations and habitat patches? And (4) Can organisms shift their geographic ranges in response to climate change?

To evaluate how fragmentation can affect each of these types of movements we need to understand that fragmentation is not a monolithic phenomenon. It exists as a continuum for which Richard Forman (1995) has identified four major stages: First, dissection occurs when a large, natural ecosystem (the landscape's matrix ecosystem) is divided by linear intrusions such as roads. Next, perforation occurs when isolated patches of a new type of ecosystem appear as holes in the overall matrix, such as clearcuts in a forested landscape. Third, fragmentation occurs as the new patches coalesce and become the matrix, leaving the original ecosystem as isolated patches. Finally, attrition occurs as the original patches become smaller and more isolated.

In conclusion, the answer to the overarching question, Is fragmentation inhibiting the biological interactions that link a landscape? depends on the species of organism, the type of movement, and the degree of fragmentation involved. A narrow road through a forest may be a barrier to the daily movements of a small mammal, but even a landscape that has suffered severe attrition may not be a barrier to the geographic range shift of a bird or a wind-dispersed plant. When considering all the complexities, it is obvious that our current level of understanding of fragmentation and landscape-scale movements is rather limited. This should encourage us to be conservative and minimize anthropogenic fragmentation as much as possible.

Acknowledgments

I thank Lisa Petit for her comments on the manuscript. Figures 4.1 and 4.2 were created by Andrea Sulzer, Figure 4.2 by Phillip DeMaynadier, and Figure 4.3 by Chris Halstead. Maine Agricultural and Forest Experiment Station Publication 1975.

Literature Cited

Baker, R. R. 1978. *The evolutionary ecology of animal migration.* New York: Holmes and Meier.

Boorman, S. A., and P. R. Levitt. 1973. Group selection on the boundary of a stable population. *Theoretical Population Biology* 4:85–128.

Brown, J. H., and A. Kodric-Brown. 1977. Turnover rates in insular biogeography: Effect of immigration on extinction. *Ecology* 58:445–449.

Case, R. M. 1978. Interstate highway road-killed animals: A data source for biologists. *Wildlife Society Bulletin* 6:8–13.

Chen, J., J. F. Franklin, and T. A. Spies. 1992. Vegetation responses to edge environments in old-growth Douglas-fir forests. *Ecological Applications* 2:387–396.

Chepko-Sade, B. D., and Z. T. Halpin. 1987. *Mammalian dispersal patterns.* Chicago: The University of Chicago Press.

Dunning, J. B., B. J. Danielson, and H. R. Pulliam. 1992. Ecological processes that affect populations in complex landscapes. *Oikos* 65:169–175.

Forman, R. T. T. 1995. *Land mosaics.* Cambridge, England: Cambridge University Press.

Frankel, O. H., and M. E. Soulé. 1981. *Conservation and evolution.* Cambridge, England: Cambridge University Press.

Franklin, J. F., and R. T. T. Forman. 1987. Creating landscape patterns by forest cutting: Ecological consequences and principles. *Landscape Ecology* 1:5–18.

Fuller, T. K., W. E. Berg, G. L. Radde, M. S. Lenarz, and G. B. Joselyn. 1992. A history and current estimate of wolf distribution and numbers in Minnesota. *Wildlife Society Bulletin* 20:42–55.

Griffith, B., J. M. Scott, J. W. Carpenter, and C. Reed. 1989. Translocation as a species conservation tool: Status and strategy. *Science* 245:477–480.

Hanski, I., and M. Gilpin. 1991. Metapopulation dynamics: Brief history and conceptual domain. *Biological Journal of the Linnean Society* 42:3–16.

Harrison, S. 1991. Local extinction in a metapopulation

context: An empirical evaluation. *Biological Journal of the Linnean Society* 42:73–88.

Hunter, M. L., Jr. 1990. *Wildlife, forests, and forestry: Principles of managing forests for biological diversity.* Englewood Cliffs, NJ: Prentice-Hall.

———. 1992. Paleoecology, landscape ecology, and the conservation of neotropical migrant passerines in boreal forests. In *Ecology and conservation of neotropical migrant landbirds,* ed. J. Hagan and D. Johnston. Washington, DC: Smithsonian Institution Press.

———. 1993. Natural disturbance regimes as spatial models for managing boreal forests. *Biological Conservation* 65:115–120.

Hunter, M. L., Jr., and A. Calhoun. 1996. A triad approach to land-use allocation. In *Biodiversity in managed landscapes,* ed. R. Szaro and D. Johnston. New York: Oxford University Press.

Knopf, F. L. 1992. Faunal mixing, faunal integrity, and the biopolitical template for diversity conservation. *Transactions of the North American Wildlife and Natural Resources Conference* 57:330–342.

Langton T. E. S., ed. 1989. *Amphibians and roads.* Bedfordshire, England: ACO Polymer Products Ltd.

Mader, H. J. 1984. Animal habitat isolation by roads and agricultural fields. *Biological Conservation* 29:81–96.

Noss, R. F. 1987. Corridors in real landscapes: A reply to Simberloff and Cox. *Conservation Biology* 1:159–164.

Noss, R. F., and L. D. Harris. 1986. Nodes, networks, and MUM's: Preserving diversity at all scales. *Environmental Management* 10:299–309.

Oxley, D. J., M. B. Fenton, and G. R. Carmody. 1974. The Effects of roads on populations of small mammals. *Journal of Applied Ecology* 11:51–59.

Paton, P. W. 1994. The Effect of edge on avian nest success: How strong is the evidence? *Conservation Biology* 8(1): 17–26.

Peters, R. L., and T. E. Lovejoy, eds. 1992. *Global warming and biological diversity.* New Haven: Yale University Press.

Prins, H. T. T., and G. R. Iason. 1989. Dangerous lions and nonchalant buffalo. *Behaviour* 108:262–296.

Ranney, J. W., M. C. Bruner, and J. B. Levenson. 1981. The importance of edge in the structure and dynamics of forest islands. In *Forest island dynamics in man-dominated landscapes,* ed. R. L. Burgess and D. M. Sharpe. New York: Springer-Verlag.

Ratti, J. T., and K. P. Reese. 1988. Preliminary test of the ecological trap hypothesis. *Journal of Wildlife Management* 52:484–491.

Rudnicky, T. C., and M. L. Hunter Jr. 1993. Avian nest predation in clearcuts, forests, and edges in a forest-dominated landscape. *Journal of Wildlife Management* 57(2): 358–364.

Schonewald-Cox, C. M., S. M. Chambers, B. MacBryde, and W. L. Thomas, eds. 1983. *Genetics and conservation.* Menlo Park, CA: Benjamin/Cummings.

Sepik, G. F., R. B. Owen Jr., and M. L. Coulter. 1981. *A landowner's guide to woodcock management in the northeast.* Publication 253. Orono: Maine Agricultural Experiment Station.

Seymour, R., and M. L. Hunter Jr. 1992. *Principles and applications of new forestry in spruce-fir forests of eastern North America.* Publication 716. Orono: Maine Agricultural Experiment Station.

Shaffer, M. L. 1981. Minimum population sizes for species conservation. *BioScience* 31:131–134.

Simberloff, D., and J. Cox. 1987. Consequences and costs of conservation corridors. *Conservation Biology* 1:63–71.

Simberloff, D., J. A. Farr, J. Cox, and D. W. Mehlman. 1992. Movement corridors: Conservation bargains or poor investments? *Conservation Biology* 6:493–504.

Temple, S. A. 1987. Predation on turtle nest increases near ecological edges. *Copeia* 1987:250–252,

Templeton, A. R., K. Shaw, E. Routman, and S. K. Davis. 1990. The genetic consequences of habitat fragmentation. *Annals of the Missouri Botanical Garden* 77:13–27,

Williamson, D., J. Williamson, and K. T. Ngwamotsoko. 1988. Wildebeest migration in the Kalahari. *African Journal of Ecology* 26:269–280.

Woods, K. D., and M. B. Davis. 1989. Paleoecology of range limits: Beech in the upper peninsula of Michigan. *Ecology* 70:681–696.

5 Riparian Management in the 21st Century

Stanley V. Gregory

Current Policies and Practices 70
Federal Policies 72
State Policies 74
The FEMAT Report 75
Future Directions 76
An Emphasis on Ecological Function and Natural Forest Pattern 76
Adoption of a Landscape Perspective of River Networks 78
Development of Ecologically Sound Systems for Restoring Ecosystem Properties 79
Attention to Social Needs for Riparian Resources 81
Conclusion 82
Literature Cited 83

Management of streams, lakes, and wetlands in forest ecosystems represents one of the most revolutionary changes in forestry in the latter half of the 20th century. Prior to 1950, forest harvesting along streams and rivers differed little from upslope harvesting: forests were cut from the ridge to the stream's edge. Logging operations dragged logs down stream channels to landings at the bottom of harvesting units. From the late 1800s until World War II, lower reaches of northwestern watersheds were subject to log drives—artificial floods created to run logs down the rivers to mills (Sedell and Luchessa 1982, Sedell and Frogatt 1984). These practices delivered large amounts of sediment to streams, lakes, and estuaries, removed forest canopies and warmed water temperatures, altered habitats associated with wood and greatly decreased future sources of wood inputs, and simplified and narrowed floodplains. After 1950, resource managers in the Pacific Northwest and other regions increasingly expressed concerns over effects of logging on streams and anadromous salmonids. Today, there is widespread agreement that historical forest practices negatively altered the structure of aquatic ecosystems and decreased their productivity (Gregory et al. 1987, Hicks et al. 1991, Bisson et al. 1992, McIntosh et al. 1993, Botkin et al. 1995). However, there are still many unanswered questions about what forest practices are appropriate in riparian areas. These are central issues in ecosystem management.

This chapter addresses future directions in riparian management for the next century. Current trends are based on the momentum of policies that have emerged over the last four decades of the 20th century. Thus, the first part of the chapter describes current policies for managing riparian areas on private, state, and federal forests to provide a context for understanding future trends in the field. The second part of the chapter then explores elements of future riparian management.

Current Policies and Practices

Regional managers have recognized the importance of ecosystem perspectives for riparian practices, but the nature of riparian areas as interfaces between ecosystems has been obscured in many applications. Ecosystems are unique assemblages of communities and their environments: riparian areas are "ecotones" or interfaces between terrestrial and aquatic ecosystems. Sharp gradients in environmental conditions, ecological processes, and species across the transitional zone between terrestrial and aquatic ecosystems make riparian areas one of the most diverse and dynamic portions of forested landscapes (Naiman et al. 1988, Gregory et al. 1991). Inaccurate designation of the land–water interface as a riparian ecosystem has obscured the ecological importance of these gradients. Riparian areas are broad interfaces with no discrete boundaries. The term "riparian management zone," however, represents distinct spatial boundaries that are designated to achieve specific management goals (Figure 5.1). Such management designations incorporate inherent tradeoffs between proportions of riparian functions included within and outside their boundaries.

Current federal and state riparian regulations are designed to minimize erosion, maintain stream shading, protect habitat conditions, maintain food resources, avoid detrimental environmental conditions, and maintain water quality. They are based on regional studies that have demonstrated that forest practices—including timber harvesting, yarding, and road building—alter many components and processes of aquatic ecosystems and the land–water interface. Forest harvest practices cause alterations in aquatic ecosystems or riparian processes, including changes in sedimentation and mass failure, stream temperatures, hydrologic regimes, channel structures, floodplain processes, amounts of woody debris, aquatic plant production, terrestrial litter inputs, and invertebrate, fish, and wildlife populations. These interactions have been evaluated and synthesized in several major symposia, reports, and books (Krygier and Hall 1971, Iwamoto et al. 1978, Newbold et al. 1980, Schlosser and Karr 1981, Harmon et al. 1986, Murphy et al. 1986, Salo and Cundy 1987, Raedeke 1988, Murphy and Koski 1989, Meehan 1991, Naiman 1992, Peterson et al. 1992, FEMAT 1993, Murphy 1995). These works provide detailed reviews of the effects of forest practices on aquatic ecosystems and variation in watershed and ecological responses observed across the Pacific Northwest.

Reviews of riparian management practices frequently focus on existing rules and regulations. They evaluate current practices in terms of protecting aquatic resources and incorporating fundamental ecological processes. While this perspective is relevant to analyses of future trends in riparian management, it poorly reflects the current state of forested landscapes in the region, the influence of riparian practices, and the absence of streamside protection prior to 1970. The emphasis on timber harvesting on federal lands in Oregon and Washington largely emerged after 1960 (Figure 5.2). The total amount of timber harvested and the proportion of timber coming from public and private lands changed as much or more from 1940 to 1960 as in recent decades. These historical trajectories form the landscape basis for future trends. For example, more than half of Oregon's private forest lands were harvested prior to 1972 (as indicated by 20- to 100-year-old age classes in 1988) and the first requirements for any form of

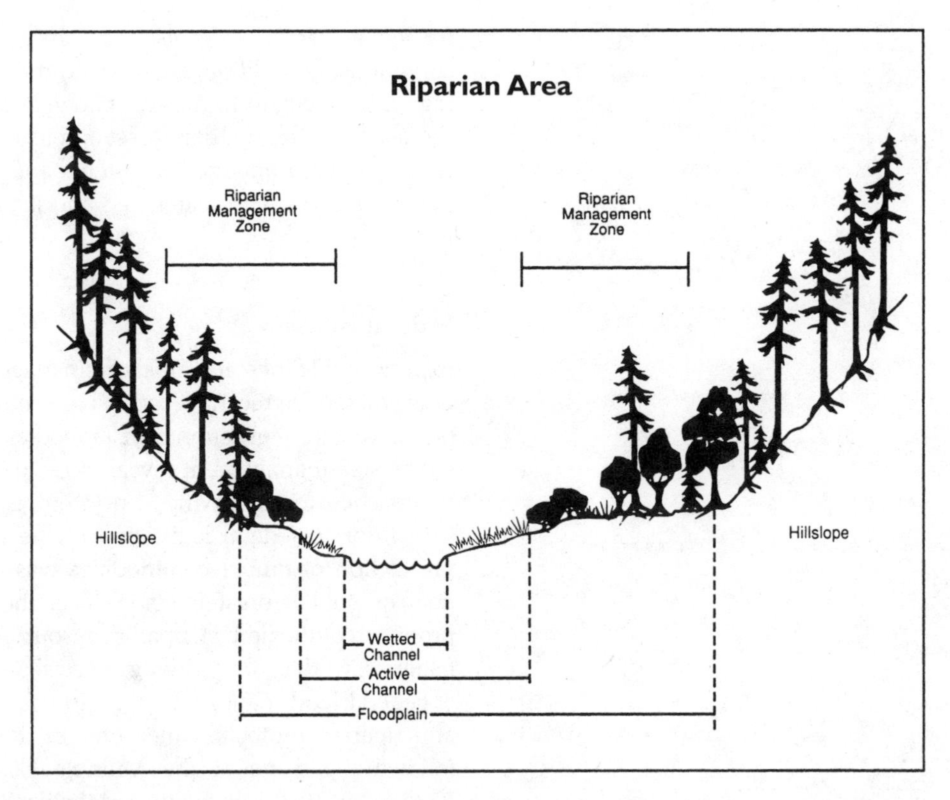

Figure 5.1 Cross-section of a riparian area and adjacent upslope forest. The riparian area may extend upslope, encompassing zones of influence for shading, leaf inputs, and wood delivery. The riparian management zone may be substantially narrower than the riparian area (from Gregory and Ashkenas 1990).

streamside protection (Gedney 1988). While recent changes in riparian management represent important advances, the landscape reflects more than a century of harvesting on all forest lands with little or no protection of riparian resources and aquatic ecosystems.

Current riparian management policies focus on several issues: (1) widths of riparian management zones, (2) retention of live trees and snags within the riparian zone, (3) the extent of shade cover, (4) floodplain protection, (5) yarding corridors, (6) culvert dimensions, (7) road crossings, (8) felling techniques, and (9) erosion protection. Riparian management practices for private and federal lands in the Pacific Northwest represent diverse approaches that have evolved for management of different types of land ownership since 1970 (Table 5.1). Most of the guidelines reported in this table have been established within the last five years; earlier riparian rules were substantially less protective.

The first question to emerge in riparian management is how wide should the riparian management zone be? This immediately raises questions of scale: What is the size of the stream, the height of the trees, and the lateral extent of riparian processes and habitats? Traditional approaches have (1) explored relevant ecological issues, (2) identified operational constraints, (3) debated tradeoffs between conflicting values, and (4) arrived at negotiated boundaries and harvest practices that fuel debate until the next revision of riparian rules. Frequently, the basis for selection of specific numerical criteria is not documented

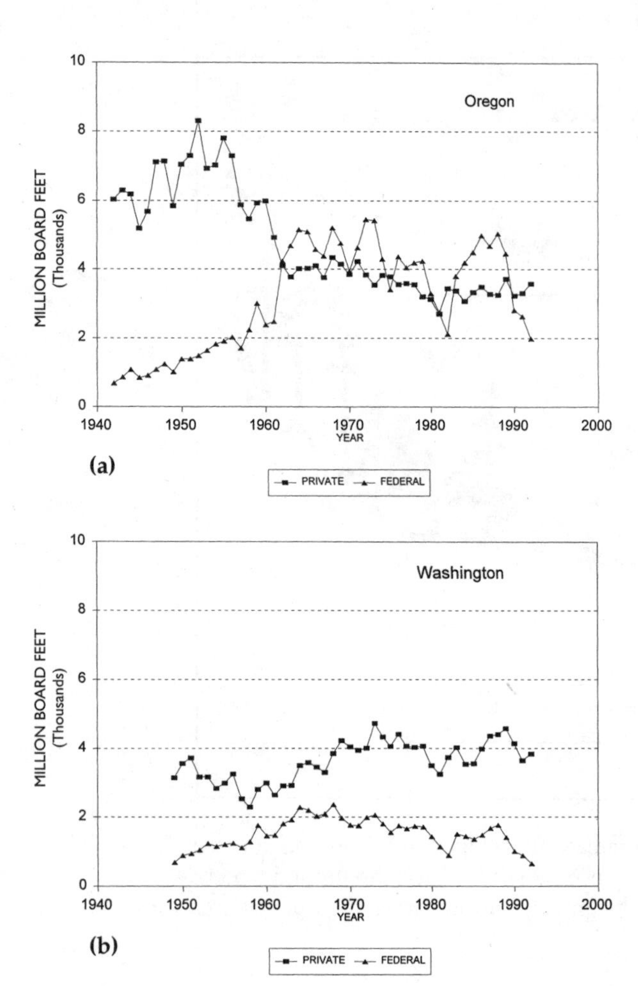

Figure 5.2 (a) Timber harvest from private and public forests in Oregon from 1940 to 1992 (from Oregon Department of Forestry). (b) Timber harvest from private and public forests in Washington from 1940 to 1992 (from Washington Department of Natural Resources).

and soon forgotten, thus insuring the starting point of the next revision—argument over the validity of numerical criteria.

Most riparian guidelines lack explicit determination of the reference conditions that represent the goals of future management. Identification of reference systems allows all parties to determine the validity of numerical standards and directs discussion over future revisions to the ecological intentions of management. Recent riparian management rules for the state of Oregon and the Northwest Forest Plan for federal lands set goals based on riparian forest conditions. In addition, federal land managers in the region have begun to match riparian zone widths to the nature of the riparian forest and its ecological interactions with aquatic ecosystems and to place riparian reserves with watershed and landscape contexts.

Federal Policies

Explicit guidelines for riparian management have only emerged in recent decades. The sequence of federal and state legislation that embodies changes in streamside management reveals a rapid evolution of approaches for achieving ecosystem goals (Figure 5.3). Prior to federal legislation in the early 1960s, production of timber commodities was the primary goal of public forest lands. Hence, the burden of proof for protection of riparian resources rested on fisheries, wildlife, or hydrology specialists.

Federal land managers began to use buffer strips and riparian protection measures in the late 1960s following passage of the Multiple Use–Sustained Yield Act of 1960, which directed the Forest Service to address forest uses other than production of timber commodities, and the National Wilderness Act and the Wild and Scenic Rivers Act, which removed specific federal lands from the timber base for other resource values. Few specific guidelines governed the management of federal riparian areas, however, until passage of the National Forest Management Act and the Federal Land Policy and Management Act of the mid-1970s. These laws required the U.S. Forest Service and the Bureau of Land Management to conduct forest planning with specific attention to relevant federal laws, such as the National Environmental Policy Act, the Clean Water Act, and the Endangered Species Act. At this point, federal forest management agencies identified protection of riparian resources as the primary objective of riparian management. Moreover, they specified that managers should "give preferential consideration to riparian-dependent resources when conflicts among land-use activities occur" (USDA and USDI 1994). This represented a major shift of the burden of proof from ecologists to timber managers, and, in theory, it required timber harvesting to be justified because of its inherent

Table 5.1 Riparian management regulations for state and federal forests in the Pacific Northwest

Agency Class	Definition	Width (ft.)	Linear Density (Trees/1,000 ft.)	Site Goal (Stand)	Harvest	Floodplain	Bank Protection
Oregon							
Type F I	> 10 cfs, with fish	100	40 conifers	Mature 159 ft.²/acre	Partial	None	20-ft. No harvest
Type F II	2–10 cfs, with fish	70	30 conifers	Mature 159 ft.²/acre	Partial	None	20-ft. No harvest
Type F III	< 2 cfs, with fish	50	Deciduous	Mature 159 ft.²/acre	Partial	None	20-ft. No harvest
Type N-D	< 2 cfs, no fish	0	Understory	0	Complete	None	N/A
Washington							
Type 1 & 2	> 75 ft.	100	50 trees		Partial	None	
Type 1 & 2	< 75 ft.	75	100 trees		Partial	None	
Type 3	> 5 ft.	50	75 trees		Partial	None	
Type 3	< 5 ft.	25	25 trees		Partial	None	
Type 4	> 2 ft.	25	0		Partial	None	
California							
Class I	Fish present	150	25% of conifers		Partial	None	
Class II	Fish within 1,000 ft.	100	25% of conifers		Partial	None	
Class III	No fish	Site-based	50% understory		Partial	None	
Idaho							
Class IA	> 20 ft., with fish	75	67 trees		Partial	None	
Class IB	10–20 ft., with fish	75	63 trees		Partial	None	
Class IC	< 10, with fish	75	42 trees		Partial	None	
Class II	No fish	5	0				
Alaska							
Type A	Anadromous fish, Unconstrained	66	All trees		No harvest	None	
Type B	Anadromous fish, Constrained	100	BMP		BMP	None	
Type C	No anadromous fish	50	BMP		BMP	None	
FEMAT							
Class I	Fish-bearing	300/2 spt	All trees	Old growth	No harvest	Protected	No harvest
Class II	Permanent, no fish	150/1 spt	All trees	Old growth	No harvest	Protected	No harvest
Class III	Seasonally flowing	100/1 spt	All trees	Old growth	No harvest	Protected	No harvest
PACFISH							
Class I	Fish- bearing	300	All trees	Old growth	No harvest	Protected	No harvest
Class II	Permanent, no fish	150	All trees	Old growth	No harvest	Protected	No harvest
Class III	Seasonally flowing	100	All trees	Old growth	No harvest	Protected	No harvest

Note: Many states have multiple standards based on regional, stand, topographic, morphological, and biotic criteria. Representative standards are presented to illustrate the general characteristics of the state and federal approaches to riparian management. All states and federal agencies include various provisions for alternative practices, waivers, and experimental applications.

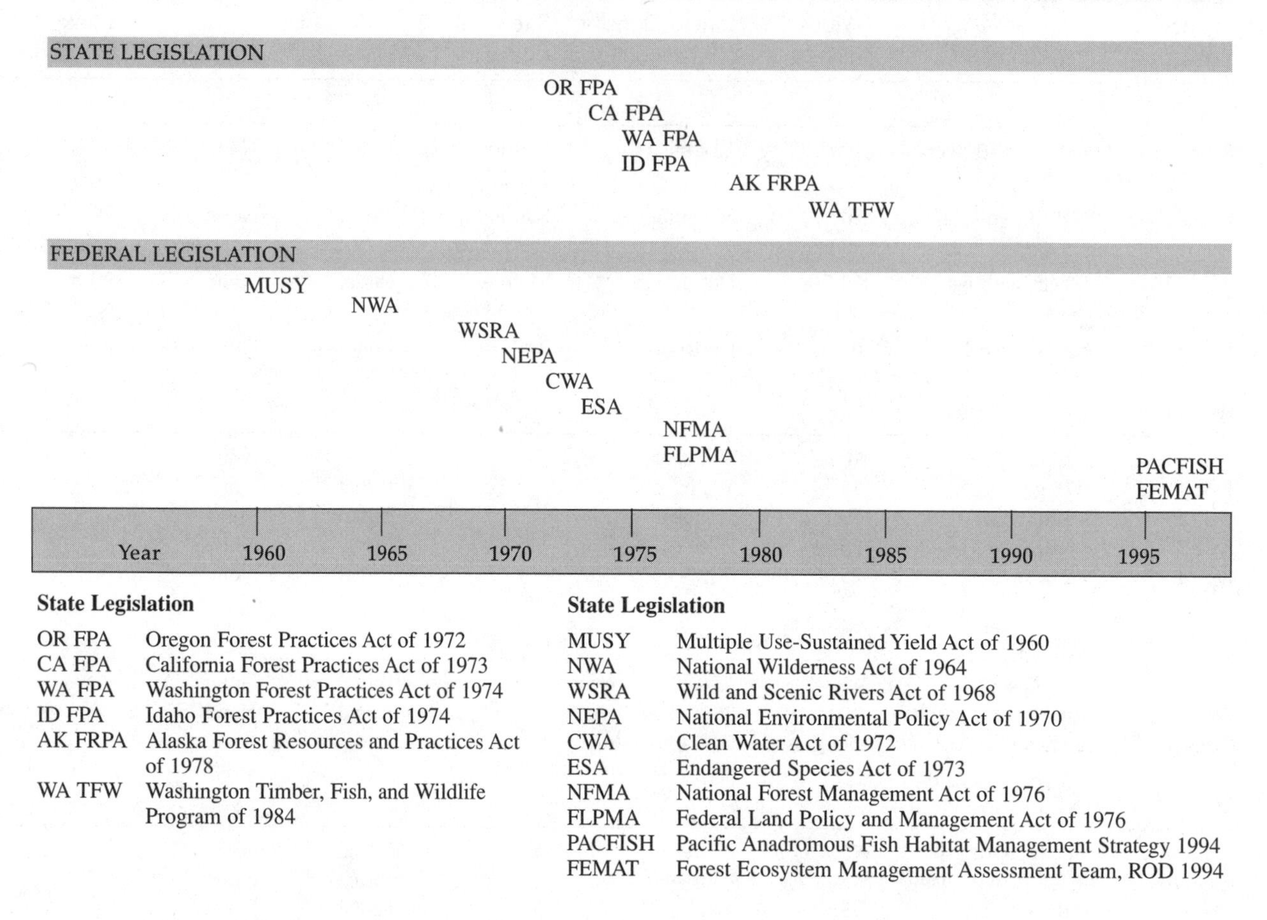

Figure 5.3 History of state and federal legislation related to riparian management practices in the Pacific Northwest.

modification of riparian and aquatic systems. However, in reality, riparian protection and harvesting practices in forest plans varied widely between national forests. Consistent guidelines for riparian practices still do not exist at a federal level. Regionally consistent riparian practices have been established only since 1994 on federal forests (USDA and USDI 1994, PACFISH 1994).

State Policies

Federal land-management policies established an emerging set of environmental considerations in riparian areas, and the water quality standards of the Clean Water Act created the motivation for defining appropriate management practices on private and state lands in the Pacific Northwest. Oregon implemented the first Forest Practices Act for state and private lands in 1972. Other western states quickly developed similar riparian regulations (Figure 5.3). Early state laws contained few requirements for the maintenance of larger size classes of either coniferous or deciduous trees and were designed primarily to provide stream shading and erosion control. Few other environmental or ecological functions of riparian areas were even considered in the early determination of riparian management guidelines.

All states in the Pacific Northwest define a range of appropriate riparian management practices for private and state forests. These standards are modified for local conditions based on a variety of criteria related to regional location, stand types, topography, channel or valley morphology, and fish communities. Approaches vary from state to state, and flexibility for site-specific management is provided in several ways. Some states, such as Oregon, have developed extensive criteria for management practices, although landowners can ask for site-specific exceptions. Washington uses an innovative system of basin analyses and coordinated development of basin-specific practices based on local consensus (known as the Timber, Fish, and Wildlife Program; see Smith, Chapter 27). It is still not known whether specific rule-based approaches or consensus models will be most effective. As such, the diversity of riparian practices on state and federal lands in the Pacific Northwest may be one of the strongest assets for future development of effective riparian management.

The FEMAT Report

In 1993 the Forest Ecosystem Management Assessment Team (FEMAT), convened by President Clinton, developed a range of options and a recommended alternative for ecosystem management of federal forests within the range of the northern spotted owl (FEMAT 1993). The FEMAT report proposed several major advances in the management of aquatic ecosystems. First, it called for the creation of key watershed and riparian reserves designed to fulfill upslope as well as riparian functions—in contrast to previous strategies geared toward the protection of selected habitats or species of interest. Second, it provided a spatial context for delineating riparian reserves based on forest–stream interactions rather than fixed riparian widths. And finally, the FEMAT report recommended that watershed analyses be conducted to identify conditions that influence the ability of a particular landscape to meet established ecosystem management goals.

The FEMAT report's designation of key watersheds was one of the first attempts to incorporate a network of watersheds into an ecosystem management plan at a landscape scale (Pacific Rivers Council 1993). The physical and ecological processes in watersheds serve as a foundation for more-site-specific conditions and selection of management alternatives. Watersheds that contain critical aquatic resources and habitats were selected to provide refuges and sources of colonists at a scale relevant to river networks and components of upland terrestrial ecosystems.

Riparian reserves were established to (1) protect riparian-dependent resources and aquatic ecosystems and (2) provide habitat for and fulfill the environmental requirements of upslope communities of plants and animals. An innovative and ecologically sound feature of these riparian reserves was the scaling criterion developed for delineating their boundaries. Instead of lapsing into the traditional debate over the width of the riparian management zone, boundaries were based on "site-potential tree heights." A site-potential tree height is the average height of trees that have attained the maximum height possible given the site conditions. This delineation of riparian widths is based on functional interactions between riparian forests, stream processes, and microclimate. Unlike previous definitions of riparian management zone boundaries, this definition is transferable to other forest types and locations and is linked directly to ecological function.

The term "riparian reserves," however, only conveys a portion of their intended purpose—they were designed to fulfill both upslope and riparian functions in ecosystem management. To fulfill both of these functions, riparian reserve widths of two site-potential tree heights were established along fish-bearing streams, and widths of one site-potential tree height were established along either perennial non-fish-bearing streams or along ephemeral streams. Riparian reserves outside lands withdrawn for other purposes represent an estimated 2.628 million acres of the 24.455 million acres of federally administrated forest lands within the range of the northern spotted owl—or 11 percent of the land base (USDA and USDI 1994). Total reserved or withdrawn lands (i.e., wilderness, national parks, late successional reserves, other administratively withdrawn lands, riparian reserves) account for 18.856 million acres or 77 percent of the total land base. Following watershed analysis, forest practices within riparian reserves can be modified to attain the overall goals of the Northwest Forest Plan. In concept, these practices could include preservation, silvicultural restoration, modification of boundaries, and testing of alternative forest practices—as long as the status of populations and communities in these forest ecosystems is not jeopardized.

The FEMAT report defined a new context for riparian management at the landscape scale. Future consideration of alternative riparian practices will be scaled by links between ecological functions and adjacent forests as well as the roles of key watersheds within the larger landscape identified in the Northwest Forest Plan.

Future Directions

Over the last 30 years, many unprecedented advances have been made in riparian forest practices in the Pacific Northwest that far exceed those of other regions of the United States and Canada. Changes in riparian management represent trends rather than arrival at an endpoint—an issue raising strong feelings on both sides in the political debate over the future of forest management in the region. Where are these changes leading? Why will riparian management change even more in the next century than the abrupt changes that we have witnessed at the end of the 20th century? Part of the answer lies in the diversity of riparian management approaches that have emerged as we move into the next century. Resource managers and planners in the Pacific Northwest are simultaneously pursuing several fundamentally different social and ecological models rather than a single uniform path. Though these paths may evolve in parallel, the wide array of land ownerships, ecosystem types, and changing demographic patterns will translate into complex goals and management practices for riparian resources within the region.

Future trajectories of change in riparian practices reflect several emerging characteristics of ecosystem management:

- An emphasis on ecological function and natural forest pattern
- Adoption of a landscape perspective of river networks
- Development of ecologically sound systems for restoring ecosystem properties
- Attention to social needs for riparian resources

These trends in riparian management are dynamic—no one can accurately predict the outcome of ecological, environmental, social, economic, and political interactions. Though the endpoint is not clear, the influences of these factors will shape the course of riparian management in the 21st century.

An Emphasis on Ecological Function and Natural Forest Pattern

In contrast to early riparian management systems, which focused on one or two characteristics of concern (e.g., sediment or stream temperature), newer systems on both public and private lands have evolved rapidly to address the full range of ecological functions that occur at the land–water interface (Salo and Cundy 1987, Raedeke 1988, Gregory et al. 1991, Naiman et al. 1992, Thomas et al. 1993). Revision of state forest rules after 1990 throughout the region include analyses of shade, food resources, woody debris, channel dynamics, sedimentation, mass failure, hydrologic regimes, and invertebrate, fish, and wildlife populations. These ecological functions are used to set management boundaries and determine alternative practices. One result of this shift toward a functional definition of riparian zones is the development of functionally based scalers for riparian boundaries, such as the site-potential tree height criteria in the Northwest Forest Plan. This is not to suggest that federal and private forest lands will adopt identical standards; rather, a common conceptual basis will evolve for guidelines for riparian forest management

Development of a functional basis for delineation of riparian management zones will necessitate explicit identification of reference conditions that represent the goals of riparian practices. Oregon recently established mature forest conditions at the midpoint of rotation as a target or reference condition for its revised riparian rules (Oregon Department of Forestry 1994). An accepted reference system for desired management outcomes provides a common basis for interpretation of the appropriate numerical criteria, potential areas for future modification of the rules, and a standard against which to measure management success. Numerical standards without such a context fuel controversies and hinder widespread development of alternative approaches. Future reviews will determine whether an adopted reference system is consistent with the goals and objectives of a particular management agency or with new legislation; revisions can then be made to reflect the changing base of scientific information.

Floodplains are a fundamental ecological component of stream ecosystems that are vital to the survival, recolonization, and productivity of aquatic communities (Junk et al. 1989, Bayley 1995). Only federal riparian policies specifically protect floodplain functions (Gregory and Ashkenas 1990). None of the forest practices acts of the northwestern states specifically identifies floodplain areas within riparian management zones, and floodplain processes are not directly addressed in riparian practices. All states have provided for protection of adjacent wetlands, but a floodplain has to be identified as a wetland before any level of protection is required. Although the fundamental role of floods on aquatic ecosystems (Junk et al. 1989), hydrologic regimes, floodplain structure, and human communities (Booth 1991) has been incorporated in floodway management for decades, private forest managers have actively avoided this aspect of forest ecosystem management. Recurrence intervals of major floods range from 50 to 200 years, well within the rotation ages of managed forests; thus consideration of flood processes and floodplains is essential for sound forest management.

Development of approaches that provide riparian functions for terrestrial wildlife presents several challenges for riparian management (Raedeke 1988, Thomas et al. 1993). Assemblages of terrestrial wildlife within a watershed exhibit varying degrees of association with riparian areas, ranging from obligate dependence to almost complete avoidance. Moreover, these relationships may shift seasonally based on moisture availability and life history requirements (Raedeke 1988, McComb et al. 1993). Wildlife dependence on riparian resources extends over diffuse gradients extending well into upslope terrestrial environments. Simple riparian interactions such as shading, inputting of woody debris, and inputting of leaf material have few parallels in wildlife–riparian habitat relationships. As a result, ecological studies of terrestrial wildlife tend to focus on species-specific patterns and ecological requirements. More general functional group or guild approaches used for aquatic ecosystems (Cummins 1974) have not been applied effectively to wildlife, in part because of the variability of wildlife–habitat relationships. Large networks of riparian reserves—such as those recommended in the Northwest Forest Plan that have widths of one site-potential tree height along ephemeral streams—reflect the commitment of larger land areas than commonly found in a riparian delineation for basin-wide wildlife protection. Alternative approaches (e.g., intensive survey, dynamic modeling, species-specific management, experimental monitoring, and long-term monitoring) may shift the emphasis away from riparian reserves. However, the time and personnel required to build adequate information bases are enormous.

One of the major limits of current riparian management is the lack of information on the structure and stand dynamics of riparian forests and their role in larger landscapes (Agee 1988; see also Spies, Chapter 2). Almost all considerations of riparian forest dynamics—regeneration, growth, survival, mortality, snag development, down wood delivery, community succession, and rates of succession—are based on assumptions derived from upslope forests. The unique gradients of light, moisture, microclimate, and flood disturbance across riparian areas make it highly unlikely that these assumptions are valid. Studies of riparian forests have demonstrated that their composition differs greatly from that of upslope forests and often exhibits a much greater richness of terrestrial plant species and a lower volume of conifers (Schoonmaker and McKee 1988, Gregory et al. 1990, Ursitti 1991, Emmingham and Maas 1994). Until forest ecologists develop a more regionally and ecologically extensive body of knowledge on riparian forest dynamics, our management systems and silvicultural approaches will be limited by a high degree of uncertainty.

Current forest management is based on generating economically desirable forest stands as rapidly as possible within the constraints of relevant ecological goals. This strategy is applied to both upslope and riparian forests, either for commercial goals of the forest industry or for ecological goals of aquatic and terrestrial ecologists. Our ability to accelerate development of short-term economic goals is well demonstrated, but our ability to generate forest structures that are recognized and used effectively by species and communities is poorly developed.

Regardless of our ability to manipulate stand structure for economic or ecological goals, we face a fundamental long-term question: Are differences in rates of forest succession, either between riparian and upslope forests or across forest landscapes, important for long-term ecosystem function? Variation in rates of succession may have implications for the diversity of plant communities, the diversity of associated nonplant species, resilience to disturbance, habitat heterogeneity, nutrient flux, and microclimate (see Spies, Chapter 2). Riparian forests experience a broader suite of disturbances—including floods, landslides, fire, disease, insect outbreaks, and windthrow—which shape riparian patterns and create some of the most complex forest structures in the landscape. We have little documentation on rates of plant succession in natural or managed riparian forests. Riparian management in the next century will require a more robust knowledge of riparian stand dynamics and forest succession. In the meantime, our application of such essential concepts of ecosystem management will be severely limited.

Adoption of a Landscape Perspective of River Networks

A central question for managing forest ecosystems at the scale of landscapes has remained largely unasked, let alone unanswered: What patterns of future forest structure will be created by our management systems, and what are the ecological implications of these patterns? Private, state, and federal forest managers in the Pacific Northwest tend to create a "forest of walls" through their practices of both commodity harvest and ecological protection. Riparian and upslope management have been addressed as separate systems with separate goals; almost no attention has been paid to ecological links between upslope and riparian forests. Fire breaks, road systems, different levels of harvest between riparian and upslope forests, and other management practices create artificial patterns that have unique spatial characteristics of edge, patch dimensions, and connectivity not found in the natural forest landscape. Even the Northwest Forest Plan, with its extensive network of riparian reserves managed to maintain mature to old-growth forests along streams and rivers and restrict harvest to upslopes and ridgelines, will produce a forest pattern that differs markedly from the structure of forests across the landscape prior to harvest.

One of the most promising areas of riparian management in the next century will be the integration of upslope and riparian management at basin and landscape scales. In many ways, our conceptual and operational evolution of (1) recognizing the important ecological functions of riparian networks (Naiman et al 1993), (2) developing practices to maintain or restore varying degrees of those functions (Gregory and Ashkenas 1990), and (3) implementing a range of approaches across different forest ownerships has been a necessary precursor for future advances in applying riparian management to entire river basins.

There are several alternatives that are likely directions for future research and forest management (Botkin et al. 1995). First, we will need to explore the long-term consequences of our collective management systems in river basins. This will require remote sensing, spatial pattern analysis, dynamic forest models that operate at large scales, models of aquatic responses, and experimental studies to identify the mechanisms of physical and ecological responses. Second, we will need to acknowledge the degree to which specific landscapes can mimic natural forest patterns and identify more-effective approaches for different mixes of ownerships. The greatest potential for creating large-scale patterns that resemble long-term dynamic patterns like those that existed prior to timber harvesting will be in areas with extensive, contiguous public forests. Checkerboard ownership patterns of public and private lands offer the least potential because of the inherent differences in their resource management goals.

Historically, riparian management has focused on harvest units. This ignores the network properties of riparian areas within river basins and creates challenges for effective management. Fundamental decisions about riparian-dependent resources, their habitats, and physical processes must be addressed within the context of the river network (Naiman et al. 1991). Most current management strategies for commercial forest lands or national forests focus on montane portions of the landscape. As a result, lowland forests primarily occur in private nonforest ownerships. Although these riparian forests and floodplain rivers are critical components of a riverine ecosystem, they are not addressed through current riparian practices (Boule and Bierly 1987). Exploring alternatives such as land exchanges and public acquisition of these areas, which are essential to restoration of large river and aquatic productivity, is a high priority for future riparian management.

Development of Ecologically Sound Systems for Restoring Ecosystem Properties

Though forest practices have certainly improved, we are left at the end of the 20th century with a legacy of past practices and an obvious need to restore ecological characteristics of riparian forests (Pacific Rivers Council 1993). Concerns about realistic goals for restoration are certainly valid—it is virtually impossible to return forest ecosystems to the conditions that existed prior to Euro-American settlement. Even if it were possible, restoration might not reflect the changes that would have occurred over the last 200 years under natural conditions. But this is an extremely restrictive perspective of restoration that is not dynamic or appropriate for modern landscapes.

Restoration is the process of encouraging a system to maintain its function and organization without continued human intervention (NRC 1992). If a system has been shifted outside its range of performance under natural conditions, restoration attempts to move the system toward that range of performance in the future. The degree to which the system can reach that desired range of behaviors will depend on many factors—cause and degree of degradation, irreversibility of past actions or changes, viability of remaining populations, financial resources, and the time frame for desired recovery (Moyle and Yoshiyama 1994). Restoration should not be considered to be the return of a system to a fixed, pre-alteration condition (NRC 1995).

Ecosystem restoration inherently depends on a framework of broader landscape management that protects, maintains, and restores ecosystem structure and function (Wissmar and Swanson 1990, Sedell et al. 1991). Unfortunately, our recent history of riparian restoration, though well intended, has emphasized engineering approaches, which erect permanent "structures" in streams, establish administrative programs, and highlight paper trails and ledger-based accountability (Reeves et al. 1991, Frissell and Nawa 1992). Sound restoration of aquatic ecosystems requires a solid foundation of ecological principles and a clear recognition of the dynamic nature of streams, rivers, wetlands, lakes, and their adjacent riparian forests.

Several fundamental principles will guide effective restoration of watersheds, riparian areas, and aquatic ecosystems:

- The goal of ecological restoration is to reestablish the ability of the system to maintain its function and organization without continued human intervention.
- Restoration is no substitute for appropriate ecosystem management. Any restoration program should be nested within a larger program of landscape management that protects, maintains, and restores ecosystem structure and function.
- Resource analysis should precede any restoration effort. Resource evaluation should begin at least at the scale of the entire river basin, focusing down to a specific watershed, and finally addressing local reach characteristics. Specific habitat characteristics and ecological processes that have been degraded should be identified. Restoration practices should be designed to alter those factors that shape the ecological processes of concern.
- Before restoration efforts are implemented, practices that caused resource degradation must be changed to prevent or reduce continued environmental degradation. Constraints on current conditions and processes should be considered in determining the appropriate timing for restoration efforts. To the degree possible, systems should be allowed to stabilize before habitats are altered unless immediate and intensive efforts are required to save resources from extinction or prevent catastrophic habitat change.
- Restoration efforts should provide materials and organisms to reestablish natural physical and ecological processes. Material should be supplied in amounts and types that would be within the range expected for the system. Species or stocks of organisms should be native to the area and maintain the integrity of the genetic characteristics of local populations.

- The time frame for ecosystem restoration should be described explicitly, and expected patterns of recovery should be identified clearly.
- Actions should be reversible either by natural processes or by human correction, if possible.
- Restoration of streams and rivers should protect or restore floodplain and channel function and recover terrestrial plant communities.
- Natural disturbances such as floods, fire, windthrow, disease, and pest outbreaks are the major agents of restoration at the scale of river basins and forest landscapes. Resource management agencies need to develop after-the-disturbance policies that protect beneficial changes caused by natural disturbances. Disaster relief efforts that simply repeat previous resource management mistakes must be prevented.
- The success of restoration efforts should be evaluated based on ecological functions and responses within the dynamics of the system.

Understanding the current status of riparian forests, channel structures, and aquatic communities is essential to developing appropriate restoration objectives. All too often, projects are initiated without identifying specific objectives. For instance, pools are created for fish habitat without determining whether existing geomorphic and hydrologic processes would maintain additional pools over the long term, or whether existing fish populations are actually limited by lack of pools.

Conditions of existing riparian forests, stream channels, and aquatic communities must be identified both for the basin and for the restoration site (Reeves et al. 1991). Projects initiated without consideration of basin-level conditions will be more likely to fail than those based on network assessment. If possible, reference systems that represent desired future conditions should be located and examined to design restoration approaches. If natural areas cannot be found within the basin, similar basins nearby can be used to compare patterns found at the proposed restoration site. Location, distribution, configuration, and size of riparian or channel modifications should be consistent with the patch structure of the riparian vegetation and the geomorphic and hydraulic properties of the stream reach (Bottom et al. 1985). If a riparian area has been damaged by a natural or land-use-related disturbance, the probability of continued disturbance must be considered. Any attempts to restore ecological conditions can be negated by treating the symptom rather than the source of disturbance.

The majority of aquatic restoration efforts in the Pacific Northwest focus on restoring channel structure, particularly adding large woody debris, but long-term restoration of riparian functions (e.g., woody debris accumulation, shade provision, nutrient retention, food production for aquatic organisms) requires either natural riparian forest succession or silvicultural restoration. The goal of silvicultural management in riparian management zones should be to provide the natural ecological functions of riparian vegetation where past practices or natural events have diminished the diversity of riparian plant communities. All stages of silvicultural activity should encourage natural patterns of succession within the constraints of our current understanding of community composition, patch dimensions, and rates of succession (Agee 1988). Silvicultural operations in riparian areas can create diverse and structurally complex riparian plant communities (see e.g., Franklin et al., Chapter 7; Debell et al., Chapter 8; and Tappeiner et al., Chapter 9). Where short-term canopy recovery is required, hardwood species can be used to rapidly reestablish vegetative cover. Coniferous species can be used to reestablish long-term shade conditions and provide for more persistent wood in channels. Snags, green trees, and cull trees can be left in place to provide a short-term debris source. Native species that decay slowly can provide long-term sources of woody debris and snags. Salvage of trees and snags from riparian areas should be avoided unless it benefits riparian-dependent resources.

Reestablishment of shade over stream channels can be accelerated by protecting any remaining streamside vegetation, especially young trees. However, in areas dominated by shrub cover, underburning may encourage regeneration of desired tree species. Fire alters forest structure in upslope areas more frequently and more intensively than it does in riparian areas, but fire is a natural disturbance in riparian forests (Agee 1993). Exclusion of fire in riparian areas, both through overall fire suppression efforts and through riparian protection, produces riparian plant communities that do not reflect natural composition or patch structure. Incorporating a full range of natural disturbances within riparian forests may improve our ability to maintain ecosystem functions throughout river networks.

Based on analysis of riparian forest conditions within a basin, precommercial thinning may offer an opportunity to attain ecological goals of restoration and produce timber resources for human use. Stand thinning and underplanting can accelerate growth, increase structural complexity, replace native riparian species diminished by harvest, create snags, and provide large woody debris for the forest floor and stream channel. Placing thinned material directly into channels may increase short-term channel and floodplain complexity, particularly in small streams lacking debris. This small woody debris can provide structure and organic matter for 5 to 15 years. At the commercial thinning stage (40–80 years) there are two options: (1) the operation can be avoided completely in the riparian management zone, leaving natural mortality at 80 to 150 years to thin the area, or (2) commercial thinning can proceed, but pole timber and culls can be placed in the channel. Debris additions are appropriate if the stream channel contains inadequate volumes of woody debris; thus, site inventories are required to coordinate thinning operations with restoration objectives.

In degraded riparian areas, structural complexity and vertical diversity can be attained partially by leaving large, distorted, or broken trees in adjacent harvest units. These groups of trees outside the riparian management zone serve to feather the riparian forest into the adjacent younger forest and provide wildlife habitat. Group selection or single-tree selection is preferable to even-age management (Agee 1988). Thinning in riparian management zones should leave trees irregularly distributed in patches that reflect local patch dimensions and composition rather than uniformly spaced throughout the stand.

The major agent of restoration in riparian areas and stream ecosystems of the Pacific Northwest is floods. Restoration is a process of change, and most ecosystems change most dramatically during episodic disturbances (Swanson et al. 1990). Stream channels are shaped not during low flows, but during short-duration, infrequent floods. During these high flows, the stream has the power to move sediment, erode deep pools, deposit floodplain surfaces, create major debris dams, and shape the aquatic ecosystem (Gregory et al. 1991, Naiman 1992, Bayley 1995). Riparian areas serve as critical refuges and contribute large wood and boulders during floods and debris flows (Lamberti et al. 1991). Human efforts to restore streams and riparian areas will pale in comparison to the enormous forces of floods, which are the natural restoration process in stream ecosystems. Unfortunately, humans also eliminate many of the ecological benefits of floods in disaster-relief efforts and in repair programs after these basin-scale events. One of the most positive steps for future riparian restoration will be policies for systematic review of both social and ecological consequences of flood events and for maintenance of beneficial changes to the greatest degree possible.

Attention to Social Needs for Riparian Resources

People simultaneously abhor and desire regulations and legislation because of their conflicting needs for freedom and flexibility on one hand and assurance of safety and common good on the other hand. This dilemma clearly is central to riparian resource issues and future trends in riparian management over the next century. Far more regulatory requirements have been placed on the management of public and private forest lands in the Pacific Northwest than on other land-use types—a reflection primarily of the large portion of forest land in public ownership.

As our society develops laws that reflect its desires for managing its public resources, questions arise about their application to private lands. Common resources such as water, air, migratory fish, and wildlife make these questions relevant to society as a whole and encourage extension of practices from the public to the private land sector. What does this suggest for future trends in riparian management? A likely outcome of human demographics is increased regulatory guidance; concerns about fairness and equality may translate into application of riparian regulations to all land-use types.

A poll of forest land owners and managers would almost certainly call for reduced regulatory constraints. This feeling is understandable, but the outcome is unlikely. Increased human populations will have dual impacts on the forest industry. Human populations in the Pacific Northwest are increasing at 1.5 to 2.0 percent per year, resulting in a projected doubling of the population in 35 to 40 years (American Almanac 1994). First, there will be a greater demand for forest products because of the increased number of users. This effect will be amplified if recent trends in per capita consumption of wood continue; between 1970 and 1988, per capita consumption of wood products in the United States increased 30 percent from 61.1 cu. ft. to 79.5 cu. ft. (American Almanac 1994). Second, social demands for water resources, fisheries, wildlife, and recreation will increase due to the greater number of people using public land resources. Oregon and Washington currently obtain 42 percent and 35 percent, respectively, of their domestic water supply from specifically designated national forest lands (Bruce McCammon, Region 6, U.S. Forest Service, personal communication, 1994).

Though landowners would prefer the use of voluntary programs, incentive systems, user fee systems, market-based incentives ("green marketing"), and other less rigid regulatory approaches, these types of guidance have little history of success. The most direct and well-demonstrated tools for influencing collective human behavior in large systems have been regulations that define a range of appropriate actions. The major advances in riparian management in the Northwest to date have been based on such legislation, and there is little evidence that this trend will change.

One of the most glaring inconsistencies in riparian management in the Northwest is the enormous disparity between riparian management requirements on private forest lands and that on other land-use types. There are numerous examples of riparian management zones on forest lands where large merchantable trees have been retained to meet riparian rules, while on adjacent agricultural lands it is legal to plow through the stream or have livestock standing in the stream. On residential land, it is legal to cut riparian vegetation and landscape stream banks with any type of structure or vegetation. On urban lands, communities and agencies can line the entire stream channel with concrete and eliminate riparian forests. Land uses on these different types of land clearly are inconsistent and held to different standards. Creation of a general land-use practices act would coordinate management directions and provide more equitable support of society's ecosystem goals. Regulatory requirements on these land-use types would differ because of the range of social expectations, but the various land uses could be evaluated through a common set of ecosystem management questions and related ecological functions. A common commitment to goals for riparian resources would also identify the types of incentives or disincentives that are woven into the current fragmented policy landscape of the Pacific Northwest.

Conclusion

Riparian forests and stream ecosystems in forests of North America have been extensively altered over the last several centuries. One year after the formation of the Bureau of Forestry in 1901, Overton Price (1902) observed that "[i]n effective methods for the harvesting and manufacture of lumber, the American lumberman has no superior, nor is he equaled in his disregard for the future of the forest which he cuts." As we look to the next century, we have the advantage of a few decades of awakening to the need to maintain and restore riparian resources and their ecological functions in forest landscapes. The ecological and social challenges of managing riparian areas in complex landscapes are becoming more acute as human populations grow along with the demand for water, forest products, fisheries, wildlife, and recreational resources.

Resource professionals and the public continually must improve approaches to managing the world's common resources. The success of our efforts will be based not on our static performance at any point in time, but rather on our ability to deal with ecological and institutional change. In *The Influence of Forestry upon the Lumber Industry,* Overton Price (1902) noted that "it is the history of all great industries directed by private interests that the necessity for modification is not seen until the harm has been done and its results are felt." It is this characteristic of human nature and our society that necessitates awareness of historical changes, anticipation of future trends, and development of more effective approaches to maintain and restore riparian forests and aquatic ecosystems.

Literature Cited

Agee, J. K. 1988. Successional dynamics of forest riparian zones. In *Streamside management: Riparian wildlife and forestry interactions,* ed. K. J. Raedeke. Contribution number 59. Seattle, WA: Institute of Forest Resources, University of Washington.

———. 1993. *Fire ecology of Pacific Northwest forests.* Washington, DC: Island Press.

American Almanac. 1994. *American almanac 1993–1994: Statistical abstract of the United States.* 113th ed. Austin, TX: The Reference Press Inc.

Bayley, P. B. 1995. Understanding large river-floodplain ecosystems. *BioScience* 45(3):153–158.

Bisson, P. A., T. P. Quinn, G. H. Reeves, and S. V. Gregory. 1992. Best management practices, cumulative effects, long-term trends in fish abundance in Pacific Northwest river systems. In *Watershed management: Balancing sustainability and environmental change,* ed. R. J. Naiman. New York: Springer-Verlag.

Booth, D. B. 1991. Urbanization and the natural drainage system: Impacts, solutions, and prognoses. *Northwest Environmental Journal* 7:93–118.

Botkin, D., K. Cummins, T. Dunne, H. Regier, M. Sobel, L. Talbot, and L. Simpson. 1995. *Status and future of salmon in western Oregon and northern California: Status and future options.* Report no. 8. Santa Barbara, CA: The Center for the Study of the Environment.

Bottom, D. L., P. J. Howell, and J. D. Rodgers. 1985. *The effects of stream alterations on salmon and trout habitat in Oregon.* Portland, OR: Oregon State Department of Fish and Wildlife.

Boule, M. E., and K. F. Bierly. 1987. History of estuarine wetland development and alteration: What have we wrought? *Northwest Environmental Journal* 3:43–61.

Cummins, K. W. 1974. Structure and function of stream ecosystems. *BioScience* 24:631–641.

Emmingham, W. H., and K. Maas. 1994. Survival and growth of conifers released in alder-dominated coastal riparian zones. *COPE Report* (Oregon State University) 7:13–15.

FEMAT. 1993. *Forest ecosystem management: An ecological, economic, and social assessment.* Report of the Forest Ecosystem Management Assessment Team (FEMAT). 1993-793-071. Washington, DC: GPO.

Frissell, C. A., and R. K. Nawa. 1992. Incidence and causes of physical failure of artificial habitat structures in streams of western Oregon and Washington. *North American Journal of Fisheries Management* 12:182–197.

Gedney, D. R. 1988. *The private timber resource: In assessment of Oregon's forests,* ed. G. J. Lettman and D. H. Stere. Salem, OR: Oregon State University, Department of Forestry..

Gregory, S. V., and L. R. Ashkenas. 1990. *Riparian management guidelines for the Willamette National Forest.* Technical report. Eugene, OR: USDA Forest Service, Willamette National Forest.

Gregory, S. V., G. A. Lamberti, D. C. Erman, K. V. Koski, M. L. Murphy, and J. R. Sedell. 1987. Influence of forest practices on aquatic production. In *Streamside management: Forestry and fishery interactions,* ed. E. O. Salo and T. W. Cundy, 233–255. Contribution no. 57. Seattle, WA: Institute of Forest Resources, University of Washington.

Gregory, S. V., B. Beschta, F. J. Swanson, J. R. Sedell, G. Reeves, and F. Everest. 1990. Abundance of conifers in Oregon's riparian forests. *COPE Report* (Oregon State University) 3:5–6.

Gregory, S. V., F. J. Swanson, W. A. McKee, and K. W. Cummins. 1991. An ecosystem perspective of riparian zones. *BioScience* 41:540–551.

Harmon, M. E., J. F. Franklin, F. J. Swanson, P. Sollins, S. V. Gregory, J. D. Lattin, N. H. Anderson, S. P. Cline, N. G. Aumen, J. R. Sedell, G. W. Lienkaemper, K. Cromack, Jr., and K. W. Cummins. 1986. Ecology of coarse woody debris in temperate ecosystems. *Advances in Ecological Research* 15:133–302.

Hicks, B. J., J. D. Hall, P. A. Bisson, and J. R. Sedell. 1991. Responses of salmonids to habitat change. In *Influences of forest and rangeland management on salmonid fishes and their habitats,* ed. W. R. Meehan. Special publication no. 19. Bethesda, MD: American Fisheries Society.

Iwamoto, R. N., E. O. Salo, M. A. Madej, and R. L. McComas. 1978. *Sediment and water quality: A review of the literature, including a suggested approach for water quality criteria.* EPA 910/9-78-048, U.S. EPA Region X. Seattle, WA: U.S. EPA.

Junk, W. J., P. B. Bayley, and R. E. Sparks. 1989. The flood pulse concept in river-floodplain systems. In *Proceedings of the International Large River Symposium,* ed. D. P. Dodge. Ottawa: Canadian Special Publication of Fisheries and Aquatic Sciences 106, Department of Fisheries and Oceans.

Karr, J., and I. Schlosser, 1977. *Impact of nearstream vegetation and stream morphology on water quality and stream biota.* EPA 600 3-77 097. Washington, DC: U.S. EPA.

Kauffman, J. B. 1988. The status of riparian habitats in Pacific Northwest forests. In *Streamside management: Ri-*

parian wildlife and forestry interactions, ed. K. J. Raedeke. Contribution no. 59. Seattle, WA: Institute of Forest Resources, University of Washington.

Krygier, J. T., and J. D. Hall. 1971. *Forest land uses and stream environment.* Proceedings of a symposium. Corvallis, OR: College of Forestry and Department of Fisheries and Wildlife, Oregon State University.

Lamberti, G. A., S. V. Gregory, L. R. Ashkenas, R. C. Wildman, and K. M. S. Moore. 1991. Stream ecosystem recovery following a catastrophic debris flow. *Canadian Journal of Fisheries and Aquatic Sciences* 48:196–208.

Li, H., C. B. Schreck, C. E. Bond, E. Rexstad. 1987. Factors influencing changes in fish assemblages of Pacific Northwest streams. In *Community and evolutionary ecology of North American stream fishes,* ed. W. J. Matthews and D. C. Heins. Norman, OK: University of Oklahoma Press.

McComb, W. C., K. McGarigal, and R. G. Anthony. 1993. Small mammal and amphibian abundance in streamside and upslope habitats of mature Douglas-fir stands, western Oregon. *Northwest Science* 67:7–15.

McIntosh, B. A., J. R. Sedell, J. E. Smith, R. C. Wissmar, S. E. Clarke, G. H. Reeves, and L. A. Brown. 1993. Management history of eastside ecosystems: Changes in fish habitat over 50 years, 1935 to 1992. In *Eastside forest ecosystem health assessment,* ed. P. F. Hessburg. Volume III, Assessment. Washington, DC: USDA Forest Service.

Meehan, W.R., ed. 1991. *Influences of forest and rangeland management on salmonid fishes and their habitats.* Special publication no. 19. Bethesda, MD: American Fisheries Society.

Moyle, P. B., and R. M. Yoshiyama. 1994. Protection of aquatic biodiversity in California: A five-tiered approach. *Fisheries* 19(2):6–18.

Murphy, M. L. 1995. *Forestry impacts on freshwater habitat of anadromous salmonids in the Pacific Northwest and Alaska: Requirements for protection and restoration.* NOAA's Coastal Ocean Program, decision analysis series no. 1. Washington, DC: U.S. Department of Commerce, National Oceanic and Atmospheric Administration, Coastal Ocean Office.

Murphy, M. L., and Koski, K. V. 1989. Input and depletion of woody debris in Alaska streams and implications for streamside management. *North American Journal of Fisheries Management* 9:427–436.

Murphy, M. L., J. Heifetz, S. W. Johnson, K. V. Koski, and J. F. Thedinga. 1986. Effects of clearcut logging with and without buffer strips on juvenile salmonids in Alaskan streams. *Canadian Journal of Fisheries and Aquatic Sciences* 43:1521–1533.

Naiman, R. J., ed. 1992. *Watershed management: Balancing sustainability and environmental change.* New York: Springer-Verlag.

Naiman, R. J., H. Decamps, J. Pastor, and C. A. Johnston. 1988. The potential importance of boundaries to fluvial ecosystems. *Journal of the North American Benthological Society* 7:289–306.

Naiman, R. J., D. G. Lonzarich, T. J. Beechie, and S. C. Ralph. 1991. General principles of classification and the assessment of conservation potential in rivers. In *River conservation and management,* ed. P. J. Boon and G. E. Petts. Chichester, UK: John Wiley & Sons, Inc.

Naiman, R. J., T. J. Beechie, L. E. Benda, D. R. Berg, P. A. Bisson, L. H. MacDonald, M. D. O'Connor, P. L. Olson, and E. A. Steel. 1992. Fundamental elements of ecologically healthy watersheds in the Pacific Northwest Coastal Ecoregion. In *Watershed management: Balancing sustainability and environmental change,* ed. R. J. Naiman. New York: Springer-Verlag.

Naiman, R. J., H. Decamps, and M. Pollock. 1993. The role of riparian corridors in maintaining regional biodiversity. *Ecological Applications* 3:209–212.

National Research Council (NRC). 1992. *Restoration of aquatic ecosystems.* Washington, DC: National Academy Press.

———. 1995. *Upstream: Salmon and society in the Pacific Northwest.* Washington, DC: National Academy Press.

Newbold, J. D., D. C. Erman, and K. B. Roby. 1980. Effects of logging on macroinvertebrates in streams with and without buffer strips. *Canadian Journal of Fisheries and Aquatic Sciences* 37:1076–1085.

Oregon Department of Forestry. 1994. *The Oregon Forest Practices Act: Water protection rules.* Salem, OR: Forest Practices Policy Unit, Oregon Department of Forestry.

PACFISH. 1994. *Environmental assessment for the implementation of interim strategies (PACFISH) for managing anadromous fish-producing watersheds in eastern Oregon and Washington, Idaho, and portions of California.* Washington, DC: USDA Forest Service and USDI Bureau of Land Management.

Pacific Rivers Council. 1993. *Entering the watershed: An action plan to protect and restore America's river ecosystems and biodiversity.* Eugene, OR: Pacific Rivers Council

Peterson, N., A. Hendry, and T. P. Quinn. 1992. *Assessment of cumulative effects on salmonid habitat: Some suggested parameters and target conditions.* TFW-F3-92-001. Olympia, WA: Department of Natural Resources.

Price, O. 1902. The Influence of forestry upon the lumber

industry. In *Yearbook of agriculture.* Washington, DC: GPO.

Raedeke, K. J., ed. 1988. *Streamside management: Riparian wildlife and forestry interactions.* Contribution no. 59. Seattle: University of Washington, Institute of Forest Resources.

Reeves, G. E., J. D. Hall, T. D. Roelofs, T. L. Hickman, and C. O. Baker. 1991. Rehabilitating and modifying stream habitats. In *Influences of forest and rangeland management on salmonid fishes and their habitats,* ed. W. R. Meehan. Special report no. 19. Bethesda, MD: American Fisheries Society.

Salo, E. D., and T. W. Cundy, eds. 1987. *Streamside management: Forestry and fishery interactions.* Contribution no. 57. Seattle, WA: Institute of Forest Resources, University of Washington.

Schlosser, I. J., and J. R. Karr. 1981. Water quality in agricultural watersheds: Impact of riparian vegetation during base flow. *Water Resources Bulletin* 17:233–240.

Schoonmaker, P., and A. McKee. 1988. Species composition and diversity during secondary succession of coniferous forests in the western Cascade Mountains of Oregon. *Forest Science* 34:960–979.

Sedell, J. R., and J. L. Froggatt. 1984. Importance of streamside forests to large rivers: The isolation of the Willamette River, Oregon, U.S.A., from its floodplain by snagging and streamside forest removal. *Verhandlungen Internationale Vereinigen Limnologie* 22:1828–1834.

Sedell, J. R., and K. J. Luchessa. 1982. Using the historical record as an aid to salmonid habitat enhancement. In *Symposium on acquisition and utilization of aquatic habitat inventory information,* ed. N. B. Armantrout. Eugene, OR: Western Division, American Fishery Society.

Sedell, J. R., R. J. Steedman, H. A. Regier, and S. V. Gregory. 1991. Restoration of human-impacted land-water ecotones. In *Ecotones: The role of landscape boundaries in the management and restoration of changing environments,* ed. M. M. Holland, P. G. Risser, and R. J. Naiman. New York: Chapman and Hall.

Swanson, F. J., J. F. Franklin, and J. R. Sedell. 1990. Landscape patterns, disturbance, and management in the Pacific Northwest, U.S.A. In *Trends in landscape ecology,* ed. I. S. Zonneveld and R. T. Forman. New York: Springer-Verlag.

Thomas, J. W., M. G. Raphael, R. G. Anthony, E. D. Forsman, A. G. Gunderson, R. S. Holthausen, B. G. Marcot, G. H. Reeves, J. R. Sedell, and D. M. Solis. 1993. *Viability assessments and management considerations for species associated with late-successional and old-growth forests of the Pacific Northwest.* Report of the Scientific Analysis Team. Washington, DC: USDA Forest Service.

Ursitti, V. L. 1991. Riparian vegetation and abundance of woody debris in streams of southwestern Oregon. M.S. thesis, Oregon State University, Corvallis.

USDA and USDI. 1994. *Record of decision for amendments to Forest Service and Bureau of Land Management planning documents within the range of the northern spotted owl: Standards and guidelines for management of habitat for late successional and old-growth forest related species within the range of the northern spotted owl.* Washington, DC: USDA Forest Service and USDI Bureau of Land Management.

Wissmar, R. C., and F. J. Swanson. 1990. Landscape disturbances and lotic ecotones. In *Ecology and management of aquatic-terrestrial ecotones,* ed. R. J. Naiman and H. Decamps. London: Parthenon Press.

6 Biodiversity of Old Forests of the West: A Lesson from Our Elders

Bruce G. Marcot

Fungi 89
Lichens 90
Bryophytes 91
Vascular Plants 92
Invertebrates 92
Arthropods 93
Fish 94
Amphibians 95
Reptiles 95
Birds 95
Mammals 95
Maintaining the Productivity and Sustainable Use of Forests 96
Some Difficult Questions About Managing for Biodiversity 100
Biodiversity as Our New Mythos for Management 101
Acknowledgments 101
Literature Cited 102

One of the most fundamental lessons of the last several decades of ecological research is that the biological diversity of North American forests is far greater than previously thought. At the same time, much more is at risk through traditional forestry programs than ever imagined. Perhaps nowhere has this been more pronounced than in the debate over the fate of the remaining old forests of the Pacific Northwest.

At the root of this debate we find central concerns for protecting species, ecological communities and processes, and entire ecosystems (Wilcove 1988). These concerns, in turn, are linked to maintaining productivity of forest ecosystems and long-term sustainable use of forest resources (Hansen et al. 1991, Probst and Crow 1991)—much like American Indian philosophies of nature, which view air, water, mountains, forests, animals, and humans as inextricably knotted in webs of ecological relations (Heizer and Elsasser 1980, Storm 1972). This chapter summarizes what we have learned about the biodiversity of old forests in the Pacific Northwest and the lessons to draw for maintaining the productivity, diversity, and sustainability—the very ecological webs—of forests into the 21st century.

The amount of late-successional forests within the range of the northern spotted owl in western Washington, Oregon, and California has decreased at least five-fold from early historic times (FEMAT 1993). At present, there remain about 4.5 million acres of multistoried and 4 million acres of single-storied medium to large conifer stands. Most of these stands constitute late-successional forests. Over half occur at medium to high elevations, because much of the highly productive old-growth forests of the low elevations were the first to be harvested on private and state lands.

What is the biological diversity of the remaining old-growth conifer forests on public federal land, how is it related to forest land productivity, and how can it best be maintained? These questions were addressed in a series of research studies (Ruggiero et al. 1991) and in two regional land management planning analyses (Thomas et al. 1993, FEMAT 1993) culminating in a new President's Northwest Forest Plan for Management of Habitat for Late-Successional and Old-Growth Forest Related Species Within the Range of the Northern Spotted Owl (USDA and USDI 1994). The following discussion draws on the technical work done in these assessments, particularly that of the Forest Ecosystem Management Assessment Team (FEMAT), which compiled information on biodiversity from research scientists in government agencies and universities. Because much has been written on biodiversity of fish and other vertebrates, the focus here is species groups that are less popular or less understood.

Table 6.1 The many faces of forest biodiversity

Special Elements	Ecological Process Elements	Patterns of Rarity and Endemism	Paleoecological Trends of Species and Environments
• Degree to which the natural complement of species of each taxonomic group is present • Degree to which the original genetic variations at organism, deme, population, and metapopulation levels are present • Viability of individual species and populations • Conservation of key environments and habitats in geographic locations vital to species dispersal or viability* • Trends of exotic and introduced species (both desired and deleterious)	• Types, frequencies, locations, and durations of disturbance dynamics and events • The role of disturbance events in maintaining forest ecosystem health, productivity, and diversity • Nutrient flow cycles, rates, and components • Energy flow pathways, rates, and components	• Status and fate of unique, rare, or declining habitats or special ecological communities • Conservation of centers of species endemism, rarity, and richness	• The patterns and role of prehistoric climate and vegetation • Coevolution of species and systems, such as ungulates and grasslands • Prehistoric and early historic ranges of natural conditions, including disturbance regimes • Prehistoric and historic influences of human activities on other biodiversity elements

Note: These are some of the elements that can be assessed in a forest management program designed to maintain long-term productivity.

*Includes a discussion on linkages, corridors, and barriers for dispersal, and on dispersal modes and conditions for successful dispersal, particularly of endemic, threatened, endangered, candidate, and sensitive species.

Biodiversity of forests and its influence on sustainability must be perceived as much more than just trees, fish, and wildlife and the presence and fate of individual species. Table 6.1 presents a partial list of key aspects of biodiversity that should be addressed when developing forest management plans designed to promote or maintain long-term sustained resource production.

Just considering the species component of biodiversity is formidable. For example, FEMAT cataloged some 1,098 species closely associated with late-successional forests within the range of the northern spotted owl (Figure 6.1). This sum does not include thousands of other species found in late-successional forests that are not considered close associates; it also does not specifically catalog the tens of thousands of invertebrate species, principally all the bacteria, protozoa, mollusks, arthropods, and other forms associated with old forests. We quickly reach the limits of our understanding when dealing with many of these "lesser" species groups, which still are so critical to the sustained function and trophic health of forest ecosystems (Table 6.2).

Indeed, the small and inconspicuous life forms—particularly the fungi, lichens, nonvascular plants, invertebrates, and small vertebrates—contribute much to the functioning of energy pathways, food chains, and nutrient cycles. Along with the more visible vascular plants, such forest species can greatly mediate the spread of fire, erosion of soil, change of air quality, and even change in climate. In the following summaries, I list some of the results of the FEMAT analyses and discuss some of the key ecological functions of each group of forest organisms.

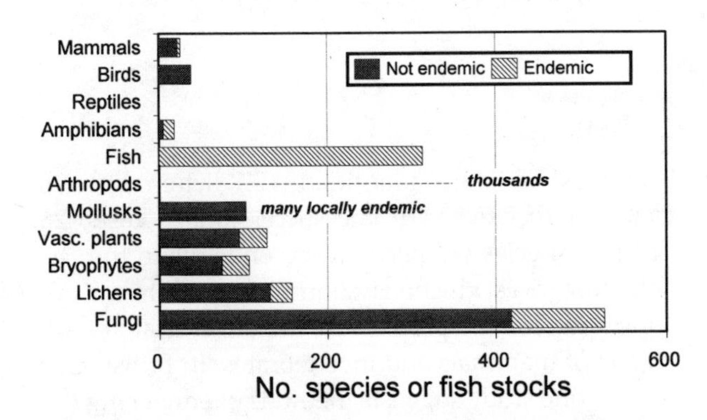

Figure 6.1 The incredible biodiversity of late-successional forests of the Pacific Northwest. Numbers of species or fish stocks that are closely associated with late-successional forests in the Pacific Northwest United States (western Washington, Oregon, and California), with numbers that are also endemic to that area (from FEMAT 1993).

Fungi

Some 527 species of fungi were identified by FEMAT as being closely associated with old-growth forests of the Pacific Northwest. This includes 109 species (21 percent) that are endemic to the Pacific Northwest.

Many forest fungi of Pacific Northwest (and many other) forests that produce fruiting bodies are coevolved in symbiosis with vascular plants. Thousands of ectomycorrhizal fungi (tiny hair-like fungi that bind to the outside of small rootlets of trees) occur within the Pacific Northwest. These inconspicuous life forms are responsible for aiding nitrogen uptake and fixation by the commercially valuable conifer tree species, and thereby directly influencing the health and productivity of conifer forests. Amaranthus and Perry (1994) have found that ectomycorrhizal fungi allow carbon and nutrients to pass among individual

Table 6.2 Ecosystem functions of late-successional forests

- Buffering of microclimate during seasonal climactic extremes
- Producing food for consumer organisms
- Storing carbon, which can act as a buffer to large-scale climate change
- Retaining high amounts of nutrients and water, including a high capacity for intercepting fog and rain (particularly by the epiphytic lichens and mosses)
- Providing sources of arthropod predators and organisms beneficial to other ecosystems or successional stages
- Maintaining low soil erosion potential

Note: These functions are for forests lacking or less well developed in younger natural forests and managed plantations. Such functions are critical to the sustainability and trophic health of forest landscapes, and are afforded by maintaining biologically diverse managed and natural forests.

Source: FEMAT (1993:IV–31).

trees, helping forests become reestablished after disturbances. Nearly 2,000 species of ectomycorrhizal fungi are associated with Douglas-fir alone, and many are host-specific (Berstein 1977). FEMAT identified some 336 species of ectomycorrhizal fungi closely associated with late-successional forests; there are more remaining to be identified, some likely for the first time.

Other species of fungi, the hypogeous forms, fruit underground. These include truffles, false truffles, and related species. They provide critical links in food chains that include ground squirrels, northern flying squirrels, red tree voles, and northern spotted owls. The small rodents consume and disperse the fungal spores, helping to inoculate forest stands with the species (Fogel and Trappe 1978, Li et al. 1986, Maser and Maser 1988). Mychorrhizae and hypogeous soil fungi in particular contribute greatly to nutrient cycling in Douglas-fir ecosystems (Fogel and Hunt 1983, Ingham and Molina 1991).

Other forms of fungi in old-growth forests include saprobes, parasites, and microfungi, and many are tightly coevolved with vascular plants. Some have future potential as medicinals, biological control agents, and possibly anticarcinogens; fungi are major sources of antibiotics. The microfungi in particular have medicinal potential, but the species are yet incompletely described. The saprobic fungi play important roles in the physical and biochemical breakdown of dead wood and in the recycling of nutrients. Some fungi are of commercial importance, such as king bolete, golden chanterelle, and matsutake. Many other fungi species play major roles in dead wood decomposition and recycling (Figure 6.2). In fact, some fungi species are specific to particular decay stages of down conifer trees (Crawford et al. 1990), which undergo their own microsuccession of organisms as they decompose.

Many species of fungi are also important indicators for monitoring forest health and stability. The disappearance of fungi species in continental Europe over the past two decades has signaled concern there for the decline in forest health (Berg et al. 1994); similar species that are still present in forests elsewhere, such as the Pacific Northwest, could serve well as signals of impending changes in forest environmental conditions.

Figure 6.2 Fungi are one of the more diverse life forms in Pacific Northwest forests and, like this conk, play vital roles in the decay of rotting wood and in the nutrient uptake by conifers (photo by B. Marcot).

Lichens

The FEMAT team identified 157 lichen species closely associated with late-successional forests; 26 of these species (17 percent) are endemic to the Pacific Northwest. Lichens contribute to nutrient cycling and biomass production and are critical in food chains of mammals and invertebrates. In British Columbia one study suggests methods of enhancing the presence and distribution of forest lichens to help provide winter food for ungulates (Stevenson and Enns 1992).

Forest lichens also accumulate biomass and carbohydrates, and arboreal lichens capture fog and retain moisture within the forest canopy (Figure 6.3). Many

Figure 6.3 Lichens provide habitat for invertebrates, food for ungulates, nutrients and moisture for forest trees, and organic material for soil. Some species, like this old-man's beard, also can be excellent indicators of atmospheric sulfur dioxide concentrations (photo by B. Marcot).

lichen species fix atmospheric nitrogen, and their litterfall contributes to soil organic material and helps increase soil moisture-holding capacity.

Like other life forms, lichen species undergo successional changes in forests. Some lichens do not appear until the forest is 200, sometimes even 500, years old. Lichens require the ecological and temporal continuity of mature trees, as they lack the ability to disperse widely, largely by vegetative fragments falling or being borne by wind. In one study in the Oregon Cascades, half the biomass of lichen litterfall occurred within only 6 m of the edge of an old-growth forest patch.

In England, many lichen species are used as indicators of woodland age and ecological continuity—temporal fragmentation—of old-growth forests (e.g., Tibell 1992). Lichens also are used for monitoring air quality, as they are sensitive to sulfur dioxide and other gases and amass heavy metals (Stolte et al. 1993). Some are potential sources for antibiotics and other medicinal uses.

Many lichens have declined worldwide. Fourteen species that are still associated with old-growth forests in the Pacific Northwest have been documented as being extirpated from parts of Europe because of declines in environmental quality.

Bryophytes

Some 106 species of bryophytes (hornworts, liverworts, and mosses) found in the Pacific Northwest are closely associated with late-successional forests; 32 of these species (30 percent) are endemic to western North America or the Pacific Northwest.

Epiphytic mosses and lichens (those that grow on other plants) can total 2.6 metric tons per hectare in old-growth Douglas-fir forests of western Oregon. Mosses alone can contribute 20 percent of the understory biomass and 95 percent of the photosynthetic tissue biomass of a forest.

Most species of bryophytes do not become established in stands until 100 years, and they are best developed in stands 400 years and older (Pike et al. 1975). Empirical studies have shown that some bryophyte species failed to reappear in regrown clearcut stands, and other species occurred in much reduced abundance; bryophytes thus are sensitive to ecological continuity of old-forest stands (Soderstrom and Jonsson 1992). Maintaining and monitoring such species can contribute to a global knowledge of effects of forest management practices and changes in environmental quality (Hallingback 1992).

Some bryophytes, such as the common canopy species *Antitrichia curtipendula,* provide food and habitat for numerous invertebrates and vertebrates. For example, marbled murrelets nest in moss mats on large limbs of inland old-growth trees. Bryophytes such as *A. curtipendula* also are key to nutrient cycling in old-growth stands. They act as sinks for nitrogen leachate; they intercept, absorb, and buffer nutrients and water; and they contribute to soil structure and stability. In riparian areas and streams, species such as *Chiloscyphus polyanthos* filter sediments and small organic material. Also, as with lichens, some bryophytes (e.g., the genus *Ulota*) are sensitive to, and can act as indicators of, air quality.

Vascular Plants

Hundreds of species of vascular plants occur in late-successional forests of the Pacific Northwest. Some 127 of these species are closely associated with such forests; 33 species of these (26 percent) are endemics or rare endemics of the Northwest, and 29 species (23 percent) have federal, state, or agency listing status.

The vegetation of the Pacific Northwest includes vascular plant representatives of both the Madro-Tertiary Geoflora (the warm-weather, southern flora, including such species as *Pacific madrone, Arbutus menzeisii,* and live oak, *Quercus chrysolepis*) and the Arcto-Tertiary Geoflora (the cold-weather, northern flora, including true firs, *Abies* spp., and big-leaf maple, *Acer macrophyllum*) in a species mix found in the United States and much of the world (Axelrod 1958, 1981). Plants of course play major roles in nearly all food chains and nutrient and energy cycles and are the primary producers of forest ecosystems. In Pacific Northwest forests, as elsewhere, there are plants that are key to nutrient accumulation, such as nitrogen-fixing species of ceanothus, alders, and legumes, including lupines.

Vascular plants also provide necessary habitats and feeding substrates for many vertebrate species. For example, anadromous fish depend on pools and riffles maintained by fallen large trees that intercept stream currents and separate sediments. Vascular plants provide important commercial products, including medicines, horticultural stocks, and foods.

Invertebrates

Invertebrates are a vast and diverse group of species that occupy nearly every conceivable niche in forest ecosystems, including the interiors of other organisms. At times considered mostly for their detrimental effects on commercial tree species, invertebrates are nonetheless essential components of ecosystem function (Murphy 1991). They perform as keystones in food webs, for pollination of flowering plants, for dispersal of seeds and spores, for decomposition of organic material, for production of chemicals and fixation of nutrients usable by other organisms, and for many other functions.

Although FEMAT considered mainly mollusks and arthropods, other invertebrate life forms are part of the biodiversity spectrum that greatly influences the productivity and function of forests. These life forms include the acellular and saprobic viruses and allies of kingdom Archetista; the anucleated autotrophic or saprobic bacteria, cyanobacteria, and spirochetes of kingdom Monera; and the nucleated species of flagellated, ameboid, spore-forming, and ciliated protozoans of kingdom Protista (Protoctista). Little has been studied or published specifically on species of these three kingdoms in forests of the Pacific Northwest. As examples of important ecological functions, bacteria produce many chemical metabolites that influence many other organisms, fix carbon dioxide into organic compounds, contribute to the biomass of the first trophic level, cycle nutrients in soil and water, and directly affect the somatic health of many host species (Muehlchen 1994). An unknown number of species of these three kingdoms occur in Pacific Northwest old-growth forests; the species have never been cataloged, nor is it known what percent and which species are endemic to the Northwest.

Mollusks of old forests of the Pacific Northwest include land snails, slugs, and aquatic snails and clams. Currently, about 350 species of mollusks are known to occur in Pacific Northwest forests. Land snails and slugs account for over 150 of these species (Figure 6.4). Some 102 of these species are closely associated with late-successional forests, and many of them are locally endemic, in some cases endemic to only one province or river drainage. With further surveys and studies, the known number of species in the Pacific Northwest may eventually double.

Many mollusk species are endemic to only one region or river drainage. Examples include members of the land snail genera *Monadenia, Trilobopsis, Megom-*

Figure 6.4 In aquatic, riparian, and terrestrial systems, mollusks, like this banana slug, help in soil formation, wood decomposition, and support of energy webs and nutrient cycles (photo by B. Marcot).

phix, and *Vespericola* and the slug genus *Hemphillia.* Many aquatic snails of the genera *Juga, Lanx,* and *Fluminicola* occur only in single permanent streams, intermittent streams, springs, or seeps. Because they are poor dispersers, land snails seldom recolonize the same areas when they become locally extirpated.

Arthropods

Arthropods are another major group of invertebrates in Northwest forests. Olson (1992) estimated that about 7,000 species of arthropods occur in late-successional forests of the Pacific Northwest (see also Mispagel and Rose 1978). Closely associated with late-successional forests of the Northwest are at least 155 insects, 25 spiders, 25 millipedes, and 1 crustacean (206 species total). The species groups are poorly studied, but a substantial number of species are likely endemic to the Pacific Northwest.

Asquith et al. (1990) described arthropods as the "invisible diversity" of forest ecosystems. The litter and soil of the forest floor are the sites of some of the greatest biological diversity found anywhere. The soil under a square meter of forest may hold up to 200,000 mites from a single taxonomic group, plus tens of thousands of other mites, beetles, centipedes, pseudoscorpions, springtails, and spiders. Lattin (in press) noted that over 3,400 different species are known from a single 6,400-ha site in Oregon. Many of the arthropod species of the Northwest (perhaps 20–30 percent), especially the arboreal forms, have yet to be described.

Some arthropods of late-successional forests are terrestrial and flightless and thus have poor dispersal capability. As hypothesized by Lattin and Moldenke (1990) for the moss lacebug in Northwest conifer forests, the evolution of flightlessness might reflect environmental stability over a long time. Many of the flightless, old-forest associates now likely occur in fragmented populations in the Northwest (Moldenke and Lattin 1990b). Those species are probably most sensitive to forest fragmentation (with mosaics of very young and old-forest patches) that are highly substrate- or host-specific, tightly coupled to other species in parasitoid or other symbiotic relations, or flightless.

Several studies of effects on invertebrates from forest management and forest fragmentation are instructive. In one study of lepidopterans in soybean fields separated by various kinds of corridors, Kemp and Barrett (1989) found that managing uncultivated corridors within the agricultural landscape could be an important way of regulating insect pests, and that establishing uncropped patches could provide net ecological and economic benefits. In dry subtropical forests of northwest Argentina, Aizen and Feinsinger (1994) reported that forest fragmentation affected native flower-visiting insects adversely and enhanced visitation by the exotic honey bee. In boreal forests of western Canada, Niemela et al. (1993) found that various species of boreal ground beetles responded differently to forest cutting. Forest generalists were not dramatically affected, open-habitat species invaded or increased, and mature-forest species disappeared or decreased. As with many other species groups, several specialist ground beetles of mature forests did not recolonize even the oldest regenerating forest stands. In Costa Rica, Roth et al. (1994) found that 109 species of ground-foraging ants were differentially affected by types and degrees of agroforestry; results suggest that it is possible to conserve diversity in a mosaic of different land uses, but the dynamics of species source populations need further investigation. More such studies as these need to be done to relate effects of forest successional stages and forest fragmentation on sundry invertebrate species in managed forest landscapes.

Arthropods are key to ecosystem function (Seastadt and Crossley 1984). They aid in nutrient cycling of down wood, contribute to organic matter of soil and litter, are herbivores of forest canopies, pollinate flowering plants, and are major decomposers, chewers, shredders, predators, and food sources in forest streams and rivers. Arthropods can serve as sensitive barometers or indicators of environmental health, including complex soil functions (Moldenke and Lattin 1990a). Olson (1992) and Lattin (in press) have suggested specific taxa that can be monitored in old forests (see examples in Table 6.3). McIver et al. (1990) suggested that litter spiders can serve as indicators of recovery of western conifer forests following clearcutting. As indicators biological diversity for inventories and conservation studies, Pearson and Cassola (1992) suggested using tiger beetles, and Kremen (1994) advocated using butterflies in Madagascar.

Fish

The FEMAT team identified 314 stocks of at-risk salmonid fishes in old-growth forest landscapes within the range of the northern spotted owl in the Pacific Northwest; all 314 stocks are endemic to the Pacific Northwest. They also described the degradation of spawning and migration habitat and prescribed methods for analyzing watersheds, delineating and protecting stream buffers, and restoring aquatic systems.

Inland and anadromous fish are significant in their role of transferring nutrients longitudinally up and

Table 6.3 Examples of arthropods that could be monitored

Species' Ecosystem Function to Monitor	Organisms to Sample	Sampling Method
Fungal and litter feeders and predators	Oribatid mites (Acarina), springtails (Collembola)	High-gradient extractors, Berlese funnels
Litter communitors (chewers)	Millipedes (Diplopoda)	Pit-fall traps
Terrestrial seed feeders	Seed bugs (Lygaedae, Hemiptera: Heteroptera, Insecta), ground beetles (Carabidae, Coleoptera, Insecta)	Pit-fall traps
Foliage feeders	Plant bugs (Miridae, Hemiptera: Heteroptera, Insecta), sawflies (Symphyta, Hymenoptera, Insecta), caterpillars (Lepidoptera, Insecta)	Beating foliage, malaise traps, branch clipping
Predators	Ground beetles (Carabidae, Coleoptera, Insecta)	Pit-fall traps
Xylophagous (sap-eating)	Bark beetles (Scolytidae, Coleoptera, Insecta)	Pheromone, intercept traps
Nectar feeders	Adult butterflies and moths (Lepidoptera, Insecta)	Blacklight traps

Source: Lattin, (in press).

down stream and river systems, which are otherwise typically nutrient poor. Populations of anadromous fish transfer nutrients and organic material inland from the ocean (Cederholm and Peterson 1985); stocks of different seasonality do so at different times of the year. Bilby et al. (1996) found that ocean-derived calcium and phosphorus from salmon carcasses were taken up by riparian vegetation. Cederholm and Peterson (1985) reported that mammals, especially raccoons, further contribute to dispersal of fish-transmitted nutrients by dragging coho salmon (*Oncorhynchus kisutch*) carcasses into riparian zones. Foerester (1968) reported that carcasses of juvenile sockeye salmon (*Oncorhynchus nerka*) are an important source of phosphorus in oligotrophic (nutrient-poor) lakes. Fish also are a main food source for bald eagles, osprey, and other birds and mammals (Cederholm and Peterson 1985).

The anadromous fish of the Northwest have had vital cultural and social values of historic importance, particularly to native cultures (Hunn et al. 1990). Historically, the large populations of fish supported in Northwest river systems may have been one of the key factors allowing native cultures to develop more sedentary habitations and to thereby advance beyond hunting and gathering (G. Reeves, personal communication, 1995).

Amphibians

There are 62 species of amphibians found in the Northwest. Eighteen are closely associated with late-successional forests and 13 of these (72 percent) are endemic to the range of the northern spotted owl in the Pacific Northwest. Old forests of the Northwest provide habitat for some relictual species (Welsh 1990) and a relatively high species diversity (Walls et al. 1992).

Amphibians, particularly salamanders, may play major roles in buffering the energy flow through forest ecosystem food webs and act as major pools of energy and substances in their large-standing crop biomass. As many as 5,000 individual amphibians can occur per acre in suitable habitat. Example species from Northwest forests are ensatinas, giant salamanders, and torrent salamanders.

Reptiles

There are no species of reptiles that seem to be closely associated with old-growth forests per se in the Pacific Northwest. However, such species as northern alligator lizard and sharp-tailed snake are dependent on down or dead wood, particularly large rotting logs in late stages of decay. Such microhabitats are found most commonly in old-growth forests and are often reduced or eliminated during more-intensive forestry operations.

Birds

Thirty-eight species of birds are closely associated with late-successional forests, but none is endemic to the Northwest. Species with federal threatened or endangered status that associate with old forests of the Northwest include the northern spotted owl, marbled murrelet, and bald eagle. Many insectivorous species, including woodpeckers, help suppress population peaks of forest insects and bark beetles (Holmes 1990, Holmes et al. 1979, Knight 1958, Loyn et al. 1983, Solomon 1969), thus conveying an economic benefit by helping to keep forests healthier (see also papers in Dickson et al. 1979). Takekawa et al. (1982) and Campbell et al. (1983) have advocated using forest insectivorous birds as cost-effective biological control agents to control outbreaks of forest insect pests.

Many bird species disperse seeds of conifers and flowering plants. At least one set of birds—the cryptic, sibling species group of crossbills—are specifically coevolved with gymnosperm cones and seeds (Benkman 1993). The beaks are specially adapted with overlapping tips, unique to the bird world, that help the birds to pry off cone bracts and extract the seeds within.

Mammals

Fifteen mammal species other than bats, and 11 species of bats, are closely associated with Northwest late-successional forests. Four species other than bats (27 percent of mammals other than bats, and 15 per-

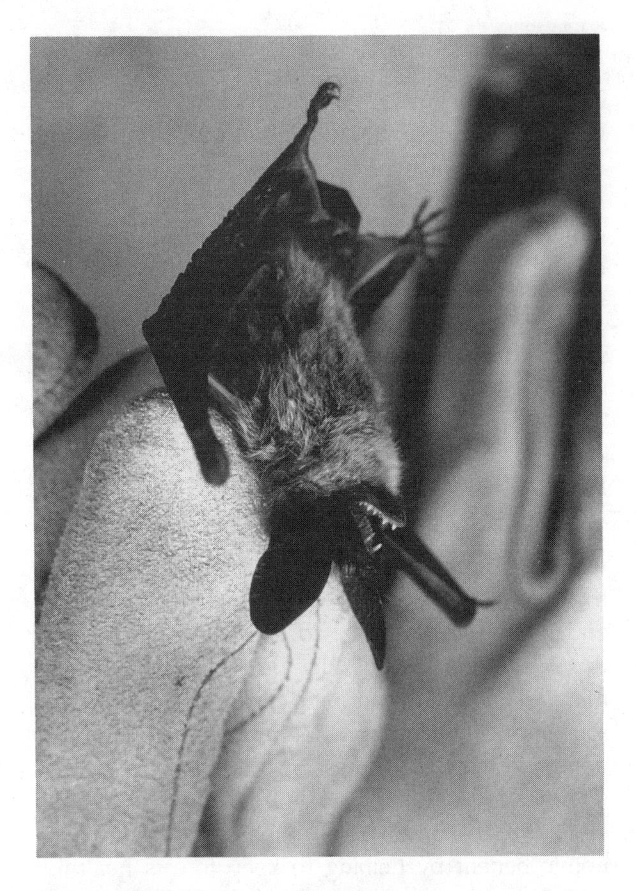

Figure 6.5 Their contribution to ecosystem function is little understood, but bats such as this long-eared myotis may contribute to redistribution of nutrients throughout the forest by foraging in open areas or over water and roosting elsewhere, as in dense tree foliage, caves, or buildings (photo by B. Marcot).

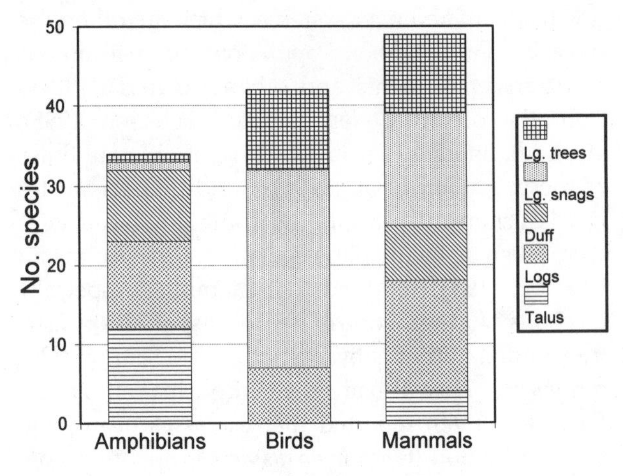

Figure 6.6 No one forest component provides for all species. Numbers of species of amphibians, birds, and mammals that are closely associated with late-successional forests in the Pacific Northwest, by forest substrate.

cent of all mammals), but none of the bats, are broadly endemic to the Pacific Northwest.

As with birds, many species of small mammals serve to distribute spores and seeds of fungi and plants throughout a forest. Small mammals such as red tree vole also help to disperse lichens. Fecal pellets of small mammals not only contain spores and seeds but also nitrogen-fixing bacteria and yeast, which add to the forest floor and contribute to nutrient cycling and substance breakdown.

Bats serve to suppress populations of flying insects, including forest insect pests; all 11 bat species are insectivorous (Figure 6.5). Large populations of bats may play important roles in nutrient cycling in forests. The species that forage in riparian areas and over water and that fly upland to roost may serve as dispersers of nutrients across a forest landscape. No one has yet estimated the quantity and geographic pattern of guano excrement deposited by foraging bats of Northwest forests, but it may likely be an important source of redistributed nutrients.

The forest microhabitats of terrestrial vertebrates closely associated with late-successional forests of the Pacific Northwest suggest associations of amphibians with large snags, a deep litter and duff layer, and talus; birds with large snags but also with large green trees and a deep litter and duff layer; and mammals with all components of large trees, large snags, a deep litter and duff layer, large down logs, and talus (Figure 6.6). Thus, no one component of late-successional forests suffices to provide for the needs of all species.

Maintaining the Productivity and Sustainable Use of Forests

What lessons can be learned from this assessment of biodiversity of Pacific Northwest forests? My list includes eight key items; others doubtless will also emerge as work continues. Although these lessons have emerged from work in the Pacific Northwest, they apply far beyond that geographic region and to other forest types as well.

1. Hidden Species Play Key Ecological Roles

We have learned much about the central role of the more furtive or hidden species and elements of biodiversity—that is, the essential ecosystem functions of many small and neglected life forms, particularly the fungi, lichens, bryophytes, and arthropods. These are central to the health and productivity of forest ecosystems (Crow 1988, 1990). They also can teach us much about impending environmental degradation by acting as early warning systems for changes in the trophic health and productivity of the soil, air, and canopy (see Table 6.3).

2. Large Reserves Alone Do Not Suffice to Ensure Conservation of the Entire Legacy of Natural Biological Diversity

Natural reserves alone—including research natural areas, botanical waysides, wildernesses, and other natural areas and land reserves—do not protect all species. Some reserves are simply set in the wrong place to ensure conservation of substrates or microhabitat conditions for some of the rare, sessile, furtive species. Of course, this does not invalidate the use of reserves for other species or purposes, however.

As a corollary, a coarse-filter approach to old forest conservation (Hunter 1991) does not necessarily ensure protection of the viability of all species closely associated with late-successional forest habitats in the Pacific Northwest (Marcot et al. 1994). The work of Thomas et al. (1993) in part was a test of the hypothesis that providing for the habitat needs of one species, such as the northern spotted owl, would provide for most needs of all other species also closely associated with similar habitats, such as the owl's late-successional forest ecosystems. As it turned out, in the Pacific Northwest the hypothesis was wrong. The specific habitat requirements and locations of only one-third of all 667 plant and animal species analyzed by Thomas et al. seemed to be met by applying guidelines for managing spotted owl forests and allocating spotted owl Habitat Conservation Area reserves. Another third of the 667 species required conditions or sites not specifically provided by the management guidelines or the reserves. And the last third were too poorly known for judgment to be drawn. As a result, Thomas et al. proposed an additional set of complementary reserves and management guidelines to provide for the second set of species, and an inventory, monitoring, and research program for gathering basic information on the last third of species. The lesson is that it is wise to check the validity of the coarse-filter approach to biodiversity conservation, lest some elements or species be excluded.

3. Providing Some Old-Forest Elements Between Reserves Is Vital

The third lesson learned from studying biodiversity of Pacific Northwest forests is that maintaining at least some old-forest components within the forest "matrix" between large reserves or conservation areas is absolutely key to conservation of furtive biodiversity (North and Franklin 1990). Tiny remnant islands of old-growth forests are important for species refugia and centers of dispersal for fungi, lichens, bryophytes, mollusks, and arthropods, as well as amphibians and the smaller terrestrial mammals. Many species of these taxa can act as indicators of the temporal continuity of old-forest ecosystems throughout a landscape. Once interrupted, many of these species would vanish even if at some later time forests are regrown to old (but more isolated) conditions. The effects of such temporal fragmentation of old forests on long-term productivity of forests of all age classes are poorly understood and deserve further investigation.

Thus, forest management should attend to conserving old-growth fragments along with clusters or patches of large remnant old trees, which collectively provide sources of large snags, sources of large down wood on the forest floor and in streams, sources of species inocula for many taxa, and connectivity among reserves for the less vagile species.

4. The Needs of Individual Species Still Need to Be Addressed, But Should Be Done So in the Context of the Ecological Community

Biodiversity is best provided by coordinating forest management across geographic and topographic conditions, and on multiple ownerships. A case in point is that large, widely spaced reserves or conservation areas might suffice for many of the vascular plants and vertebrates. However, in the Northwest, federal lands largely occur in mid to upper elevations and do not include the most productive lowland sites that are state-managed or privately owned; they may not include particular centers of biodiversity or species endemism (e.g., of mollusks and arthropods). Management standards therefore need to be developed for maintaining at least some older forest components within the forest matrix on federal lands and for coordinating forest management with states and other land management administrators and owners.

Again, this points to the conclusion that, on federal public lands of the western United States, reserves or conservation areas alone probably will not guarantee the future for all late-successional taxa. As well, the diversity of vertebrate species extends throughout all physiographic provinces of the Pacific Northwest (Figure 6.7) where differences in species composition and habitat requirements may warrant designing province-specific management guidelines for reserves and forest matrix areas.

5. Riparian Areas and Wetlands, As Well As Rare and Tiny Environments, Play Key Roles in Conservation of Biodiversity

The forest biodiversity studies in the Pacific Northwest have underscored the importance of small streams, seeps, wetlands, bogs, ponds, lakes, and other unique and rare environments for maintaining many of the smaller elements of biodiversity. Among vertebrates closely associated with late-successional forests of the Pacific Northwest, 63 percent of amphibians, 21 percent of birds, and 23 percent of mammals also were found to be closely associated with riparian habitats (Figure 6.8). Many of the more sessile endemic and scarce life forms commonly occur only on specific substrates or microhabitats. Examples include the tailed frog found mostly in cold mountain streams, the fungus *Tuber rufum* found only on oaks, and the lichens *Platismatia* spp., *Parmelia* spp., and *Cetraria* spp., which grow on tree boles and twigs. Many fungi are host- or substrate-specific as well. Local inventories and surveys, or in some cases, broad-scale regional studies, need to be conducted to determine additional habitat requirements and to ascertain the presence of such species in particular drainages or project areas.

6. How Forests Are Connected over Time and Across Space Can Greatly Affect Biodiversity

Related to the role of local microhabitats are specific elements of forest ecosystems that are essential for linking populations across space. Such elements include (1) contiguous forest canopies that help link populations of such species as the red tree vole and old-forest lichens, (2) riparian and aquatic systems that help link associated species such as anadromous fish and mink (*Mustela vison*) (Naiman et al. 1993), (3) belowground terrestrial conditions that link soil mesoinvertebrates and

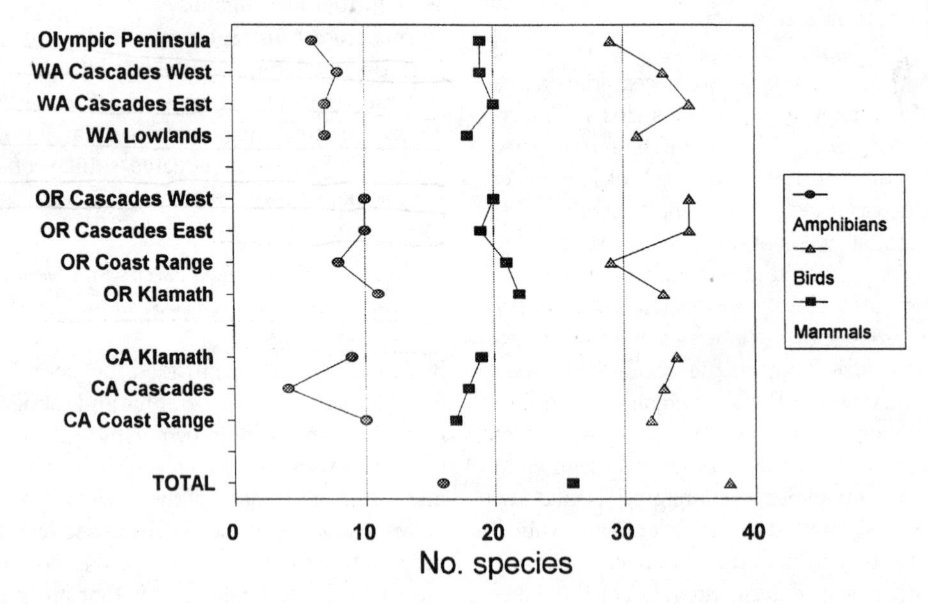

Figure 6.7 Despite minor variations in species numbers, the richness of terrestrial vertebrates extends throughout the Pacific Northwest United States. Numbers of species of amphibians, birds, and mammals that are closely associated with late-successional forests in the Pacific Northwest, by physiographic province.

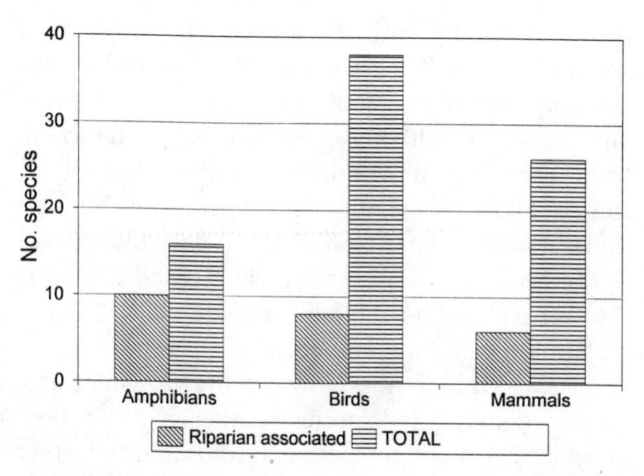

Figure 6.8 Riparian areas play key roles in providing for terrestrial vertebrate species closely associated with late-successional forests. Numbers of species of amphibians, birds, and mammals that are closely associated with riparian habitats within late-successional forests in the Pacific Northwest.

mycorrhizal fungi (Perry et al. 1990) so critical to maintaining soil productivity and tree growth, (4) hyporheic ecosystems, which are broad belowground extensions of stream and river systems that are critical to hydrologic and biotic processes (Naiman et al. 1992, 1993) and that contain species of invertebrates found only belowground (G. Reeves, personal communication, 1995), and (5) the atmosphere itself, insofar as air quality affects the survival and productivity of species sensitive to particulates and pollution, such as some arboreal lichens.

Connectivity of ecosystems and species' habitats through space must also be understood through time in dynamic, changing forest environments. As the loss of lichens in forests of Europe has shown, temporal as well as physical fragmentation of habitat patches and populations can eliminate species. As such, we need further understanding of the role of fire in leaving behind patches of live trees and dead standing and down wood, stimulating regrowth of vegetation, and changing floral succession. Management in the next century explicitly should assess and address how to re-create near-natural disturbance dynamics in forest landscapes.

7. Nonfederal Lands Can Play a Key Role in Biodiversity Conservation

Some locally endemic species, such as some fungi, arthropods, and even some vertebrates such as the Columbia torrent salamander, occur largely on nonfederal lands. Among the arthropods, centers of biodiversity or key locations of rare or endemic species on state and private lands of the Northwest include southwest Washington (Willapa Hills and coastal areas), northwest Oregon (Coast Range), and coastal northern California. As well, nonfederal lands can help maintain feeding and resting habitat for threatened and endangered vertebrates. Again, it may be to the long-term advantage of landholders to maintain slices of older forests throughout more intensively managed forest landscapes to help maintain the elements of biodiversity. Such elements include providing natural sources of mycorrhizal fungi, hypogeous fungi, and their dispersers, including red tree voles and northern flying squirrels, to maintain soil productivity; providing sources of large down wood to help stabilize soil loss on steeper slopes and to stave off soil erosion; and providing sufficiently wide stream buffers and managing both near-stream and upslope forests for sources of large down wood to maintain productive fish and aquatic habitats.

Thus, the mix of land ownerships and associated landholder objectives should be acknowledged and respected in setting broad-based goals for biodiversity and species conservation (e.g., Bird 1994, Brussard et al. 1992). And at the same time, specific management actions can be prescribed to help individual landholders play their appropriate role in ensuring the continuation of biodiversity elements through space and time. For instance, management for the furtive, small elements of biodiversity—especially for the rare endemics—could involve carefully tailored, site-specific management actions with at least some site inventories to help establish habitat relationships of the species. This should be followed by appropriate management prescriptions to maintain the hydrology and ecological characteristics of the site.

8. The Concept of "Desired Future Conditions" Should Be Replaced with "Desired Future Dynamics"

It is more than a leap of lexicon from species management to ecosystem management. Human activities can affect environmental conditions for populations of game and threatened and endangered species. But unless simplified radically far beyond natural conditions, forest ecosystems are too complex and labile to allow for very accurate management of all components. What this means is that, while we strive to maintain the very complexity inherent in natural forest ecosystems, we lose the ability to accurately predict future levels of resource production and composition. It is as if the uncertainty of

Heisenberg has extended to ecosystems: We can know the dynamics, but not the specific future composition, and when we manage for future composition the dynamics and associated biodiversity are severely compromised.

This is not a new concept. In 1978, Regier proposed managing natural resources by a means-variance approach whereby acceptable results of management would be judged as a range of potential future states (now sometimes called range of natural conditions) rather than as a single, specific condition. Thus, the concept of identifying a desired future condition in public forest planning should be replaced with the concept of attaining desired future dynamics of ecosystems. That is, management guidelines and land allocations should allow for natural disturbances, which play such a central role in determining the physical structure, layout, and mixed species communities of forests.

In concert, the above lessons underscore the need for a broad gradient of approaches to managing forest biodiversity. Assessment and management guidelines range from maintaining structures such as down wood to maintaining large late-successional forest reserves where appropriate and sustainable. To do this, ecosystem managers of the new century must consider the ecological roles and requirements of furtive biodiversity in maintaining ecosystem functions. They must recognize that many disturbance events are stochastic and that specific conditions and yields of forest products might not be predictable on specific sites or watersheds. They must also plan for the repetition of disturbance dynamics at several scales (Gavin 1991), including within stands and across the forest matrix. By addressing these multiple levels of space, time, and biological organization, they can move more efficiently toward conservation of biological diversity and long-term site productivity of forest ecosystems.

Some Difficult Questions About Managing for Biodiversity

Which biodiversity counts? Of all the questions raised in the debate over forest biodiversity, this is among the most basic and the most difficult to answer. Argermeier (1994) recently raised the question, Does biodiversity include artificial diversity? Since the distribution and abundance of many forest species have already been greatly influenced by anthropogenic effects, should the goal be to (1) maintain current conditions or dynamics, (2) return to some early historic conditions or dynamics, or (3) simply allow for continued influence of humans under the argument that humans ineluctably are part of the ecosystem too? Argermeier argued that only the native elements of biodiversity should be considered in the definition and goals.

A related question is, How can the growing number of federally listed species be managed in a more holistic, ecosystem perspective? According to Flather et al. (1994), between 1976 and 1992, on average, 34 species have been listed per year by USDI Fish and Wildlife Service as threatened or endangered, 64 percent of which are plant taxa, and the rate has been increasing. Many authors have written on expanding the application of the Endangered Species Act to focus more on ecosystems and ecological communities (e.g., Barnes 1993, Hunt 1989, LaRoe 1993, Marcot 1994, Marcot et al. 1994, Noss 1991, Salwasser 1991, Sidle and Bowman 1988) and to better consider economic forces in land allocations and species protection (Babbitt 1994, Barker 1993).

Likewise, which areas should be protected first, and why? In Idaho, Wright et al. (1994) evaluated the percentage of each vegetation type within the state that would be protected in parks under a proposal to create four new parks. They applied the criterion that at least 10 percent of each vegetation type should be in protected status within the state. However, the park proposals would provide such protection for little more than 15 percent of the vegetation types in the state. Other approaches to natural area protection elsewhere have focused variously on habitat conservation and species-specific needs.

Are there no exotic species that are desirable elements of biodiversity? Indeed, the regulations (36 CFR 219) that implement the National Forest Management Act speak to maintaining viability of all "native and desired nonnative species." Which nonnative species are desirable, which take precedence for management, and who decides this? OTA (1993) has provided a compendium on which nonnatives they consider to be harmful to native biota, but the question remains as to how to allocate finite resources for their eradication (Coblentz 1990). Ecosystem-based forest management planning could do well to include a risk assessment of potential spread of undesired nonnative organisms under different planning scenarios.

When and how is a species deemed to be absent, present, or safe in a specific planning area? Proving local extinction—particularly in dynamic landscapes allowing for at least occasional colonization by founder populations—is often difficult. As Diamond (1987) noted, should a species be considered extant unless proven extinct, or extinct unless proven extant? This has major implications for allocating finite funds for recovery activities, for jeopardy or "may affect" calls by the USDI Fish and Wildlife Service under the Endangered Species Act, for opportunity costs of land and resource use, and for real costs for habitat restoration projects. The question of demonstrating extirpation, too, can apply to any geographic scale, from nations to bioregions to land ownerships to forest patches. For example, how much of the southern pine forests should be restored to old longleaf pine (*Pinus palustris*) habitat for the red-cockaded woodpecker (*Picoides borealis*) if the species has not been seen in a particular area for several years or decades?

How should potentially conflicting objectives for maintaining all native species and their habitats be resolved (Franklin 1993)? Some authors (e.g., Hof and Raphael 1993) have advocated an optimization approach in which the appropriate mix of species and habitats is defined by balancing the needs for all species simultaneously. Again, this has ramifications for allocating scarce financial and management resources. Margules et al. (1994) based their recommendations for allocating conservation areas on expected patterns of persistence and turnover of species within the reserves. Other authors (e.g., Hansen and Walker 1985) have advocated a less reductionist approach by maintaining the general dynamics and conditions of native forests rather than attempting to micromanage specific species, habitat elements, and vegetation successional stages.

Finally, what should be monitored over time to ensure the long-term productivity and diversity of forest ecosystems? Can indicators of trophic health adequately represent the conditions of species populations and interactions? Oliver and Beattie's (1993) rapid assessment method for characterizing biodiversity in an area could be applied to forest resources. Reid et al. (1993) outlined a series of indicators that could be monitored for indexing the health of communities and ecosystems. Noss (1990) presented a classification of ecosystem structures, functions, and processes that can help as indicators for monitoring biodiversity; indeed, this was further developed and used on one national forest in the Northwest (Williams and Marcot 1991).

Biodiversity as Our New Mythos for Management

We have come full circle in many ways. The new focus on ecosystem management has returned us to the systems models of energy flow, nutrient cycles, and food webs popular in the ecology of the 1960s. Moreover, ecosystem management focuses on both individual species (including *Homo sapiens*) and ecological systems rather than merely one or the other. Interesting arguments ensue on our ethical and moral onus in conservation planning; conserving forest biodiversity for viable wildlife populations and for our long-term, sustained use of forest products truly are but different facets of the same problem with a common solution.

And in perhaps the greatest irony, the American Indian philosophy of the web of life is now becoming a part of public land management policy. As no mere quaint or anachronistic tale, the vision of the web of life provides a new mythos, a fresh root metaphor, for this era of forest ecosystem management. We must embrace the practical lessons from this new mythos and learn from scientific research, from monitoring and observation, and from the elders of our ecological household. Only this will help ensure that our management actions and lifestyles will become truly conservatory and that the diversity of the living and the nonliving will remain wed, productive, diverse, and sustainable.

Acknowledgments

Much of my discussion of the ecology of species groups in old forests of the Pacific Northwest drew from the assessments by Thomas et al. (1993) and especially by FEMAT

(1993). In particular, I gratefully acknowledge the outstanding work on fungi, lichens, bryophytes, plants, and invertebrates by John Christy, Nancy Fredricks, Robin Lesher, Roger Rosentreter, Joan Zeigltrum, and a host of species experts who kindly provided both published and unpublished information to make these two reports, Thomas et al (1993) and FEMAT (1993), possible. As well, it is my honor to report that Robert Anthony, Eric Forsman, Patricia Greenlee, A. Grant Gunderson, Ed Starkey, and Cindy Zabel helped compile information on vertebrates, and Richard Holthausen, E. Charles Meslow, and Martin Raphael helped oversee and interpret the work on compiling species information. Thanks, too, to Gordon Reeves for discussions and references on the ecological role of fish and hyporheic ecosystems. My appreciation to Kathryn Kohm for reviewing the manuscript and suggesting improvements.

Literature Cited

Aizen, M. A., and P. Feinsinger. 1994. Habitat fragmentation, native insect pollinators, and feral honey bees in Argentine "Chaco Serrano." *Ecological Applications* 4(2): 378–392.

Amaranthus, M. P., and D. A. Perry. 1994. The functioning of ectomycorrhizal fungi in the field: Linkages in space and time. *Plant and Soil* 159:133–140.

Argermeier, P. L. 1994. Does biodiversity include artificial diversity? *Conservation Biology* 8(2):600–602.

Asquith, A., L. D. Lattin, and A. R. Moldenke. 1990. Arthropods: The invisible diversity. *The Northwest Environmental Journal* 6:404–405.

Axelrod, D. I. 1958. Evolution of the Madro-Tertiary Geoflora. *Botanical Review* 24:433–509.

———. 1981. Holocene climatic changes in relation to vegetation disjunction and speciation. *American Naturalist* 117:847–870.

Babbitt, B. 1994. Protecting biodiversity. *Nature Conservancy.* 44(1):16–21.

Barker, R. 1993. *Saving all the parts: Reconciling economics and the Endangered Species Act.* Washington, DC: Island Press.

Barnes, B. V. 1993. The landscape ecosystem approach and conservation of endangered spaces. *Endangered Species Update* 10(3 & 4):13–19.

Benkman, C. W. 1993. Adaptation to single resources and the evolution of crossbill (*Loxia*) diversity. *Ecological Monographs* 63:305–325.

Berg, A., B. Ehnstrom, L. Gustafsson, T. Hallingback, M. Jonsell, and J. Weslien. 1994. Threatened plant, animal, and fungus species in Swedish forests: Distribution and habitat associations. *Conservation Biology* 8(3):718–731.

Berstein, M. E. 1977. Internal fungi in old-growth Douglas-fir foliage. *Canadian Journal of Botany* 55:644–653.

Bilby, R. E., B. R. Fransen, and P. A. Bisson. 1996. Incorporation of nitrogen and carbon from spawning coho salmon into the trophic system of small streams: Evidence from stable isotopes. *Canadian Journal of Fisheries and Aquatic Sciences* 53:164–173.

Bird, J. 1994. The Nature Conservancy and biodiversity. *Nature Conservancy* 44(1):22–27.

Brussard, P. F., D. D. Murphy, and R. F. Noss. 1992. Strategy and tactics for conserving biological diversity in the United States. *Conservation Biology* 6(2):157–159.

Campbell, R. W., T. R. Torgersen, and N. Srivastava. 1983. A suggested role for predaceous birds and ants in the population dynamics of the western spruce budworm. *Forest Science* 29(4):779–790.

Cederholm, C. J., and N. P. Peterson. 1985. The retention of coho salmon (*Oncorhynchus kisutch*) carcasses by organic debris in streams. *Canadian Journal of Fisheries and Aquatic Sciences* 42:1222–1225.

Coblentz, B. E. 1990. Exotic organisms: A dilemma for conservation biology. *Conservation Biology* 4:261–265.

Crawford, R. H., S. E. Carpenter, and M. E. Harmon. 1990. Communities of filamentous fungi and yeast in decomposing logs of *Pseudotsuga menziesii. Mycologia* 82(6): 759–765.

Crow, T. R. 1988. Biological diversity: Why is it important to foresters? In *Managing north central forests for non-timber values.* Duluth, MN: Society of American Foresters.

Crow, T. 1990. Old growth and biological diversity: A basis for sustainable forestry. In *Old growth forests: What are they? How do they work?* Toronto, Canada: Canadian Scholars' Press.

Diamond, J. M. 1987. Extant unless proven extinct? Or, extinct unless proven extant? *Conservation Biology* 1:77–79.

Dickson, J. F., R. N. Conner, R. R. Fleet, J. C. Kroll, and J. A. Jackson, eds. 1979. *The role of insectivorous birds in forest ecosystems.* Nacagdoches, TX: Academic Press.

FEMAT. 1993. *Forest ecosystem management: An ecological, economic, and social assessment.* Report of the Forest Ecosystem Management Assessment Team (FEMAT). 1996-793-071. Washington, DC: GPO.

Flather, C. H., L. A. Joyce, and C. A. Bloomgarden. 1994. *Species endangerment patterns in the United States.* General technical report RM-241. Washington, DC: USDA Forest Service.

Foerester, R. E. 1968. The sockeye salmon, *Oncorhynchus nerka. Bulletin of Fisheries Research Board of Canada* 162:

Fogel, R., and G. Hunt. 1983. Contribution of mycorrhizae and soil fungi to nutrient cycling in a Douglas-fir ecosystem. *Canadian Journal of Forest Research* 13:219–232.

Fogel, R., and J. M. Trappe. 1978. Fungus consumption (mycophagy) by small animals. *Northwest Science* 52:1–31.

Franklin, J. F. 1993. Preserving biodiversity: Species, ecosystems, or landscapes? *Ecological Applications* 3(2):202–205.

Franklin, J. F., and T. A. Spies. 1984. Characteristics of old-growth Douglas-fir forests. In *New forests for a changing world.* Bethesda, MD: Society of American Foresters.

Franklin, J. F., F. Hall, W. Laudenslayer, C. Maser, J. Nunan, J. Poppino, C. J. Ralph, and T. Spies. 1986. Interim definitions for old-growth Douglas-fir and mixed-conifer forests in the Pacific Northwest and California. Research note PNW- 447. Washington, DC: USDA Forest Service.

Gavin, T. A. 1991. New approaches in managing biodiversity: A matter of scale. In *Challenges in the conservation of biological resources,* ed. D. J. Decker, M. E. Krasny, G. R. Goff, C. R. Smith, and D. W. Gross. Boulder: Westview Press.

Hallingback, T. 1992. Sveriges boreala mossflora i ett internationellt perspektiv. (The boreal bryophyte flora of Sweden in an international perspective.) *Svensk Botanisk-Tidskrift* 86:117–184.

Hansen, A. J., and B. H. Walker. 1985. The dynamic landscape: Perturbation, biotic response, biotic patterns. *South African Institute of Ecology Bulletin* 4:5–14.

Hansen, A. J., T. A. Spies, F. J. Swanson, and J. L. Ohmann. 1991. Conserving biodiversity in managed forests. *BioScience* 41(6):382–392.

Heizer, R. F., and A. B. Elsasser. 1980. *The natural world of the California Indians.* Berkeley: University of California Press.

Hof, J. G., and M. G. Raphael. 1993. Some mathematical programming approaches for optimizing timber age class distributions to meet multispecies wildlife population objectives. *Canadian Journal of Forest Research* 23:828–834.

Holmes, R. T. 1990. Ecological and evolutionary impact of bird predation on forest insects: An overview. *Studies in Avian Biology* 13:6–13.

Holmes, R. T., J. C. Schultz, and P. J. Nothnagle. 1979. Bird predation on forest insects: An exclosure experiment. *Science* 206:462–463.

Hunn, E. S., with J. Selam and family. 1990. *Nch'i-Wana "The Big River": Mid-Columbia Indians and their land.* Seattle, WA, and London, England: University of Washington Press.

Hunt, C. E. 1989. Creating an endangered ecosystem act. *Endangered Species Update* 6:1–5.

Hunter, M. L. 1991. Coping with ignorance: The coarse-filter strategy for maintaining biodiversity. In *Balancing on the brink of extinction,* ed. K. A. Kohm. Washington, DC: Island Press.

Ingham, E. R., and R. Molina. 1991. Interactions among mycorrhizal fungi, rhizosphere organisms, and plants. In *Microbial mediation of plant-herbivore interactions,* ed. P. Barbosa, V. A. Krischik, and C. G. Jones. New York: John Wiley & Sons.

Kemp, J. C., and G. W. Barrett. 1989. Spatial patterning: Impact of uncultivated corridors on arthropod populations within soybean agro-ecosystems *Ecology* 70:114–128.

Knight, F. B. 1958. The effects of woodpeckers on populations of the Engelmann spruce beetle. *Journal of Economic Entomology* 51:603–607.

Kremen, C. 1994. Biological inventory using target taxa: A case study of the butterflies of Madagascar. *Ecological Applications* 4(3):407–422.

Kremen, C., R. K. Colwell, T. L. Erwin, D. D. Murphy, R. F. Noss, and M. A. Sanjayan. 1992. *Arthropod assemblages: Their use as indicators in conservation planning.* Portland, OR: The Xerces Society.

Kuusipalo, J., and J. Kangas. 1994. Managing biodiversity in a forestry environment. *Conservation Biology* 8(2):450–460.

LaRoe, E. T. 1993. Implementation of an ecosystem approach to endangered species conservation. *Endangered Species Update* 10(3 & 4):3–6.

Lattin, J. D. In press. Arthropod diversity and conservation in old-growth Northwest forests. *American Zoologist.*

Lattin, J. D., and A. R. Moldenke. 1990. Moss lacebugs in northwest conifer forests: Adaptation to long-term stability. *The Northwest Environmental Journal* 6:406–407.

Li, C.Y., C. Maser, Z. Maser, and B. A. Caldwell. 1986. Role of three rodents in forest nitrogen fixation in western Oregon: Another aspect of mammal–mycorrhizal fungus–tree mutualism. *Great Basin Naturalist* 46:411–414.

Loyn, R. H., R. G. Runnalls, and G.Y. Forward. 1983. Territorial bell miners and other birds affecting populations of insect prey. *Science* 221:1411–1413.

Marcot, B. G. 1994. Integrating environmental management through conservation strategies for threatened and endangered wildlife species. In *Implementing integrated environmental management,* ed. J. Cairns Jr., T. V. Crawford, and H. Salwasser. Blacksburg, VA.: Virginia Polytechnic Institute and State University.

Marcot, B. G., M. J. Wisdom, H. W. Li, and G. C. Castillo. 1994. *Managing for featured, threatened, endangered, and sensitive species and unique habitats for ecosystem sustainability.* General technical report PNW-GTR-329. Portland, OR: USDA Forest Service, Pacific Northwest Research Station.

Margules, C. R., A. O. Nicholls, and M. B. Usher. 1994. Apparent species turnover, probability of extinction and the selection of nature reserves: A case study of the Ingleborough limestone pavements. *Conservation Biology* 8(2):398–409.

Maser, C., and Z. Maser. 1988. Interactions among squirrels, mycorrhizal fungi, and coniferous forests in Oregon. *Great Basin Naturalist* 48:358–369.

McIver, J. D., A. R. Moldenke, and G. L. Parsons. 1990. Litter spiders as bio-indicators of recovery after clearcutting in a western coniferous forest. *The Northwest Environmental Journal* 6:410–412.

Mispagel, M. E., and S. D. Rose. 1978. Arthropods associated with various age stands of Douglas-fir from foliar, ground, and aerial strata. *Coniferous Forest Biome Bulletin* 13. Seattle: University of Washington.

Moldenke, A. R., and J. D. Lattin. 1990a. Density and diversity of soil arthropods as "biological probes" of complex soil phenomena. *The Northwest Environmental Journal* 6:409–410.

———. 1990b. Dispersal characteristics of old-growth soil arthropods: The potential for loss of diversity and biological function. *The Northwest Environmental Journal* 6:408–409.

Muehlchen, A. M. 1994. Eastside Ecosystem Management Project: Functional groups of bacteria. Unpublished contract report to USDA Forest Service, Walla Walla, WA.

Murphy, D. D. 1991. Invertebrate conservation. In *Balancing on the brink of extinction,* ed. K. A. Kohm. Washington, DC: Island Press.

Naiman, R. J., T. J. Beechie, L E. Benda, D. R. Berg, P. A. Bisson, L. H. MacDonald, M. D. O'Connor, P. L. Olson, and E. A. Steel. 1992. Fundamental elements of ecologically healthy watersheds in the Pacific Northwest coastal ecoregion. In *Watershed management: Balancing sustainability and environmental change,* ed. R. J. Naiman. New York: Springer-Verlag.

Naiman, R. J., H. Decamps, and M. Pollock. 1993. The role of riparian corridors in maintaining regional biodiversity. *Ecological Applications* 3(2):209–212.

Niemela, J., D. Langor, and J. R. Spence. 1993. Effects of clear-cut harvesting on boreal ground beetle assemblages (*Coleoptera carabidae*) in western Canada. *Conservation Biology* 7(3):551–561.

Norse, E. A., K. L. Rosenbaum, D. S. Wilcove, B. A. Wilcox, W. H. Romme, D. W. Johnston, and M. L. Stout. 1986. *Conserving biological diversity in our national forests.* Washington, DC: The Wilderness Society.

North, M., and J. Franklin. 1990. Post-disturbance legacies that enhance biological diversity in a Pacific Northwest old-growth forest. *The Northwest Environmental Journal* 6:427–429.

Noss, R. F. 1990. Indicators for monitoring biodiversity: A hierarchical approach. *Conservation Biology* 4:355–364.

———. 1991. From endangered species to biodiversity. In *Balancing on the brink of extinction,* ed. K. A. Kohm. Washington, DC: Island Press.

Oliver, I., and A. J. Beattie. 1993. A possible method for the rapid assessment of biodiversity. *Conservation Biology* 7 (3):562–568.

Olson, D. M. 1992. *The northern spotted owl conservation strategy: Implications for Pacific Northwest forest invertebrates and associated ecosystem processes.* Final report submitted for the Northern Spotted Owl EIS Team. Portland, OR: USDA Forest Service.

OTA. 1993. *Harmful non-indigenous species in the United States.* OTA-F-565 (2 vols.). Washington, DC: U.S. Congress, Office of Technology Assessment.

Pearson, D. L., and F. Cassola. 1992. World-wide species richness patterns of tiger beetles (*Coleoptera cicindelidae*): Indicator taxon for biodiversity and conservation studies. *Conservation Biology* 6(3):376–391.

Perry, D. A., J. G. Borchers, S. L. Borchers, and M. P. Amaranthus. 1990. Species migrations and ecosystem stability during climate change: The belowground connection. *Conservation Biology* 4:266–274.

Pike, L. H., et al. 1975. Floristic survey of epiphytic lichens and bryophytes growing on old-growth conifers in western Oregon. *Bryologist* 78:389–402.

Probst, J. R., and T. R. Crow. 1991. Integrating biological diversity and resource management. *Journal of Forestry* 2:12–17.

Regier, H. A. 1978. *A balanced science of renewable resources.* Washington Sea Grant publication. Seattle: University of Washington.

Reid, W. V., and K. R. Miller. 1989. *Keeping options alive: The scientific basis for conserving biodiversity.* Washington, DC: World Resources Institute.

Reid, W. V., J. A. McNeely, D. B. Tunstall, D. A. Bryant, and

M. Winograd. 1993. *Biodiversity indicators for policy-makers.* Washington, DC: World Resources Institute.

Roth, D. S., I. Perfecto, and B. Rathcke. 1994. The effects of management systems on ground-foraging ant diversity in Costa Rica. *Ecological Applications* 4(3):423–436.

Ruggiero, L. F., K. B. Aubry, A. B. Carey, and M. H. Huff. 1991. Wildlife and vegetation of unmanaged Douglas-fir forests. General technical report PNW-GTR-285. Portland, OR: USDA Forest Service.

Salwasser, H. 1991. In search of an ecosystem approach to endangered species conservation. In *Balancing on the brink of extinction,* ed. K. A. Kohm. Washington, DC: Island Press.

Sandlund, O. T., K. Hindar, and A. H. D. Brown, eds. 1992. *Conservation of biodiversity for sustainable development.* New York: Oxford University Press.

Seastadt, T. R., and D. A. Crossley. 1984. The influence of arthropods on ecosystems. *BioScience* 34:157–161.

Sidle, J. C., and D. B. Bowman. 1988. Habitat protection under the endangered species act. *Conservation Biology* 2:116–118.

Soderstrom, L., and B. G. Jonsson. 1992. Naturskogarnas fragmentering och mossor par temporara substrata (Fragmentation of old-growth forests and bryophytes on temporary substrates). *Svensk Botoanisk Tidskrift* 86:185–198.

Solomon, J. D. 1969. Woodpecker predation on insect borers in living hardwoods. *Annals of the Entomological Society of America* 62:1214–1215.

Stevenson, S. K., and K. A. Enns. 1992. *Integrating lichen enhancement with programs for winter range creation.* Part 1: Stand/lichen model. Publication IWIFR-41.Victoria, BC, Canada: Ministry of Forests.

Stolte, K., D. Mangis, R. Doty, and K. Tonnessen, tech. coords. 1993. *Lichens as bioindicators of air quality.* General technical report RM-224. USDA Forest Service, Rocky Mountain Forest and Range Experiment Station.

Storm, H. 1972. *Seven arrows.* New York: Harper and Row.

Takekawa, J. Y., E. O. Garton, and L. A. Langelier. 1982. Biological control of forest insect outbreaks: The use of avian predators. *Transactions of the North American Wildlife and Natural Resources Conference* 47:393–409.

Thomas, J. W., M. G. Raphael, R. G. Anthony, E. D. Forsman, A. G. Gunderson, R. S. Holthausen, B. G. Marcot, G. H. Reeves, J. R. Sedell, and D. M. Solis. 1993. *Viability assessments and management considerations for species associated with late successional and old-growth forests of the Pacific Northwest.* 1993-791-566. Washington, DC: GPO

Tibell, L. 1992. Crustose lichens as indicators of forest continuity in boreal coniferous forests. *Northwest Journal of Botany* 12(4):427–450.

USDA and USDI. 1994. Standards and guidelines for management of habitat for late successional and old-growth forest related species within the range of the northern spotted owl. In *Record of decision for amendments to Forest Service and Bureau of Land Management planning documents within the range of the northern spotted owl.* Washington, DC: GPO.

Walls, S. C., A. R. Blaustein, and J. J. Beatty. 1992. Amphibian biodiversity of the Pacific Northwest with special reference to old-growth stands. *The Northwest Environmental Journal* 8(1):53–69.

Welsh, H. H. 1990. Relictual amphibians and old-growth forest. *Conservation Biology* 4:309–319.

Wilcove, D. S. 1988. *National forests: Policies for the future.* Vol. 2, *Protecting biological diversity.* Washington, DC: The Wilderness Society.

Williams, B. L., and B. G. Marcot. 1991. Use of biodiversity indicators for analyzing and managing forest landscapes. *North American Wildland Natural Resource Conference* 56:613–627.

Wright, R. G., J. G. MacCracken, and J. Hall. 1994. An ecological evaluation of proposed new conservation areas in Idaho: Evaluating proposed Idaho national parks. *Conservation Biology* 8(1):207–216.

Section

II Silvicultural Systems and Management Concerns

There is no technical area more central to forestry than silviculture: It is where the natural sciences, engineering, and management objectives converge and are hopefully successfully integrated. There is also no other area in forestry as steeped in tradition and as conservative.

Silviculturists are challenged as never before by a multiplicity of management objectives and by recent scientific insights into forest ecosystems and landscapes. Society and landowners are asking for silvicultural systems that integrate multiple and often conflicting objectives, including provisions for organisms, structures, and processes about which we have limited understanding. Basic premises underlying traditional practices have been rendered obsolete. Trees continue to perform important ecological roles when they die and become snags and logs on the forest floor; traditional silviculture did not recognize the values of the dead tree. Spatial context does matter; forest stands and proposed silvicultural activities must be considered as elements in landscapes, not as isolated patches.

Ultimately silviculturists must develop and apply approaches that produce stand structures and landscapes that incorporate much more of the complexity—and irregularity—of natural forests and forest landscapes. The agricultural paradigm of forestry adopted in this century—simplification and uniformity in structure, pattern, and product—and the regulated landscape—fully occupied by an ordered age sequence of managed stands—no longer suffices. The simplistic notion that four regeneration harvest practices, designed with the knowledge and objectives of the 19th century, can meet the objectives of the 21st century must be given up. Some major responses that will occur in silviculture in the 21st century are already clear: long rotations, structural retention, and structural restoration.

Long Rotations

Long rotations involve the use of rotation ages that are significantly longer, often much longer, than the economic rotations typically adopted on industrial forestlands or even the semibiological measure adopted for federal lands, culmination of mean annual increment. Long rotations can address important environmental challenges in forestry, including cumulative effects issues, which are a consequence of having too large a percentage of a landscape in a recently harvested condition. Long rotations can be coupled with a series of silvicultural treatments to produce complex managed forests. Recent research on forest growth and effects of intensive stand management practices show that such a regime can dramatically delay culmination of mean annual increment and increase yields of commodities from forest landscapes (see Curtis, Chapter 10).

Structural Retention

Structural retention—silvicultural prescriptions which retain significant structural elements from a harvested stand for incorporation into the new stand—offers a nearly infinite array of possible alternatives to clearcutting and other traditional regeneration harvest systems (see Franklin et al., Chapter 7, and Tappeiner et al., Chapter 9). Acceptance of structural retention has not come easily, particularly in regions committed by tradition to view alternatives to clearcutting as neither technically feasible nor socially desirable. Yet, the opportunities inherent in structural retention are increasingly apparent to forest stakeholders and, however reluctantly, being adopted by silviculturists.

Objectives in structural retention include maintenance of refugia for organisms and processes on harvested areas and structural enrichment of the next forest stand (see Franklin et al., Chapter 7). Variables involve the type and number of structures that are retained and the spatial pattern of the retention. Large, old, and decadent trees, standing dead trees (snags), and logs on the forest floor are examples of individual structures that are foci of many retention harvest prescriptions. There is increasing interest in the potential for aggregated retention—maintenance of small forest patches—on harvested units as an alternative to dispersion of retained structures.

Structural Restoration

Restoration involves silvicultural treatments to speed development of structural complexity in young forests lacking such complexity either as a result of past management practices or natural events (see DeBell et al., Chapter 8, and Tappeiner et al., Chapter 9). A rich array of possibilities exists, including precommercial and commercial thinning, creation of snags, enhancement of woody debris on the forest floor, active creation of cavities and other habitat niches, underplanting, and development or maintenance of shrub and herbaceous layers. Many nontraditional approaches can be taken in thinnings, including prescriptions that enhance development of multiple-species, multistoried stands. Similarly, use of spatially varying prescriptions—rather than the traditional uniform treatments—can assist in creation of structurally heterogeneous stands.

Forest Protection Issues

The expanded array of intermediate and harvest stand treatments often lead to concerns over the impacts that such treatments and structures will have on the protection and genetic composition of the future forests (see Schowalter et al., Chapter 11, Agee, Chapter 12, and Friedman, Chapter 13). The effects of silvicultural treatments on the potential for wildfire are nearly always a management concern: What effect will the new practices have on fuels and traditional approaches to fuel reduction, such as slash burning? Fire can also be an important tool for restoring species and desired structural conditions in stands, particularly in forest types traditionally subject to frequent, low-intensity burns; an important unanswered question in many regions is the degree to which mechanical treatments can substitute for fire.

Regardless of silvicultural practice, foresters are challenged to view insects, fungi, and similar organisms in new ways (see Schowalter et al., Chapter 11). For example, invertebrate detritivores are major elements in energy and nutrient cycles, and fungi are half of the essential absorbing structures known as mycorrhizae. Phytophagous insects, diseases, and parasitic plants can play important roles as well in maintaining diverse and productive forest stands. In attempting to control such "undesirable" organisms, foresters have sometimes failed to appreciate their role and, more often, the fact that practices designed to control such organisms are likely to have drastic impacts on the much larger set of invertebrates and fungi that carry out critical ecological functions.

Perhaps the most fundamental contribution of forest ecosystem science to silviculture has been to help us appreciate the limitations of existing knowledge. The numerous "surprises" of the last several decades make clear that much remains to be learned about how forests work. Traditional approaches to silvicultural research have contributed significantly to the problem by focusing almost exclusively on trees—and living trees at that—rather than the interplay of all components of the ecosystem. Experiments have focused on comparisons of the classical silvicultural systems, such as various forms of even-age management, rather than on taking more fundamental approaches, such as investigating how organisms or processes respond to removal or, alternatively, retention of increasing numbers of trees.

In the 21st century, silvicultural prescriptions must be viewed as the working hypotheses that they are rather than as treatments with determinate and predictable ends. Adoption of the principles of adaptive management are essential. Hopefully, silviculturists will be leaders in creating the infinite array of silvicultural prescriptions that will be needed to achieve the complex multiple objectives, to abandon the straightjacket of the traditional regeneration harvest systems, and to embrace the view that silviculture is the art and science of manipulating forest stands, regardless of objectives.

7 Alternative Silvicultural Approaches to Timber Harvesting: Variable Retention Harvest Systems

Jerry F. Franklin, Dean Rae Berg, Dale A. Thornburgh, and John C. Tappeiner

Long Rotations and Structural Retention at Harvest 113
Ecological Advantages of Long Rotations 114
Limitations of Long Rotations 114
The Variable Retention Harvest System 115
Lifeboating: Refugia and Inocula 115
Structural Enrichment of Established Forest Stands 116
Enhancing Connectivity in the Managed Landscape 117
Design Elements in a Variable Retention Harvest System 118
Box: Plum Creek Timber Company's Approach to Environmental Forestry 121
Examples of Variable Retention of Silvicultural Prescriptions 124
Low Retention, Mix of Aggregated and Dispersed 124
Box: Menominee Sustained-Yield Management 125
Retention of Dispersed Large-Diameter Cohorts 126
Group Harvest with Low Retention 126
Management Issues in the Application of Variable Retention 127
Forest Protection 127
Forest Harvest and Management 131
Rotations, Entries, and Age Classes 132
Scientific Issues Raised by Variable Retention 133
Evidence for Effectiveness of Structural Retention 133
Research Needs 134
Conclusions 136
Acknowledgments 136
Literature Cited 136

Silviculture is the art and science of manipulating forest stands to achieve human objectives, including the production of various goods and services. As a discipline, silviculture has very strong traditions, most of which are rooted in European forest practices. Basic concepts underlying the establishment, tending, and harvest of forest stands were established by the beginning of this century.

Nowhere are traditions more firmly established than in the approaches to regeneration harvesting of forest stands. There are four recognized regeneration harvest methods (Smith 1986): clearcut, seed-tree, and shelterwood methods for use as a part of an even-age management system, and selection for use in uneven-aged systems. While rather precisely defined, there are recognized variations on these several methods, variations that have been defined and described in textbooks.

All regeneration harvest methods were created with a singular objective: regeneration and subsequent growth of a commercially important tree species (Smith 1986). Management objectives for forest harvesting have become increasingly complex during the last several decades however—a trend that will certainly continue into the 21st century. We are no longer seeking simply to create a free-growing replacement forest while safely and efficiently harvesting the mature stand. Today, multiple objectives typically include maintenance of specific levels of ecosystem processes, including habitat for elements of biological diversity. Tree regeneration and its subsequent growth are often still concerns, although these objectives—especially for rapid growth of the regeneration—often are subordinated to other goals. Harvest cutting may include such diverse goals as maintaining tree root strength; providing for specified levels of snags of various species, sizes, and conditions; and fulfilling specific aesthetic criteria. Although there is substantial flexibility in application of existing harvest methods (Smith 1986), foresters sometimes are forced to either take liberties with the technical definitions of the four harvest methods or to adopt awkward and confusing terminology, such as "clearcut with reserves." Most important, even with substantial modifications, the four recognized methods provide a very limited set of choices.

Recent research on forest ecosystems has clarified the importance of structural complexity to forest ecosystem functioning and the maintenance of biological diversity (Franklin 1993, Bormann and Likens 1979, Swank and Crossley 1988, Franklin et al. 1987, Maser et al. 1988, Harmon et al. 1986, Spies, Chapter 2). Important structural features include snags, woody debris on the forest floor, multiple canopy layers, varied sizes and conditions of live trees, and presence of canopy gaps. Research has also made clear the dramatic impacts that clearcutting and other management activities can have on biological diversity and ecosystem function; for example, in Sweden clearcutting is the major factor threatening endangered forest organisms (Berg et al. 1995).

Investigation of the effects of natural disturbances on forest ecosystems and their subsequent recovery also have dramatically altered our understanding of these events (see Perry and Amaranthus, Chapter 3). Results from these studies emphasize the importance of biological legacies—surviving organisms and organically derived structures, such as snags, logs, and soil organic layers—to the rapid reestablishment of ecosystems that have high levels of structural, functional, and compositional diversity. Similar patterns of extensive legacies emerge from disturbances as diverse as wildfires (Christensen et al. 1989, Schullery 1989, Knight and Wallace 1989), hurricanes (Foster and Boose 1992, Walker et al. 1991) and other storm events (Peterson and Pickett 1995), and volcanic eruptions (Franklin et al. 1995, Franklin et al. 1985). These natural patterns contrast sharply with low levels of biological legacies associated with even-aged regeneration harvest practices, particularly clearcutting, even when treatments do not involve intensive site preparation (see, e.g., Keenan and Kimmins 1993).

As a result of this new knowledge, the creation and maintenance of structurally complex managed stands is being developed as the primary approach to managing forests for multiple, complex objectives, including production of wood products. Indeed, such approaches have emerged independently in many countries and on several continents (see, e.g., Arnott et al. 1995; Lunney 1991; Larsen 1995; Squire 1990; National Board of Forestry Sweden 1990; Swanson and Franklin 1992; Scientific Panel for Sustainable Forest Practices in Clayoquot Sound 1995; Ciancio and Nocentini 1994a, 1994b; Ciancio, Iovino, and Nocentini 1994; Watanabe and Sasaki 1993).

Proposed approaches to creation of structurally complex managed stands include the use of long rotations, retention of structural features at the time of harvest, and silvicultural treatment of established stands to create specific structural conditions. None of these are mutually exclusive, although each has specific circumstances where it is particularly appropriate. For example, silvicultural treatments to achieve specific structural features often are proposed to "restore" structurally simplified stands created using traditional even-aged systems (see Carey et al. submitted, Debell et al., Chapter 8, and Tappeiner et al., Chapter 9). Both restoration and retention approaches have analogies in traditional practices, although there are significant differences, as will be seen. Retention is of course focused most heavily on harvest practices in mature and old stands, while restoration addresses the challenging issue of what can be done in young stands.

Silvicultural methods based on significant structural retention at the time of harvest are the subject of this chapter. Such approaches may involve retention of individual trees, snags, logs, or small patches of forest on the harvest unit, usually for at least the next rotation. Such cuttings are conducted in stands that are at least at economic, if not biological, maturity. Because long rotations are often proposed as an alternative to retention harvest methods, we begin by contrasting the relative advantages of the two approaches. Next we illustrate the flexibility of using a retention harvest philosophy and identify the important variables in retention silvicultural prescriptions: which structures, how much of each, and in what spatial patterns. Current evidence for the effectiveness of structural retention in achieving objectives such as maintenance of wildlife habitat is reviewed, and important research needs are identified. We conclude by proposing that traditional regeneration harvest methods and terminology be supplemented by a more flexible and ecumenical approach based upon a continuum of retention levels.

Although this chapter is focused on structure, we acknowledge the important contribution that tree species diversity (stand composition) can also have on maintenance of ecological functions, including habitat for specific organisms. Compositional diversity can be extremely important, as illustrated by the simple example of including some representation of hardwoods in conifer-dominated stands. Furthermore, many specific structural conditions are associated with only one or a few species. Hence, compositional diversity is commonly implicit in structural goals.

Long Rotations and Structural Retention at Harvest

Long rotations, structural restoration, and structural retention at harvest are approaches that can be combined effectively. However, long rotations and structural retention generally address different environmental issues and have different economic implications. Furthermore, long rotations are often proposed as an alternative to reservation of late-successional forest areas as well as to structural retention at harvest. Hence, it is important to contrast these approaches.

Long rotations involve management of forests on rotation ages that are longer, often much longer, than those currently in use for the forest property in question (Weigand, Haynes, and Wikowski 1994). Rotations are typically based upon either economic or biological criteria. In the Douglas-fir region, economic rotations of 40 to 60 years are common on private lands where good growing conditions exist; such rotations are driven primarily by traditional investment economics. Biological rotations, based on culmination of mean annual increment (MAI), are required on national forest lands with typical rotation ages of 80 to 120 years, depending upon site conditions. There is the potential for considerable flexibility, however, as culmination often extends over several years or even decades and typically is delayed by intensive forest management (see Curtis 1994, 1995, Chapter 10). Although the appropriateness of culmination of mean annual increment as an index of stand biological maturity is debatable from an ecological point of view, it is the traditional measure in forestry.

Proposals for long rotations take many forms, depending upon management objectives (Weigand, Haynes, and Wikowski 1994). Possible objectives can include (from Curtis 1995) reduced land area in regeneration and early development stages, hence reduced visual impacts; low annual regeneration costs and less need for herbicides and slash burning; higher quality wood and larger trees; improved habitat for some wildlife species; hydrological and long-term soil productivity benefits; increased carbon storage; opportunity to adjust present unbalanced age distributions toward a regulated forest; and maintenance of options to allow adaptation to future changes and to correct errors stemming from incomplete knowledge.

One generic proposal involves the use of long rotations to develop structurally complex managed forests that include large-diameter trees (Weigand, Haynes, and Wikowski 1994). Such proposals usually include a series of silvicultural treatments during development of the stand to ensure creation of specific structural elements. Rotations may be extended by 50 to 300 percent—for example, sites traditionally managed on an 80-year rotation would be extended to 120 to 240 years.

Ecological Advantages of Long Rotations

The primary ecological application of long rotations is in places where area-based ecological effects are of primary concern, such as with the cumulative impacts of timber harvest on watershed conditions. For example, clearcutting can increase peak flows associated with rain-on-snow storm events (Harr 1986, Harr et al. 1989); recovery to hydrologic conditions comparable to those in the preharvest forest may take many decades. Hence, the percentage of a watershed in a particular condition may have to be limited to reduce the potential for adverse effects.

Long rotations can effectively address the issue of cumulative effects, since they reduce the area of a watershed that is harvested in any given year. If rotations are increased from 50 to 100 years, then the percentage of a watershed harvested is reduced from 2 to 1 percent. In terms of cumulative effects, if 20 years is used as the recovery period, then the area potentially contributing to peak flows is changed from 40 percent of the watershed (with a 50-year rotation) to 20 percent of the watershed (with the 100-year rotation).

Long rotations also may make it possible to reduce the density of permanent transportation systems, an important consideration in reducing impacts of harvesting (see, e.g., Keenan and Kimmins 1993). While this is possible under other management scenarios, greater use of temporary roads and harvest systems that utilize fewer roads are likely where there are much longer time intervals between final harvests. Of course, if the management scenario calls for repeated silvicultural entries throughout the rotation, this advantage is much less likely to be realized.

Lengthening rotations from those based on discounted present net worth to those based on culmination of mean annual increment also will increase wood production. For example, Curtis (1994) has noted that harvest ages of 40 to 50 years in Douglas-fir reduces volume production relative to potential. Furthermore, intensive management, such as systematic thinning, generally delays culmination, which makes it possible to utilize even longer rotations without penalties in mean annual levels of wood production.

Limitations of Long Rotations

Long rotations have important ecological limitations if carried out without structural retention at harvest. First, some structural elements and related species and processes are completely lost from the harvested site until such structures can be re-created. This means that a much smaller percentage of the landscape will have key structural components than would be the case if such structures were retained at the time of harvest.

Large-diameter, moderately decayed snags provide an example. Using clearcutting, all such structures are removed at harvest. Re-creation of snags of this diameter and decay state would take at least 100 years. Under a 120-year rotation, the harvested site will have such structures in place for 20 years; in a fully managed landscape only one-sixth of the land area will have such structures. In contrast, retention harvest could maintain either a population of (1) large moderately decayed snags or (2) large-diameter green trees for postharvest conversion to snags. Thus, most or all of the managed landscape would have such structural features.

Large differences in the percentage of the managed landscape with such structural features (between 17 and 100 percent in this example) can be of great importance for associated species and processes. In part, this relates to the absolute number of such structures present; for example, research has shown that both diversity and density of cavity- or snag-dependent vertebrates is related directly to density of snags or trees with cavities (e.g., Lindemayer and Franklin submitted). Contributing factors include the territorial nature of some species and their need to move among several snags or trees.

Long rotations also have important limitations as alternatives to reserves for maintenance of values associated with intact old-growth forests. It is very doubtful that a forest ecosystem can be re-created by silvicultural treatments that is compositionally, functionally, and structurally complete, even over long rotations. Some elements of late-successional forests require very long periods of time for reestablishment (e.g., Henderson 1994). More important, we do not even know many of the organisms and processes that occur in natural late-successional forests, nor is even rudimentary information available on the temporal and spatial patterns associated with such organisms and processes, especially in soils and canopies.

The Variable Retention Harvest System

The variable retention harvest system is based upon the concept of retaining structural elements of the harvested stand for at least the next rotation in order to achieve specific management objectives. Variable retention is extremely flexible in application since it utilizes a continuum of structural retention options (Figure 7.1a) in creating silvicultural prescriptions to meet specific management objectives.

Development and maintenance of structurally complex managed forests is the overall rationale for retaining structural elements of the harvested stand. Unlike traditional regeneration harvest systems, the objective of regeneration and growth of a new crop of trees may not be a primary or even a secondary objective. Variable retention harvesting is also flexible with regard to age class and may lead to even-aged, multi-aged, or uneven-aged stands.

Variable retention harvest prescriptions are appropriate where management objectives include maintenance or rapid restoration of environmental values associated with structurally complex forests. At least three major purposes should be recognized: (1) "lifeboating" species and processes immediately after logging and before forest cover is reestablished, (2) "enriching" reestablished forest stands with structural features that would otherwise be absent, and (3) "enhancing connectivity" in the managed landscape.

Lifeboating: Refugia and Inocula

A primary objective of structural retention is to provide refugia for elements of biological diversity that might otherwise be lost from the harvested area—lifeboating. Lifeboating is achieved in at least three ways: (1) by providing structural elements that fulfill habitat requirements for various organisms, (2) by

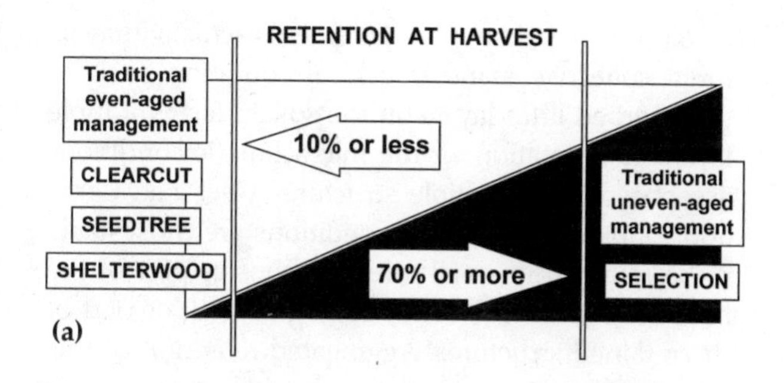

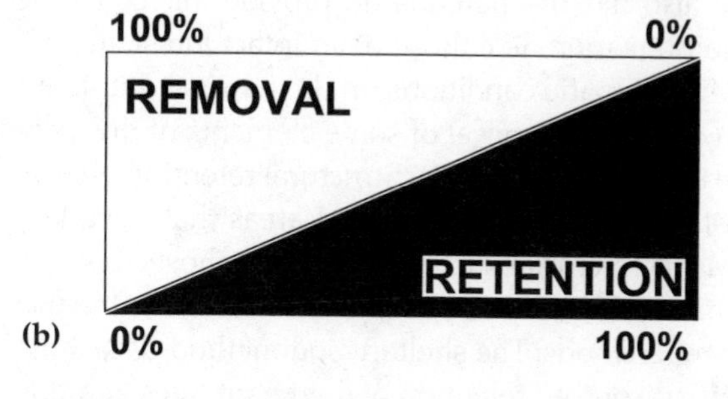

Figure 7.1 (a) The variable retention harvest system utilizes the full spectrum of structural retention or, conversely, removal that is available to silviculturists to achieve the complex and varied objectives typical of modern forestry. (b) Traditional regeneration harvest systems utilize a limited portion of this spectrum.

ameliorating microclimatic conditions in relation to those that would be encountered under clearcutting, and (3) by providing energetic substances to maintain nonautotrophic organisms.

Structural retention strategies may focus on many different types of organic structures—living, dead, or both—and individual or various combinations of structures. Individual structural features include living trees of various species, sizes, and conditions and their derivatives, such as standing dead trees or snags and logs on the forest floor. Such structures are critical habitat elements for many species (see, e.g., Berg et al. 1995, Carey and Johnson 1995, Lindemayer and Franklin submitted) and ecosystem functions. These species can be eliminated from the harvested stand when all of the structures on which they depend are removed; conversely, many of these species and processes can tolerate conditions on a harvested area provided that the structures are still present.

Necessary structural requirements actually may be for a collective stand structural condition such as undisturbed litter layers on the soil surface, multiple layers of vegetation, or the microclimatic conditions associated with multiple structures (see, e.g., Carey and Johnson 1995). Such conditions are most likely to be retained in small forest patches as part of the harvested unit—that is, by aggregating all or part of the retained structures. Aggregated retention of this type also has the potential to provide microclimatic conditions more like those of an intact forest stand.

Microclimatic conditions on the harvested unit are also critical for survival of some elements of diversity (see e.g., Berg et al. 1995). Structural retention will almost always result in harvested areas that have less stressful microclimatic regimes than those that are found on clearcuts. A well-known example is the use of shelterwoods. The shelterwood method uses temporary dispersed retention of dominant trees at moderate density to alleviate climatic stresses such as frost and high temperatures, thereby improving the prospects for successful tree regeneration. Aggregated retention can produce habitat patches on the harvest unit which have microclimates that are even more forest-like than shelterwoods (Jiquan Chen, personal communication, 1995). However, the microclimate within these patches will still be very different from those in the interior of large, intact forest patches, often known as forest interior environments.

Interactions between structures and microclimatic conditions are also important to the lifeboating function. Some species will not persist in a clearcut microclimate even if the necessary structures are present. Their persistence depends on at least some level of climatic protection, along with the required structures.

Provision of critical substrate to maintain populations of heterotrophs is a third aspect of lifeboating. For example, live trees are needed to function as host plants and energy sources. This is particularly important in the case of soil organisms, such as mycorrhizal fungal symbionts. The soil community is very dynamic with high turnover rates in populations and structures such as mycorrhizae. The persistence of many elements of the soil community depends upon a continuing source of readily available, high-quality energy, which is provided by vascular plants, especially trees (Perry 1994). Loss of this energy source following clearcutting may result not only in loss of species, but also of entire critical functional elements of the soil community (Perry 1994).

It is important to note in this regard that there are two distinct patterns of mycorrhizal association: endo- and ectomycorrhizae. These two patterns are associated with different groups of vascular plants and of fungi. Many angiosperms and the gymnosperm families Cupressaceae, Taxaceae, and Taxodiaceae form endomycorrhizae, while Pinaceae and the angiosperm family Ericaceae form ectomycorrhizae; hence, mixtures of tree and shrub species are important in maintaining a full complement of fungal associates.

Many elements of biological diversity that can be maintained by structural retention are essential to the sustained productivity and health of a harvested area. A common perspective is that lifeboating of biological diversity is primarily intended to sustain species of esoteric or peripheral interest in managed stands, such as officially listed rare or endangered species. In fact, much of the diversity that is sustained by structural retention, such as fungal species capable of forming mycorrhizae, play important functional roles.

Use of structural retention to sustain biological diversity assumes that refugia will provide the inocula for reestablishing species in the harvested area once the new forest stand and other suitable habitat conditions are reestablished.

Structural Enrichment of Established Forest Stands

Many forest species are displaced or eliminated even with significant structural retention. Except under light partial cutting, closed forest stand conditions typically are lost for significant periods of time. Factors responsible for displacement or loss include logging disturbances, absence or reduced levels of key structures, loss of forest integrity, and creation of more extreme microclimatological conditions. With tree establishment and growth, closed forest cover is reestablished, correcting several of these conditions. However, important structural features, such as large and decadent trees, snags, and logs, may be absent either permanently or for the extended periods of time required for their re-creation. Retention of some of these structures at the time of harvest can result in stands with much higher levels of structural diversity and therefore habitat carrying capacity.

Structural retention is a technique for enriching the structural complexity of managed forest stands for an entire rotation. As such, suitable conditions for species can be reestablished much earlier in the rotation than would otherwise be possible. In some cases, where rotation ages or management practices do not provide for re-creation of specific structures, they are lost entirely from the managed forest if there is no retention. Furthermore, structural retention can be used to restore structures that cannot be maintained during the harvest period. For example, large and highly decayed snags often are completely eliminated for safety reasons; retention of appropriate sizes and species of live trees at harvest can provide the material for managers or nature to quickly reestablish such structures following harvest.

Numerous examples illustrate how structural retention can enrich subsequent stands and thereby provide suitable habitat for species that are generally rare or absent in young stands of simple and homogeneous structure. These examples are of both natural and human origin. There are many forest stands 80 to 200 years of age in the Douglas-fir region of the Pacific Northwest that provide suitable nesting and foraging habitat for northern spotted owls (Figure 7.2) (e.g., North 1993). Such stands also sustain populations of other species associated with late-successional forest habitats (e.g., Carey 1995), even though old-growth forests are generally considered to exceed 200 years of age. These younger stands typically incorporate a component of large old trees and snags that survived the natural disturbance or partial harvest of the preceding stand; hence, they are multiaged rather than truly even-aged young stands.

Such examples provide a model whereby retention of some old-growth Douglas-fir trees can create managed stands that provide suitable nesting and foraging habitat for spotted owls within 50 or 60 years of harvest. Without retention, it may take 120 years or more to create the necessary structural elements, even with intensive silvicultural efforts. As noted earlier, desired habitat conditions exist for most of a rotation under retention approaches; whereas they are present for only a limited period under a long-rotation strategy.

Figure 7.2 Natural mixed-age stands resulting from legacies of large, living trees, snags, and logs provide models for structural retention. Large surviving trees and snags are components of this dominantly 80-year-old Douglas-fir stand developed following wildfire. Consequently, the stand provides habitat for species that would otherwise be absent, such as the northern spotted owl.

Retention is particularly critical where rotation ages or other conditions exist that prevent the predictable re-creation of some particular structure such as very large, old, and decadent live trees and large, highly decayed snags.

Enhancing Connectivity in the Managed Landscape

A third value of structural retention is enhancing the movement of organisms within a managed landscape. Conditions in the matrix or dominant patch type are the most important factor controlling connectivity in that landscape, including dispersion and migration of most organisms (Franklin 1993). Traditional conservation biology approaches fail to recognize this fact by fixing on intact corridors of specific habitat conditions as the primary technique to facilitate organismal movement; this limited perspective probably originated from an early focus on vertebrate organisms. In fact, most organisms probably do not respond to a corridor-based strategy. Rather, they are influenced most strongly by the conditions of the matrix. The importance of the matrix is obvious in many forest landscapes where it is composed almost entirely of managed stands.

Structural retention in managed stands can be designed to facilitate dispersion of organisms. In terms of traditional island biogeographical theory, the objective is to make the "sea" (i.e., the matrix) a less hostile environment for dispersion. The "sea" can effectively be made shallower and be provided with stepping stones by furnishing, for example, well-spaced logs, trees, and shrub patches for protective cover or transient habitat on cutover lands.

Retained forest aggregates—small forest patches—provide larger, more structurally diverse, and microclimatologically moderate habitat islands for dispersing organisms than do dispersed individual structures such as trees, snags, or logs, but they also tend to be more widely spaced. In designing strategies for improved matrix connectivity, a variety of issues needs to be considered, including the necessary size and spacing of various retained structural elements. For example, in coastal British Columbia one scientific team concluded that retained aggregates should be spaced no more than four tree heights apart (Scientific Panel for Sustainable Forest Practices in Clayoquot Sound 1995).

Structural retention also can work to the disadvantage of some organisms, such as when it creates favorable conditions for predation of a specific organism. Concerns have been raised, for example, that retention of trees and snags in cutover areas may create a "killing ground" for prey species dispersing through the cutover, such as predation by great horned owls on northern spotted owls.

Design Elements in a Variable Retention Harvest System

There are three major issues in development of harvest prescriptions based upon the variable retention concept: (1) what structures to retain on the harvested site, (2) how much of each of these structures to retain, and (3) the spatial pattern for the retention—that is, dispersed or aggregated or in some combination. Decisions regarding each of these questions is, of course, dependent upon management objectives and specific stand conditions.

Given the multiple and complex management objectives typical in modern forestry, standardized prescriptions are not likely to be sufficient. The variable retention concept allows silviculturists to be aware of and to utilize a broad array of harvest prescriptions (Figure 7.1b).

What to Retain?

A wide variety of individual and stand-level structural features can be conserved during harvest depending upon management objectives. Exemplary structural elements include (Figure 7.3) (1) live trees, especially large-diameter trees and trees with distinctive features such as rot pockets, cavities, and large limbs or clusters of limbs, (2) snags in varying states of decay, including snags of larger diameter, (3) logs and other woody debris in varying states of decay, (4) undisturbed layers of forest floor, and (5) forest understory species, including moss, herb, shrub, and small tree components. Following is a brief review of considerations with regards to these elements.

Large-diameter, decadent trees are particularly important features to consider for retention because they provide critical habitat for many organisms and will otherwise be absent from many managed stands. For example, in the mountain ash forests of southeastern Australia, large trees with hollows are essential habitat for over 400 species of vertebrates (Lindemayer and Franklin submitted). Trees with extensive, large-diameter branch systems are important to species such as marbled murrelets and northern spotted owls in northwestern North America (FEMAT 1993). Live trees also provide habitat for many other organisms, including invertebrates, epiphytes (e.g., mosses, lichens, and liverworts), and microbial organisms.

Large live trees are important hosts and energy sources for a wide variety of soil organisms, including fungal species that form mycorrhizae (Perry 1994). As the major photosynthesizing organisms in a forest, trees produce and transfer immense amounts of high-quality carbohydrates from leaves to root systems. A large proportion of these carbohydrates is utilized in the maintenance of mycorrhizae and fine

Figure 7.3 Structural features of old-growth stands having high value for retention include large old trees, snags, and down logs. The silvicultural prescription on this unit is for low (10–15 percent) levels of dispersed structural retention of large trees, snags, and logs (by volume) to meet minimal long-term goals for coarse woody debris (Blue River Ranger District, Willamette National Forest).

root systems. Eventually, they fuel most of the dynamic and complex belowground energy web.

Large live trees are also the sources of large-diameter snags and logs. The importance of snags to a large variety of animal species is well known for temperate forest environments throughout the world (Harmon et al. 1986, Maser et al. 1988).

Logs and other coarse woody debris on the forest floor and in associated aquatic ecosystems fulfills a wide variety of ecological functions (Harmon et al. 1986, Maser et al. 1988). These include habitat for a large variety of vertebrate, invertebrate, plant, fungal, and microbial species; sites for biological fixation of nitrogen; and long-term sources of organic matter and nitrogen. Coarse woody debris, including logs, plays similar roles in providing habitat in freshwater and marine ecosystems and by influencing geomorphic processes such as erosion and sediment retention. These influences are particularly well known for stream and river ecosystems where large logs are often critical structural elements for retentive and diverse stream reaches.

Similarly, understory plants are often critical resources that may require long periods of time to reestablish once eliminated by logging (Halpern and Spies 1995). Herbs, shrubs, and small trees may provide important resources and habitat for animal species. For example, in Australian mountain ash forests, small trees found in the lower canopy belonging to the genus *Acacia* are important as a foraging substrate as well as in facilitating movement of arboreal marsupials (Lindemayer and Franklin submitted). In the same forests, tree ferns (*Dicksonia antartica* and *Cyathea australis*) act not only as nursery sites for other plants, but also as substrate for fungi that are a food resource for some marsupials. The importance of the diverse herbaceous and shrubby understories in coastal Alaskan Sitka spruce–western hemlock forests to deer is well known as is the very long periods of time required to reestablish such understories following clearcutting (see, e.g., papers in Meehan et al. 1984).

In addition to individual structural features, foresters may want to consider some aspects of stand structure, such as provision of multiple canopy layers and maintenance of areas of intact forest floor. Multiple canopy layers provide a diversity of habitat conditions for bird and invertebrate species. Undisturbed areas of forest floor can provide important refugia for many species of ground-dwelling invertebrates and fungal species; maintaining areas with deeper layers of organic matter can also be important, as indicated by North (1993), who found strong relationships between occurrence of truffles and forest floor depth. As will be seen, retention of small forest patches or aggregates is one strategy to provide for some stand-scale structural elements.

How Much to Retain?

Answering the question of how much to retain is conceptually very simple—it depends upon the management objectives for the harvest unit, which of course includes landscape-level considerations. As a beginning point, the silviculturist can decide whether to maintain essentially closed forest conditions, which will require high levels of retention, or to sacrifice closed forest conditions for some period of time, which will allow for low levels of retention.

Detailed decisions about actual levels of retention are complex, however, and data are limited. There is increasing evidence that retention is effective in maintaining biological diversity. However, there is very little quantitative information available on how specific ecological objectives respond to various levels of structural retention. For example, there are no quantitative studies on the effect of various amounts and spatial patterns of logs on movement of small mammals through cutover areas. Similarly, there have been no studies of the numbers of trees that are needed to effectively maintain the hydrologic behavior of an intact forest stand during rain-on-snow storm events. Finally, although live tree retention is known to be effective in maintaining certain bird (e.g., Hansen et al. 1995b), invertebrate (Schowalter in press, Berg et al. 1995), and lichen species (Hunter 1995, Berg et al. 1995, Sillett 1995), there is little quantitative data for most groups of organisms on how species diversity and population levels respond to levels of retention.

Resource managers have begun to develop guidelines for retention of some structural features, such as snags, logs, and live trees. But these guidelines are based upon limited scientific data. Hence, managers have had to rely on inferences based upon knowledge of species and processes of interest and upon practical field experience. Many agencies have developed guidelines for retention of wildlife trees (e.g., Washington State Department of Natural Resources 1992). Earlier wildlife habitat guides provided general information on types, levels, and distribution of snags and logs that is valuable in addressing questions of how much to leave (e.g., Thomas 1979, Brown 1985). Some of the most detailed information that has been developed is for woody debris in aquatic ecosystems, where the objective is often to maintain natural levels of such structures (Maser et al. 1988). This contrasts with objectives in harvested terrestrial areas, where it is understood that levels of specific structures will be substantially below that of natural stands. Individual forest units, such as national forests, have used a variety of information sources, including expert panels, to develop standards for structural retention.

What Spatial Pattern for Retention?

Dispersed and aggregated retention are two contrasting spatial models of structural retention. Each approach has advantages (Table 7.1); moreover, combinations of the two can be designed to gain the ecological benefits of both approaches.

Under dispersed retention, structures selected for retention are evenly distributed over a harvest unit (Figure 7.3). The tree overstory in a classical shelterwood harvest unit provides a model of this spatial pattern. Familiarity with this shelterwood model made dispersed retention the first approach adopted by foresters when challenged to develop alternative silvicultural prescriptions that call for significant retention.

A common application of dispersed retention focuses on dominant and strong codominant trees, since these are likely to be the most wind-firm and stress-tolerant individuals. Such trees can function as refugia for many organisms as well as provide for well-distributed sources of soil energy. They also provide a component of well-distributed large-diameter trees for the new stand once it becomes reestablished. Ultimately, these green trees will also become a well-distributed source of snags, logs, and woody debris incorporated into the forest floor and soil.

Aggregated retention focuses on small patches of forest within a harvested unit (see "Plum Creek Timber Company's Approach"). Patches may be of varied size and shape, but as currently applied in northwestern North America, aggregates are typically 0.05 to 1.0 ha in size. Patch size and placement, initial conditions, and treatment at the time of harvest can vary widely based upon management objectives. Objectives often include provision of patches that are representative of initial stand conditions in terms of composition and diameter distribution and that provide intact forest understories and soil organic layers (e.g., Scientific Panel for Sustainable Forest Practices in Clayoquot Sound 1995). Under such prescriptions,

Plum Creek Timber Company's Approach to Environmental Forestry

Plum Creek Timber Company L.P., a Seattle-based company, is the second largest owner of private timberland in the Pacific Northwest. The company owns 1.2 million acres of forest land in Washington, Idaho, and Montana. In the early 1990s, Plum Creek decided to change both the perception and the reality of their operations by experimenting with variable retention harvesting methods.

In an effort to redress a negative public image, Plum Creek began to transform their corporate identity by first adopting a set of environmental principles. Among other things, these principles include the enhancement of ecological and structural diversity where feasible, meeting or exceeding state and federal standards for protecting water quality and fisheries, enhancing soil and site productivity, and protecting wildlife habitat. Working from this base, company foresters began to develop alternatives to standard clearcutting. Their goal has been to improve upon the aesthetics and ecological functionality of harvest units—particularly with regard to the provision of habitat for the northern spotted owl and other old-growth-dependent species. Two harvest units inhabited by spotted owls are particularly noteworthy:

Frost Meadows is a 183-acre harvest unit located on the east side of the Cascades in Washington State. In 1990, when an active spotted owl nest was discovered just prior to harvesting, Plum Creek modified the harvest prescription to incorporate variable retention harvest strategies rather than the planned, traditional clearcut.

First a minimum 400-foot-wide, no-cut corridor was established along the creek as a travel and dispersal route for owls. Second, on the unit closest to the nest site, approximately 80 percent of the residual stand was left intact to maintain old-growth characteristics, such as large-diameter trees and decayed, dead, and downed timber. Yet, nearly 50 percent of the merchantable timber volume was removed from the stand. The remainder of the sale was changed to marked leave-tree units, reflecting the dispersed retention strategy. About 18 to 20 old-growth overstory trees per acre and the associated understory trees in their immediate vicinity were retained. The objective was to accelerate the development of future spotted owl habitat by providing for multilayered tree canopies and interspersed overmature trees.

Since the harvest, the adult spotted owls at Frost Meadows have been banded, and radio transmitters have been attached to monitor their movements throughout the year. This work has revealed that the female owl discovered just prior to harvest in 1990 remained in the area. In 1991, the male with whom she previously had nested was replaced by a sub-adult. The new pair did not mate in 1991. In 1992, however, this pair successfully raised two young owls, which have been banded and fitted with radio transmitters.

Cougar Ramp is a harvest unit in southwest Washington State on which Plum Creek has pioneered the development of aggregated retention of Douglas-fir forests. The Cougar Ramp unit was designed to meet three management objectives: (1) retain patches of representative green trees, snags, and downed logs with as little disturbance as possible for wildlife habitat; (2) address aesthetic concerns for travelers along a nearby highway; and (3) allow for future blowdown salvage and site preparation for reforestation, should the need arise.

Using uphill cable logging, approximately 15 percent of the 73-acre unit was set aside in contoured patches. The patches were designed, mapped, and marked within each tower setting. Timber was directionally felled away from the patches. Cutting of nonhazardous snags and understory vegetation was avoided in the harvest areas.

The contoured-patch design at Cougar Ramp provided a diverse mix of trees and shrubs representative of the pre-harvest stand. Douglas-fir was the primary tree species retained in the unit, although red alder, maple, and western hemlock were also represented.

Following harvest, researchers at the University of Washington evaluated residual vegetation and microclimate conditions in the unit. Wildlife habitat conditions have been studied by comparing bird, small mammal, and amphibian use of cut areas, uncut patches, and adjacent forest. Preliminary results of the wildlife research indicate that bird species diversity and abundance are highest in the contoured retention patches. The 55 bird species using the Cougar Ramp unit represent groups normally associated with forest canopies as well as species commonly found in clearcuts and openings. Research and monitoring will continue at Cougar Ramp, especially as the harvested areas regenerate to provide additional forest structure.

For more information on these and other environmental forestry projects underway on Plum Creek Timber Company lands, contact Lorin Hicks, Wildlife Biologist, 999 Third Avenue, Suite 1900, Seattle, WA 98104.

Table 7.1 Contrasts between dispersed and aggregated structural retention

	Pattern of Retention	
Objective on Harvest Unit	Dispersed	Aggregated
Microclimate modification	Less, but generalized over harvest area	More, but on localized portions of harvest area
Influence on geohydrological processes	Same as above	Same as above
Maintenance of root strength	Same as above	Same as above
Retain diversity of tree sizes, species, and conditions	Low probability	High probability
Retain large-diameter trees	More emphasis	Less emphasis
Retain multiple vegetation (including tree) canopy layers	Low probability	High probability
Retain snags	Difficult, especially for soft snags	Readily accomplished, even for soft snags
Retain areas of undisturbed forest floor and intact understory community	Limited possibilities	Yes, can be as extensive as aggregates
Retain structurally intact forest habitat patches	Not possible	Possible
Distributed source of coarse woody debris (snags and logs)	Yes	No
Distributed source of arboreal energy to maintain belowground processes and organisms	Yes	No
Carrying capacity for territorial snag- and/or log-dwelling species	More	Less
Windthrow hazard for residual trees	Average wind firmness greater (strong dominants), but trees are isolated	Average wind firmness less, but trees have mutual support
Management flexibility in treating young stands	Less	More
Harvest (e.g., logging) costs	Greater increase over clearcutting	Less increase over clearcutting
Safety issues	More	Less
Impacts on growth of regenerated stand	More, generalized over harvest area	Less, impacts are localized

Note: Contrasts various ecological and operational objectives. Comparable overall levels of retention are assumed.

selected aggregates are not entered during harvest or subsequent silvicultural treatments, such as slash treatment and site preparation.

Retention of patches of intact forest or aggregates as an integral part of a harvest unit is a relatively new concept for foresters and biologists. It is not analogous to any traditional harvest cutting practice, nor has it received much attention from wildlife managers or conservation biologists. Hence, aggregated retention is often misunderstood as an effort to create small forest reserves. In fact, aggregates are intended to be functional elements of a harvested unit—lifeboating and ultimately enriching a managed stand. As such, they are not intended to provide habitat for interior forest species that require large areas because of the edge influences that are experienced with small residual forest patches (see, e.g., Chen et al. 1992, 1993a, 1993b), home range requirements, or other factors.

Although dispersed and aggregated retention both broadly address maintenance of structurally complex forest stands, each has its own set of ecological advantages and disadvantages (Table 7.1) and therefore specific applications in retention harvesting.

Dispersed retention is obviously most appropriate where ecological objectives require that structures be well distributed (Table 7.1). Objectives for which relatively uniform distribution is desirable include provision of coarse woody debris (from logs and snags) to the forest floor and soil and provision of well-distributed sources of arboreal energy (from vigorous living trees) to maintain belowground organisms and processes. Dispersed retention can also provide habitat for wildlife that are strongly territorial or incompatible or that require high levels of some structure; vertebrates dependent upon cavity-bearing trees in Australian mountain ash forests provide an example of this circumstance (Lindemayer and Franklin submitted).

Dispersed retention is also appropriate where the objective is to broadly mitigate some condition over the entire harvest unit, such as modification of microclimate or hydrological processes, or maintenance of root strength to stabilize soils (Table 7.1). Although aggregated retention provides for conditions that are more forestlike, these effects are confined to the immediate vicinity of the aggregate.

Aggregated retention provides opportunities to maintain a broader variety of stand structural elements than dispersed retention (Table 7.1). For example, it is easier to maintain a variety of tree species, sizes, and conditions in aggregates. Some species and sizes of trees have low survival rates in open areas due to poor wind resiliency or low tolerance of moisture, temperature, or insolation stress. Retaining a diversity of trees is important because many organisms, such as invertebrates, are associated with only a limited set of host species (see, e.g., Schowalter 1996). It is also easier to retain a diversity of snag sizes and conditions in aggregates; this is particularly true for soft, highly decayed snags, which are likely to collapse when disturbed by logging activity.

Aggregate retention also affords opportunities to maintain multiple canopy layers, understory plant species and communities, and intact forest floor layers. Retaining these features provides refugia for many species, such as invertebrates and fungi, and processes that would decline or disappear in their absence.

There are many questions about appropriate sizes, shapes, and distributions for aggregates. Applications of the approach should be flexible and fitted to specific management objectives and stand conditions. In many applications, it will be desirable to incorporate representative areas of the preharvest forest stand.

Selective placement of aggregate areas of lower productivity or stand density may reduce their ability to provide desired structures or conditions following harvest. Integrating aggregates with protection of aquatic ecosystems may be appropriate in some areas. However, ecological goals for retention of the aggregates, along with economic and operational issues, are primary considerations.

Isodiametric shapes are often desirable for aggregates, but there are also good reasons for using linear shapes. Compact designs will be more effective than linear designs in modifying microclimatic conditions for a given aggregate size. Linear designs can provide efficient visual screening, corridors for movement, and protection for linear features, such as streams. Special shapes may be designed to achieve specific objectives; teardrop-shaped aggregates are being explored, for example, to reduce risk of windthrow.

Selecting an appropriate size for aggregates involves a variety of considerations, including tradeoffs between the number, potential distribution, and size of aggregates. In most applications, a well-distributed system of aggregates is preferred. Research on edge effects suggests that many microclimatic benefits can be achieved with relatively small aggregates (e.g., less than 1 ha in size). Achieving true interior forest conditions is impossible, of course, in the context of heavily harvested areas due to the extent of edge influences in most forest types (Chen et al. 1992, 1993a, 1993b).

Finally as will be noted below, aggregated retention generally provides fewer operational constraints and has less impact on growth of regeneration than dispersed retention.

Other Considerations in Designing Prescriptions

Although the preceding sections have emphasized stand-level considerations, it is extremely important to recognize that landscape- or larger spatial-scale concerns will be very influential in development of variable retention harvest prescriptions. What is prescribed for an individual stand will very much depend upon its immediate and long-term relationship to conditions and activities in surrounding areas. While such viewpoints are implicit in the notion of objectives such as "enhancing connectivity," this is sufficiently important to warrant specific mention.

Ecological objectives will rarely be resolved by activities on small individual tracts of land. Furthermore, current and planned conditions on surrounding areas may well mitigate much of the potential impact of a harvest unit, reducing the level of structural retention required to achieve some landscape-level objectives. As an example, riparian protection zones may provide for much of the retention required in some landscapes. This occurred under some options developed by FEMAT (1993); stream systems and associated riparian reserves occupied such a high percentage of some coastal regions that retention requirements could be relaxed on some of the matrix available for harvest.

Examples of Variable Retention Silvicultural Prescriptions

While there are essentially infinite possibilities, some generalized variable retention harvest prescriptions already are emerging. These reflect some broad similarities in management objectives as well as the early stages in learning about retention harvesting. Three examples are reviewed below.

Low Retention, Mix of Aggregated and Dispersed

Common objectives for many public and some private timberlands include provision of moderate to high levels of timber production, regeneration of shade-intolerant tree species, and maintenance of minimal structural levels to fulfill basic lifeboat and stand-enrichment functions. Silvicultural prescriptions to achieve this mix of objectives typically involve relatively low retention levels (10 to 20 percent) using both aggregated and dispersed approaches.

The standards and guidelines for structural retention during regeneration harvesting on "matrix" lands in the Northwest Forest Plan (USDA and USDI 1994b) are a generic silvicultural prescription of this type. In brief, the guidelines call for permanent retention of at least 15 percent of green trees on each cutting unit: "Seventy percent of the total area to be retained should be in aggregates of moderate to larger size (0.2 to 1 ha or more) with the remainder as dispersed structures (individual trees, and possibly including smaller clumps less than 0.2 ha)." This direction assumed that a mixture of dispersed and aggregated retention was most likely to achieve the full array of ecological objectives incorporated into the plan. Flexibility was provided to allow silviculturists to fit the mix of dispersed and aggregated retention and size of aggregates to specific site conditions and objectives (USDA and USDI 1994b).

Adoption of these guidelines reflected the strong sentiment of biologists working as a part of FEMAT (1993) that aggregated retention is likely to be more successful in conserving elements of biological diversity than comparable levels of dispersed retention. Indeed, a subsequent team preparing the final environmental impact statement (USDA and USDI 1994a) favored total reliance on larger (1 ha or greater) aggregates. The final wording in the Record of Decision was only adopted after energetic debate; absence of information about the effectiveness of various sizes of aggregates contributed to the difficulty in arriving at a decision.

Similar retention harvest guidelines were provided for cutting units "without significant values for resources other than timber, or without sensitive areas" in the Clayoquot Sound region of British Columbia (Scientific Panel for Sustainable Forest Practices in Clayoquot Sound 1995). These recommendations provided for retention of at least 15 percent of the forest, primarily as aggregates of 0.1 to 1 ha that are well dispersed throughout each cutting unit. Regardless of retention level, the scientific panel recommended that all portions of a cutting unit be within two tree heights of an existing aggregate or stand edge. The panel also advised that aggregates should be representative of forest conditions in a cutting unit—that is, not disproportionately located on sites of lower timber volumes or productivity.

The Scientific Panel for Sustainable Forest Practices in Clayoquot Sound (1995) did recommend high levels of retention on cutting units with significant values for resources other than timber. They rec-

Menominee Sustained-Yield Management

The Menominee Indian Reservation in northeastern Wisconsin stretches over 235,000 acres of land, 220,000 of which are forested. To the casual observer, the Menominee forest looks pristine. Large-diameter trees in a natural setting belie the fact that it is one of the most intensively managed tracts in the Great Lakes states. Over 2 billion board feet of lumber have been removed from the forest in the last 140 years—yet the volume of sawtimber currently is greater than when the reservation was established in 1854.

The 140-year history of forest resource use and management on the Menominee forest is a practical example of sustainable forestry—forestry that is ecologically viable, economically feasible, and socially desirable. The Menominee concept of sustained-yield management refers not only to forest products and social benefits, but also to wildlife, site productivity, and other ecosystem functions.

The Menominees' approach to forestry is based on a simple, farsighted management objective: to maximize the quantity and quality of sawtimber grown under sustained-yield management principles while maintaining a diversity of native species. To the Menominees, quality and quantity are concepts that favor growing those tree species most suitable to a particular site for as long as they remain healthy and vigorous. This concept is based upon the direction chosen by earlier Menominee leaders, who recognized the need to harvest trees for economic survival, but only at a speed or intensity under which the forest could replace itself. They promoted a timber harvesting system that removed timber according to vigor rather than merchantable size alone. The Menominee tradition of harvesting according to tree vigor (the ability to grow and regenerate itself) retains more larger and older trees compared to adjacent timber lands. As such, the Menominee forest is a mixture of older, larger trees with ample younger regeneration. This has provided the tribe with a diversity of forest plants and animals not seen on surrounding forests managed under short-term economic formulas.

A cornerstone of Menominee forestry is its monitoring program. Monitoring is accomplished through two inventory systems called the continuous forest inventory (CFI) and the operations inventory (OI). Using a systematic grid of permanent plots, the CFI monitors forest health (including the area, volume, and condition of the timber) to determine how much of the forest can be harvested annually or over a longer period. Observers thus can track changes in the forest due to management practices or natural occurrences. The OI system monitors all forest land to determine where the timber types described in the CFI occur. Data common to both the CFI and OI systems (such as cover type) are collected with the same specifications, allowing information from both inventories to be merged. This detailed stand information provides the basis for planning when and where to cut.

The annual allowable cut on the Menominee forest is determined based on the CFI. The forest management plan specifies the minimum stocking level necessary for each cover type before any green standing timber can be harvested. In this way, harvest prescriptions are based on the excess stocking of fully stocked stands—not on the net growth of all stands. Silviculture, rather than market forces, determines how much timber is harvested. Understocked acreage is allowed to grow and develop for future harvest.

The forest is divided into 109 compartments. Compartment management activities are based on a 15-year cutting cycle for timber types subject to all-age management—approximately 65 percent of the forest. The remaining forest types, which are managed under an even-age system, are harvested as closely as possible to the compartment schedule.

The compartment cutting schedule was initially determined by looking at the harvest history of specific areas and combining these areas into units of roughly equivalent volume or acreage. Planners have also tried to balance areas of dominant species composition throughout the 15-year cycle. The current schedule reflects the best combination of acres and volume that would produce a reasonably even flow of sawtimber and pulpwood to the mills. It may be revised as new information becomes available through the CFI.

The Menominee sustained-yield management program predates the ecosystem management concepts currently being debated among natural resource professionals. The Menominee tribe has inhabited the forests of this region for thousands of years. They have understood that the whole resource was needed to protect any individual part. This is a heritage that has been passed from generation to generation.

This case study was adapted from the Menominee Tribal Enterprises Forest Management Plan 1995-2005, as well as from the following articles which provide further information on the Menominee forestry program:

Bristol, T. 1992. Edge of the woods: Forestry for the seventh generation. *Turtle Quarterly* Fall.

Pecore, M. 1992. Menominee sustained yield management: A successful land ethic in practice. *Journal of Forestry* 90(7):12–16.

ommended retention of "at least 70 percent of the forest in a relatively uniform distribution . . . [including] some large-diameter, old, and dying trees; snags; and downed wood throughout the forest."

Retention of Dispersed Large-Diameter Cohorts

Prescriptions directed toward management of multiple cohorts are also emerging as an approach to maintaining structurally diverse stands, especially when maintaining a component of large-diameter trees is a major objective. Such approaches are appropriate in the fire-prone forests of western North America. In many of these forests, maintaining a large-diameter old-growth cohort in perpetuity is an important objective for wildlife and fire resiliency objectives, while managing the small- and medium-diameter component for wood production and reduction of catastrophic fire potential.

An example of such a silvicultural prescription of this type for the mixed-conifer forests common in the Sierra Nevada and interior mountains of eastern Oregon might be one that has as one of its objectives maintaining a population of 6 to 10 large-diameter trees and the snags and logs created through the periodic death of these trees. Definition of the diameter objective would probably vary with site productivity; a range might be from 75 to 100 cm d.b.h. (diameter at breast height). The stand would be managed to insure that replacements are available for losses from the large-diameter tree population. No salvage of the dead trees would occur in order to insure that there is also a continuing population of large snags and logs.

The interim California spotted owl harvest guidelines provide a starting point for this kind of system (Verner et al. 1992) (Figure 7.4). In forest stands of a type "selected" by California spotted owls (primarily mixed-conifer forest dominated by large trees), the guidelines call for retention of 40 percent of the basal area of large old trees, including all trees 30 or greater in d.b.h. In other strata that might be used for nesting, prescribed retention levels are 30 percent and at least 50 sq. ft. per acre of large old trees, including all trees 30 or greater in d.b.h. Retention of large snags (to a maximum of eight per acre) and the largest down logs (to at least 10 to 15 tons per acre) is also a part of the interim recommendations. These guidelines are currently being used on national forest lands within the range of the California spotted owl. Additional desirable developments for a long-term strategy might include refinement of the large-diameter tree population goals (i.e., numbers and species) and providing for replacements as mortality occurs.

Figure 7.4 Silvicultural prescription designed to maintain a dispersed population of large-diameter trees on the harvest unit following the California spotted owl interim guidelines. All trees greater than 30 in. d.b.h. have been retained with complete removal of smaller merchantable stems (Plumas National Forest, California).

Silvicultural prescriptions designed to produce multiple cohorts as outlined above contrast sharply with traditional selection-cutting approaches. Traditional selection prescriptions focus on creation and maintenance of a particular tree diameter distribution. Larger, older trees are systematically removed as a part of this process.

Group Harvest with Low Retention

Group selection is often proposed as a technique to mitigate impacts of timber harvesting on biological diversity since it generally involves clearing of relatively small areas within a forest matrix. Smith (1986) describes a maximum size for selected groups as an opening two tree heights in diameter, about 0.7 and 2.9 acres for trees 100 and 200 feet tall, respectively. The harvest of these small areas cycles through the stand, eventually resulting in harvest of the entire area. As traditionally practiced, group selection does not provide for retention of structural features within the harvested patch. Consequently, any structural features that have development periods longer than the rotation will be lost from an area subjected to classical group selection. Groups that are completely cleared obviously have much simpler structures than most natural openings, which have a structural legacy of living or dead trees or both.

Hence, group selection combined with structural retention has been proposed as an alternative approach to maintaining structures that have very long development times or are required in large numbers. Such a modification has been proposed by the California Spotted Owl Technical Group (Verner et al. 1992) for Sierra Nevada pine and mixed-conifer types. Incorporating structural retention within selected groups is very straightforward in such timber types.

One of the few examples of group selection with structural retention known to the authors is in second-growth coast redwood stands located in the Arcata, California, city forest (Figure 7.5). This approach was developed by the third author in 1982 after an attempt to apply a uniform, single-tree selection system proved unsatisfactory. Prescriptions provided for retention of large live trees (especially dominant coast redwoods), snags, and down logs within selected groups up to 4 acres in size. Typical retention levels are 25 percent of the merchantable volume.

Another alternative for maintaining structural features under a group selection approach would be to permanently reserve an appropriate percentage of the subject stand from harvest.

Figure 7.5 Silvicultural prescription for second-growth stand involving harvest of selected tree groups up to 4 acres in size with 25 percent retention (by volume) of coast redwood and other trees and large snags and logs (watershed for City of Arcata, California).

Management Issues in the Application of Variable Retention

Conceptually, structural retention is as old as forestry. But, at the same time, it is revolutionary because none of the traditional regeneration harvest systems truly has utilized the concept, as will be discussed below. Consequently, practical experience—let alone designed scientific studies—is very limited. Numerous questions are associated with forest protection, regeneration and growth, and the effectiveness of retention prescriptions in achieving management objectives. Some of these issues are discussed below along with current scientific information providing valuable insights for understanding and evaluating retention harvest approaches. The Cascade Center for Ecosystem Management has produced a very useful review of much of the existing scientific and anecdotal information in its *Residual Trees as Biological Legacies* (Hunter 1995).

Forest Protection

Protection of forests from a variety of factors—including wind, wildfire, insect pests, and diseases—has always been an important element of any forest management program. Indeed, justifications of even-age management often are based, at least partially, upon control or exclusion of some pest or pathogen, such as parasitic mistletoes. In particular, clearcutting—which eliminates all existing trees—is viewed as a technique to reduce or eliminate pathological legacies and start over with a fresh slate.

In any case, retention harvest prescriptions must address forest protection concerns if they are to be successful.

Wind

Wind is always an important consideration in timber harvest because of possible impacts of residual stand and landscape conditions on the potential for major windthrow events. Hence, wind is a key consideration in designing silvicultural prescriptions. All harvest cutting practices can increase the potential for windthrow. The only way to ensure that there will be no windthrow is to cut all of the trees. Yet even clearcutting affects windthrow potential in adjacent unharvested stands. Increases over endemic levels of windthrow will depend upon the amount (e.g., Franklin and Forman 1987) and topographic location of boundaries or edges between uncut and cut forest (e.g., Gratkowski 1956).

Retention harvest systems are of particular concern because isolated trees and tree groups are much more vulnerable to wind damage than trees in intact stands. In the Douglas-fir region, an early study found very high rates of mortality in residual seed trees (e.g., Isaac 1943). This led to the conclusion that such trees were ineffective as seed sources. On this basis, large tree retention on harvest units was dropped until the 1960s when Roy R. Silen, working at H. J. Andrews Experimental Forest in the western Oregon Cascades, resurrected seed tree and shelterwood cutting to address problems in natural regeneration of Douglas-fir forests (Franklin 1963). Silen hypothesized that if residual trees were sound dominant trees—rather than culls or defective trees as had previously been the case—survival would be much better. He was proven correct by his initial trials (Franklin 1963) and by a much more extensive series of Douglas-fir shelterwood cuttings stimulated by his research (Williamson 1973).

Trees will be lost to wind-related causes. However, retention harvest prescriptions can be designed to minimize windthrow. Selection of species and individuals that have a high likelihood of surviving windstorms is an important element in minimizing potential losses; such distinctions are generally understood in most forest types. For example, in old-growth western hemlock–western red cedar stands along the Pacific Coast of North America, red cedar have a much higher probability of surviving as residual trees in partially cut stands than western hemlock. As noted earlier, sound dominant trees are more likely to survive than trees that are subdominant or have advanced states of decay. Deep-rooted species, such as Douglas-fir and many pines, are more likely to survive than are shallow-rooted species such as hemlocks and spruces.

With regard to windthrow, aggregated patterns of retention are likely to be superior to dispersed patterns of retention. A group of trees provides at least some mutual support. It is also possible to site aggregates in more wind-firm locations within a harvest unit. Likewise, the shape of aggregates can minimize windthrow potential; for example, teardrop shapes tend to be more aerodynamic than linear designs. The advantage regarding windthrow may not be entirely with aggregated retention, however, if trees selected for dispersed retention are generally sound dominants of more wind-firm species. Furthermore, the windthrow susceptibility of retained trees can be reduced by removing a portion of the canopy by topping or branch pruning, thereby reducing the sail area.

Topographic and soil conditions are also important variables in the design of retention harvest prescriptions. Sites that are exposed to frequent, intense windstorms are not prime candidates for retention harvest if long-term survival of residual green trees is the objective. Note, however, that retention harvesting may be appropriate if development of snags and logs on the harvested area is the objective. Similarly, sites with restricted rooting depths as a result of shallow or poorly drained soils are not good candidates for retention harvest.

The preceding comments apply primarily to harvest prescriptions involving low levels of structural retention and high levels of exposure for residual trees. Silvicultural prescriptions with very high levels of partial retention—which basically retain the conditions of an intact forest—present different circumstances. For example, experience suggests that selective harvest of up to 30 percent of stand volume may be possible in alluvial Sitka spruce–western hemlock stands in coastal British Columbia without seriously increasing windthrow in the residual stand (Scientific Panel for Sustainable Forest Practices in Clayoquot Sound 1995). Of course, much depends upon the local situation (e.g., species), specific silvicultural prescription, and the damage sustained by a stand in such a selective harvest operation. Of course, potential wind damage should not be ignored at high levels of retention. Rather, prescriptions with a high level of retention offer different opportunities and problems than those with low levels of retention.

Levels of mortality in residual trees following retention harvest will vary widely depending upon such variables as retention level, age, condition, and species of residual trees; geographic region; topographic location; and soil conditions. Some early data are already available for retention units in northwestern North America (Adler 1994, Hunter 1995). Variability in mortality is high: Zero to 58 percent of residual trees were windthrown on 44 units 1 to 10 years of age, while average mortality was 15.9 percent (Adler 1994). Losses are generally viewed as being within acceptable levels (Hunter 1995). Rates of loss of retained trees appear to be highest in the first few years following harvest (Adler 1994, Hunter 1995).

High rates of mortality in retained trees may necessitate retention of larger numbers at harvest in order to maintain minimal levels. This is a potential problem in Australia, for example, where trees with cavities experience accelerated rates of collapse following harvesting (Lindemayer and Franklin submitted).

Of course, windthrown trees have not lost all of their functional value. As noted earlier, objectives for structural retention often include provision of logs on the forest floor and in streams. Generally, however, resource managers prefer not to have the majority of retained live trees converted to down logs within the first 5 to 10 years of harvest.

Fire

Fire is an important element in the management of most forest types for both its potential negative and positive roles. Protection of forests from destructive, uncontrolled wildfires is typically one of the first issues addressed in forest management. For this reason, treatment of slash and fuels generated during harvest and other management activities is typically a part of most silvicultural systems. Slash burning and other activities are often used to reduce or redistribute fuels.

Fire also can have important and positive effects on ecological processes within forests; as such, it is an important silvicultural tool. Prescribed fire appears to be especially important in forest types which evolved under frequent light to moderate fire regimes. Such types are widespread in western North America and Australia. Hence, the effects of silvicultural practices on opportunities for use of prescribed fires is an important consideration.

Retention harvesting—in comparison with clearcutting—does introduce some complexities to the treatment of slash and other forest fuels. However, forest managers have adapted traditional practices—such as broadcast burning or piling and burning—for slash treatment on shelterwood and on retention harvest areas. When treating retention harvest units, there are the duel concerns of keeping the location and intensity of the fire within desired bounds and insuring the survival of the retained vegetation and other structural elements. There is, of course, the potential to use prescribed fire to convert retained live trees to snags; such an approach might have some ecological advantages over creation of snags by topping.

In the Douglas-fir region, fire has been used to treat slash on the majority of retention harvest units cut to date. A variety of techniques has been used, including broadcast burning and burning of concentrations and piles. The presence of retained vertical structures does not appear to create insurmountable problems, although slash treatment costs are generally greater than after clearcutting. Even though most existing cuttings involve dispersed retention, slash treatments have generally been achieved without causing major damage to retained green trees. Much remains to be learned, however, regarding long-term effects of varying fire intensities on survival of residual trees. It is also clear that very intense slash fires can cause unacceptable rates of mortality in residual trees, as has been reported for Eucalyptus forests, for example (Lindemayer and Franklin submitted).

Fire is generally less of a problem with aggregated retention if managers are trying to limit losses of retained structures while treating fuels. The problem is simply one of keeping slash fires out of the aggregates, rather than trying to manage an intricate mosaic of slash and dispersed structures. Of course, in some forest types subject to frequent low intensity fires, managers may specifically wish to burn within aggregates as part of a management strategy.

Using prescribed fire to manage structurally diverse stands developed under retention harvest prescriptions does not appear to pose any unique problems. In fact, it is possible to create stands that are not only well suited to prescribed fire, but also are more resistant and resilient to intense wildfire. For example, large-diameter dominant trees—such as those maintained in the dispersed, large-diameter cohort prescription described earlier—are most likely to survive intense wildfire.

Insects and Diseases

Insects and diseases are important considerations in forest management; often they provide the rationale for specific silvicultural prescriptions. For example, clearcutting is sometimes justified based on the notion that all potential hosts need to be removed from an area to eliminate sources of infection. Conversely, the existence of insect or disease infestations or the potential for their development are often identified as factors that preclude retention harvest systems.

Conceptually, there is a basis for concerns over retention harvest prescriptions providing opportunities for persistence or intensification of pathological problems (see, e.g., Shaw et al. 1994). If structural retention can be used to provide refugia and inocula for desirable insects, fungi, and similar organisms, then it also can provide refugia and inocula for pathogens. Note that the converse is also true—drastic treatments to completely eliminate pathogens, hosts, or critical substrate have the real potential to eliminate many desirable organisms, such as detritivores and symbionts. Some criticisms by pathologists are based on the assumption that retention systems involve more-frequent stand entries than even-aged systems (e.g., Shaw et al. 1994), but this is not necessarily the case.

From a managerial perspective, the key is specificity—identification of the pathogens of specific concern and development of silvicultural prescriptions that balance those concerns against other objectives. Even-age management can accentuate pest problems, especially when it involves monocultures; similarly, long rotations can result in increased problems when they involve numerous intermediate stand entries (Shaw et al. 1994). Dwarf mistletoe, root rots, and bark beetles provide examples of differing challenges and potential silvicultural responses.

Dwarf mistletoe is a parasitic plant that is very common in forests of western North America. Often it is the rationale for rejecting retention harvest prescriptions. If a mistletoe-infected overstory is retained, it will infect susceptible tree species in the understory. Aggregated retention can reduce the conflict between structural retention and containment of problem areas, although it does not totally eliminate the problem. The potential for intensification of mistletoe infections within aggregates remains along with the potential for spread to adjacent areas. An alternative in such situations is dispersed retention in which only nonsusceptible tree species or mistletoe-free individuals of susceptible species are retained.

Root diseases are among the most difficult pathological problems faced in forest management (Shaw et al. 1994). With regards to root diseases, it is not clear where the advantages, if any, would lie between clearcut and retention harvest systems. Even-aged approaches allow for removal of all host species and, possibly, replacement with nonsusceptible or less susceptible species, but this is also possible with retention harvest systems. The retention approach would provide for greater structural complexity in the subsequent forest, but might also result in higher levels of inoculum than under clearcutting. Complete elimination of the root disease from a site is unlikely under either approach. Retention approaches are more likely to provide for a more complex soil ecosystem, which may help hold pathogens in check.

Forest insect pests are generally not likely to be a problem with retention harvest systems. Different species of bark beetle attack different age classes and species of trees as well as different sizes and conditions of material. Typically, retained trees are much older and have a different set of insect pests than the younger, managed component of a stand. Perhaps the greatest problem is the potential for excessive mortality in the retained tree component, especially if wind or fire create opportunities for increased populations of specific pests. Again, choosing various combinations of aggregated and dispersed retention and appropriate retained tree species on dispersed retention sites may minimize insect problems.

Emerging evidence suggests that retention harvest will maintain greater varieties and numbers of insect predators and parasites than occur under clearcutting regimes (Schowalter 1989, 1996). This could be an important factor in maintaining natural controls on insect pests.

Forest Harvest and Management

Operational issues associated with retention harvest strategies include safety; transportation and logging issues, including costs; impacts on management practices and costs; and impacts on forest product receipts.

Worker safety is a critical issue in all forest operations, particularly logging operations. Retention of structural features, especially decadent trees and snags, has the potential to create major hazards for workers who are felling, bucking, and yarding logs. Indeed, removal of all snags and any hazardous trees in and adjacent to work areas is standard practice for most forest regions. Until recently, all logging contracts on federal lands in the Pacific Northwest called for removal of all snags within 200 ft. of a cutting boundary or road. Safety issues are not only confined to harvest operations, but also extend to workers in subsequent operations, such as tree planting and thinning.

Developing safe approaches to structural retention is a major challenge that must be addressed. From a safety perspective, aggregated retention may be particularly appropriate since it is consistent with creation of no-work zones around areas where snags and hazardous trees have been retained.

Worker safety was a major concern of the Scientific Panel for Sustainable Forest Practices in Clayoquot Sound (1995). The panel, which included a highly qualified representative of the Worker's Compensation Board, concluded that "[s]afety concerns are inherent in any silvicultural system. The hazards of clearcutting are better understood than those of other systems. . . . Safety concerns are much easier to address using aggregated [than dispersed] retention . . . principles, which must be observed regardless of the silvicultural system used: procedures must be developed and implemented to minimize risk to workers; and workers must have the right to refuse to carry out procedures that place them at risk."

Retention harvest practices will generally result in higher logging costs than clearcutting, and they may or may not require selection of alternative logging technology. Experiences with shelterwood harvesting are probably very relevant to low to moderate levels of dispersed retention. In general, logging costs for dispersed retention are likely to be significantly greater than for clearcutting, but costs will be only slightly greater for aggregated retention than for clearcutting. Some data from actual retention harvest operations (e.g., Zielke 1993) support these hypotheses.

Loggers can provide some very useful insights into the relative merit of various retention harvest prescriptions and logging methods under specific conditions of access and topography (Berg 1995). These ideas address a variety of concerns, including worker productivity and safety. Incorporating worker input into the design of silvicultural prescriptions needs to be greatly expanded. Training programs, including basic education in ecological concepts, clearly is another important part of implementing new and complex silvicultural procedures, including retention harvest prescriptions (Scientific Panel for Sustainable Forest Practices in Clayoquot Sound 1995).

Retention harvest methods also can introduce complexities into the management of the subsequent stands. One of the early objections to dispersed retention was the potential for interference with aerial applications of pesticides, fertilizers, or herbicides. Aggregated retention can essentially eliminate this problem since it is possible to manage intervening areas essentially as even-aged stands. In such circumstances, aggregates should be laid out so as not to interfere with flight paths.

Concerns have also been raised about the compatibility between retention harvesting and the amount and genetic composition of tree regeneration. Retention harvesting is likely to result in high levels of natural tree regeneration since significant seed sources and protective cover are retained on the site. This does not preclude planting or seeding of desirable species or genetic strains not already present on the site. Indeed, there are many options for multiple-species management using retention approaches utilizing a mixture of regeneration practices and rotation periods on the same site. Genetic impacts of retention harvest techniques on tree populations will depend, of course, upon the nature of the silvicultural prescriptions and are not necessarily more likely to have negative consequences than traditional silvicultural systems (see Friedman, Chapter 13).

Excessive stocking of tree regeneration is likely to be the most common outcome using retention harvest prescriptions. Hence, precommercial thinning and other activities that reduce stand densities will generally be the most important intermediate stand-level silvicultural treatment.

Retention harvesting clearly will reduce wood yields relative to even-aged systems, especially clearcutting. These reductions take two forms: (1) wood volume in the structures (trees, snags, and logs) permanently retained on a site and (2) reduced growth of a regenerated stand due to effects of the residual overstory.

Calculation of volumes and values associated with retained structures is straightforward, assuming that no subsequent removal of these structures is planned. For example, Weigand and Burditt (1992) found that the potential value left behind ranged from $102 to $1,114 per acre depending upon the prescription that was used. The key variables are the amount and type of material retained and the per unit value of such material.

The impact of retained structures on growth of subsequent stands is much more difficult to calculate. Few empirical data exist on the effects that residual trees have on growth of a regenerated stand. Similarly, there are few growth models that allow realistic simulation of growth under such complex stand structures. Two general hypotheses are (1) that growth effects will be related to level of overstory retained and (2) that dispersed retention should have greater impacts than aggregated retention.

Some empirical data on effects of residual trees on growth of associated younger cohorts do exist for forests in northwestern North America (Zenner 1995, Rose 1994). In young to mature stands of Douglas-fir and western hemlock in the Cascade Range, Zenner (1995) found that total understory volume reduction was 22 and 45 percent with residual tree densities of 5 and 50 per acre, respectively; this converts to a growth reduction of 2.4 and 1.5 percent per residual tree. Douglas-fir volume and basal area declined more rapidly than that of western hemlock when residual tree density exceeded 15 trees per acre, although average size and growth rates of dominant Douglas-fir were not reduced by residual trees. Zenner (1995) hypothesized that growth impacts can be reduced by thinning in the understory. Rose (1994) compared 70- to 110-year-old stands with and without an overstory of large old trees. Young stand densities declined when remnant tree densities exceeded about 15 per hectare. Total stand basal area was relatively constant regardless of remnant densities.

Rotations, Entries, and Age Classes

There is often confusion about such issues as rotation lengths, numbers of regeneration harvest entries per rotation, and age class structures (even-, multiple-, or uneven-aged) under retention harvest strategies.

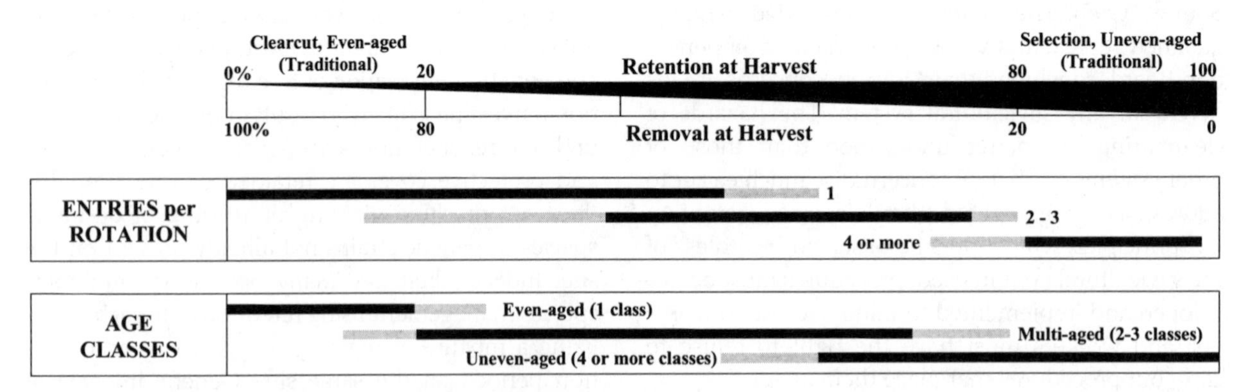

Figure 7.6 Hypothetical relationships between retention levels and entries per rotation and stand age classes.

Foresters often assume that retention harvest approaches necessarily involve more than one entry per rotation or involve an uneven-aged strategy. Some hypothetical relationships between rotation, entries per rotation, and age class structures are illustrated in Figure 7.6. There is substantial overlap among variables over the retention gradient depending upon the particular silvicultural prescription, which of course reflects specific management objectives for the stand.

Strategies involving low to moderate levels of retention (10–30 percent) typically involve only a single entry per rotation and creation of a two-aged stand. After initial harvest operations and regeneration of a new stand, entries probably can be confined to precommercial and commercial thinning operations in the young stand. Strategies involving high levels of retention may involve multiple regeneration harvest entries per rotation and creation of multi-aged (two to three age classes) or uneven-aged (four or more age classes) of trees.

Scientific Issues Raised by Variable Retention

There are numerous important scientific and technical issues regarding the variable retention harvest system. These include fundamental questions about which and how many structures need to be retained and what spatial patterns are desirable to achieve specific objectives. Current working hypotheses are largely based upon inferences from ecosystem science, conditions in multi-storied stands created by nature or past human activities, and short-term studies of recent harvest cuttings that used some form of retention. Problems with using traditional silvicultural systems, particularly clearcutting, to achieve many objectives are clear, along with the need for greater variety in harvest prescriptions. Nevertheless, the knowledge base for implementation of retention harvest is limited.

Evidence for Effectiveness of Structural Retention

Questions are frequently raised about the effectiveness of retention harvest prescriptions in achieving either the lifeboating or structural enrichment objectives outlined earlier. Is there greater species diversity in stands with structural retention than in those that were clearcut?

Studies of young and mature stands which incorporate an old-growth tree cohort provide evidence of greater biodiversity than stands of comparable age lacking such a cohort. In the Pacific Northwest, catastrophic wildfires in the 1800s and early 1900s produced numerous stands that contained residual old-growth Douglas-fir trees; portions of the 1902 Yacholt burn in southwestern Washington State provide a good example of these stands (Figure 7.2). A 1921 windstorm on the Olympic Peninsula produced stands with similar structural complexity. In both cases the resulting mixed-aged stands provide suitable habitat for northern spotted owls when the dominant age class is only 70 to 90 years.

Vertebrate research in such stands provides further evidence. Carey (1995) found that northern flying squirrel populations in young stands with old-growth legacies and well-developed overstories were equivalent to those in old-growth stands. Flying squirrel abundance could be predicted by the density of large snags and abundance of ericaceous shrubs; at least seven large snags per hectare were required to achieve high abundances. North (1993) examined ecological features of 41 non-old-growth stands utilized by northern spotted owls on the Olympic Peninsula and in the northern Cascade Range of Washington. These stands originated following partial destruction of old-growth forests 60 to 70 years ago by wind, fire, or selective cutting. Snag volume and tree height diversity were important predictors of owl use intensity in these stands.

Recent harvest units have been studied for the effects of retention on microclimate, birds, small mammals, lichens, and invertebrates. Unpublished microclimate studies indicate significant amelioration of environment under either dispersed or aggregated retention (Jiquan Chen, personal communication, 1995). Of course, the extensive body of data on microclimatic conditions under shelterwood cutting is relevant in assessing the effects of dispersed retention.

Results of avian studies on retention harvest units and in other stand types clearly indicate that retention enhances structural complexity and provides habitat for many native bird species, including species which are characteristic of late-successional forests but absent from clearcuts (Hansen and Hounihan 1995, Hansen et al. 1995a, 1995b, Vega 1993). Hansen and Hounihan (1995) found higher bird species richness and diversity in retention stands than in clearcuts in the High Cascades of central Oregon. In a more comprehensive analysis, Hansen et al. (1995b) compared bird abundance and habitat functions for forest birds across a wide range of natural and managed stand structures and ages and found that several species of birds benefit from retention of canopy trees, including four species that were characteristic of late-successional forests. Model simulations (Hansen et al. 1995a) also predicted favorable effects of structural retention on several species of birds.

Retrospective surveys of mixed-cohort stands have shown that structural retention can provide effective refugia for invertebrates and lichens. Schowalter (in press) found substantially greater arthropod diversity in partially harvested stands than in plantations regenerated following clearcutting. Similarly, Sillett (1995) finds that many lichen species, including the nitrogen-fixing cyanolichens, will survive on old-growth trees retained on cutover areas and will inoculate the young trees.

Research Needs

Implementation of new silvicultural prescriptions must proceed even in the face of limited existing knowledge in order to meet the complex, multiple objectives of modern forest management. A great deal of relevant scientific knowledge does exist on which to base alternative harvest prescriptions.

An adaptive management approach is required, however, which recognizes that all silvicultural prescriptions are effectively working hypotheses, as Smith (1986) emphasizes. Retention harvest systems should incorporate a strong monitoring component. Furthermore, harvest units that allow for informal comparisons of alternative prescriptions can be established and observed.

Formal research on retention harvest approaches is also critical, however. Particularly critical are experiments designed to provide quantitative information about the types and levels of structures and spatial patterns of retention to achieve various objectives. For example, what are the tradeoffs between levels of coarse woody debris on cutover areas and persistence and movement of small mammal populations? How does avian diversity respond to increasing numbers of retained snags and green trees?

The relative merits of dispersed and aggregated retention currently rank as one of the most important silvicultural research questions associated with retention harvest approaches. This is because of the important economic and operational tradeoffs between the two approaches as well as issues of ecological effectiveness. Currently there are no empirical data comparing the two approaches.

For aggregated retention, important questions center on the appropriate size of aggregates and tradeoffs between aggregate distribution and size under a specific level of retention. These two issues were intensely debated during development of the Northwest Forest Plan (FEMAT 1993, USDA and USDI 1994b) as noted earlier. Few empirical data exist on environmental and organismal responses to increases in the size of isolated forest aggregates (e.g., between .05 and 4 ha). Studies of microclimatic gradients at clearcut-forest boundaries (Chen et al. 1993a, 1993b) provide a basis for inferring that the environmental changes in aggregates are rapid with increased aggregate size initially, but quickly slow after an aggregate size of 0.5 to 1 ha is attained. Creation of interior forest conditions requires very large forest patches (e.g., 25 to 40 ha) and cannot be achieved in aggregates.

Tradeoffs between aggregate size and distribution at a given level of retention is similar to the SLOSS (single large or several small) debate over reserve strategies at the landscape and regional scale (Noss and Cooperrider 1994). Are more ecological objectives achieved by having more small aggregates well distributed over a harvest unit or by having a few large aggregates? The Scientific Panel for Sustainable Forest Practices in Clayoquot Sound (1995) emphasized distribution—no portion of the harvest area was to be greater than four tree heights from an aggregate or patch. At one stage in the development of the Northwest Forest Plan (USDA and USDI 1994a), the strategy using a few large aggregates was emphasized, although this was changed in the final document (USDA and USDI 1994b). In any case, aggregate size and distribution are very important topics for future scientific work.

Major silvicultural experiments to answer basic questions about retention levels and patterns are difficult and expensive undertakings. This is why so few experiments have been conducted, and those that are undertaken are often abandoned after a short time (see, e.g., the Silvicultural Systems Project in Australia [Squire 1990]). There has never been a large-scale, replicated scientific study of any regeneration harvest system in the Douglas-fir region, for example, including clearcutting. Such experiments are made even more challenging when relatively large areas need to be treated to allow evaluations of at least some vertebrate responses. Some harvest trials, such as at Plum Creek Timber Company's Cougar Ramp unit (Zielke 1993) and at several innovative retention harvest units created by the city of Seattle at the Cedar River Watershed have been very instructive. They have provided sites for initial studies of wildlife responses, tree mortality, regeneration and growth, and microclimate. However, most trials of retention harvest approaches to date do not qualify as statistically valid scientific experiments.

Replicated experiments are now being planned and implemented at several locations in the United States, Canada, and other countries. One of the more notable is a project known as DEMO (Demonstration of Ecosystem Management Options), subtitled "A Study of Green Tree Retention Patterns and Levels in Western Oregon and Washington" (USDA Forest Service 1996). DEMO involves six different treatments (Figure 7.7), each replicated at eight locations:

- 15 percent retention in dispersed pattern
- 15 percent retention in aggregated pattern
- 40 percent retention in dispersed pattern
- 40 percent retention in aggregated pattern
- 75 percent retention with harvest in small groups
- 100 percent retention for control

Response variables include small mammal, bird, and fungal populations; forest understory communities, including vascular plants and ground-layer cryptogams; understory response; regeneration and growth of tree regeneration; and growth and mortality of retained trees.

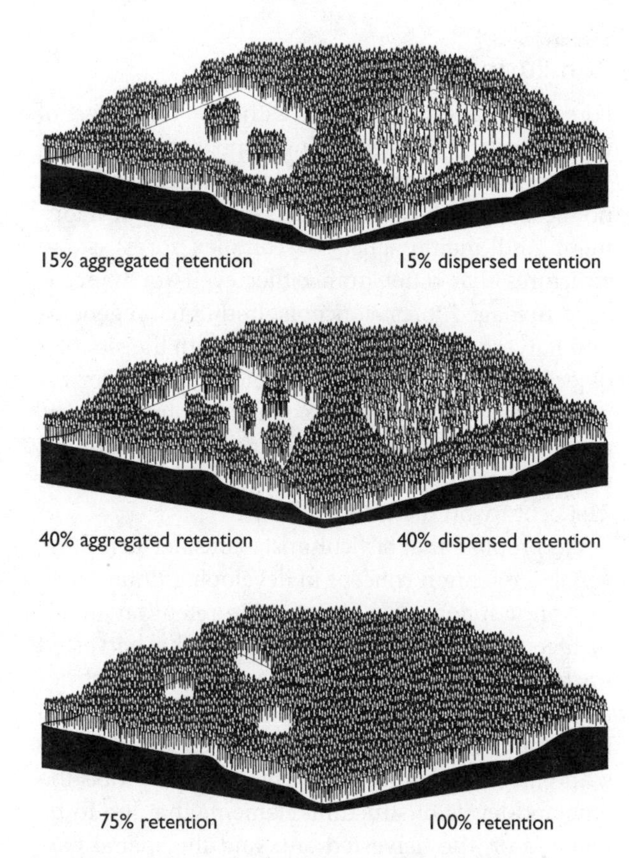

Figure 7.7 The primary focus of the DEMO (Demonstration of Ecosystem Management Options) experiment underway in Oregon and Washington is comparison of ecological and silvicultural responses to three different retention levels (15, 40, and 75 percent) and two retention patterns (dispersed and aggregated) replicated in eight locations. The six treatments are illustrated here.

Another example of an important large-scale harvest-cutting experiment is the Montane Alternative Silvicultural Systems (MASS) project being conducted on Vancouver Island, British Columbia (Arnott et al. 1995). This experiment is designed to address regeneration, wildlife habitat, and aesthetic concerns in managing forests at higher elevations, and it considers both biologic and economic issues. Prescriptions under study are small patch cuts, green tree retention, and shelterwood.

Conclusions

Forest managers are faced with the challenge of designing and implementing timber harvest prescriptions that address multiple ecological and economic objectives. These often call for the development and maintenance of complex forest stand structures that differ from either even- or uneven-aged managed forest concepts. Traditional regeneration harvest systems were designed with the singular objective of harvesting trees while providing for regeneration and growth of commercial tree species, and they do not readily accommodate the complex management objectives that will be typical of the 21st century on many forest lands.

We propose that silviculturists consider using the variable retention concept in developing timber harvest prescriptions. The objectives in retention are (1) to lifeboat species and processes on the harvested tract immediately following harvest, (2) to structurally enrich the subsequent forest stand, and (3) to improve connectivity in the managed forest landscape. In this approach, silviculturists prescribe the type and levels of structural elements that are to be retained on the harvested area and the spatial patterns for the retention.

Variable retention harvest prescriptions are emerging as a major strategy for integrating ecological and economic objectives throughout the temperate forest regions of the world. This approach is extremely flexible and provides benefits that cannot be achieved simply by using long rotations with traditional clearcutting.

Variable retention harvest is not simply a modification or adaptation of the traditional harvest systems—clearcut, seed tree, shelterwood, and selection. Variable retention harvest prescriptions typically focus on ecological objectives, such as maintaining biological diversity, as well as economic management objectives. Regeneration of commercial tree species or the rapid growth of this regeneration may not be primary objectives of such harvest prescriptions.

We feel that by stipulating the kind, amount, and distribution of retained structures, silviculturists can clearly communicate their objectives and proposed treatment. Confusion associated with efforts to adapt to old terminology, including such oxymorons as "clearcut with reserves," could thereby be avoided.

Acknowledgments

The Pew Charitable Trusts and The Nature Conservancy provided partial support for preparation of this chapter through a grant, Ecology and Management of Forests As Ecosystems, provided to the senior author. USDA Forest Service Pacific Northwest Forest and Range Experiment Station also provided significant support through ecosystem grants to the University of Washington.

Literature Cited

Adler, P. B. 1994. New forestry in practice: A survey of mortality in green tree retention harvest units, western Cascades, Oregon. B.A. thesis, Harvard University, Cambridge.

Arnott, J. T., W. J. Beese, A. K. Mitchell, and J. Peterson, eds. 1995. *Montane alternative silvicultural systems (MASS): Canada–British Columbia partnership agreement on forest resource development.* FRDA report 238. Victoria, BC: British Columbia Ministry of Forestry.

Berg, A., B. Ehnstrom, L. Gustafsson, T. Hallingback, M. Jonsell, and J. Weslien. 1995. Threat levels and threats to red-listed species in Swedish forests. *Conservation Biology* 9:1629–1633.

Berg, D. R. 1995. Forest harvest-setting design evaluation incorporating logger's preference. Ph.D. dissertation, University of Washington, Seattle.

Bormann, F. H., and G. E. Likens. 1979. *Pattern and process in a forested ecosystem.* New York: Springer-Verlag.

Brown, E. R., tech. ed. 1985. *Management of wildlife and fish habitats in forests of western Oregon and Washington.* Portland, OR: USDA Forest Service .

Carey, A. B. 1995. Sciurids in Pacific Northwest managed and old-growth forests. *Ecological Applications* 5:648–661.

Carey, A. B., and M. L. Johnson. 1995. Small mammals in managed, naturally young, and old-growth forests. *Ecological Applications* 5:336–352.

Carey, A. B., C. Elliott, B. R. Lippke, J. Sessions, C. J. Chambers, C. D. Oliver, J. F. Franklin, and M. J. Raphael. Submitted. A pragmatic ecological approach to small-landscape management: Final report of the biodiversity pathways working group of the Washington Landscape Management project.

Cascade Center for Ecosystem Management. 1995. *Residual trees as biological legacies.* Communique no. 2. Corvallis, OR: Department of Forest Science, Oregon State University.

Chen, J., J. F. Franklin, and T. A. Spies. 1992. Vegetation responses to edge environments in old-growth Douglas-fir forests. *Ecological Applications* 2:387–396.

———. 1993a. An empirical model for predicting diurnal air temperature gradients from edge into old-growth Douglas-fir forest. *Ecological Modeling* 67:179–198.

———. 1993b. Contrasting microclimates among clearcut, edge, and interior of old-growth Douglas-fir forest. *Agricultural and Forest Meteorology* 63:219–237.

———. 1994. Growing-season microclimatic gradients from clearcut edges into old-growth Douglas-fir forests. *Ecological Applications* 5(1):74–86.

Christensen, N. L., et al. 1989. Interpreting the Yellowstone fires of 1988. *BioScience* 39:678–685.

Ciancio, O., and S. Nocentini. 1994a. Gurnaud's control method and silviculture on natural basis: A forest management and silvicultural question. *L'Italia Forestale e Montana* 49(4):336–356.

———. 1994b. Problems and perspectives in forest management. *L 'Italia Forestale e Montana* 49(6):550–566.

Ciancio, O., F. Iovino, and S. Nocentini. 1994. The theory of the "normal forest."*L'Italia Forestale e Montana* 49(5): 446–462.

Curtis, R. O. 1994. *Some simulation estimates of mean annual increment of Douglas-fir: Results, limitations, and implications for management.* Research paper PNW-RP-471. Washington, DC: USDA Forest Service.

———. 1995. *Extended rotations and culmination age of coast Douglas-fir: Old studies speak to current issues.* Research paper PNW-RP-485. Washington, DC: USDA Forest Service.

Curtis, R. O., and D. D. Marshall. 1993. Douglas-fir rotations: Time for reappraisal? *Western Journal of Applied Forestry* 8(3):81–85.

FEMAT. 1993. *Forest ecosystem management: An ecological, economic, and social assessment.* Report of the Forest Ecosystem Management Assessment Team(FEMAT). 1993-793-071. Washington, DC: USDA Forest Service.

Foster, D. R., and E. R. Boose. 1992. Patterns of forest damage resulting from catastrophic wind in central New England, U.S.A. *Journal of Ecology* 80:79–98.

Franklin, J. F. 1963. *Natural regeneration of Douglas-fir and associated species using modified clear-cutting systems in the Oregon Cascades.* Research paper PNW-RP-3. Portland, OR: USDA Forest Service Pacific Northwest Forest and Range Experiment Station.

———. 1993. Preserving biodiversity: Species, ecosystems, or landscapes? *Ecological Applications* 3:202–205.

Franklin, J. F., and R. T. T. Forman. 1987. Creating landscape patterns by forest cutting: Ecological consequences and principles *Landscape Ecology* 1:5–18.

Franklin, J. F., J. A. MacMahon, F. J. Swanson, and J. R. Sedell. 1985. Ecosystem responses to the eruption of Mount St. Helens. *National Geographic Research* 1985: 196–215.

Franklin, J. F., H. H. Shugart, and M. E. Harmon. 1987. Tree death as an ecological process. *BioScience* 37:550–556.

Franklin, J. F., P. M. Frenzen, and F. J. Swanson. 1995. Recreation of ecosystems at Mount St. Helens: Contrasts in artificial and natural approaches. In *Rehabilitating damaged ecosystems,* 2nd ed., ed. J. Cairns, Jr. Boca Raton: Lewis Publishers.

Gratkowski, H. J. 1956. Windthrow around staggered settings in old-growth Douglas-fir. *Forest Science* 2:60–74.

Halpern, C. B., and T. A. Spies. 1995. Plant species diversity in natural and managed forests of the Pacific Northwest. *Ecological Applications* 5:913–934.

Hansen, A. J., and P. Hounihan. 1995. A test of ecological forestry: Canopy retention and avian diversity in the Oregon Cascades. In *Biodiversity in managed landscapes: Theory and practice,* ed. R. Szaro. London, England: Oxford University Press.

Hansen, A. J., S. L. Garman, J. F. Weigand, D. L. Urban, W. C. McComb, and M. G. Raphael. 1995a. Alternative silvicultural regimes in the Pacific Northwest: Simulations of ecological and economic effects. *Ecological Applications* 5:535–554.

Hansen, A. J., W. C. McComb, R. Vega, M. G. Raphael, and M. Hunter. 1995b. Bird habitat relationships in natural and managed forests in the West Cascades of Oregon. *Ecological Applications* 5:555–569.

Harmon, M. E., J. F. Franklin, F. J. Swanson, P. Sollins, S. V. Gregory, J. D. Lattin, N. H. Anderson, S. P. Cline, N. G. Aumen, J. R. Sedell, G. W. Lienkaemper, K. Cromack Jr., and K. W. Cummins. 1986. Ecology of coarse woody debris in temperate ecosystems. *Advances in Ecological Research* 15:133–302.

Harr, R. D. 1986. Effects of clearcutting on rain-on-snow

runoff in western Oregon: A new look at old studies. *Water Resources Bulletin* 22:1095–1100.

Harr, R. D., B. A. Coffin, and T. W. Cundy. 1989. *Effects of timber harvest on rain-on-snow runoff in the transient snow zone of the Washington Cascades: Interim final report submitted to Timber, Fish, and Wildlife(TFW) sediment, hydrology, and mass wasting steering committee for Project 18 (rain-on-snow).* Portland, OR: USDA Forest Service Pacific Northwest Forest and Range Experiment Station.

Henderson, J. A. 1994. The ecological consequences of long-rotation forestry. In *High quality forestry workshop: The idea of long rotations,* ed. J. F. Weigand, R. W. Haynes, and J. L. Mikowski. Special paper no. 15. Seattle: Center for International Trade in Forest Products, University of Washington.

Hunter, M. 1995. *Residual trees as biological legacies.* Management communique no. 2. Corvallis, OR: Cascade Center for Ecosystem.

Isaac, L. A. 1943. *Reproductive habits of Douglas-fir.* Washington, DC: Charles Lathrop Pack Forestry Foundation.

Keenan, R. J., and J. P. Kimmins. 1993. The ecological effects of clear-cutting. *Environmental Review* 1:121–144.

Knight, D. H., and L. L. Wallace. 1989. The Yellowstone fires: Issues in landscape ecology. *BioScience* 39:700–706.

Larsen, J. B. 1995. Ecological stability of forests and sustainable silviculture. *Forest Ecology and Management* 73:85–96.

Lindemayer, D. B., and J. F. Franklin. Submitted. Managing forest structure as part of ecologically sustainable temperate forestry: A Case study from Australia. *Conservation Biology.*

Lunney, D., ed. 1991. *Conservation of Australia's forest fauna.* Chipping Norton, Australia: Surrey Beatty and Sons.

Maser, C., R. F. Tarrant, J. M. Trappe, and J. F. Franklin, tech. eds. 1988. *From the forest to the sea: A story of fallen trees.* General technical report PNW-GTR-229. Washington, DC: USDA Forest Service.

Meehan, W. R., T. R. Merrell Jr., and T. A. Hanley. 1984. *Fish and wildlife relationships in old-growth forests.* Morehead City, NC: American Institute of Fishery Research Biologists.

National Board of Forestry Sweden. 1990. *A Richer Forest.* Jonkoping, Sweden.

North, M. 1993. Stand structure and truffle abundance associated with northern spotted owl habitat. Ph.D. thesis, University of Washington, Seattle.

Noss, R. F., and A. Y. Cooperrider. 1994. *Saving nature's legacy: Protecting and restoring biodiversity.* Washington, DC: Island Press.

Perry, D. A. 1994. *Forest ecosystems.* Baltimore, MD: Johns Hopkins University Press.

Peterson, C. J., and S. T. A. Pickett. 1995. Forest reorganization: A case study in an old-growth forest catastrophic blowdown. *Ecology* 76:763–774.

Rose, C. R. 1994. Relationships of green-tree retention following timber harvest to forest growth and species composition in the western Cascade Mountains. M.S. thesis, Oregon State University, Corvallis.

Schowalter, T. D. 1989. Canopy arthropod community structure and herbivory in old-growth and regenerating forests in western Oregon. *Canadian Journal of Forest Research* 19:318–322.

Schowalter, T. D. In press. Canopy arthropod communities in relation to forest age and alternative harvest practices in western Oregon. *Forest Ecology and Management.*

Schullery, P. 1989. The fires and fire policy. *BioScience* 39:686–694.

Scientific Panel for Sustainable Forest Practices in Clayoquot Sound. 1995. *Sustainable ecosystem management in Clayoquot Sound planning and practices.* Victoria, BC: Cortex Consultants Inc.

Shaw, D., R. Edmonds, W. Littke, J. Browning, K. Russell, and C. Driver. 1994. Influence of forest management on annosus root disease in coastal western hemlock, Washington state, USA. In *Proceedings of eighth meeting of IUFRO working party* s2.06.01, Root and Butt Rots, August 9–16, 1993, Sweden and Finland.

Sillett, S. C. 1995. Canopy epiphyte studies in the central Oregon Cascades: Implications for the management of Douglas-fir forests. Ph.D. dissertation, Oregon State University, Corvallis.

Smith, D. M. 1986. *The practice of silviculture.* New York: John Wiley & Sons.

Squire, R. O. 1990. *Report on the progress of the Silvicultural Systems Project: July 1986–June 1989.* Melbourne, Australia: Victoria Department of Conservation and Environment.

Swank, W. T., and D. A. Crossley Jr., eds. 1988. *Forest hydrology and ecology at Coweeta.* New York: Springer-Verlag.

Swanson, F. J., and J. F. Franklin. 1992. New forestry principles from ecosystem analysis of Pacific Northwest forests. *Ecological Applications* 2:262–274.

Thomas, J. W., tech. ed. 1979. *Wildlife habitats in managed forests: The Blue Mountains of Oregon and Washington.* Agricultural Handbook 553. Washington, DC: USDA Forest Service.

USDA Forest Service. 1996. DEMO (demonstration of

ecosystem management options): A study of green tree retention levels and patterns in western Oregon and Washington. General study plan. Olympia, WA: USDA Forest Service Pacific Northwest Research Station.

USDA and USDI. 1994a. *Final supplemental environmental impact statement on management of habitat for late successional and old-growth forest–related species within the range of the northern spotted owl.* Vol. 2, appendices. Portland, OR: USDA Forest Service.

———. 1994b. *Record of decision for amendments to Forest Service and Bureau of Land Management planning documents within the range of the northern spotted owl.* Portland, OR: USDA Forest Service.

Vega, R. 1993. Bird communities in managed conifer stands in the Oregon Cascades: Habitat associations and nest predation. Ph.D. thesis, Oregon State University, Corvallis.

Verner, J., K. S. McKelvey, B. R. Noon, R. J. Gutierrez, G. I. Gould Jr., and T. W. Beck. 1992. *The California spotted owl: A technical assessment of its current status.* General technical report PSW-GTR-133. Washington, DC: USDA Forest Service.

Walker, L. R., N.V. L. Brokaw, D. J. Lodge, and R. B. Waide. 1991. Ecosystem, plant, and animal responses to hurricanes in the Caribbean. *Biotropica* 23:313–521.

Washington State Department of Natural Resources. 1992. *Guidelines for selecting reserve trees.* Olympia, WA.

Watanabe, S., and S. Sasaki. 1993. The silvicultural management system in temperate and boreal forests: A case history of the Hokkaido Tokyo University Forest. *Canadian Journal of Forest Resources* 24:1176–1185.

Weigand, J. F., and A. L. Burditt. 1992. *Economic implications for management of structural retention on harvest units at the Blue River Ranger District, Willamette National Forest, Oregon.* Research note PNW-RN-510. Washington, DC: USDA Forest Service.

Weigand, J. F., R. W. Haynes, and J. L. Wikowski, eds. 1994. *High-quality forestry workshop: The idea of long rotations.* Special paper no. 15. Seattle: Center for International Trade in Forest Products, University of Washington.

Williamson, R. L. 1973. *Results of shelterwood harvesting of Douglas-fir in the Cascades of Western Oregon.* Research paper PNW-RP-161. Washington, DC: USDA Forest Service.

Zenner, E. K. 1995. Effects of residual trees on growth of young to mature Douglas-fir and western hemlock in the western central Oregon Cascades. M.S. thesis, Oregon State University, Corvallis.

Zielke, K. 1993. *Environmental forestry: Plum Creek Timber Company's approach to forest management.* Victoria, BC, Canada: British Columbia Ministry of Forests.

8 Shaping Stand Development Through Silvicultural Practices

Dean S. DeBell, Robert O. Curtis, Constance A. Harrington, and John C. Tappeiner

Social Expectations and Silvicultural Opportunities 142
Silvicultural Practices 143
- Early Density Control 143
- Thinning in Older Stands 143
- Nutrient Management 145
- Pruning 146
- Managing Dead Wood 146

Implementation 147
Literature Cited 147

Regeneration systems and harvest cutting patterns, including retention of live and dead trees, determine the nature of a forest after harvest and influence early growth (see Tappeiner et al., Chapter 9, and Franklin et al., Chapter 7). Subsequent silvicultural treatments are the tools for channeling stand development to meet specific or multiple objectives, which may change over the lifetime of the stand.

During the past half century, silvicultural practices such as spacing control, fertilization, and pruning have been applied extensively throughout the world wherever soils, climate, social policies, and markets combine to favor and sustain forestry enterprises. Research, development, and implementation of these practices were justified primarily on the basis of quantity or quality of wood that could be grown or harvested. The basic knowledge and experience acquired, however, can be used to meet additional objectives. Expansion of social desires and concerns for forest values and products may provide the justification for further silvicultural investments.

In this chapter, we first provide some background

on changing societal expectations and their influence on silvicultural opportunities. Next, we describe several silvicultural practices, how conventional application influences stand development, and how they might be applied or modified to meet other or additional objectives. Finally, we discuss considerations related to implementation throughout broader forest landscapes.

Social Expectations and Silvicultural Opportunities

Public expectations and support for the multiple benefits of forests have been kindled for decades through forestry advertisements by both public and private organizations. Increased attention to such values is now demanded by various segments of society and required by public law.

Although managed forest landscapes appear more natural and more diverse than landscapes associated with most human endeavors, extensive areas of young managed plantations do contrast markedly with natural mature and old-growth forests. Many of the features associated with older natural forests (Franklin et al. 1981) are minimized with conventional management. Such features may include large trees, snags, down woody debris, ragged edges, within-stand structural complexity, and diversity in age, size, shape, and distribution stand patterns across the landscape (Hansen et al. 1991). Many of these features contribute to wildlife habitat (Thomas 1979, Hunter 1990) or are otherwise considered related or essential to other forest values and general forest health. The desire to retain some of these features in managed forests has stimulated interest in various modifications of conventional silvicultural practice (Franklin et al. 1986, DeBell 1989).

Existing young natural stands (or plantations) will not necessarily develop into stands comparable to present old growth in the absence of human intervention. Our present younger natural stands were established after severe fires (or logging and fire) in the 19th and early 20th centuries. They were established and are developing under considerably different and milder climatic conditions than did present old growth that originated in the more severe climate of the Little Ice Age (Henderson and Brubaker 1986).

The general nature of silvicultural practices needed to foster features similar to those of natural old growth is fairly apparent—although the specifics of implementation are not. It may be obvious, for example, that a given feature provides important habitat for one or more species. But little is known about the response of the forest ecosystem in general—or even of specific species—to various levels or distribution patterns of that feature. Even less is known about other biological or economic costs associated with such management. The answers will vary with forest type, landscape, and specific stand and site conditions. These uncertainties should not be used to justify reluctance or failure to modify conventional practices, however. But they should make us wary of widespread implementation or legislation requiring specific untested modifications, and they should be a stimulus for experimenting with a *range* of approaches in a designed and controlled manner.

The greatest near-term opportunity to develop such management knowledge and experience and to provide more diverse habitat lies in millions of acres of existing forest plantations, which range in age from 1 year to 40 or 50 years. Plantations dominate the landscape on industrial lands, and scattered smaller plantations are a major feature of public lands. Most in the Pacific Northwest were planted as pure Douglas-fir. There are, however, some mixed-species plantings—and many plantations acquire a considerable mixture of other conifers, hardwoods, and shrubs through natural seeding or sprouting.

Plantations are sometimes disparaged as uniform monocultures of minimal or even negative value for purposes other than wood production. This superficial and shortsighted view arises in part from the fact that many plantations in the Pacific Northwest are now in the early stem exclusion stage. In this stage, stands are relatively uniform, and dense canopies have shaded out understory vegetation. Even without further management, this uniformity is gradually modified by natural processes and by agents such as root diseases and snow breakage. With management, stand differentiation will occur much more rapidly. Most of these plantations are now in a very plastic stage of development, and there are major opportunities to mold them—individually and in a landscape context—toward a variety of objectives.

Silvicultural Practices

Objectives for forest lands will vary by ownership, geographic location, and condition of surrounding stands and landscapes. Continued pursuit of rapidly growing uniform stands focused primarily on wood production will be a major or sole objective of some plantations. Accelerated development of habitat for threatened or endangered species may be a primary objective for others. Judicious modification of silvicultural practices can produce managed forest landscapes that meet a wide range of objectives. A number of modifications of conventional silvicultural practices merit consideration, and some are now being applied by several organizations.

Early Density Control

Conventional early (pre-commercial) thinning is widely applied near the end of the stand initiation stage to enhance the survival, growth, and value of residual trees. Thinning specifications usually are aimed at leaving the most valuable larger trees at relatively even spacing (Reukema 1975). This increases stand uniformity, but the reduced stand density also accelerates tree growth and promotes development of a shrub and herbaceous understory. It frequently leads to early establishment of tolerant tree species such as western hemlock and western red cedar.

Modifications could involve selection criteria for residual trees, spacing distances, and intentional creation of small openings. Trees to be left after thinning could be selected to increase size and species diversity, thus accelerating stand differentiation and increasing structural complexity. Spacing could be widened to enhance development of understories. Moreover, spacing might be varied in patches throughout the plantations; small openings or gaps might be created to retain some components of the early initial stage and ultimately develop patches of younger trees. The root rot diseases common throughout the Douglas-fir region often produce such openings in abundance as stands grow older, even without intervention.

Thinning in Older Stands

Thinning in older stands has long been a generally accepted practice in European countries, and it rests on over a century of experience and research embodied in an enormous literature. Yet, it has not been widely applied in the Pacific Northwest until very recently because markets for small material were poor and large volumes of old timber were available. But the economic and social context has changed, and thinning is assuming increasing importance. A considerable body of research on thinning practices dates back to about 1950 (Worthington and Staebler 1961, Reukema 1972, Reukema and Pienaar 1973, Reukema and Bruce 1977, Oliver et al. 1986, King 1986, Curtis and Marshall 1986). Thinning combined with extended rotations can maintain forest cover for long periods while still producing wood products.

The traditional purposes of thinning are to maintain growth rates of residual trees and promote stem quality and vigor during the stem exclusion stage. Over time, thinning produces larger trees and visually more attractive stands (Figures 8.1a and 8.1b). One recent study in the Pacific Northwest (Marshall et al. 1992) showed that over a 20-year period, thinning in a young stand produced increases of 33 to 56 percent in diameter growth of the largest 80 stems per acre compared to the unthinned condition. Thinning also provides income and timber flow during the intermediate stages of stand development. Thinning may or may not promote vertical stratification, depending on how it is done. It usually accelerates understory development and succession and movement of the stand into the understory re-initiation stage. If begun early, changes can be striking over comparatively short time periods (Figures 8.2a and 8.2b)—although on some sites, thinning can also produce dense shrub layers that inhibit establishment of desired conifers.

Thinning encourages seedling establishment of conifers (Del Rio and Berg 1979), hardwoods (Fried et al. 1988), and shrubs (Tappeiner and Zasada 1993, Huffman et al. 1994, O'Dea et al. 1995), but also results in vegetative expansion of shrubs by rhizomes

(Tappeiner et al. 1991) and layering. By reducing overstory density and providing a seedbed, thinning results in invasion of plants not previously in the stand and the vegetative spread of those already established. This generally produces a dense, diverse understory of shrubs, herbs, and tree seedlings and saplings.

Regeneration of understory conifers after thinning will often enhance the development of stands with structures similar to old-growth stands. On some sites, however, dense understories of salal, Oregon grape, salmonberry, vine maple, and other shrubs

(a)

(b)

Figure 8.1 (a) Unthinned portion of an 80-year-old Douglas-fir stand in the Black Rock Forest Management Research Area near Fall City, Oregon. (b) Nearby portion of the same 80-year-old stand, heavily thinned at about age 45.

(a)

(b)

Figure 8.2 (a) Unthinned area in 45-year-old plantation at Iron Creek study of levels of growing stock near Randle, Washington. (b) Adjacent area in same 45-year-old plantation, with early and repeated thinning.

may develop and prevent establishment of a second layer of conifers. Conifer seeds will germinate under such shrub layers, but seedlings may not survive. These shrub layers are quite persistent. They produce new aerial stems (ramets) annually that replace older stems as they die, maintaining a dense cover (Huffman et al. 1994). Thus, on some sites (particularly in the Oregon Coast Range), some disturbance of these brush layers and possibly underplanting of conifers may be necessary to establish multilayered conifer stands.

Modifications might include favoring trees of diverse species and sizes to foster crown stratification and understory development. Small openings or gaps could be created, which ultimately will be occupied by younger trees. Root rots often do this independent of human intervention. Additional species can be introduced through underplanting, if seed source of desired species is lacking. An example of such modification is irregular thinning in the Forest Ecosystem Study now being conducted in Fort Lewis, Washington (Carey 1993). As stands grow older, some trees will die and provide the snags needed for wildlife. If these are judged insufficient, additional snags can be created by intentionally killing or topping selected trees.

A less obvious but important effect of repeated thinning is that trees and stands maintain rapid growth to older ages than without thinning. Culmination age for repeatedly thinned Douglas-fir in the Pacific Northwest has yet to be determined—but it is greater than that for dense unthinned stands (Worthington and Staebler 1961, Curtis 1994, 1995). If terrain and harvesting methods permit thinning, larger trees can be grown on longer rotations with no loss and perhaps with even an increase in wood production per unit area. Stem quality and value per unit of volume may be increased. Repeated thinnings and longer rotations may also provide additional opportunities and flexibility to capture benefits from stands composed of mixed species with differing tolerances and growth patterns.

Nutrient Management

Nutrient management is an important consideration, particularly in the Douglas-fir region where soils are young and considerable nitrogen is immobilized in organic matter. Inadequate supplies of available nitrogen limit natural productivity and rates of stand development on many sites. Relatively small and infrequent applications of nitrogen fertilizers have been used to increase wood production in conventionally managed forests. These applications often produce striking responses in tree growth and stand development (Chappell et al. 1992a.). The largest and most long-lasting responses occur when nutrient deficiencies are severe (Miller and Tarrant 1983) and when fertilizer application is combined with thinning (Chappell et al. 1992b).

In newly established and very young plantations, broadcast application of fertilizer results in earlier canopy closure and greater stand uniformity (DeBell et al. 1986); in older stands where trees have begun to differentiate into crown classes, fertilizer application accelerates the stand development process of crown differentiation and competition-related mortality (Miller and Pienaar 1973, Miller and Tarrant 1983). Other properties, including soil organic matter, may be enhanced, and site resources such as amount and nutritional value of wildlife forage (Sullivan and Rochelle 1992) may be increased.

A modification of conventional practice that could be considered is small-scale individual tree or group application. This could be used to promote development of a layered canopy, in contrast to broadcast applications, which simply increase the overall rate of stand development. It also could be used to enhance the growth and nutritional value of forage in openings created for that purpose. Individual tree application was deemed practicable and profitable more than 15 years ago by scientists at the Washington Department of Natural Resources (Anderson and Hyatt 1979). Coupled with selective thinning and pruning, individual tree applications could also increase production of high-quality wood.

Introduction or favoring of nitrogen-fixing plants also merits consideration. This may provide benefits similar to those of fertilization on some nitrogen-deficient sites (Miller and Murray 1978, Binkley 1983, Tarrant et al. 1983). In addition, it may have additional benefits in terms of economic and ecological diversity. Selected red alder trees are now favored in early thinnings of young conifer plantations on nitrogen-deficient soils of the Siuslaw National Forest (Turpin 1981).

Pruning

Pruning has been advocated and used primarily to increase the amount of clear wood produced in young stands. It also has been recommended as part of a program to control blister rust in western white and sugar pines (Russell 1988). Most pruning for improved wood quality has been done during the stem exclusion stage. Current practice is to begin pruning much earlier than was common in the past.

If pruning were begun prior to stand closure, removal of lower branches could increase the amount of light reaching the forest floor and favor development of the woody and herbaceous understory. Other modifications could include altering the number of trees pruned and the portion of the crown that is removed. In regimes with repeated thinnings on relatively long rotations, removal of some pruned trees during thinning could increase financial returns and perhaps offset costs of other silvicultural activities.

Managing Dead Wood

In stands managed for wood production, dead trees have been viewed primarily as problems to be avoided or minimized: as such they often are harvested before they die or deteriorate. Thus, spacing guidelines are based on the number of trees that can be grown to some desired size prior to the onset of significant competition-related mortality. Trees killed by insects, diseases, and fire are commonly harvested immediately if economics and accessibility permit. Yet dying, dead, and down trees are important components of many forest ecosystems, and a variety of organisms are associated with them. These organisms range from cavity-dwellers like bears, squirrels, and woodpeckers to amphibians and invertebrates to diverse vascular and nonvascular plants and microorganisms. Consideration should therefore be given to dead wood management in multipurpose forests, particularly those where conservation of biodiversity is a primary management objective.

Plans for management of snags and down wood must consider numbers, sizes, species, and dynamics of decay and persistence. Current ideas regarding desired snag numbers are speculative, but based on assumptions derived from basic information on woodpecker populations and snag use. Forest managers and wildlife biologists from nearly every region have suggested that an average of 5 to 10 snags per hectare (two to four snags per acre) is adequate (Hunter 1990). Generally, there is a minimum size (diameter and height) suitable for each species, but larger is always better because more species will make greater use of the snag. Biologists seem to prefer slowly decaying species because they will persist longer. However, both hard and soft snags are needed. New snags of suitable characteristics must be recruited periodically to replace those that have fallen and are providing values associated with down wood. Lacking information to the contrary, managers can assume that management efforts that provide adequate snag populations will also result over time in adequate amounts of down wood.

Three matters specific to dead wood management should be considered in silvicultural efforts to shape stand development and vegetative habitat:

1. Protection or retention of standing dead trees (snags) and down trees that have developed naturally. At least some and perhaps all of these should be retained during silvicultural operations where they are consistent with management objectives and safety considerations.

2. Creation of snags and down trees in stands where they are scarce or absent due to past human or natural disturbances. Trees can be killed by girdling or injection with herbicides—yet the biological effectiveness of such methods is uncertain because decay proceeds from the outside in rather than from the inside out. Some biologists and managers have tried other techniques, including blowing off tops with explosives, cutting off tops with saws, inoculating trees with fungi, and attracting beetles with sex pheromones (Bull and Partridge 1986, Conner 1983). Operational and safety considerations may make it more feasible to create and preserve snags within small uncut groups of trees, rather than distributed over an area. Because of the importance of snags to wildlife, and the high economic cost of devoting large trees to such purposes, it is critical that uncertainties be resolved and methods for snag creation be developed

that are both biologically effective and economically acceptable.

3. Creation of cavities. In general, retention of existing snags, perhaps supplemented with snag creation, will provide cavities sufficient to meet wildlife objectives. In specific instances, however, installation of nest boxes (e.g., for wood ducks and bluebirds) or creation of cavities by den-routing (Carey and Sanderson 1981) or by cutting a hole with a chainsaw and covering it with a faceplate (Carey and Gill 1983) may be appropriate. These approaches currently are being evaluated for effects on flying squirrel populations in young-growth Douglas-fir forests (Carey 1993).

In general, many standard silvicultural practices, such as thinning, fertilizing, and pruning, can be applied to favor the growth and development of trees of desired characteristics (size, species, form, age) in specific locations. With appropriate rotation lengths (see Curtis, Chapter 10, and Franklin et al., Chapter 7) and an understanding of the dynamics of snag populations, it should be relatively easy to provide standing dead and down wood in amounts and locations consistent with forest management objectives.

Implementation

Selecting and applying silvicultural measures requires managers to identify specific objectives. What is the desired future stand condition? What are the relative values to the landowner and to society of the forest outputs involved—wood, wildlife (what kinds of wildlife?), water, aesthetics. And what are the associated costs?

Decisions also require that options be considered in the context of the surrounding landscape, rather than on a stand-by-stand basis. Will productivity for any of these values be markedly influenced by treatment of the stand? Can a particular treatment regime enhance or provide conditions that are needed and now lacking in the larger unit? Applicable measures and reasonable objectives will differ with stand type.

It must also be recognized that there is much uncertainty, both in objectives and current judgments, of treatment effects on values other than wood production. Management objectives of a few years ago were markedly different from those of today, and those a few years hence will differ from current ideas. For example, it is generally recognized that some snags and down wood are important, but we do not know how much is necessary or how it should be distributed in different forest landscapes, and we will not know without long-term experimentation and monitoring. Incomplete knowledge and uncertainty about the future are facts of life.

Despite uncertainties, we have a large body of knowledge based on research and experience showing that stand characteristics and rates of stand development can be markedly altered within relatively short time periods through silvicultural treatment—for whatever future objective may be selected.

Literature Cited

Anderson, H. W., and M. Hyatt. 1979. Feasibility of hand application of urea to forest land in western Washington. In *Proceedings of Forest Fertilization Conference,* eds. S. P. Gessel, R. M. Kenady, and W. A. Atkinson. Institute of Forest Resources contribution no. 40. Seattle: University of Washington, College of Forest Resources.

Binkley, D. 1983. Ecosystem production in Douglas fir plantations: Interaction of red alder and site fertility. *Forest Ecology and Management* 5:215–227.

Bull, E. L., and A. D. Partridge. 1986. Methods of killing trees for use by cavity nesters. *Wildlife Society Bulletin* 14:142–146.

Carey, A. B. 1993. *The Forest Ecosystem Study: Experimental manipulation of managed stands to provide habitat for spotted owls and to enhance plant and animal diversity: A summary and background for the interagency experiment at Fort Lewis, Washington.* Olympia, WA: Forestry Sciences Laboratory.

Carey, A. B., and J. D. Gill. 1983. Direct habitat improvements: Some recent advances. In *Snag habitat management: Proceedings of a symposium,* tech. coord. J. W. Davis, G. A. Goodwin, and R. A. Ockerfells. General technical report RM-99. Washington, DC: USDA Forest Service.

Carey, A. B., and H. R. Sanderson. 1981. Routing to accelerate tree-cavity formation. *Wildlife Society Bulletin* 9:14–21.

Chappell, H. N., G. F. Weetman, and R. E. Miller, eds. 1992a. *Forest fertilization: Sustaining and improving nutrition and growth of western forests.* Institute of Forest Resources contribution no. 73. Seattle: University of Washington, College of Forest Resources.

Chappell, H. N., S. A. Y. Omule, and S. D. Gessel. 1992b. Fertilization in coastal northwest forests: Using response information in developing stand-level tactics. In *Forest fertilization: Sustaining and improving nutrition and growth of western forests,* ed. H. N. Chappell, G. F. Weetman, and R. E. Miller. Institute of Forest Resources contribution no. 73. Seattle: University of Washington, College of Forest Resources.

Conner, R. N., J. G. Dickson, and J. H. Williamson. 1983. Potential woodpecker nest trees through artificial inoculation with heart rots. In *Snag habitat management: Proceedings of a symposium,* tech. coords. J. W. Davis, G. A. Goodwin, and R. A. Ockerfells. General technical report RM-99. Washington, DC: USDA Forest Service.

Curtis, R. O. 1994. *Some simulation estimates of mean annual increment of Douglas fir: Results, limitations, and implications for management.* Research paper PNW-RP-471. Portland, OR: USDA Forest Service.

———. 1995. *Extended rotations and culmination age of coast Douglas fir: Old studies speak to current issues.* Research paper PNW-RP-485. Portland: USDA Forest Service .

Curtis, R. O., and D. M. Marshall. 1986. *Levels-of-growing-stock cooperative study in Douglas fir.* Report no. 8, *The LOGS study: Twenty-year results.* Research paper PNW-PNW-356. Portland, OR: USDA Forest Service.

DeBell, D. S. 1989. Alternative silvicultural systems—West, a perspective from the Douglas-fir region. In *Silvicultural challenges and opportunities in the 1990s,* Proceedings of the national silviculture workshop, Petersburg, AK, July 10–13, 1989. Washington, DC: USDA Forest Service.

DeBell, D. S., R. S. Silen, M. A. Radwan, and N. L. Mandel. 1986. Effect of family and nitrogen fertilizer on growth and foliar nutrients of Douglas fir saplings. *Forest Science* 32(3):643–652.

Del Rio, E., and A. Berg. 1979. *Growth of Douglas fir reproduction in the shade of a managed forest.* Corvallis, OR: Forest Research Lab, Oregon State University.

Franklin, J. F., K. Cromack Jr., W. Denison, A. McKee, C. Maser, J. Sedell, F. Swanson, and G. Juday. 1981. *Ecological characteristics of old-growth Douglas fir forests.* General technical report PNW-GTR-118. Portland, OR: USDA Forest Service.

Franklin, J. F., T. Spies, D. Perry, M. Harmon, and A. McKee. 1986. Modifying Douglas fir management regimes for nontimber objectives. In *Douglas-fir: Stand management for the future,* ed. C. D. Oliver, and J. A. Johnson. Institute of Forest Resources contribution no. 55. Seattle: University of Washington, College of Forest Resources.

Fried, J., J. Tappeiner, and D. Hibbs. 1988. Bigleaf maple seedling establishment and early growth in Douglas fir forests. *Canadian Journal of Forest Research* 18:1226–1233.

Hansen, A. J., T. A. Spies, F. J. Swanson, and J. L. Ohman. 1991. Conserving biodiversity in managed forests: Lessons from natural forests. *BioScience* 41(6):382–392.

Henderson, J., and L. Brubaker. 1986. Response of Douglas fir to long-term variations in precipitation and temperature in western Washington. In *Douglas fir: Stand management for the future,* ed. C. D. Oliver, D. P. Hanley, and J. A. Johnson. Institute of Forest Resources contribution no. 55. Seattle: University of Washington, College of Forest Resources.

Huffman, D. W., J. C. Tappeiner, and J. C. Zasada. 1994. Regeneration of salal in the central Coast Range forests of Oregon. *Canadian Journal of Botany* 72:39–51.

Hunter, M. L., Jr. 1990. *Wildlife, forests, and forestry: Principles for managing forests for biological diversity.* Englewood Cliffs, NJ: Prentice-Hall.

King, J. E. 1986. Review of Douglas fir thinning trials. In *Douglas fir: Stand management for the future,* ed. C. D. Oliver, D. P. Hanley, and J. A. Johnson. Institute of Forest Resources contribution no. 55. Seattle: University of Washington, College of Forest Resources.

Marshall, D. D., J. F. Bell, and J. C. Tappeiner. 1992. *Levels-of-growing-stock cooperative study in Douglas fir.* Report no. 10, *The Hoskins study,* 1963–83. Research paper PNW-RP-448. Portland, OR: USDA Forest Service.

Miller, R. E., and M. D. Murray. 1978. The effects of red alder on growth of Douglas fir. In *Utilization and management of alder,* ed. D. G. Briggs, D. S. DeBell, and W. A. Atkinson. General technical report PNW-GTR-70. Portland, OR: USDA Forest Service.

Miller, R. E., and L. V. Pienaar. 1973. *Seven-year response of 35-year-old Douglas fir to nitrogen fertilizer.* Research paper PNW-RP-165. Portland, OR: USDA Forest Service.

Miller, R. E., and R. F. Tarrant. 1983. Long-term growth response of Douglas fir to ammonium nitrate fertilizer. *Forest Science* 29:127–137.

O'Dea, M., J. C. Zasada, and J. C. Tappeniener. 1995. Vine maple clone growth and reproduction in managed and unmanaged coastal Douglas fir forests. *Ecological Applications* 5:63–73.

Oliver, C. D., K. L. O'Hara, G. McFadden, and I. Nagame.

1986. Concepts of thinning regimes. In *Douglas fir: Stand management for the future,* ed. C. D. Oliver, D. P. Hanley, and J. A. Johnson. Institute of Forest Resources contribution no. 55. Seattle: University of Washington, College of Forest Resources.

Reukema, D. L. 1972. *Twenty-one-year development of Douglas fir stands repeatedly thinned at varying intervals.* Research paper RP-PNW-141. Portland, OR: USDA Forest Service.

———. 1975. *Guidelines for precommercial thinning of Douglas fir.* General technical report PNW-30. Portland, OR: USDA Forest Service.

Reukema, D. L., and D. Bruce. 1977. *Effects of thinning on yield of Douglas fir: Concepts and some estimates obtained by simulation.* General technical report GTR-PNW-58. Portland, OR: USDA Forest Service.

Reukema, D. L., and L. V. Pienaar. 1973. *Yields with and without commercial thinnings in a high-site-quality Douglas fir stand.* Research paper RP-PNW-155. Portland, OR: USDA Forest Service.

Russell, K. W. 1988. Management of western white pine in nurseries and plantations to reduce white pine blister rust. In *Proceedings at Western Forestry and Conservation Association Annual Meeting, Dec. 4–7, 1988.* Reprint available from Washington State DNR Forest Health, Box 47048, Olympia, WA 98504-7048.

Sullivan, T. P., and J. A. Rochelle. 1992. Forest fertilization and wildlife. In *Forest fertilization: Sustaining and improving nutrition and growth of western forests,* ed. H. N. Chappell, G. F. Weetman, and R. E. Miller. Institute of Forest Resources contribution no. 73. Seattle: University of Washington, College of Forest Resources.

Tappeiner, J. C., and J. C. Zasada. 1993. Establishment of salmonberry, salal, vine maple, and bigleaf maple seedlings in the coastal forests of Oregon. *Canadian Journal of Forest Research* 23(9):1775–1780.

Tappeiner, J. C., J. C. Zasada, P. Ryan, and M. Newton. 1991. Salmonberry clonal and population structure: The basis for a persistent cover. *Ecology* 72:609–618.

Tarrant, R. F., B. T. Bormann, D. S. DeBell, and W. A. Atkinson. 1983. Managing red alder in the Douglas fir region: Some possibilities. *Journal of Forestry* 81:787–792.

Thomas, J. W., tech. ed. 1979. *Wildlife habitats in managed forests: The Blue Mountains of Oregon and Washington.* Agriculture handbook 553. Washington, DC: USDA Forest Service.

Turpin, T. C. 1981. Managing red alder on the Siuslaw National Forest. In *Proceedings of a National Silviculture Workshop: Hardwood Management.* Washington, DC: USDA Forest Service.

Worthington, N. P., and G. R. Staebler. 1961. *Commercial thinning of Douglas fir in the Pacific Northwest.* Technical bulletin no. 1230. Washington, DC: USDA Forest Service.

9 Silvicultural Systems and Regeneration Methods: Current Practices and New Alternatives

John C. Tappeiner, Denis Lavender, Jack Walstad, Robert O. Curtis, and Dean S. DeBell

Development of Regeneration Practices 151
Early Logging Practices and Regeneration Methods 152
Studies of Natural Regeneration Processes 153
Artificial Conifer Regeneration 154
Producing Stands of Diverse Structures and Habitats 156
At Harvest 156
Young Stand Establishment 158
Regeneration in Young to Mature Stands 158
Natural Succession 159
Conclusion 160
Literature Cited 160

In this chapter, we discuss silvicultural systems and regeneration methods to meet the needs of society over the next several decades. We begin with a brief history of silvicultural systems and what we have learned about forest regeneration in the Pacific Northwest. We then discuss how regeneration methods might evolve over the next several decades.

We believe that the practice of silviculture generally will be applied to two different forest management philosophies and objectives, providing (1) old-forest characteristics and (2) wood production. Significant amounts of wood can be produced under objective 1, and considerable wildlife habitat and other values can be provided while producing relatively high yields of wood under objective 2. The knowledge and skills are available to pursue both objectives effectively. Moreover, many owners will likely manage their forests under both approaches.

Development of Regeneration Practices

Traditional methods for regenerating forests as part of a timber harvest fall into two broad categories: (1)

even-age management systems, which include clearcutting, shelterwood, and seed-tree methods, and (2) uneven-age systems, which include single-tree and group selection methods. As part of these methods, regeneration can be obtained by natural seeding or planting, by release of advanced regeneration (i.e., seedlings established in the previous stand), or by coppice from sprouting tree species.

These methods all have been used successfully in western North America, and all will have their place in future forest management. They are the foundation upon which we will build new strategies to meet society's desire for sustaining forests with old-growth characteristics as well as its demand for wood.

Early Logging Practices and Regeneration Methods

Generally, early logging in the late 1800s and early 1900s was done to harvest high-quality commercial timber at the least cost with little concern for reforestation, protection of soil or water, or provisions for aquatic wildlife habitat. Unmerchantable trees were left standing, logged areas often were burned, and cutting frequently began at the bottom of a watershed and continued to the uplands until an entire basin was logged. Ironically, this pattern of cutting may have more closely mimicked natural disturbance by large fires than the staggered-setting, dispersed clearcutting approach that followed. Natural regeneration of woody plants following early logging or intense fires readily occurred because mineral soil was exposed and seed was available from residual trees or adjoining stands. Regeneration, however, varied from well-stocked, vigorous young conifer stands to dense stands of red alder or sprouting hardwoods to dense covers of shrubs with occasional conifers. A high priority for early research was to provide methods for consistently regenerating forests after fires and timber harvests.

The early efforts at developing silvicultural systems in the Pacific Northwest included both clearcutting (Hoffman 1924, Isaac 1956) and partial cutting (Kirkland and Brandstrom 1936). Kirkland and Brandstrom proposed a method for managing Douglas-fir and hemlock forests that partitioned the forest into relatively small tracts that were planned for timber yield, logging systems, and regeneration. Their ideas drew heavily on European experience and called for intensive practices and detailed stand analyses. The skills and techniques needed to implement this system were probably unrealistic for those times; however, it was used on federal lands. Sales were designed to remove less than 35 percent of the volume per stand (Munger 1950). Artificial regeneration and tending of conifer seedlings to ensure adequate conifer regeneration in unstocked parts of the stand were not part of this partial cutting system.

Debate over the use of partial cutting for Douglas-fir was quite lively (Munger 1950, Smith 1970). However, its use ended in the late 1940s. Munger (1950) reported that all new Bureau of Land Management and Forest Service timber sales called for clearcutting, except for the use of the shelterwood regeneration method and some selective cutting in southwestern Oregon.

Evaluation of partial cutting practices and studies of natural regeneration helped establish the use of clearcutting. Isaac (1956) evaluated a series of partially cut old stands 5 to 10 years after cutting. Because of residual tree damage and mortality, windthrow, change in species composition from shade intolerant to tolerant species, and lack of Douglas-fir regeneration, he recommended abandoning widespread use of this system. He acknowledged that not enough time had elapsed to determine if uneven-age management would eventually work on his study sites, but felt that in all probability there would be loss of growth and Douglas-fir stocking would be reduced. He suggested that partial-cutting or uneven-age management might be appropriate for drier sites in southwestern Oregon, gravel soils of the Puget Sound region, and severe southerly exposures elsewhere in the region (where moisture and shade are critical factors). Smith (1970) suggested that the uneven-age system did not work in this region because "theoretical ecological considerations were not verified, thereby making the system inappropriate for the future."

In retrospect, use of partial-cutting, uneven-age management was probably discontinued for several reasons:

- It was difficult and probably not appropriate to implement a single policy or approach over such a broad range of forest stand conditions.

- Inadequate attention was given to creating environments and making use of treatments that would regenerate Douglas-fir and other conifer species.
- Insufficient thought was given to leaving vigorous, undamaged trees to provide adequate growing stock.
- Logging planning and technology were inadequate to implement Kirkland and Brandstrom's (1936) ideas, especially on steep slopes.
- Unfortunately, partial-cutting, uneven-age management ended abruptly. Its long-term use, even on some sites, could have provided useful information to design other silvicultural systems.

Studies of Natural Regeneration Processes

Although some of the very early work on reproduction of western conifers was with planted seedlings (Munger 1911), most of it focused on natural regeneration, in both undisturbed forests and clearcuts. Hoffman (1924) and Isaac (1930, 1955) studied the dispersal of seed and the possibility of storage of conifer seed in the forest floor. Later work by Isaac (1938, 1940, 1943) focused on determining the environmental variables that control natural regeneration and identifying microsites that favor seedling establishment. For example, Lavender (1958) studied the effects of seeding date and ground cover on the germination and survival of Douglas-fir in the Tillamook burn. Hooven (1958) studied the effects of rodents and other predators on seed supply. Hermann and Chilcote (1965) simultaneously studied the effects of seed bed, shade, and insect predation on conifer seedling establishment, while Christie and Mack (1984) and Harmon and Franklin (1989) compared dead wood and mineral soil as a substrate for hemlock regeneration. Parallel studies were underway throughout the West (Haig 1936, Haig et al. 1941, Dunning 1923), including the classic work by Pearson (1923) in Arizona.

Similar work was done on the regeneration of hardwoods (Tappeiner et al. 1986; Fried et al. 1988; Haeussler and Tappeiner 1993, 1995; Tappeiner and Zasada 1993). Regeneration of forest shrubs was studied by Gratkowski (1961), Zavitkowski and Newton (1968), Hughes et al. (1987), Tappeiner and Zasada (1993), Huffman et al. (1994), and O'Dea et al. (1995). These studies provide information on forest plant autecology and regeneration and help define the regeneration niche (Grubb 1977). They provide a basis for understanding the response to silvicultural practices.

Evaluation of Natural Regeneration Practices

Studies of natural regeneration were integrated with evaluation of reforestation projects. This work evaluated applied regeneration practices, identified problem sites, and helped to develop alternative regeneration practices (Roeser 1924). Lavender et al. (1956) examined natural regeneration on staggered settings, and Franklin (1963) assessed natural regeneration on strip cuts, small patch cuts (0.25 to 4 acres), and staggered cuttings. Considerable work has been done on regeneration using the shelterwood method (Tesch and Mann 1991; Laacke and Fiddler 1986; McDonald 1983; Seidel 1983; Laacke and Tomascheski 1986; McDonald 1976b; Gordon 1970, 1979; Williamson 1973). McDonald (1976a) studied natural regeneration in the Sierra Nevada of mixed conifers in all five principal regeneration methods, ranging from single-tree selection to clearcutting. Minore (1978) and Stein (1981, 1986) examined regeneration results following harvesting on sites considered difficult to regenerate in southwestern Oregon. For example, Minore (1978) found that shelter prevented forest damage to seedlings and saplings on the Dead Indian Plateau in southwestern Oregon, and both Minore (1978) and Stein (1981) and Williamson and Minore (1978) pointed out the value of advanced natural regeneration and natural seeding among planted seedlings on sites with extreme variation in temperatures, such as those described by Holbo and Childs (1987) or on rocky soils that are difficult to plant.

Effectiveness of Advanced Regeneration

Use of advanced regeneration, established naturally prior to logging, is a very effective way to regenerate forest stands and should be a common practice on sites that are difficult to regenerate and in uneven-age systems (Minore 1978, Stein 1981). Helms and Standiford (1985), Oliver (1986), Gordon (1973), and Tesch and Korpela (1993) developed methods for assessing potential vigor of advanced regeneration following logging. Tesch et al. (1993) found that damaged seedlings often recovered within three to six years. Growth of advanced regeneration was similar to that of planted seedlings (Korpela and Tesch 1992).

The following generalizations emerge from studies of seedling establishment and regeneration methods:

- Seed production is variable, often with six to eight years or more between adequate seed crops for some species.
- Seed predation rates of Douglas-fir, ponderosa pine, and hardwoods such as bigleaf maple and tanoak are high both on the tree and on the forest floor.
- Mortality rates during the first two to three years following germination are high. Causes are high soil temperatures, pathogens and insects in the forest floor, competition for light and soil water, litterfall, and frost.
- Douglas-fir, red alder, ponderosa pine, and true fir seedling survival is usually highest on bare mineral soil; spruce, hemlock, and large-seeded hardwoods survive well on both mineral soil and organic seed beds.
- Moderate shade often aids seedling survival—even for intolerant species— because during the summer it reduces soil temperatures and the evaporative capacity and temperature of the air near the ground, while in the winter and early spring it reduces the chance of frost.
- Shady conditions that foster early seedling survival in the understory of forest stands do not always favor later growth. Seedlings of most species tend to be shade tolerant when very young, but less so as they grow older and larger.
- Intense competition from established populations of grasses, shrubs, and herbs may cause high rates of natural seedling mortality.
- Advanced regeneration can successfully regenerate a new stand after logging, especially on hard-to-regenerate sites. It may have to be augmented by seeding or planting.
- Group selection and single-tree selection methods generally favor shade-tolerant species such as true firs, western hemlock, western red cedar, tanoak, bigleaf maple, and Douglas-fir on dry and warm sites in southern Oregon and northern California.
- Shelterwood methods and small openings or clearcuts (less than 10 acres) appear to be most suitable for the regeneration of true firs at high elevations (Gratkowski 1958, Gordon 1970, 1973, 1979), although they will work for other species (Worthington 1953).
- Natural reproduction is often patchy and variable. It may maintain a sparse forest cover, but does not ensure desired species composition, stocking, and distribution or timely stand reestablishment.

Artificial Conifer Regeneration

Need for Artificial Regeneration Practices

Studies on planting western conifers began in the early 1900s (Munger 1911) and Show (1929). Reforestation of the Cispus and Yacolt burns led to establishment of the Wind River Nursery and extensive planting programs. Also, the decision to use clearcutting on federal lands (Munger 1950) and the Tillamook burn (a series of fires in 1933, 1939, 1945, and 1951 that burned over 460,000 acres and created the area now known as the Tillamook Forest) stimulated artificial reforestation work in the Pacific Northwest. The Columbus Day storm of 1962 was also a catalyst for artificial regeneration efforts. Large acreages of windthrown timber were salvaged, and the areas were planted. Prior to the Tillamook burn, the emphasis had been on either natural regeneration or direct seeding. Both of these methodologies were strongly limited by seed predation by small mammals (Hooven 1958, 1970; Schubert and Adams 1971) and by the lack of seed trees following the intensive fire.

Early in the reforestation of the Tillamook burn, the emphasis was upon direct seeding with annual projects of 10,000 to 15,000 acres that were seeded with Douglas-fir seed treated with rodenticides such as endrin. Such treatments were only partially successful in their goal of preventing seed predation. Also, direct seeding was limited by seed supplies and location of large contiguous areas suitable for aerial seeding. Direct seeding accounted for about 50 percent of the burn reforested, but as reforestation progressed, there was an increasing emphasis on planting.

State forest practices acts in Washington, Oregon, and California also affected reforestation practices. The Oregon State Forest Practices Act (1941), which required leaving seed trees or planting, was changed in 1971 to require that reforestation efforts on clearcuts must begin within 12 months, planting must be completed within two planting seasons, and at least 200 trees per acre must be "free to grow" within five growing seasons of planting.

Producing Planting Stock

Tree mortality was often high in many of the early plantations, and research was done to increase seedling survival and growth. It had two major facets: (1) studies of seedling size and morphology (Iverson 1984, Jenkinson 1980) and (2) studies of seedling physiology with particular reference upon the effects of nursery practices such as lifting date, storage, and fertilization upon seedling vigor (Lavender 1964; Hermann 1967; Lavender et al. 1968; Lavender and Hermann 1970; Lavender and Wareing 1972; Hermann et al. 1972; Lavender 1984, 1985, 1988, 1990a, 1990b; Ritchie 1984).

As a result of this research, methods for producing high-quality seedlings are available (Duryea and Dougherty 1991, Margolis and Brand 1990). Now, foresters are able to prescribe stock types best suited to various microsites on harvest areas, the average survival of seedlings has increased to 85 percent or better, and there is a physiological basis for understanding seedling growth potential and stress from planting or planting site conditions.

Managing Young Plantations

Competition also affects regeneration success. Consequently, an extensive program of research (Walstad and Kuch 1987) was designed to evaluate effects of competition and to develop methods for controlling it when needed. Control of grasses, forbs, and shrubs is generally used to ensure adequate soil moisture for seedling survival and growth. On moist sites, competition for light by tall shrubs and hardwoods is generally of more concern.

Hardwood Plantations

The role of hardwoods—ecologically and economically—has received increasing recognition during the past 20 years. Red alder, black cottonwood, and hybrid poplars are now intensively managed. All are rapid-growing, shade-intolerant pioneer trees that require bare or new soil to regenerate naturally. Natural stands of alder and cottonwood commonly occur along streams and other areas where soil has been exposed and moisture conditions are favorable. Thus, plantations are established on moist sites or are irrigated.

The primary objective of pure red alder plantings is raw material for solid wood products or fine papers. Mixed plantings are established mostly for enhancing site productivity (through nitrogen fixation) and forest biodiversity. Also, red alder is immune to root rots that affect conifers and is often planted in areas severely infected with *Phelinus.* Although competing vegetation must be controlled, trees can be planted successfully without the extensive exposure of bare soil needed for natural seeding.

Hybrid poplar plantations have been established on marginal agricultural land along the lower Columbia and on irrigated, sagebrush steppe land in the Columbia basin of eastern Washington and Oregon. These plantations are managed similarly to agricultural crops and are harvested about five to seven years after planting.

Genetics

For several decades basic and applied research in forest tree genetics has been an important part of artificial regeneration. The programs have two major phases: (1) the identification of large numbers of site-adapted trees from breeding zones and (2) planting or grafting these genotypes in progeny test sites and seed orchards. Programs are designed to increase yield while maintaining the genetic variability to ensure long-term stability of artificially regenerated forest stands (Hermann and Lavender 1968, Campbell 1979, Silen 1982).

Guidelines for Artificial Regeneration

Successful artificial regeneration requires careful attention to the details of seed source and nursery and planting practices, as well as a thorough evaluation of environmental conditions (Hobbs et al. 1992). Microclimate, competition from herbs, shrubs, and hardwoods, and animal browsing affect seedling survival and growth.

Cafferata (1986) has provided an excellent overview of the application of current reforestation practices. Guidelines for conifers (DeYoe 1986, Strothman and Roy 1984, Schubert and Adams 1971) and hardwoods (Ahrens et al. 1992) are available. In summary, capable professionals and technicians must be involved at all stages, including the following:

- Proper seed source and seed handling
- Nursery procedures that optimize seedling root regeneration potential, and storage and handling procedures in the nursery and field that minimize seedling dehydration and respiration
- Careful planting, including onsite inspection of planting procedures
- Site-specific prescriptions for site preparation and for weed and pest control
- Monitoring for three or more years to ensure seedling survival and growth
- Early thinning to control stocking and species composition

The ability to regenerate forests is demonstrated in the annual reforestation reports of the Oregon State Board of Forestry for 1992. It shows that of the 85,689 acres requiring reforestation by the end of 1992, 82,034, or 96 percent, were in compliance.

Producing Stands of Diverse Structures and Habitats

Habitat and biodiversity goals can be stated best in terms of forest stand structure and species composition. Stand structure includes the vertical and horizontal arrangement of trees, shrubs, herbs, grasses, and nonvascular plants, as well as such things as snags, down logs, and forest floor depth. Thus many, but not all, components of stand structure are affected by or can be produced by silvicultural practices.

There are several periods in the life of a forest stand during which its structure and composition can be altered by silvicultural practices (Table 9.1). Below we present examples of some possible stand structures, and practices to produce them, that should be fairly easy to implement given today's technology.

At Harvest

Retaining Trees and Wood

Retaining large trees, snags, and down logs in some ways mimics the results of natural disturbance by fire or wind agents (Spies and Franklin 1991). Natural disturbances usually do not kill all the trees in a stand; they do, however, produce large pieces of dead wood in the form of snags or logs lying on the forest floor. Regeneration can be accomplished by planting, by use of advanced reproduction, or by natural seeding, providing proper seed trees are left. The method must be determined site-by-site. In a recent study on MacDonald Forest near Corvallis, Oregon, natural regeneration of Douglas-fir was plentiful when 10 to 12 large trees per acre were left after logging (Ketchum 1995).

Retention of large trees—especially those with large limbs and cavities—as well as large snags and logs will help ensure that a stand with diverse structure develops after harvest. This is very similar to the irregular shelterwood method described by Smith (1986). After regeneration is established, shelterwood trees are retained to produce large overstory trees and a multilayered stand. Leaving groups of large trees (small patches) rather than scattered individuals may be easier from a logging and reforestation standpoint. Also, leaving undisturbed groups of trees may allow some plants and organisms in the forest floor to survive from one stand to the next. Surveys of plant communities in areas that have been clearcut and burned indicate that most herbaceous plants that grow in old forests are adapted to disturbance (Franklin and Dyrness 1971, Dyrness 1973). They also are commonly found in clearcuts five or more years of age.

Small Openings

Making small openings, as in group shelterwood or group selection methods (Smith 1986), is similar to small-scale disturbance by wind, insects, or root dis-

Table 9.1 Methods of producing mixed-species stands

Harvest Considerations	Stand Establishment	Managing Young to Mid-Age Stands	Introducing Conifers in Riparian Zones
• Thin and/or defer for longer rotation • Retain green trees in groups or singly—especially trees with large limbs, cavities, or broken tops • Retain snags and logs on forest floor • Save advanced regeneration of seedlings, saplings, poles • Use irregular shelterwoods to produce two-story stands • Regenerate a stand over time using group selection, group shelterwoods, strip shelterwoods • Protect carryover of herbaceous and shrubby plants • Protect within-stand variability, such as seeps and rock outcrops	• Plant mixed species at varying spacings • Save advanced regeneration of seedlings, saplings, poles • Leave parts of stand undisturbed by site preparation or slash disposal • Vary treatments to consider within-stand variability such as seeps, rock outcrops, etc. • Save patches of shrubs and hardwoods to increase future stand variability • Encourage establishment of natural seedlings among planted ones by leaving seed trees	• Thin to produce or maintain large trees with deep crowns • Thin around hardwoods to encourage mast production • Make snags and large logs • Release advanced tree and shrub regeneration by thinning overstory trees • Protect within-stand variation—lichens, rock outcrops • Under-plant with shade-tolerant species	• Concentrate along certain reaches of streams • Release advanced conifer regeneration • Use large planting stock • Consider all variables that affect conifer establishment (browsing, flooding, and overstory and understory competition) • Avoid frequently flooded sites • Manage riparian zones in conjunction with the upland part of the stand • Use intensive site preparation in small, strategic areas

Source: 1992 reforestation accomplishment report, Oregon Department of Forestry.

ease. Experience over four years in regenerating small openings (0.5 acre) on MacDonald Forest indicates that the same reforestation methods used in larger clearcuts are applicable to small openings. Growth and survival in the openings was not different from that of the clearcuts (Ketchum 1995). This trend may not continue unless openings are widened. Just as with clearcuts, animal browsing and shrub competition affect seedling survival and growth.

There is a great deal of variability among openings: Some with high light intensity developed covers of grass or low shrubs; others with low light levels became dominated by tall shrubs. Based on fourth-year results, grand fir appears to be better suited to regeneration in small openings than Douglas-fir because it is browsed much less and is more shade tolerant.

The size of an opening, its aspect, and the height of surrounding trees are all likely to affect regeneration success. On north aspects or where surrounding trees are tall, widening openings or thinning around them may be necessary to increase light and growing space for the young conifers. Natural Douglas-fir regeneration was not plentiful in these small openings, most likely due to lack of soil disturbance. It was plentiful in adjoining stands where 10 to 12 trees per acre were left.

In this study we used only 0.5-acre openings for

experimental purposes. In practice, larger openings or a variety of opening sizes might be more appropriate for biological, administrative, or economic reasons.

Young Stand Establishment

Use of Advanced Reproduction and Mixed Species Planting

Current methods of planting and tending young stands can be altered to produce stands of diverse structures and species composition after fire, harvesting, or other disturbances. Poles, saplings, and seedlings (from the previous stand) will help accelerate the regeneration of the next stand. In addition, they will probably encourage more patchy stands and a variety of tree sizes and species (Tesch and Korpela 1993).

Mixed species planting will produce multilayered stands because of differential species growth patterns. For example, Douglas-fir and western red cedar planted together may form a two-story stand. Cedar grows slower than Douglas-fir, it is more susceptible to browsing, and it's shade tolerance enables it to survive in the understory.

Mixed stands of red alder and conifers have been shown to be more productive than pure conifer stands in soils with low nitrogen levels (Tarrant and Miller 1963, Tarrant 1961). These mixed species stands potentially can benefit some wildlife species. Professors William Emmingham and Denis Lavender have established mixed plantations of these species at the research forest of Oregon State University's College of Forestry. Their purpose was to use alder's nitrogen-fixing ability to increase Douglas-fir growth and to produce more diverse tree and herbaceous layers than might occur in pure Douglas-fir. Because of alder's rapid juvenile height growth rate and its ability to overtop Douglas-fir within three to four years after planting, alder was planted when the Douglas-fir were over 15 feet tall. If alder is to be used only as a source of nitrogen, Tarrant and Miller (1963) suggest that an off-site alder seed source might be used so that frost damage would keep it from overtopping the Douglas-fir.

Shrub and Hardwood Management

Managing shrub and hardwood density at the time of regeneration will affect the species composition and structure of the next stand. The models developed by Harrington et al. (1991a, 1991b) for tanoak and Pacific madrone and by Knowe et al. (1995) for bigleaf maple estimate the amount of cover produced by sprouting hardwoods, the effects on conifer survival and growth rates, and the effects on the stocking of understory shrubs and herbs. Such models will help forest managers forecast the development and influence of hardwoods during early stages of stand establishment.

Shade-tolerant hardwoods like tanoak and bigleaf maple can be managed in groups to produce a second layer in parts of the new stand while not shading out understory shrubs or herbs or substantially reducing the growth of conifer regeneration in the rest of the stand. Typically, they are overtopped by conifers at about 40 to 50 years of age. Large overtopped hardwoods provide cavities as large branches die and decay.

On coastal sites, red alder natural regeneration often is abundant in conifer plantations. Mixed alder-conifer stands could be established by spacing alder during precommercial thinning. Like tanoak and maple, alder's early height growth is much greater than that of new conifers. Therefore, it would have to be spaced to enable conifers to grow among it. Unlike other hardwoods, alder is not likely to survive beneath conifer stands.

Regeneration in Young to Mature Stands

Thinning and Tree Regeneration

In the Douglas-fir region there are many well-stocked young stands (10–50+ years of age) that were established following fire or timber harvest. About 30 to 60 percent of most watersheds on federal land are stocked with these young stands. For the most part, they have been regenerated and managed at high densities (150 to 200+ trees per acre) to produce wood, not to develop diverse structures. In contrast, old-growth stands often have only 10 to 30 trees per acre (Spies and Franklin 1991). Thus, to help develop old-growth characteristics in these younger stands, considerable reduction in stocking is needed to produce large trees with deep crowns and provide a more open environment for understory development. Seedlings can be established under higher overstory densities (100–150 trees per acre), but canopy densities need to be reduced to ensure understory development.

Thinning and regulation of overstory density can produce large trees quickly, develop stand structure, and generally aid the development of old-growth characteristics (Newton and Cole 1987, Curtis and Marshall 1993). In addition, thinning to improve wood yields can release advanced conifer and hardwood regeneration in the understory and ultimately produce multilayered stands. There are often numerous hardwood and conifer seedlings (Tappeiner and McDonald 1984, Fried et al. 1988) in the understory of stands 50 or more years of age that will respond to a reduction in overstory density. In dense stands with no tree understory, increased light and some soil disturbance favor establishment of both conifer seedlings (Del Rio and Berg 1979) and hardwood seedlings (Fried et al. 1988, Tappeiner et al. 1986).

In a western Oregon study that compared understory characteristics in thinned and unthinned Douglas-fir stands, one of the most striking differences between the stands was the stocking of natural conifer seedlings in the understory of the thinned stands (J. D. Bailey, personal communication, 1996). Additional thinning to mimic natural stand development could release conifers and hardwoods and leave the overstory at variable densities to encourage patchy understory development.

Shade-tolerant conifers can be planted in the understory following thinning to develop multilayered stands. Professor Alan Berg at Oregon State University thinned a 40-year-old Douglas-fir stand to 50 trees per acre and planted western hemlock in the understory. Now, approximately 40 years later, there is a well-developed two-storied stand (Curtis and Marshall 1992, 1993). At 80 years of age, the overstory of 50 trees per acre is probably too dense for continued understory growth. In this example, the average diameter of the Douglas-fir trees is about 30 inches. In the unthinned stand, tree diameters average 15 inches, and there is practically no understory development.

Pure red alder stands are common in riparian areas and on sites with northerly exposure, especially in coastal forests. However, some conifer component often is desirable in these stands to provide large logs for stream channel structure and to produce a more diverse forest for wildlife. Emmingham et al. (1989) successfully regenerated western hemlock under thinned red alder stands. Both overstory density reduction and intensive salmonberry control were needed to establish hemlock. Release of advanced conifer regeneration from red alder also can be used to grow large conifers in some riparian areas.

Shrub Regeneration

Thinning also will favor regeneration of shrub understories. Salal (Huffman et al. 1994) and salmonberry (Tappeiner et al. 1991) clonal development and rhizome extension increase with reduction of overstory density. Vine maple clones are spread by "layering" as a result of thinning. Slash from natural disturbance of the overstory or from commercial thinning pins the vine maple crowns to the forest floor where the branches often root and form a dense understory of new sprouts (O'Dea et al. 1995). Establishment of salal, vine maple, and salmonberry seedlings also is favored by thinning—however, their rate of expansion and the development of a dense cover probably is slower than that of vegetative clonal expansion. (Huffman et al. 1994, Tappeiner and Zasada 1993, Tappeiner et al. 1991). Because of the potential for rapid clonal expansion of shrubs, there may be a relatively narrow window for establishment of new plants by natural seeding or planting without the need for vegetation control.

Natural Succession

Our studies of the ecology and development of forest stands and our observations of forests in the Coast Range of Oregon lead us to the hypothesis that many young stands in these forests will not naturally develop cohorts of multistoried conifers that are considered to be typical of old growth (Spies and Franklin 1991). Stands in the western hemlock zone often have well-developed understories of shrubs (e.g., salal and vine maple) and little conifer regeneration in the understory. Studies of natural regeneration suggest that these shrub layers will continue to prevent the establishment of conifers. Similarly, alder stands that were established on many acres following logging frequently have well-developed understories of salmonberry, sword fern, and elderberry (Carlton 1988, Henderson 1970). As the short-lived alder dies, it is likely that many of these stands will be dominated by a dense cover of salmonberry that may persist and prevent the establishment of conifers for many decades.

Thus we believe that in many cases natural succession in today's forests will not produce the same kinds of stands and habitats as it has in the past. Reasons for this may be a combination of the following:

- The lack of fire in these forests over the past 75 years or more has resulted in development of dense shrub understories.
- Exotic species have become established.
- Logging has changed species composition and seed supply and has favored the development of shrubs and hardwoods with the potential for vigorous sprouting.
- Stands that are established after logging are often more heavily and uniformly stocked with Douglas-fir than the original natural stands.
- Climate or weather patterns are different today than they were when the present old-growth forests developed.

Consequently, treatment to facilitate understory conifer establishment and reduce shrub density and overstocking of the conifers in the overstory is likely to benefit the development of old-forest characteristics on many sites.

Conclusion

Continued research and practical experience are critical to the successful implementation of the stand management treatments suggested above. Key issues requiring further investigation include the following:

- Information on the reproduction and growth of hardwoods, conifers, and shrubs in the understory of conifer stands, as well as the effects of understory and overstory density on other components of the ecosystem
- Use of different types of stand structures by wildlife species—for example, use of snags, wood on the forest floor, and shrub and tree understory layers
- Growth and development of mixed species stands and old stands over 100 years old
- Practicality of implementing these different types of treatments
- Landscape evaluations of a range of silvicultural systems and treatments (through space and time) to evaluate their effect on wildlife populations
- Effects of stand density on insects, pathogens, windthrow, etc., on the morphology and structure of conifer trees.

The art and science of silviculture is continuing to evolve. Fortunately, there is a good foundation of research information and practical experience on which to build new practices for the future. The success of silvicultural systems and regeneration methods will depend to a large extent upon public perception and acceptance. Increasingly, societal pressures, not always based on reliable information, constrain the use of practices that would yield positive long-term results in terms of habitat, wood production, and biodiversity. Therefore, forest managers and researchers need to involve the public in the development of alternative silvicultural systems.

Literature Cited

Ahrens, G. R., A. Dobkowski, and D. E. Hibbs. 1992. *Red alder guidelines for successful regeneration.* Special publication 24. Corvallis, OR: Forest Research Lab, Oregon State University.

Cafferata, S. L. 1986. Douglas-fir stand establishment overview: Western Oregon and Washington. In *Douglas-fir stand management for the future,* ed. C. D. Oliver, D. P. Hanley, and J. A. Johnson. Seattle: College of Forest Resources, University of Washington.

Campbell, R. K. 1979. Genecology of Douglas-fir in a watershed in the Oregon Cascades. *Ecology* 60(5):1036–1050.

Carlton, G. C. 1988. The structure and dynamics of red alder communities in the central coast range of western Oregon. M.S. thesis, Oregon State University, Corvallis.

Christie, J. E., and R. N. Mack. 1984. Variation in demography of juvenile *Tsuga heterophylla* across the substrate mosaic. *Journal of Ecology* 72:75–91.

Curtis, R. O., and D. D. Marshall. 1992. A new look at an old question: Douglas-fir culmination age. *Western Journal of Applied Forestry* 7:97–99.

———. 1993. Douglas-fir rotations: Time for reexamination. *Western Journal of Applied Forestry* 8:81–85.

Daniel, T. W., J. A. Helms, and F. S. Baker. 1979. *Principles of silviculture.* New York: McGraw-Hill.

Del Rio, E., and A. Berg. 1979. A growth of Douglas-fir reproduction in the shade of a managed forest. Research paper 40. Corvallis, OR: Forest Research Lab, Oregon State University.

DeYoe, D. R. 1986. Guidelines for handling seeds and seedlings to ensure vigorous stock. Special publication 13. Corvallis, OR: Oregon State University.

Dunning, D. 1923. Some aspects of cutting in the Sierra Nevada forests of California. USDA bulletin 1176. Washington, DC: USDA.

Duryea, M. L., and P.M. Dougherty. 1991. *Forest regeneration manual.* Hingham, MA: Kluwer Academic Publishers.

Dyrness, C. T. 1973. Early stages of plant succession following logging and burning in the western Cascades of Oregon. *Ecology* 54(1):57–69.

Emmingham, W. H., M. Bondi, and D. E. Hibbs. 1989. Underplanting western hemlock in a red alder thinning: Early survival, growth, and damage. *New Forest* 3:31–43.

Franklin, J. F. 1963. *Natural regeneration of Douglas-fir and associated species using modified clear-cutting systems in the Oregon Cascades.* Research paper PNW-RP-3. Portland, OR: PNW Research Station, USDA Forest Service Pacific Northwest Forest and Range Experiment Station.

Franklin, J. F., and C. T. Dyrness. 1971. *A checklist of vascular plants on the H. J. Andrews Experimental Forest.* Research paper PNW-RN-138. Portland, OR: USDA Forest Service.

Franklin, J. F., K. Cromack Jr., W. Denison, A. McKee, C. Muser, J. Sedell, F. Swanson, and G. Juday. 1981. *Ecological characteristics of old-growth forests.* General technology report PNW-118. Washington, DC: USDA Forest Service.

Fried, J. J., J. C. Tappeiner, and D. Hibbs. 1988. Bigleaf maple seedling establishment and early growth in Douglas-fir forests. *Canadian Journal of Forest Research* 18:1226–1233.

Gordon, D. T. 1970. *Natural regeneration of white and red fir: Growth, damage, mortality.* Research paper PSW-RP-58. Berkeley, CA: USDA Forest Service.

———. 1973. *Released advance reproduction of white and red fir: Growth, damage, mortality.* Research paper PSW-RP-95. Berkeley, CA: USDA Forest Service.

———. 1979. *Successful natural regeneration cuttings in California true firs.* Research paper PSW-RP-140. Berkeley, CA: PSW Research Station, USDA Forest Service.

Gratkowski, H. J. 1958. Natural reproduction of Shasta red fir on clear cuttings in southwestern Oregon. *Northwest Science* 32(1):9–18.

———. 1961. Brush seedlings after controlled burning of brushlands in southwestern Oregon. *Journal of Forestry* 59:885–888.

Grubb, P. J. 1977. The maintenance of species richness in plant communities: The importance of regeneration niches. *Biology Review* 52:107–145.

Haeussler, S., and J. C. Tappeiner. 1993. Effect of light environment on seed germination of red alder. *Canadian Journal of Forest Research* 23:1487–1491.

———. 1995. Germination and first-year survival of red alder seedlings in the central Oregon Coast Range. *Canadian Journal of Forest Research* 25:1639–1651.

Haig, I. T. 1936. *Factors controlling initial establishment of western white pine and associated species.* Bulletin 41. New Haven, CT: Yale University School of Forestry.

Haig, I. T., K. R. Davis, and R. H. Weidman. 1941. *Natural regeneration in the western white pine type.* Technical bulletin 767. Washington, DC: USDA.

Hallin, W. E. 1959. *The application of unit area control in the management of ponderosa Jeffrey Pine at Black Mountain Experimental Forest.* Technical bulletin 1191. Washington, DC: USDA.

Harmon, M. E., and J. E. Franklin. 1989. Tree seedlings on logs in *Picea tsuga* forests of Washington and Oregon. *Ecology* 70:48–59.

Harrington, T. B., J. C. Tappeiner, and T. F. Hughes. 1991a. *Planning with PSME: A growth model for young Douglas-fir and hardwood stands in southwestern Oregon.* Special publication 21. Corvallis, OR: Forest Research Lab, Oregon State University.

———. 1991b. Predicting average growth and size distributions of Douglas-fir saplings competing with sprouts of tanoak or Pacific madrone. *New Forests* 5:109–130.

Helgerson, O. T., K. A. Wearstler Jr., and W. K. Bruckner. 1982. *Survival of natural and planted seedlings under a shelterwood in southwest Oregon.* Research note 69. Corvallis, OR: Oregon State University.

Helms, J. A., and R. B. Standiford. 1985. Predicting release of advance reproduction of mixed conifer species in California following overstory removal. *Forest Science* 31(1): 3–15.

Henderson, J. A. 1970. Biomass and composition of the understory vegetation in some *Alonus ruvia* stands in Western Oregon. M.S. thesis, Oregon State University, Corvallis.

Hermann, R. K. 1967. Seasonal variation in sensitivity of Douglas-fir seedlings to exposure of roots. *Forest Science* 13:140–149.

Hermann, R. K., and W. W. Chilcote. 1965. *Effect of seedbeds on germination and survival of Douglas-fir.* Research paper 4. Corvallis, OR: Oregon State University.

Hermann, R. K., and D. P. Lavender. 1968. Early growth of Douglas-fir from various altitudes and aspects in southern Oregon. *Silvae Genetica* 17(4):141–153.

Hermann, R. K., D. P. Lavender, and J. B. Zaerr. 1972. *Lifting and storing western conifer seedlings.* RP 17. Corvallis, OR: Forest Research Lab, Oregon State University.

Hobbs, S. D., S. D. Tesch, P. W. Owston, R. E. Stewart, J. C. Tappeiner, and G. E. Wells, eds. 1992. *Reforestation practices in southwestern Oregon and northern California.* Corvallis, OR: Forest Research Lab, Oregon State University.

Hoffman, J. V. 1924. *Natural regeneration of Douglas-firs in the Pacific Northwest.* Bulletin 1200. Washington, DC: USDA.

Holbo, H. R., and S. W. Childs. 1987. Summertime radiation balances of clearcut and shelterwood slopes in southwest Oregon. *Forest Science* 33(2):504–516.

Hooven, E. 1958. *Deer mouse and reforestation in the Tillamook burn.* Research note 37. Corvallis: Oregon Forest Lands Research Center.

Hooven, E. F. 1970. Animal damage to seeds and seedlings. In *Regeneration of ponderosa pine,* ed. R. K. Hermann. Corvallis, OR: Oregon State University.

Huffman, D. W., J. C. Tappeiner, and J. C. Zasada. 1994. Regeneration of salal in the central coast range forests of Oregon. *Canadian Journal of Botany* 72:39–51.

Hughes, T. F., C. R. Latt, J. C. Tappeiner, and M. Newton. 1987. Biomass and leaf area estimates for varnishleaf ceanothus, deer brush, and white leaf manzanita. *Western Journal of Applied Forestry* 2:124–128.

Isaac, L. A. 1930. Seed flight in the Douglas-fir region. *Journal of Forestry* 28:492–299.

———. 1938. *Factors affecting the establishment of Douglas-fir seedlings.* Circular 486. Washington, DC: USDA.

———. 1940. Vegetative succession following logging in the Douglas-fir region, with special reference to fire. *Journal of Forestry* 38:716–721.

———. 1943. *Reproductive habits of Douglas-fir.* Washington, DC: Lathrop Pack Forestry Foundation.

———. 1955. Where do we stand with Douglas-fir natural regeneration research? In *Proceedings of the Society of American Foresters Meeting.* Washington, DC: Society of American Foresters.

———. 1956. *Place of partial cutting in old-growth stands of the Douglas-fir region.* Portland, OR: PNW Research Station, USDA Forest Service.

Iverson, R. D. 1984. Planting-stock selection: Meeting biological needs and operations realities. In *Forest nursery manual: Production of bareroot seedlings,* ed. M. L. Duryea and T. D. Landis. The Hague, The Netherlands: Martinus Nijhoff/Dr. W. Junk.

Jenkinson, J. L. 1980. *Improving plantation establishment by optimizing growth capacity and planting time of western yellow pines.* Research paper PSW-154. Berkeley, CA: USDA Forest Service.

Ketchum, J. S. 1995. Douglas-fir, grand fir, and plant community regeneration in three silvicultural systems in Western Oregon. M. S. thesis, Oregon State University, Corvallis.

Kirkland, B. P., and A. J. F. Brandstrom. 1936. *Selective timber management in the Douglas-fir region.* Washington, DC: USDA Forest Service.

Knowe, S. A., B. D. Carrier, and A. Dobkowski. 1995. Effects of bigleaf maple sprout clumps on diameter and height growth of Douglas-fir. *Western Journal of Applied Forestry* 10:5–11.

Korpela, E. J., and S. D. Tesch. 1992. Plantations vs. advance regeneration: Height growth comparisons for southwestern Oregon. *Western Journal of Applied Forestry* 7(2): 44–47.

Laacke, R. J., and G. O. Fiddler. 1986. *Overstory removal: Stand factors related to success and failure.* Research paper PSW-RP-183. Berkeley, CA: USDA Forest Service.

Laacke, R. J., and J. H. Tomascheski. 1986. *Shelterwood regeneration of true fir: Conclusions after 8 years.* Research paper PSW-RP-184. Berkeley, CA: USDA Forest Service.

Lavender, D. P. 1958. *Effects of ground cover on seedling germination and survival.* Research note 34. Salem, OR: Department of Forestry.

———. 1964. *Date of lifting for survival of Douglas-fir seedlings.* Research note 49. Corvallis, OR: Forest Research Lab, Oregon State University.

———. 1984. Plant physiology and nursery environment: Interactions affecting seedling growth. In *Forest nursery manual: Production of bareroot seedlings,* ed. M. L. Duryea and T. D. Landis. The Hague, The Netherlands: Martinus Nijhoff/Dr. W. Junk.

———. 1985. Bud dormancy. In *Evaluating seedling quality principles, procedures, and predictive abilities of major tests,* ed. M. L. Duryea. Corvallis, OR: Forest Research Lab, Oregon State University.

———. 1988. Characterization and manipulation of the physiological quality of nursery stock. In *Proceedings of the tenth North American forest biology workshop,* ed. J. Worrall, J. Loo-Dinkins, and D. P. Lester. Vancouver, BC, Canada: University of British Columbia.

———. 1990a. Measuring phenology and dormancy. In *Techniques and approaches in forest tree ecophysiology,* ed. J. P. Lassoie and T. M. Hinckley. Boca Raton, FL: CRC Press.

———. 1990b. Physiological principles of regeneration. In *Regenerating British Columbia's forests,* ed. D. P. Lavender et al. Vancouver, BC, Canada: University of British Columbia.

Lavender, D. P., and R. K. Hermann. 1970. Regulation of growth potential of Douglas-fir seedlings during dormancy. *New Phytologist* 69:675–694.

Lavender, D. P., and P. F. Wareing. 1972. Effects of day length and chilling on the responses of Douglas-fir (*Pseudotsuga menziesii* [Mirb] *Franco*) seedlings to root damage and storage. *New Phytologist* 71:1055–1067.

Lavender, D. P., M. H. Bergman, and L. D. Calvin. 1956. *Natural regeneration on staggered settings.* Research bulletin. Oregon State University, Corvallis: Oregon State Board of Forestry.

Lavender, D. P., K. K. Ching, and R. K. Hermann. 1968. The effect of environment on the development of dormancy and growth of Douglas-fir seedlings. *Botanical Gazette* 1 (129):70–83.

Lord, C. M. 1938. Natural reproduction in Douglas-fir stands as affected by the size of opening. M.S. thesis, Oregon State University, Corvallis.

Margolis, H. A., and D. G. Brand. 1990. An ecophysiological basis for understanding plantation establishment. *Canadian Journal of Forest Research* 20(4):375–390.

Mathews, J. D. 1989. *Silvicultural systems.* London: Oxford University Press.

McDonald, P. M. 1976a. *Forest regeneration and seedling growth from five major cutting methods in north-central California.* Berkeley, CA: PSW Research Station, USDA Forest Service.

———. 1976b. *Shelterwood cutting in a young-growth, mixed-conifer stand in north-central California.* Research paper PSW-RP-117. Berkeley, CA: PSW Research Station, USDA Forest Service.

———. 1983. *Clearcutting and natural regeneration: Management implications for the northern Sierra Nevada.* General technical report PSW-GTR-70. Berkeley, CA: PSW Research Station, USDA Forest Service.

Minore, D. 1978. *The Dead Indian Plateau: A historical summary of forestry observations and research in a severe southwestern Oregon environment.* General technical paper PNW-GTR-72. Portland, OR: USDA Forest Service.

———. 1986. *Germination, survival, and early growth of conifer seedlings in two habitat types.* Research paper PNW-RP-348. Portland, OR: USDA Forest Service.

Munger, T. T. 1911. *Growth and management of Douglas-fir in the Pacific Northwest.* Circular 175. Washington, DC: USDA Forest Service.

———. 1950. A look at selective cutting in Douglas-fir. *Journal of Forestry* 48:97–99.

Newton, M. 1978. *Test of western hemlock wildlings in brushfield regeneration.* Research paper 39. Corvallis, OR: Oregon State University School of Forestry.

Newton, M., and E. C. Cole. 1987. A sustained yield scheme for old-growth Douglas-fir. *Western Journal of Applied Forestry* 2:22–25.

O'Dea, M., J. C. Zasada, and J. C. Tappeiner. 1995. Vine maple clonal development in coastal Douglas-fir forests. *Ecological Applications* 5:63–73.

Oliver, W. W. 1986. *Growth of California red fir advance regeneration after overstory removal and thinning.* Research paper PSW-RP-180. Berkeley, CA: USDA Forest Service.

Pearson, G. A. 1923. *Natural reproduction of western yellow pine in the Southwest.* Bulletin 1105. Washington, DC: USDA.

Ritchie, G. A. 1984. Assessing seedling quality. In *Forest nursery manual: Production of bareroot seedlings,* ed. M. L. Duryea and T. D. Landis. The Hague, The Netherlands: Martinus Nijhoff/Dr. W. Junk.

Roeser, J., Jr. 1924. A study of Douglas-fir reproduction under various cutting methods. *Journal of Agricultural Research* 28:1233–1242.

Schubert, G. H., and R. S. Adams. 1971. *Reforestation practices for conifers in California.* Sacramento: California State Board of Forestry.

Seidel, K. W. 1983. *Regeneration in mixed conifer and Douglas-fir shelterwood cuttings in the Cascade Range of Washington.* Research paper PNW-RP-314. Portland, OR: USDA Forest Service.

Show, S. B. 1929. *Forest nursery and planting practice in the California pine region.* Circular 92. Washington, DC: USDA.

Silen, R. R. 1982. *Nitrogen, corn, and forest genetics: The agricultural yield strategy implications for Douglas-fir manage-*

ment. General technical report PNW-GTR-137. Washington, DC: USDA Forest Service.

Smith, D. M. 1970. Applied ecology and the new forest. In *Joint Session Proceedings of Western Forest Fire, Pest, and Reforestation Coordinating Committee.* Vancouver, BC, Canada: Western Forestry and Conservation Association.

———. 1986. *The practice of silviculture.* New York: John Wiley & Sons.

Spies, T. A., and J. F. Franklin. 1991. The structure of natural young and old-growth Douglas-fir forests in Oregon and Washington. In *Wildlife and vegetation of unmanaged Douglas-fir forests,* ed. L. F. Ruggiero et. al. General technical report PNW-GTR-285. Portland, OR: USDA Forest Service.

Stein, W. I. 1981. *Regeneration outlook on BLM lands in the southern Oregon Cascades.* Research paper PNW-RP-284. Portland, OR: USDA Forest Service.

———. 1986. *Regeneration outlook on BLM lands in the Siskiyou Mountains.* Research paper PNW-RP-349. Portland, OR: USDA Forest Service.

———. 1995. *Ten-year development of Douglas-fir and associated vegetation after site preparation on Coast Range clearcuts.* Research paper PNW-RP-473. Portland OR: Pacific Northwest Research Station, USDA Forest Service.

Strothmann, R. O., and D. F. Roy. 1984. *Regeneration of Douglas-fir in the Klamath Mountains region, California and Oregon.* General technical report PSW-GTR-81. Berkeley, CA: USDA Forest Service.

Tappeiner, J. C., and P. M. McDonald. 1984. Development of tanoak understories in conifer stands. *Canadian Journal of Forest Research* 14:271–277.

Tappeiner, J. C., and J. C. Zasada. 1993. Establishment of salmonberry, salal, vine maple, and bigleaf maple seedlings in the coastal forests of Oregon. *Canadian Journal of Forest Research* 23:1775–1780.

Tappeiner, J. C., P. M. McDonald, and T. F. Hughes. 1986. Survival of tanoak (*Lithocarpus densiflorus*) and Pacific madrone (*Arbutus menziesii*) seedlings in forests of southwestern Oregon. *New Forests* 1:43–45.

Tappeiner, J. C., J. C. Zasada, P. Ryan, and M. Newton. 1991. Salmonberry clonal and population structure: The basis for a persistent cover. *Ecology* 72:609–618.

Tarrant, R. F. 1961. Stand development and soil fertility in a Douglas-fir–red alder plantation. *Forest Science* 7:238–246.

Tarrant, R. F., and R. E. Miller. 1963. Accumulation of organic matter and soil nitrogen beneath a plantation of red alder and Douglas-fir. *Soil Science Society of America Proceedings* 27:231–234.

Tesch, S. D., and E. J. Korpela. 1993. Douglas-fir and white fir advanced regeneration for renewal of mixed conifer forests. *Canadian Journal of Forest Research* 23:1427–1437.

Tesch, S. D., and J. W. Mann. 1991. *Clearcut and shelterwood reproduction methods for regenerating southwest Oregon forests.* Research bulletin 72. Corvallis, OR: Oregon State University.

Tesch, S. D., K. B. Katz, and E. J. Korpela. 1993. Recovery of Douglas-fir seedlings and saplings wounded during overstory removal. *Canadian Journal of Forest Research* 23:1684–1694.

Walstad, J. D., and P. J. Kuch. 1987. *Forest vegetation management for conifer production.* New York: John Wiley & Sons.

Williamson, D., and D. Minore. 1978. *Survival and growth of planted conifers on the Dead Indian Plateau east of Ashland, OR.* Research paper PNW-RP-242. Portland, OR: USDA Forest Service.

Williamson, R. L. 1973. *Results of shelterwood harvesting of Douglas-fir in the Cascades of western Oregon.* Research paper PNW-RP-161. Portland, OR: USDA Forest Service.

Worthington, N. P. 1953. *Reproduction following small group cuttings in virgin Douglas-fir.* Research note 84. Portland, OR: Pacific Northwest Forest and Range Experiment Station.

Zavitkowski, J., and M. Newton. 1968. Ecological importance of snowbrush in the Oregon Cascades. *Ecology* 49:1134–1145.

10 The Role of Extended Rotations

Robert O. Curtis

Extended Rotations 167
Wildlife and Biodiversity Values 169
The Timber Supply Problem 169
Conclusion 169
Literature Cited 170

The theme of this chapter is that progressive shortening of rotations in recent decades has been a factor in the genesis of current forest resource management controversies, and that a shift to extended rotations on some part of the land base—combined with certain other measures—can be a valuable component of any overall strategy to deal with these problems.

Forest management practices in the Pacific Northwest evolved from beginnings in the 1850s. During the next three-quarters of a century, most cutting took place on private lands. Early cutting practices were essentially liquidation, at rates determined by market forces, of what was then regarded as a wasting asset. With the gradual adoption of planned forest management over the period from about 1920 to 1950, systematic planning for long-term production was introduced. A basic principle, accepted by all at the time, was that old stands were essentially static and unproductive and should be replaced by young rapidly growing stands as quickly as possible. With dwindling old growth and a growing appreciation of the inherent high productivity of northwestern forests, it soon became apparent that the future was in managed second growth. In the 1950s, many foresters, still thinking in terms of relatively long rotations, thought that commercial thinning was the wave of the future. But markets for small material were poor, and thinning costs were high. Most owners abandoned commercial thinning in favor of early harvest of second-growth stands, which today is often as early as 40 to 50 years of age.

Similar developments took place on public lands, but at a much slower pace. Harvests were relatively small until after World War II. Because public land managers still had large amounts of old timber and were concerned about even flow of timber and about other forest values, harvest ages were not reduced to the same extent.

The net effect we see today is landscapes dominated by recent clearcuts and stands in the stand initiation and stem exclusion stages. Much of the public finds this aesthetically distasteful and tends to regard it as irresponsible destruction of nature. These visual impressions create horrendous public relations and political problems, with huge indirect costs to landowners and the public.

Many scientists also have concerns about possible effects associated with frequent drastic disturbance and regimes in which most of the landscape is occupied by young stands that are usually quite uniform, with few snags and little understory vegetation or coarse woody debris. Questions include effects on wildlife, hydrology, species diversity, site productivity, timber quality, and sustainability of yield. Concerns have also been raised about landscapes where stand harvest and regeneration are done in large units (with few edges or corridors) and at ages at which volume and value growth are substantially less than the potential.

We need regimes that will reduce the visual impacts of forestry operations and combine maintenance of species diversity, wildlife, and other environmental values with continued timber production at a relatively high and sustainable level. One approach with potential advantages is a shift to extended rotations, combined with increased use of commercial thinning and—where feasible—a shift to regeneration systems other than the large clearcuts of the recent past.

Definitions:

Rotation—the planned number of years between stand regeneration (or removal of the previous stand) and final harvest.

Yield—the sum of standing volume plus the cumulative amount cut since stand establishment at a specified age.

Mean annual increment (MAI)—volume produced (standing volume plus thinnings) divided by stand age; that is, average production rate from establishment to the age in question.

Current annual increment (CAI)—growth rate in volume per year at a specified age.

Periodic annual increment (PAI)—the difference in stand volume at two successive measurements, divided by the number of years between measurements. PAI is an approximation to current annual increment, which is not directly measurable.

MAI and PAI trends over time have characteristic patterns shown schematically in Figure 10.1. The MAI and PAI curves intersect at the maximum point (culmination) of the MAI curve. The stand development patterns represented by the MAI and PAI curves are important both in relation to timber yields and rotations and because they are related to other stand and species characteristics.

There are three basic theoretical concepts of optimum rotation. Each differs in assumptions and ob-

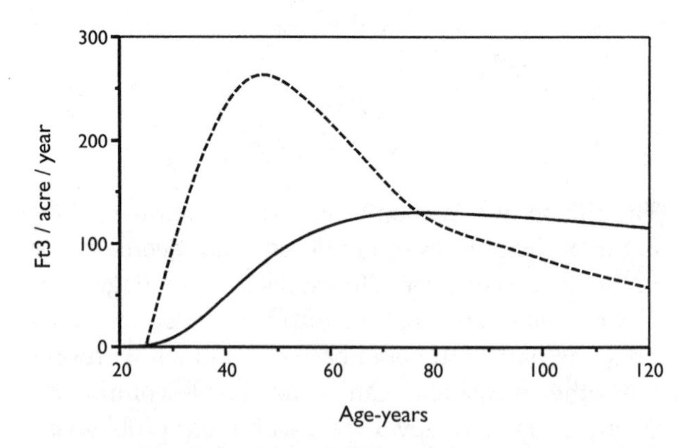

Figure 10.1 Mean annual increment and periodic annual increment curves for a medium quality site (site III) from McArdle et al. 1961. Values adjusted to merchantable volume to 6-inch top.

jectives. In traditional presentations, these are based on timber volume or value production only (Chang 1984, Newman 1988, Pearse 1967).

1. The first is known in older forestry jargon as "the rotation of forest rent." This maximizes the value of timber produced per year and corresponds to age of culmination of mean annual value increment.

2. The second is that which maximizes timber volume produced per year and corresponds to the age of culmination of mean annual volume increment.

3. The third, or financial rotation, is known in traditional forestry literature as "the rotation of soil rent." This maximizes the net value of all costs and returns, discounted to year zero, using some specified interest rate to give a "soil expectation value."

In general, the first is greater than the second because of the increased value per unit volume of large trees. The third is the shortest of the three, much shorter if the assumed interest rate is high. Foresters have argued about these for a century and a half and will probably continue to do so. Conventional economists usually advocate the third option as the theoretically "correct" answer. In real life, many other considerations enter into the rotation decision. A major factor today is the existence of multiple management goals, not restricted to or necessarily emphasizing timber production.

Extended Rotations

As rotations become shorter, a greater percentage of forest land is cut annually. Consequently, most of the life of an individual stand, as well as most of the total forest area, is in the least-attractive stand initiation stage and the least-diverse stem exclusion stage. With the short rotations in common use today, trees are smaller, wood quality and value are lower, and wood productivity and productivity of other forest values are reduced in comparison with their potentials.

Long rotations are biologically reasonable in the Pacific Northwest because Douglas-fir and its associates are very long-lived and can maintain rapid growth to rather advanced ages. Possible advantages of extended rotations (combined with commercial thinning) include the following:

- Reduced land area in regeneration and early development stages, hence—
  - — Reduced visual impacts
  - — Lower regeneration and respacing costs
  - — Less need for herbicides, slash burning, etc.
  - — Reduced frequency of drastic disturbance affecting biodiversity
- Larger trees and higher-quality wood
- Opportunity to adjust present unbalanced age distributions
- Higher-quality wildlife habitat associated with reduced area in the early, uniform, stem exclusion stage of stand development
- Hydrological and long-term site productivity benefits
- Increased carbon storage associated with larger growing stock
- Preservation of options for future adaptive management

Extended rotations also imply a major expansion of commercial thinning. This can provide both intermediate income and employment and a means of maintaining tree and stand vigor and growth rates. Thinning can be used to promote stand health and stability, forestall or salvage major mortality, and develop stand structures favorable to certain species of wildlife.

The feasibility of extended rotations depends in part on their effect on timber yields, both long-term and short-term. There is a widespread perception that high yields and intensive management imply short rotations. This is not necessarily so. Recent evidence indicates that extended rotations would not necessarily reduce long-term yields, and they might even increase them in comparison to rotations now in common use (Figure 10.2) (Curtis and Marshall 1993, Curtis 1994).

Maximum volume production is attained at culmination of MAI. This is the rotation age specified for national forest lands under the National Forest Management Act of 1973. In principle this seems a reasonable compromise for public lands managed for multiple objectives, which include aesthetic and

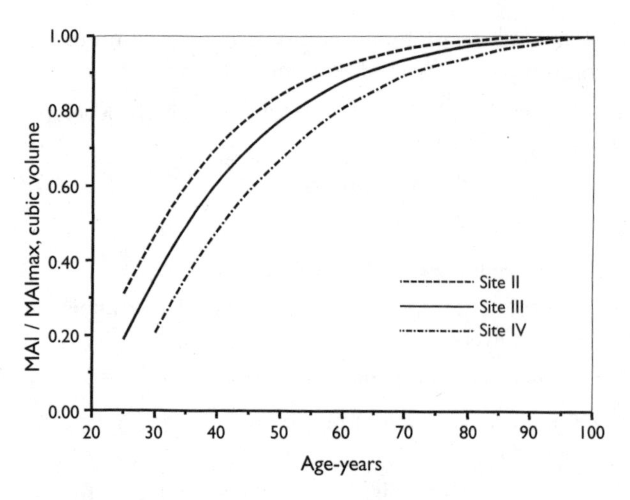

Figure 10.2 Fraction of potential volume yield attained at various ages on good (II), medium (III), and poor (IV) sites as predicted by DFSIM (Curtis et al. 1982) for stands established at 300 stems per acre and receiving repeated commercial thinning.

wildlife benefits and do not give high priority to maximizing returns on capital value of timber.

But culmination age is not a fixed known quantity. It is influenced by management and obscured by differences among available estimates. For many years, estimates were based on the work of McArdle et al. (1930, 1961) (Figure 10.1). It now appears that these estimates were seriously in error, in part because of errors in the height development curves used and in part because systematic density control affects patterns of stand development. A recent analysis of results from 17 long-term thinning trials, some extending to maximum ages of about 70 years on the best sites to 117 years on a poor site, showed that MAI had not culminated in any of the stands examined (Curtis 1995). To the upper limits of the data, PAI remained nearly constant or only slowly declined over an extended period of years. Development patterns were very different from those portrayed in Figure 10.1 for heavily stocked unmanaged stands. The MAI curve is relatively flat in the vicinity of culmination age, and there is a considerable range of possible rotation ages that will produce approximately the same MAI. Moderate extension of rotations would not decrease long-term volume production, might well increase value production, and would certainly increase aesthetic and some wildlife and biodiversity values.

The traditional economic arguments lead to the rotation that maximizes soil expectation value—the discounted value of all future costs and returns. Except at very low interest rates, this gives great weight to short-term results and negligible weight to the long term. A noneconomist might well argue that no individual or society bases decisions solely on discounted monetary values; each determines goals on other bases and then uses economics as an aid in deciding how best to reach those goals.

More concretely, one can point out shortcomings—perhaps not inherent—that are characteristic of the most common and simplistic presentations. They commonly ignore wildlife, water, fish, and amenity values, which usually do not accrue to the landowner. They usually do not recognize that public policy considerations include maintaining productivity of all values and that public perceptions—justified or otherwise—are the driving force behind costly and increasing regulatory constraints and pressures for land withdrawals for special uses. We have no satisfactory way to assign dollar values to most nontimber outputs, although costs of specified management modifications to provide them can be estimated.

Estimates of future returns often do not adequately account for the fact that longer rotations with thinning are expected to produce stands with higher values per unit volume because of the higher value of large trees and quality improvement from selection of leave trees. With unknown future markets and future quality premiums, estimates are necessarily uncertain, and the natural tendency is to be conservative.

These factors are real even though expression in monetary terms is imprecise or impossible. Nontimber values and public perceptions have become dominant forces influencing management of national forest and other public lands. The political and regulatory consequences of these changes cannot be ignored by any owner.

The short rotations now used by many owners produce considerably less volume and value than the potential. But rotations are influenced by financial and supply considerations and owner objectives. Rotations of maximum MAI in volume or in value are probably not appropriate for many private owners. Continuation of the recent trend toward very short rotations on many nonfederal lands, however, can only mean sharply reduced productivity, restricted future management options, reduced nontimber benefits, and exacerbation of antiforestry attitudes among major segments of the public.

Wildlife and Biodiversity Values

In recent years, a great deal of emphasis has been placed on nongame wildlife and on protection of endangered species and conservation of biodiversity. We have learned that attempts to rescue individual species can be extremely disruptive and expensive and are often too late. It is not possible to provide for each of the myriad of species individually. The only viable course is to develop forest management regimes that provide both commodity production and support for most forest-dwelling species (Carey et al. 1996). This will be politically and economically feasible only if severe conflicts with the economic and social well-being of rural communities and forest-based industries can be avoided.

Several aspects of extended rotations and the associated thinning and regeneration options will be generally favorable to biodiversity and production of wildlife (Curtis and Carey 1996):

- Extended rotations allow development of a wide range of age classes, tree sizes, and structures, as well as a more balanced stand age distribution.
- The combination of extended rotations, thinning, and alternative regeneration systems will (1) minimize the influence of the stem exclusion stage and clearcuts on dispersal and colonization processes, (2) promote development of biologically rich forest floors and complex trophic pathways, and (3) reduce spatial isolation of high-quality forest habitat. The combination can also reduce or eliminate the unfavorable effects often thought to be associated with forest fragmentation into disconnected units of regeneration and old stands.
- Thinning entries will usually be infrequent during the latter part of the rotation. Therefore, there will be some natural unsalvaged mortality providing snags and coarse woody debris additional to that from thinning slash. If these are thought insufficient, additional trees can be killed or treated to provide snags, cavities, and coarse woody debris.
- Thinning provides flexibility in control of stand density and structure that can be used to promote wildlife values through development of understory and creation of small openings and within-stand density variation. Systematic thinning begun at a relatively early age can produce large trees and markedly alter stand structures over relatively short periods of time.

Emphasis will obviously be strongly influenced by existing land ownership and owner management objectives.

The Timber Supply Problem

Extended rotations would, if anything, increase long-term timber supply. Unfortunately, changing rotations is a one-way street. Rotations are easy to shorten, but difficult to lengthen. The practical implementation problem is maintaining an acceptable level of timber supply during the transition period. This cannot be ignored by public owners, and it is a critical obstacle for others. Reductions in short-term supply could be offset to some degree by increased thinning and decreased pressures to remove land from the timber base.

A number of examples (Curtis 1995) suggest that thinning regimes can be designed that combine high intermediate yields and relatively infrequent entries with development of stands acceptable for both timber and nontimber values.

Many currently perceived problems reflect the fact that existing stand age distributions are highly unbalanced both regionally and locally. A move toward more balanced distributions implies both that some stands should be carried to advanced ages and that others should be harvested at ages substantially less than might otherwise be desirable. Spatial distribution, as well as total areas involved, will be important.

Conclusion

Silviculture consists of the techniques for manipulating the composition, structure, and rates of development of forest trees and stands. These techniques are the product of a long history of accumulated research and experience. Although many were originally developed for the purpose of enhancing wood production, they can be extended with little difficulty to the broader problems of developing forests and forest stands with characteristics desired for multiple forest objectives (DeBell et al., Chapter 8, Tappeiner et al., Chapter 9).

The choice of rotation is an integral part of a management regime and has reciprocal ties to the nature and appropriateness of other silvicultural measures. A considerable part of present forest management controversies can be attributed to the visual and ecological effects of the short-rotation management that has become common in recent decades. A shift to longer rotations on some portion of the land base should mitigate these problems and conflicts. There is much evidence that this need not reduce long-term timber outputs and that it might even increase them while at the same time increasing production of aesthetic, wildlife, and other nontimber values.

Literature Cited

Carey, A. B., C. Elliott, J. R. Sessions, C. J. Chambers, J. F. Franklin, C. D. Oliver, and M. J. Raphael. 1996. *Washington landscape management project report no. 2.* Olympia, WA: Washington Department of Natural Resources. In press.

Chang, S. J. 1984. *Determination of the optimal rotation age: A theoretical analysis. Forest Ecology and Management* 8(1984):137–147.

Curtis, R. O. 1994. *Some simulation estimates of mean annual increment of Douglas-fir: Results, limitations, implications for management.* Research paper PNW-RP-471 Portland, OR: USDA Forest Service.

———. 1995. *Extended rotations and culmination age of coast Douglas-fir: Old studies speak to current issues.* Research paper PNW-RP-485. Portland, OR: USDA Forest Service.

Curtis, R. O., and A. B. Carey. 1996. Managing for economic and ecological values in Douglas-fir forests. *Journal of Forestry.* In press.

Curtis, R. O., and D. D. Marshall. 1993. Douglas-fir rotations: Time for reappraisal? *Western Journal of Applied Forestry* 8(3):81–85.

Curtis, R. O., G. W. Clendenen, D. L. Reukema, and D. J. DeMars. 1982. *Yield tables for managed stands of coast Douglas-fir.* General technical report PNW-135. Portland, OR: USDA Forest Service.

McArdle, R. E., and W. H. Meyer. 1930. *The yield of Douglas-fir in the Pacific Northwest.* Technical bulletin 201. Washington, DC: USDA.

McArdle, R. E., W. H. Meyer, and D. Bruce. 1961. *The yield of Douglas-fir in the Pacific Northwest.* Technical bulletin 201. Washington, DC: USDA.

Newman, D. H. 1988. *The optimal forest rotation: A discussion and annotated bibliography.* General technical report SE-48. New Orleans: Southeastern Forest Experiment Station, USDA Forest Service.

Pearse, P. H. 1967. The optimum forest rotation. *Forestry Chronicle* 43:178–195.

11 Integrating the Ecological Roles of Phytophagous Insects, Plant Pathogens, and Mycorrhizae in Managed Forests

Timothy Schowalter, Everett Hansen, Randy Molina, and Yanli Zhang

Insect and Fungus Biology 172
Functional Roles 173
Determinants of Forest Structure, Composition, and Succession 174
Maintaining Fitness and Productivity 176
Nutrient Capture and Recycling 177
Pollination 178
Biological Control 179
Legacy of Past Practices 179
Conditions That Cause Problems 181
Introduced Organisms 182
Monocultures 182
Fire Control, Overcrowding, and Thinning 183
Fertilization 184
Fragmentation 184
What Changes Are Necessary? 185
Literature Cited 186

Insects, fungi, other arthropods, and bacteria often are overlooked, both in catalogs of forest dwellers and in discussions of forest ecosystem function. Though small, these organisms are numerous and diverse. In most forests, they comprise the greatest portion of biological diversity and even a surprising portion of the biomass, especially belowground. They impact many ecosystem functions, sometimes controlling them, and their populations can fluctuate dramatically, responding quickly to changes in resource availability and environment. Some are considered pests when they interfere with management objectives, usually by killing or damaging valuable trees, and others are valued for their essential contribution to survival and growth of planted trees.

Much has been written about the economic losses caused by tree-killing insects and plant pathogenic fungi in forests. Similarly, the benefits of mycorrhizal fungi to tree growth are well documented. In this chapter, we aim to broaden the traditional view of forest pests and symbionts in order to consider them as part of functioning ecosystems, and thus to provide a better basis for predicting their responses to changes in forest management and other disturbances. To reach this goal, we provide a very brief review of insect and fungus biology as background to a discussion of the contributions of insects and fungi, especially herbivores, pathogens, and mycorrhizae, to ecosystem function. From this base, we discuss the responses of phytophagous insects, tree pathogens, and mycorrhizal fungi to changes in forest conditions as a basis for integrating these organisms into the "new" practices of ecosystem-based forest management. One of the keys to successful forestry is understanding and accepting the species and site specificity of the knowledge base. Our examples come largely from the forests of the western United States with which we are most familiar. Different organisms in different environments will respond differently, but we hope to establish underlying principles to allow predictions in new situations.

Insect and Fungus Biology

Fungi and insects are important in forest ecosystems in terms of numbers of species and ecological and economic impact. A given site will harbor thousands of species of insects and fungi, many undescribed, compared to a few dozen species of plants and vertebrates. Insects and fungi have diversified through evolution to fill a broad array of often very narrow niches in forest ecosystems.

The Fungi compose a separate kingdom of organisms, very different from plants, animals, and bacteria. The insects are the most numerous and diverse of the groups in the animal kingdom (Wilson 1992). Both groups are heterotrophic—that is, they gain their energy for growth and reproduction indirectly from carbohydrates manufactured first by plants through photosynthesis. Many species are detritivores or saprotrophs, feeding on dead organic matter. These are essential to ecosystem function for their roles in nutrient cycling. Another large group of species are predaceous or parasitic and represent important mechanisms for regulating prey or host populations. Our focus, however, is on a relatively small subset of species that have evolved nutritional strategies based on feeding on living plants, especially the trees that dominate forests. These phytophagous insects and pathogenic fungi, together with the mutualistic mycorrhizal fungi, have evolved with the trees and other organisms of the forests and are normal components of all forests. The diversity of nutritional strategies among these species utilizing living trees for energy is great. Some canker fungi and bark beetles, for example, are only successful on weakened or dying trees or parts of trees and may spend much of their life on dead material, but other species, such as rust fungi and some aphids, prefer vigorously growing plants and die when their host plant dies. Many live on leaves and needles, but because trees have many leaves, their impact on whole tree growth is insignificant, except in years when populations are very large. Others kill trees, singly or in large groups, and can alter the structure and composition of the entire forest.

The reproductive potential of insects and fungi is immense. Some species produce hundreds or thousands, even millions, of offspring each generation, and generation times may be as short as two weeks. They often possess effective means of dispersal as well, carried through the air on their own wings or on air currents. New genotypes, produced through mutation or recombination, appear frequently and spread widely. Beneficial changes to the insect or fungus may be favored through natural selection. Asexual reproduction is found in both groups, and the fungi have haploid genomes. Both life cycle adaptations favor rapid selection and increase of favorable mutations, such as pathogenicity to new host genotypes. These reproductive advantages of the insects and fungi are countered in the ecosystem by the continuing evolution of host defenses and by limitations imposed by the environment.

The success of both insects and fungi is dependent on access to suitable hosts and a favorable environment. The interplay between tree resistance and environmental limitations serves to limit the frequency and extent of damage from insects and fungi. Trees have evolved a diverse array of mechanisms of defense against attack. Most tree species are resistant to attack by most insects and fungi. Susceptibility is the special case, and phytophagous insects and pathogenic fungi have evolved the necessary specialized behaviors and physiology to bypass host resistances.

Most, perhaps all, plant-feeding insect and fungus species show strong host preferences; many are host specific. Even a broad host-ranging species, such as the laminated root rot fungus found on most western conifers, spreads much more rapidly on Douglas-fir than on western hemlock and does not attack angiosperms at all (Thies 1984). The mountain pine beetle kills all western pines, but allozyme and morphometric analyses of beetles indicate that sympatric populations from lodgepole pine are distinct from populations from limber pines, with populations from ponderosa pine intermediate (Sturgeon and Mitton 1986).

Insects and fungi often are protected from adverse environments by their host trees, but during critical times of reproduction and dispersal they are exposed and vulnerable. Environmental and host conditions normally limit the opportunities for rapid population growth to particular microsites or especially favorable years.

Mycorrhiza translates literally as "fungus-root" and defines the common association of specialized soil fungi with the fine roots of nearly all forest plants. Mycorrhizal associations represent one of the more widespread forms of mutualistic symbioses in terrestrial ecosystems. Indeed, these plant–fungus associations have coevolved over the millennia such that each partner depends on the other for survival. The mycorrhizal fungus basically serves as an extension of the plant root system, exploring soil far beyond the roots' reach and transporting water and nutrients to the roots. The uptake of phosphorus and nitrogen is an especially critical function of mycorrhizal fungi, which can release bound forms of these nutrients otherwise unavailable to the roots. In return, the plant is the primary energy source for the fungus, providing simple sugars and vitamins produced through photosynthesis and transported to the roots and then the fungus.

Although mycorrhizal fungi compose a functional group, individual species differ strongly in their biology, ecology, and host range. Three primary classes of mycorrhizae occur in forest ecosystems: ectomycorrhizae, vesicular-arbuscular mycorrhizae, and ericoid mycorrhizae. Ectomycorrhiza is pervasive in temperate coniferous forests because tree species in the Pinaceae, Betulaceae, and Fagaceae families form this type. Fungi are primarily basidiomycetes and ascomycetes (mushroom, truffle, and cup fungi); many of the diverse mushroom species seen on the forest floor during autumn are the reproductive structures of ectomycorrhizal fungi. Some ectomycorrhizal fungi have broad host ranges, while others are restricted to certain tree genera (Molina et al. 1992). Vesicular-arbuscular mycorrhiza is the most common type worldwide. and many trees and understory plants form this type. Vesicular-arbuscular mycorrhizal fungi are zygomycetes and reproduce primarily as large soil-born spores. Ericoid mycorrhiza is restricted to the Ericales order and is widespread in forest ecosystems because many ericaceous species are dominant understory components. Ericoid mycorrhizal fungi are mostly ascomycetes that form small cuplike reproductive sporocarps. This diversity in fungal life histories and host associations among mycorrhizal fungi leads to a variety of functional interdependencies between forest plants and fungi in space and time (Harley and Smith 1983).

Functional Roles

Until relatively recently, forest insects and fungi, to the extent that they were noticed by forest managers and researchers at all, were regarded largely as pests, interfering with commodity production or experimental design. Beginning in the mid-1970s, this view began to change. A rapidly developing ecosystem concept of forests was fueled by multidisciplinary studies of functioning forest communities initiated by programs such as the International Biological Program (IBP). These programs raised awareness of the integration of biotic and abiotic ecosystem components and the coupled ecological processes underlying ecosystem stability and productivity (e.g., Bormann and Likens 1979, Edmonds 1982, Swank and Crossley 1987).

Mattson and Addy (1975) first challenged the narrow view of phytophagous insects and pathogens as pests by suggesting that these organisms could actually increase primary productivity through pruning, thinning, and stimulating nutrient cycling. Wickman (1980) and Alfaro and Shepard (1991) subsequently showed that trees often compensate over long time periods for short-term growth losses during defoliator outbreaks. A number of authors have presented an integrated view of insects and pathogen roles within forest ecosystems (Castello et al. 1995; Hansen 1977; Schowalter 1985, 1994; Van der Kamp 1991).

Much has been written from an economic perspective on the losses caused by insects and pathogenic fungi in the forest and on the benefits of mycorrhizal fungi to tree seedling success (Castellano 1994, Castellano and Molina 1989), and these impacts will not be reviewed again here. The impacts of these organisms on ecosystem structure and function have received much less attention. They kill big trees and in the process change the character of the forest, and they kill small, young, and weak trees, maintaining the fitness of the ecosystem. They are important in nutrient capture and cycling, and many of the insects are pollinators. Species of both insects and fungi are predators and parasites on other insects, fungi, and small organisms, helping to regulate populations in the forest.

Determinants of Forest Structure, Composition, and Succession

Insects and pathogens affect forests most dramatically by killing trees. They are agents of disturbance with pattern and periodicity different from physical disturbances such as fire and wind. In some ecosystems, they are the most important disturbance agents, particularly in the long intervals between stand-replacing events such as fire or harvest, and they thus determine the character of the forest (Dickman and Cook 1989). The range of effects is illustrated with observations from the Douglas-fir forest ecosystem, where extensive tracts of wild forest are still available for observation.

The Douglas-fir forests of western North America are largely in the western hemlock zone of Franklin and Dyrness (1984). They are maintained in the seral stage by periodic stand-replacing wildfire or, in recent decades, by clearcut forest harvest. The late-successional western hemlock regenerates in the shade of the seral Douglas-fir, but extensive late-successional forests, with hemlock replacing itself in a dynamic equilibrium, are seldom encountered. Instead, hemlock trees remain in the understory until mortality begins to create light gaps in the fir overstory. Pathogens, particularly root and stem decay fungi and the associated Douglas-fir beetle, are the principal gap formers in these forests and the agents that drive succession between wildfires.

The two decay fungi that cause brown cubical butt rot and red ring rot in tree boles are responsible for much of the single-tree mortality of Douglas-fir in old forests. Butt rotted trees typically snap near the base in wind storms and may destroy or damage adjacent trees as they fall. Butt rot–caused mortality is often ascribed to wind, but in most storms few sound trees are toppled. In one study, windthrow resulting from a winter storm on Mary's Peak in the Oregon Coast Range was almost exclusively related to root rot or to exposure along recent clearcut margins (Hansen unpublished data). Red ring rot kills trees as they stand or weakens them sufficiently to make them susceptible to attack by the Douglas-fir beetle. It is often difficult to assign cause of death to standing dead trees, since both decay indicators and signs of beetle attack may be far above ground. Death of single trees gradually reduces the dominant overstory, but canopy openings may fill with the expanding crowns of surrounding trees before understory trees can reach the upper levels.

Larger gaps in the Douglas-fir canopy are frequently caused by the laminated root rot fungus. This pathogen spreads slowly from tree to tree across root contacts. It takes decades to kill large trees, and they either fall from root decay or die standing, mortality often hastened by the Douglas-fir beetle. A large portion of the Douglas-fir beetle mortality in the Northwest is intimately associated with root rot (Goheen and Hansen 1993). The result is slowly expanding openings that develop through the life of the stand and may reach several acres in size (Figure 11.1). In the Douglas-fir forest, fir is most susceptible, hemlock is tolerant of infection, cedars are resistant, and hardwoods are immune. If hemlock or cedar seed sources are available in the adjacent stand, root rot openings fill with these species, and succession is advanced toward the late-successional condition. If seed sources are too distant, gaps fill with hardwood brush and trees and succession is reversed to earlier seral stages (Holah et al. 1993). Regional surveys suggest that

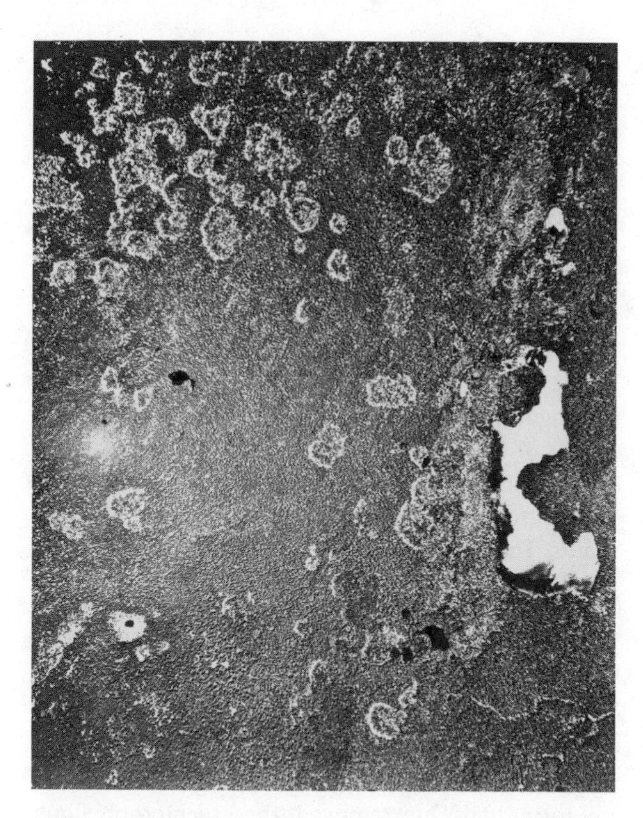

Figure 11.1 Patchiness in forest structure resulting from multi-tree gaps formed by Douglas-fir mortality because of laminated root rot near Waldo Lake, Oregon.

about 5 percent of the land area is affected by laminated root rot, with individual stands having 30 percent or more of their area affected (Figure 11.1).

Death of dominant trees from pathogens also affects forest structure and diversity. As dying trees are replaced with smaller individuals, two- and three-level canopies, characteristic of the old-growth forest condition, are created. Dead trees, as standing snags or as down logs, are ecosystems in themselves, providing food and shelter for a succession of microorganisms, insects and other arthropods, animals and plants (Harmon et al. 1986, Schowalter et al. 1992). Gaps created in the canopy allow light to reach the forest floor, creating growth opportunities for a different suite of herbaceous and shrubby plants and the animals that feed on them. If hemlock colonizes the gap, the long-term consequence may be a reduction in local diversity of herbaceous and shrub species beneath the dense shade of the developing hemlock canopy (Holah et al. 1993). Temperature and moisture relations are changed in the gaps, sometimes greatly increasing rates of nutrient release and cycling from killed trees and litter (Waring et al. 1987).

While pathogens and insects change the character of stands, they usually do so slowly, across the decades. Some bark beetles can change entire stands within a few years. The mountain pine beetle, western pine beetle, and southern pine beetle, for example, often act like a slow-moving fire through sufficiently dense pine stands, killing all host pines of particular size classes or physiological condition in a matter of months and either accelerating successional transition toward nonhost species or fueling stand-replacing fire that facilitates regeneration of pines (e.g., Amman et al. 1988, Schowalter and Turchin 1993, Schowalter et al. 1981a).

Mycorrhizae can be primary regulators in the successional interactions between plants and influence community development (Allen 1991). Successional outcomes, especially following disturbance, are often affected by the presence or abundance of mycorrhizal fungus propagules. For example, the lack of mycorrhizal fungus propagules was a primary cause of afforestation failures around the world, particularly in establishing pine plantations where pines had not occurred previously (Mikola 1973). Many exotic plantations were established successfully only after mycorrhizal fungi were purposely, or accidentally, introduced. Mine spoils and other reclamation sites often need inoculation with mycorrhizal fungi to reestablish plants because the sites are devoid of fungus propagules. Likewise, severely disturbed forest sites can experience low mycorrhizal inoculum potential that impedes forest recovery and development (Perry et al. 1987). Hence, it is essential that anthropogenic disturbances of forest ecosystems fall within the threshold tolerances of mycorrhizal fungi so that adequate viable inoculum remains on site.

Successional patterns will also be influenced by the degree of host dependence on mycorrhizal functions. Plants range from obligate, to facultative, to lacking dependence on mycorrhizal fungi for survival and growth. Janos (1980) and Allen and Allen (1990) present an interactive model of successional outcomes as determined by relative plant dependence on mycorrhizae and environmental conditions. Plants less dependent on mycorrhizal fungi often invade disturbed habitats quickly, while obligately mycorrhizal plants invade more slowly, especially if fungal inoculum is low. Allen (1991) notes examples that fit this model in several ecosystems. The microhabitat scale of microbial interactions in soil, however, often hides the successional dynamics influenced by mycorrhizae, such that they are overlooked by plant ecologists. For example, ectomycorrhizae of forest trees are often abundant in buried wood (Harvey et al. 1991), and many ectomycorrhizal fungi specialize in this soil habitat. Because many plant species, such as western hemlock and several ericaceous plants, often use logs and coarse woody debris as primary colonization sites (i.e., nurse logs), the presence of active, ecologically adapted mycorrhizal fungi in those substrates directly affects plant survival and final community composition. Similarly, live legacies such as the scattered trees and shrubs that survive disturbance (or purposely retained living trees) maintain active mycorrhizal fungi on their roots, thereby directly benefiting the next generation of trees and other forest plants (Molina et al. 1992). Hence, fungus and plant community dynamics are intricately linked through mycorrhizal connections in space and time.

Maintaining Fitness and Productivity

The stem decay and laminated root rot fungi attack and parasitize trees with little regard for tree vigor. If anything, the larger, more rapidly growing trees are more likely to be infected. Opportunistic pathogens, including most canker fungi, some root-rotting organisms, and wound decay fungi, on the other hand, successfully attack only weakened or injured trees. They are usually found on understory, suppressed trees or trees wounded or weakened by some other agent such as lightning or severe drought. The bark beetles preferentially attack trees weakened by drought, competitive stresses, or pathogens. By removing weaker individuals from the stand, these organisms reduce competition for limiting resources and help maintain fitness of the entire population.

Canker fungi and opportunistic bark beetles are often particularly active at the edges of a tree species's range. The mountains and valleys of southwest Oregon enjoy hot, dry summer conditions. Douglas-fir dominates the forest at mid and upper elevations, where winter rainfall is higher. The drier valleys are grassland-savanna, grading into ponderosa pine woodland. In many years, Douglas-fir can establish and grow at lower elevations, but in periodic drought cycles, the encroaching fir is killed by various canker fungi, Douglas-fir beetle, and flathead borers.

The absolute requirement for mycorrhizal functioning, especially at seedling establishment, characterizes the profound influence of mycorrhizae on plant fitness. Such is true not only for well-studied woody trees and shrubs, but for the smallest of forest plants. The notoriously tiny seeds of orchids, for example, require mycorrhizal colonization from the earliest stage of germination to begin their life cycle (Harley and Smith 1983). Chlorophyll-lacking plants in the Ericaceae family likewise depend completely on the mycorrhizal fungi that they share with neighboring overstory trees to obtain carbon (Furman and Trappe 1971); when the trees (the original sources of photosynthate) are removed, these achlorophyllous plants do not survive. After plant establishment, the many functional benefits of mycorrhizae continue to affect plant fitness directly and indirectly. Some mycorrhizal fungi produce antibiotics that protect fine roots from pathogens (Goldfarb et al. 1989, Marx 1973, Pfleger and Linderman 1994). The enhanced uptake of nutrients and water via mycorrhizae improves plant vigor and natural defenses against insects and pathogens.

The functional diversity of mycorrhizal fungi also contributes to plant fitness. Most plants form mycorrhizae simultaneously with several fungal species, and mycorrhizal fungi differ in their seasonal activity as well as in their dominance in different seres of community development (Deacon and Fleming 1992). Some mycorrhizal fungi are host-specific and provide unique benefits or exclusive access to a particular resource to their co-adapted host (Molina et al. 1992). Perry et al. (1989) found that competitive interactions between Douglas-fir and ponderosa pine seedlings were influenced by the presence of mycorrhizal fungi host-specific to either of the conifer hosts. Such diversity in mycorrhizal associations and function allows mycorrhizal fungi to partition use of the soil resource in space and time, thereby maximizing plant fitness over the life of the stand.

Phytophagous insects and plant pathogens also contribute to forest productivity over the long-term through defoliation and pruning. By killing foliage and twigs, they reduce plant competition for water and nutrients by reducing water use, increasing water penetration through the canopy, and stimulating nutrient turnover. Schowalter et al. (1991) reported that 20 percent defoliation of young Douglas-fir doubled the amount of precipitation and litter reaching the forest floor and increased nitrogen, potassium, and calcium input to the litter by 25 to 30 percent. Defoliator outbreaks associated with drought conditions (Mattson and Haack 1987) could improve forest conditions by reducing water use and increasing water supply to the forest floor. Parks (1993) found that defoliation of drought-stressed grand fir seedlings reduced susceptibility to Armillaria root rot by improving the internal moisture status of the seedlings. Trees often compensate for short-term growth losses during defoliation events by increasing productivity for decades following defoliator outbreaks (Alfaro and Shepard 1991, Trumble et al. 1993, Wickman 1980).

Mycorrhizal fungi make their contribution to productivity directly by enhancing nutrient and water uptake and plant fitness, as noted above. However, their effects on total ecosystem productivity go beyond enhancing accumulation of plant biomass. Mycorrhizae and mycorrhizal fungi also are major sinks for photosynthate in forest ecosystems. Fogel and Hunt (1983) estimated that fungi accounted for 28 percent of the total organic matter throughput in a Douglas-fir forest; Vogt et al. (1982) estimated that mycorrhizal fungi consume 15 percent of the net primary productivity in Pacific silver fir forests. These data indicate the range in carbon cost for reciprocal plant benefits in mycorrhizal mutualisms. Productivity cannot be measured simply as carbon allocation to various forest components. Use of carbon (energy to fuel organisms) must also be considered within the soil to understand the broad effects of fungi on forest productivity. Organic molecules from live and dead hyphae feed the microbial food web in forest soils. Some are exuded from fungal hyphae and, together with fungal mycelium, bind soil particles, producing the microaggregates critical to soil porosity, hence to aeration and moisture retention (Amaranthus et al. 1989, Perry et al. 1987). Fungi also use their stored energy to develop reproductive structures such as mushrooms and truffles. Although the amount of carbon spent on reproductive structures is small compared to total net primary production, the sporocarps are essential to fungal life cycles and are significant dietary components for many animals. Finally, the recent large-scale commercial harvest of wild edible forest mushrooms in the Pacific Northwest (nearly 4 million pounds were collected in Oregon, Washington, and Idaho in 1992, contributing $41 million to the regional economy [Schlosser and Blatner 1995]) exemplifies the value of this fungal portion of forest productivity to the public.

Nutrient Capture and Recycling

Nutrient capture and recycling are essential to the long-term stability and health of ecosystems. When nutrient inputs no longer balance nutrient loss (such as following disturbance or climate change), nutrients become limiting and vegetation changes, initially as increased mortality of the most-sensitive species, followed by reduced stature of dominant vegetation.

Nitrogen is often limiting in forest ecosystems because of high demands for protein and nucleotide synthesis by living tissues coupled with losses from denitrification (under anaerobic conditions, such as from soil compaction and flooding) and volatilization during fires. The rate of nitrogen-fixation by some bacteria, blue-green algae, and lichens often is substantial, but depends on conditions. For example, free-living nitrogen-fixing bacteria require suitable habitats, such as decomposing logs (Schowalter et al. 1992, Silvester et al. 1982). Nitrogen fixation rates in more-decayed wood are substantially higher than rates in less-decayed wood or soil. Symbiotic nitrogen fixation depends on host plant productivity. In this case, nitrogen-fixing bacteria provide nitrogen to the host plant and receive photosynthates in return. Symbiotic hosts, such as alder and ceanothus, typically are fast-growing plants with the high resource requirements that characterize earlier successional ecosystems and frequently are viewed as competitors with the more valuable tree species. Suppression of the symbiotic hosts through management or eventual overtopping by later successional trees reduces photosynthate allocation to root nodules and suppresses nitrogen fixation. Canopy lichens may fix substantial amounts of nitrogen, but are limited in occurrence to canopies of old-growth trees. Forest management practices have only begun to address the importance of nitrogen-fixing organisms to forest productivity and the sensitivity of these organisms to silvicultural manipulations.

Mycorrhizal fungi play a critical role in nutrient acquisition and plant nutrition. These symbiotic fungi acquire nutrients from decomposing litter and soil pools and exchange nutrients for photosynthates from their host plants, thereby facilitating nutrient and water uptake by their host plants and minimizing nutrient loss from the ecosystem. Carbon and nutrients also can be exchanged among plants via mycorrhizal connections. Such linkages influence interplant interactions and successional patterns in forest development (Molina et al. 1992, Amaranthus and Perry 1994).

The direct allocation of photosynthate from plant to mycorrhizal fungi sets mycorrhizal fungi apart in ecosystem function from soil saprobes that cycle carbon from organic matter. Host photosynthate not only fuels the physiological activity of the symbiotic fungi, but contributes directly to the soil processes and functions performed by mycorrhizal fungi. The large biomass of mycorrhizal fungi stores significant quantities of nutrients in the soil. Some ectomycorrhizal fungi form dense, perennial mats that occupy up to 28 percent of the forest floor (Cromack et al. 1979). Numbers of microbes and microarthropods are greater in these physiologically active fungal mats than in immediately surrounding soil (Cromack et al. 1988). The mycelium of mycorrhizal fungi also exudes organic molecules that not only support rhizosphere microbes, but also bind soil particles.

Mycorrhizal fungi also compose a key functional group in forest food webs. Approximately 80 percent of soil microarthropods are fungivores. Ectomycorrhizal fungal sporocarps (mushrooms and truffles) are major food sources for many forest animals, especially small mammals. Because small mammals are the primary prey for predators such as the endangered northern spotted owl, understanding these functions is vital to integrated conservation efforts.

Detritivores and saprophytes have a widely recognized role in nutrient cycling (Seastedt 1984). These organisms are instrumental in the decomposition and mineralization of organic matter. Larger invertebrates fragment ingested litter and infuse it with saprophytic fungi and bacteria, increasing decomposition rate. Wood-boring insects (including many bark beetles) penetrate bark and inoculate wood with symbiotic saprophytic fungi and bacteria that enhance nutritional quality of wood. Bark penetration also facilitates colonization by a diverse community of detritivores, saprophytic fungi, and bacteria, including nitrogen-fixing bacteria. The activities of these organisms increase wood porosity and water-holding capacity and nutrient availability, making logs "hot spots" of limited resources (Harmon et al. 1986, Schowalter et al. 1992).

Litter organic compounds are decomposed by action of extracellular enzymes secreted by saprophytic fungi and bacteria. The nutrients released become available for uptake by the saprophytes, but also by mycorrhizal fungi and tree roots infusing the litter.

Mycorrhizal fungi are typically viewed as being limited in saprophytic capabilities. Recent research, however, has shown that several species are able to decompose organic substrates, particular organic nitrogen (Read 1992). Mycorrhizal fungi of ericaceous plants, for example, produce proteases that allow them to mobilize protein-nitrogen. Because ericaceous plants such as salal, rhododendron, and huckleberry are widespread dominant understory forest plants and often colonize woody substrates, their mycorrhizal associates directly benefit not only their hosts, but contribute to nitrogen cycling in many forest ecosystems.

Pollination

Pollination is the transport of pollen to the ovary. For most trees, pollination is accomplished by air currents. However, some important forest components require insect pollinators, which transport pollen more efficiently than wind. Whereas wind transport is largely random, many insects orient toward specific flowers and thereby target pollen to conspecific flowers where fertilization is more likely. Given insect capacity for long-distance flight, outcrossing is more efficient with insect pollinators. Trees such as maples and dogwoods, as well as many shrubs, including ceanothus and huckleberry, and understory herbs, including many rare or endangered forest floor plants, require insect pollinators for adequate seed development.

Pollinators represent a number of insect groups, including butterflies and moths; flies, bees, wasps, and ants; and beetles. Most of these insects pollinate only as adults, when they can fly. Immature stages may be defoliators (butterflies and moths), wood borers (some beetles), fungivores (many flies), or predators (beetles, wasps, and flies). Whereas pollination is recognized as a beneficial role, these other roles often seem to conflict with forest management objectives, requiring that we address tradeoffs between the roles represented by different life stages of the same organism.

Biological Control

Much of the diversity of invertebrates and microorganisms in forest ecosystems is represented by predators and parasites (Parsons et al. 1991, Schowalter 1995a), including predaceous insects; arachnids (spiders, mites, centipedes, etc.); nematodes; entomopathogenic, mycopathogenic, or endophytic fungi; bacteria; and viruses. While the importance of many predators and parasites has been recognized and their populations have been augmented for purposes of pest suppression, other groups remain relatively unknown. Examples include endophytic fungi, whose mutualistic association with host foliage and production of mycotoxins may limit feeding by herbivorous insects (Carroll 1988, McCutcheon and Carroll 1993). Mycorrhizal or other saprophagous fungi also may inhibit root infection by pathogenic fungi (Goldfarb et al. 1989, Marx 1973, Pfleger and Linderman 1994).

Biological control research has often focused on predator or parasite species that specifically target pest species, thereby maximizing suppression efficiency. However, specialist species must track their host population in time and space, limiting their ability to respond quickly to incipient pest outbreaks. Abundances of generalist predators and parasites, on the other hand, may be relatively more stable, making these species more responsive to increased prey populations. The diversity of predaceous and parasitic organisms is important in the regulation of prey populations because different species attack different life stages of their prey, thereby limiting prey escape in time or space.

Legacy of Past Practices

Forest management practices, even decisions to do nothing, affect the forest environment in many ways. The responses of various invertebrate and microbial species in the community can be quite different, depending on species tolerances or adaptations to particular conditions. Some species will be favored by a particular silvicultural practice and become "pests," whereas others may decline or disappear as conditions change, leaving healthier forests or eliminating important ecological functions, depending on the roles of the organisms involved.

Predicting the response of insects, fungi, and other microorganisms to changed forest practices is an uncertain art, but comparison with natural disturbance regimes and successional pathways in the environment in which the forest evolved may be useful. Contrasting the conditions in east-side and west-side forests in Oregon and Washington illustrates the point.

A forest health crisis has been declared in many forests of interior western North America as the combined effects of past management practices, bark beetles and defoliating insects, pathogens, drought, and catastrophic wildfire threaten to change the forests more rapidly and more completely than previously recorded. Managed coastal forests have also been changed dramatically in recent decades, but epidemics of insects and fungi are few, and any regional health crisis remains a future concern, not a present reality.

Typically, wild forests west of the Cascade Mountains are relatively even-aged and dominated by Douglas-fir. The intervals of stand-replacing wildfire range from 100 to 300 or more years (Agee 1993). The light-demanding Douglas-fir is often the first tree species to colonize the fire-disturbed sites, and its rapid growth, large stature, and long life span allow it to maintain dominance for centuries. Clearcutting, often followed by slash burning, is the preferred harvest method in most west-side forests, and sites are planted with Douglas-fir. The disturbance pattern and successional trajectory roughly mimic the wild forest, but on an accelerated timeline. Insects and fungi of the Douglas-fir forest evolved in an environment of periodic catastrophic disturbance and Douglas-fir dominance, and they generally play the same roles in today's managed forests. There are exceptions, and they are instructive.

Several studies show that mycorrhizal fungi and the organisms important in nutrient cycling are more abundant and diverse in old forests than in recent clearcuts or young plantations (e.g., Amaranthus et al. 1994, Perry et al. 1987). This comparison of the ends of the successional spectrum is interesting, but these organisms are adapted to an environment that includes periodic stand-replacing wildfire. More relevant to possible changes in forest health would be a comparison of early successional microbial and insect populations following wildfire and following clearcutting. Removal during harvest of the large woody material (valuable logs), so important to mycorrhizal fungi and free-living nitrogen fixers alike, is cause for concern. Although effects of wood removal were mitigated somewhat by utilization practices through the 1950s that removed only the most perfect logs from the site, unmerchantable timber often was yarded or piled and burned, leaving many regenerated stands with insufficient woody residues. In recent years changes to the forest practices acts of the states have mandated leaving specified numbers of logs following harvest.

Reduction in the average age of managed west-side forests has led to decreases in populations of species that are more abundant in or even dependent on large trees or their residues or old-growth conditions and to increases in populations of organisms that thrive in or are dependent on young forests. Many decay fungi are much less common in young managed forests than in forests with bigger older trees. The fungus that causes red ring rot, for example, was often recorded in more than 50 percent of the trees in early surveys of mature Douglas-fir stands (Boyce and Wagg 1953), but in stands less than 100 years old, incidence seldom exceeds 4 percent. Old western hemlock, the potential climax species in much of the west side, generally has much more decay than comparable Douglas-fir because it is thin-barked and more easily wounded by animals and falling trees and because the heartwood is much less resistant to decomposition than is Douglas-fir wood.

The fungus that causes black stain root disease is one of very few pathogenic fungi that have increased measurably in young managed west-side forests. This fungus moves from tree to tree by direct growth across adjacent root systems or is carried by root-feeding bark beetles. Young plantations, with regularly spaced susceptible trees, are ideal for spread of the pathogen, especially in stands where pre-commercial thinning has created stumps (breeding opportunities) for the insect vectors and removed non-host trees. The disease is most common on sites where the soil was compacted by tractors used to harvest the previous stand (Hansen et al. 1988).

Populations of some herbivorous insects are higher in young Douglas-fir plantations than in adjacent mature forests (Figure 11.2) (Schowalter 1995a). In west-side forests, however, this has not led to any increase in damaging outbreaks of defoliators. Several needle diseases have caused local epidemics in plantations, and they serve as warning of the importance of using appropriate tree species and genotypes in reforestation. Pine needle cast fungus damages east-side ponderosa pine seed sources planted west of the Cascades in several locales in the southern Oregon Cascades (Harvey 1976), and Swiss needle cast fungus damages off-site Douglas-fir planted on spruce and hemlock sites near Tillamook on the Oregon coast.

Forests east of the Cascade Mountains grow in generally much drier environments than on the west side and are more variable in tree composition, depending largely on local precipitation. On many sites on the drier end of the spectrum, ponderosa pine is seral and Douglas-fir and the true firs are typical of mid to late-successional forests. The historic fire pattern was typically frequent (10- to 20-year interval) with light underburns that often killed young trees, especially Douglas-fir and grand fir, and kept overall

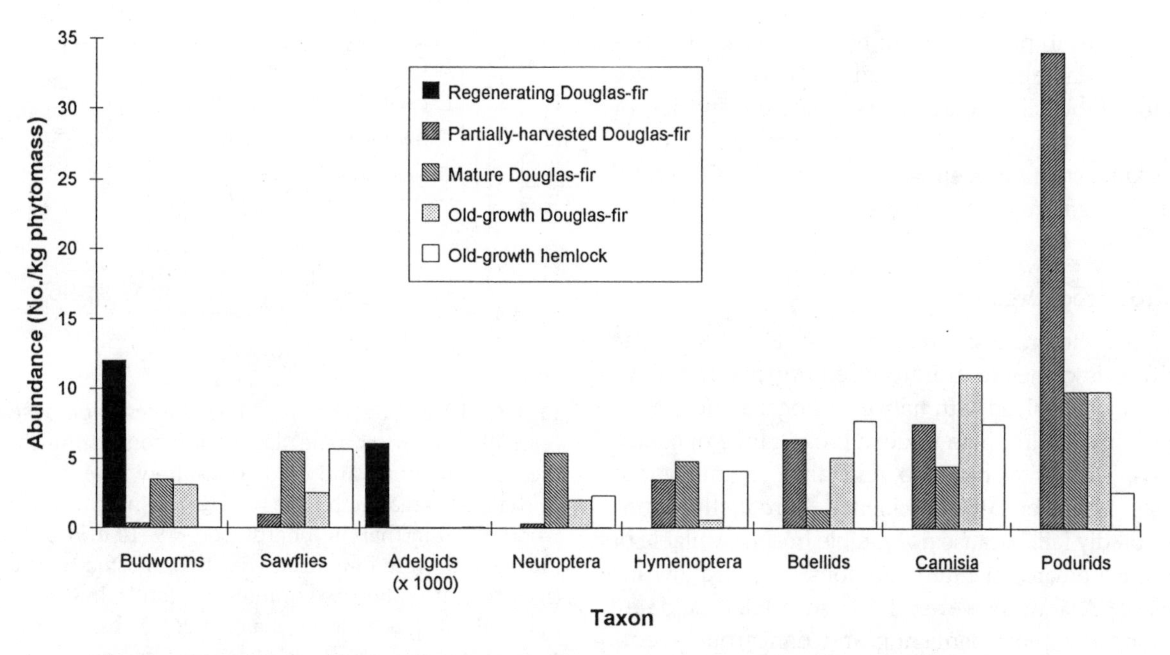

Figure 11.2 Eight canopy arthropod taxa showing significant differences in abundance in forests of contrasting age or harvest practices in western Oregon. Budworms and sawflies are defoliators; adelgids are sap-sucking herbivores; Neuroptera, Hymenoptera, and bdellids (mites) are predators, and *Camisia* (an oribatid mite) and podurids are detritivores. Data represent six replicates of each treatment intermixed in a 15,000-hectare area on and around the H. J. Andrews Experimental Forest (from Schowalter 1995a, and based on a figure from Schowalter 1995b).

stocking levels low. Forests were patchy, with mixed age classes, and often dominated by pine.

Forest management has brought dramatic changes to many sites. Harvest traditionally was by selective cutting, with the goal of maintaining an all-age forest, mimicking the natural fire disturbance regime. Economics and fire-control practices conspired against this outcome, however. Large ponderosa pine and other valuable species were selectively removed, shifting the species mix to the firs. Fire control accelerated this trend, and the result is thousands of acres of very densely stocked stands highly susceptible to insect and pathogen epidemics. Dead wood on the forest floor was consumed in the frequent ground fires of earlier times, but has now accumulated to dangerous levels. The next fire will be a stand-replacing event.

Some phytophagous insects and root pathogens have increased in response to these changes. The Douglas-fir tussock moth and spruce budworm prefer fir species for food, and the frequency and extent of epidemic outbreaks has increased, favored by prolonged drought. Drought conditions have been exacerbated by the overstocked stands, increasing competition for what water is available. The selective harvest created stumps, interspersed among the living trees, and Armillaria and annosus root rot fungi have taken advantage of the new food sources to increase their activity. Again, the firs are more susceptible than the pines. Several species of bark beetle have also responded with increased populations, resulting in increased mortality. Trees weakened by drought, competition, and root rot are especially vulnerable. The killed trees add to the fuel loads and the danger of catastrophic fire.

Conditions That Cause Problems

Changes in forest environments, composition, or structure, beyond the range of conditions that guided development of the ecosystem in question, may lead to changes in populations of both beneficial and potentially damaging insects and fungi. Major changes induced by silviculture include altered species or genotypic composition, fire suppression and overstocking, creation or elimination of food sources, and forest fragmentation.

Introduced Organisms

Perhaps the greatest threat that insects or fungi pose to forests comes from introduced organisms. Native trees have evolved with native pathogens and insects and coexist with them under a broad range of conditions. They may have no resistance to introduced pests, however. Most accidental introductions undoubtedly fail, because no suitable host is available or because environmental conditions do not allow increase. A few, however, have succeeded and are among the most damaging and dangerous agents threatening our forests. In the West, familiar examples include the larch casebearer, Port Orford cedar root rot fungus, and white pine blister rust. There is no practical way to predict which foreign organisms will be damaging in a particular environment; most are unknown or innocuous in their native forests. Effective quarantines, with constant vigilance at ports of entry and in the forest, and determination to respond quickly to early reports of introductions are necessary.

Monocultures

The limited genetic diversity within monocultures of commercially valuable trees, especially when coupled with intensive breeding programs or genetic engineering to increase growth rates, provides a concentrated resource for adapted herbivores and pathogens (Kareiva 1983, Schowalter and Turchin 1993). This resource inevitably will become more vulnerable to herbivores and pathogens as populations grow through time, especially during adverse conditions that stress plants or limit defensive capability. Localized outbreaks occur where isolated stands become vulnerable. These scattered outbreaks can be restricted in a diverse landscape that provides barriers to herbivore population spread (Figure 11.3). Region-wide outbreaks may occur where a sufficient proportion of the landscape is occupied by suitable hosts (Schowalter and Turchin 1993). Large populations can expand into otherwise resistant stands and affect resistance of sparsely distributed hosts.

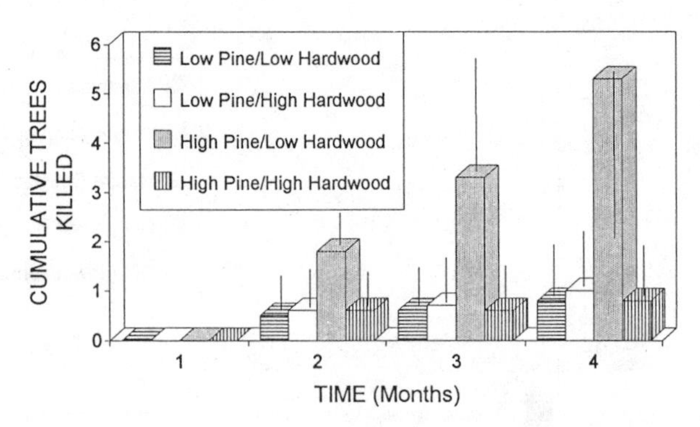

Figure 11.3 Southern pine beetle infestation development in southern pine stands with contrasting basal areas of pine hosts and hardwoods. Low pine = 11–14 m^2/ha basal area; high pine = 23–29 m^2/ha; low hardwood = 0–4 m^2/ha; high hardwood = 9–14 m^2/ha. All infestations were initialized with two infested trees introduced into otherwise uninfested stands. Infested material was placed around a central tree baited with attractant to ensure attack. Data combine four replicated treatment blocks each from the Homochitto National Forest in Mississippi in 1989 and the Kisatchie National Forest in Louisiana in 1990 (from Schowalter and Turchin 1993).

In some ecosystems, seral conifers, such as Douglas-fir, lodgepole pine, ponderosa pine, and the southern pines, develop naturally as extensive single-species stands, especially following stand-replacing fire. It is important to note that these natural populations of a single species are genetically very diverse and adapted to local conditions, including insect and pathogen populations. In the absence of thinning (through disturbance or management), these trees become vulnerable to competitive stress and eventual insect and pathogen outbreaks that accelerate (1) the succession to late-successional species or (2) fuel accumulation and the return of fire (e.g., Hagle and Schmitz 1993).

Silvicultural practices that favor development of one vegetation type over another (e.g., planting monocultures of tree species and eliminating competing vegetation) can also have consequences for the soil microbial community and for ecosystem resilience. After plant community disturbance, the succession of vegetation and soil microbes are interrelated. When natural successional processes are eliminated, the biotic and functional diversity of soil microbes may likewise be altered. For example, many species of hardwoods, such as madrone, manzanita, chinkapin, tan oak, and oaks, quickly reoccupy disturbed sites in forest ecosystems in southwestern Oregon. These pioneering hardwoods form mycorrhizae with many of the same fungal species as the climax conifer species (Molina and Trappe 1982, Molina et al. 1992). Suppression of these hardwoods to promote conifer regeneration thus eliminates a refuge for the diverse array of mycorrhizal fungi and rhizosphere microbes critical to conifer productivity. Recolonizaton must then come from dispersal, via airborne spores or animals, from adjacent stands.

Planting exotic tree species or nonadapted genotypes (off-site planting) or otherwise altering the natural species composition can lead to increased losses. For example, foliage diseases seldom create serious problems in the forest, except where trees are planted off-site in environments that favor fungus infection—for example, pine needle cast suppression of east-side ponderosa pine planted on the west side of the Cascades (Harvey 1976).

Fire Control, Overcrowding, and Thinning

Most forests of the West suffer from too many trees. Fire suppression has had dramatic effects on forest conditions, especially in ecosystems adapted to frequent burning. Fire is a primary mechanism for thinning and reducing competition and also for mineralizing accumulated litter and preventing bottlenecks in nutrient supply. Canopy closure and altered tree species composition as a result of fire suppression have led to serious declines in forest health and consequent insect and pathogen epidemics in forests of the interior West and elsewhere (e.g., Hagle and Schmitz 1993). In stands where fire suppression has lead to an increase in Douglas-fir and true firs in the understory of pine stands, both pines and firs are at increased risk to insects and pathogens. The firs are more susceptible to several root rots and defoliating insects than are the pines, and the increased competition for moisture with the firs places the large overstory pines at greater risk from bark beetles due to moisture stress. Similarly, the replacement over large portions of the southern United States of fire-tolerant longleaf pine savannas with dense commercial plantations of less tolerant loblolly pine, a species more susceptible to the southern pine beetle and fusiform rust, has contributed to recent widespread outbreaks of these organisms (Schowalter and Turchin 1993, Schowalter et al. 1981a).

Crowding leads to intense competition for limited light, water, and nutrients and to pest outbreaks. Thinning is the usual means of preventing overcrowding. Thinning is an effective way to manage some insects and pathogens sensitive to tree spacing or tree competitive stress. The southern pine beetle and the mountain pine beetle can be controlled by spacing trees at least 6 m (20 ft.) apart, the effective limit to pheromone communication (Larsson et al. 1983, Lorio 1980, Mitchell et al. 1983, Sartwell and Stevens 1975, Schowalter and Turchin 1993, Schowalter et al. 1981b). Some defoliators also are sensitive to tree spacing. Host-switching by gypsy moth larvae can be reduced by increased spacing between hosts (Lance 1983).

However, thinning or partial harvest systems that leave stumps or slash among crop trees may promote insects and pathogens that colonize stumps or woody debris and allow them to spread to surrounding trees. Stumps, created by cutting down a living tree, have no close counterpart in the natural forest. They represent a new infection pathway and a large and suddenly available food source for some insects and pathogens. Black stain root disease is vectored from tree to tree by several bark beetles and weevils that are attracted to and breed in fresh stump roots. These beetles introduce the fungus into healthy roots during feeding prior to breeding in fresh stump roots or the roots of diseased trees. Witcosky et al. (1986) reported that thinning at any time increased colonization of Douglas-fir stumps and crop trees by three species of root-colonizing beetles in western Oregon, but that beetle populations increased most in stands thinned during fall and winter. The stumps remained suitable at the time of beetle dispersal in spring, whereas stands thinned during spring and summer were less attractive due to drying of stumps and loss of volatiles by the time of the next insect dispersal.

Armillaria root rot fungus attacks and kills several species of conifers, especially when they are growing adjacent to previously infected stumps. If the fungus is present, it quickly colonizes the stump and root system, and with the extra energy it derives from decomposition, hyphal rhizomorphs are able to spread to, colonize, and kill surrounding trees. Annosus root rot fungus relies primarily on spore spread instead of vegetative growth to colonize new trees. Fresh stumps represent a new infection opportunity for the fungus. Spores germinate free from competition from saprophytic decay fungi, grow down through the stump and roots, and pass into the roots of surrounding trees. Tree mortality from each of these root diseases increases in thinned or partially harvested stands as compared to unharvested or clearcut situations (Goheen and Hansen 1994, Hagle and Schmitz 1993 and references therein).

Fertilization

Fertilization is practiced to increase growth and to replace nutrient capital removed from the forest as a result of harvest, fire, or soil disturbance. However, while fertilization may be necessary in some cases, the form and balance of nutrients most appropriate to accomplish silvicultural goals remain poorly understood. The form of subsidized nutrients can have significant effects on plant growth and insect-microbial-plant interactions. Addition of selected nutrients can create relative shortages of other nutrients, limiting realized tree growth.

Fertilization can profoundly affect soil microbial populations. Nitrogen fertilization, for example, typically reduces ectomycorrhizal development and sporocarp formation, and also alters the composition of fungal species (Menge and Grand 1977, 1978). For these reasons, high nitrogen input from pollution is considered partly responsible for ectomycorrhizal fungus decline in European forests (Arnolds 1991, Schafer and Schoeneberger 1994). Fertilization effects may be short-lived, but nonetheless should be considered within the overall objectives of forest management.

Effects of forest fertilization on herbivory are poorly understood and apparently contradictory, perhaps because few studies have included examination of plant allocation of subsidized nutrients (see references in Schowalter et al. 1986). Tuomi et al. (1984) suggested that plants reallocate carbon and nutrients in response to nutrient limitations. If nutrients such as nitrogen are limiting, carbon cannot be used for growth and may be diverted to the production of plant defenses (Haukioja et al. 1985). If other nutrients necessary for plant growth are limiting, nitrogen cannot be used for growth and may increase the nutritional suitability of the plant for herbivores or be used with excess carbon in production of additional defenses (Mattson 1980, Waring 1987).

Examples of changed insect and pathogen behavior in fertilized stands include increased losses to pitch canker in southern pines (Dwinell et al. 1985), increased incidence of fusiform rust in fertilized southern pine stands (Powers et al. 1993), and increases in spruce budworm populations and defoliation (but also net increases in tree growth) (Mason et al. 1992, Wickman et al. 1992). In commercial coastal Douglas-fir forests, operational fertilization has not led to any alarming changes in insect or fungus activity. High levels of nitrogen fertilizer were applied experimentally to a Douglas-fir plantation to test the hypothesis that increased nitrogen availability would lead to increased populations of soil microorganisms antagonistic to laminated root rot fungus (Nelson 1989). There was no effect on the root rot, but black bears showed a strong preference for the fertilized trees, resulting in a dramatic increase in wounding and tree mortality.

Fragmentation

Forest fragmentation can have serious consequences for many species. Fragmentation increases the proportion of stands subject to edge effects such as wind, drying, and temperature extremes. These conditions are unsuitable for organisms that require the moderate temperatures and humidities provided by extensive, intact forests. Soil and litter organisms and canopy lichens are especially vulnerable to warming and drying (Amaranthus and Perry 1994, Perry et al. 1987, Seastedt 1984, Schowalter and Sabin 1991).

Schowalter (1995a) found that arboreal predators and detritivores also are sensitive to canopy opening and virtually disappear from exposed sites (Figure 11.2). Sunscald and wind breakage damage trees directly and also create infection courts for wound decay fungi.

What Changes Are Necessary?

Fungi and insects are integral parts of forest ecosystems. By acquiring and releasing nutrients, promoting growth, and killing trees, they help to shape the forest. Humans are also part of forest ecosystems and have altered forest structure and controlled fire as means for maximizing production of certain forest commodities. However, some anthropogenic changes have induced undesirable changes in species associated with forest ecosystems, causing destructive outbreaks of some and reduced populations and impaired ecosystem functions for others. Successful ecosystem management will recognize and manage the needs and effects of all forest components—insect, fungus, and human—to assure sustainable productivity of the whole ecosystem.

Populations of phytophagous insects, fungal pathogens, and humans can increase to levels that threaten ecosystem processes and values. A forest without insects and fungi probably couldn't exist and would certainly look and function very differently from its wild counterpart. These organisms must be maintained and managed with as much attention as any other forest component. The key is to prevent problems, because available direct remedies for crisis situations are usually at least as disruptive to the ecosystem as, for example, an insect outbreak or a temporary loss of mycorrhizal inoculum.

A real need is to recognize thresholds. Populations of insects and fungi are sensitive to normal variation in the local environment, and they change through the course of stand development. We need to differentiate the normal fluctuations from trends that are leading toward damaging outbreaks or disappearances. Dangerous population levels, high or low, must be defined according to the specific management objectives for particular stands.

A better understanding of insect, pathogen, and mycorrhizal population responses to environmental changes is needed. We know that some populations track some environmental conditions closely, quickly, and dramatically, making these organisms potentially useful indicators of subtle changes in environmental factors. However, we do not yet know if observed high populations of defoliating insects in young plantations or decreased numbers and changed species of decomposing fungi in recent clearcuts are well within the range of normal successional changes or indicate long-term changes in resource availability or microclimate. The challenge, then, is to understand relationships between natural ecosystem processes and populations of insects and microorganisms in order to evaluate the use of these organisms as bioindicators or to anticipate future undesirable trends in populations.

We will never have complete understanding in a timely way, so certain conservative principles of management are needed. We do not know all the functional roles of the various species nor how future environmental changes will affect their populations and ability to continue critical ecosystem functions. Actions that mimic natural disturbances and successional changes are less likely to have unforeseen consequences, but the rate of disturbance and forest fragmentation may exceed the ability of some populations to adjust or to detect and colonize suitable habitats. New silvicultural approaches should be tried on a small scale and in diverse environments. Ecosystem management must be adapted locally to site-specific conditions. Regular monitoring of populations and ecological conditions is important.

Changing forest management practices to meet objectives of sustainable resource production requires an integrated ecosystem framework for managing forests that incorporates the various roles of insects, pathogens, mycorrhizae, and other species, especially the functional importance of interactions among various species. We will need to accept tradeoffs among species with differing responses to changes in forest conditions. For example, partial harvests will reduce stand susceptibility to most bark beetles and provide refuges for many species, including mycorrhizal fungi, but will likely increase the incidence of stump-colonizing beetles that vector black stain root disease. However, we can maximize protection of biodiversity and ecosystem function while at the same time minimizing risk of pest epidemics by maintaining a greater variety of options (e.g., diversity of forest types and ages, clearcuts, partial cuts, etc.) across landscapes, rather than applying a favored silvicultural system uniformly. This will ensure availability of refuges and stable populations for most species and limit the ability of potential pests to spread unimpeded across the landscape.

Literature Cited

Agee, J. K. 1993. *Fire ecology of Pacific Northwest forests.* Washington, DC: Island Press.

Alfaro, R. I., and R. F. Shepard. 1991. Tree-ring growth of interior Douglas-fir after one year's defoliation by Douglas-fir tussock moth. *Forest Science* 37:959–964.

Allen, E. B., and M. F. Allen. 1990. The mediation of competition by mycorrhizae in successional and patchy environments. In *Perspectives on plant competition,* ed. J. B. Grace and G. D. Tilman. New York: Academic Press.

Allen, M. F. 1991. *The ecology of mycorrhizae.* Cambridge, England: Cambridge University Press.

Amaranthus, M. P., and D. A. Perry. 1994. The functioning of ectomycorrhizal fungi in the field: Linkages in space and time. *Plant and Soil* 159:133–140.

Amaranthus, M. P., J. M. Trappe, and R. J. Molina. 1989. Long-term forest productivity and the living soil. In *Maintaining the long-term productivity of Pacific Northwest forest ecosystems,* ed. D. A. Perry et al. Portland, OR: Timber Press.

Amaranthus, M. P., J. M. Trappe, L. Bednar, and D. Arthur. 1994. Hypogeous fungal production in mature Douglas-fir forest fragments and its relation to coarse woody debris and animal mycophagy. *Canadian Journal of Forest Research* 24:2157–2165.

Amman, G. D., M. D. McGregor, R. F. Schmitz, and R. D. Oakes. 1988. Susceptibility of lodgepole pine to infestation by mountain pine beetles following partial cutting of stands. *Canadian Journal of Forest Research* 18: 688–695.

Arnolds, F. 1991. Decline of ectomycorrhizal fungi in Europe. *Agriculture, Ecosystems and Environment* 35:209–244.

Bormann, F. H., and G. E. Likens. 1979. *Pattern and process in a forested ecosystem.* New York: Springer-Verlag.

Boyce, J. S., and J. W. B. Wagg. 1953. *Conk rot of old-growth Douglas-fir in western Oregon.* Bulletin 4. Oregon State University, Corvallis: Oregon Forest Products Laboratory.

Carroll, G. C. 1988. Fungal endophytes in stems and leaves: From latent pathogen to mutualistic symbiont. *Ecology* 69:2–9.

Castellano, M. A. 1994. Current status of outplanting studies using ectomycorrhiza-inoculated forest trees. In *Mycorrhizae and plant health,* ed. F. L. Pfleger and R. G. Linderman. St. Paul, MN: American Phytopathology Press.

Castellano, M. A., and R. Molina. 1989. Mycorrhizae. In *The container tree nursery manual.* Vol. 5 of *Agricultural handbook 674,* ed. T. D. Landis et al. Washington, DC: USDA Forest Service.

Castello, J. D., D. J. Leopold, and P. J. Smallidge. 1995. Pathogens, patterns, and processes in forest ecosystems. *BioScience* 45:16–24.

Cromack, K., Jr., B. L. Fichter, A. R. Moldenke, and E. R. Ingham. 1988. Interactions between soil animals and ectomycorrhizal fungal mats. *Agriculture, Ecosystems and Environment* 24:161–168.

Cromack, K., Jr., P. Sollins, W. C. Graustein, K. Speidel, A. W. Todd, G. Spycher, C. Y. Li, and R. L. Todd. 1979. Calcium oxalate accumulation and soil weathering in mats of the hypogeous fungus, *Hysterangium crassum. Soil Biology and Biochemistry* 11:463–468.

Deacon, J. W., and L. V. Fleming. 1992. Interactions of ectomycorrhizal fungi. In *Mycorrhizal functioning: An integrative plant-fungal process,* ed. M. F. Allen. New York: Chapman and Hall.

Dickman, A., and S. Cook. 1989. Fire and fungus in a mountain hemlock forest. *Canadian Journal of Botany* 67:2005–2016.

Dwinell, L. D., J. B. Barrows-Broaddus, and E. G. Kuhlman. 1985. Pitch canker: A disease complex of southern pines. *Plant Disease* 69:270–276.

Edmonds, R. L. 1982. *Analysis of coniferous forest ecosystems in the western United States.* Stroudsburg, PA: Hutchinson & Ross.

Fogel, R., and G. Hunt. 1983. Contributions of mycorrhizae and soil fungi to nutrient cycling in a Douglas-fir ecosystem. *Canadian Journal of Forest Research* 13:219–232.

Franklin, J. F., and C. T. Dyrness. 1984. *Natural vegetation of Oregon and Washington.* Corvallis, OR: Oregon State University Press.

Furman, T. E., and J. M. Trappe. 1971. Phylogeny and ecology of mycotrophic achlorophyllous angiosperms. *Quarterly Review of Biology* 46:219–225.

Goheen, D. J., and E. M. Hansen. 1993. *Effects of pathogens and bark beetles on forests.* In *Beetle-pathogen interactions in conifer forests,* ed. T. D. Schowalter and G. M. Filip. New York: Academic Press.

———. 1994. Tree vigor and susceptibility to infection by *Phellinus weirii:* Results of field inoculations. In *Proceedings of Eighth International Conference on Root and Butt Rots, August 9–16, 1993.* Uppsala: Swedish University of Agricultural Sciences.

Goldfarb, B., E. E. Nelson, and E. M. Hansen. 1989. *Trichoderma* species from Douglas-fir stumps and roots infested with *Phellinus weirii* in the western Cascades of Oregon. *Mycologia* 81:134–138.

Hagle, S., and Schmitz, R. 1993. Managing root disease and bark beetles. In *Beetle-pathogen interactions in conifer forests,* ed. T. D. Schowalter and G. M. Filip. New York: Academic Press.

Hansen, E. M. 1977. Forest pathology: Forests, fungi, and man. In *Mushrooms and man,* ed. A. C. Waters. Albany, OR: Linn Benton Community College.

Hansen, E. M., D. J. Goheen, P. F. Hessburg, J. J. Witcosky, and T. D. Schowalter. 1988. Biology and management of black stain root disease in Douglas-fir. In *Leptographium root diseases on conifers,* ed. T. C. Harrington and F. W. Cobb Jr. St. Paul, MN: APS Press.

Harley, J. L., and S. E. Smith. 1983. *Mycorrhial symbiosis.* London: Academic Press.

Harmon, M. E., J. F. Franklin, F. J. Swanson, P. Sollins, S. V. Gregory, J. D. Lattin, N. H. Anderson, S. P. Cline, N. G. Aumen, J. R. Sedell, G. W. Lienkaemper, K. Cromack Jr., and K. W. Cummins. 1986. Ecology of coarse woody debris in temperate ecosystems. *Advanced Ecological Research* 15:133–302.

Harvey, A. E., D. S. Page-Dumroese, R. T. Graham, and M. F. Jurgensen. 1991. Ectomycorrhizal activity and conifer growth interactions in Western-montane forest soils. In *Proceedings management and productivity of Western-montane forest soils.* General technical report INT 280. Ogden, UT: USDA Forest Service.

Harvey, G. J. 1976. Epiphytology of a needle cast fungus, *Lophodermella morbida,* in ponderosa pine plantations in western Oregon. *Forest Science* 22:223–230.

Haukioja, E., P. Niemela, and S. Siren. 1985. Foliage phenols and nitrogen in relation to growth, insect damage, and ability to recover after defoliation in the mountain birch, *Betula pubescens* ssp. *tortuosa. Oecologia* 65:214–222.

Holah, J. C., M. V. Wilson, and E. M. Hansen. 1993. Effects of a native forest pathogen, *Phellinus weirii,* on Douglas-fir forest composition in western Oregon. *Canadian Journal of Forest Research* 23:2473–2480.

Janos, D. P. 1980. Mycorrhizae influence topical succession. *Biotropica* 12:56–64.

Kareiva, P. 1983. Influence of vegetation texture on herbivore populations: Resource concentration and herbivore movement. In *Variable plants and herbivores in natural and managed systems,* ed. R. F. Denno and M. S. McClure. New York: Academic Press.

Lance, D. R. 1983. Host-seeking behavior of the gypsy moth: The influence of polyphagy and highly apparent host plants. In *Herbivorous insects: Host-seeking behavior and mechanisms,* ed. S. Ahmad. New York: Academic Press.

Larsson, S., R. Oren, R. H. Waring, and J. W. Barrett. 1983. Attacks of mountain pine beetle as related to tree vigor of ponderosa pine. *Forest Science* 29:395–402.

Lorio, P. L., Jr. 1980. Loblolly pine stocking levels affect potential for southern pine beetle infestation. *Southern Journal of Applied Forestry* 4:162–165.

Marx, D. H. 1973. Mycorrhiza and feeder root diseases. In *Ectomycorrhizae: Their ecology and physiology,* ed. G. C. Marks and T. T. Kozlowski. New York: Academic Press.

Mason, R. R., B. E. Wickman, R. C. Beckwith, and H. G. Paul. 1992. Thinning and nitrogen fertilization in a grand fir stand infested with western spruce budworm. Part I: Insect response. *Forest Science* 38:235–251.

Mattson, W. J. 1980. Herbivory in relation to plant nitrogen content. *Annual Review of Ecology and Systematics* 11:119–161.

Mattson, W. J. and N. D. Addy. 1975. Phytophagous insects as regulators of forest primary production. *Science* 190:515–522.

Mattson, W. J., and R. A. Haack. 1987. The role of drought in outbreaks of plant-feeding insects. *BioScience* 37:110–118.

McCutcheon, T. L., and G. C. Carroll. 1993. Genotypic diversity in populations of a fungal endophyte from Douglas-fir. *Mycologia* 85:180–186.

Menge, J. A., and L. F. Grand. 1977. The effect of fertilization on growth and mycorrhizae numbers in 11-year-old loblolly pine plantations. *Forest Science* 23:37–44.

———. 1978. Effect of fertilization on production of epigeous basidiocarps by mycorrhizal fungi in loblolly pine plantations. *Canadian Journal of Botany* 56:2357–2362.

Mikola, P. 1973. Application of mycorrhizal symbiosis in forestry practice. In *Ectomycorrhizae: Their ecology and physiology,* ed. G. C. Marks and T. T. Kozlowski. New York: Academic Press.

Mitchell, R. G., R. H. Waring, and G. B. Pitman. 1983. Thinning lodgepole pine increases tree vigor and resistance to mountain pine beetle. *Forest Science* 29:204–211.

Molina, R., and J. M. Trappe. 1982. Lack of mycorrhizal specificity by the ericaceous hosts *Arbutus menziesii* and *Arctostaphylos uva-ursi. New Phytology* 90:495–509.

Molina, R., H. Massicotte, and J. M. Trappe. 1992. Specificity phenomena in mycorrhizal symbioses: Community-ecological consequences and practical implications. In *Mycorrhizal functioning: An integrative plant-fungal process,* ed. M. F. Allen. New York: Chapman and Hall.

Nelson, E. E. 1989. Black bears prefer urea fertilized trees. *Western Journal of Applied Forestry* 4:13–15.

Parks, C. G. 1993. The influence of induced host moisture stress on the growth and development of western spruce budworm and *Armillaria ostoyae* on grand fir seedlings. Ph.D. thesis, Oregon State University, Corvallis.

Parsons, G. L., G. Cassis, A. R. Moldenke, J. D. Lattin, N. H. Anderson, J. C. Miller, P. Hammond, and T. D. Schowalter. 1991. *Invertebrates of the H. J. Andrews Experimental Forest, western Cascade Range, Oregon: An annotated list of insects and other arthropods.* General technical report PNW-GTR-290. Portland, OR: USDA Forest Service.

Perry, D. A., R. Molina, and M. P. Amaranthus. 1987. Mycorrhizae, mycorrhizospheres, and reforestation: Current knowledge and research needs. *Canadian Journal of Forest Research* 17:929–940.

Perry, D. A., M. P. Amaranthus, J. G. Borchers, S. L. Borchers, and R. E. Brainerd. 1989. Bootstrapping in ecosystems. *BioScience* 39:230–237.

Pfleger, F. L., and R. G. Linderman. 1994. *Mycorrhizae and Plant Health.* St. Paul, MN: APS Press.

Powers, H. R., T. Miller, and R. P. Belanger. 1993. Management strategies to reduce losses from fusiform rust. *Southern Journal of Applied Forestry* 17:146–149.

Read, D. J. 1992. The mycorrhizal mycelium. In *Mycorrhizal functioning: An integrative plant-fungal process,* ed. M. F. Allen. New York: Chapman and Hall.

Sartwell, C., and R. E. Stevens. 1975. Mountain pine beetle in ponderosa pine: Prospects for silvicultural control in second growth stands. *Journal of Forestry* 73:136–140.

Schafer, S. R., and M. M. Schoeneberger. 1994. Air pollution and ecosystem health: The mycorrhizal connection. In *Mycorrhizae and plant health,* ed. F. L. Pfleger and R. G. Linderman. St. Paul, MN: American Phytology Press.

Schlosser, W. E., and K. A. Blatner. 1995. The wild edible mushroom industry of Washington, Oregon, and Idaho. *Journal of Forestry* 93:31–36.

Schowalter, T. D. 1985. Adaptations of insects to disturbance. In *The ecology of natural disturbance and patch dynamics,* ed. S. T. A. Pickett and P. S. White. New York: Academic Press.

———. 1994. An ecosystem-centered view of insect and disease effects on forest health. In *Sustainable ecological systems: Implementing an ecological approach to land management,* ed. W. W. Covington and L. F. DeBano. Technical report RM-247. Ft. Collins, CO: USDA Forest Service.

———. 1995a. Canopy arthropod community response to forest age and alternative harvest practices in western Oregon. *Forest Ecology Management* 78:115–125.

———. 1995b. Canopy invertebrate response to disturbance and consequences of herbivory in temperate and tropical forests. *Selbyana* 16:41–48.

Schowalter, T. D., and T. E. Sabin. 1991. Litter microarthropod responses to canopy herbivory, season, and decomposition in litterbags in a regenerating conifer ecosystem in western Oregon. *Biology and Fertility of Soils* 11:93–96.

Schowalter, T. D., and P. Turchin. 1993. Southern pine beetle infestation development: Interaction between pine and hardwood basal areas. *Forest Science* 39:201–210.

Schowalter, T. D., R. N. Coulson, and D. A. Crossley Jr. 1981a. Role of southern pine beetle and fire in maintenance of structure and function of the southern pine beetle. *Environmental Entomology* 10:821–825.

Schowalter, T. D., D. N. Pope, R. N. Coulson, and W. S. Fargo. 1981b. Patterns of southern pine beetle (*Dendroctonus frontalis Zimm*) infestation enlargement. *Forest Science* 27:837–849.

Schowalter, T. D., W. W. Hargove, and D. A. Crossley Jr. 1986. Herbivory in forested ecosystems. *Annual Review of Entomology* 31:177–196.

Schowalter, T. D., T. E. Sabin, S. G. Stafford, and J. M. Sexton. 1991. Phytophage effects on primary production, nutrient turnover, and litter decomposition of young Douglas-fir in western Oregon. *Forest Ecology Management* 42:229–243.

Schowalter, T. D., B. A. Caldwell, S. E. Carpenter, R. P. Griffiths, M. E. Harmon, E. R. Ingham, R. G. Kelsey, J. D. Lattin, and A. R. Moldenke. 1992. Decomposition of fallen trees: Effects of initial conditions and heterotroph colonization rates. In *Tropical ecosystems: Ecology and management,* ed. K. P. Singh and J. S. Singh. New Delhi: Wiley Eastern Ltd.

Seastedt, T. R. 1984. The role of microarthropods in decomposition and mineralization processes. *Annual Review of Entomology* 29:25–46.

Silvester, W. B., P. Sollins, T. Verhoeven, and S. P. Cline. 1982. Nitrogen fixation and acetylene reduction in decaying conifer boles: Effects of incubation time, aeration, and moisture content. *Canadian Journal of Forest Research* 12:646–652.

Sturgeon, K. B., and J. B. Mitton. 1986. Allozyme and morphological differentiation of mountain pine beetles *Dendroctonus ponderosae* Hopkins (*Coleoptera: Scolytidae*) associated with host trees. *Evolution* 40:290–302.

Swank, W. T., and D. A. Crossley Jr., eds. 1987. *Forest hydrology and ecology at Coweeta.* New York: Springer-Verlag.

Thies, W. G. 1984. Laminated root rot: The quest for control. *Journal of Forestry* 82:345–356.

Trumble, J. T., D. M. Kolodny-Hirsch, and I. P. Ting. 1993. Plant compensation for arthropod herbivory. *Annual Review of Entomology* 38:93–119.

Tuomi, J., P. Niemela, E. Haukioja, S. Siren, and S. Neuvonen. 1984. Nutrient stress: An explanation for plant antiherbivore responses to defoliation. *Oecologia* 61:208–210.

Van der Kamp, B. J. 1991. Pathogens as agents of diversity in forested landscapes. *Forestry Chronicle* 67:353–354.

Vogt, K. A., C. C. Grier, C. E. Meier, and R. L. Edmonds. 1982. Mycorrhizal role in net production and nutrient cycling in *Abies amabilis* ecosystems in western Washington. *Ecology* 63:370–380.

Waring, R. H. 1987. Characteristics of trees predisposed to die. *BioScience* 37:569–573.

Waring, R. H., K. Cromack Jr., P. A. Matson, R. D. Boone, and S. G. Stafford. 1987. Responses to pathogen-induced disturbance: Decomposition, nutrient availability, and tree vigour. *Forestry* 60:219–227.

Wickman, B. E. 1980. Increased growth of white fir after a Douglas-fir tussock moth outbreak. *Journal of Forestry* 78:31–33.

———. 1992. *Forest health in the Blue Mountains: The influence of insects and disease.* General technical report PNW-GTR-295. Washington, DC: USDA Forest Service.

Wickman, B. E., R. R. Mason, and H. G. Paul. 1992. Thinning and nitrogen fertilization in a grand fir stand infested with western spruce budworm. Part II: Tree growth response. *Forest Science* 38:252–264.

Wilson, E. O. 1992. *The diversity of life.* Cambridge, MA: Harvard University Press.

Witcosky, J. J., T. D. Schowalter, and E. M. Hansen. 1986. The influence of precommercial thinning on the colonization of Douglas-fir by three species of root-colonizing insects. *Canadian Journal of Forest Research* 16:745–749.

12 Fire Management for the 21st Century

James K. Agee

Historical Themes 191
Recent Policy Developments 192
Integrating Fire Management with Ecosystem Management 194
Predicting Fire Behavior and Effects 194
Fire Management Strategies 195
Fire at the Landscape Level 198
Conclusion 200
Literature Cited 201

Current debates about forest management sometimes obscure the tremendous progress we have made in understanding and managing forest ecosystems. At the same time, forests change at predictable and somewhat slow rates, so mistakes we have made may be with us for a while. As we look toward the 21st century and its implications for fire management, we need to first reflect on the balance of management decisions—in retrospect good or bad—that we made in the 20th century.

Historical Themes

In 1900 there was essentially no organized fire protection outside of volunteer cooperatives, and in many places there was virtually no need for protection: Fire ran free in the forest. Two major themes in fire management occurred during this century. The first was the development of organized fire protection from 1900 to 1950—that is, the institution building necessary to combat a common "enemy" (see, e.g., Cowan 1961). The second was the evolution of these institutions to reflect varied goals of land management, making our response to fire more complex over the second half of the century.

The evolution of fire management over the past century has been analyzed elsewhere (Pyne 1982, Agee 1993) and will not be summarized in detail here. Suffice it to say that, like insect- and disease-protection issues, initial approaches to fire management focused on simplistic and eventually unsuccessful attempts at total fire exclusion. We suppressed fire in landscapes where it was historically frequent, as well as where it was not. Successful fire prevention messages featuring Smokey Bear, in addition to continually improving fire-fighting technology, reduced wildfire acreage to historic low levels by the mid-1950s (Figure 12.1).

Many decades later, we have come to realize the paradox inherent in our noble efforts. The more intensely we have protected the forest from fire, as well as from insects and disease, the worse many of these problems have become. U.S. fire statistics show an alarming trend in wildfire acreage, primarily attributable to fuel buildups in western forests (Figure 12.1).

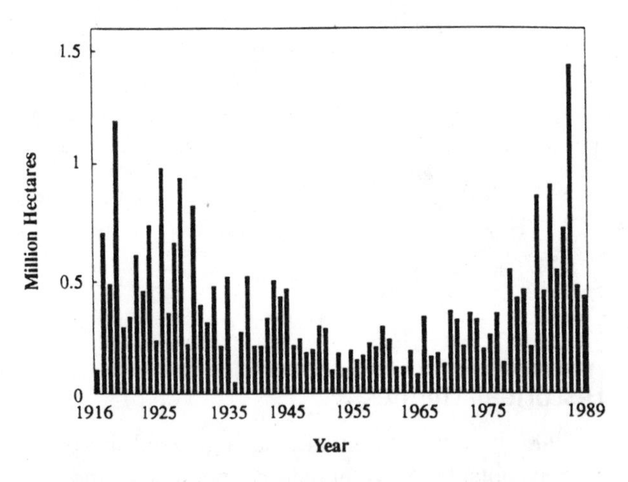

Figure 12.1 Wildfire acreage in the 11 western states between 1915 and 1990. Data before 1931 omit burned areas in national parks and Native American reservations, and before 1926 only show forested land burned (data compiled by B. Mitchell, USDI BLM, Boise, ID; provided by J. K. Brown and S. F. Arno, USDA Forest Service, Missoula, MT; and reproduced from Agee [1993]).

The fuel buildups were substantially in place by 1950, as clearly shown by photographic (Gruell et al. 1982) and age class information (McNeil and Zobel 1980). We have been sitting on a time bomb with little idea of how long the fuse is. Are we at the worst case scenario now, or will it get worse?

Furthermore, protection issues concerning fire, wind, disease, and insects were managed separately as if there were no links between them. For example, we have found that the absence of one disturbance can increase another. A good example is the insect defoliator problems on shade-tolerant trees in the Blue Mountains of Oregon and Washington. Much of the current high tree density in those species is a direct result of overprotection from historic, low-intensity surface fires. Removal of fire from these systems has allowed defoliator-susceptible trees to increase, making insect epidemics longer and more damaging than they were in previous centuries. There may also be synergistic effects between disturbances—the presence of one disturbance can increase another. Wind disturbance, or blowdown, in the coastal forests of the Pacific Northwest often may be followed by Douglas-fir beetle attack on residual trees.

Many of our forests are in much less sustainable condition than they were a century ago, in part due to mismanagement and in part due to a lack of a good information about the long-term effects of management actions. Because we have been more concerned about outputs, as opposed to taking a more holistic look at the system, the state of the forest ecosystem has in many cases deteriorated.

Recent Policy Developments

Major shifts in fire management paradigms occurred in the latter part of this century. The most significant shift was associated with recognition of the role of fire in maintaining natural ecosystems and the adoption of monitored natural fires as a management strategy for parks and wilderness areas (Kilgore and Heinselman 1990). Although fire was recognized as a significant influence on natural ecosystems for many decades, particularly in the southern United States (Komarek 1974), it was in the West that natural fire policies were tested and implemented. At first, these were called let-burn fires (Kilgore and Briggs 1972), but this phrase more appropriately described what the fire did, not what the management strategy entailed. It was later changed to "prescribed natural fire" to reflect the monitoring and decision processes associated with these naturally ignited fires in backcountry areas. As the early successes with this policy multiplied, the application of truly wild fire expanded past its socially acceptable limit at Yellowstone in 1988. The Yellowstone fires of 1988 burned over a million acres in the national park and environs. Almost half of the area in the park was burned by some of the most intense fires ever recorded. Often understated by both proponents and opponents of the fires of 1988 was the fact that about half the burned area resulted from human-associated fire (downed power lines, campfires, or equipment) on which suppression action was taken from the very beginning. In addition to a year of policy change, it was a bad fire year. In the end, the natural fire policy was declared appropriate. However, very conservative implementation procedures were adopted, such that fire may not play a very natural role in most areas due to greatly expanded checklists that in most instances result in a decision to suppress the fire.

Other paradigm shifts were driven by external factors such as air quality. Slash burning began to be controlled for air quality purposes by 1970 (Agee 1989) with continued pressure to burn less near urban areas. Further pressure to burn less slash was applied with federal visibility protection regulations for wilderness and park areas (called Class I). The transition from old-growth to second-growth harvesting was associated with less slash creation too, so there was less need to apply an increasingly expensive site preparation tool.

Prescribed fires and natural fires have become part of the working tools of today's fire managers, who have an ever-expanding set of land management goals, including protection of commercial timber, managing fuels around rural developments (also called the urban interface), and reintroducing or maintaining the natural role of fire in park and wilderness ecosystems. As a result, numerous strategies have emerged to achieve these goals, including combinations of fire prevention, suppression, prescribed fire, and prescribed natural fire (allowing natural fires to burn with monitoring). Fire has moved from being a suppression-oriented goal in and of itself to a subset of land management planning. To some extent, the pendulum may have shifted a little too far because some recent land management strategies have not incorporated wildfire risks to the extent perhaps needed (e.g., the northern spotted owl recovery plan), even though some risk has been recognized.

In the last half of the 20th century, major strides were made in predicting the behavior of wildfires given the inputs of fuels, weather, and topography (Figure 12.2). Improved and standardized fire danger ratings (Deeming et al. 1977) and predictions of fire spread and intensity (Rothermel 1983) have improved our ability to forecast and predict fire behav-

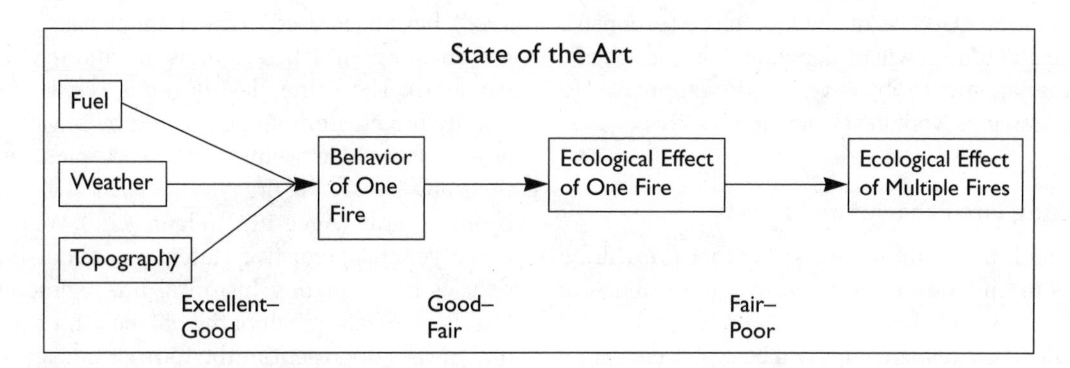

Figure 12.2 The prediction of fire behavior (rate of spread and fire-line intensity) from fuels, weather, and topography is good to excellent, while the extension of that prediction to ecological effects is weaker, particularly those effects associated with smoldering or large log burnout. The prediction capability for effects of multiple fires is poor.

ior. These developments have greatly aided fire managers in suppression situations. Either standardized fuel models or site-specific, customized fuel models can be used in these applications. Our current ability to take the next step and apply the prediction outputs of fire behavior to ecological effects is less developed (Figure 12.2). Some effects, like crown scorch, are accurately predicted from the fire behavior prediction outputs (Van Wagner 1973). Cambial and root heating predictions have been attempted (Peterson and Ryan 1986, Frandsen 1991), but are not yet predictable with any stated accuracy. The landscapes we see are often the result of many fires of differing character: myriad frequencies, unpredictable intervals, variable intensities, different seasons, and different interactions. Our understanding of how multiple fires interact remains poor, so that future fire management strategies are likely to be hampered by this lack of information.

Integrating Fire Management with Ecosystem Management

Fire management in the 21st century will be an evolving combination of strategies, most of which were developed in some form in the 20th century. During the 20th century, we experienced institution building on a massive scale. Those institutions lost track of the overall goals as they focused on narrow fire control objectives. We will see a broader mix of prescribed fire and thinning and prescribed natural fire strategies mixed with fire prevention and suppression strategies. The mix will be more responsive in space and time; it will be dependent on the specific area management objectives and responsive to which strategies work best over time.

Predicting Fire Behavior and Effects

As we look to the future, we know that fire will be with us in one form or another. To some extent, we have the ability to choose what that relationship will be. None of the relationships will be cost-free. Large, catastrophic wildfires will still occur at times in some places. Even with continued and improved safety measures, disasters such as the 1994 Storm King fire in Colorado, with 14 firefighter deaths, will occasionally happen—although hopefully not more frequently than the interval between the last disaster in which the Mann Gulch fire of 1949 killed 13 firefighters. Prescribed fires will escape due to poor planning or execution or because of intangible factors that combine to move the fire out of prescription. If fire approaches even a fraction of its range of historic variation, air quality may significantly decline, affecting visibility and perhaps human health and welfare.

The challenges are great, but so are the opportunities. Much of the natural diversity we see on the landscape today is the result of disturbance by fire. The presence of some tree species on Pacific Northwest landscapes, such as knobcone pine, are the result of past fires opening their serotinous cones. Shrubs such as ceanothus are post-fire nitrogen fixers. The mix of hemlock, Pacific silver fir, and Douglas-fir we see in the Cascades and Olympics reflects a complex history of fire and subsequent windthrow. In fact, almost all the superlative old-growth Douglas-fir forests are first-generation forests borne of fire.

We need to recognize that fire is not a constant in space or time, but that fire behavior and effects are not so unique that generalizations are impossible. The patterns of fire frequency, predictability, intensity, extent, synergism, and timing that together are known as the fire regime (Agee 1993) are capable of being classified and categorized in relatively understandable ways. Vegetation classifications, such as potential vegetation or habitat typing (Daubenmire 1968), have been used to order and explain the physical environment (such as moisture and temperature variation). The same classifications can be used to classify fire environments, ordered by physical characters of fire (Heinselman 1973), by plant associations and interacting fire effects (Davis et al. 1980), or by fire severity (Agee 1993)(Figure 12.3). We then can make general predictive statements about fire tendencies by evaluating historical fire regime characters. Fortunately, much of the database for such evaluations is still present in the form of tree age classes or fire scars on live trees or stumps. It is, however, a slowly disappearing database being erased from the landscape by natural decay, wildfire, or prescribed

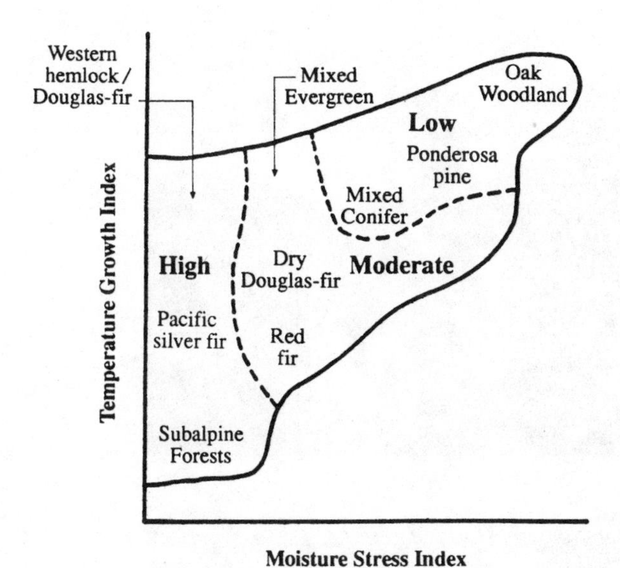

Figure 12.3 Fire severity can be defined as the effects of fire on the dominant vegetation of a site. In this diagram, three fire severity regimes (high, moderate, and low) illustrate the historical fire severity levels for the forest types of the Pacific Northwest.

fire. The fire scar record on stumps is preserved by the resinous nature of the stump cross sections with the scars, even though much of the rest of the stump is decayed. Fire frequency and its variation can still be reconstructed from stumps created in the 1920s. Major salvage efforts are needed to recover site-specific fire history information, or it will be lost forever.

The development of prediction capability for the behavior of surface fires over the last 20 years has increased our ability to predict the behavior of wildfires, prescribed fires, and prescribed natural fires. Future research will likely concentrate on the behavior of crown fires and ground fires as well as improvements in the surface fire model (Rothermel and Andrews 1987). At a landscape level, the co-evolution of geographic information systems containing vegetation and fuel layers and fire behavior models will allow real-time simulations of projected fire behavior—prototypes are already available (Dr. M. Finney, Systems for Environmental Management, Missoula, MT, personal communication, 1995). The key remaining variables will be weather inputs that drive fire across the landscape, and these will remain somewhat unpredictable.

Fire Management Strategies

Fire management strategies include all appropriate actions related to fire—its prevention, suppression, and use. Fire prevention efforts will shift from the rather stuffy image of Smokey Bear for two reasons. First, Smokey's message, "Only you can prevent forest fires," is too simple for what will be a more complex fire management job. Second, the fire prevention message will be focused at specific sources and targets, requiring awareness and education beyond the simple and, in many respects, inaccurate message of Smokey: For example, lightning, the major cause of fires on western national forests, is not preventable by "you." As population continues to increase and merge with wildland environments, fire prevention will be less generic and more site- and population-specific. Fire control technology will continue to evolve, and the costs of fire control will continue to increase (e.g., Hoover 1993). Decisions will have to be made about what kind of suppression actions are cost-effective.

Expansion of prescribed fire use will occur, primarily in drier forest types, probably not without problems. There will be a need to apply fire at a landscape level, and developing that expertise will require a learning curve. Aerial ignition, common in dry Australian forests, will likely emerge as a preferred ignition technique to allow coverage of large areas in short "prescription windows" or at times when acceptable fire behavior will occur. If prescribed fire use is expanded, regional-scale air quality analyses will be necessary to avoid transport of large volumes of smoke to other regions, and substantial interaction with local communities will be necessary. Without local community involvement, prescribed burning does not have a future.

Prescribed natural fire use is currently rebounding from the Yellowstone year, in which a number of initially allowed fires eventually required millions of dollars of fire control efforts. Most of these fires were in backcountry and subalpine and alpine areas (Figure 12.4) where their results will be evident into the 22nd century. The development of geographic infor-

Figure 12.4 (a) Subalpine fires are high-severity events that may transform forests to non-forest vegetation for a century or more. (b) A 55-year-old burn (center) has paltry regeneration and substantial snag longevity. (c) At the edge of the same 55-year-old burn, the seed source is more plentiful and takes advantage of a favorable climate. (d) A 90-year-old burn still has an open landscape after nearly a century of post-fire regeneration.

mation system models incorporating fire behavior will allow gaming strategies to be employed. This will be a powerful addition to current decision models and may increase the confidence of managers that a fire allowed to burn will remain within predetermined limits. At the same time, there may be increased use of prescribed fire or fuel breaks to increase fire control probability at or near the margin of prescribed natural fire zones (Agee 1995).

Every resources management issue we will face will have a fire component as part of the solution. The solution may be aggressive fire suppression or the use of fire in some form. An instructive example of the need for site- and population-specific strategies is the case of the northern spotted owl in the Pacific Northwest. Fire and the northern spotted owl seem strange bedfellows, and to some extent they are. But across the range of the threatened owl, natural fire cycles vary from greater than 500 years to less than 20 years. The northern spotted owl lives in old-growth forest over much of its range, preferring multilayered canopies typical of forests exceeding 150 to 200 years of age. In the coastal portion of its range, the owl's habitat develops as natural thinning occurs and small canopy gaps are filled with trees released from the understory. In these areas, where annual precipitation exceeds 80 inches, natural fire return intervals begin to exceed 100 to 200 years, implying that fire occurs under infrequent conditions. Many of these fires begin as lightning strikes that smolder for days or weeks before weather needed for significant fire spread occurs (typically east-wind events in the Pacific Northwest). In this wetter end of the northern spotted owl range, the best fire strategy is an aggressive fire suppression response (Agee and Edmonds 1992). Prescribed fire has little place because it would be difficult to apply and fuel management is not needed due to generally moist conditions. The multilayered old-growth structure, which the owls prefer, should be best preserved under such a fire strategy.

In the drier part of the owl range, multilayered canopy conditions are an artifact of fire suppression (Figure 12.5). Such multilayered structure was rare and limited to riparian areas. Fires burned frequently, but were rarely stand-replacing events; these forests experienced the classic low-severity fire regimes which had little impact on overstory trees. The 1908 photo (Figure 12.5a) shows the result of these benign forest fires. In those days, fire suppression commonly occurred when individual forest rangers tied pine boughs to the tails of their horses and walked them along the edge of the flickering flames, spreading the needles to the side and creating a successful fire line. Fire exclusion over the 20th century has resulted in multilayered forests (Figure 12.5b) composed of shade-tolerant species. Habitat for northern spotted owls may well have increased, but these forest structures are not sustainable. First, the shade-tolerant understory of Douglas-fir and grand fir is much more susceptible to insect and disease problems (Everett et al. 1994). Second, when wildfires burn these forests now, more continuous, heavier, and three-dimensional fuel loads cause these fires to be of high severity. Much of the east-side spotted owl habitat will be lost over the next century if we pretend we can draw a line around these areas and leave them alone.

These areas need to be broken up with fuel treatment, either by prescribed underburning (Figure 12.6a) or by fuel break construction (Figure 12.6b). In some cases commercial thinning can be integrated with the operation to reduce site biomass and smoke impacts resulting from fire. At present, there is the

(a)

(b)

Figure 12.5 (a) A ponderosa pine forest in western Montana in 1908 showing low fuel conditions, substantial herb and shrub cover, and wide-spaced trees. Limited selective logging has recently occurred, but appears to have had little impact on the condition of the forest. (b) The same stand in 1948, showing invasion of Douglas-fir with fire exclusion in this Douglas-fir/snowberry plant association. The multi-storied canopy presents a much greater fire hazard. Photographs showing the succession to Douglas-fir on this and other sites are reproduced in Gruell et al. (1982) and in Agee (1993). Note that this photo is from a forest outside the range of the northern spotted owl, but the structural changes within the owl's range are similar.

(a)

(b)

Figure 12.6 (a) Two prescribed fires have reduced tree density and opened up this mixed-conifer stand in southern Oregon. With thinning, one prescribed fire might show the same effect. (b) A fuelbreak in southern Oregon, where small groups of understory trees were left, illustrates that these corridors can be visually pleasing and yet break up continuous blocks of heavy fuel on the landscape.

potential of decreasing owl habitat if owls are not adapted to more open forest structure in these areas. Clearly, experimental areas need to be established first with careful monitoring to guide more extensive application of techniques. The hopeful result will be a balance between sustainable forest structure and sustainable owl populations. Our ability to predict and manage fire behavior has advanced incredibly in the last two decades. Our understanding of fire effects is increasing, but it is not yet as well advanced as our ability to control and use fire.

Our choice in forest protection is a partial Hobson's choice. Hobson's Choice was to accept what was given or get nothing at all. If we ignore forest protection in owl reserve planning, we will get something—but not what we want. But if we incorporate forest protection into planning, we probably will also not get all of what we want because other objectives cannot be maximized if forest sustainability is to be achieved.

Fire at the Landscape Level

As we begin to grasp the ecological effects of fire at the stand level, we are faced with the new challenge of understanding how fire on one piece of the landscape interacts with the other pieces. Several emerging issues will require a better understanding of fire effects at larger scales.

Retention Harvesting Strategies

New forestry has been presented as a conceptual approach to light-on-the-land harvesting, particularly in the reservation of biological legacies such as downed logs, snags, and green tree retention. The basic idea is that the transition to young growth will be more gradual and, in theory at least, have less impact on forest flora and fauna and the processes that sustain them. Such units may provide a less hostile landscape matrix for wildlife around core areas of protected landscapes. In the past, the more traditional clearcutting approach often involved slash burning after harvest for site preparation—although there has been a trend away from slash burning on many lands since the 1970s. It is not yet clear how fire and new forestry will evolve. For example, it is possible to underburn units with residual trees, snags, and logs. Many logs are wet enough that only a thin shell of wood will be burned off the exterior (Ottmar 1983). Snags can be treated with foam to reduce the probability of ignition, and large live trees of some species (e.g., Douglas-fir, ponderosa pine, western larch) are relatively resistant to fire. However, the logistics involved are expensive and the prescription windows for such burns are narrow, so opportunities for burning will be limited. Where Sitka spruce, western hemlock, or true firs comprise the green tree component, burning will result in almost total residual tree mortality, negating the purpose of their retention. Together with smoke and air quality constraints, it is unlikely that substantial burning will accompany new forestry cuts of the future, particularly in west-side forests.

The presence of these unburned harvest units on the landscape will increase the continuity of slash and, with it, the risk of extensive wildfires. For example, the Falls fire east of Mount Rainier in 1987 burned through large units of untreated slash and all but skipped intervening forested areas, spotting from plantation to plantation. Even though the units were not contiguous, the presence of untreated activity fuels helped generate fire spread over many miles that probably would not have occurred if those fuels had been treated. The significance of such risk will vary quite a bit depending on the forest type, the degree of recent activity across that landscape, and the degree of biomass removal by the operation. Risk by forest type for unburned units will be lowest in spruce-hemlock and western hemlock plant associations because of rapid natural decay. It will be highest in the drier forest types where partial harvest (often by high grading of ponderosa pine) has been in place for many decades. However, in the drier forest types, prescription windows are wider and trees are more fire tolerant so that fire is more easily applied. Where chipping of residue is possible and economical, it can reduce unusual fuel loads created by new forestry harvesting operations.

As long as fire risk is in the equation for new forestry prescriptions, new forestry and fire need not be incompatible. There will clearly be the need for more precise prescriptions and control of fire-line intensity throughout the units chosen for treatment.

Global Climate Change

The 21st century is likely to be a century of global climate change. Vegetation change, which is likely to occur under global warming, is usually associated with massive insect outbreaks or forest fires (Franklin et al. 1991). Where drier climates are hypothesized, fire frequency may increase, with decreases in fire frequency in wetter climates (Romme and Turner 1991). Yet the evidence to support such cause-and-effect scenarios is lacking. It is clear that more than just annual or seasonal precipitation is associated with fire activity (Agee 1993). In the coastal Pacific Northwest, short-term summer drought, lightning storms, and east wind episodes interact with longer-term drought to affect fire occurrence and spread (Agee and Flewelling 1983). Large-scale fires (millions of acres) that have been reconstructed from forest age class data in the coastal forests of the Pacific Northwest appear to have occurred during global cooling episodes (Agee 1993). The cooling may have been associated with changes in one or more of the four contributing fire factors listed above, with a resulting alteration of the fire regime. Interior forests may react somewhat differently to the same regional changes depending on the net changes in the drivers of the fire regime. Global change and fire must be dealt with at a subregional level—in other words, global warming might cause opposite directions of change in the disturbance regimes of moist west-side forests compared to drier east-side forests.

The Outlier Event

The catastrophic fire that may be unexpected and unmanageable is an event for which few managers plan. They hope that it will not occur on their watch of the resource. In some cases, the occurrence of such events are manageable to a degree. Nevertheless, the threat of unusual weather and fire will always exist whether or not human-induced climate change occurs. Fire strategies for old-growth west of the Cascade Mountains assume that aggressive fire suppression will be effective in protecting owl and old-growth reserves. Fuel management strategies are not likely to have major effects. Yet a Yellowstone type of event is neither out of the probability range nor absent from the historical record. Large-scale landscape fires were likely about 300 years ago in the Olympics (Figure 12.7) and are thought to have covered 3 to 10 million acres in western Washington. Many areas

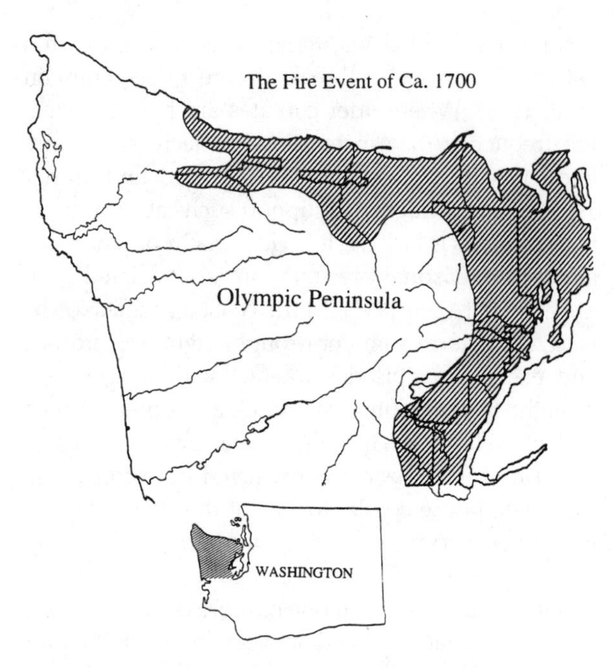

Figure 12.7 Estimated extent of a single fire or series of fires about A.D. 1700 in the eastern Olympic Mountains of Washington (from Henderson et al. 1989). Other widespread events in the western Olympics appear to have occurred ca. A.D. 1230 and 1480.

have 500- and 750-year age classes of Douglas-fir, suggesting very widespread fires at those earlier dates. These events were probably weather-driven rather than fuel-driven—large lightning storms on the west side followed by strong east winds—but we really do not know.

In the eastern Cascades, current catastrophic fire potential is the result of past management. Most of the historical fires in that area were of low severity. While we can implement fuel management strategies, significant reduction in catastrophic fire potential at a landscape level will take many decades (Agee 1993). Events such as the 1994 Wenatchee fires in Washington State, which were initiated by one human-caused and three lightning-caused wildfires, burning over 200,000 acres, will continue to occur. We simply do not know how much of the landscape needs fuel treatment to make the entire landscape more fire safe. So the unpredictable severe fire has potential to occur almost anywhere in the western United States. Such fires in the 21st century will depend on the mitigative effects of fuel reductions interacting with the unknown effects of global climate changes on fire regimes.

Conclusion

Forest protection issues will force us to put confidence limits on predictions about sustainable stand structures across the landscape. Although we know much more about fire ecology and fuel issues than we did in the first half of the 20th century, we need to incorporate that knowledge more into management plans than we have in the past. In so doing, adaptive management strategies will be crucial—planning on an ecosystem level and monitoring the effect of various strategies. Incorporating the possibility of the unusual disturbance will be needed, but we need more information about what these outliers are and under what conditions they may have occurred. Effective forest protection will require hands-on management, even in reserves, whether they are owl reserves, national parks, or wilderness. It will only occur with increased public confidence in our ability to manage forests and with our courage to admit that much of what we do is an experiment.

One of my favorite cartoons has a balding, skinny guy peering out an apartment window with the caption, "With the bodybuilding contest only two hours away, Larry wondered what was keeping Vincent with the steroids." It illustrates a good lesson about advance planning: We will not be able to create desirable landscape structures overnight. Trees cannot respond that fast and watersheds may not be resilient enough to absorb substantial restoration disturbance in a short period of time. Today's standard dogma reiterates one of Aldo Leopold's statements about landscapes: "Intelligent tinkering requires keeping all the pieces." Decisions we make about our forest landscapes will be with us for a long time. We do not want to lose the pieces, but at the same time if we continue to delay action, we may indeed lose the pieces through catastrophic loss from wildfire. Such losses by fire, insects, or disease can ruin the best landscape management plans. We need to incorporate what we do know about forest protection to minimize those undesirable changes in our forest landscape planning for the century to come.

Literature Cited

Agee, J. K. 1989. A history of fire and slash burning in western Oregon and Washington. In *The burning decision: Regional perspectives on slash,* ed. D. P. Hanley et al. Seattle: University of Washington, College of Forest Resources.

———. 1993. *Fire ecology of Pacific Northwest forests.* Washington, DC: Island Press.

———. 1995. Alternatives for implementing fire policy. In *Symposium on fire in wilderness and park management: Past lessons and future opportunities.* General technical report INT-GTR-320. Ogden, UT: USDA Forest Service.

Agee, J. K., and R. L. Edmonds. 1992. Forest protection guidelines for the northern spotted owl. Appendix F in *Recovery plan for the northern spotted owl.* Washington, DC: USDI Fish and Wildlife Service.

Agee, J. K., and R. Flewelling. 1983. A fire cycle model based on climate for the Olympic Mountains, Washington. *Fire and Forest Meteorology Conference* 7:32–37.

Cowan, C. S. 1961. *The enemy is fire.* Seattle: Superior Publishing Company.

Daubenmire, R. 1968. *Plant communities: A textbook of plant synecology.* New York: Harper & Row.

Davis, K. M., B. D. Clayton, and W. C. Fischer. 1980. *Fire ecology of Lolo National Forest habitat types.* General technical report INT-79. Ogden, UT: USDA Forest Service.

Deeming, J. E., R. E. Burgan, and J. D. Cohen. 1977. *The national fire-danger rating system.* General technical report INT-39. Ogden, UT: USDA Forest Service.

Everett, R., P. Hessburg, M. Jensen, and B. Bormann. 1994. *Executive summary.* Vol. I in *Eastside Forest ecosystem health assessment.* General technical report PNW-GTR-317. Portland, OR: USDA Forest Service.

Frandsen, W. 1991. Burning rate of smoldering peat. *Northwest Science* 65:166–172.

Franklin, J. F., F. J. Swanson, M. E. Harmon, D. A. Perry, T. A. Spies, V. H. Dale, A. McKee, W. K. Ferrell, J. E. Means, S. V. Gregory, J. D. Lattin, T. D. Schowalter, and D. Larsen. 1991. Effects of global climatic change on forests in northwestern North America. *Northwest Environmental Journal* 7:233–254.

Gruell, G. E., W. C. Schmidt, S. F. Arno, and W. J. Reich. 1982. *Seventy years of vegetative change in a managed ponderosa pine forest in western Montana: Implications for resource management.* General technical report INT-130. Ogden, UT: USDA Forest Service.

Heinselman, M. L. 1973. Fire in the virgin forests of the Boundary Waters Canoe Area, Minnesota. *Quaternary Research* 3:329–382.

Henderson, J. A., D. H. Peter, R. D. Lesher, and D. C. Shaw. 1989. *Forested plant associations of the Olympic National Forest.* R6 ECOL Technical Paper 001-88. Portland, OR: USDA Forest Service.

Hoover, K. L. 1993. The State of Washington view. In *The future: Landowners, loggers, and protection.* Northwest Fire Council proceedings, Portland, Oregon. Portland, OR: Northwest Fire Council.

Kilgore, B. M., and G. S. Briggs. 1972. Restoring fire to high elevation forests in California. *Journal of Forestry* 70:266–271.

Kilgore, B. M., and M. L. Heinselman. 1990. Fire in wilderness ecosystems. In *Wilderness management,* ed. J. Hendee, G. H. Stankey, and R. C. Lucas. Golden, CO: North American Press.

Komarek, E. V. 1974. Effects of fire on temperate forests and related ecosystems: Southeastern United States. In *Fire and ecosystems,* ed. T. T. Kozlowski and C. E. Ahlgren. New York: Academic Press.

McNeil, R. C., and D. B. Zobel. 1980. Vegetation and fire history of a ponderosa pine–white fir forest in Crater Lake National Park. *Northwest Science* 54:30–46.

Ottmar, R. D. 1983. Predicting fuel consumption by fire stages to reduce smoke from slash fires. In *The future: Landowners, loggers, and protection.* Northwest Fire Council proceedings, Olympia, Washington. Portland, Oregon: Northwest Fire Council.

Peterson, D. L., and K. Ryan. 1986. Modeling post-fire conifer mortality for long-range planning. *Environmental Management* 10:797–808.

Pyne, S. 1982. *Fire in America: A cultural history of wildland and rural fire.* Princeton, NJ: Princeton University Press.

Romme, W. H., and M. G. Turner. 1991. Implications of global climate change for biogeographic patterns in the Greater Yellowstone Ecosystem. *Conservation Biology* 5:373–386.

Rothermel, R. C. 1983. *How to predict the spread and intensity of wildland fires.* General technical report INT-143. Washington, DC: USDA Forest Service.

Rothermel, R. C., and P. L. Andrews. 1987. Fire behavior system for the full range of fire management needs. In *Proceedings of the Symposium on Wildland Fire 2000,* ed. J. B. Davis and R. E. Martin. General technical report PSW-101. Washington, DC: USDA Forest Service.

Van Wagner, C. E. 1973. Height of crown scorch in forest fires. *Canadian Journal of Forest Research* 3:373–378.

13 Forest Genetics for Ecosystem Management

Sharon Friedman

The Conundrum of Parentage and Proximity 204
Genetic Changes During Seed and Stand Development 205
Tradeoffs Between Genetic Diversity Among and Within Populations 207
Strategies for Coping with Uncertainty 208
Acknowledgments 210
Literature Cited 210

Evolution is the process that has led to the incredible diversity of life that makes earth a spiritually and physically nourishing place to call home. It is a process that is never static, and in itself is value-free. Organisms will continue to evolve whether or not humans are here to observe them. Yet there are specific organisms that humans value for a variety of reasons. Through time, as human and other forces change the environment, humans have an opportunity to influence the evolution of species—and, in fact, will by their very existence.

In the past, genetics was used to develop populations of forest trees with traits of value to humans—ranging from the ability to survive when planted to rust resistance to low wood specific gravity. As such, genetics was the domain of breeders and tree improvement specialists. In ecosystem management, however, genetic considerations include and extend far beyond breeding. The impacts of different management practices on the genetics of a wide variety of species needs to be addressed, as well as broader questions about what our desired goals for genetics should be.

The conceptual difficulty in dealing with the

process of evolution is the same as that for another dynamic process—fire. As with vegetation and fire, attempting to keep the current genetic composition and structure through time is expensive at the minimum, impossible at the maximum, and runs counter to the very nature of the process. Genetic diversity and evolutionary fitness (i.e., relative reproductive success) at the scale of the individual, the population, the metapopulation, and the species are interrelated in a constantly shifting dynamic. Unlike Aldo Leopold's intelligent tinkerer, who saves all the pieces, evolution is unintelligent and operates more like a game of musical chairs—whoever is in the right place at the right time with the right characteristics continues to play, while the others leave the game. To try to hold this constant—to simulate what we have today or had yesterday—would be like listening to a pleasant note in a symphony and then playing it over and over. Just as a melody is beautiful as notes and rhythms change, so too is evolution in the dance of organisms, populations, and species as they modulate their numbers, characteristics, and interactions with each other and the environment. As Namkoong (1994) states, "We can, we have, and we will alter the directions of change, either as a consequence of inaction or by accident or, if we choose, by design. The pattern that emerges will be influenced by us whether we wish to accept that responsibility or not. The only questions are who chooses, and what forces will guide future evolution."

Considering this complexity and dynamism, a goal for genetics in ecosystem management might be what Namkoong calls a conservation objective: "The next evolutionary step will enable trees to adapt as well as they have thus far." We might term this "evolutionary sustainability." Within this framework of sustainability, there is also room for influencing genetics in directions that meet human needs and values. In this sense, genetics is a microcosm of the overall sustainability issue. In the words of Jack Ward Thomas, the current chief of the U.S. Forest Service, "We must collect the golden eggs without killing the goose."

The basic question that presents itself in managing forest tree populations is what situations, changes, or rates of change might indicate we are in the process of positively or negatively influencing evolution? What are the situations that might impact the ability of species to adapt? When and where is action needed, and when might it be effective in improving ecosystem services and health? These are difficult questions. The answers will be found by focusing research in these areas and by taking extant research and blending it with human values and practicality in a host of judgment calls. In many of these areas, data are lacking, but conjecture is not.

This chapter examines concerns raised about silvicultural practices in the context of some of the key lessons from the past 20 years of forest genetics research. The single, overriding conclusion is that ecosystem genetics are inherently dynamic. Currently there is no evidence that one regeneration system (e.g., clearcutting vs. partial cutting) is better than the other in terms of genetic impacts. In the future, we must learn to lift our eyes from current genetic configurations to better understand the evolutionary processes that led to those configurations.

The Conundrum of Parentage and Proximity

An important lesson to come out of the study of naturally regenerated forests is that pollen and seed travel much further than previously thought. This raises one of the most fundamental and difficult to answer questions in forest genetics: Who are likely to be the parents of the next generation? Even for widely studied forest trees, we do not know the answer to this question.

Forest trees can be pollinated by wind, insects, or mammals (such as bats). Seed can be dispersed by gravity, wind, water, birds, and mammals (including humans), both intentionally and unintentionally. Due to the commercial importance of wind-pollinated conifers, pollen dispersal in these genera have received a great deal of study.

In 1962, Silen measured Douglas-fir pollen density at the center of a 3- by 14.5-mile treeless area and found 769 grains per square inch. Silen also cites studies by Sarvas (1955) which indicate that pollen fell on a lightship 20 km from the coast of Finland in amounts approaching or even exceeding that beneath a stand on the shore. Since the 1950s, pollen has been known to travel far, but thinking in the genetics community was that pollen which traveled long distances wouldn't be viable. Due to the difficulties with experimentation, this hypothesis could not be tested. Nevertheless, the idea found a home in mainstream forest genetics thinking, and today is still found in some genetics and silviculture textbooks. There is some population biology literature based on studies of smaller plants that suggests that dispersal in plants is limited (Levin 1979).

With the advent of biochemical genetic markers in the late 1970s, it was finally possible to test hypotheses about the origin of pollen effective in fertilizing ovules. Friedman and Adams (1985) first used protein biochemical markers (isozymes) to monitor pollen coming into a seed orchard and found that 36–60 percent of the pollen coming into a 10-acre orchard complex of loblolly pine was from outside the orchard. Neale (1985) used allozyme markers in natural stands of Douglas-fir and found that in a shelterwood at least 22 percent of the seed and 28 percent of the pollen came from outside a 6-acre plot. Due to the estimation procedures, this is likely to be an underestimate of the true proportion. In Yazdani et al. (1995), the authors estimate that more than 90 percent of seedlings do not have the closest trees as one of their parents. They also estimate that 0–35 percent of genes come from trees over 50 m away from the seedling. In these studies, we can only say where the pollen or seed does not come from—the identified trees that are close. What we don't know is whether the pollen or seed is from a couple of kilometers away or with the pollen at least, from tens or hundreds of kilometers away, nor how the location of the successful parent varies from year to year.

Information on seed dispersal also can be found in silvicultural literature. For example, Franklin and Smith (1974) tracked amounts of seed coming into clearcuts. For Shasta red and white fir, both heavy-seeded species, it was found that large amounts of seed fell even on central portions of a 40-acre clearcut. They assumed that if only 1 percent of the sound seed produced established seedlings, this would be more than adequate to restock the site. Although more seed fell close to the trees, so many trees produced so much seed that the tail of the distribution was more than enough to restock the site.

In addition, seeds can be dispersed by birds and mammals. For example, corvid seed dispersers can fly up to 22 km (Vander Wal 1990). Even without invoking human actions, many species of trees are capable of long-distance seed and pollen dispersal. Genetically this is significant, as even rare migrants can decrease population differentiation. Further discussion of the complexity of seed dispersal within the genus *Pinus* can be found in Lanner (1995).

The take-home message from this literature is that we do not know how much of the successful natural regeneration on a site will come from the seed or pollen of whatever trees are left. If the trees cut were reproductively active, there is likely to be some change in parents represented, but a restriction on the number of parents represented does not seem likely given the studies mentioned above.

Genetic Changes During Seed and Stand Development

As our knowledge of the genetics of seed and stand development has increased, we have gained new awareness and respect for the adaptive strategies of forest trees, which have dominated the continent since long before the advent of humans. While geneticists have wondered whether negative genetic changes such as inbreeding, a decrease in genetic diversity, or dysgenic selection could occur due to human influence, trees have a substantial variety of mechanisms that have ensured their survival during difficult times—including the ice ages. We are only beginning to recognize the beauty and subtlety of some of these mechanisms, which encompass a mix of stochastic and selective factors.

In the past, we have often considered female plants as passive recipients of pollen. Better understanding of their reproductive biology has led to a more complex picture of what happens between the time when pollen grains land and an embryo germinates. A thorough discussion of this subject can be found in Willson and Burley (1983). For example, delayed fertilization is common among gymnosperms and some angiosperms. Short-term delays can be interpreted as female tactics to increase the number of males available—more pollen is accumulated before fertilization occurs. Extremely long delays, such as a year or more in conifers, may allow the females to improve mate choice and abort zygotes from poor pollen years. In simple polyembryony in conifers, a varied array of zygote genotypes within an ovule (the same maternal haploid genotype, potentially different male haploid genotypes) presents the female with choice among them. This may permit the elimination of inbred zygotes without the loss of the ovule, as suggested by forest geneticists (Bramlett and Popham 1971). A similar phenomenon may occur in multiovulate carpels in angiosperms. Female reproductive organs may contain various numbers of ovules; in some species all can develop, in others all but one are aborted. For example, in *Quercus* six ovules typically develop, of which five abort early (including some that get fertilized), and in *Betula alleghenienis* only one of four ovules per fruit develops into a seed (Clausen 1973). In these cases, diversity and fitness are being selected for before a seed even falls to the ground.

Given that the inbreds, or genetically inferior ovules, would survive the competition to become embryos, what is the likelihood that they would survive through germination and stand development? We now know that in several different stand types that have been studied, inbreds become suppressed during stand development. Thus, the less-inbred individuals survive to reproduce (Plessas and Strauss 1986, Tigerstedt et al. 1982, Yazdani et al. 1985, Hawley et al. 1989). There has been some discussion about whether genetically variable (heterozygous) individuals survive due to the heterozygosity being a marker for inbreeding (more heterozygous are less inbred) or whether allozyme heterozygosity has intrinsic selective value. For most people's purposes, the bottom line is the same: Selection occurs during stand development and less inbred trees are winners (at least for the conifers that have been studied). Unless trees are protected from competition, it would be difficult to argue that inferior growers and inbreds are able to survive and produce offspring, and in fact, the literature supports inbreds being weeded out through stand development.

Knowing that the male and female parents of progeny have been changed by a given silvicultural practice, or knowing that the proportion of inbred embryos has increased, does not tell us whether these changes will have an impact in stands. In many cases, stands are planted, and the genetic arrays found at the seedling stage are different from those resulting from natural regeneration. In fact, genetic arrays from natural regeneration vary from year to year. Evidence shows that genetic changes, some selective, take place during stand development. Therefore, one cannot predict genetic changes at the adult stage without knowing both the original genetic raw material and the selective and stochastic forces operating through stand development. Unfortunately, this is the meaningful kind of research that is difficult to do and has received relatively little attention.

This simple example illustrates two important aspects of applying genetics to forest ecosystem management. Knowledge of basic reproductive biology (who is mating with whom, whose offspring survive and reproduce) is the first and most essential information needed to understand genetic implications of management. The second principle is that the timing and spatial scale of potential impacts always need to be explicit. For example, is there a concern with inbreeding depression leading to slow growth this generation, or to a lack of fertility? Or is there a concern with some lowering of within-population variation over several generations? The only way to address these concerns is to think through the cycles of mating and selection, cycle by cycle, through time.

Unfortunately, for many organisms, the basic questions of reproductive biology must remain unanswered. In fact, in some cases we are only now finding out where organisms go, let alone whom they mate with when they get there. For example, resident stream fishes have been perceived as relatively sedentary, but recent studies suggest that they can move as far as 96 km (Young in press). Current interest in stream fishes is focused on the possibility that many stream fishes compose metapopulations consisting of subpopulations linked by immigration and emigration (Fausch and Young in press). These "big picture" questions are critical but difficult to answer; while more restrictive questions (such as, Does apparent population A differ from apparent population B?) are easier to answer, but relatively little help in developing management strategies.

Tradeoffs Between Genetic Diversity Among and Within Populations

A statement that a certain management practice is good or bad for genetic diversity should always be suspect. Although the term "genetic diversity" has often been oversimplified—particularly in popular literature—it can incorporate many complex concepts. Most important, genetic diversity is not an entity, but varies by trait, by area, and over time.

An organism is a genetic unit that can, in general, be easily defined and observed. A population, however, is more difficult to pinpoint. When geneticists refer to a population, they are generally referring to a group of organisms that have an equal chance of interbreeding with each other. Yet, in reality, whether a population is a gene pool represented by one big pool or a series of interconnected small pools, or whether individuals who are located in the same vicinity breed with each other or not, is often unknown. Nevertheless, the use of the term "population" is essential for describing the array of genes across the landscape.

Conceptually, genetic diversity can be described as a matrix with four columns and thousands of rows. The columns can be labeled "diversity within an individual," "diversity within populations," "diversity among populations within regions," and "diversity among regions within species." The rows are the myriad of different traits that describe an organism—for example, needle color, seed size, frost resistance, etc. There are literally thousands of traits, and there is no particular reason to think that the distribution of these traits across individuals, populations, and regions would be the same. Information in one cell of the matrix, or even one row, cannot be expected to summarize genetic diversity for a species.

One of the key issues that should be considered in forest resource management is the tradeoff in genetic diversity within and among populations. Comparison of plantings from seed orchards with that from natural regeneration provides an instructive example. Some people argue that seed orchards lead to a reduction of genetic diversity because many thousands of acres will be planted from the same number of clones. Others reply that the clones in an orchard are selected to be from throughout a zone and hence represent more genetic diversity than would occur if the area were naturally regenerated. If we go back to the above matrix, we can see that both arguments are correct, but that each is referring to different cells in the matrix. It is true that the diversity among populations (considering each planting unit as potentially a different population) is reduced by planting seed orchard stock. This is simply due to the fact that seed orchard seedlots reflect the clones in the orchard: More or less the same genetic material will be planted on many plantation sites across space and time. At the same time, for a specific plantation site, the diversity of traits important to adaptation, which we assume varies across the zone, is likely to be greater from seed orchard material than if the area were naturally regenerated. So in our matrix with seed orchard material, the within-tree and within-population diversity is likely to increase for adaptive traits, and the among-population diversity is reduced compared to natural regeneration. In other words, when actions are taken to divide populations into smaller groups, among-population diversity goes up and within-population diversity goes down; when actions are taken to combine populations, the within-population diversity goes up and the among-population diversity goes down.

Genetic diversity within populations is important to adaptation through time. If a trait has diverse values within a population and is adaptively important (e.g., frost resistance), it enables the population to survive under a broader range of environments than if there were less diversity for that trait. However, there is a tradeoff with fitness in any one environment. For a simple example, imagine a certain drought-resistance mechanism in a forest tree that is an asset on dry sites and a liability on moist sites. Populations that do not have the mechanism will do the best on moist sites; populations with the mechanism will do the best on dry sites; and populations that carry both mechanisms will do the best when they are exposed to both environments. Since future environments may not be the same as current ones, diversity today is important for survival tomorrow. At its simplest, this is the argument for within-population diversity.

So it is a good thing to mix different genetic types to increase diversity—acknowledging that flexibility may be enhanced at the expense of adaptation to a particular site. Yet there is another tradeoff in addition to the diversity/adaptation tradeoff. When populations are mixed to increase within-population diversity, among-population diversity is reduced; genetic "homogenization" is a value-laden but descriptive term for this. Among-population diversity is also important to the long-term survival of a species. Dividing a species into small, isolated populations can increase diversity in the species as a whole (Wright 1932). If populations are small, random genetic fluctuations can enhance genetic diversity. Isolation is important so that populations can adapt to their own unique environments and become increasingly differentiated from each other. Thus, a species divided into small, isolated populations is likely to have a greater total diversity than one large, randomly mating population. The advantages of using multiple populations in breeding forest trees have been discussed by Namkoong et al. (1988). In discussing the concept of fragmentation, Simberloff (1988) points out that when separated populations are genetically distinct, the spread of diseases and insects may be inhibited. So, among-population diversity has many advantages as well.

Hence, we come to the conclusion that both within-population and among-population diversity are desirable for maintaining the capacity of the species to adapt. Yet, one can only be attained at the expense of the other. Combining and separating populations is a dance that goes on in nature that we can influence directly or indirectly. How do we honor this natural process without attempting to freeze it at a specific point in time? We cannot and should not attempt to return diversity to a certain state. The only answer is to carefully consider the tradeoffs and appreciate the complexity when management decisions are made.

Strategies for Coping with Uncertainty

Perhaps the most important contribution that geneticists can make to the forestry of the future is a philosophical contribution in the form of humility. We know that populations and species change through time, they changed in the past, and they will continue to change in the future. Yet, we do not know how our practices will influence that change, nor can we really delineate an "important" change from an "unimportant" change, except in terms of our own values (e.g., how we feel about species loss).

The preceding discussion underscores how little we know about the genetics and evolution of forest organisms. We will never have the funding to fully understand the genetics of a wide sampling of organisms. Nevertheless, in the absence of knowledge, experience has led to some basic conservative, commonsense guidelines that apply across a wide variety of organisms.

First, for species that appear to be doing fine, or for which genetic problems are unknown, the best general approach is to maintain current populations and provide special protection for populations that appear to be unique—such as those growing in an unusual area or having unusual features.

Second, for species that are to be planted or stocked, additional care needs to be taken. The first step is to develop some kind of zoning scheme and to carefully manage the adaptation/diversity balance within and among populations. It is important to acknowledge that there is no truth here, but rather a set of biological, social, and practical judgment calls. Special care needs to be taken with both intentional and unintentional selection throughout the process from selecting sources of propagules to the planting operation itself. Care should be taken so that human-induced selection does not reduce the genetic base below a desired level, nor select against traits important for reproductive fitness in the wild. For a plantation or stocking program, there are some key questions to be answered: Where should the parental organisms come from? How do you choose them? How different from each other genetically will they be? After leaving the originating site, will selection be applied to the parents or offspring and, if so, how strongly and for what traits? Are there characteristics of seed and propagule storage and early growth in a hatchery or nursery that might foster the death of certain genetic combinations that might contribute to increased diversity or fitness in the wild? (An organism must be fit enough to survive—dead organisms don't help diversity.)

Third, for threatened or endangered species, greater research investment and technological latitude are warranted. More-intense management techniques such as captive breeding need, at the very least, to be informed by state-of-the-art genetic knowledge. For example, better genetic knowledge of the dusky seaside sparrow could have contributed to a more successful crossing program (Avise and Nelsen 1989). Advanced technologies such as genetic engineering of forest trees are more likely to be socially acceptable in cases such as the American chestnut where a species, life or death is at issue (Friedman and Foster in press).

For research in forest genetics, the challenge is to focus attention in areas where new information will lead to improved management. The first common question is whether the genetics of a given species or population is in good shape or whether there are genetic problems. Is "who is mating with whom" the reason undesirable outcomes are occurring? Is inbreeding the reason for observed viability and fertility problems? Is the reason for lack of adaptation or loss of desirable characteristics some form of hybridization among different populations or species (for example spotted owl and barred owl)? These types of questions are best answered by observing reproductive behavior and biology, followed by genetic analysis. As mentioned above, the most basic question is, Who is mating with whom? And does that contribute to undesirable outcomes? If this is found to be the case, some management actions (introducing new material or removing the unwanted hybridizer) can be undertaken.

Another common question is, Will a given management practice have a negative impact on the genetic fitness or diversity of a population or species? This is a conceptually more difficult question to answer than the first. The second question has two elements—How much of an impact occurs? and How much would it take to be important? In short, what is a negative genetic impact? Is removing one organism, 1 population, or 20 populations removing enough to have a negative impact? How do we measure the negative impact? What traits do we use to measure decreases in diversity? How do we choose how to calculate estimates of diversity? What is the sensitivity of our findings to the trait and to the estimator used? What are the spatial and temporal scales that we are looking at and how do we weigh each scale in deriving the overall answer to genetic fitness and diversity? The effect of a given management practice on different kinds of genetic diversity is a vector of pluses and minuses. Different people might weight the values of the different elements of the vector differently in arriving at a conclusion on whether there is an overall negative effect.

In addition, there are fundamental questions concerning timing. When do we measure the genetic changes? Do we measure trees standing on a site? Do we measure embryos from seed collected from the trees standing on the site? Do we measure successfully regenerated offspring of those trees as seedlings, as saplings, or when they have their own first reproductive success? Since it takes a long time for trees to go through a generation, management decisions must often be made before final information is available.

To complicate matters even further, the traits and estimators of diversity one chooses can influence the answers to the above questions. For example, one could argue that we should focus on adaptive traits so that populations can survive and thrive under changing environmental conditions. Yet, much of our current information is based on isozyme and DNA traits, which tend not to be correlated with adaptive traits nor with each other. Nevertheless, there is a tendency among researchers to reify the variation patterns based on different traits.

This is reminiscent of an old story about blind men and an elephant. One man, touching the elephant's trunk, describes the elephant as long and skinny, like a snake. Another, feeling a leg, describes it as thick and straight like a column, and so on. Many authors describe their results as being about the elephant (genetic variation of the species) rather than about a foot or an ear (genetic variation in a certain trait of a given species). In reality, each trait is likely to show its own pattern of genetic variation, depending on evolutionary history, natural selection, and stochastic factors. There is no reason to think that studies of different traits will agree; in fact, empirical evidence shows that they do not (Beer et al. 1993). Infinitely many studies could be done for a given species based on different traits. The key question is why different traits are of interest, and whether at some point more information on the pH of the elephant's saliva, the growth rate of its toenails, or the color of its stomach lining, exceeds what we need to know to have an adequate understanding of the elephant.

The question that we really want to answer is, How can we tell if the elephant is basically well or sick? but our snapshot studies of population variation cannot answer it. We need to study the elephant through time and watch as the cycles wax and wane. This will be difficult—and yet for forest trees there have been a few empirical studies that crossed generations (Roberds and Conkle 1984). The dynamics of change also can be approached by modeling (Namkoong and Bishir 1989).

The third question is, Can genetics be used as a tool to meet some management goals? For example, when seed is collected for riparian restoration projects, should there be selection and breeding done to ensure that the new trees perform needed ecological services as quickly as possible? In many cases, genetics can be used to move more quickly toward ecosystem management goals.

Ultimately, we must ask different types of questions than we have in the past. We need more understanding of the basic reproductive biology of organisms and the importance of genetic and stochastic factors in determining reproductive success. We need genetics research to be more integrated with other research on management options. For example, the basis for understanding the response of species to climate change is within-population levels of phenotypic and genetic diversity for adaptive traits. Unfortunately, the basic information necessary for predicting vegetative change and for modeling plant populations under climate change—the phenotypic and genetic diversity within stands—is unknown and so far unstudied. Finally, and perhaps most important, we should carefully think through the concepts and paradigms underlying forest genetics research. For example, if populations change every generation, what would be the kind of genetic change we would think of as problematic? This up-front thinking before research is initiated is perhaps the single most critical task in developing a genetics for the 21st century.

Acknowledgments

The author would like to express her appreciation to Carol Aubry for many productive discussions on these topics and to Gene Namkoong and Dean DeBell for their helpful reviews.

Literature Cited

Avise, J. C., and W. S. Nelson. 1989. Molecular genetic relationships of the extinct dusky seaside sparrow. *Science* 243:646–648.

Beer, S. C., J. Goffreda, T. D. Phillips, J. P. Murphy, and M. E. Sorrells. 1993. Assessment of genetic variation in *Avena sterilis* using morphological traits, isozymes, and RFLPs. *Crop Science* 33:1386–1393.

Bramlett, D. L., and T. W. Popham. 1971. Model relating unsound seed and embryonic lethal alleles in self-pollinated pines. *Silvicultural Genetics* 20:192–193.

Clausen, K. E. 1973. *Genetics of yellow birch.* Research paper WO-18. Washington, DC: USDA Forest Service.

Fausch, K. D., and M. K. Young. In press. Movement of resident stream fishes and evolutionarily significant units: A cautionary tale. In *Evolution and the aquatic ecosystem: Defining unique units in population conservation,* ed. J. L. Nielsen. Bethesda, MD: American Fisheries Society.

Franklin, J. F., and C. E. Smith. 1974. *Seeding habits of upper-slope tree species III: Dispersal of white and shasta red fir seeds on a clearcut.* Research note PNW-215. Portland, OR: USDA Forest Service.

Friedman, S. T., and W. T. Adams. 1985. Estimation of gene flow into two seed orchards of loblolly pine (*Pinus taeda* L.). *Theoretical and Applied Genetics* 69:609–615.

Friedman, S. T., and G. S. Foster. In press. Concerns about genetic diversity on public forest lands. *Canadian Journal of Forest Research.*

Hawley, G. J., D. H. DeHayes, and S. F. Gage. 1989. The relationship between genetic diversity and stand viability: A case study with jack pine. In *Proceedings: 31st Northeastern Forest Tree Improvement Conference and the 6th Northcentral Tree Improvement Association.* University Park, PA: Pennsylvania State University.

Lanner, R. 1995. Seed dispersal. In *Pinus: Ecology and Biogeography of Pines,* ed. D. L. Richardson. New York: Cambridge University Press.

Levin, D. A. 1979. The nature of plant species. *Science* 204:381–3841.

Namkoong, G. 1994. An evolutionary concept of breeding. Lecture by the 1995 Wallenberg Prize winner, September 22, 1994, at the Marcus Wallenberg Foundation, Stockholm, Sweden.

Namkoong, G., and J. Bishir. 1989. The interaction of genetics and ecology in forest ecosystems. In *Proceedings: 31st Northeastern Forest Tree Improvement Conference and the 6th Northcentral Tree Improvement Association.* University Park, PA: Pennsylvania State University.

Namkoong, G., H. C. Kang, and J. S. Brouard. 1988. *Tree breeding principles and strategies.* New York: Springer-Verlag.

Neale, D. B. 1985. Genetic implications of shelterwood regeneration of Douglas-fir in southwest Oregon. *Forest Science* 31:995–1005.

Plessas, M. E., and S. H. Strauss. 1986. Allozyme differentiation among populations, stands, and cohorts in Monterey pine. *Canadian Journal of Forest Research* 16:1155–1164.

Roberds, J. H, and M. T. Conkle. 1984. Genetic structure in loblolly pines stands: Allozyme variation in parents and progeny. *Forest Science* 30(2):319–329.

Sarvas, R. 1955. Ein Beitrag zur Fernverbreitung des Blütenstaubes einiger Walbäume. Zeitschrift. *Forestgenetik* 4 (4/5):137–142.

Silen, R. 1962. Pollen dispersal considerations for Douglas-fir. *Journal of Forestry* 60(11):790–795.

Simberloff, D. 1988. The contribution of population and community biology to conservation science. *Annual Review of Ecological Systems* 19:473–511.

Tigerstedt, P. M. A., D. Rudin, T. Niemela, and J. Tammisola. 1982. Competition and neighboring effect in a naturally regeneration population of Scots pine. *Silvae Fennica* 16:122–129.

Vander Wal, S. B. 1990. *Food hoarding in animals.* Chicago: University of Chicago Press.

Willson, M. F., and N. Burley. 1983. Mate choice in plants: Tactics, mechanisms, and consequences. In *Monographs in population biology,* vol. 19. Princeton, NJ: Princeton University Press.

Wright, S. 1932. The roles of mutation, inbreeding, crossbreeding, and selection in evolution. *Proceedings of the Sixth International Congress of Genetics* 1(1932):356–366.

Yazdani, R., D. Lindgren, and D. Rudin. 1995. Gene dispersion and selfing frequency in a seed-tree stand of Pinus sylvestris L. In *Population Genetics in Forestry,* Lecture notes in biomathematics 60, ed. H. R. Gregorious. New York: Springer Verlag.

Yazdani, R., O. Muona, D. Rudin, and A. E. Szmidt. 1985. Genetic structure of a *Pinus sylvestris L.* seed-tree stand and naturally regenerated understory. *Forest Science* 31(2):430–436.

Young, M. K. 1994. Mobility of brown trout in south-central Wyoming streams. *Canadian Journal of Zoology* 72:2078–2083.

Section

III Approaches to Management at Larger Spatial Scales

All major management issues facing foresters in the next century will include issues at large spatial scales—landscapes and regions. Maintenance of biological diversity and protection of wetlands and riparian ecosystems are only two among many examples. FORPLAN, the computer planning model that the U.S. Forest Service used in its planning efforts in the 1980s, fell short of expectations because of spatial constraints. In some cases FORPLAN's harvesting schedules were unworkable because they conflicted with other requirements, such as limitations on the cumulative size of adjacent clearcuts. Experiences with this and other planning models have made clear that spatially explicit project planning is essential to credible projections—and that much-expanded knowledge of ecological responses to larger spatial patterns is needed (see Crow and Gustafson, Chapter 14).

Understanding the environmental consequences of larger spatial patterns is the objective of landscape ecology. Although its roots are old, modern landscape ecology—with its dynamic context and functional focus—is relatively new. Furthermore, a great deal of learning actually has occurred as a result of dysfunctional landscapes, which in turn arose out of ignorance of such phenomena as cumulative effects and habitat fragmentation.

At the most fundamental level, landscape ecology is about patches. It is about (1) how patchworks evolve and persist at larger spatial scales as a consequence of the interactions between the geophysical template and disturbances (see Swanson et al., Chapter 15), (2) the importance of patch size and context, including edge effects, the key interactions that occur at boundaries between patches, and (3) the collective consequences of different patch mosaics for various resources, including habitat for organisms and the production of goods and services of humankind.

The lessons are surprising only in prospect, not retrospect. All parts of the landscape are not created equal with regards to function. Areas differ dramatically in their contribution to various services and goods, whether for wildlife, hydrologic regulation, or wood production; problems arise when important conflicting values overlap. For example, in the valleys of coastal British Columbia and Alaska, the alluvial valley bottoms turn out to be critical sites for timber production, deer, bears, fish, eagles, and other species.

Patch context matters. Where is the required habitat or the proposed timber sale? Projections must be spatially explicit to be credible. Simplistic traditional approaches to landscape patterns fail to succeed in producing credible projections of habitat availability—whether it is the forester's fully regulated forest (or its modernized version, the shifting mosaic) or the conservation biologist's division of the landscape into suitable habitat (reserves) and unsuitable habitat (matrices).

Integrating knowledge at larger spatial and temporal scales is one of the major challenges of the 21st century. Fortunately, powerful new tools have emerged, just in the nick of time. Geographic information systems (GIS) allow us to manage large amounts of spatially explicit data; advancements in this technology continue to make management and use of such data easier and more broadly accessible. The major spatial data sources provided by remote imagery have expanded dramatically from black-and-white aerial photography of the mid-century to multispectral and radar images, often of high resolution. In addition, geographic positioning satellites (GPS) provide incredible capabilities for on-the-ground geographic locating. New models are being developed to take advantage of such databases and provide assessments of alternative management strategies (see Sessions et al., Chapter 18).

We are at the beginning of the learning curve in dealing with large spatial and temporal scales, however. Our level of ignorance is immense with regard to both natural science and social issues. For example, what landscape designs are desirable to achieve various objectives? What kind of guidance do historical landscapes provide for future management (see Diaz and Bell, Chapter 17)?

Designing large spatial-scale strategies for conservation of biological diversity is a critical area in which we need fresh perspectives. What are the limitations of conservation strategies focused primarily upon biological reserves and corridors? How do conditions in the matrix ultimately affect conservation of biological diversity and connectivity? What level of particular habitats, such as late-successional forests, is needed to achieve conservation objectives?

Many of the challenges relate to assessing and planning activities on larger multiowner watersheds (see Naiman et al., Chapter 16). How do we allocate obligations at the scale of multiowner watersheds? What technical approaches and social incentives can be provided to deal with such difficult issues? What are the relative merits of dispersing management activities such as timber harvest in time and space, as opposed to concentrating such activities, particularly with regard to impacts on aquatic ecosystems? How rapidly do hydrologic conditions recover in harvested landscapes?

14 Ecosystem Management: Managing Natural Resources in Time and Space

Thomas R. Crow and Eric J. Gustafson

Landscape Ecology and Forest Management 216
Landscape Structure 217
Box: Restaging an Evolutionary Drama: Thinking Big on the Chequamegon and Nicolet National Forests *by Linda Parker* 218
Temporal Dynamics and Spatial Patterning in Managed Landscapes 220
Projecting in Time and Space 222
Results 223
Implications for Management 225
Literature Cited 226

When asked to visualize themselves relative to a forest, most managers are likely to see themselves standing on the forest floor viewing trees from below with an umbrella of foliage extending overhead. Fewer are likely to select another perspective—a view from above the forest, but both perspectives are equally valid for considering spatial and temporal patterns created through forest management. In this chapter, we explore the implications of the alternative perspective, the view from above the forest, in considering the spatial and temporal patterns created through forest management.

Different patterns become apparent when the observer moves from within to above the forest. The details of individual trees that are so important within the stand become less obvious. The broad, general patterns of landscape ecosystems that reflect changes in the physical environment, along with the related human imprint on the land, become more obvious. A mosaic of patches is seen with different compositions, including forests, agricultural lands, and urban lands, each with a different size and shape. This broad, comprehensive view, in which the forest is considered within the context of all land uses, is necessary when dealing with the complex problem of integrating commodity production with other values and benefits derived from the forest. A landscape perspective is especially helpful in the analysis of cumulative effects (Risser 1988), as a basis for spatial planning and resource analysis (Crow 1991, Mladenoff et al. 1994), for the assessment of biological diversity (Probst and Crow 1991), and as a framework for adaptive management (Walters and Holling 1990). In short, a landscape perspective is critical to practicing integrated resource management.

So far, we have looked at a number of points in space, but only a single point in time. However, because landscape ecosystems are dynamic, changing in time, we need to put this picture in motion. So the patches begin to shift, changing in their cover types, age classes, sizes, shapes, and relative positions on the land. Now we have combined the two elements that are of concern in this chapter—time and space. These elements define variation, or heterogeneity. Kolasa and Rollo (1991) provide a formal definition of spatial heterogeneity: "[An] environment is heterogeneous if the rate of a process varies over space in relation to structural variation in the environment." In other words, an environment can be considered homogeneous if a process has a uniform rate across space. Heterogeneity also applies to time. As the term suggests, temporal heterogeneity refers to variation in time that can be expressed in a deterministic, random, or chaotic fashion (Kolasa and Rollo 1991).

Our objective in this chapter is to consider temporal and spatial landscape patterns created through forest management and other land uses. We also explore some of the potential ecological implications of these patterns, as well as their possible impacts on commodity outputs and other values and benefits derived from the forest. First we consider how ecological science has altered our viewpoint on spatial and temporal patterns in forest management, and then we apply a spatially explicit model to an actual landscape to evaluate alternative management strategies.

Landscape Ecology and Forest Management

Our understanding of ecosystems has increased dramatically during the past two or three decades. Pioneering ecosystem research, often conducted over long periods within watersheds—such as that at Hubbard Brook Experimental Forest in New Hampshire, at the Coweeta Hydrologic Laboratory in North Carolina, and at H. J. Andrews Experimental Forest in Oregon—has greatly improved our knowledge about stand dynamics, forest productivity, decomposition, and biogeochemical processes. Research conducted in these watersheds exemplifies a large-scale experimental approach to studying natural and altered landscapes. To develop appropriate solutions to a variety of complex and interrelated ecologic, social, and economic problems, we need to study large landscapes and regions as well as understand the significance of the spatial and temporal variability embedded in these ecosystems.

Landscape ecologists seek to understand the ecological structure and function of large areas and to relate landscape patterns to ecological processes. Landscape ecology is not a distinct discipline or simply a branch of ecology, but rather a synthetic intersection of many related disciplines that focus on the spatial-temporal pattern of the landscape (Risser et al. 1984). It extends ecosystem analysis to the interactions among ecosystems and involves both natural and managed ecosystem attributes (Risser 1985).

A rich literature in landscape ecology has developed in a relatively short period of time. Among the key topics that characterize the science are landscape fragmentation (Whitcomb et al. 1981, Johnson 1988, Saunders et al. 1991, Fore et al. 1992, Bratton 1994, Esseen 1994, Kruess and Tscharntke 1994); spatial-temporal scale (Delcourt et al. 1983, Meentemeyer and Box 1987); hierarchy theory (O'Neill et al. 1986, Allen and Starr 1982); analysis of landscape pattern (Turner 1990, Ripple et al. 1991); spatial models (Gardner et al. 1987, Milne 1992); relations of pattern and process (Wiens 1976); boundaries and edges (Wiens 1992); corridors, networks, and connectivity (MacClintock et al. 1977, Fahrig and Merriam 1985, Forman 1991); fluxes of matter and energy (Wiens 1992), and applications to planning and resource management (Crow 1990, 1991; Mladenoff et al. 1994).

Landscape Structure

Patches and corridors are ubiquitous features in the landscape. Their composition, size, shape, and arrangement in time and space characterize landscape structure and dynamics. Forman and Godron (1986) define a patch as a "nonlinear surface area differing in appearance from its surroundings." Often, these differences are expressed in terms of composition or vegetation, although delineations can be based on other factors, such as physiography, soil, human land use, or differences in age class for forests. Furthermore, patches can be classified according to their causative mechanisms or origins. Windthrow and forest harvesting, examples of disturbance patches, create gaps in the forest canopy. A disturbance that is both extensive and intensive can convert the landscape matrix, but small, isolated patches of the former matrix often remain in place. Deforestation and conversion fragment the forest, leaving remnant forest patches embedded in an agricultural or urban matrix (e.g., Curtis 1956, Sharpe et al. 1987). If farmland remains fallow, regenerating patches of forest can resemble remnant patches, but with a different origin. Reforestation also can create new forest islands on lands that were previously nonforested. If reforestation is allowed to continue, these emerging forest islands will expand in size and eventually coalesce with other forest patches to form the landscape matrix (Nyland et al. 1986, Zipperer et al. 1990).

The patch types discussed above owe their origin to either natural or human disturbances. Patches also exist because of differences in the physical environment. Forman and Godron (1986) call these "environmental resources patches." One example is the common landscape patterning created by the interspersion of lakes and bogs with upland forests in the glaciated Great Lakes region. Compared to disturbance and regenerating patches, environmental resources patches are likely to have lower patch turnover rates and thus higher persistence time. In summary, variation in landscape structure is the result of both heterogeneity in the physical environment as well as differences in disturbance patterns, including human land use. Because land use is related in part to the physical environment, these factors creating variation in the landscape are interrelated.

A corridor is a linear strip of land that differs from the landscape elements on either side. Forman and Godron (1986) identify at least two different corridor types: A line corridor is a narrow strip that is totally dominated by edge environment, and a strip corridor is sufficiently wide to contain interior environment. Corridor width is considered the most important variable affecting all ecological functions.

The elements that create heterogeneity in the landscape—patches and corridors—are defined by their boundaries, which represent spatial and temporal discontinuities in the physical environment or land cover (Wiens 1992). At the interface, or ecotone, between two landscape elements, rates and magnitudes of ecological processes—such as fluxes of energy, flows of materials, and interactions between plants and animals—are likely to change abruptly compared to those processes within patches (Wiens et al. 1985). Abiotic factors such as temperature, wind, and moisture are strongly affected by the structural features associated with boundaries. In comparing the growing season microclimates of adjacent clearcut, edge, and old-growth Douglas-fir forest in the Pacific Northwest, Chen et al. (1993) found the highest variability for temperature and moisture in the edge environment, not in the clearcut. This trend was due primarily to the influences of edge orientation. Their results emphasize that edges have a different microclimate from either the interior forest or the recent clearcut, and that the edge serves as a climatic mediator between the clearcut and the forest. Much research has been conducted about edge effects, including the importance of edge habitat for some wildlife species (e.g., Thomas et al. 1979, Yahner 1988) and the effect of these boundaries on species interactions, such as competition, predation, and parasitism (e.g., Kroodsma 1984, Andren and Angelstam 1988, Gustafson and Crow 1994). Research is now focusing more on the dynamics that occur at boundaries between landscape ecosystems (Wiens et al. 1985).

Both Wiens et al. (1985) and Forman and Moore (1992) compare a landscape boundary to a cellular membrane. Like membranes, boundaries vary in their permeability to flows and fluxes. Boundaries may be highly impermeable to the movement of some materials across their interface, but allow other transfers and fluxes to occur unimpeded. The rates and magnitudes of flows outward across a boundary may not be the same as the rates and magnitudes inward for the same material. Given these properties, boundaries separating landscape ecosystems will obviously have important influences on system properties both within homogeneous patches and among components of a landscape (Wiens et al. 1985). It is also important to recognize that the flows and fluxes across boundaries can, in turn, act to change the location and structure of the boundary. Most landscape ecosystems, however, are open systems in which species, matter, and energy move across boundaries with relatively little resistance; that is, ecosystems do not generally exist in isolation from their landscape matrix, and thus context becomes important in understanding the function and structure of these systems.

Still another term mentioned above, "landscape matrix," needs to be defined. It is the dominant cover class or land use within the landscape. Common classes for the landscape (or waterscape) matrix are

Restaging an Evolutionary Drama: Thinking Big on the Chequamegon and Nicolet National Forests

Linda Parker

Working under the notion that "endangered spaces" result in endangered species, the Chequamegon and Nicolet National Forests have launched a strategy in part to identify and protect representative, natural area-quality examples of all the ecosystems of the forests. However, no reserve system on its own can represent and sustain all the variety of life. Therefore, a critical part of this strategy is to integrate this network into the managed landscape (or matrix). This strategy, entitled Landscape Analysis and Design (LAD) after the model described in Chapter 17 of this volume, has been adapted to produce two products for the forests: a landscape-scale assessment of ecological capabilities and a redesigned landscape based on this new information. The National Hierarchy of Ecological Units is an essential component of this strategy, integrated at multiple spatial scales. Overall, this case study illustrates how protection and restoration can be integrated into the forest matrix using principles of zoning and aggregation of harvest units. Specifically, it describes efforts to restore pine barrens, a globally rare ecosystem, in a core reserve area while conducting complementary, sustainable forest management through large block harvesting in the forest matrix.

The Moquah Barrens are a vast area of rolling sand hills on the Chequamegon National Forest in northwest Wisconsin. They are one of the last remnants in a peninsula of pine barrens that once stretched to the prairies of Minnesota, reaching across six counties in Wisconsin. Pine barrens are a type of savanna—a community in which tree density is low enough to allow grasses, forbs, and shrubs to dominate. Reoccurring fire maintained this structure prior to European settlement. Although pine barrens once covered over 2 million acres in Wisconsin, they are now considered to be imperiled in the state and to be rare globally. Around 1900, Moquah was logged, plowed, and burned until its soils were depleted and abandoned. Then, in the 1930s, they were "reforested" in jack pine and red pine plantations by the Civilian Conservation Corps (CCC).

Today, the 8,000-acre Moquah Barrens Wildlife Area is one of the largest and best examples of a restored barrens. This area is now a core zone of open-land habitat in what has since become a landscape dominated by pine plantations and second-growth forests.

The first step in the restoration process was clearcutting very large (500- to 1,000-acre) jack pine plantations to re-create the savanna-like structure of this system. After harvest, fire was reintroduced to the system; the intensity, frequency, and seasonality of the prescribed burns were designed to mimic the natural disturbance regime.

Since 1991, restoration efforts have expanded to a 2,800-acre area adjacent to the Moquah Barrens Wildlife Area. The focus of this new effort is to restore a savanna generally broad categories such as agricultural land, forest, grassland, water body, or urban land. By definition, the matrix exceeds the total area of any other landscape element.

Patches, boundaries, and the heterogeneity they create vary with the resolution and extent at which they are observed. Increasing the extent of observation tends to homogenize variation, and increasing the resolution or grain of observation transforms a variable into a constant (Kolasa and Rollo 1991). Shifts in scale therefore create homogeneity where heterogeneity existed, or vice versa, depending on the direction of change. In addition to a change in perspective, new properties often emerge with shifts in scale. Thus an understanding of processes at one scale does not necessarily translate directly to another scale.

Various indices and metrics have been applied to characterize landscape structure (e.g., O'Neill et al. 1988, Turner 1990). The theory of fractal geometry (Mandelbrot 1983), for example, involves simultaneous measurements over multiple spatial scales. Thus, it provides a basis for determining the scale or scales at which an ecological hypothesis is valid (Milne 1988). The effect of pattern on process has been the central theme for many ecological studies. Because pattern is a function of scale, a multiscale approach to understanding heterogeneity in time and space, or

system more typical of the pre-European settlement condition, which included red pine savanna—another community which, although once common, no longer exists. These restored acres, combined with the Moquah Barrens, will result in one of the largest, most contiguous pine barrens in the state. Already, the barrens are home to 7 of the 11 critical grassland bird species in Wisconsin.

Perhaps the factor that most distinguishes this project is the integration of the entire landscape into the reserve design. When the old jack pine plantations outside the restoration areas were harvested, two basic questions were debated: Should we harvest in the traditional 40-acre blocks that produce a patchwork pattern? and Do we replant all these acres and maintain a monoculture?"

When ice storms hit the Bayfield Peninsula in the spring of 1991, they helped answer these questions and made one thing perfectly clear—natural jack pine systems withstood the forces of nature better than plantations. So, instead of converting to red pine plantations, the Wisconsin Forest Service launched a successful program of natural regeneration of jack pine with the use of anchor chaining. The sandy, droughty soils produce a vegetation mosaic of jack pine mixed with oak and aspen that will always have open pockets and shrubby patches. This community structure is very similar to a post-fire pine barrens ecosystem. More significantly, the Forest Service abandoned past cutting patterns, which had created a fragmented, patchwork landscape. Harvest units were aggregated into blocks as large as 1,000 acres; size and shape were dependent upon the ecological unit of the site. From an ecological perspective, the large temporary openings complement the nearby pine barrens. The large, naturally regenerated patches create a landscape pattern more typical of the dynamic, fire-driven system that persisted here for thousands of years.

At a community scale, these large temporary openings have greater structural and compositional diversity than plantations and are surrogate barrens for open-lands species. This dramatically increases the effectiveness of the restored barrens as refugia for these species. Many of the bird species that are typically associated with pine barrens also nest in these large, naturally regenerated clearcuts. At a large scale, the landscape structure is much more typical of the natural system—a mosaic of large open forested patches. In addition, these large patches may serve as corridors for species dispersal to newly restored areas. These units will eventually be harvested again in about 40 years—the approximate return interval for catastrophic fire in this system. As they get older, and thus more forested, they eventually will become less suitable for barrens species. However, planning and scheduling efforts can help to ensure that there is an appropriate distribution of large blocks of open habitat across the landscape over time.

Although one would expect that 1,000-acre clearcuts would create a firestorm of opposition, the reaction has been overwhelmingly supportive. The Forest Service's extensive public involvement—including hilltop discussions on the barrens—has been successful. There seems to be an understanding that these efforts are an attempt to work with the land instead of against it.

more specifically, a multiscale approach to managing natural resources in time and space, is critical to achieving a more comprehensive, integrated approach to forest management.

Temporal Dynamics and Spatial Patterning in Managed Landscapes

Landscape patterns created by forest harvesting vary with the size of harvest units, the rates of harvest, and the spatial arrangement of harvest units. At the landscape scale, concerns about timber harvesting often center around forest fragmentation. Lord and Norton (1990) define fragmentation as the loss of continuity in either time or space. More directly, fragmentation is the process of increasing the number of landscape patches, increasing the amount of edge environment, decreasing the amount of interior habitat, and increasing the isolation of remnant patches in the landscape matrix.

Dispersing small harvest units throughout the landscape, such as in the checkerboard (staggered-setting) model used to harvest Douglas-fir in the Pacific Northwest, rapidly reduces the amount of forest interior and increases the amount of forest edge. In the checkerboard model, no forest interior environment remains when 50 percent of the forest has been harvested using 10-hectare harvest units, assuming an edge width of 160 m (Franklin and Forman 1987). If cutting units are dispersed regularly, the last forest patch that is 100 ha in size will disappear from the landscape when only 1 percent of the forest is harvested using 1-hectare cutting units (Franklin and Forman 1987). To further explore the relationship between cutting pattern and forest fragmentation, Li et al. (1993) simulated five cutting patterns—random patches, maximum dispersion, staggered setting, partial aggregation, and progressive cutting—in a hypothetical landscape. They then measured edge density (length of edge per unit area), patch shape, relative patchiness (a measure of structural complexity of the landscape), and amount of forest interior as a function of the percent of landscape harvested. Levels of fragmentation as measured by these variables increased rapidly with increased harvest levels for the maximum-dispersion, random-patch, and staggered-setting harvest patterns, increased slowly with partial aggregation, but remained virtually unchanged when a system of progressive harvesting was applied. With a progressive-cutting strategy, harvesting is initiated at a point in the landscape and then moves outward in a concentric pattern.

In addition to patch size, patch shape has important ecological implications. Patch shape affects the amount of interior habitat. Harvest units of equal size but with different perimeter-to-area ratios will vary in the amount of forest interior. When patch area is held constant, units with a higher ratio of perimeter to area will have less interior environment; for example, a long, linear cutting unit will have less area greater than 100 m from an edge than will a circular area of the same size. Likewise, irregular patches with long lobes and sinuous boundaries increase the perimeter-to-area ratios and thus decrease the area of effective habitat for interior species. Although less is known about the relation between ecological process and patch shape, the movement and fluxes of organisms, matter, and energy across the landscape are likely related to patch shape and the related landscape heterogeneity (Forman and Godron 1986).

Finally, the spatial arrangement of patches is important for many ecological interactions and processes. Landscape pattern and the demographic behavior of individual species combine to shape population dynamics in habitat patches. Population-level processes, including competition, predation, parasitism, and mutualism, are affected by the spatial arrangement of organisms and their habitats (Roff 1974, Lefkovitch and Fahrig 1985). A number of studies demonstrate the importance of proximity and connectivity in the persistence of populations (e.g., Fahrig and Paloheimo 1988). Patchiness can enhance persistence of populations if the spatial pattern facilitates the recolonization of patches in which local extinction has occurred. In studies of small mammals associated with woodland patches surrounded by cropland, local extinction within the woodland patches was a frequent phenomenon, but so was rapid recolonization (Merriam 1990). The network of fencerows connecting the woodlots allowed the movement of small mammals between forest patches.

Applying the principles of proximity and connectivity, Mladenoff et al. (1994) designed a landscape to maintain and increase the value of small, scattered patches of old growth embedded in a matrix dominated by second-growth forest. A 100-meter "restoration zone" adjacent to the old growth provides a "soft edge" that effectively removes the edge environment outside the old-growth patch. A secondary buffer adjacent to the restoration zone is subject to uneven-age management for timber, but serves to maintain a canopy cover and to connect the otherwise isolated old-growth patches. Most of the landscape, including the secondary buffer and the remaining forest, is available for commodity production.

At landscape and regional scales, human activities (e.g., producing food, building roads, harvesting forests), combined with physical features (e.g., lakes, streams, wetlands), contribute to landscape heterogeneity. Few of these factors remain constant in time. In comparing landscape dynamics on public and private lands on a 2,589-square-kilometer landscape in western Oregon, Spies et al. (1994) found that large remaining patches (greater than 5,000 ha) of contiguous, closed-canopy forest were restricted to public lands protected from timber production, such as wilderness areas and research natural areas. When public and private lands were combined, the percentage of closed-canopy forest declined from 71 percent in 1972 to 58 percent in 1988; the amount of edge increased and the amount and percentage of interior forest (greater than 100 m from edge) decreased during the same time period. All these trends suggest increasing forest fragmentation.

In contrast to the patterns associated with deforestation in western Oregon, Turner and Ruscher (1988) documented changes in the composition and structure of the Georgia landscape during a period in which reforestation was the prevailing trend. During the half century starting in 1930, fragmentation declined in the Georgia landscape, in part because of the increase in forest lands and the aggregation of small patches of forest into larger patches. The complexity of patch shape, as measured by the fractal dimension of patches, however, decreased for croplands and pine forests during the same period of time, although hardwood forests showed a slight increase in complexity of patch shape. The contrasting patterns in Oregon and Georgia illustrate the importance of the initial conditions when assessing landscape change. The current state of a landscape, as measured by its composition and structure, profoundly affects future conditions. A good example of this legacy of past land-use practices is the difficulty in altering landscape patterns created by dispersed harvesting, such as those created in the Pacific Northwest. Rapid changes in landscape patterns will not result from changing from a dispersed- to an aggregated-cutting pattern without significant reductions in harvest rates and significant reductions in the age at which stands can be harvested (Wallin et al. 1994).

What are the ecological consequences of changes in patch size, shape, and configuration relating to cutting patterns? Many concerns, including decreased biological diversity, increased susceptibility to disturbance, and increased vulnerability to predation and parasitism, have been related to fragmentation. The propagation of disturbance in landscapes depends on the structure of the landscape as defined by the composition, size, shape, and distribution of patches as well as on the intensity and frequency of the disturbance event (Franklin and Forman 1987, Turner et al. 1989). Susceptibility to windthrow, for example, increases with the amount of forest edge, especially with high-contrast edges or along narrow corridors of remnant forests. Frequency of fire in a landscape is higher with the interspersion of cutover patches in a closed-canopy forest compared to a matrix dominated by closed-canopy forest (Franklin and Forman 1987). As with other species, the response of pests and pathogens to changing landscape patterns is likely to be highly specific to the interaction between the host and the life-history features of the pest or pathogen. At least with one insect, the forest tent caterpillar, the amount of forest edge per kilometer (i.e., edge density) was the best predictor of the duration of tent caterpillar outbreaks in northern Ontario (Roland 1993). It is known that natural enemies of the forest tent caterpillar drive its population dynamics. A fragmented boreal forest may limit the dispersal of parasitoids and pathogens of tent caterpillar, and the population processes determining the duration of outbreaks may be directly modified by the physical environment associated with forest edges.

Designing a landscape involves explicit consideration of the composition, size, shape, and distribution of patches in both time and space. Desired future conditions will depend in large part on existing conditions, the regional and landscape context in which a management unit exists, and the specific set of values and benefits that are desired from the land. Despite these variables, general guidelines can be obtained by using natural landscapes as guides for developing appropriate managed landscapes. In northern Wisconsin and upper Michigan, Mladenoff et al. (1993) compared a managed landscape dominated by second-growth northern hardwoods to an unmanaged-forest (old-growth) landscape with a similar climate, soil, and landform to determine the effect of human activity on landscape structure. The managed-forest landscape had significantly more small forest patches and fewer large patches than the unmanaged-forest landscape. Furthermore, forest patches were more regular in shape (having a lower fractal dimension) in the managed landscape compared to the unmanaged area. The baseline landscape in this study was a hemlock–sugar maple forest in which small-scale windthrows were frequent, but catastrophic disturbance was infrequent (greater than a 1,000-year return time). Others (e.g., Ripple et al. 1991, Spies et al. 1994) found similar trends in forested landscapes with fire-dominated disturbance regimes. The results of these and many other studies suggest that simplifying the composition and structure of landscape ecosystems is a common cumulative effect of human land use. Economic efficiency, at least in the short term, may require simplification of ecosystems used for the production of commodities. However, ecologic efficiency, as defined by maintaining the integrity of self-organizing systems such as ecosystems, may well require embracing complexity (Kay and Schneider 1994).

Strategies to maintain forest contiguity and minimize fragmentation should be based on landscape analysis and other ecological considerations. One strategy in the search for appropriate baselines for designing landscapes is to consider the range of variability in disturbance regimes before massive alteration of the landscape (e.g., prior to European settlement in North America). In the Pacific Northwest, such a strategy has resulted in recommendations to create a network of large (20,000–50,000 ha) reserves (Noss 1993), extend rotation lengths, harvest larger but fewer units, abandon the strategy of dispersing clearcuts, and adopt a more aggregated distribution of harvest units (Franklin and Forman 1987, Li et al. 1993).

In summary, the following generalizations about the effects of forest harvesting on fragmentation seem reasonable: (1) Large cutting units cause less fragmentation than small units. (2) Aggregated and progressive cutting patterns create less fragmentation than dispersed cutting patterns. (3) Low harvest rates produce less fragmentation than high cutting rates. The interactions among size of harvest units, their spatial distribution, and harvest rate, however, require further investigation. What is their relative importance in creating forest fragmentation? What are the most effective strategies to minimize fragmentation in real landscapes? These and other related questions are explored in the next section.

Projecting in Time and Space

Linking geographic models with geographic information systems (GISs) can increase a manager's ability to evaluate the effects of alternative management strategies on landscape structure. One such model, HARVEST, allows the application of specific rules for even-age management (clearcuts and shelterwoods) and group selection to actual landscapes within a GIS environment (Gustafson and Crow 1994, 1996). The model produces landscape patterns that have spatial attributes resulting from the initial landscape conditions and the planned management actions. The goal is not to optimize timber production, but instead to simulate alternative management strategies by incorporating decisions typically made by resource managers. Model outputs are displayed visually as maps that allow for easy comparisons among different management scenarios through time.

HARVEST was constructed to run within the ERDAS GIS software (produced by Erdas, Inc., Atlanta, GA) and was written using ERDAS Toolkit routines for inputs and outputs. ERDAS is a grid-cell (raster) GIS that allows flexible display and manipulation of digital maps. Timber harvest allocations were made by the model using a digital stand map, where grid-cell values reflect the age of each timber stand. The model allows control of the size distribution of harvests, the total area of forest to be harvested, the rotation length (as given by the minimum age on the input stand map in which harvest may be allocated), and the width of buffers left between adjacent harvests and between harvests and nonforest habitats.

Applying HARVEST to a 23,000-hectare landscape in southern Indiana, we designed a factorial experiment to test the interaction between the size and distribution of harvest units and the rate of harvest. The starting point was a digital land-cover map derived from LANDSAT Thematic Mapper satellite imagery collected in April 1988. The area is largely forested, with some patches of nonforest cover. For the simulation, all forests in the study area were assumed to be commercial stands. Two levels of each factor were tested: 18-hectare and 0.18-hectare harvest units, 1 percent and 7 percent of the forest harvested per decade, and randomly dispersed and aggregated harvests. The simulation period was 80 years, with output provided in 10-year increments. Three replications were run for each factorial combination. The variability of the spatial pattern of the simulations was very low (less than 0.05 percent), and three replicates were adequate to ensure robust results. We used a GIS proximity function to produce maps of forest interior (defined as closed-canopy forest at least 90 m from an opening or edge) for each of the eight decades. Harvested stands were assumed to return to a closed canopy condition 20 years after harvest. From these maps, we calculated the total area of forest interior and the length of linear forest edge.

To determine the relative impact of the three sources of variation (size of timber harvest openings [*size*], harvest intensity [*intensity*], and the spatial dispersion of openings [*dispersion*]) on forest spatial pattern, we conducted a repeated-measures analysis of variance (ANOVA) to test for treatment effects on the total area of forest interior and total linear forest edge. We included the simulated time periods (*decade*) in the analysis to account for the potential of autocorrelation among spatial patterns measured in successive decades.

Results

Forest interior was greatest when harvest opening size was large, harvest intensity was low, and openings were clustered; it was lowest when openings were small, harvest intensity was high, and openings were dispersed (Figure 14.1a). The total area of forest interior maintained on the study area was primarily influenced by the size of harvest openings and harvest intensity, although the influence of the spatial dispersion of openings was also significant. Examination of the ANOVA sums of squares revealed that *size* explained 46.9 percent of the total variance in forest interior, and *intensity* explained 40.5 percent of the variance, while *dispersion* explained only 3.7 percent (Table 14.1). The most significant interaction was be-

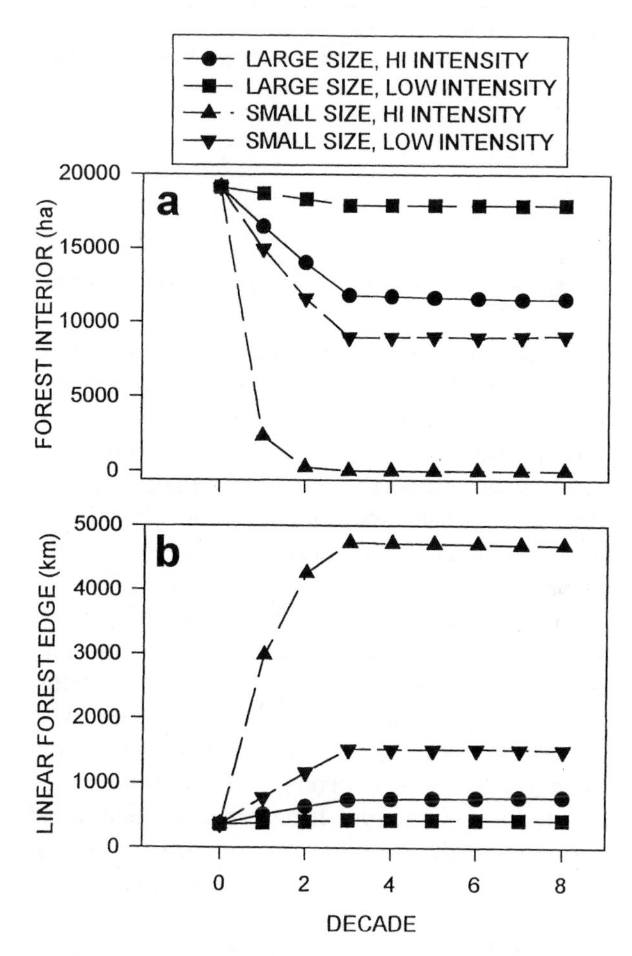

Figure 14.1 Changes over time in the amount of forest interior (forest >90 m from an edge) and linear forest edge under alternative harvest strategies. These data represent a random dispersion of harvest openings. A clumped dispersion produced the same trends with slightly higher amounts of forest interior and lower amounts of edge. The standard deviation at all points is less than the width of the symbols used, and is not shown.

Table 14.1 Treatment effects on the total area of forest interior

Source	Degrees of Freedom	Sums of Squares	F	Probability >FR2
Size	1	3.6138+09	1152.67	0.0001
Intensity	1	3.1213E+09	995.58	0.0001
Dispersion	1	1.2301E+08	39.23	0.0001
Decade	7	2.8465E+08	12.97	0.0001
Error	181	5.6746E+08		
TOTAL	191	7.7102E+09		0.93

Note: Using repeated measures analysis of variance, this table compares the effects of harvest size, harvest intensity, and the spatial dispersion of harvest openings on the total area of forest interior habitat (forest > 90 m from the nearest opening) on the landscape. Analysis based on three replicates.

tween *size* and *intensity,* explaining 4.6 percent of the variance.

We would expect that the difference between harvesting 1 percent and 7 percent of the forest would have a greater effect than the difference in harvest opening size on the amount of forest interior and edge. However, note that harvesting only 1 percent of the forest each decade using small openings leaves less forest interior than harvesting 7 percent of the forest using large openings (Figure 14.1a). The primary consequence of varying harvest-opening size is to change the number of openings produced to achieve a given level of harvest. In these simulations, 100 of the small harvest openings were necessary to harvest the same area of forest as one of the large openings. Since each opening perforates forest interior and introduces edge habitat in a 90-meter band surrounding the opening, the cumulative effect of small opening size becomes very large. These results are consistent with the findings of Gustafson and Crow (1994), who found that forest interior declines sharply with reductions in cutting-unit size below about 20 ha.

The amount of linear forest edge was lowest when harvest opening size was large, harvest intensity was low, and openings were clustered; it was highest when openings were small, harvest intensity was high, and openings were dispersed (Figure 14.1b). Here again, even a low rate of harvest using small openings produces more edge than a high rate of harvest using large openings. *Size* explained 49.7 percent of the variance in forest edge, while *intensity* explained 27.1 percent of the variance. *Dispersion* and *decade* each explained less than 2.2 percent of the variance (Table 14.2). The most significant interaction was between *size* and *intensity,* explaining 18 percent of the variance.

These results are intuitive because perimeter-area ratios vary with the size of a two-dimensional object such as a harvest opening. Also, since each opening introduces edge into the landscape, changes in the intensity of harvest directly impact the amount of edge produced.

The results of these and other simulations (Gustafson and Crow 1994, 1996) clearly show that when retention of forest interior habitat and the reduction of edge are management goals, timber extraction using large, clustered openings is much more effective than using small, dispersed harvest openings. Single-tree selection removes trees without necessarily creating openings, and for some management goals might be preferred. However, single-tree selection favors late-successional species and requires a dense network of roads for frequent access. Regardless, each management option needs to be considered within the framework of time and space. Our study illustrates that the complex interaction of the spatial and temporal implementation of timber management and forest habitat conditions can be better understood using spatial modeling tools.

Table 14.2 Treatment effects on the total linear forest edge

Source	Degrees of Freedom	Sums of Squares	F	Probability >FR2
Size	1	2.2035E+14	436.55	0.0001
Intensity	1	1.2026E+14	238.26	0.0001
Dispersion	1	2.1576E+12	4.27	0.0401
Decade	7	9.4934E+12	2.69	0.0114
Error	181	9.1360E+13		
TOTAL	191	4.4362E+14		0.79

Note: Using repeated measures analysis of variance, this table compares the effects of harvest size, harvest intensity, and the spatial dispersion of harvest openings on the total linear forest edge on the landscape. Analysis based on three replicates.

Implications for Management

Habitat fragmentation, increases in edge environments (especially hard or high-contrast edge), and simplification of patch shape are common characteristics associated with the transformation to human-dominated landscapes. Forest harvesting often contributes to these changes. Spatially explicit simulation models such as HARVEST and others (Li et al. 1993, Wallin et al. 1994), as well as analysis of structural changes in actual landscapes (e.g., Mladenoff et al. 1993, Spies et al. 1994), illustrate the ease with which a forest can be fragmented when small harvest units are dispersed across the landscape. A combination of larger harvest units, the aggregation of those units, and lower harvest rates may be necessary to reduce forest fragmentation.

Relationships among the size of harvest units, their spatial distribution, and harvest rates can be visualized using GISs in combination with remote sensing and mathematical models. A rule-based model such as HARVEST allows the application of realistic management alternatives to be applied to actual landscapes. The initial landscape patterns and landscape patterns projected after 80 years—using a large harvest unit and high cutting rates, a large harvest unit and low cutting rates, a small harvest unit and high cutting rates, or a small harvest unit and low cutting rates—can be compared easily as maps (Figure 14.2). The reduction in forest interior by dispersing small harvest units across the landscape is obvious in this format (Figures 14.2d and 14.2e). Even at a low har-

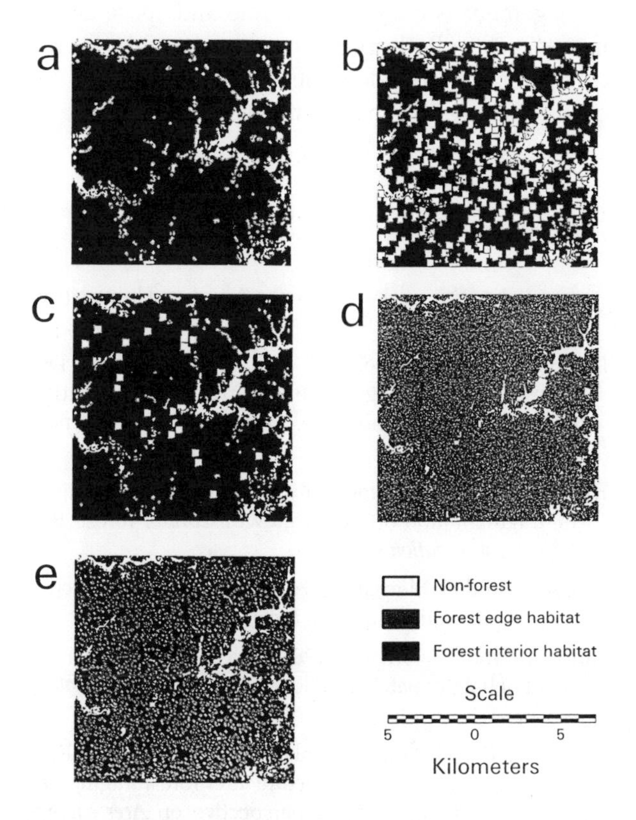

Figure 14.2 Visual outputs from the HARVEST simulation. Two harvest unit sizes (18 ha and 0.18 ha with dispersed distributions) and two harvest rates (1 percent and 7 percent of the forest per decade) are compared. The maps show the forest area > 90 m from an edge, forest edge, and nonforest at the beginning (a), and after 80-year simulations with large unit, high harvest rate (b), large unit, low harvest rate (c), small unit, high harvest rate (d), small unit, low harvest rate (e).

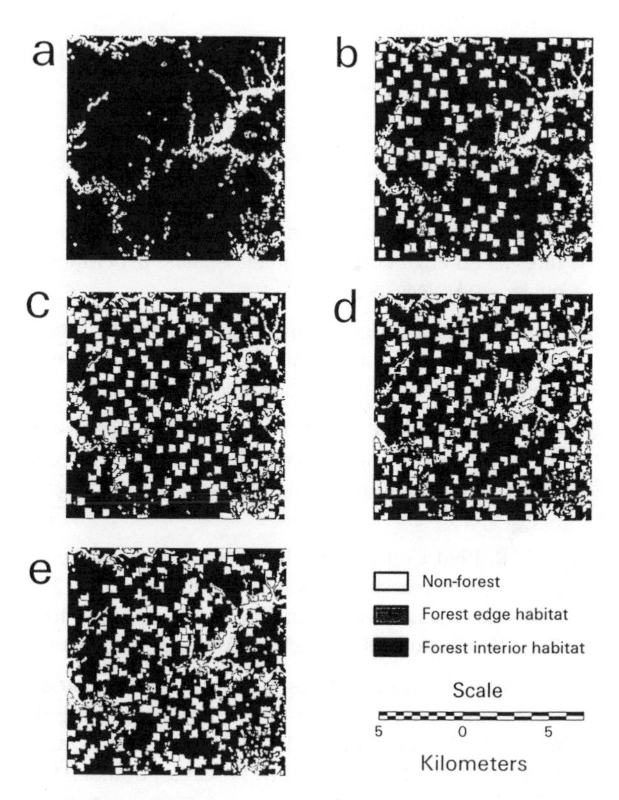

Figure 14.3 Initial conditions and simulation results at 20-year increments for the management alternative using large harvest units with a dispersed distribution and a high harvest rate.

vest rate, little forest interior remains (Figure 14.2e). The output at year 80 provides only a static picture. The dynamic nature of the landscape under different management scenarios is obvious when plotting the landscape at various points in time. In Figure 14.3, the landscape in southern Indiana is shown at year 0 and subsequently plotted at 20-year increments for the treatment with dispersed harvest units that are 18 ha in size and with a 7 percent harvest rate per decade. The shifting mosaic of harvest units reflects our assumption that closed-canopy conditions develop 20 years after harvest in this Midwestern forest.

Decisions about the size, shape, distribution, and timing of resource management actions should be based on a sound understanding of the relation between pattern and process. Landscape ecology, by defining responses to management apparent at the landscape scale together with appropriate characterizations of heterogeneity, can provide a unifying framework for developing consistent predictive models of use in resource management (Risser et al. 1984). The characterization of heterogeneity is only the first step. More important is knowing the ecologic and economic significance of the patterns measured. The tools for assessing patterns, such as GIS and remote sensing, are widely available and have been widely applied by resource managers. Our understanding of the significance of the landscape patterns created through management is far from complete, but is increasing rapidly. There is growing recognition that informed resource management decisions cannot be made exclusively at the stand level. A shift to regional and national decisions will place greater emphasis on landscape ecology concepts. Ecosystem management is managing in time and space across many scales. Both perspectives, from above the canopy as well as below the canopy, are needed for effective ecosystem management.

Literature Cited

Allen, T. F. H., and T. B. Starr. 1982. *Hierarchy: Perspectives for ecological complexity.* Chicago: University of Chicago Press.

Andren, H., and P. Angelstam. 1988. Elevated predation rates as an edge effect in habitat islands: Experimental evidence. *Ecology* 69:544–547.

Bratton, S. P. 1994. Logging and fragmentation of broadleaved deciduous forests: Are we asking the right ecological questions? *Conservation Biology* 8:295–297.

Chen, J., J. F. Franklin, and T. A. Spies. 1993. Contrasting microclimates among clearcut, edge, and interior of old-growth Douglas-fir forest. *Agricultural and Forest Meteorology* 63:219–237.

Crow, T. R. 1990. Conservation biology and landscape ecology: New perspectives for resource managers. In *Proceedings: Society of American Foresters National Convention.* Washington, DC: Society of American Foresters.

———. 1991. Landscape ecology: The big picture approach to resource management. In *Challenges in the conservation of biological resources,* ed. D. J. Decker, M. E. Kransy, G. R. Goff, C. R Smith, and D. W. Gross. Boulder, CO: Westview Press.

Curtis, J. T. 1956. The modification of mid-latitude grasslands and forests by man. In *Man's role in changing the face of the earth,* ed. W. L. Thomas. Chicago: University of Chicago Press.

Delcourt, H. R., P. A. Delcourt, and T. Webb III. 1983. Dynamic plant ecology: The spectrum of vegetation change in space and time. *Quatenary Science Review* 1:153–175.

Esseen, P. A. 1994. Tree mortality patterns after experimental fragmentation of an old-growth Conifer forest. *Biological Conservation* 68:19–28.

Fahrig, L., and G. Merriam. 1985. Habitat patch connectivity and population survival. *Ecology* 66:1762–1768.

Fahrig, L., and J. Paloheimo. 1988. Effect of spatial arrangement of habitat patches on local population size. *Ecology* 69:468–475.

Fore, S. A., R. J. Hickey, J. L. Vankat, S. I. Guttman, and R. L. Schaefer. 1992. Genetic structure after forest fragmentation: A landscape ecology perspective on *Acer saccharum. Canadian Journal of Botany* 70:1659–1668.

Forman, R. T. T. 1991. Landscape corridors: From theoretical foundations to public policy. In *Nature conservation 2: The role of corridors,* ed. D. A. Saunders and R. J. Hobbs. Chipping Norton, Australia: Surrey Beatty and Sons.

Forman, R. T. T., and M. Godron. 1986. *Landscape ecology* New York: John Wiley & Sons, Inc.

Forman, R. T. T., and P. N. Moore. 1992. Theoretical foundations for understanding boundaries in landscape mosaics. In *Landscape boundaries: Consequences and principles,* ed. A. J. Hansen and F. di Castri. New York: Springer-Verlag.

Franklin, J. F., and R. T. T. Forman. 1987. Creating landscape patterns by forest cutting: Ecological consequences and principles. *Landscape Ecology* 1:5–18.

Gardner, R. H., B. T. Milne, M. G. Turner, and R. V. O'Neill. 1987. Neutral models for the analysis of broad-scale landscape pattern. *Landscape Ecology* 1:19–28.

Gustafson, E. J., and T. R. Crow. 1994. Modeling the effects of forest harvesting on landscape structure and the spatial distribution of cowbird brood parasitism. *Landscape Ecology* 9:237–248.

———. 1996. Simulating the effects of alternative forest management strategies on landscape structure. *Journal of Environmental Management* 46:77–94.

Johnson, W. C. 1988. Estimating dispersibility of *Acer, Fraxinus,* and *Tilia* in fragmented landscapes from patterns of seedling establishment. Landscape Ecology 1:175–187.

Kay, J. J., and E. Schneider. 1994. Embracing complexity: The challenge of the ecosystem approach. *Alternatives* 20:32–39.

Kolasa, J., and C. D. Rollo. 1991. Introduction: The heterogeneity of heterogeneity, a glossary. In *Ecological heterogeneity,* ed. J. Kolasa and S. T. A. Picket. New York: Springer-Verlag.

Kroodsma, R. L. 1984. Effect of edge on breeding forest bird species. *Wilson Bulletin* 96:426–436.

Kruess, A., and T. Tscharntke. 1994. Habitat fragmentation, species loss, and biological control. *Science* 264:1581–1584.

Lefkovitch, L. P., and L. Fahrig. 1985. Spatial characteristics of habitat patches and population survival. *Ecological Modelling* 30:297–308.

Li, H., J. F. Franklin, F. J. Swanson, and T. A. Spies. 1993. Developing alternative forest cutting patterns: A simulation approach. *Landscape Ecology* 8:63–75.

Lord, J. M., and D. A. Norton. 1990. Scale and the spatial concept of fragmentation. *Conservation Biology* 4:197–202.

MacClintock, L., R. F. Whitcomb, and B. L. Whitcomb. 1977. Evidence for the value of corridors and minimization of isolation in preservation of biotic diversity. *American Birds* 31:6–16.

Mandelbrot, B. B. 1983. *The fractal geometry of nature.* New York: W. H. Freeman.

Meentemeyer, V., and E. O. Box. 1987. Scale effects in landscape studies. In *Landscape heterogeneity and disturbance,* ed. M. Turner. New York: Springer-Verlag.

Merriam, G. 1990. Ecological processes in the time and space of farmland mosaics. In *Changing landscapes: An ecological perspective,* ed. I. S. Zonneveld and R. T. T. Forman. New York: Springer-Verlag.

Milne, B. T. 1988. *Measuring the fractal geometry of landscapes. Applied Mathematics and Computation* 27:67–79.

———. 1992. Spatial aggregation and neutral models in fractal landscapes. *American Naturalist* 139:32–57.

Mladenoff, D. J., M. A. White, J. Pastor, and T. R. Crow. 1993. Comparing spatial patterns in unaltered old-growth and disturbed forest landscapes. *Ecological Applications* 3:294–306.

Mladenoff, D. J., M. A. White, T. R. Crow, and J. Pastor. 1994. Applying principles of landscape design and management to integrate old-growth forest enhancement and commodity use. *Conservation Biology* 8:752–762.

Noss, R. F. 1993. A conservation plan for the Oregon Coast Range: Some preliminary suggestions. *Natural Areas Journal* 13:276–290.

Nyland, R. D., W. C. Zipperer, and D. B. Hill. 1986. The development of forest islands in exurban central New York. *Landscape and Urban Planning* 13:111–123.

O'Neill, R. V., D. L. DeAngelis, J. B. Waide, and T. F. H. Allen. 1986. *A hierarchical concept of ecosystems.* Princeton: Princeton University Press.

O'Neill, R. V., J. R. Krummel, R. H. Gardner, G. Sugihara, B. Jackson, D. L. DeAngelis, B. T. Milne, M. G. Turner, B. Zygmunt, S. W. Christensen, V. H. Dale, and R.L. Graham. 1988. Indices of landscape pattern. *Landscape Ecology* 1:153–162.

Probst, J. R., and T. R. Crow. 1991. Integrating biological diversity and resource management. *Journal of Forestry* 89:12–17.

Ripple, W. J., G. A. Bradshaw, and T. A. Spies. 1991. Measuring forest landscape patterns in the Cascade Range of Oregon, USA. *Biological Conservation* 57:73–88.

Risser, P. G. 1985. Toward a holistic management perspective. *BioScience* 35:414–418.

———. 1988. General concepts for measuring cumulative impacts on wetland ecosystems. *Environmental Management* 12:585–589.

Risser, P. G., J. R. Karr, and R. T. T. Forman. 1984. *Landscape ecology: Directions and approaches.* Special publication no. 2. Champaign, IL: Illinois Natural History Survey.

Roff, D. A. 1974. Spatial heterogeneity and the persistence of populations. *Oecologia* 15:245–258.

Roland, J. 1993. Large-scale forest fragmentation increases the duration of tent caterpillar outbreak. *Oecologia* 93:25–30.

Saunders, D. A., R. J. Hobbs, and C. R. Margules. 1991. Biological consequences of ecosystem fragmentation: A review. *Conservation Biology* 5:18–32.

Sharpe, D. M., G. R. Guntenspergen, C. P. Dunn, L. A. Leitner, and F. Stearns. 1987. Vegetation dynamics in a southern Wisconsin agricultural landscape. In *Landscape heterogeneity and disturbance,* ed. M. G. Turner. New York: Springer-Verlag.

Spies, T. A., W. J. Ripple, and G. A. Bradshaw. 1994. Dynamics and pattern of a managed coniferous forest landscape in Oregon. *Ecological Applications* 4:555–568.

Thomas, J. W., C. Maser, and J. E. Rodick. 1979. Edges. In *Wildlife habitats in managed forest: The Blue Mountains of Oregon and Washington,* ed. J. W. Thomas. Agricultural handbook no. 553. Washington, DC: USDA Forest Service.

Turner, M. G. 1990. Spatial and temporal analysis of landscape pattern. *Landscape Ecology* 3:153–162.

Turner, M. G., and C. L. Ruscher. 1988. Changes in landscape patterns in Georgia, USA. *Landscape Ecology* 1:241–251.

Turner, M. G., R. H. Gardner, V. H. Dale, and R. V. O'Neill. 1989. Predicting the spread of disturbance across heterogeneous landscapes. *Oikos* 55:121–129.

Wallin, D. O., F. J. Swanson, and B. Marks. 1994. Landscape pattern response to changes in pattern generation rules: Land-use legacies in forestry. *Ecological Applications* 4:569–580.

Walters, C. J., and C. S. Holling. 1990. Large-scale management experiments and learning by doing. *Ecology* 71:2060–2068.

Whitcomb, R. F., C. S. Robbins, J. F. Lynch, B. L. Whitcomb, M. K. Klimkiewicz, and D. Bystrak. 1981. Effects of forest fragmentation on avifauna of the eastern deciduous forest. In *Forest island dynamics in man-dominated landscapes,* ed. R. L. Burgess and D. M. Sharpe. New York: Springer-Verlag.

Wiens, J. A. 1976. Population responses to patchy environments. *Annual Review of Ecology and Systematics* 7:81–120.

———. 1989. Spatial scaling in ecology. *Functional Ecology* 3:385–397.

———. 1992. Ecological flows across landscape boundaries: A conceptual overview. In *Landscape boundaries: Consequences for biotic diversity and ecological flows,* ed. A. J. Hansen and F. di Castri. New York: Springer-Verlag.

Wiens, J. A., C. S. Crawford, and J. R. Gosz. 1985. Boundary dynamics: A conceptual framework for studying landscape ecosystems. *Oikos* 45:421–427.

Yahner, R. H. 1988. Changes in wildlife communities near edges. *Conservation Biology* 2:333–339.

Zipperer, W. C., R. L. Burgess, and R. D. Nyland. 1990. Patterns of deforestation and reforestation in different landscape types in central New York. *Forest Ecology and Management* 36:103–117.

15 The Physical Environment as a Basis for Managing Ecosystems

Frederick J. Swanson, Julia A. Jones, and Gordon E. Grant

Landscape Patterns and Processes 231
History of Landscape Change in the Pacific Northwest 232
Regional Ecological Assessments 233
Future Landscape Management 233
Conclusions 236
Acknowledgments 237
Literature Cited 237

The physical environment sets the stage on which ecological phenomena operate. Our understanding of the physical environment, especially disturbances in forest and stream systems, has evolved through time. How we have used what we have learned has changed in tandem with changes in management focus. Early in the 20th century, when resource extraction was preeminent, hydrology and geomorphology were used to support activities such as repair of landslide-damaged roads. In the mid–20th century, environmental concerns grew, expressed in part by a series of laws—the Clean Water Act, the National Environmental Policy Act, the National Forest Management Act, and the Endangered Species Act—which directed public land managers to analyze and disclose the effects of proposed actions and to protect species and ecosystem productivity. Information on the physical environment was used to minimize undesired environmental effects through actions such as surveying proposed road locations to avoid potential landslide sites.

Late in the 20th century, growing concerns about

the effects of management actions dispersed over large watersheds (considered under the rubric of cumulative watershed effects) has led to more holistic views of watershed function. Understanding of the physical environment dominates watershed analyses designed to identify and mitigate cumulative watershed effects (Washington Forest Practice Board 1992, FEMAT 1993).

A parallel development during the late 1980s and 1990s has been concern with potential loss of species, especially those associated with old-growth forests. This has led to conservation strategies focused on recovery of forest habitat for wildlife species. Knowledge about the physical environment has played a significant role in conservation strategies for fish species, but less so for terrestrial species.

What will the 21st century bring? We expect understanding of the physical environment to gain a more prominent position in land-use planning, forming the starting point for ecosystem management on public lands. Emphasis will shift from commodity production and single species to sustaining the dynamic ecosystems that support life. Development of the concept of managing within a range of natural variability is a harbinger of this shift. This concept recognizes that ecosystems are naturally very dynamic and subject to periodic disturbances. As such, they have exhibited a range of conditions (e.g., successional states) in the past that resulted in important controls on geographic patterns of disturbances and biotic resources.

The management implication of this perspective is that ecosystem patterns and processes should be managed with adequate similarity to those of the past so that diversity and productivity of assemblages of native species can be maintained in conformance with federal laws governing public forest land where some cutting and active fire management are permitted. Such an approach contrasts with past landscape management, which was largely a de facto consequence of numerous stand-level decisions. In contrast, an approach considering ecosystem dynamics is based on an understanding of landscape history and natural disturbance regimes. It is intended to fit the managed biotic landscape with the natural biological and physical character of landscapes, including disturbance regimes. The rationale is that native species have adapted to the natural disturbance regime over thousands of years. The survival potential of native species would be expected to be reduced if their environment is pushed outside the range of conditions that existed over this time period. Karr and Freemark (1985:167), for example, argue that "disturbance regimes . . . must be protected to preserve associated genetic (Frankel and Soulé 1981), population (Franklin 1980), and assemblage (Karr 1982a, 1982b; Kushlan 1979) dynamics."

Biotic landscapes include the patterns of vegetation patches of various ages and compositions and the fauna that find habitat in such vegetation patterns. Natural biotic patterns reflect, in part, the physical landscape of landforms, climate, and soil patterns. Managed biotic landscapes may be forced into poor fits with physical landscapes, creating biological and social problems. Fire suppression in ecosystems with naturally high fire frequency, for example, has led to fuel buildups that increased fire severity when it occurred. Forest cutting has reduced the extent of old-growth forest conditions, thus reducing habitat for a host of species. On the other hand, efforts to create areas of old-growth forest where it would not naturally occur because of frequent fire, wind, or other disturbances, have either failed or created unnatural habitat patterns. Midslope roads, a notably unnatural feature in landscapes, can contribute to increased peak flows in downstream areas (Wemple 1994, Jones and Grant 1996) and be major sources of debris slides (Sidle et al. 1985).

In this chapter, we consider some implications of the physical environment and ecosystem dynamics in land management. We begin with a brief description of pattern-process relations in landscapes as a basis for managing ecosystems. We follow with a discussion of the history of landscape management in the Pacific Northwest and its relation to our understanding of the physical environment. To conclude, we consider implementation of these concepts in the 21st century.

These perspectives and the examples considered are derived from the Pacific Northwest, an area of strong interactions between biota, physical environment, and instructive pre- and post-settlement land-use histories. Rich vegetation and fire histories can be described, for example, from dendrochronological data (records extending back to 500–800 years before the present), paleoecological data (records extending back to 40,000 years before the present), and satellite remote-sensing data (records extending back to 1972 to the present).

Landscape Patterns and Processes

To see the value of using the physical environment as a basis for ecosystem management, it is necessary to have a conceptual framework. Such a framework considers forest landscapes and riverscapes in terms of a hierarchy of spatial and temporal scales. Questions regarding evolution of landscape patterns in natural and managed systems must be posed and answered at the correct temporal and spatial scale (Delcourt et al. 1983, O'Neill et al. 1986, Gregory et al. 1991). Many terms are useful for identifying scale, depending on subject matter—site, landscape, province, and region are common geographic scales. The specific definition of these terms depends on objectives: A region, for example, may be defined by the ranges of critical species of interest. Many tough environmental problems today are a result of management decisions made at fine spatial scales, which had undesired, aggregated consequences at coarser scales.

Interactions between biotic and physical patterns and processes (Figure 15.1) operate at each scale. Landscape patterns are dynamic as a result of interactions between vegetation succession and disturbance processes. Biotic patterns can take the form of patchworks of vegetation in various stages of development. Vegetation patchworks partly reflect pattern-creating (disturbance) processes and geophysical patterns, such as landform controls on disturbance or variations in soil properties that influence vegetation. Vegetation patterns in natural landscapes include both the indelible imprint of patches created by resource limitations (*sensu* Forman and Godron 1986), such as lack of soil on rock outcrops, and ephemeral patches initiated by biotic disturbances. Landform influences vary between disturbance processes that follow local gravitational flow paths (e.g., landslides, floods) and those that do not

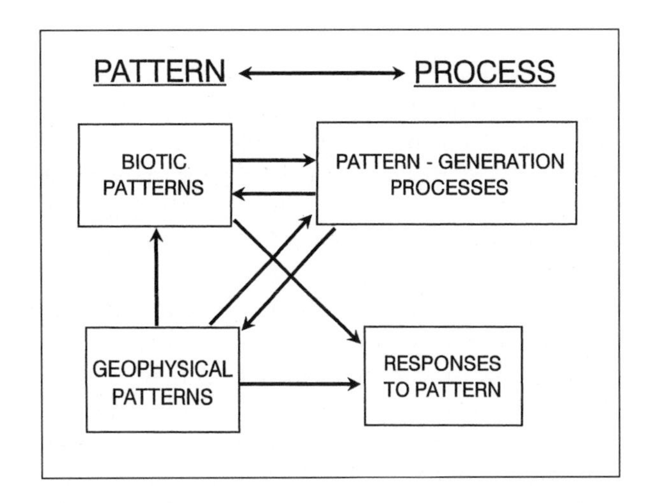

Figure 15.1 Dominant pattern-process relations. Geophysical patterns (e.g., soils, topography) underlie biotic patterns (e.g., vegetation type, productivity) and may influence landscape-related processes. Disturbance processes generate patterns; other processes may simply respond to patterns without modifying existing patterns.

(e.g., wildfire, windthrow). Human activities (e.g., cutting units) may overprint natural patterns, creating intentional or unintentional patterns (e.g., those arising from escaped slash fires). Some processes, such as generation of low streamflows, respond to biotic and landform patterns without creating new landscape patterns.

Landscapes are composed of patchwork and network structures, each with characteristic dynamics and functions. In forest landscape studies, patchwork structures are typically a collection of patches of varied vegetation type and seral stage. Natural network structures include streams, riparian areas, and ridgelines. Some disturbance processes, such as floods and debris flows, propagate through networks, while others, such as wildfire and windthrow, move through patchworks.

Most landscape ecology studies have examined patchwork structures; few have considered network components of landscapes; and still fewer have dealt with network–patchwork interactions. Understanding interactions between landscape structures of the same or different type is critical to forest and watershed management. Road segments, for example, may follow ridges and large streams (fourth-order and larger). But roads in steep terrain are likely to intersect groundwater flow paths and first- through third-order streams at right angles. This may accentuate the effects of roads in increasing the density of stream channels and thereby the potential of watersheds to generate peak flows (Wemple 1994, Jones and Grant 1996). An extensive network of road ditches and stream flow paths may help to transport increased runoff from clearcut patches (Harr 1981, 1986) to downstream areas where peak flows may be increased (Jones and Grant 1996). Such interactions between network and patchwork structures are significant but often overlooked because of disciplinary instincts; stream ecologists tend to see landscapes as networks, while forest and wildlife ecologists see them as patchworks.

History of Landscape Change in the Pacific Northwest

Postglacial landscape patterns in the Pacific Northwest have developed sequentially through three stages: (1) a wild landscape dominated by natural disturbances and actions of native people, spanning most of the Holocene until the early 1800s; (2) the landscape managed by European settlers through much of the 19th and 20th centuries; and (3) the present period of regional ecological assessments and associated management plans focused on species conservation on public lands. The physical environment dominated ecosystem change in the first stage; human forces attempted to override some influences of the physical environment in the second stage; and the physical environment will be an increasing consideration in management through the present and future.

The transition from the wild landscape of the first stage to the present, managed landscape created during the second stage has occurred over a century of piecemeal land-use decisions. A simple, visual expression of this is revealed in maps of closed-canopy forest and of disturbance patterns (mainly harvest cutting) from 1972 to 1988 in an area of mixed ownership and varied land-use designations in the central Cascade Mountains of Oregon (Figure 15.2) (Spies et al. 1994). Here we see patterns created by

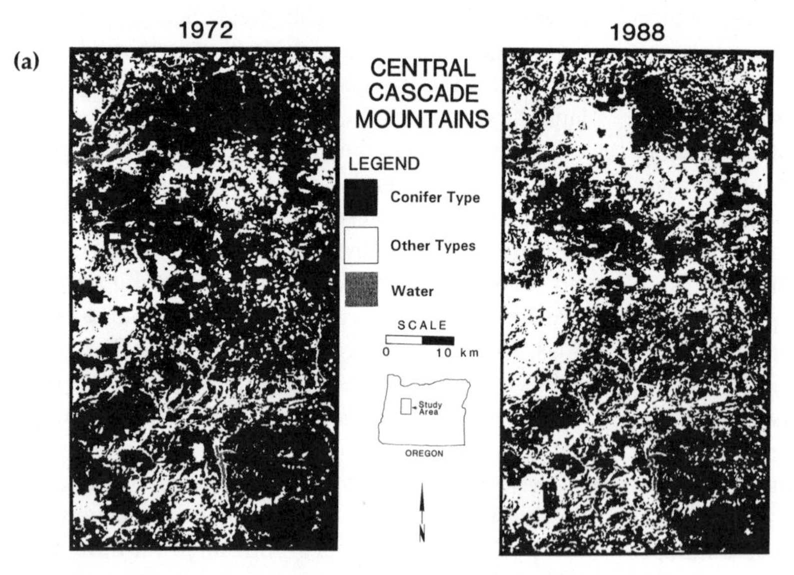

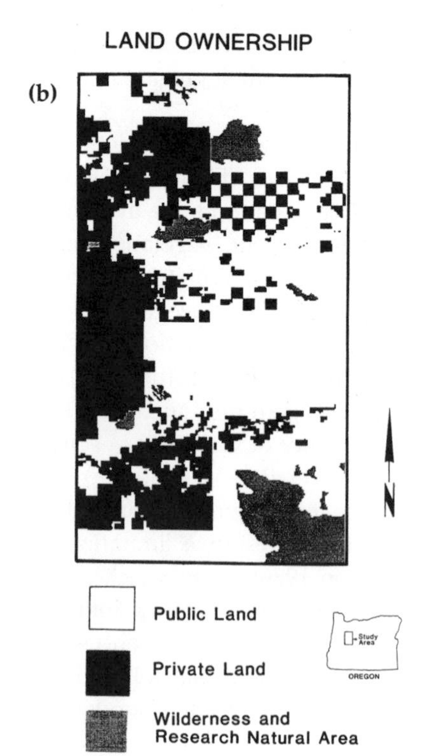

Figure 15.2 (a) Maps of closed-canopy conifer (>40 years in age) forest type in 1972 and 1988 for a 259,000-ha study area in the Oregon Cascade Range (from Spies et al. 1994). (b) Land ownership and land-use designations (modified from Spies et al. 1994).

large, aggregated cutting units on private industrial lands cut on approximately 50-year rotations, smaller, dispersed cutting units on public lands where the cutting cycle was 80 to 100 years, and very low levels of disturbance in Wilderness, Research Natural Areas, and other reserves where logging does not occur and fire is suppressed. Note how the reserves and other areas of limited forest cutting emerge as distinctive landscape features over this 16-year period.

Management of landscape patterns on public lands in the Pacific Northwest has changed dramatically in the past few years. After 40 years of dispersed-patch clearcutting, brief consideration was given to more aggregation of cutting units, assuming that the rate of cutting was unlikely to change (Franklin and Forman 1987, Swanson and Franklin 1992). Yet, from the early 1990s, cutting rates have dramatically decreased, and concepts about how we manage landscape patterns have changed. For example, there is growing interest in allowing more wildfire in the wilderness areas, and state regulations are limiting the size of cutting units on private lands. The most profound change will probably be in areas formerly designated as general forest lands of the U.S. Forest Service and Bureau of Land Management. Management in those areas will be altered by the Northwest Forest Plan developed by the Forest Ecosystem Management Assessment Team (FEMAT 1993).

Regional Ecological Assessments

Recently for the first time, broad-scale patterns of forest vegetation in the Pacific Northwest were subject to conscious design. Regional ecological assessments, such as FEMAT (1993), the Columbia River Basin/East-Side Ecosystem Project, and the Sierra Nevada Ecosystem Project, are defining broad templates for ecosystem management on major pieces of public lands in the United States. Some of these assessments derive from legal and other imperatives to conserve threatened and endangered species and old-growth forest ecosystems.

These regional management plans use a variety of land designations to define future patterns of forest development. Existing reserve lands, such as congressionally designated wilderness areas, are combined with new designations to create an interacting network of reserves. Guidance for management of lands between reserves and for designation of riparian corridors is intended to provide dispersal habitat between reserves. In these and other ways, reserves established in the past as independent entities are intended to fit within a larger design to meet ecological objectives. Physical processes and landscape features have minor roles in the design.

The focus on species conservation and the cumulative effects of past ecosystem alterations severely limit opportunities to reintroduce features of natural disturbance regimes in some areas. This is particularly acute in areas characterized by infrequent, stand-replacement wildfire and where management is focused on conserving old-growth forest habitat. Reserve placement and design in the Northwest Forest Plan, for example, are structured to provide old-growth habitat for the northern spotted owl and to accommodate aquatic resources, but with little reference to landscape patterns created by historic, natural disturbance regimes. Under this regional plan, only a small fraction of the forest landscape of public land is subject to cutting—but the forest patterns created on those lands are highly fragmented and unlike natural patterns. Perhaps once certain levels of ecosystem recovery have been achieved, it will be possible to more nearly match management patterns with natural patterns.

In other regions, ecological assessments and plans may provide for a somewhat better fit of managed biotic patterns to patterns controlled by natural disturbance regimes and landforms. Where wildfire was frequent and has been suppressed for several natural fire recurrence intervals (with resulting problems of insect outbreaks and buildup of fuels), there is greater attention to matching management practices with natural disturbance regimes. This may be the case in the Columbia River Basin assessment and plan.

Future Landscape Management

We expect that management of forest and riverine ecosystems in the 21st century, at least on public lands, will depart substantially from earlier and current approaches. These traditional approaches have been based either on site-specific best management practices, which lack long-term, large-scale perspectives, or on the habitat-restoration emphasis of conservation strategies for threatened and endangered species. Emerging new approaches will define a range of desired ecosystem states based on natural states and disturbance regimes interpreted from analysis of past events and controls of climate, landforms, soils, etc., on process location and rates. Hence, management interventions are treated as modifications of the natural disturbance regime. They can be applied at the site or landscape level—the first is most suitable to sustaining productivity and the latter to sustaining elements of the larger ecosystem.

Characterization of landscape dynamics is essential to planning ecosystem management. This can be approached in several complementary ways. One method is to define the range of past ecosystem conditions, such as extent of old-growth forests, early successional habitat, or unstable stream banks (Figure 15.3)(Caraher et al. 1992). A second approach focuses on disturbance processes in terms of the range of frequency, severity, and geographic pattern (such as size distribution) of patches created by these processes before and during periods of intensive forestry management (Figure 15.4)(Swanson et al. 1993). Mapping forest conditions or the extent of individual disturbance events through time can help define either the range of vegetation conditions or the disturbance regime.

An even more useful, spatially explicit characterization can be developed by mapping disturbance regime units across landscapes (Figure 15.5). This is accomplished by interpreting event histories for individual sites and then compiling site histories across a landscape. Disturbance regime patterns are interpreted by considering effects of topography, vegetation type, soil, climate, or other relevant factors that exercise long-term controls on the frequency and severity of disturbance processes.

Mapping units for characterizing wildfire disturbance regimes may describe the frequency and severity of disturbances—such as infrequent, stand-replacement wildfires or 60-year rotation clearcutting on private land holdings of the forest products industry. In some areas, landscapes can be stratified by vegetation types (Barrett and Arno 1991, Agee 1993) to provide a basis for mapping fire regimes. Where vegetation types are not highly differentiated with respect to fire regimes, topography may be more useful in interpreting fire regimes over landscapes. In either case, information on landform effects on fire patterns assists in interpreting long-term landscape

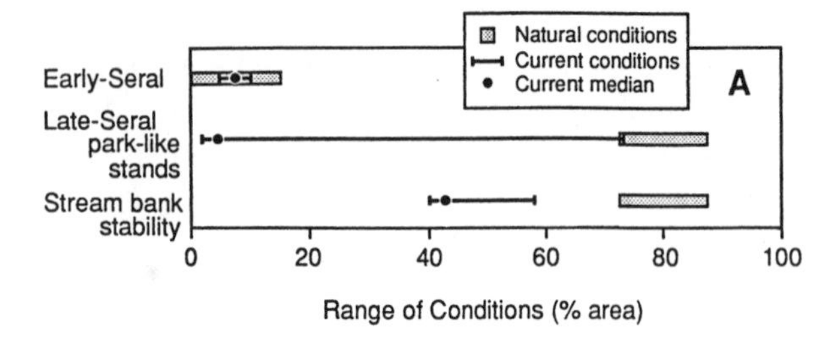

Figure 15.3 Example of range of interpreted natural and present ecosystem conditions for the Silvies River area (from Caraher et al. 1992, p. 22).

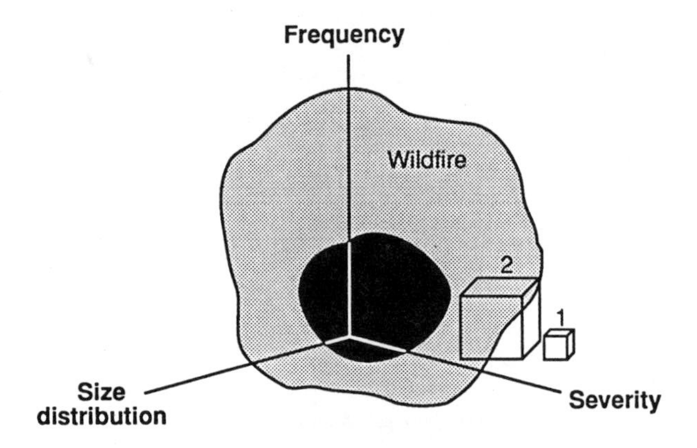

Figure 15.4 Hypothetical representation of a natural disturbance regime as the large, irregular "cloud" showing a probability distribution of wildfire events. Box 1 represents a management system of dispersed clearcuts with broadcast burning, assuming no landscape disturbance by wind or fire. Box 2 represents the disturbance regime resulting from the interaction of the management system (box 1) with disturbance processes that could not be suppressed (e.g., windthrow at stand edges and wildfire). The black area represents a range of conditions resulting from a management system within the range of natural variability, but not a mimic of the full natural range (Swanson et al. 1993).

dynamics beyond the time scale reflected in current vegetation (Figure 15.5).

Disturbance regimes imposed by geomorphic processes are another important aspect of fitting managed biological landscapes with the physical environment. Examples of mapable disturbance regimes include landslide hazards and the potential for lateral change in river channels (Grant and Swanson 1995). Some elements of watershed analysis are designed to provide such information (Washington Forest Practice Board 1992, FEMAT 1993, Montgomery et al. 1995). Information on geomorphic disturbance regimes is used not only to minimize effects of management on accelerated erosion but also to identify parts of the landscape where natural disturbances should not be impeded—such as stream processes that create and maintain secondary channels on wide valley floors.

There are, however, some important challenges to defining natural ranges of variability and disturbance regimes: First, interpretation of past disturbance regimes has limitations. Second, native people influenced fire regimes to varying extents for much or all of the period that the present dominant forest species have been significant components of the flora of the region. And finally, exotic species, such as undesired

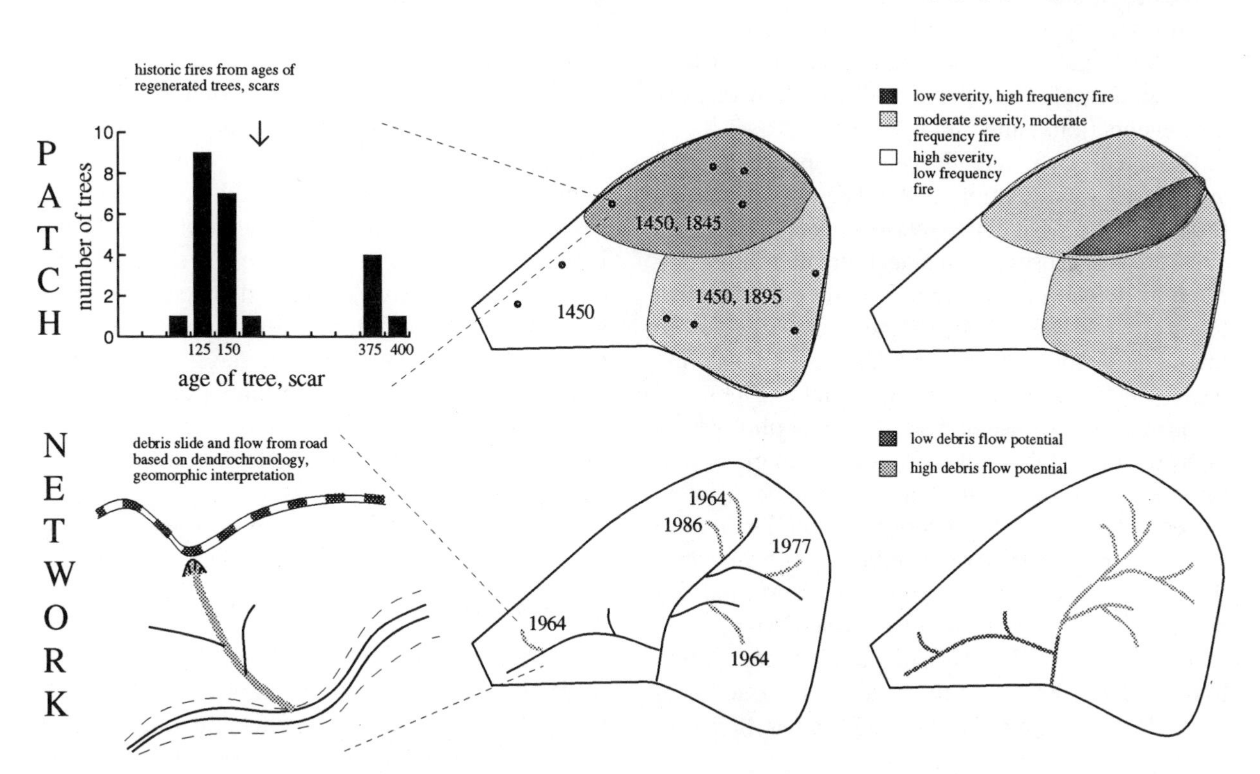

Figure 15.5 Schematic example of steps for going from site-level interpretation of disturbance history to landscape history of disturbances to disturbance regime mapping. Fire patterns are an example of patch-creating disturbances, and debris flows down stream channels are an example of disturbances in networks.

weeds and planted fish stocks, and engineered structures, such as roads and dams, are unnatural parts of the ecosystem that are likely to stay in most cases; hence, they must be accommodated in managing ecosystems at large scales (Swanson et al. 1993). There are concerted efforts to improve understanding of each of these factors. Although some uncertainty surrounds each of these considerations, we believe that the range of natural variability concept for management remains important and viable.

Adopting a management approach based on disturbance regimes has rather different implications in terrestrial habitats of uplands areas than it does in stream and riparian networks. Stream and riparian networks are potentially affected not only by direct local disturbances, but also by disturbances in upland areas that may have effects transported downslope by streamflow and other processes. Furthermore, disturbances that may be limited in uplands areas may have substantial effects on the disturbance regime of stream and riparian networks. Roads, for example, may affect less than 5 percent of uplands, but they may significantly alter the size distribution of peak flow events (Jones and Grant 1996).

These concepts actually can be applied to real landscapes. One example is the Augusta Landscape Project on the Willamette National Forest in Oregon, a test area for ecosystem management planning (Cissel et al. 1994, in preparation, Wallin et al. in press). Planning for this 7,600-hectare area began with mapping disturbance event histories for wildfire, based on dendrochronology (A.D. 1500–1900) and landslides and the use of field and aerial photograph interpretation (1950–1994). To develop maps of disturbance regimes, event histories have been interpreted in light of landform influences on disturbance patterns. This information, combined with information on present landscape conditions and land-use designations, has been used to develop a landscape management plan that sustains aquatic and terrestrial habitat more within the range of historic variation. Moreover, the plan has been modeled for several centuries into the future.

Key features of this disturbance-based Augusta landscape design include variable rotation lengths (100–300 years), varying levels of tree removal at each cutting, and size distribution of cutting units. Selection of silvicultural prescriptions is based on wildfire frequency, severity, and geographic patterns of the past. The range of cutting rotations, levels of live trees retained, and unit sizes are much more variable under this management system than they were under earlier patch clearcut systems or the Northwest Forest Plan.

Approaches to riparian zone management also differ significantly. The Willamette National Forest Plan and the Northwest Forest Plan prescribe streamside buffer strips throughout the stream and riparian network. The Augusta landscape design uses a small number of large reserves to protect aquatic and riparian species. These are connected by wide buffers along only the major channels. Some cutting is permitted within other riparian areas.

The result of the Augusta landscape design is a future landscape with forest habitat patterns more similar to the historic patterns than would be the case under the previous patch clearcut system or the Northwest Forest Plan (Wallin et al. in press, Cissel et al. in preparation). This conclusion has been examined in a variety of respects, including the extent of interior forest habitat in mature and old-growth classes and the density of edge between forest and open areas. Edge density is a useful measure of landscape pattern because it may serve as an index of blowdown potential—that is, freshly exposed forest edges are susceptible to wind damage. Also, some species that favor interior forest habitat, such as northern spotted owls, may be more susceptible to predation in landscapes with higher edge densities. The Augusta landscape design, therefore, has a higher likelihood of sustaining terrestrial and aquatic species.

Conclusions

The use of understanding of the physical environment, including natural disturbance regimes, has evolved through the stages of helping to repair damage to commodity production systems, to minimizing effects of further management, to using the physical environment and ecosystem dynamics as a basis for management design. In the 21st century, we expect land management to better fit the managed biological landscape with the physical landscape and its historic disturbance processes and patterns. The major impetus for doing so is that past poor fits have proven socially untenable; consequences of poor fits include the actual or potential listing of threatened or endangered species as a result of habitat alteration well outside the historic range of conditions.

Prototypes of landscape management designs are being developed that illustrate use of natural disturbance regimes and other aspects of the physical environment to plan ecosystem and watershed management. In some regions, however, the need to provide for habitat recovery following a prolonged period of intensive land use may slow implementation of an ecosystem-based approach to management. This seems to be the case in the range of the northern spotted owl. In areas where landscapes have been strongly influenced by suppression of wildfire, a focus on restoring the disturbance regime through management may be more immediately acceptable.

Viewed from the mid-1990s, it is interesting to consider how difficult changing to a more ecosystem-based approach to management may be. The recent growth of understanding about how ecosystems and watersheds function and the major revamping of natural resource management approaches would seem to set the stage for improving the fit of managed landscapes to natural landscape patterns and functions. However, it may be well into the 21st century before much progress is made. Social changes are critical—such as increased public understanding of the role of disturbances in ecosystems. Changes in policy also are needed to direct planning to regional and watershed scales, rather than restricting it to the scales of projects (e.g., timber sales) and agency management units (e.g., an individual national forest) where cumulative ecological and watershed effects may be obscured.

Acknowledgments

We thank John Cissel and David Wallin for many thoughtful discussions of these topics and leadership in development of the Augusta Landscape Project.

Literature Cited

Agee, J. K. 1993. *Fire ecology of Pacific Northwest forests.* Washington, DC: Island Press.

Barrett, S. W., and S. F. Arno. 1991. Classifying fire regimes and defining their topographic controls in the Selway-Bitterroot Wilderness. In *Proceedings of the 11th Conference on Fire and Forest Meteorology, April 16–19, 1991,* Missoula, Montana, ed. P. L. Andrews, and D. F. Potts. Boise, ID: USDA Forest Service, Boise National Forest.

Caraher, D. L., J. Henshaw, F. Hall, et al. 1992. *Restoring ecosystems in the Blue Mountains: A report to the regional forester and the forest supervisors of the Blue Mountain Forests.* Portland, Oregon: USDA Forest Service, Pacific Northwest Research Station.

Cissel, J. H., F. J. Swanson, W. A. McKee, and A. L. Burditt. 1994. Using the past to plan the future in the Pacific Northwest. *Journal of Forestry* 92(8):30–31, 46.

Cissel, J., F. Swanson, D. Olson, G. Grant, S. Gregory, S. Garman, L. Ashkenas, M. Hunter, J. Kertis, J. Mayo, M. McSwain, D. Wallin. In preparation. *A disturbance-based landscape design in a managed forest ecosystem: The Augusta Creek study.*

Delcourt, H. R., P. A. Delcourt, and T. Webb III. 1983. Dynamic plant ecology: The spectrum of vegetational change in space and time. *Quarterly Science Review* 1:153–175.

FEMAT (Forest Ecosystem Management Assessment Team). 1993. *Forest ecosystem management: An ecological, economic, and social assessment.* Report of the Forest Ecosystem Management Team (FEMAT), 1993-793-071. Washington, DC: GPO.

Forman, R. T. T., and M. Godron. 1986. *Landscape ecology.* New York: John Wiley & Sons.

Frankel, O. H., and M. E. Soulé. 1981. *Conservation and evolution.* New York: Cambridge University Press.

Franklin, I. R. 1980. Evolutionary change in small populations. In *Conservation biology: An evolutionary-ecological perspective,* ed. M. E. Soule and B. A. Wilcox. Sunderland, MA: Sinauer.

Franklin, J. F., and R. T. T. Forman. 1987. Creating landscape patterns by cutting: Ecological consequences and principles. *Ecology* 1(1): 5–18.

Grant, G. E., and F. J. Swanson. 1995. Morphology and processes of valley floors in mountain streams, western Cascades, Oregon. In *Natural and anthropogenic influences in fluvial geomorphology: The Wolman volume,* ed. J. E. Costa, A. J. Miller, K. W. Potter, and P. R. Wilcock. Geophysical monograph no. 89. Washington, DC: American Geophysical Union.

Gregory, S.V., F. J. Swanson, W. A. McKee, and K.W. Cummins. 1991. An ecosystem perspective of riparian zones. *BioScience* 41(8):540–551.

Harr, R. D. 1981. Some characteristics and consequences of snowmelt during rainfall in western Oregon. *Journal of Hydrology* 53:277–304.

———. 1986. Effects of clearcutting on rain-on-snow runoff in western Oregon: A new look at old studies. *Water Resources Research* 22(7):1095–1100.

Jones, J. A., and G. E. Grant. 1996. Peak flow responses to clearcutting and roads in small and large basins, western Cascades, Oregon. *Water Resources Research* 32:959–974.

Karr, J. R. 1982a. Avian extinctions on Barro Colorado Island, Panama: A reassessment. *American Naturalist* 119:220–239.

———. 1982b. Population variability and extinction in the avifauna of a tropical land bridge island. *Ecology* 63:1975–1978.

Karr, J. R., and K. E. Freemark. 1985. Disturbance and vertebrates: An integrative perspective. In *The ecology of natural disturbance and patch dynamics,* ed. S. T. A. Pickett and P. S. White. New York: Academic Press, Inc.

Kushlan, J. S. 1979. Design and management of continental wildlife reserves: Lessons from the Everglades. *Biological Conservation* 15:281–290.

Montgomery, D. R., G. E. Grant, and K. Sullivan. 1995. Watershed analysis as a framework for implementing ecosystem management. *Water Resources Bulletin* 31(3): 369–386.

O'Neill, R.V., D. L. DeAngelis, J. B. Waide, and T. F. H. Allen. 1986. *A hierarchical concept of ecosystems.* Princeton: Princeton University Press.

Sidle, R. C., A. J. Pearce, and C. L. O'Loughlin. 1985. *Hillslope stability and land use.* Water resources monograph 11. Washington, DC: American Geophysical Union.

Spies, T. A., W. J. Ripple, and G. A. Bradshaw. 1994. Dynamics and pattern of a managed coniferous forest landscape in Oregon. *Ecological Applications* 4(3):555–568.

Swanson, F. J., and J. F. Franklin. 1992. New forestry principles from ecosystem analysis of Pacific Northwest forests. *Ecological Applications* 2(3):262–274.

Swanson, F. J., J. A. Jones, D. O. Wallin, and J. H. Cissel. 1993. Natural variability: Implications for ecosystem management. In *Eastside forest ecosystem health assessment.* Vol. II, *Ecosystem management: Principles and applications,* ed. M. E. Jensen and P. S. Bourgeron. Portland, OR: USDA Forest Service, Pacific Northwest Research Station.

Wallin, D., F. Swanson, and B. Marks. In press. Comparison of managed and pre-settlement landscape dynamics in forests of the Pacific Northwest, U.S.A. *Forest Ecology and Management.*

Washington Forest Practice Board. 1992. *Washington Forest Practice Act Board manual: Standard methodology for conducting watershed analysis.* Ver 1.10. Olympia, WA: Department of Natural Resources, Forest Practices Division.

Wemple, B. C. 1994. *Hydrologic integration of forest roads with stream networks in two basins, western Cascades, Oregon.* Corvallis, OR: Oregon State University.

16 Approaches to Management at the Watershed Scale

Robert J. Naiman, Peter A. Bisson, Robert G. Lee, and Monica G. Turner

Management Perspectives at the Watershed Scale 240
The Natural System: Variability in Time and Space 240
A Holistic Perspective: Persistence and Invasiveness 240
Connectivity and Uncertainty 241
Human Cultures and Institutions 241
Quantitative Approaches for Implementing Watershed Management 242
Watershed Analysis 243
Quantitative Measures 245
Integrated Socioenvironmental Models 245
Indices of Socioenvironmental Conditions 246
Addressing Institutional Organization 247
Formulating Shared Socioenvironmental Visions 247
Public Stewardship in Watershed Management 249
Monitoring 249
Public Outreach 250
Conclusion 251
Literature Cited 251

Freshwater and freshwater ecosystems lie at the heart of the challenge of ecosystem management (Naiman 1992, Lee 1993, Naiman et al. 1995a, 1995b). Because they integrate natural resource and socioeconomic systems, freshwater issues embody the complexity that will characterize natural resource management as we move into the 21st century. Changes in human demography, resource consumption, cultural values, institutional processes, technological applications, and information all contribute to that complexity. If we are to achieve long-term social stability as well as ecological vitality, we must understand the abilities and limits of freshwater ecosystems to respond to human-generated pressures. Yet, even though human actions and cultural values drive environmental issues, few holistic approaches for watershed management offer effective resolution.

In the current debate over the scope of ecosystem management (Grumbine 1994, Montgomery et al. 1995), it is widely recognized that there are significant technical and cultural constraints to effective implementation. These constraints are related to such issues as identifying appropriate spatial and temporal scales, monitoring and assessment, developing an adaptive management process, and developing cultural values and philosophies that allow ecosystem management to be successful (Levin 1993, Grumbine 1994). Nonetheless, the ability of a rapidly increasing human population to dramatically impact local, regional, and global ecosystems makes it essential to incorporate an ecological perspective into watershed management if we are to leave a healthy resource base for future generations.

The first part of this chapter suggests several features that are fundamental to contemporary watershed management. The second part then presents several practical approaches for implementing effective watershed management programs.

Table 16.1 Special goals fundamental to ecosystem management

- Maintain viable populations of all native species in situ
- Represent, within protected areas, all native ecosystem types across their natural range of variation
- Maintain evolutionary and ecological processes (i.e., disturbance regimes, hydrological processes, nutrient cycles, and so forth)
- Manage over periods of time long enough to maintain the evolutionary potential of species and ecosystems
- Accommodate and balance human use and occupancy within these constraints

Source: Grumbine (1994).

Management Perspectives at the Watershed Scale

Four watershed-scale features provide the foundation for effective management: variability in time and space, persistence and invasiveness of species, system connectivities and uncertainties, and the role of human cultures and institutions. These features are closely related to specific goals frequently endorsed as being fundamental to ecosystem management (Grumbine 1994) (Table 16.1).

The Natural System: Variability in Time and Space

Natural processes, such as climate, soil formation, and geological disturbances, structure the diversity, productivity, and availability of natural resources that human societies depend upon. The challenge is to understand how naturally variable systems operate and to predict the environmental consequences of human activities in these systems (Naiman 1992, Naiman et al. 1995a, 1995b).

The vitality of natural ecosystems is created and maintained by substantial variation in time and space (Reice et al. 1990, Reice 1994, Turner 1990). For example, the ecological characteristics of riparian forests are structured by a complex array of dynamic and spatially variable hydrological processes that erode and deposit materials, deliver nutrients, and remove waste products (Gregory et al. 1991)(Figure 16.1). Variability in time and space results in the biological diversity and productivity characteristically found in riparian environments (Fetherston et al. 1995, Naiman et al. 1993). A key managerial challenge is balancing human needs with variations in physical and chemical characteristics so that significant declines or losses of species and ecological attributes do not occur.

A Holistic Perspective: Persistence and Invasiveness

The persistence of ecological attributes for the long term (decades to centuries) requires maintenance of naturally variable environmental regimes as well as isolation from invading organisms that can alter those regimes. When the natural environmental regime is altered, adjustments occur within the ecosystem producing new combinations of biophysical environments susceptible to the invasion of exotic organisms and the establishment of non-native ecological processes and structures (Drake et al. 1989). Understanding and quantifying persistence and invasiveness of species and their ecological processes are important for watershed management because these components are sensitive to change, integrate change over broad spatial and temporal scales, and can be used as measures of change. Moreover, many species are linked to a cultural identity, and many ecological processes are essential for sustaining human populations (Botkin 1990).

There are a variety of quantitative approaches and

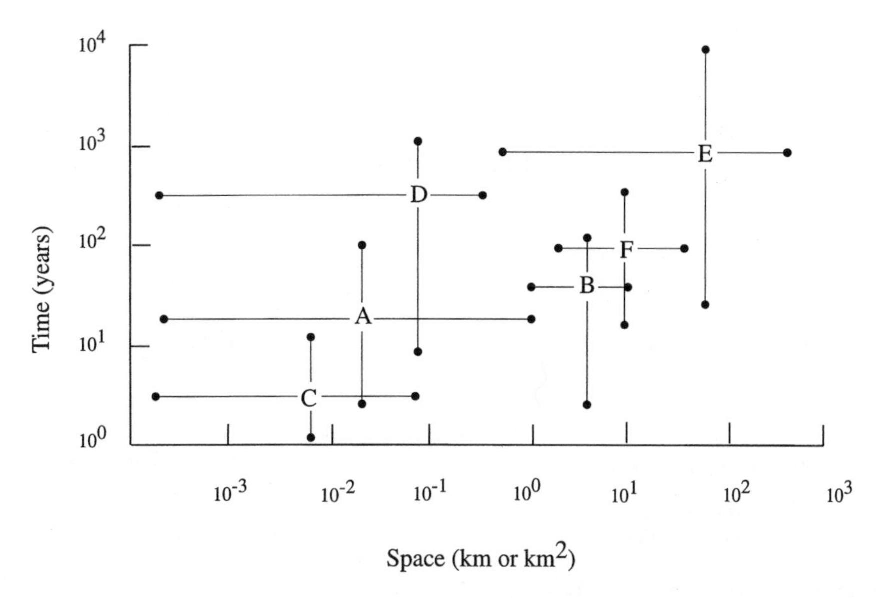

Figure 16.1 Illustration of the diversity of spatial and temporal scales influencing the creation and maintenance of riparian forests in the coastal temperate rain forest of North America. (A) Colonization surfaces created by flooding, (B) Colonization surfaces created by debris flow, (C) Seedling germination and establishment, (D) Longevity and size of species patches, (E) Persistence and movement of dead wood in channel, and (F) Impact of herbivores.

technical tools for analyzing persistence and invasiveness at the watershed scale. Many other techniques are in the design and testing stages. Existing techniques include new approaches to statistical analyses, patch and boundary analyses, modeling cumulative effects, developing indices of biotic integrity, and creating knowledge-based land-use analysis systems (Risser 1993, Fortin and Drapeau 1995, Karr 1991, Berry et al. 1996, Turner et al. 1996). These techniques are especially useful for setting goals related to desired future conditions and for preliminary examinations of the long-term effects of new or anticipated institutional regulations and policy (Turner et al. 1996, Wear et al. in press).

Connectivity and Uncertainty

The goal of watershed management is to let all components of human and nonhuman communities exist in a relative but dynamic state of balance (Naiman 1992). This goal explicitly recognizes strong connections between social and environmental components at multiple scales. This means giving serious consideration to diverse components such as water, fish, soils, forests, education, resource extraction, and cultural values, as well as managing the strong interactions between them (Stanford and Ward 1992).

Unfortunately, quantitative approaches for managing connectivity are not well formulated. There remains considerable uncertainty among scientists and decision makers as to how to proceed. This means accepting risk since the magnitude of current socioenvironmental issues requires decisions that cannot wait until all information is available (Figure 16.2). How can this be accomplished at the watershed scale? There is no definitive answer or one right way. However, we will discuss approaches that are being used by (1) small organizations to address risk, (2) groups that are helping to define social and environmental viewpoints for future conditions, and (3) researchers and managers struggling to monitor and assess change at regional scales.

Human Cultures and Institutions

In human-dominated watersheds, the land mosaic is created by a mixture of cultural practices, traditions, myths, and institutions (Lee et al. 1992, Décamps et

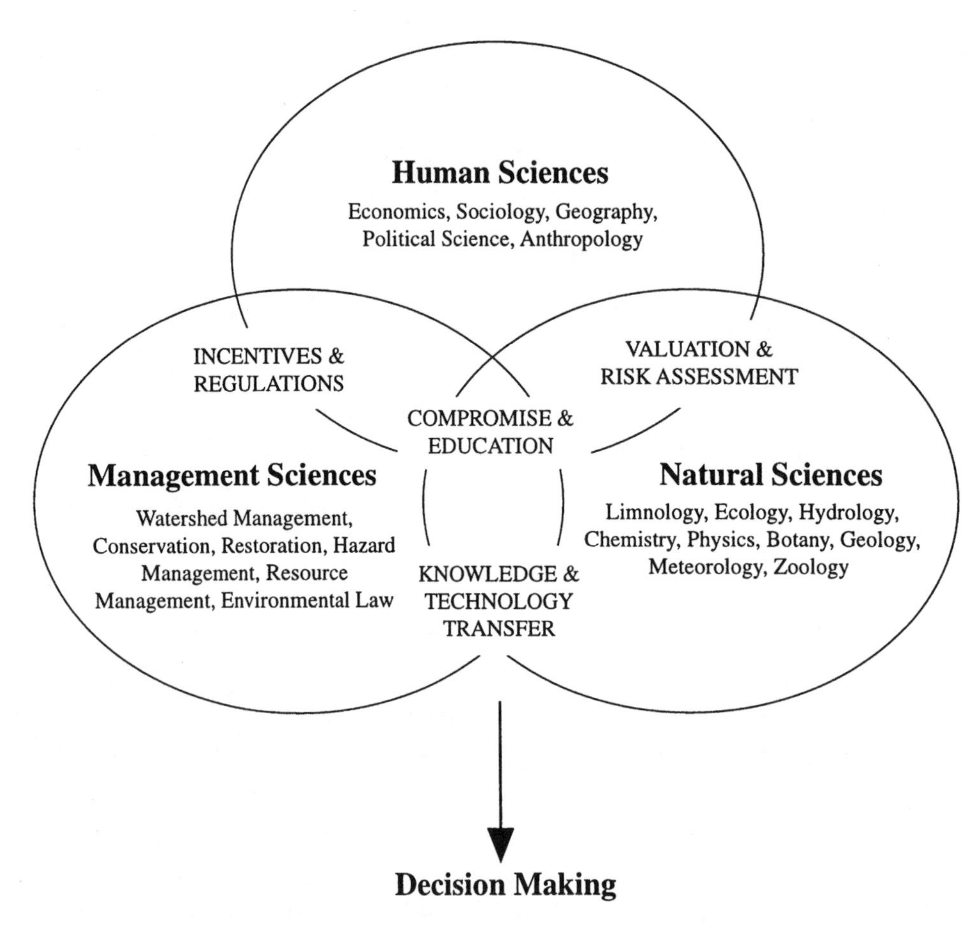

Figure 16.2 Advances in watershed management come at the interfaces between natural, human, and management sciences (from Naiman et al. 1995a).

al. 1997). The spatial extent and temporal duration of each patch and boundary type is ultimately determined by laws, regulations, taxation, technologies, cultural values and beliefs, and traditional land-use practices (Turner et al. 1990).

Developing an integrated socioenvironmental system means confronting and resolving important issues related to social and ecological literacy, the role and accommodation of changing cultural values, the increasing migration of people away from traditional homelands and cultures, balancing consumption rates and population growth, weathering political change, and establishing knowledge-based cooperative institutions (Lee 1993). These issues are closely interrelated and cannot be resolved separately. How to implement an integrated program that addresses these and related issues may not be immediately apparent since each watershed has a unique set of issues to resolve. There are, however, basic principles and practical approaches to guide the development of effective watershed management (Montgomery et al. 1995).

Quantitative Approaches for Implementing Watershed Management

Attempts to manage watersheds with more than one demand on the principal resources have been ineffective for the most part. Well-known examples include the Columbia River, the Sacramento and San Joaquin Rivers, and the Colorado River. These and other examples have been hampered by the inherent difficulties of identifying appropriate spatial and temporal scales for management, by the cumulative effects from multiple users, and by conflicting management goals. In addition, lack of accepted statistical or realistic modeling approaches and a dearth of indices for evaluating a dynamic socioenvironmental system have confounded problems (Lee 1993, Volkman and Lee 1994). Fortunately, as public awareness of watershed-level issues has improved, so has the array of quantitative approaches for assessing complex issues which have several causes and competitive solutions. Watershed analysis techniques, quantitative measures, assessing risk with integrated socioenvironmental models, and development of socioenvironmental indices are but a few of the empirical approaches available (Table 16.2).

Watershed Analysis

Quantitative approaches to documenting the status and dynamics of entire watersheds are still in the early stages of development (Montgomery et al. 1995). Most of the techniques developed have been concentrated in Western states heavily impacted by forest management (Table 16.2). The intent of watershed analysis is to provide a scientifically based understanding of the environmental processes and their interactions occurring within a watershed (Washington Forest Practices Board 1994, USDA Forest Service 1994). This understanding, which focuses on specific issues, values, and uses within the watershed, is essential for making sound management decisions (Table 16.3). Protecting beneficial uses, such as those identified by state and federal environmental laws (e.g., the Clean Water Act and the Endangered Species Act), is a fundamental objective for watershed analysis. Watershed analysis encompasses the entire watershed because of the strong fluvial linkages between headwater areas, valley floors, and downstream users.

Watershed analysis generally is carried out by an interdisciplinary team of resource professionals who are already experts in their fields and who are familiar with the area to be evaluated (Table 16.3). Different methods apply to different areas, and teams must use their professional judgment to select or design appropriate methods. Because it is an iterative and evolving process, watershed analysis should explicitly allow for methods to be improved and replaced as experience and knowledge accumulate.

Although watershed analysis may identify potentially conflicting objectives and uses, it is not a decision process per se. Rather, it is a process that generates information to assist in decision making. To the degree that it brings factual information to the decision-making process, watershed analysis is a substantial advancement over past management approaches.

Watershed analysis assumes there are demonstrable linkages between physical patches and biological processes, and that human values and perceptions do not change. This is a flawed assumption that contradicts the discussion on risk, to follow. Despite the promise of watershed analysis, it does have features that impede effective watershed management. Local and regional political influences; nonbinding agreements; the lack of long-term accountability for institutions, decision makers, and land managers; and the avoidance of an interactive synthesis of information are potentially fatal flaws in the concept. Further, to date, there has been no scientific validation of the approach, which was developed primarily by physical scientists.

Watershed analysis is now a part of the regulatory framework for managing state and privately owned commercial forests in Washington (Washington Forest Practices Board 1994). The Washington Department of Natural Resources, the agency charged with implementing watershed analysis, has identified a number of forested subbasins (fifth- and sixth-order river systems) termed watershed analysis units (WAUs). Within these units, watershed analysis forms a basis for local forest practice decisions. The first WAU to be analyzed was the Tolt River drainage, located in the Puget Sound basin of western Washington. The Tolt River watershed includes mixed ownership dominated by private forestland, as well as a reservoir that supplies water to Seattle. Because the Tolt River contained valuable fishery resources (salmon, trout, and steelhead) as well as an important drinking water supply, many interest groups participated in the analysis process.

The Tolt watershed analysis procedure identified

Table 16.2 Summary of practical approaches for watershed management

Approach	Description	Applicability	Advantages	Disadvantages	Case Studies
Watershed Analysis	Provides a scientifically based understanding of environmental processes and their interactions	Largely limited to forested watersheds of 50 to 500 km^2, although it can be adapted to other situations	Provides a spatially explicit description of resources, hazards, environmental variation, and potentials, as well as potential conflicts over resource use; adaptable to new technological methodologies	Requires highly trained, interdisciplinary teams, familiar with the terrain; assumes demonstrable linkages between physical patches and biological processes	Tolt Watershed Analysis Prescriptions, 1993, Montgomery et al. 1995
Quantitative Measures	Inventory of the abundance and spatial arrangement of vegetation land cover, or habitat characteristics	All watersheds	Provides a resource inventory for establishing spatial and temporal trends; takes advantage of existing GIS databases; acts to centralize storage of information; requires personnel with only moderate levels of training	Database development often requires a substantial investment; data availability is often incomplete; requires long-term monitoring and analyses to be useful	Turner and Gardner, 1991; Turner et al. 1996
Integrated Socioenvironmental Models	Models explicitly combining the social, economic, and environmental factors influencing watershed characteristics	Still in an experimental stage; best applied to watersheds with few, direct human influences on resources	Allows a holistic (and more realistic) perspective to be developed where human activities and values are a central component of the ecosystem; allows evaluation of a wide range of social choices	Database development is expensive and time consuming; essential data are often incomplete; requires a moderate-to-high level of technical expertise	Le Maitre et al. 1993; Warwick et al. 1993; Berry et al. 1996; Wear et al. in press
Indices of Socioenvironmental conditions	Components contributing to the long-term vitality of a social-economic-environmental system	Watersheds with a significant human population	Provides a regular report to the citizens; improves literacy about watershed-scale issues; develops stewardship for the long-term; easily maintained	Requires regular monitoring and analysis of data, some of which may be difficult to obtain	Willapa Alliance and Ecotrust 1995

Table 16.3 The fundamental steps involved in watershed analysis, and some of the basic products to be expected from the process.

Steps

1. Identify issues, describe desired conditions, and formulate key questions
2. Identify key processes, functions, and conditions
3. Stratify the watershed
4. Assemble analytic information needed to address the key questions
5. Describe past and current conditions
6. Describe condition trends and predict effects of future land management
7. Integrate, interpret, and present findings
8. Manage, monitor, and revise information

Products

1. A description of the watershed, including its natural and cultural features
2. A description of the beneficial uses and values associated with the watershed and, when supporting data allow, statements about compliance with water quality standards
3. A description of the distribution, type, and relative importance of environmental processes
4. A description of the watershed's present condition relative to it's associated values and uses
5. A map of interim riparian reserves

areas where salmonid habitat features, such as proper stream temperature and abundance of large woody debris, were degraded, as well as areas where delivery of sediment to streams would be likely from unpaved logging roads and geologically unstable slopes. Prescriptions for preventing or mitigating these problems (Tolt Watershed Analysis Prescriptions 1993) were developed by a team that included six foresters representing the Department of Natural Resources and the Weyerhaeuser Company, a forest road engineer, a tree physiologist, an environmental analyst from the Washington Department of Ecology, two aquatic biologists from the Tulalip Indian tribe, and a forest hydrologist. Their prescriptions for future forestry operations are not voluntary; a landowner who does not comply is subject to civil and criminal prosecution.

Over 40 people officially participated in the Tolt watershed analysis. The five-month process was at times contentious. This was perhaps to be expected given the diversity of interests in the watershed. Members of the analysis team generally agreed that the process of working together was at least as important as the process of using available data to guide management decisions.

Quantitative Measures

Watersheds can be characterized by a variety of quantitative measures when digital data are available. Most simply, the total area and proportion of the watershed occupied by each cover type (vegetation or habitat) can be identified and its area and perimeter recorded. Analyses of the total number of patches, of arithmetic mean patch size, of standard deviation of mean patch size, of size of the largest patch, of weighted-average patch size, of amount of interior habitat, of total edge, and of mean patch shape are easily computed. In addition to metrics describing individual cover types, edges between habitats (which are sensitive measures of habitat fragmentation) can be tabulated as the length of edge between each pair of land-cover classes (e.g., forest-grassy, forest-unvegetated, grassy-unvegetated) or as edge-to-area ratios.

While the development of quantitative measures of watershed conditions has proceeded rapidly, empirical studies that test for significant relationships between watershed metrics and ecological conditions (e.g., presence or abundance of species or water quality) are still few in number (Johnston et al. 1990). There is a clear need to identify the most important watershed metrics to monitor as well as the levels beyond which socioenvironmental conditions change significantly. In addition, it is essential to be aware of the assumptions and constraints that are implicit in the metrics. For example, selection of land-cover categories can determine the results of an analysis. Similarly, the spatial scale of the data—both the total extent of the area and the resolution, or grid cell, size—can strongly influence the numerical results (Turner et al. 1989a, 1989b).

Integrated Socioenvironmental Models

The risk of undesirable future conditions can be assessed by using integrated socioenvironmental models to explore alternative land management scenarios

(Le Maitre et al. 1993, Warwick et al. 1993, Flamm and Turner, 1994a, 1994b). An example of such a model is LUCAS, the Land-Use Change and Analysis System (Berry et al. 1996, Wear et al. 1996). LUCAS is a spatial simulation model in which the probability of a land being converted from one land-cover type to another depends on a variety of social, economic, and ecological factors (Figure 16.3). Conditional transition probabilities are estimated empirically by comparing land cover at different times (Turner et al. 1996).

Simulations begin with an initial map of land cover. Next, equations are used to generate a transition probability for each grid cell in the watershed map based on ownership type, elevation, slope, aspect, distance to roads, distance to markets, and population density (Flamm and Turner 1994a, 1994b, Wear et al. 1996). An integrated modeling approach permits the effects of a wide range of alternatives to be evaluated. For example, one can examine the effects of residential development in different locations within the basin or the effects of moving a large parcel of land into or out of intensive timber production. Linking projected land-cover maps with effects on ecological indicators (such as species persistence or water quality) allows comparison of the potential long-term implications of alternative human decisions.

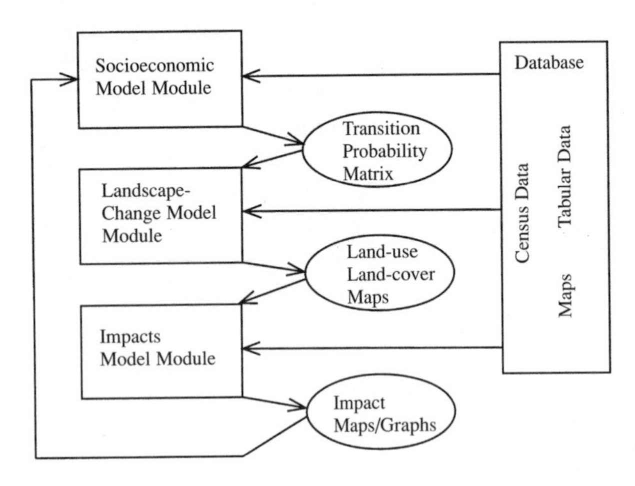

Figure 16.3 Integration of social, economic, and environmental aspects of watershed management can be accomplished with the use of the Land Use Change Analysis System (LUCAS), a modeling environment (after Berry et al. in press).

Indices of Socioenvironmental Conditions

Methods for separately examining the status and trends of environmental, social, and economic factors are well established (Finstenbusch and Wolf 1981, Burch and DeLuca 1984, Karr 1991). However, watershed management requires integrated socioenvironmental indices that provide a holistic understanding of watershed condition (Table 16.2). In a broad sense, a socioenvironmental index is a report to citizens, resource users, and government agencies on the vitality of a space they hold in common. Ideally, a socioenvironmental index should provide usable information on the important aspects of a watershed's environment, economy, and communities.

Components of a socioenvironmental index are chosen to reflect the unique characteristics of a watershed. For example, in the Willapa Bay watershed of western Washington, shellfish harvest, timber production, fishing, and agriculture are important in maintaining the local economy and culture. The Willapa Alliance, a consortium of concerned citizens and resource users, has developed an index based on indicators of environmental quality, economic vitality, and community health (Table 16.4). Environmental quality is indicated by oyster condition, which re-

Table 16.4 Socioenvironmental index

Natural resource–based industries: shellfish harvest, timber production, fishing, and agriculture	
Environmental quality	• Oyster condition • Vegetation cover • Salmon escapement • Bird abundance
Economic conditions	• Finfish, shellfish, and timber harvest • Income distribution • Unemployment rates • Bank loans per capita
Community health	• Healthy birth-weight babies • High school graduation rates • Voter turnout

Note: This socioenvironmental index was developed for the Willapa River watershed, Washington State (Willapa Alliance and Ecotrust 1995).

flects water quality; by changes in vegetation cover, which reflect terrestrial condition; by escapement of wild and hatchery salmon, which reflects ecological conditions in the rivers and streams; and by counts of wetland and riparian birds, which reflect habitat condition. Economic conditions are gauged by annual finfish, shellfish, and timber harvests, income distribution, local unemployment rates, and bank loans per capita. Community health is measured by the percentage of babies born with healthy birth weights, high school graduation rates, and voter turnout in county elections. Each category has alternate candidates that could have been used in developing the socioenvironmental index. Adult literacy rates could replace high school graduation rates, vegetative biodiversity could replace bird abundance, and so forth. Citizens and resource users within the watershed can use this index to develop a holistic understanding of the watershed and to create the stewardship necessary for long-term sustainability (The Willapa Alliance and Ecotrust 1995).

Addressing Institutional Organization

Watershed-scale activities must ultimately be integrated across a larger region, including the landscapes that make up that region. The challenge of integration involves a paradox of scale: Large-scale (regional) ecological systems can be most effectively regulated by small-scale (local) social organizations (Lee and Stankey 1992). Since people's interests, commitments, and knowledge generally are localized, bottom-up approaches that aggregate the local initiatives of citizens may be the most likely to succeed in achieving regional goals (Dryzek 1987). Experience throughout the world has shown that regional ecological stability is more likely to be achieved by permitting variability in land-use practices and, within limits of critical biological thresholds, by allowing and encouraging localized fluctuations in management practices (Ostrom 1990, Wheatley 1993, Korten 1987).

The paradox of scale is a general principle found to apply to systems as diverse as business organizations, chemical and physical processes, and ecological systems (Wheatley 1993). Stability in large-scale processes arises when the small-scale processes are allowed freedom to operate. This is illustrated by business organizations when individual and small-group initiatives respond to prices or other incentives by developing resources within the limits set by large-scale organizations.

Maintaining local initiatives, commitments, and knowledge also helps promote sustainability by insulating the management of ecological processes from the political cycles that affect large-scale organizations. National and state (or provincial) policies for regulating ecological systems generally are affected by the policies and preferences of political elites currently in power. Since political elites cycle in and out of power, especially in democratic systems, top-down control becomes a source of substantial instability. Hence, ecological regulations that rely on top-down control become highly unstable. Moreover, they may result in levels of unpredictability that discourage local initiatives requiring long-term commitments and investments of time and money.

The continuity of commitments and knowledge embodied in small-scale local organizations also helps foster effective adaptive management (Lee 1993, Pinkerton 1993). Political cycles in top-down administrations make it difficult to sustain long-term data gathering and monitoring. But even more difficult for large-scale organizations are commitments to take experimental actions for purposes of monitoring results. And when commitments to experiments are made, they often lack the diversity of trials necessary for eliminating multiple rival hypotheses about the operation of complex systems. Local initiatives, when supported by the generalized commitment of large-scale organizations, can be far more effective in implementing adaptive management. Experimental practices are insulated from the influence of political elites, fostering long-term commitments and ensuring that a diversity of trials will be put in place (McLain and Lee submitted).

Formulating Shared Socioenvironmental Visions

Loss of species, destruction of habitat, declining productivity, unstable social systems, and the disintegration of cultures are occurring on regional to global scales. How might these trends be reversed? Can it be accomplished by accepting risk at individual to institutional scales? One approach is to develop a shared socioenvironmental vision of future conditions (Figure 16.4). In an ideal sense, this may prove to be a nearly impossible task—although the process of trying to identify socioenvironmental endpoints for the short and long terms is an exercise that aids communication and acts as an effective form of education about the diverse cultural beliefs and values embedded in a watershed. For example, environmental endpoints may be related to the extent and condition of riparian forests, to acceptable levels of water quality and aquatic habitat, or to the persistence of viable populations of ecologically or culturally valuable plants and animals. Social endpoints may relate to the level of socioenvironmental literacy, to the development of adaptive institutions, to the formulation of unique public and private partnerships, and to the realization of levels of personal stewardship and responsibility that allow for the long-term maintenance of a balanced socioenvironmental system.

Successful examples of the development of shared socioenvironmental visions and the methods used to attain those visions can be found in British Columbia (Fraser River), Florida (Kissismee River), New England (Connecticut River), northern California (Metolius River), and western Washington (Willapa Bay), among many others. There are two fundamental traits inherent in these successful attempts: (1) a long-term commitment by the citizens who initially provide much of the vision and leadership and (2) an empowerment of citizens with the responsibility for their own future (Lee 1993).

Identify Criteria

- Environmental Endpoints
  - Riparian forest condition
  - Species persistence
  - Water and habitat quality
- Social Endpoints
  - Literacy
  - Adaptive Institutions
  - Partnerships
  - Stewardship and Responsibility

Personal Responsibility and Stewardship

- Long-term commitments by leadership
- Empowerment of citizens
- Communication of vision
- Education about value of vision
- Active monitoring
- Continued learning

Shared socioenvironmental vision

Figure 16.4 Components in developing and attaining a shared socioenvironmental vision for watersheds.

Public Stewardship in Watershed Management

Watershed management requires thoughtful stewardship that cannot be attained solely by government regulations or the work of technical specialists. Thus, concerned and educated citizens are essential. They represent a reservoir of human resources whose involvement can benefit management organizations and increase the overall awareness of socioenvironmental conditions. Citizens particularly can play an important role in monitoring socioenvironmental conditions. However, to do so successfully, they require continuing education to keep abreast of scientific and cultural advances.

Monitoring

Public involvement in coordinated monitoring activities instills a sense of ownership. Through monitoring programs, citizens can provide inputs to decision makers based on first-hand observations. As such, monitoring can become a learning opportunity for those setting policy as well as for those seeking to influence watershed management decisions.

There are unique advantages to including citizens in monitoring programs. First, due to limited budgets and staff availability, funds for collecting information about watershed features are usually directed to severely degraded sites. Many watersheds in need of monitoring are ignored unless volunteer efforts are undertaken. Thus, monitoring by volunteer groups or networks of individuals provides valuable information on watershed conditions that may not be high on political priority lists. Second, public involvement in monitoring projects helps ensure data continuity. Staff turnover in public agencies and large landowner organizations often results in discontinuities in data collection or undocumented changes in techniques. Local citizens, working with public and private organizations, fill gaps inevitably created when monitoring staffs change. They also provide insights to new staff members that otherwise might not be obtained within existing organizational structures. Third, the sheer number of citizens available to assist with monitoring makes it possible to conduct large-scale adaptive management experiments that would be impossible with limited agency or landowner resources.

Understanding changes in watershed conditions requires distinguishing between localized and large-scale effects, assessing system responses that separate human-related impacts from uncontrolled environmental factors, and having institutional agreements that provide for decades-long measurements (Walters and Holling 1990). These requirements generally go beyond the capabilities of individual organizations; thus, cooperative monitoring programs must become the rule instead of the exception.

In most cases, monitoring tasks need to focus on measurements readily understandable and not requiring specialized skills. Often this precludes the collection of biological samples. However, there are a number of monitoring activities well within the abilities of average citizens. These include the following:

Photographs

The importance of time-series photographs cannot be overstated. Some of the most valuable information about historical conditions is derived from photographs, particularly those where locations can be clearly identified. In addition, reference photo-points within watersheds are helpful in tracking long-term trends in vegetative structure and stream conditions. Reference photo-points also can be used to display the effects of seasonal changes and of large disturbances such as floods and fires. Important photographs often exist in family albums or businesses, and public involvement can bring these historical records to light.

Water Samples

Long-term trends in water quality require regularly scheduled sampling. However, the number of sites that can be routinely monitored by agencies is limited by the availability of automated sampling equipment and staff time. The U.S. Geological Survey monitored water quality parameters in many watersheds after passage of federal water laws in the 1960s and 1970s. Yet, it was forced to abandon many sites in the late 1970s when funding for monitoring programs expired. A network of water sampling locations at which local volunteers obtain periodic samples is an especially effective means of monitoring easily preserved parameters such as suspended sediment. Likewise, maximum-minimum thermometers placed throughout a watershed and checked at regular intervals by citizen's groups or individual landowners provide an indication of temperature fluctuations over time. These easily measured parameters have immediate significant effects on aquatic ecosystems and provide important information about erosion and upstream riparian conditions.

Habitat Measurements

Stream morphology is an integrative measurement of overall watershed condition. Pools, for example, which are very sensitive to change, provide important habitat for certain types of aquatic organisms, including many fish species. Citizen participation in simple habitat measures, such as counting the number of large pools, increases the area for which inventory information is available. Sportsman's clubs and conservation organizations (including adopt-a-stream groups) are especially suited to this type of project.

Riparian Forest Surveys

Riparian forests are critical to watershed health, yet insufficient attention is paid to their condition. Riparian plots in which surveyors periodically identify and count the number of trees, measure changes in species composition and growth, and note causes of mortality provide integrated long-term information about watershed characteristics. Plots do not have to be revisited every year, as long as their locations are well documented; they can be resurveyed by the same group or rotated among several groups over longer periods. Information generated by these surveys is useful for verifying remote sensing data, providing riparian vegetation information for watershed analysis, and teaching citizens about the dynamic nature of watershed processes.

Socioeconomic Conditions

Although socioeconomic conditions are already well monitored, data are seldom compiled for conditions within watershed boundaries. Annually collected socioeconomic indices, such as annual capital investment, resource exports, unemployment, high school literacy, healthy births, and so forth, provide essential information about human conditions in a larger community that includes the watershed (Table 16.4). Commitments to maintain and enhance the biological and physical conditions of watersheds are most likely to arise when local economies and societies are healthy and the population is well informed.

Public Outreach

Effective watershed management requires that scientists and managers provide knowledge about watershed processes and management techniques to citizens on a regular basis. Although citizens and local groups usually act with good intentions, they do not always have the benefit of current professional insights into human and environmental processes. The result may be that restoration and enhancement projects fail to achieve their objectives—or worse, that they actually impair socioenvironmental functions. For example, stream-cleaning projects largely have been discontinued by public agencies but are occasionally sponsored by citizen groups. Many citizens continue to be unaware of the ecological functions of woody debris or to regard these functions as secondary to providing unimpeded fish passage—even though this activity diminishes fish production in the long term.

How can educational and scientific communities maintain socioenvironmental literacy and instill a sense of stewardship among citizens? First, scientists need to explain concepts such as the importance of watershed connectivity, the role of natural disturbances in maintaining productivity, and the need to view watershed management in terms of large landscape units. They also need to articulate how social and environmental components work as an integrated system. Agricultural and forestry extension services, where citizens turn for advice from local specialists familiar with the region, serve as models for the establishment of an integrated watershed extension service. Watershed extension specialists, serving as local sources of the latest information, can act as liaisons between small and large landowners, natural resource consumers, and management agencies.

Second, colleges and universities must do more to educate citizens about important watershed management issues. Although educational institutions sponsor many meetings, the presentations are often too technical for citizens. Weekend or evening workshops aimed at communicating applied watershed science to a general audience are needed to increase public understanding of management options. Workshops featuring a combination of university faculty, research scientists, resource managers, citizens, and environmental policy makers are essential if we are to develop effective watershed management based on an integrated socioenvironmental perspective.

Conclusion

Watershed management is an ongoing experiment guided by fundamental principles and a common vision of the future. It utilizes a multitude of approaches to achieve an integrated and balanced socioenvironmental system (Naiman 1992, Lee 1993). There is no universal methodology for achieving effective watershed management. However, fundamental principles related to cooperation, balance, fairness, integration, communication, and adaptability can help guide the process:

1. It is important to recognize that watershed management demands unparalleled cooperation between citizens, industry, governmental agencies, private institutions, and academic organizations. In most situations, the complexity of information processing and the scope of socioenvironmental change exceed the capacity of any single group to manage a watershed effectively.

2. Technical solutions (such as fish hatcheries and waste management) to specific human-generated problems need to be balanced with the wide-scale maintenance of appropriate environmental components that provide similar ecological services.

3. Data-driven policy and management decisions need to become the standard for resolving issues; decisions based on conceptualization and perception should be minimized.

4. Regulations guiding the structure and behavior of socioenvironmental systems need to be evenly and fairly applied throughout a watershed. For example, basic regulations in areas such as riparian protection and chemical applications should not differ between forestry, agricultural, and urban areas. Regulations should encourage citizen initiatives and offer landowners incentives to provide greater protection and reduce chemical applications.

5. Human activities need to be accepted as fundamental elements of the watershed, as are the structure and dynamics of the environmental components. Both have inherent rights to exist for the long term.

These principles, when combined with approaches outlined in this chapter, provide initial steps toward achieving effective watershed management. Cultural values, social behavior, and environmental characteristics will continue to evolve. Unfortunately, critical evaluation of the approaches for watershed management outlined here will not be possible for several decades. Will the evaluation be positive? If so, it will be because citizens, regulators, educators, and industries shared a common long-term vision and adapted to change by adopting appropriate actions to meet that vision.

Literature Cited

Bella, D. A. 1987. Organizations and systematic distortion of information. *Journal of Professional Issues in Engineering* 113:360–370.

Berry, M. W., R. O. Flamm, B. C. Hazen, and R. M. MacIntyre. In press. The land-use change and analysis system (LUCAS) for evaluating landscape management decisions. *IEEE Computations Science and Engineering.*

Botkin, D. B. 1990. *Discordant harmonies: A new ecology for the twenty-first century.* New York: Oxford University Press.

Burch, W. R., and D. R. DeLuca. 1984. *Measuring the social impact of natural resource policies.* Albuquerque: University of New Mexico Press.

Campbell, D. T. 1969. Reforms as experiments. *American Psychologist* 24:409–429.

Décamps, H., R. J. Naiman, and J.-C. Lefeuvre. 1997. Landscape ecology and regional development. In *Integrating conservation, development, and research,* ed. M. Hadley. Paris, France: UNESCO.

Drake, J. A., H. A. Mooney, R. diCastri, R. H. Groves, F. J.

Kruger, M. Rejmánek, and M. Williamson, eds. 1989. *Biological invasions: A global perspective.* New York: John Wiley & Sons.

Dryzek, J. S. 1987. *Rational ecology: Environment and political choice.* Oxford, England: Basil Blackwell.

Fetherston, K. L., R. J. Naiman, and R. E. Bilby. 1995. Large woody debris, physical process, and riparian forest development in montane river networks of the Pacific Northwest. *Geomorphology* 13:133–144.

Finstenbusch, K., and C. P. Wolf. 1981. *Methodology of social impact assessment.* Stroudsburg, PA: Hutchison and Ross.

Firey, W. A. 1960. *Man, mind, and land: A theory of resource use.* Glencoe, IL: Free Press.

Flamm, R. O., and M. G. Turner. 1994a. Multidisciplinary modeling and GIS for landscape management. In *Forest ecosystem management at the landscape level: The role of remote sensing and integrated GIS in resource management planning, analysis, and decision making,* ed. V. A. Sample. Washington, DC, and Covelo, CA: Island Press.

———. 1994b. Alternative model formulations of a stochastic model of landscape change. *Landscape Ecology* 9:47–46.

Fortin, M. J. and P. Drapeau. 1995. Delineation of ecological boundaries: Comparison of approaches and significant tests. *Oikos* 72:323–332.

Gregory, S. V., F. J. Swanson, and W. A. McKee. 1991. An ecosystem perspective of riparian zones. *BioScience* 40:540–551.

Grumbine, R. E. 1994. What is ecosystem management? *Conservation Biology* 8:27–38.

Johnston, C. A., N. E. Detenbeck, and G. J. Niemi. 1990. The cumulative effect of wetlands on stream water quality and quantity: A landscape approach. *Biogeochemistry* 10:105–141.

Karr, J. R. 1991. Biological integrity: A long-neglected aspect of water resource management. *Ecological Applications* 1:66–84.

Korten, D. 1987. *Community management: Asian experience and perspectives.* West Hartford, CT: Kumarian Press.

Le Maitre, D. C., B. W. van Wilgen, and D. M. Richardson. 1993. A computer system for catchment management: Background, concepts, and development. *Journal of Environmental Management* 39:121–142.

Lee, K. N. 1993. *Compass and gyroscope: Integrating science and politics for the environment.* Washington, DC, and Covelo, CA: Island Press.

Lee, R. G., and G. S. Stankey. 1992. Major issues associated with managing watershed resources. In *Balancing environmental, social political, and economic factors in managing watershed resources,* ed. P. W. Adams and W. A. Atkinson. Corvallis, OR: Oregon State University.

Lee, R. G., R. O. Flamm, M. G. Turner, C. Bledsoe, P. Chandler, C. M. DeFerrari, R. R. Gottfried, R. J. Naiman, N. H. Schumaker, and D. N. Wear. 1992. Integrating sustainable development and environmental vitality: A landscape ecology approach. In *Watershed management: Balancing sustainability and environmental change,* ed. R. J. Naiman. New York: Springer-Verlag.

Levin, S. A., ed. 1993. Science and sustainability. *Ecological Applications* 3:550–589.

Ludwig, D., R. Hilborn, and C. Walters. 1993. Uncertainty, resource exploitation, and conservation: Lessons from history. *Science* 260:17–36.

McLain, R. J., and R. J. Lee. Submitted. Adaptive management: Promises and pitfalls. *Journal of Environmental Management.*

Meyer, J. W., and R. W. Scott, eds. 1983. *Organizational environments: Ritual and rationality.* Beverly Hills: Sage.

Montgomery, D. R., G. E. Grant, and K. Sullivan. 1995. Watershed analysis as a framework for implementing ecosystem management. *Water Resources Bulletin* 31(3): 1–18.

Naiman, R. J., ed. 1992. *Watershed management: Balancing sustainability and environmental change.* New York: Springer-Verlag.

Naiman, R. J., H. Décamps, and M. Pollock. 1993. The role of riparian corridors in maintaining regional biodiversity. *Ecological Applications* 3:209–212.

Naiman, R. J., J. J. Magnuson, D. M. McKnight, and J. A. Stanford. 1995a. *The Freshwater imperative: A research agenda.* Washington, DC: Island Press.

———. 1995b. Freshwater ecosystems and management: A national initiative. *Science* 270:584–585.

Ostrom, E. 1990. *Governing the commons: The evolution of institutions for collective action.* New York: Cambridge University Press.

Pinkerton, E. W. 1993. Co-management efforts as social movements. *Alternatives* 19:34–38.

Reice, S. R. 1994. Nonequilibrium determinants of biological community structure. *American Scientist* 82:424–435.

Reice, S. R., R. C. Wissmar, and R. J. Naiman. 1990. Disturbance regimes, resilience, and recovery of animal communities and habitats in lotic ecosystems. *Environmental Management* 14:647–659.

Reich, R. 1991. *The work of nations: Preparing ourselves for 21st century capitalism.* New York: A. A. Knopf.

Risser, P. G. 1993. Ecotones. *Ecological Applications* 3:367–368.

Schrader-Frechette, K. S. 1991. *Risk and rationality: Philosophical foundations for popular reforms.* Berkeley: University of California Press.

Selznick, Philip. 1966. *TVA and the grassroots: A study on the sociology of formal organizations.* New York: Harper & Row.

Skoal, R. R., and R. J. Rohlf. 1981. *Biometry.* San Francisco: W. H. Freeman and Company.

Slovic P. 1984. Low-probability/high consequence risk analysis and the public. In *Low-probability high-consequence risk analysis,* ed. R. A. Waller and V. T. Covello. New York: Plenum Press.

Stanford, J. A., and J. V. Ward. 1992. Management of aquatic resources in large catchments: Recognizing interactions between ecosystem connectivity and environmental disturbance. In *Watershed management: Balancing sustainability and environmental change,* ed. R. J. Naiman. New York: Springer-Verlag.

Suter, G.W., II. 1992. *Ecological risk assessment.* Boca Raton: Lewis Publishers.

Tolt Watershed Analysis Prescriptions. 1993. *Tolt Watershed Analysis Prescriptions.* Tacoma, WA: Weyerhaeuser Company.

Turner, B. L., II, W. C. Clark, K. W. Kates, J. F. Richards, J. T. Mathews, and W. B. Meyer, eds. 1990. *The earth as transformed by human action.* Cambridge, England: Cambridge University Press.

Turner, M. G. 1990. Spatial and temporal analysis of landscape patterns. *Landscape Ecology* 4:21–30.

Turner, M. G., and R. H. Gardner, eds. 1991. *Quantitative methods in landscape ecology.* New York: Springer-Verlag.

Turner, M. G., V. H. Dale, and R. H. Gardner. 1989a. Predicting across scales: Theory development and testing. *Landscape Ecology* 3:245–252.

———. 1989b. Effects of changing spatial scale on the analysis of landscape pattern. *Landscape Ecology* 3:153–162.

Turner, M. G., G. J. Arthaud, R. T. Engstrom, S. J. Hejl, J. Liu, S. Loeb, and K. McKelvey. 1996. Usefulness of spatially explicit animal models in land management. *Ecological Applications* 5:12–16.

Turner, M. G., D. N. Wear, and R. O. Flamm. In press. Influence of land ownership on land-cover change in the Southern Appalachian Highlands and the Olympic Peninsula. *Ecological Applications.*

USDA Forest Service. 1994. *A federal agency guide for pilot watershed analysis.* Vers. 1.2. Portland, OR: Pacific Northwest Research Station.

Volkman, J. M., and K. N. Lee. 1994. The owl and Minerva: Ecosystem lessons from the Columbia. *Journal of Forestry* 92:48–52.

Walters, C. J., and C. S. Holling. 1990. Large-scale management experiments and learning by doing. *Ecology* 71:2060–2068.

Warwick, C. J., J. D. Mumford, and G. A. Norton. 1993. Environmental management expert systems. *Journal of Environmental Management* 39:251–270.

Washington Forest Practices Board. 1994. *Standard methodology for conducting watershed analysis.* Vers. 2.1. Olympia, WA: Washington Department of Natural Resources.

Wear, D. N., M. G. Turner, and R. O. Flamm. In press. Ecosystem management in a multi-ownership setting: Exploring landscape dynamics in Southern Appalachian watershed. *Ecological Applications.*

Wheatley, M. J. 1993. *Leadership and the new science: Learning about organization from an orderly universe.* San Francisco: Barrett-Koehler Publishers, Inc.

The Willapa Alliance and Ecotrust. 1995. Willapa indicators for a sustainable community. Unpublished report. P. O. Box 278, South Bend, WA.

Elements of Ecosystem Management

A Photo Essay by Jerry F. Franklin

A vastly increased appreciation for the complexity of ecological and organizational systems is a central theme in development of 21st century forestry. Discovery and recognition of complexity has been especially dramatic in regard to the composition, function, and structure of natural forest ecosystems.

Natural forests have proven very heterogeneous in terms of individual structures and their spatial arrangements within stands.

Dead standing trees and boles on the forest floor have proven to be as important as living trees to ecosystem function, challenging foresters to view the dead trees and coarse woody debris in new ways, and creating an immense set of new questions about how much is needed to fulfill specific functions.

Functional capabilities of natural forests clearly are related to the inherent complex architecture, and these capabilities differentiate natural forests from structurally simplified, young managed forests. For example, snow interception in old-growth forests of the Pacific Northwest (see photo on next page) is a significant factor in reducing potential streamflow during rain-on-snow storm events.

Structurally complex 500-year-old western hemlock stand on the Ashael Curtis Nature Trail, Mt. Baker-Snoqualmie National Forest, Washington.

Rotting Jeffrey pine logs, Stanislaus National Forest, California.

Douglas-fir snags at Wind River Canopy Crane Research Facility, Gifford Pinchot National Forest, Washington.

Snow in old-growth Douglas-fir–western hemlock forest canopy at Wind River Canopy Crane Research Facility, Gifford Pinchot National Forest, Washington.

Sulfur fungus on rotting wood, Sister Rocks Research Natural Area, Gifford Pinchot National Forest, Washington.

Nitrogen-fixing lichen (Lobaria) *growing on a branch of an old-growth Douglas-fir tree, Gifford Pinchot National Forest, Washington.*

Banana slug feeding on devilsclub, Olympic National Park, Washington.

Small elements of biodiversity—such as lichens, fungi, and invertebrates—provide essential functions in natural forests. Strategies for their maintenance need to be a part of forest management plans. Many of these are inhabitants of the underground portion of ecosystems, which rivals or exceeds the canopy as the most diverse and biologically active part of the forest.

Recent clearcut, H.J. Andrews Experimental Forest, Willamette National Forest, Oregon.

Windthrown old-growth forest, Bull Run Watershed, Mount Hood National Forest, Oregon.

Sierra Nevada mixed-conifer forest killed by crownfire, Yosemite National Forest, California.

Old-growth conifer forest killed by eruption of Mount St. Helens in May 1980, Bear Creek, Mount St. Helens National Volcanic Monument, Washington.

Research on the effects of forest disturbances and natural recovery processes has revealed previously unappreciated complexities within and between disturbance types. Almost all natural disturbances leave immense "biological legacies" of living organisms and organically derived structures such as snags and logs. These legacies, which are extremely important in the recovery process, contribute to the resulting compositional and structural diversity in young, natural ecosystems. Levels of legacies, such as numbers of living trees, also show wide spatial variation where disturbances are large, providing additional large-scale complexity.

There are immense differences between even-aged silvicultural systems (especially clearcutting) and natural disturbances such as windthrow, wildfire, and even volcanic eruptions. For example, clearcutting does not resemble wildfire except in regard to the light environment; levels of biological legacies typically are extremely limited under traditional clearcutting methods, in contrast to natural disturbances.

Forested landscape composed of a mosaic of mature Douglas-fir forest (smooth canopy) regenerated after a fire in 1902 and old-growth Douglas-fir–western hemlock forest (rough canopy), North Siouxon Creek drainage, Washington.

Highly fragmented landscape created by dispersed patch clearcutting; Warm Springs Indian Reservation (previously Mount Hood National Forest), Oregon.

Extensively clearcut landscape, Cascade Range of southwestern Washington.

Most major forest management issues involve planning and conducting activities at larger spatial scales—that is, landscapes and watersheds. Past failures in landscape planning have included forest fragmentation. Fragmentation creates small forest patches and large amounts of high-contrast edge as well as cumulative effects, such as undesirable impacts resulting from rapid and extensive forest harvest. Natural landscapes provide useful models for the design of managed landscapes in terms of the size, condition, and amount of patches as well as their shape and degree of edge contrast.

Logs (including root wads) along Queets River, Olympic National Park, Washington.

Brown bear and salmon, Katmai National Park, Alaska.

Riparian areas associated with streams and rivers are typically the most ecologically important elements of forested landscapes. Forests have many functional links to associated waterbodies, including providing energy, nutrients, and coarse woody debris. In the case of smaller streams, forests have direct control over light and temperature conditions. Forestry in the 21st century must be designed to maintain these important ecological functions. In addition, managers must think in terms of larger spatial scales—such as providing for coarse woody debris from headwater streams through rivers and estuaries, and eventually to marine ecosystems. Conflicts between resource values can be intense because riparian areas often provide critical wildlife habitat as well as large timber and the most productive forest sites.

Riparian zone of Suwannee Creek, Sequoia-Kings Canyon National Park, California.

Development of harvest systems that provide for structural and other biological legacies is critical to creating managed stands. These stands, in turn, provide for higher levels of ecological function and biological diversity than has been characteristic of traditional even-aged harvest systems such as clearcutting. Particularly important are legacies of large decadent live trees, large snags, and large down logs, which are difficult to produce in managed forests except under very long rotations.

The objective of such legacies is to initially "lifeboat" many elements of biological diversity and then to enrich the young forest structurally so that additional species can return to inhabit the area. Such structural legacies also assist in movement of many organisms through harvested "matrix" lands and help buffer reserved areas.

Important questions that silviculturalists must address in developing prescriptions using variable retention harvest systems include which structures to retain, how many of each, and whether to disperse or aggregate the structures.

Old-growth Douglas-fir trees, snags, and logs permanently retained in harvest unit, Blue River Ranger District, Willamette National Forest.

Dispersed live tree retention in mature Douglas-fir forest, Cedar River watershed (City of Seattle), Washington.

Aggregated retention (retention of small islands and strips of forest), Big Dog Unit, Plum Creek Timber Co., Carbon River drainage, Washington. Photo courtesy of Plum Creek Timber Co.

Prescribed burn in longleaf pine forest, Joseph E. Jones Ecological Research Center, Georgia.

Gaging station to monitor streamflow on a small watershed, H. J. Andrews Experimental Forest, Willamette National Forest, Oregon.

Navajo tribal representatives discussing forestry issues with a scientific team, Navajo Indian Reservation, Arizona.

Forestry in the 21st century must effectively link stakeholders, managers, and scientists to provide integrated solutions to social and biological issues. New tools, organizational relationships, and interdisciplinary assessments can all contribute to these linkages. More emphasis on nonharvest tools, such as prescribed burning, also is required. Basic and applied research are important parts of the adaptive management approach.

Management prescriptions must be recognized as working hypotheses, the predictability of which is limited—a reflection of the always limited state of our knowledge and the iterative nature of learning. Consequently, we should adopt humility as a basic attitude and use adaptive management for the continued systematic accumulation and incorporation of knowledge in resource management.

Remeasuring a permanent sample plot in old-growth forest, Olympic National Park, Washington.

Construction crane providing access to 5.6 acres of old-growth Douglas-fir–western hemlock forest at Wind River Canopy Crane Facility, Gifford Pinchot National Forest, Washington.

Montane Alternative Silvicultural Systems (MASS) harvest cutting experiment on MacMillan-Bloedel Company lands near Campbell River, British Columbia, Canada.

17 Landscape Analysis and Design

Nancy M. Diaz and Simon Bell

Ecosystem Management Applied at Larger Spatial Scales 255
Creating "Intentional" Landscapes—Landscape Analysis and Design 256
Landscape Analysis and Design Process—Analysis Phase 256
Landscape Analysis and Design Process—Design Phase 259
Design Part 1—Narrative Objectives 259
Design Part 2—Spatial Design 259
Design Process Case Study—The West Arm Demonstration Forest 261
Literature Cited 269

The purpose of this chapter is to present a means of understanding and organizing ecological and social information about larger-scale ecosystems and for using that information to plan and design future landscapes. The focus is on public forests and grasslands in the Pacific Northwest (in British Columbia, Washington, and Oregon), but the framework of ideas and methods could be (and has been) applied globally to a wide range of ecosystem types.

Ecosystem Management Applied at Larger Spatial Scales

In the United States, recent controversies over the effects of human manipulations on public lands, as well as skepticism about the ability of these lands to sustain anticipated levels of commodity production, have lead to a significant realignment of resource management objectives. In fact, on federal lands the concept of resource management (in the sense of managing the production of individual resources like timber, minerals, forage for livestock, and scenery) has virtually given way to the more systemic view of ecosystem management—managing the patterns and processes in a holistic manner to provide for sustained character and function, as well as for benefits and commodities for humans. This has not yet happened in British Columbia, which shares the same types of ecosystems, but there are signs that the same pressures will lead that way, as will be shown in the example below.

The focus of much of the controversy over U.S. federal land management has been at the larger spatial scales, particularly over the serious consequences of human activities (e.g., habitat loss and fragmentation) on species that require large areas to sustain healthy populations. Until relatively recently, there has been little systematic study of the ways in which human activities interact with and affect various ecological processes that operate at the landscape scale in Pacific Northwest forests and grasslands. And there have been even fewer attempts actually to try to create sustainable landscapes. However, with recent conceptual and technological developments, we increasingly have the ability to understand (albeit in a rudimentary fashion) how landscapes function and the role humans can have in them, both to interfere with ecological functions and to promote improvements.

To implement ecosystem management, we must understand the workings of landscape ecosystems and use this knowledge to develop policies and plans that protect them. A key task is the articulation of intended future conditions that create and maintain sustainable landscapes, for if you don't know where you're going, any road will take you there.

At a variety of spatial scales, it has therefore become necessary to describe conditions that will allow forest and grassland landscapes to sustain their characteristic biological diversity, viability of ecological processes, and productive potential (USDA Forest Service 1993). This resulting picture of sustainable conditions provides a backdrop for developing realistic expectations about the ability of ecosystems to produce things people want. The question is, How?

Creating "Intentional" Landscapes—Landscape Analysis and Design

A process for describing sustainable landscape pattern and process objectives has been presented by Diaz and Apostol (1992). The landscape analysis and design process is intended to be used by forest and grassland managers to synthesize information about landscape character and resources with issues and policies that direct land management. The goal is to be able to describe and depict, spatially and temporally, sustainable landscape patterns—not just conceptually, but for real places—that can be created and maintained through various management activities ranging from preservation to intensively managed.

The landscape analysis and design process has two parts (Figure. 17.1). The first, *analysis,* consists of gathering and interpreting information about the structure and function within the landscape such that the sustainable conditions for a given landscape can be understood.

The second phase, *design,* uses the information from the analysis phase—along with policies, local issues, and desires for resources—to sort out realistic objectives for what the landscape can provide and for the spatial patterns that will sustain them. These objectives involve not only direct human uses and benefits, but also the viability of other living and nonliving landscape features and processes. A key step of the design phase is to spatially represent the location of different pattern types on the actual landforms—in other words, to make maps and three-dimensional drawings.

In this chapter, we present a conceptual discussion of the analysis phase and give a more detailed account of the use of the design process in a case study.

Landscape Analysis and Design Process—Analysis Phase

A purposeful approach to the creation of sustainable landscapes must be informed from the start by an understanding of the interactions of patterns and processes at larger spatial scales. The central idea is

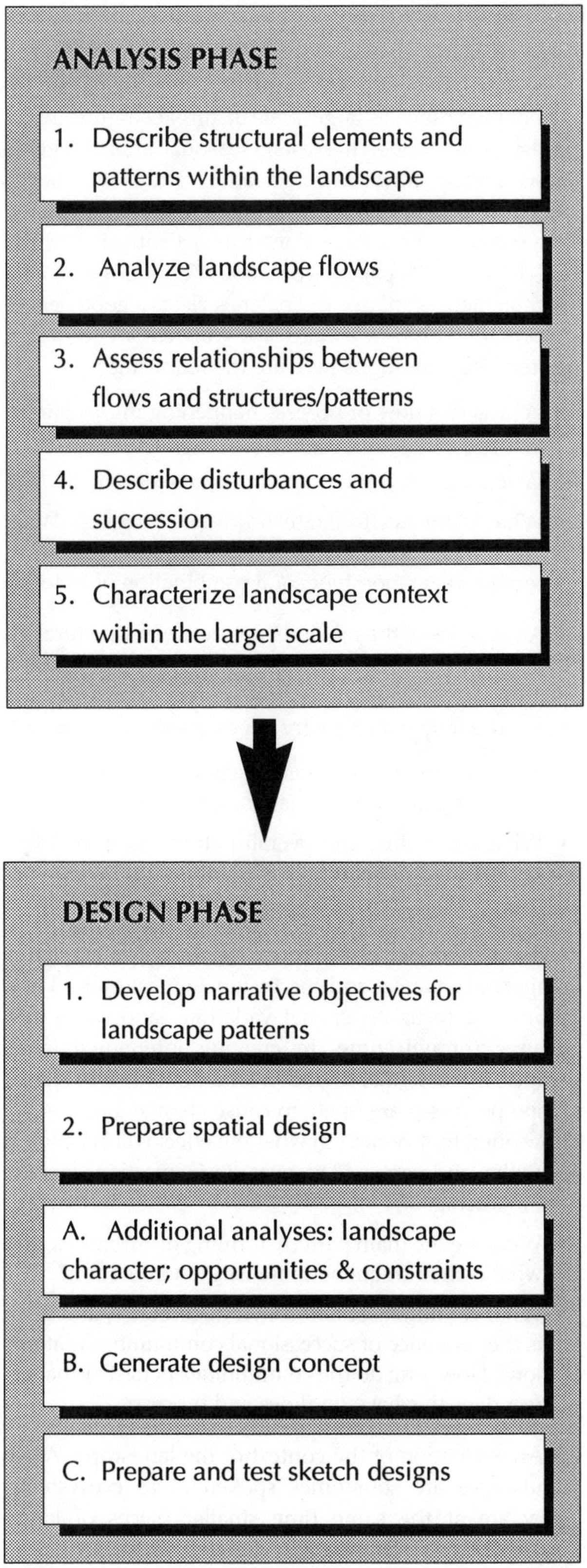

ANALYSIS PHASE

1. Describe structural elements and patterns within the landscape
2. Analyze landscape flows
3. Assess relationships between flows and structures/patterns
4. Describe disturbances and succession
5. Characterize landscape context within the larger scale

DESIGN PHASE

1. Develop narrative objectives for landscape patterns
2. Prepare spatial design

A. Additional analyses: landscape character; opportunities & constraints

B. Generate design concept

C. Prepare and test sketch designs

Figure 17.1 Landscape analysis and design process.

that landscape patterns (the spatial arrangement of matrices and patches)(Forman and Godrun 1986) and processes affect each other. For example, in western Cascades landscapes, wildfire has created a mosaic of forest stands of varying ages and sizes. At the same time, the spatial patterns of this mosaic affect the way that wildfire will behave subsequently due to the varying resistance of different-aged stands to the ignition and spread of fire. Thus, the pattern and the process, like the chicken and the egg, exist in a loop of mutual influences.

As humans, we manipulate features of both the vegetation pattern (by changing the sizes, shapes, ages, and juxtaposition of vegetation patches) and the action of processes like fire (by excluding or controlling it). We also introduce new processes, like logging, which have different consequences than any natural process to which an ecosystem has become adapted. Sorting out the numerous effects of these and other ways humans affect landscapes, and setting clear objectives for our activities, requires a complex analysis process. To understand a landscape to its fullest extent is beyond what our current knowledge can support. However, there are many aspects of the problem that can and must be addressed and that give us a large part of the picture. These include the following:

1. A description of the structural elements present in the landscape. Pertinent questions are the following:

- Is there a matrix (a "most connected" portion of the landscape) that exerts influence over various ecological processes? What is the character (species composition, stand structure) of the matrix?
- What kinds of nonmatrix patches are present? How do they contrast (in composition and physical structure) with the matrix?
- What patterns of matrix and patches occur? Is the matrix's ability to control ecological processes affected by fragmentation? How much edge relative to core areas exists? How are the sizes and shapes of patches characterized? To what degree is the pattern self-organized, and to what degree is it the result of human interference?

2. An analysis of the landscape "flows" and processes that occur (or occurred in the past or will occur in the future), where they occur, and what structural elements (vegetation patterns, landforms, etc.) they are related to. Phenomena such as large-scale disturbances (fire, insect infestations, timber harvests), seasonal animal migrations, human travel and recreation, and water movement are some examples of landscape flows. A detailed assessment of how these flows interact with (1) individual matrix and patch types and (2) a given spatial arrangement of matrix and patches yields a good deal of information on how a landscape works as an ecological system. Key questions here are the following:

- Why is this flow or process happening in this particular place?
- Where does it come from, and where does it go?
- What pathways facilitate flows or processes? What are the pathways composed of: Are they related to terrain, vegetation type, or a combination of factors?

3. An analysis of the relationships between structural elements, patterns, and flows. It is this step that gives a picture of the landscape as a system of interrelated parts. The following primary questions should be asked:

- What role do the various structural elements play in facilitating or hindering landscape flows?
- What effect does the overall pattern have in determining the kind and location of flows that are occurring?

4. An assessment of the forces of change in the landscape: disturbances and succession. Items number 1 and number 2 focus on spatial variation; landscapes also change through time. In creating intentional landscapes, it is essential to understand not only what landscape processes are likely to cause change, but also (1) how often they occur, (2) what the effect will be, and (3) how the landscape will recover. Basic questions include the following:

- What are the main pattern-forming disturbances, and what is their frequency, intensity, and duration?
- What conditions result from a disturbance, and what is the sequence of successional communities that follow? How long do the communities persist? What effect does this have on flows and processes?

5. An evaluation of the context of the landscape. While landscapes are sometimes spoken of as ecosystems, they are at the same time smaller pieces of larger ecosystems. No landscape completely circumscribes all the processes that occur within it; many things cross the

borders as parts of processes going on at scales larger than the landscape. Human travel, air pollution, and migration of animals with very large home ranges are examples of issues that often transcend landscape boundaries. A larger-scale look is thus needed to complete the picture:

- What flows and processes present in the landscape are actually operating at a larger scale?
- Is the landscape under consideration acting as a source or destination for flows or processes that cross the boundaries? Does the landscape contain a key piece of a flow (e.g., elk winter range) or is it simply part of a general pathway?

Answering these kinds of questions helps establish the nature of the linkages between a landscape and surrounding areas.

Landscape Analysis and Design Process—Design Phase

If carefully done, the above analysis phase should yield the best picture possible (given current knowledge) of how a landscape works as an ecological system and what aspects of structure and process are key to sustaining its character, function, and productivity. Subsequent steps of the landscape analysis and design system assume that there are (1) policies, guidelines, or other direction regarding resource objectives for the landscape and (2) an understanding of the issues, concerns, opportunities, and constraints unique to the particular area (commonly acquired in a "scoping" process and from inventories and assessments by various resource specialists).

Design Part 1—Narrative Objectives

The next task of landscape analysis and design is to translate policies and direction and resource constraints and opportunities into narrative statements of landscape patterns—that is, spatial arrangements of matrix and patches. A challenge at this point is to integrate the usually numerous (and often competing) considerations for a given piece of the landscape. A good starting point is to ask this question:

- What features must be present in this landscape to provide for maintenance of biological diversity, viable ecological processes, and productivity?

A good template to answer that question is to look at what might have been present in a natural, self-organized landscape. Once that has been determined, a simple but critical question can be asked:

- What are the resource objectives in different parts of the landscape, and what landscape patterns are necessary to accomplish them?

In other words, for those resources whose presence or use is in some way dependent on landscape patterns—late-successional forest inhabitants, clean water and healthy streams, timber production, scenery, certain types of recreational experiences, big game habitat, and so on—what patterns need to be in the landscape, and where do they need to be?

For example, in winter range areas of big game on national forest lands, there are commonly well-documented guidelines for a desirable mix of mature forest cover with open patches and for the size and dispersal of openings. It is relatively simple to convert such standards into the landscape ecology terms of matrix, patch, and pattern and to map where this type of pattern should occur on the landscape in order to provide the desired habitat and to see how it might change through time. For other resource objectives, the relationships to landscape patterns may be less well studied, but they are equally important to describe.

Design Part 2—Spatial Design

How do the ecological analysis, forest structural objectives, and pattern concept described above fit into the social, economic, and operational context? How is the broad level of understanding translated into realistic prescriptions for action at the stand or site level? This is where design comes in. So far, a model has been described for understanding the landscape as an ecological system. Now this has to be applied to the real landscape, with all its variations and subtleties of topography, microclimate, soil, drainage, and vegetation types. These combine to create patterns that are unique to each landscape area and that should be harnessed by the designer.

The spatial design step assumes the following tasks have been completed:

Collect Relevant Information About the Site

Besides the data collected in the analysis process, other resource information about the area will likely have been gathered. At this stage, available information about physical, biological, social, and cultural resources within the area should be compiled. Information should be relevant to the task at hand and not be overly detailed for the scale of the work.

Set Narrative Management Objectives

In addition to the landscape pattern objectives developed in design part 1, management objectives should be articulated for each resource category and should include indicators of success in meeting the objectives, if possible. For example, an objective to "protect and enhance fish and wildlife resources" might be successful when "all critical or sensitive habitats have been protected by design of buffer zones, and all seral stages for each site type are present at suitable levels at any point in time." Such objectives should come both from existing policies and from knowledge gained in the analysis steps.

At this point, the designer is ready to proceed with the process of spatial design. It consists of the following tasks:

1. Develop a design concept.
2. Prepare sketch design proposals.
3. Evaluate the sketch design to see how it meets the objectives.
4. Iterate or modify the sketch design until it is satisfactory; test it through time.

The ecological analysis has already been described above as the main imperative driving the plan. There are two other equally important analyses required, though they are far less complicated to prepare.

The first of these is a landscape character analysis (Lucas 1991, Bell 1993, B.C. Ministry of Forests 1994). This is a description of how the landscape appears to us as we see it in real life as opposed to a map or geographical information system model. Panoramic photographs should be taken from suitable viewpoints. These, or copies of them, can then be annotated with a description of the patterns, shapes, colors, textures, and other visual aspects that together make the landscape what it is. Past management activities such as logging may have introduced geometric or other unnatural shapes into the landscape that conflict visually with the natural patterns of vegetation types. This aspect is particularly important when naturalness is a specific aim of ecologically led design. It also helps to ensure that the aesthetic result of the design, usually a significant factor and often a resource category in its own right, emerges from a deep reading of the landscape rather than from the application of a preconceived style. Each design will then be special to each landscape, not to a standard solution.

A particularly valuable and interesting part of the landscape character analysis is that of the underlying landform. This greatly reflects the way in which we perceive a landform visually, but also relates to many of the vegetation patterns superimposed on it. For both plan (from contour lines) and perspective (over photographs), an analysis should be compiled with arrows, varying in strength, running down ridgelines, spurs, and convexities and, conversely, running up valleys, hollows, and concavities. Shapes that rise up into hollows and descend on ridges tend to look more comfortable and frequently reflect real vegetation patterns (British Forestry Commission 1989, Lucas 1991, Bell 1993).

The next analysis sifts out those factors from the inventory that constrain certain possibilities or that provide opportunities. Many of the factors relate to practical concerns important in the shorter term, such as accessibility, road construction options, or operability.

One key aspect of this analysis is to separate those factors that are "real" constraints, such as soils prone to mass wasting, cliffs preventing road construction, and sites of rare and endangered species, from "artificial" constraints such as maximum clearcut unit sizes, minimum widths of stream buffer zones, and minimum time requirements for logging adjacent logging units. The former are always present; the latter vary with policy and may be more flexible in practice.

Constraints may also be opportunities: A constraint on logging a streamside old-growth area may provide an opportunity to develop a recreation trail.

Once the three analyses (ecological, landscape character, and constraints and opportunities) have been completed, the objectives should be checked to see if new information or understanding will require them to be modified. This is part of the key to ensuring that the plan will be sustainable in the longer term.

The perspective should not be used only for the landscape character analysis: The ecological pattern objectives and the constraints and opportunities should also be depicted in three dimensions ready for the next design steps.

1. *Design concept generation.* This is where analysis turns into synthesis. The concept comprises a broad pattern for the landscape that combines the spatial organization of the forest (different stands or management units) with the processes at work (human-controlled or natural). The concept will be no more than a rough perspective sketch or bubble diagram annotated as to what might be done and where it might be done in the landscape.

The creative ideas that are developed at this stage do not come out of thin air. The landscape is the designer's canvas and it already gives structure, pattern, and context. It reduces the range of possibilities into a limited number of probabilities. Landform, vegetation patterns, the ecological zones, past natural disturbance patterns, and the degree of diversity can all be used for inspiration. This is easier in relatively intact landscapes compared with fragmented areas where it might be more difficult to see beyond the present disturbed pattern to the underlying potential.

2. *Sketch design.* Gradually the loose ideas and shapes coalesce, are refined, and become more detailed. They turn into real shapes that will eventually be given a management prescription and be laid out on the ground. Therefore, they need to be tested for their practicality against the constraint-and-opportunity analysis. All the design so far should be carried out in perspective. Simple sketches can be used to show how the landscape might change over time. This enables some of the possible processes to be evaluated, and allows the rate of change, its scale, and its ecological effect to be checked against the indicators identified at the objective-setting stage. The design is iterated and modified until it works as well as possible, meets the objective, is workable, and looks right.

Design Process Case Study—The West Arm Demonstration Forest

The best way in which to demonstrate the process is to use a case study. The West Arm Demonstration Forest (WADF) on the north side of the west arm of Kootenay Lake near Nelson in British Columbia was subject during early 1994 to a comprehensive, integrated design following the process described above. While not yet an agreed-upon design, it is sufficiently developed to be able to demonstrate the efficacy of the approach.

The area as a whole (Figure 17.2) comprises some 15,000 ha (36,000 acres), so only a section (one design subunit) will be shown here. The design was assembled by a team from the Ministry of Forests and Ministry of Environment of British Columbia, with the chief landscape architect of the British Forestry Commission, Simon Bell, acting as facilitator and designer of the group. All resource values were represented in the group. Two interested members of the public also attended and offered valuable input.

At the objective-setting stage the group compiled a wide range of objectives related to each resource category. Indicators showing when each would be met were defined. An example of one of these is water: The objective was to maintain and enhance water quality and quantity over the year, every year. The indicators were (1) having a minimum buffer zone on all streams, (2) maintaining appropriate equivalent clearcut areas at all times, and (3) maintaining natural drainage patterns.

A review of each resource category was undertaken by the entire group to flesh out the main factors at work and to understand how these presented opportunities or constraints. The factors included biogeoclimatic zones, forest cover, forest health, hy-

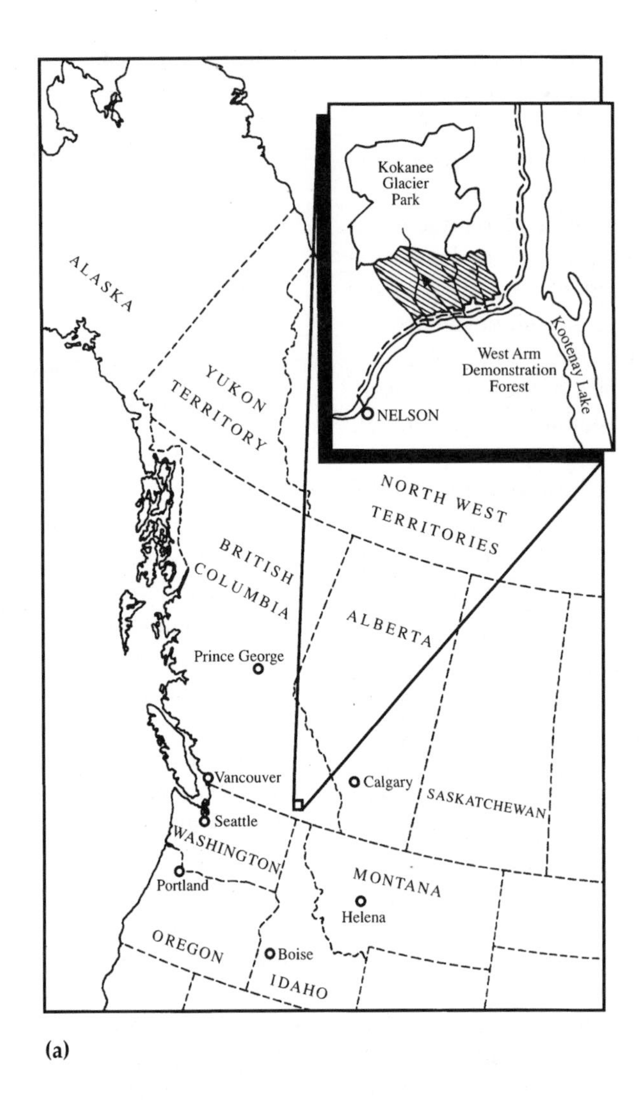

(a)

(b)

Figure 17.2 (a) Location of the West Arm Demonstration Forest. (b) A view of the section of the West Arm Demonstration Forest used in the case study.

drology, mass-wasting hazard, old-growth areas, recreation, riparian zones, sediment-yield hazard, silviculture, soils, timber and operability, topography, possible road locations, visual-quality objectives, and wildlife habitat.

Following this, the group prepared the landscape character analysis (Figure 17.3a and b) and constraints and opportunities analysis (Figure 17.3c and d) in plan and perspective forms. The perspectives were first manually transferred from the plan and then more accurately overlaid using computer-aided design (CAD) on a digital terrain model (DTM).

The ecological analysis worked through the process outlined above. A spreadsheet matrix of structures and flows was compiled which showed that a variety of structures was important for a majority of flows. This emphasized the importance of having each seral stage represented in the forest. It also showed that a number of seral stages are now largely absent from the forest as a result of regrowth over the last 90 years following large fires at the beginning of the 20th century.

One of the most significant aspects of the ecological analysis was an understanding of the main natural agents of disturbance and how they affected stand composition. Fire proved to be the main agent over most of the forest, although pathogens, wind, and snow also contributed. However, fire was so important that a spreadsheet was developed that listed biogeoclimatic sections along one axis and fire characteristics (periodicity, size range, type of burn) along the other. This naturally developed into the forest pattern objectives (Table 17.1).

The landscape was then divided into zones representing each stand type from Table 17.1. These zones were digitized so that they could be draped over the DTM (Figure 17.4). Following this, detailed stand structural objectives were prepared for each of the zones to determine how to maintain ecological functioning by choosing the appropriate intervention or silviculture, by harvesting to mimic natural disturbances, and by determining sustainable amounts of harvested timber. A typical stand structural objective description is shown in Table 17.2.

The next stage of sketch design was undertaken using the photos and three-dimensional CAD printouts. Units were designed for each of the ecological zones, taking account of landform and natural patterns in shape and scale and using the constraints and opportunities to ensure their practicality of the unit delineations (see Figure 17.5). The CAD system allowed the units to be translated from perspectives into plans, and each unit was assigned relevant stand structural objectives. This enabled the design to be evaluated over time using both perspective sketches and computer analyses—GIS and ATLAS (a timber-flow modeling program). Adherence to various operational guidelines, such as harvesting proximity rules and equivalent clearcut areas, could then be guaranteed.

There is considerable interest in forest planning, especially for logging and road construction, by members of the public who can be encouraged to participate in the planning and design process. This helps take social values into account when setting objectives and rating their relative importance. Design presented in 3-D using photographs and sketch perspectives can help to communicate management proposals in a much more meaningful and realistic way than can maps. Such sketches should be as accurate and lifelike as possible. During the WADF workshop session, two interested members of the local community joined in and helped the professionals determine whether their ideas would likely command public support.

A plan like the WADF design will undoubtedly become an important part of sustainable forest management. Without the strategic intentions and long-term vision contained in the design and fitted to the real landscape, managers have very little chance of knowing how their work on the ground fits into the bigger picture in space and time. The design, while quite detailed in many respects, remains highly flexible.

One of the criticisms leveled at designers producing 100-year plans is that policies and attitudes change frequently and plans will require constant reworking. A response to this argument is that, while policies and attitudes do change, together with the relative values placed on different resources, landforms, soils, and the vegetation potential of forest sites are unlikely to change over that period. If we base our design on the landscape, then the design elements will relate to the most constant aspects. Fu-

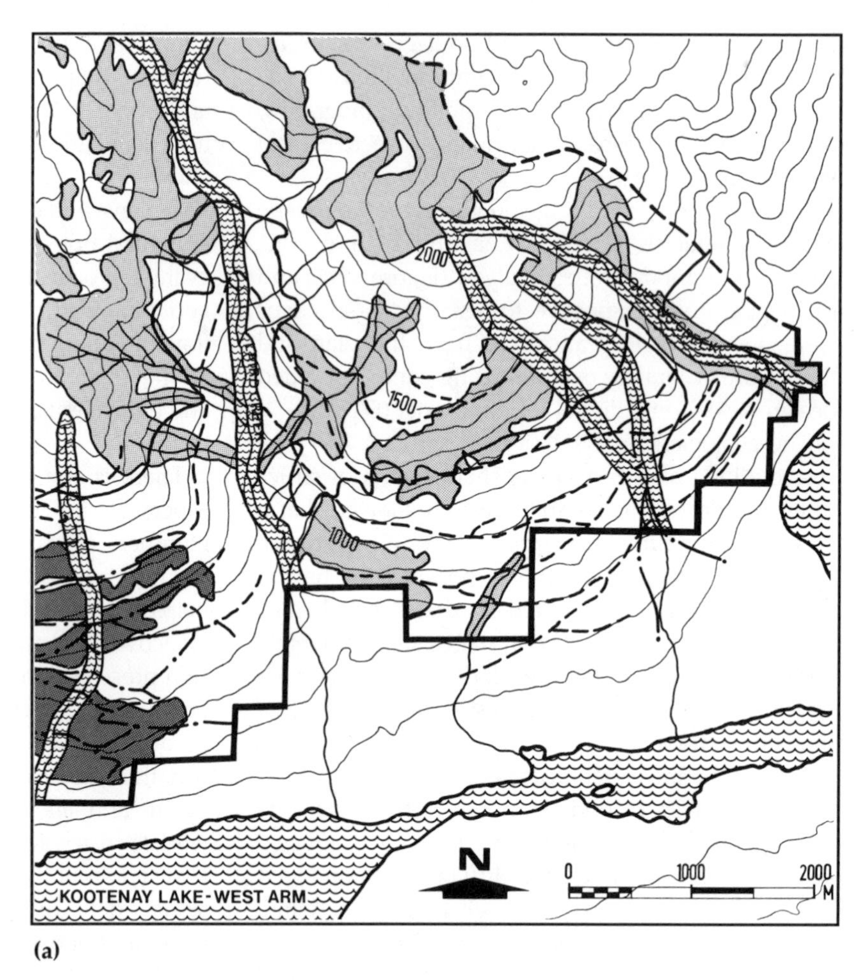

(a)

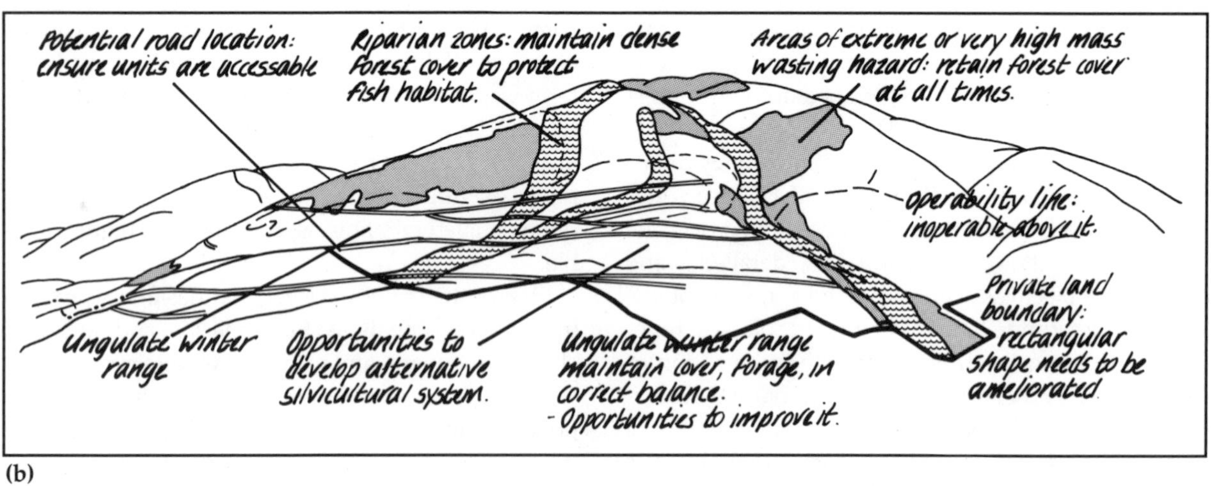

(b)

Figure 17.3 (a) The landscape character analysis for the West Arm Demonstration Forest in plan view, showing landforms, with a description of major features. (b) The perspective view of the landscape character analysis. (*Figure continues.*)

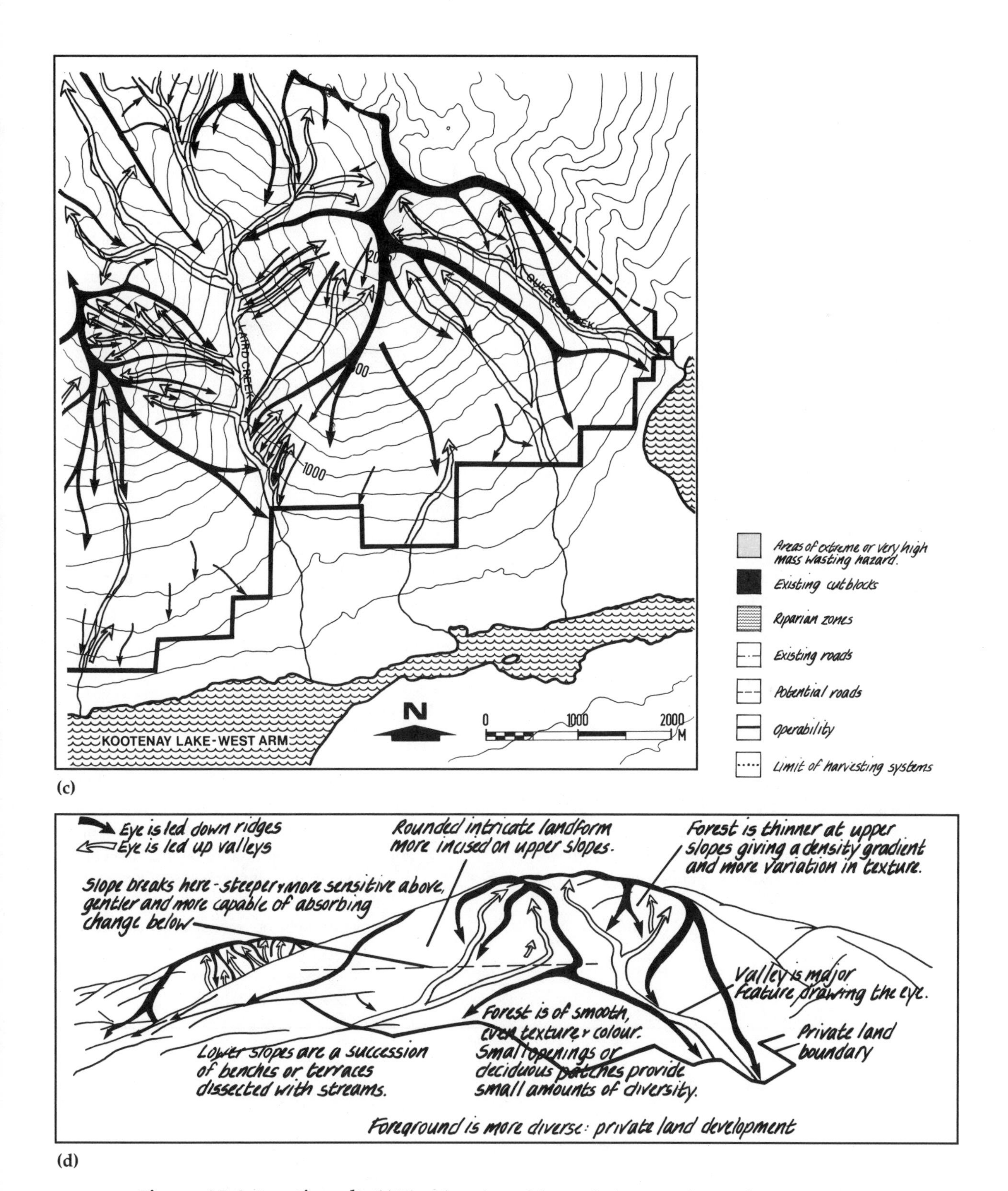

Figure 17.3 (*continued*) (c) The plan view of the analysis constraints and opportunities. (d) The constraints and opportunities in perspective.

Table 17.1 Ecological pattern objectives

Zone				
Alpine	Krummholz and parkland fading to Tundra; no intervention: climatically controlled dynamics.			
Engelmann spruce, sub-alpine fir	**Inoperable** ···			··· **Operable**
	100- to 500-ha natural burns in higher, more remote locations: control or let burn? (En)		Approximately 50-ha openings in lower sections; opening with residual patches, irregular shapes, larger on N&E slopes vs. S&W slopes; 150-year rotation. (E50)	
Interior cedar, hemlock (moist, warm)	**Low fire hazard** ···			··· **High fire hazard**
	Valley bottoms old-growth areas. Low or no intervention (Mwn)	Mature forest, root rot, pockets, insects. Moist N&E slopes. Group selection (Mwg)	Zone on Gradient of NW & SW slopes. 20–40 ha openings with trees left; islands or single (Mwi)	SW ridges similar to ESSF burn patches (50 ha) with residual areas. (Mwb)
Interior cedar, hemlock (dry, warm)	**Moister** ···			··· **Drier**
	Fairly open stands, mature with root-rot pockets; runs into drainage with west and east aspects; groups selection (very patchy). (Dwm)		South and west faces and coarse soils; open stands with short return period; small fires and large trees surviving above; two-story group selection with denser patches left in moist pockets; occasional larger clearances with large trees left. (Dwd)	

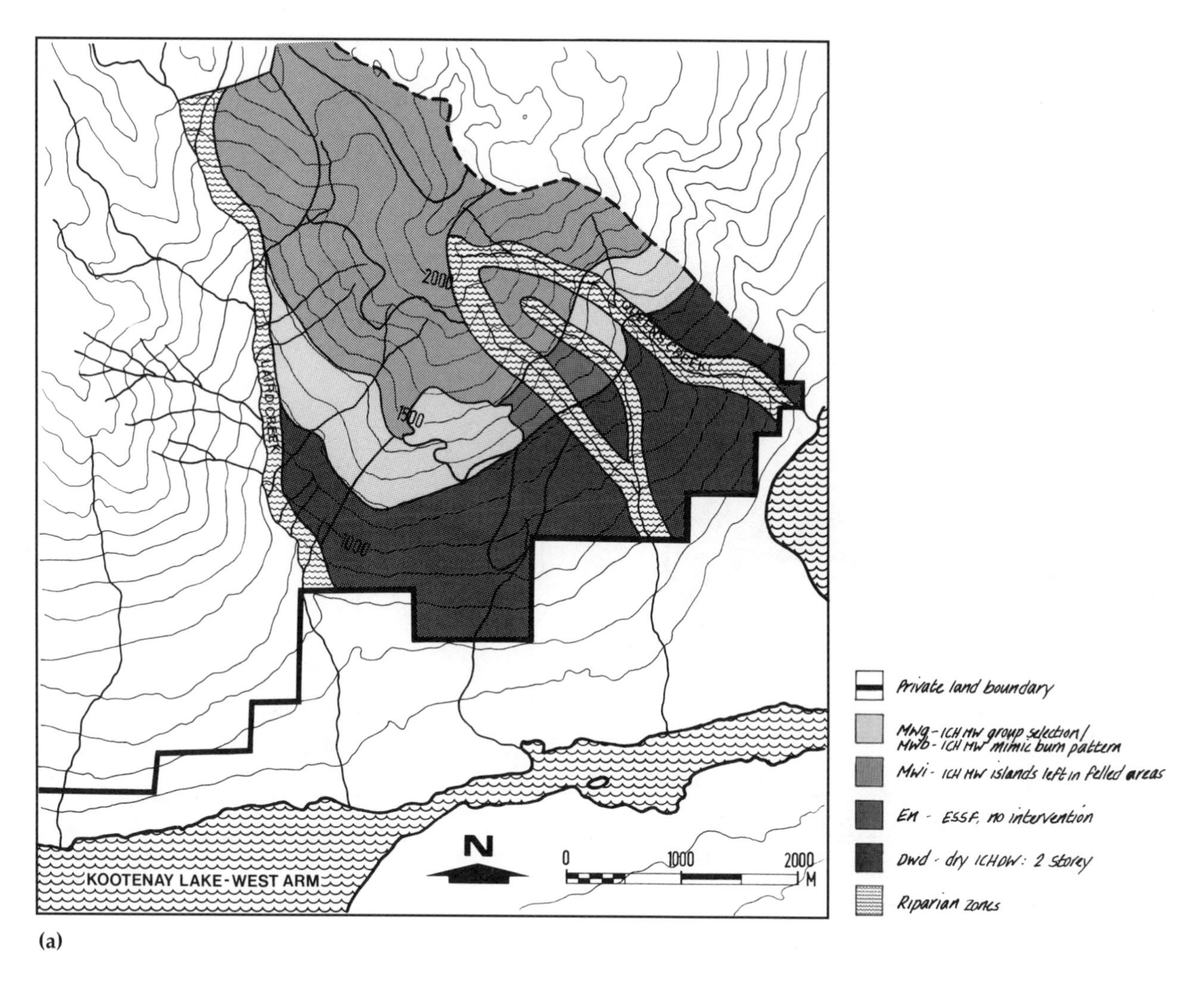

(a)

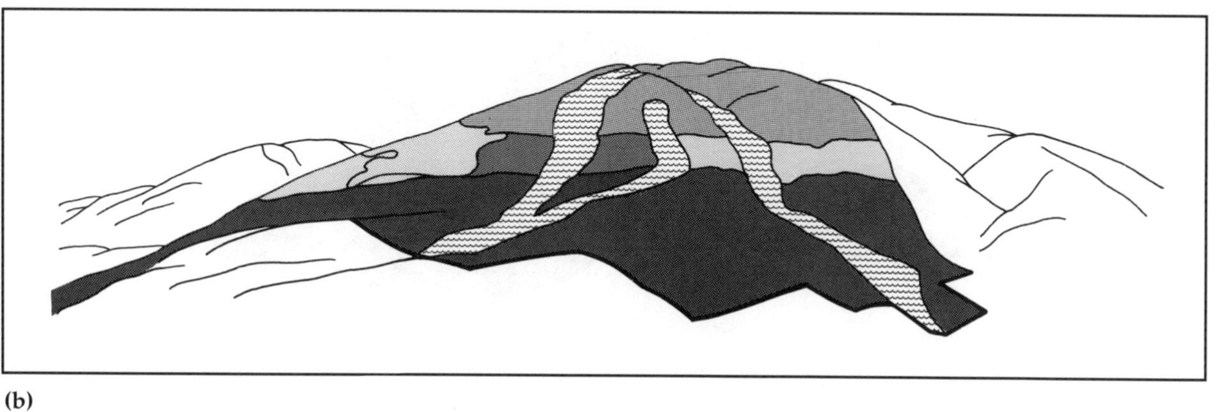

(b)

Figure 17.4 (a) The ecological units in plan view. The codes refer to the stand structural objectives in Table 17.1. (b) The ecological units seen in perspective. Note how they are layered following the main biogeoclimatic types.

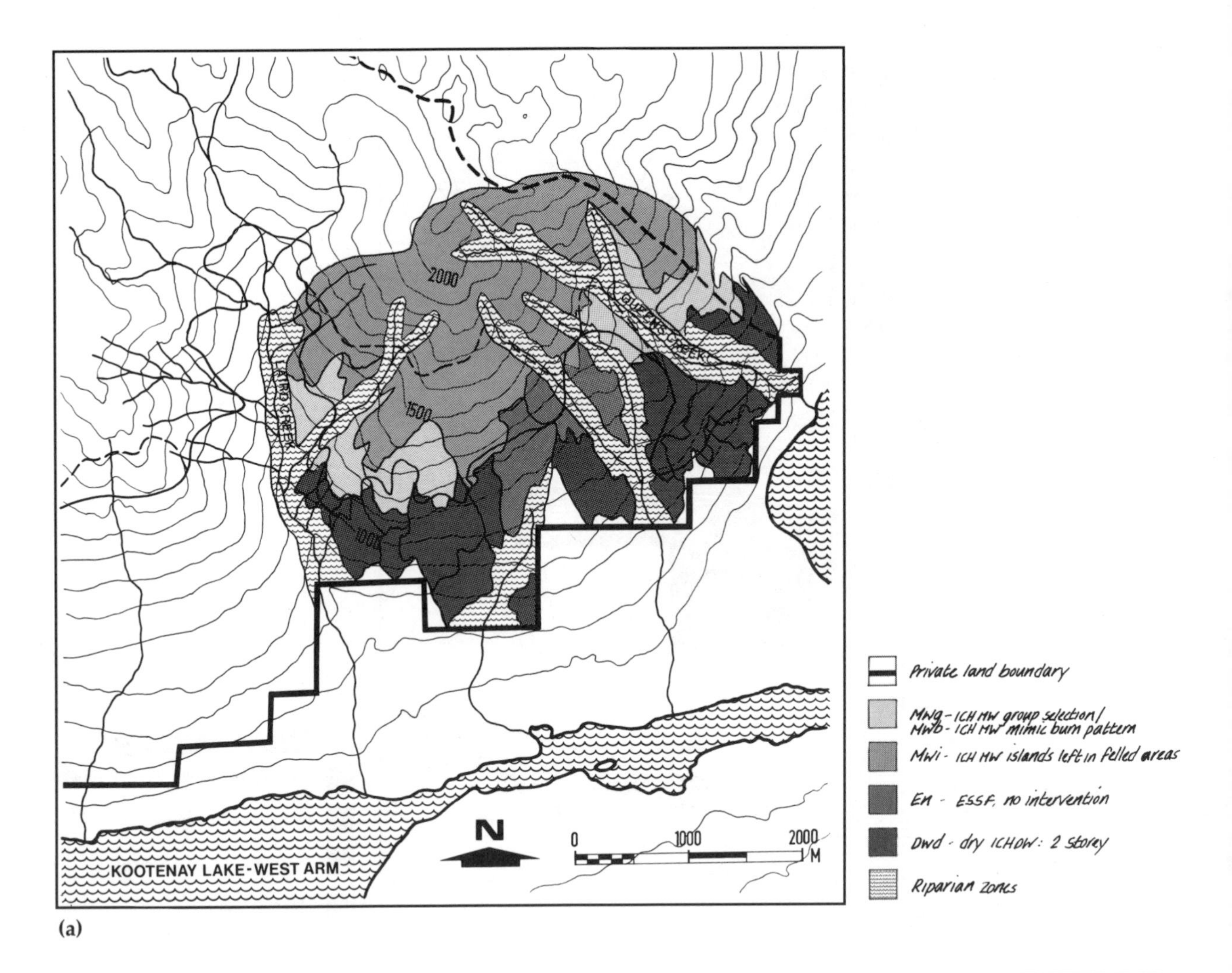

(a)

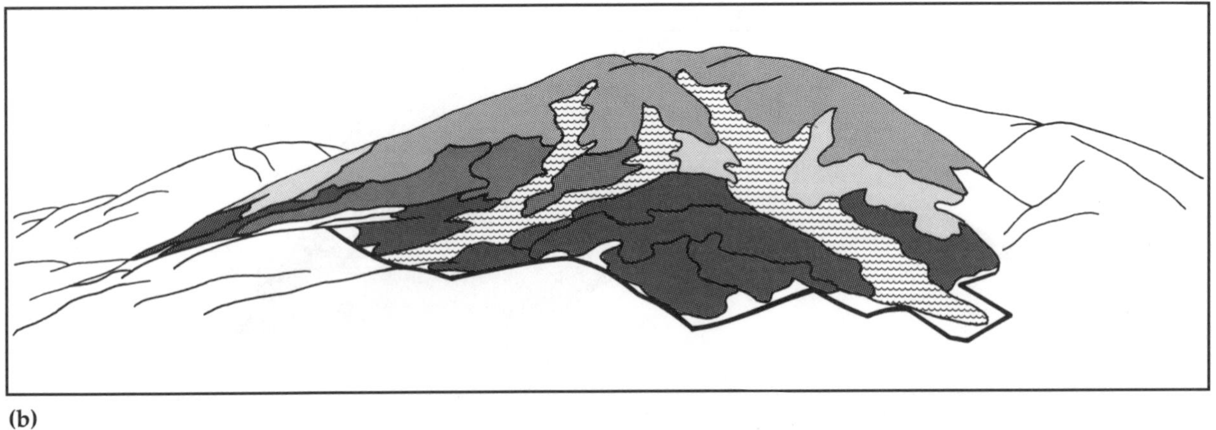

(b)

Figure 17.5 (a) Plan view of the designed units coded to refer to the appropriate stand structural objectives. (b) The perspective view of the designed units. Note how the shapes are irregular, interlocked and borrowed from the types of shapes found in the landscape. They vary in scale, being larger higher up and progressively smaller lower down.

Table 17.2 Typical stand structural objective description

ICH dw = interior cedar and hemlock; dry, warm zone
Dwd = dry, warm zone; drier site

Typical Characteristics

- Found on south and west aspects
- Trees tend to be widely spaced; predominantly yellow pine, Douglas-fir, and western larch; numerous large veterans present (escaped frequent fires)
- Multi-storied in structure with some denser mature patches on wet sites or fire exclusions
- Heavy disease and insect attacks result in hardwood patches
- Surface fires every 30 to 50 years resulting in small lower-story clearings

Type of Silviculture

- Units of 20 to 50 ha managed as two-story group selection retaining older veterans and recruiting younger trees as replacements; intervention every 30 to 50 years, over four to five passes, taking 15 to 20 percent of the unit area per pass; natural regeneration

ture managers may have other—and better—ideas. Our short-term actions should not reduce their options. Preparing long-term designs will set short-term actions in a safe context and at a level of intensity that does not diminish the future capacity of the site to yield all the possible benefits for future generations to enjoy.

Literature Cited

Bell, S. 1993. *Elements of visual design in the landscape.* London: L. E. and F. N. Spon Press.

British Columbia Ministry of Forests. 1994. *Visual landscape design training manual.* Victoria, BC, Canada: Ministry of Forests.

British Forestry Commission. 1989. *Forest landscape design guidelines.* Edinburgh: British Forestry Commission.

Diaz, N., and D. Apostol. 1992. *Forest landscape analysis and design: A process for developing and implementing land management objectives for landscape patterns.* R6 ECO-TP-043-92. Portland, OR: USDA Forest Service, Pacific Northwest Region.

Forman, R. T. T., and Godrun, M. 1986. *Landscape ecology.* New York: John Wiley & Sons.

Lucas, O. W. R. 1991. *The Design of forest landscapes.* Oxford: Oxford University Press.

USDA Forest Service. 1993. *Measuring sustainable forest management in the United States.* International forestry brief issue no. 5. Washington, DC: USDA Forest Service.

18 Implementing Spatial Planning in Watersheds

John Sessions, Gordon Reeves, K. Norman Johnson, and Kelly Burnett

Watershed Assessments 272
Problem Description 272
Generation of Alternatives 273
Spatial and Temporal Relationships 273
Linking Goals and Activities 273
Analysis 275
Conclusion 279
Literature Cited 279

More than ever before, forest managers and planners are faced with demands for a diverse array of products and services—including timber and other forest products, wildlife habitat, recreational opportunities, and the protection of ecological processes and aesthetic resources. Moreover, many of their activities are constrained by legal and administrative mandates. Planners and managers therefore face immense challenges attempting to meet multiple needs spatially and temporally.

Increasingly, watersheds are recognized as an important spatial unit for planning. The Scientific Agency Team (SAT)(Thomas et al. 1993), the Forest Ecosystem Management Assessment Team (FEMAT 1993), and the State of Washington Forest Practice Rules (Washington DNR 1993) refer to the need for planning at a watershed level. The spatial unit described by a watershed captures many of the attributes needed to define ecosystem functions. For those attributes expressed at a larger scale, watersheds are appropriate building blocks that can be aggregated. In addition, the plan for a watershed is an important bridge between broad strategic plans and site-specific project plans.

Spatial planning in watersheds requires three elements: (1) an assessment of the current situation, (2) a statement of goals for the watershed linked to forest-wide objectives, and (3) a means of identifying spatially feasible alternatives.

Decision models can be used to create alternatives for a broad range of management activities, including the timing and location of silvicultural treatments, the choice of harvest methods, the siting of roads, and the initiation of watershed restoration projects. Watershed restoration projects include in-channel work, road and culvert reconstruction or relocation, and riparian restoration through silviculture.

Watershed Assessments

Assessments are an integral component of watershed planning in that they provide a systematic procedure for characterizing a watershed and its associated ecological characteristics. Specifically, the objectives of an assessment are to (1) define the existing condition of a watershed, (2) define the structure and processes in a watershed and their linkage to management activities, and (3) stratify land by environmental effect. Assessments are an important part of the watershed analysis process envisioned by SAT, FEMAT, and the Timber, Fish, and Wildlife (TFW) program of Washington State. The National Council of the Paper Industry for Air and Stream Improvement (NCASI) also is developing watershed assessment procedures for use on private lands.

Assessments provide information on active processes in watersheds (e.g., landslides, debris floods, etc.). They also include current conditions of uplands and riparian areas, how these factors influence riparian and in-channel habitat and other attributes of the aquatic ecosystems, and the distribution of aquatic and riparian-dependent organisms. The information gathered at this initial stage is used to help design management alternatives, to meet objectives compatible with watershed and ecosystem function, and to guide site-level planning.

Watershed planning is based on information collected during the watershed assessment. It operates at the interface between management actions and ecosystems. Planning combines an understanding of the existing ecological state of the watershed with the means to project, control, and manipulate vegetation over time and space to attain management objectives.

In this chapter, we discuss the mechanics of spatial and temporal planning in watersheds, beginning with a typical planning problem faced by many public land managers in the Pacific Northwest. We illustrate many of our ideas through use of a decision model called SNAP (the Scheduling and Network Analysis Program).

Problem Description

On public lands, strategic plans have been developed to establish broad direction for management. The suitable timber base and a ceiling on sustainable harvest levels usually are established as part of this direction. While strategic plans often have a life expectancy of 10 years, 150- to 200-year planning horizons often are employed to ensure that harvest levels are sustainable and to provide opportunities to review future effects of current management actions.

In the Pacific Northwest, strategic plans, such as Option 9 of the FEMAT report, contain limited spatial definition due to their broad scope. Many spatial details—for example, the location, timing, type, and extent of harvesting—are left to be resolved during implementation of the plans. Resource planners therefore must determine whether it is possible to create an activities plan (with harvest schedules, road programs, restoration projects) for a subarea of the forest that achieves the desired goals in the overall strategic plan. Goals may include output levels of timber and other forest products and protection for wildlife, fisheries, soils, and other ecological concerns.

Operability is an important question to be addressed during watershed planning. The rules for stream crossings by roads, the location of roads in or near riparian zones, buffer widths along streams, the location of cable ways across riparian zones, and the sizes and distribution of harvest units must be considered when determining the operability of an activity plan.

Activity plans for subareas of the forest usually focus on the short term and are referred to as tactical plans. The size of the tactical problem varies both in area and potential treatment units. Various authors have suggested subareas between 20 sq. mi. and 200 sq. mi. with 500 to 5,000 potential treatment units to be scheduled over 10 to 20 time periods (FEMAT 1993).

The basic spatial unit, referred to as a polygon in this chapter, is the building block for activity planning for watershed plans. Its definition depends on the issues involved. It is the smallest land area to be treated that is considered homogeneous for the purpose of simulating environmental and economic effects. Its definition can be based upon ecological, physical, or economic factors—or a combination of factors. If harvesting is an option, the polygon's boundaries often conform to a unit that can logically be harvested. Groups of treatment units can be scheduled simultaneously (as will be discussed later), but for the purpose of analysis, a treatment unit cannot be subdivided.

Since harvest systems, road systems, and restoration work can affect stream habitat and other resources, it is necessary to consider transportation systems and restoration work along with harvest schedules. Restoration work may include road reconstruction, road removal, placing of coarse woody debris in streams, and riparian silviculture. A typical transportation system consists of 1,000 to 3,000 existing and potential road segments.

Generation of Alternatives

Identifying feasible alternatives is at the core of watershed planning. A feasible alternative is an activity schedule that achieves the goals of the plan. Often, there is more than one way to meet goals—sometimes there is no way. Identification of feasible alternatives provides two important functions: First, it provides planners with options that can be weighted with factors that are difficult to quantify. Second, it can help identify the most efficient way to meet goals.

Spatial and Temporal Relationships

A specific requirement of tactical plans is that they contain sufficient spatial and temporal detail to define when and where the polygons will be treated relative to key ecological, social, and economic variables. Examples of ecological variables include stream flow (peak and low), stream sediment and temperature, coarse woody debris, and wildlife habitat structure and distribution. Social and economic variables include visual quality, timber harvest levels, timber revenue projections, and harvest, road, and restoration costs. The tactical plan must provide the detail to show:

1. where to locate the harvest units with respect to one another,
2. when the harvest units will be treated relative to each other,
3. where the units are located with respect to streams and the transportation system, and
4. when they will be harvested relative to road construction and reconstruction.

Linking Goals and Activities

One of the most important aspects of implementing spatial planning in watersheds is relating ecological goals to management activities (Table 18.1). Management activities to reach ecological goals may require:

1. control of the maximum or minimum treatment area,
2. control of re-entry periods,
3. control of rate of harvest and road construction near streams, and
4. mitigation projects such as species conversion or road reconstruction or removal.

In the following examples we illustrate several relationships between ecological and social goals and the management activities that could be scheduled during watershed planning to reach them.

Table 18.1 Relationship between goals and activities in watershed planning

Activities	Road Location/Std.	Opening Size	Silvicultural Treatment	Mitigation/ Restorer	Buffer Width
Reduce stream sedimentation	X	X	X	X	X
Reduce stream temperature			X		X
Increase coarse woody debris			X	X	X
Increase wildlife habitat	X	X	X	X	X
Maintain or increase visual quality	X	X	X		
Reflect natural disturbance		X	X		
Produce timber harvest	X	X	X	X	X
Produce income	X	X	X	X	X
Produce employment	X	X	X	X	X

Increase Fish Habitat

Fish habitat quality is a function of stream flow, fine sediment, temperature, coarse woody debris, pool frequency, substrate characteristics, and bank configuration.

Timing, Magnitude, and Duration of Stream Flow

Peak flow may be affected by different methods of logging, seral-stage distribution, and the location and density of a road system. Within a watershed plan, management activities to control peak flow might include: (1) limiting the types of logging systems that can be used, (2) specifying the maximum number of acres that can be in critical seral stages, (3) defining the maximum patch size (especially in the rain-on-snow zone), (4) identifying how many miles of roads can exist on lands with certain physical attributes (such as midslope and ridge top), and (5) controlling the size of disturbances. Low flows, to the extent that they are connected with these variables, can be controlled in a similar way.

Decrease Stream Sediment

Logging and harvest methods, as well as road systems, can affect stream sediment. As such, sediment can be managed by controlling factors such as the seral-stage distribution on lands with critical attributes (e.g., steep, erodible soils), logging methods, road density, road standards, and stream buffers. Considerations also should be given to the placement of activities relative to fish-bearing channels—recognizing the debris-scouring relationship between entry angles of lower-order streams. If the linkage between roads, harvest units, and stream sediment is well defined, stream sediment reduction goals might explicitly include tradeoffs between sediment delivery and mitigation actions such as road improvements or removals.

Decrease Stream Temperature

Stream temperature can be affected by the width of buffer strips, activities permitted within buffer strips, and the species and seral-stage distribution of trees along streams. Information on tree height and shade density are linked to individual species and the seral stage of vegetation.

Increase Coarse Woody Debris

The same types of planning choices that affect stream temperature affect coarse woody debris. In addition, adding woody debris to streams might be used as a mitigation measure. Likewise, long-term restoration strategies might include species conversion.

Increase Wildlife Habitat

Wildlife habitat can be affected by seral-stage quantity and distribution as well as road density. Hence, increasing wildlife habitat may involve (1) limiting the number of acres in certain seral stages in a given time period on land areas with special attributes, (2) limiting the size of treatment areas, (3) controlling the distribution of seral stages, and (4) controlling road densities. The definition of seral stages plays an important role in managing wildlife habitat—for example, it can be defined by age, species, or the structure of both the upper and lower canopy, including snags. Maintenance of landscape connections also can affect wildlife habitat.

Maintain or Improve Visual Quality

Visual quality can be affected by the same factors as wildlife habitat. Management activities include (1) stratification of the watershed into visual zones that specify the maximum number of acres which can be in each seral stage in a time period, (2) control of opening size and green-up period, and (3) control of acceptable road locations.

Reflect Natural Disturbance

To reflect a natural disturbance pattern, planners may want to require that a group of polygons forming some type of ecological unit be treated simultaneously—although not necessarily with the same treatment. Furthermore, they may not want adjacent ecological units to be treated in the same time period. Grouping all polygons in a subwatershed in order to concentrate treatment activities while other areas recover can mimic natural asynchronous disturbance patterns of subwatersheds throughout a watershed. This would involve scheduling between ecological units, scheduling of treatments within ecological units, and development of transportation systems within and between ecological units.

Analysis

There are two key pieces of technology that forest planners need at their disposal to develop management alternatives that incorporate the spatial and temporal relationships outlined in the previous section: (1) geographic information systems (GISs) and (2) spatial- and temporal-based decision models. The following example uses the Scheduling and Network Analysis Program (SNAP)(Sessions and Sessions 1992), a decision model being used in a number of locations in the western United States.

The project area is Snap Creek (Figure 18.1a), which is composed of 175 polygons, each representing a unique seral stage and each subject to three possible types of initial silvicultural treatments. Each of the polygons or land parcels is also a unique combination of attributes, such as slope class (high, medium, low), visual class (foreground, background), and risk class (high and low risk soils). Other key information includes a potential road network of 214 road segments that might be constructed, reconstructed, or left unbuilt (Figure 18.1b) and the stream network (Figure 18.1c) identified by 100 individual stream reaches and their attributes.

Information used to construct these types of planning problems can be extracted from a GIS. The management unit and stream attributes, for example, can be retrieved from different layers in a GIS. The Department of Natural Resources in the state of Washington has constructed linkages between a GIS program called ARCINFO and SNAP to facilitate data transfer between the two systems.

Among other things, planners may want to know if it is possible to reach a 4.5-million-board-foot (4,500-MBF) timber harvest goal from Snap Creek during the next four time periods, given other goals. Their initial set of ecological requirements for the area includes: (1) a late-successional seral-stage wildlife corridor must be maintained on the area through time, (2) at least 30 percent of Snap Creek must be in a mature or late-successional seral stage at any given point in time, (3) created openings in the foreground and background areas cannot exceed 10 acres and 40 acres in size, respectively, (4) a maximum of 25 percent of the area can be in a treated or disturbed state at any given time, (5) regeneration harvesting cannot occur on high-risk soils, and (6) harvest buffer zones of 150-foot slope distance must be maintained along all perennial streams.

With these objectives and requirements in hand, the model is run to determine if harvesting can take place, where it can occur, when it will occur, what type of silvicultural treatments are appropriate, and

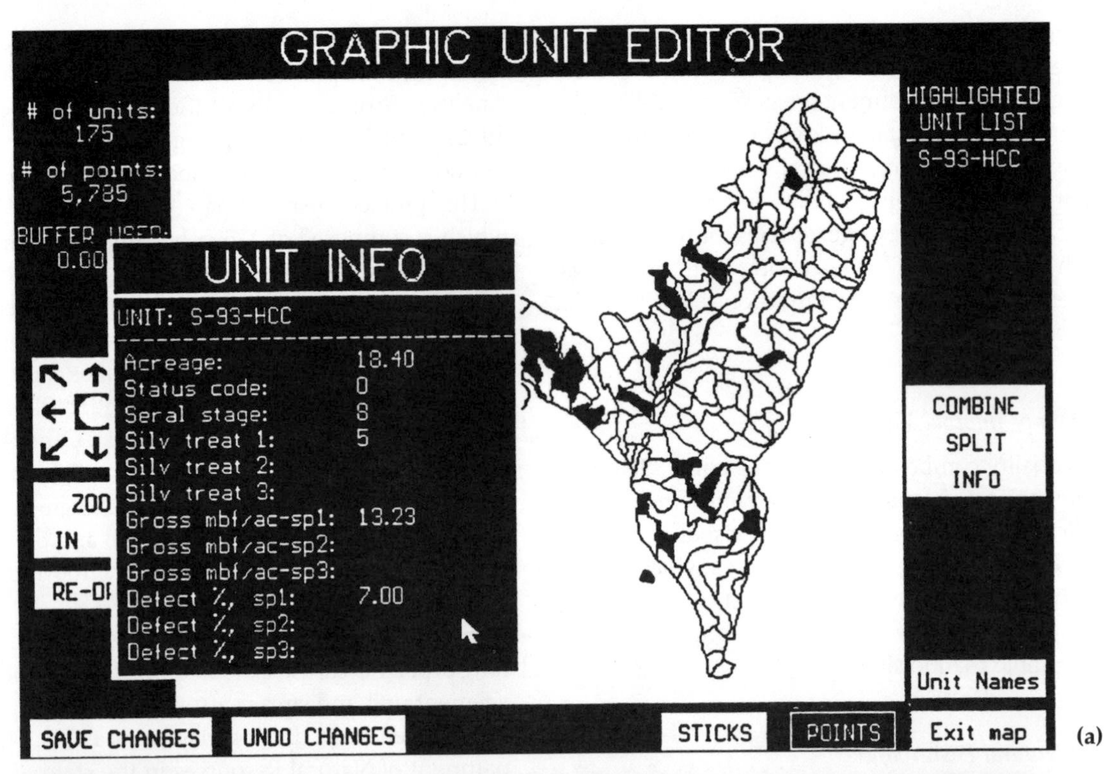

(a)

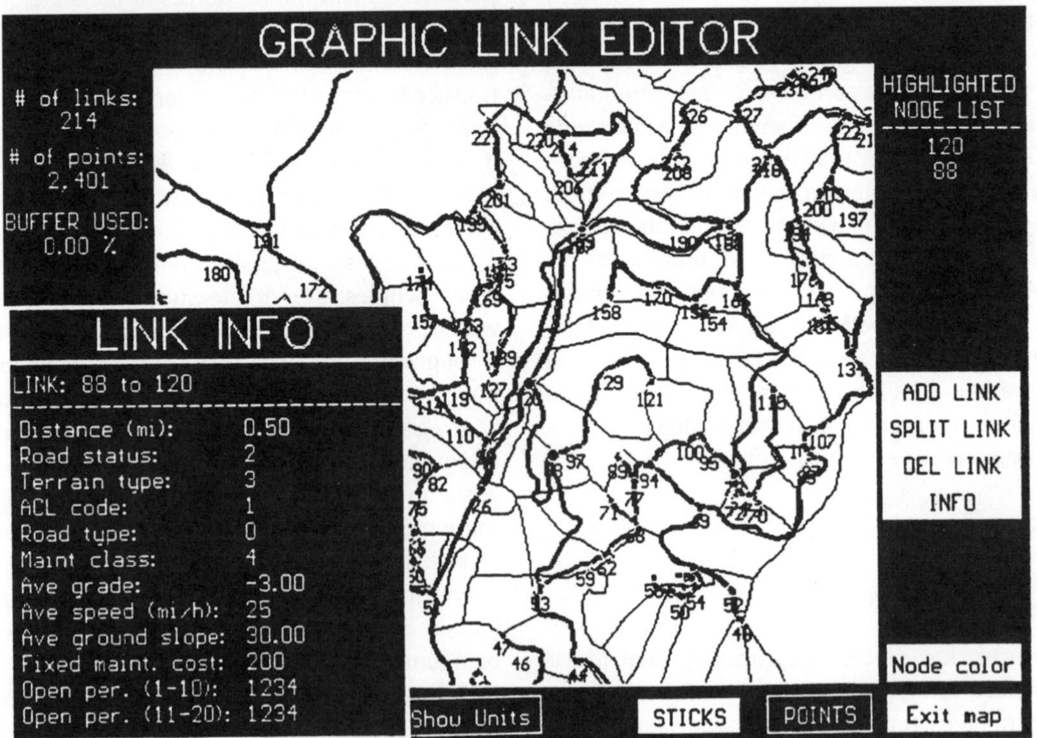

(b)

Figure 18.1 (a) Snap Creek Project Area potential harvest units, (b) Snap Creek Project Area potential road network, (c) Snap Creek Project stream network.

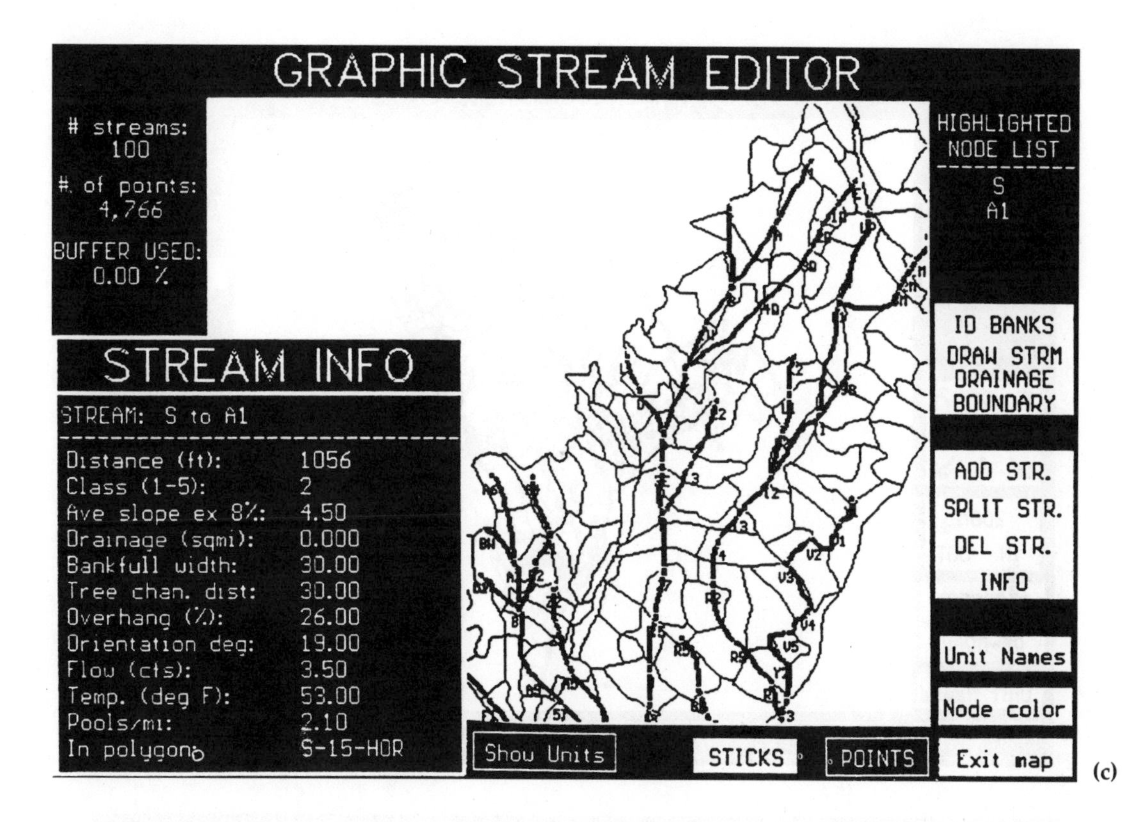

(c)

the set of associated logging systems and road activities that are needed. Five feasible harvest patterns are generated, each producing close to 4,500 MBF in each of the first three time periods. In the fourth time period, it is not possible to produce 4,500 MBF due to cumulative spatial constraints, even though a substantial volume of mature timber remains. Taking a look at one particular pattern, we can see the harvest units and transportation system selected (Figure 18.2a), the silvicultural treatments employed (Figure 18.2b), and the location of the wildlife corridor (Figure 18.2c).

At this point, resource planners can take a solution and export it back out to GIS for further analysis. We have found, however, that before planners export solutions to a GIS environment, they often want to perform further analysis directly in the decision model software—that is, they want to be able to recycle solutions quickly without switching analysis environments.

Post-analyses can provide valuable information to wildlife biologists, landscape architects, fisheries biologists, and others. For example, the area and effectiveness of acceptable big game cover at various distances from forage can be calculated for any solution and time period using techniques similar to those described by Wisdom et al. (1986). Similarly, the area of effectiveness of forage at various distances from acceptable cover can be reported. The sensitivity of habitat effectiveness to road traffic can also be tested through specification of road closure policies.

To assess visual objectives, the planning area can be viewed in perspective from user-defined points (inside or outside the planning area) after each time period for any solution identified by SNAP.

Post-analysis evaluations also can be done on aquatic resources using the stream temperature model based on techniques developed by Park (1993). In addition to looking at temperature loadings, fishery biologists can evaluate the proximity of harvest activity to individual stream reaches. If the biologists feel, for example, that particular harvest units are too close to a stream segment rich in aquatic structure, they can make these land parcels off-limits to harvesting and quickly generate new solutions that take these additional requirements into account.

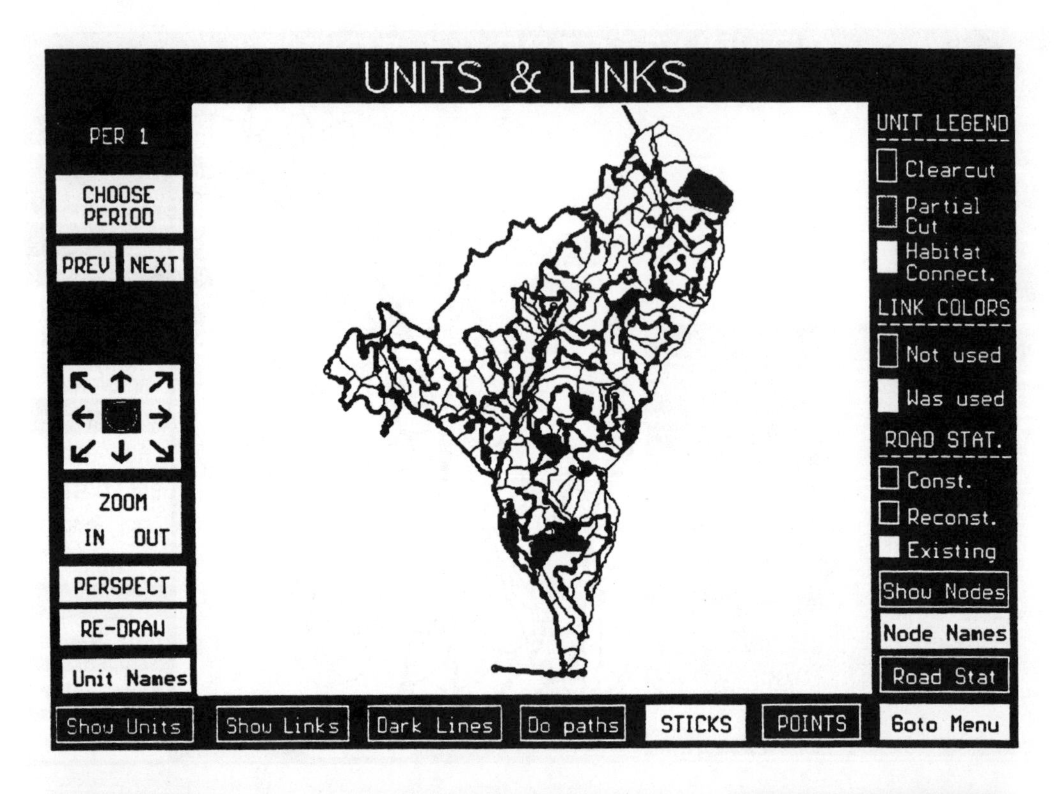

(a)

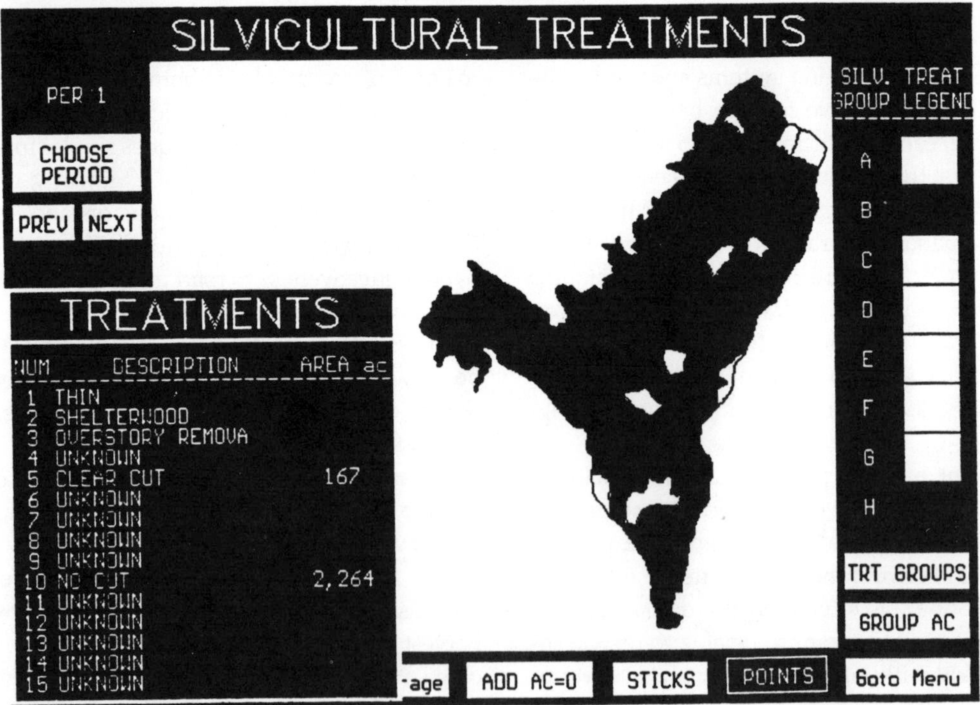

(b)

Figure 18.2 (a) Solution showing selected harvest units and road links, (b) Solution showing selected silvicultural treatments, (c) Solution showing the selected wildlife corridor during time period 4.

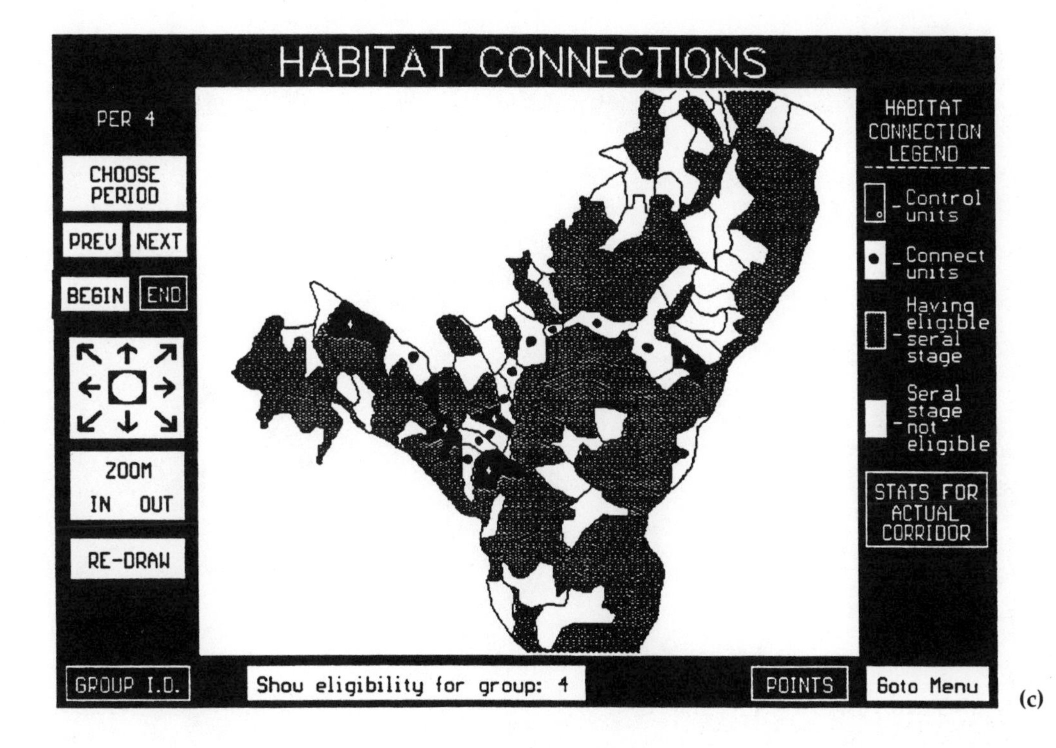

(c)

Conclusion

As the above discussion demonstrates, decision models can be used to assist resource specialists in exploring alternative ways to reach ecological and social goals. A key to effectively using decision models is understanding how to relate ecological goals to management actions. If ecological concerns are to be considered simultaneously with social concerns, defining these relationships is essential. Once this is done, a decision model becomes a valuable way to develop spatially and temporally explicit management alternatives that capture the complex, dynamic nature of the forests and their treatment. As our understanding of the linkage between watershed processes and management activities grows, decision models will become even more useful.

Literature Cited

FEMAT. 1993. *Forest ecosystem management: An ecological, economic and social assessment.* Report of the Forest Ecosystem Management Assessment Team (FEMAT). 1993-793-071. Washington, DC: GPO.

Park, C. 1993. *SHADOW: Stream temperature management program.* Portland, OR: USDA Forest Service.

Sessions, J., and J. B. Sessions. 1992. Tactical forest planning using SNAP 2.03. In *Proceedings: Computer-Supported Planning of Roads and Harvesting Workshop, August 26–28.* Munich, Germany: International Union of Forest Research Organizations.

Thomas, J. W., M. G. Raphael, R. G. Anthony, E. D. Forsman, A. G. Gunderson, R. S. Holthausen, B. G. Marcot, G. H. Reeves, J. R. Sedell, and D. M. Solis. 1993. *Viability assessments and management considerations for species associated with late successional and old-growth forests of the Pacific Northwest.* Washington, DC: GPO.

Washington DNR. 1993. *Washington Forest Practices Act.* Chapter 222-22 WAC. Olympia, WA: Department of Natural Resources, Forest Practices Division.

Wisdom, M. J., L. Bright, C. Caret, W. Hines, R. Pedersen, D. Smithey, J. Thomas, and G. Witmer. 1986. *A model to evaluate elk habitat in western Oregon.* Portland, OR: USDA Forest Service, Region Six.

Section

IV Forest Economics: Products and Policies

Traditionally, forest economists have focused on a relatively narrow set of questions pertaining to the production, sale, and distribution of wood fiber and wood products. While timber economics remains important, it is fast becoming only one part of a much larger picture. With the emergence of ecosystem management, the field of forest economics has expanded dramatically. New economic concerns range from developing system frameworks to modeling complex economic-ecologic interactions to encouraging economic development for natural resource–dependent communities to redefining forest product markets. The chapters in this section touch upon some of economic issues that have emerged in relation to ecosystem management.

Reinventing the Wood Products Industry

One of the stark realities of ecosystem management and the new concepts of forestry is that forests (at least those in public ownership) are no longer simply sources of commodities. The operation of the forest products industry always has been based on the amount of raw material available. What was available was determined largely by natural processes. Increasingly, however, availability will be determined by social and political processes as much as by nature. To remain viable in the 21st century, the forest products industry is faced with first realizing, then accepting, and finally adjusting to this new reality.

In the short term, questions center on harvesting and administrative costs relative to new silvicultural strategies such as retention harvesting. Clearly, the shift to ecosystem management raises the costs of forest operations and the competitive position of some forests (Haynes and Weigand, Chapter 19). Determining what those costs are and mitigating associated low stumpage values is essential to the continued development of new silvicultural practices.

Long-term issues are of a more fundamental nature. Ecosystem management raises some basic questions about how we define timber commodities and timber values. For example, Haynes and Weigand (Chapter 19) point out that old-growth forests might be managed as a renewable resource—rather than something to be depleted to make room for more regulated stands. This would require us to think in terms of rotations of several centuries and to forego current supplies so that future generations have sustainable supplies. Providing the diverse structures and habitats associated with old growth, as well as a sustainable supply of high-quality saw timber, could prove to be an important ecological and economic niche for public forests. Plantations grown on shorter rotations for pulp and smaller logs would be the domain of private industry.

From a manufacturing perspective, ecosystem management challenges the wood products industry to get more value out of less timber (Whittenbury, Chapter 20). The increasing emphasis on protecting ecological values and services translates into reduced timber harvests for commercial purposes and new incentives for the wood products industry. Whereas performance of the industry traditionally has been based on the volume of material produced, in the future it will be based on value. The solid two-by-four may become a thing of the past.

To cope with changes in the quantity and quality of available commercial timber, we need to move beyond the narrowly defined debate over allowable sale quantities (ASQs). The industry will have to explore new technologies in engineered wood products and added-value manufacturing, as well as alternative sources of fiber that is produced agriculturally. We must begin asking fundamental questions such as: What are wood products? What are their basic characteristics? Can they be provided through alternative means that still offer the environmental advantages of biomaterials but that exist in harmony with the forests of the 21st century (Whittenbury, Chapter 20)?

Special Forest Products

Another modern shift in production on public lands is the increasing recognition of a class of forest products known as special forest products. These products range from floral greens to foods to medicinal herbs. Although many are not new, managers and the public increasingly are recognizing that special forest products are important components of forest ecosystems and local economies (Molina et al., Chapter 21). In the Pacific Northwest alone, special forest products account for over $200 million in revenue.

Like timber, the issues related to managing special forest products as part of an overall ecosystem management strategy include developing guidelines for sustainable harvests and protection of ecosystem functions. But as we move into the 21st century, these products may also be an important component of diversifying formerly timber-dependent communities. To that end, we need to develop reliable inventory and monitoring programs, gather information on changing demands and market niches, and gain a better understanding of management costs and returns from investment (Molina et al., Chapter 21).

Economic Incentives for Ecosystem Management

Achieving ecosystem management objectives on public as well as private lands will depend upon creating economic incentives—and eliminating disincentives—for ecologically sensitive practices. There are at least three sources of such incentives. First, government programs have long been used to influence forest practices (Cubbage, Chapter 22). Regulations, tax incentives, and cost-sharing programs were first created to provide for regeneration of sites after harvest and to control fire, insects, and disease. Many have been expanded to mitigate negative environmental effects associated with harvesting and have the potential to incorporate the broader objectives of ecosystem management. For example, regulations might require large buffer zones around riparian or other sensitive areas, or technical assistance programs might provide private nonindustrial forest landowners with expertise to custom fit retention harvest systems to their land and ecological objectives.

In addition, disincentives for ecosystem management need to be identified and, where possible, changed. Property tax laws that tax standing timber and assess old-growth at a higher value than younger trees encourage premature harvesting. Similarly, taxing land at its "highest and best use" (particularly near urban areas) can create excessive tax burdens that encourage forest landowners to convert their property to more developed uses.

But it is also clear given the antiregulatory mood in government and the lack of resources to enforce prohibitive policies, that we must begin to explore a range of nontraditional forest policy tools to encourage ecologically sensitive management practices.

These might include purchasing or exchanging development rights and easements for timber harvesting or developing private landowner cooperative associations (see Cubbage, Chapter 22).

A second source of incentives for ecosystem management is the changing nature of forest product markets. Just as changes in timber availability will drive changes in the wood products manufacturing industry, manufacturing changes can influence forest management practices. In other words, we must build stronger links between the forest and forest products. If wood products of the future use far less wood fiber from forests than they do today, the pressure to produce large volumes may be replaced by incentives for high-quality material. Once relieved of pressures to "get the cut out," forest managers can focus on producing high-quality hardwoods, large old-growth saw timber, and other products that are more compatible with ecological objectives.

Finally, incentives for moving toward more ecologically based forest practices will come from consumers. Forest industry spokesmen often have decried environmentally conscious urbanites who want to build new wood homes, but condemn cutting down trees. But these attitudes may very well spur the development of innovative, new products that make use of veneers and core materials not made from trees. Growing consumer consciousness and green marketing are just beginning to have an impact on the way forests are managed. For example, industry associations have begun to design certification programs to mark products that are produced in an ecologically and socially responsible manner. The influence of green parties in European countries has produced a market that is demanding more attention to ecological values; changes in forest practices in Sweden, Finland, Chile, and Canada have been strongly influenced by the environmental emphasis of the European market.

Regardless of whether they are "carrots" rewarding sound stewardship or "sticks" excluding products produced at high environmental cost, market-based stimuli hold great promise for influencing forest practices. Certification programs are likely to be one important element. There is even the exciting prospect of thoroughly educating consumers regarding not just the sources of their products, but also ecological and social costs and benefits.

19 The Context for Forest Economics in the 21st Century

Richard W. Haynes and James F. Weigand

Challenges to Economic Analysis 286
Systems Frameworks for Economic Analysis of Ecosystem Management 286
The Challenges of Scale to Economic Analyses 287
The Changing Nature of Forest Products 290
Timber Supply and Demand: Traditional Forest Economics in a New Light 291
Valuing Nontraditional Outputs 294
Economic Implications of Old Growth as a Renewable Resource 295
Social Forestry 296
Strategies for Rural Development and Employment 296
Community Stability 297
Conclusion 298
Acknowledgments 299
Literature Cited 299

Forest stewardship on public lands in North America has become a balancing act to meet societal expectations for the human well-being derived from forests and the need for ecosystem health that sustains forest function to satisfy human needs. In this chapter, we discuss economic concepts that are central to achieving that balance as society expands the scope of forest ecosystem management. Forest management continues to be a societal response to meet human demands for goods, services, and desired ecosystem conditions. Traditional concepts in neoclassical economic theory, such as supply and demand, efficiency, and equity, remain vital in any discussion of ecosystem management. Economic analysis will continue to examine management options and focus on the flows and distribution patterns of costs and benefits to society over time. Analysis will ascertain the degree of congruence of these flows and distributions with social values and expectations for economic development.

Forest economics has traditionally concentrated on economic issues of timber production (see Duerr 1993). Indeed, the phrase "forest products" continues to mean timber products in common North American parlance. Increasingly, however, forest economists are exploring a spectrum of resources and public policy issues seldom addressed in North American society. Current forest economics research challenges traditional economic assumptions and methodologies that appear not to model current societal preferences and economic behavior adequately.

New economic concerns about forest ecosystems include, but are not limited to, valuation methods of nontraditional ecosystem products, discount rates that reflect societal preferences to preserve options for future generations, the human economic role in ecosystem sustainability, appropriate systems constructs to model complex economic-ecologic interactions by way of scenario planning, methods to weigh the benefits and costs of ecosystem restoration, redefinition of old growth as a sustainable resource, the significance of scale in economic analysis, and links between forest management and economic development for natural resource–dependent communities. Also, the range and magnitude of goods, services, and ecosystem states demanded are growing as knowledge of ecosystems and human population expands. Hence, economists find themselves challenged to be more inventive in their efforts to provide adequate models and explanations of economic interactions between people and forest ecosystems.

This chapter is divided into four broad sections. The first three deal with selected economic issues: specific challenges to economic analysis as they relate to ecosystem management, the changing nature of forest products, and the emerging discipline of social forestry. In the last section, we weave these issues together.

Challenges to Economic Analysis

Ecosystem management poses two challenges to economic analysis. The first is the need for greater use of systems constructs that make risks and uncertainty manifest. The second is working at multiple organizational levels that vary over time and space.

Systems Frameworks for Economic Analysis of Ecosystem Management

In North America, economic and ecological concerns have often been depicted as ideologically irreconcilable. Yet, recent resource management policy issues in the Pacific Northwest involving the northern spotted owl, marbled murrelet, and numerous wild salmon stocks underscore the immediacy with which economic behavior and ecosystem function interact. Recognizing the need for more comprehensive and integrative frameworks to address these issues, the U.S. Forest Service has embarked on a major policy initiative aimed at managing ecosystems (Overbay 1992). This policy requires national forest managers to plan over the long term for ecosystem integrity and for the benefit of multiple generations of people. Moreover, it requires decision makers to be acquainted with a wide range of biological, physical, social, and economic consequences of each alternative decision under consideration.

Not surprisingly, practical adoption of this policy is problematic. Real-world management strategies entail development of more-complex and explicit models that show the outputs of joint production of multiple resources and their effects on ecosystems and society. To this end, a number of analysts have proposed new models based on a systems approach to economic analysis for ecosystem management. For example, Hof (1993) suggests that mathematical programming can be used to analyze the complexity of multiple ecological processes simultaneously with joint production of ecosystem products for various interest groups. Another alternative focuses on cost-benefit analysis combined with input-output data, but makes use of the qualitative links where quantitative data are lacking or inappropriate. These adaptive models mirror iteratively generated strategies for ecosystem management (Bormann et al. 1994b) with attention to consistent model construction, comparison to real-world information, and resolution through hypothesis generation and research (Richmond and Peterson 1994). These systems models provide information about outputs and insights into the joint workings of ecosystems and socioeconomic systems in the form of feedback loops, continually revised economic goals based on new information about the joint systems, and revised ecosystem and societal impacts based on revised economic goals.

Ecosystem managers develop models of constituent parts of an ecosystem using input variables such as timber yield functions, habitat indicator relationships linked to attributes of the forests, mushroom yield relationships, and available labor force, for example. Solution variables might be timber and mushroom production, habitats, stand conditions (i.e., seral stages), jobs, community stability, and patterns of stand disturbance. Some of these variables would define the boundaries to acceptable combinations of management options for individual resources.

Modeling systems allow managers and the interested public to explore possibilities for economic development of forest landscapes over time under alternative versions of future management strategies. They differ from traditional approaches by directly treating uncertainty in the analysis and not attempting to program a constrained economic optimization. In the modeling literature, this has been called scenario planning (Wack, 1985). Scenario planning does not attempt to predict the future, but rather postulates a set of plausible futures, each dependent on the assumptions that underlie a future vision. This way, techniques focus on what might happen or go awry and how people can respond effectively under multiple events. Planning to achieve a single predetermined "optimal" future is avoided.

In addition, scenario modeling can model the ripple of economic effects of management decisions and ecological processes through a spatial hierarchy, through societal organization, and across time. Most forestry applications (e.g., USDA Forest Service 1983, Haynes 1990) take a classical sensitivity analysis approach in which a number of predetermined variables are varied in advance. Then, key projected results from scenarios are examined for differences. These differences point to emerging problems and measure the effectiveness of possible solutions. Use of such a framework has been absent in the recent large-scale ecosystem management efforts. The Forest Service adopted management for sustainable ecosystems as policy, assuming that both society and ecosystems benefit in the long term. Without a framework for planning and decision making, evaluating ecosystem management strategies remains an ideological exercise marred by value-laden, all-or-nothing choices.

Systems modeling for risk assessment is an essential element of ecosystem management because virtually all decisions made in ecosystem management are decisions made under uncertainty. Stochastic and chaotic events and incomplete information make accurate prediction of outcomes about biophysical supply and human demand fundamentally uncertain.

With scenario planning, managers experiment with decision weights based on initially subjective assessments of probabilities when information is incomplete. The richer the model in its ecological and economic verisimilitude, the more a manager gains in approaching an objective estimate of risk likelihood. Of particular importance to management is recognition of human bias in decision making under risk and uncertainty. Montgomery (1994) points out the differences between allocation of private and of public risk (i.e., at different human organizational scales), especially in an individual's predilection for risk adversity to make economic gain and risk taking to avoid economic loss. Public land managers can protect the long-term interests of society by being risk-neutral in both regards. Model building with scenarios provides a tool for seeing through bias.

The Challenges of Scale to Economic Analyses

Like ecosystems, economic systems are heterogeneous and evolutionary. Processes such as market forces, policy decisions, and management practices occur at different spatial and temporal scales and influence the course of human economic behavior. Disturbances in these economic processes at critical scales can disrupt markets and have far-reaching consequences (see, e.g., Sohngen and Haynes 1994) that significantly modify ecosystem development.

Analysis at Multiple Spatial Scales

Economic analysis at multiple spatial scales is complex. There are six general scales for relevant economic analyses in ecosystem management: the stand level, the local level, watersheds, regions, and national and global levels. Traditional forest economic methods, however, have been limited to the stand level, based on the Faustmann determination of optimal timber rotation length. The stand remains detached from its surrounding environment, and outside influences are considered external to the economic analysis and the scope of management. The shortcomings of this perspective have become increasingly obvious over the past several years, as has the need for simultaneous consideration of different scales. Many resources do not appear until a sufficiently broad geographic scale is considered. Analysis of the economic importance of elk management cannot occur only at the stand level. Elk may have an economic effect on timber at the stand level, resulting from browsing and antler rubbing, but the effect of trees and other vegetation on elk is first seen at a local or watershed level where the pattern of elk forage and shelter habitats first can be evaluated.

In addition, calculation of economic benefits can be positive at one geographic scale and negative at another. Decisions about large-scale economic policy may have far-reaching outcomes on local or stand-level conditions. Likewise, forest management policies drawn up at lower scales initiate cumulative economic effects at broader scales. This was starkly demonstrated in the spotted owl controversy. The Forest Ecosystem Management Assessment Team (FEMAT 1993) concluded that federal policy about forests in the Pacific Northwest would have different economic effects on at least three scales. Financial costs resulting from substantial changes in management of federal forestlands would be fairly negligible at the level of the national economy, which comprises multiple industrial sectors. At the regional level (the Pacific Northwest), the rapid expansion of other economic sectors would offset the failure of timber products producers to maintain historical harvesting and manufacturing levels. The severest impacts would be in small rural communities whose local economies depended on a vital timber products industry (Niemi and Whitelaw 1994).

Considerations of intratemporal equity enter into economic policy formulation at this point. The greater welfare of society, expressed in the affirmation of social and economic values in ecosystem management, sacrifices the economic welfare of a minority. Political decisions such as the Northwest Forest Plan have sought to mitigate the disadvantages to rural communities by implementing incentives to guide rural economic development in affected regions in new ways.

Even at the local scale, it is difficult to assess what society values. Questions constantly arise: What do people expect from ecosystem management? What are people willing to accept in tradeoffs? What are people willing to forego? What level of cost is acceptable? and What type and amount of benefits will prompt people to invest in ecosystems given the multitude of opportunities elsewhere? These classical economic questions achieve relatively easy solutions in the idealized model world of *certeris paribus* (i.e., where all other variables are held constant with the exception of the variable of interest), but frustrate applications of economic analysis to real policy formulation, which typically addresses dynamic situations with multiple variables. The more-relevant scale for economic impacts may be regional rather than local because of the opportunities for substitution among local places within ecosystems.

At the national and international levels, an important question is how alternative management in one country influences management in other countries. Specifically, the United States, the world's largest importer of softwood timber, will need to assess its influence on the Canadian forest sector. The shift to ecosystem management with less emphasis on timber harvests may be accompanied by higher product prices and greater dependence on lumber imports from Canada. Harder to define are the ripple economic effects of implementation of ecosystem management in the Pacific Northwest over other ecosystems, whether they be in Canada, the southeastern United States, Malaysia, or Siberia.

Finally, policy makers are assessing the need for ecosystem management and the state of its technology while global environmental policies come to be driven increasingly by nonforestry concerns. For example, policies to manage the terrestrial biosphere to sequester atmospheric carbon dioxide may greatly influence forest management in the coming decades.

Analysis at Different Temporal Scales

A key issue in forest management is finding the socially appropriate balance between the wants and needs of future generations and those of the present generation. Subjective philosophical and ethical values determine attitudes of the current generation and define its willingness to forego a portion of its potential consumption so that some set of capital stocks and resource production capacities of ecosystems is guaranteed for future generations. The current generation decides whether its population should gain at the possible expense of the well-being of future generations. Our culture is imbued with an ethos of transmitting to successive generations ecosystems that are capable of producing goods, services, and ecosystem states equivalent to those enjoyed by the current generation. But there is no consensus about how we can most efficiently and equitably transmit the economic benefits of forest ecosystem patrimony to a succeeding generation.

There are two general ways to view the minimum conditions for sustainable ecosystem patrimony (Pearce and Turner 1990, Ervin and Berrens 1994). The first outlook maintains that the total ecosystem resource capital and production capacity bequeathed ought to be equal to or greater than the sum of the nonrenewable and renewable components of the ecosystem received by the current generation. The second view allows for the total capital available to the current generation to be greater than or equal to the renewable and nonrenewable capital of the future ecosystem plus physical human-made capital and human capital. The latter view allows substitution of human capital (through advances in technology, for example) for depleted ecosystem capital of nonrenewable natural resources (Solow 1991) or of renewable resources (Ervin and Berrens 1994). Stricter views on the moral and practical constraints for sustainability of ecosystem patrimony require that people maintain all the parts of an ecosystem without any depletion (Leopold 1970). Past history has shown that the ideas about sustainability have evolved as people have become more aware of ecosystem information and of the limits to economic production derived from ecosystems. Society will likely not maintain its definition of economic sustainability from forest ecosystems from one generation to the next.

In an economic sense, discount rates serve as indicators of the degree to which the current generation wishes to value ethical obligations to future generations. A high discount rate implies that people prefer to reap benefits from ecosystem management sooner rather than later. Benefits to be obtained in the future are discounted against their present value by some annual rate. Planners and other decision makers conventionally use relatively high discount rates to evaluate forest management practices. At an 8 percent discount rate, returns on investments have to be twice as high as those made at a 4 percent rate. Another way of regarding this is to consider the value of a dollar: To the current generation, the value of a dollar is one dollar—but the value of that dollar in 20 years is $0.46, assuming a discount rate of 4 percent, and $0.21 at an 8 percent discount rate. People clearly prefer to use the dollar of benefits now rather than defer receipt of benefits if the value of the benefits is perceived to drop in the future.

Concern about economic sustainability in ecosystems has prompted a reevaluation of the traditional assumptions of a positive discount rate for time preference in economic analyses. If unconditional sustainability of ecosystems and economic production from ecosystems is a societal goal, positive societal discount rates are probably inappropriate because such rates favor consumption in the near term. Greater consideration to future generations could be obtained if a zero or negative discount rate were used.

Social Scales: The Organization and Expression of Human Interests

In a diverse and variable society, cultural attitudes, experiences, education, and history shape economic behavior. The organization and expression of human interests influence the analysis and outcomes of economic actions. Common interests draw people together to act collectively at higher scales of social organization than at the level of the individual consumer on whose behavior much of neoclassical microeconomic theory is founded (Arrow 1983). In recent years, some economists have indicted the emphasis on individual consumer preference and competition at the expense of collective realities of social interaction and cooperation of groups (see, e.g., Bradley 1991).

The processes of people organizing themselves into associations of common economic interest have "emergent scalar properties." Moreover, a consumer's economic behavior changes in a societal context (Bormann et al. 1994a). Whether called social learning (Lee 1993) or civic discovery (Iverson and Allston 1994), the collective need to maintain social order during disputes over resource distribution and equity (traditionally the normative questions of neoclassic economics) transforms individual self-interested behavior into collective behavior in regard to the use of ecosystem resources needed by many people. Simple aggregation of individual unrestrained economic interests would make the public policy process of determining the uses of forest ecosystems impossible. Economic interest groups ensure that public debate precedes actual resource use.

Aggregations of societal units for the purpose of advancing common interests are apt to be highly fluid given changes in personal preferences and perceptions of self-interest and collective interests. An increasingly all-embracing or powerful scale unit of the "public interest" may in many cases supersede less-comprehensive interests. In recent years, changes in the Forest Service's perception of the public interest have greatly changed national forest management. The Forest Service has redefined national public interest from primarily the interests of the timber products sector and its many clients to the interests of a broader, multiple mix of interest groups, including such sectors as tourism and recreation, nontimber forest products, and ecological services. The meaning of public interest has become more complex as more new economic interest groups have coalesced, realized their economic and political power, and become vocal in public debate.

Human organization in response to economic uses of forest ecosystems is even more complex because the scales of human organization do not correspond neatly with the nested spatial and temporal hierarchy of ecosystems. Overlap exists among organizational units where an individual or some members of one group are part of multiple groups at different scales (for example, participation at different levels or branches in government). The increasing trend in globalization of markets has expanded the interest and involvement of people from distant places in economic choices about ecosystem products from a local site. A given site is no longer the domain of local residents. The worldwide audience for recreational experiences in national parks exemplifies the extreme scalar disjunction between global demand and site-specific resource supply. No prescription exists or probably can exist that specifies how the needs of different levels of communities of interest are to be met at any one scale of time or space. Moral questions on precedence of preferential rights for local or indigenous communities may confound power sharing or struggles among interest groups. Inherent disjunction in scales and derived power of economic interest groups will remain a source of controversy concerning the economic use of forest ecosystems.

The Changing Nature of Forest Products

Society's demand for products from forest ecosystems defines economic goals and the management policies to achieve them. In turn, basic cultural values from historical traditions and current innovations shape society's demands and economic decisions. North Americans have long expressed a philosophy favoring the use of renewable resources over nonrenewable resources (USDA Forest Service 1982). This tenet has guided policy for forest management for most of the 20th century. A more recent corollary to that belief is the notion that renewable forest products entail lower ecological cost and are more benign to the environment than substitutable products made from nonrenewable resources such as earthen materials, ores, and fossil fuels (Cliff 1973, Alexander and Greber 1991, Kershaw et al. 1993).

A second tenet for economic analysis is that efficiency determines basic supply and demand projections. On the supply side, individual producers' perceptions of profitability influence productive capacity, technological improvements, and land management actions. On the demand side, changes in prices, preferences, and income of consumers determine levels of demand.

Timber Supply and Demand: Traditional Forest Economics in a New Light

Presently, and probably at a diminishing rate in the future, timber is the primary market good obtained from North American forests. Economic theories of natural resource use often assume that resource use increases as population and income grows. Facts point to a more complex reality. Per capita consumption of all forest products in the United States was 80 cu. ft. in 1950, declined to a low of 54.1 cu. ft. in 1975, reached 78.8 cu. ft. in 1988, and currently is 80 cu. ft. (see Figures 19.1 and 19.2). Perhaps more important has been the change in the composition of product consumption (see Figure 19.1). Per capita consumption of solid wood products has fallen—although in recent years it has rebounded as Americans have spent more to repair and alter existing housing. Per capita fiber consumption has increased 45 percent throughout the past four decades, with most of the growth taking place in the 1950s and 1960s. Finally, fuelwood consumption, which fell by 82 percent between 1950 and 1970, rose rapidly in the 1970s, reaching the 1950 level by the late 1980s.

Recent research in the United States (USDA Forest Service 1988, Haynes 1990, Haynes et al. 1995) suggests that per capita wood product consumption will decline slightly over the next 50 years. This view of the future places less reliance on solid wood products manufactured from logs, but greater reliance on engineered and reconstituted products for structural applications. It assumes greater use of recycled fiber and greater demand for nontimber commodities and noncommodity benefits from forests. In spite of this, total roundwood consumption in the United States is expected to grow at about the same rate as population growth—0.6 percent per year—to 27.6 billion cubic feet by 2040 (Figure 19.2).

Stumpage prices are expected to increase in the next 15 years as the transition from harvesting natural stands to harvesting managed stands is completed and as public harvests are reduced in the West. In the longer term, prices will stabilize, reflecting a balance between supply and demand in both stumpage and wood product markets. Between 1990 and 1995, however, there were painful adjustments, particularly in the Pacific Northwest, as a consequence of cumulative Forest Service planning decisions before 1989 and subsequent actions to protect the northern spotted owl, marbled murrelet, and anadromous fish species. Since 1989, total timber harvests west of the Cascade crest fell by 40 percent. By 1995, real (inflation-adjusted) stumpage prices increased from 1989 levels an average of 118 percent in eastern Oregon

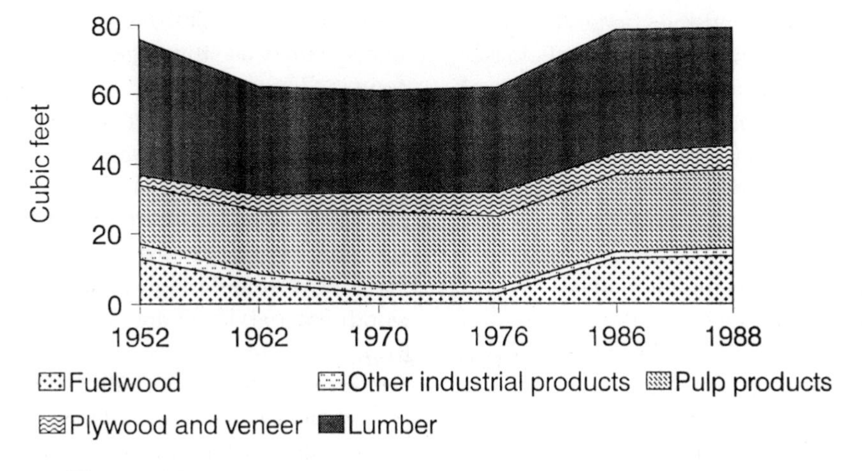

Figure 19.1 Per capita consumption of timber products 1952–1988.

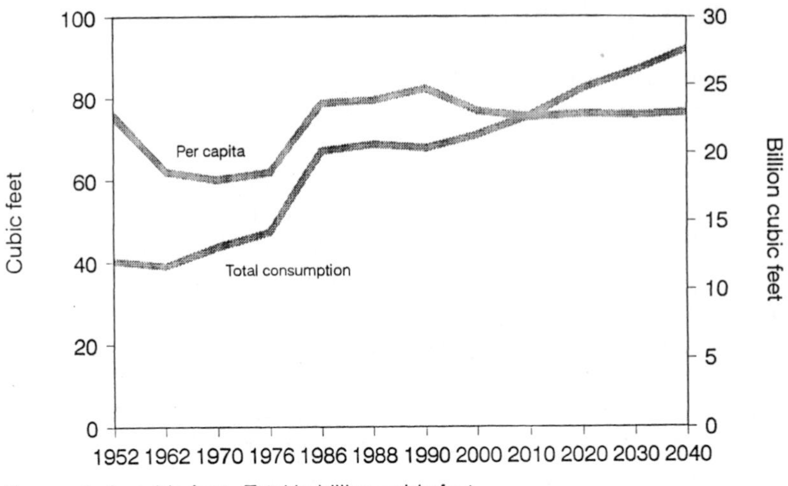

Figure 19.2 Consumption of timber products.

and Washington, 59 percent in western Oregon and Washington, and 49 percent in the south.

Resulting higher stumpage prices stimulated increased harvests on private forest lands. Much of the increase was in nonsawtimber material that replaced wood residues lost by reductions in lumber and plywood production. These increases were not sustained beyond the early 1990s—by 1994 private harvests fell back to mid-1980 levels.

Logging Costs

The shift to ecosystem management impacts the costs of forest operations and the competitive position of some forests. Logging costs from the Lolo National Forest (Keegan et al. 1993) illustrate how changing to more complex stand retention patterns can increase logging costs (Table 19.1).

Under the current system of federal timber sales, higher logging costs on national forests relative to private timberlands will lead to lower stumpage prices for national forest timber sales and result in lower payments to counties. Raising logging costs has the potential to force some sales to be appraised as below-cost sales. But, stumpage prices might not fall if the Forest Service contracted with and supervised loggers directly to cut timber to careful specifications for multiple resource production and management. The cost of the timber and vegetation management would then be variously accounted for with production of multiple resources and not just against the timber value alone. All of this raises questions about how timber is sold on federal timberlands—either as stumpage or perhaps as logs, as is done in Germany.

The recent and continuing rapid increases in sawtimber stumpage prices also indicate changing perceptions about feasible land management options in the West—although these price increases are generally consistent with past projections of rising prices for sawtimber and roughly constant prices for roundwood material. There are questions about the extent to which higher stumpage prices will change the economic feasibility of harvesting some of the small-diameter dense stands prevalent in the interior West. Using data provided by the Colville National Forest, we have calculated 1994 breakeven proportions of

Table 19.1 Logging costs for alternative silvicultural methods

Silvicultural Method	Tractor	Cable
Shelterwood	$89	164
Clear cut and seed tree	$87	155
Group selection	$98	219

chipwood (roundwood) and small sawtimber at 42 and 58 percent, respectively. Assuming current price projections and that logging costs will grow at 0.8 percent per year, the proportion will change so that by 2040 it will be 59 and 41 percent. The implication is that in the future, as now, stands to be marketed will need to have significant amounts of sawtimber.

Gainers and Losers

Economic analysis identifies groups of people who derive benefits or incur costs as the result of policies for resource management. Under assumptions of changing quantity and quality in timber supply, with resulting shifts in end products and means of production, the timber supply has different bearings on the economic livelihoods of various groups of people. Hence, equity considerations are an important component of forest management policy. Conventionally, three broad groups are affected by changes in forest product markets: stumpage owners, producers of forest products, and consumers. Changes in either property values or the regulatory climate affect stumpage owners. Changes in costs of production and changes in the prices of final wood products affect producers; changes in prices of wood products also impact consumers. Often, changes in the wood products sector do not affect the three groups equally or even in the same direction. For example, the long-term declines in harvest levels from federal timberlands in the Douglas fir region generally affect regional producers most while at the national level little affecting consumers. One estimate of consumer impacts associated with harvest reductions to protect the spotted owl showed that annual consumer expenditures for softwood lumber would increase by $12 (1989 real dollars) per household by 2010 (Haynes 1991). During the next two decades, increased production in other regions, particularly Canada, should reduce potential negative impacts on consumers. In later decades, opportunities for maintaining production levels will be exhausted, increasing total consumer burden.

There is also the issue of interregional gainers and losers. The forest sector is made up of a number of major regional markets, each with a unique set of circumstances. Forest product markets in the eastern United States typically have little involvement in western public land issues. Similar markets in the West may be dominated by the actions of a single public land management agency (e.g., the Forest Service in eastern Oregon). Many of the recent policy issues in forestry have involved the extent to which changes in one region might affect timber markets in other regions. The concern is that changes in timber supplies in one region could preempt opportunities in other regions. For example, increased land management in the South would reduce opportunities for long-term land management in some western regions. The northern spotted owl controversy, as it turns out, is relatively localized. Most of its impact is in western Oregon and Washington and in northern California.

As harvests on public lands in the West are reduced, losses are greatest in the next decade for wood product producers in the affected regions, but losses level off after the year 2000 as producers reduce mill capacity in response to lower harvest levels and higher stumpage prices. Least-affected are private timberland owners who see increases in revenues from the sale of sawtimber. Private landowners will have financial incentives to manage and harvest their lands more intensively as federal timber harvests are reduced. In the context of ecosystem management, this economic behavior may create an undesirable fragmented pattern of forest age-class distributions over the regional landscape.

Forest practices acts at the state level have in recent years increasingly regulated forestry-related practices on private timberlands. These conservation measures often benefit the public (e.g., providing watershed protection, water quality, and biological diversity) without compensation to private landowners for the costs of foregoing profitable timber harvesting over shorter rotations. Society as a whole gains, but private landowners, who are legally obligated to provide ecosystem benefits, suffer economic loss. There is a critical need to devise suitable incentive policies that assign a cost to the public as compensation to private landowners. In both Washington State and Oregon, there has been resistance to expanding forest practice laws because the question of who pays for social benefits has yet to be resolved. Greater regulation under integrated ecosystem management across legal boundaries of ownership without compensation constitutes governmental taking of property rights. Inequitable takings might build citizens' disaffection with the practice of ecosystem management and hobble efforts to achieve its societal acceptance.

Valuing Nontraditional Outputs

The essence of multiple-use management is managing ecosystems to produce a bundle of diverse goods, services, and ecosystem states simultaneously. The challenge to economists is to reflect the true, but often fluctuating and obscure, values to individuals and society of these goods, services, and states. Communicating a complete summary of economic values is important for formulation of socially acceptable policies for ecosystem management. Ideally, transactions between buyers and sellers in markets for ecosystem products reveal economic values. But market transactions exist only for a small number of goods and services (primarily timber and range allotments) produced in forested ecosystems.

Information needed for describing the economic worth of ecosystem products includes appropriate units of measure, data on the biophysical yields or human rates of use (both consumptive and nonconsumptive), and prices (including willingness to pay). Inadequate information about many "nonmarket" ecosystem resources and public demands for them leads to an incorrect allocation of resources within an ecosystem. Policy makers and resource managers charged with assessing the value of resource outputs often overlook resource values that are difficult to quantify or to perceive within the conventional methods of cost-benefit analysis and net-present-value estimation. "Market failure" occurs in such cases because incomplete information sends the wrong signal to policy makers, managers, and the public that they should use or neglect resources in ways that are ultimately socially undesirable. Government intervention represents the public's perceived willingness to pay by establishing surrogate values for nonmarket goods and services and all ecosystem states in lieu of market values. Information failure compounds if government measures to adjust access to resources for the common good do not intervene effectively. The cost of imperfect information about economic values held by society can be high if the economic values expressed in ecosystem management for resources generates societal discord.

Different resources represent different kinds of values and may be valued more accurately by one of several available methods. Turner (1991) provides a useful taxonomy of economic values and valuation methods (see Figure 19.3). For a detailed discussion of individual applications of valuation methodologies see Freeman (1993).

Unfortunately, traditional economic debate tends to be fixed on the costs of investment in natural resources while failing to quantify the economic benefits of unconventionally analyzed resources. For example, Rubin et al. (1991) suggest that the benefit of northern spotted owls on the West Coast has regional variation—a positive value for all Californians, but a negative value for residents of Oregon and Washington. A nationwide estimate of net benefits is clearly positive as well. A national expression of willingness to pay for owl and old-growth habitat conservation is balanced by the willingness to compensate local people who suffer from the abrupt shift in national policy. Failure to address shifts in perceptions of economic values and to use suitable economic tools to detect changing or emerging new resources portends great costs.

The need for empirical studies is critical. For example, Bolon (1994) modeled valuation of nonconsumptive elk-viewing and provided the data and methodology for this recently perceived emerging resource. In some studies, researchers have found that actual public values for forested landscapes are quite different from what resource managers believe society wants. In a study assessing the economic impacts of listing the Columbia and Snake sockeye salmon, the economic costs of halting various recreation, timber, range, and mineral programs for the nine national forests in the Snake River basin were measured (Haynes et al. 1992). These costs had a total net value of \$2.6 billion; the value for range was \$10.3 million, timber \$480.8 million, recreation \$2,030.4 million, and mineral programs \$72.8 million (in 1990 dollars). The cost of lost recreation was 4.3 times greater than

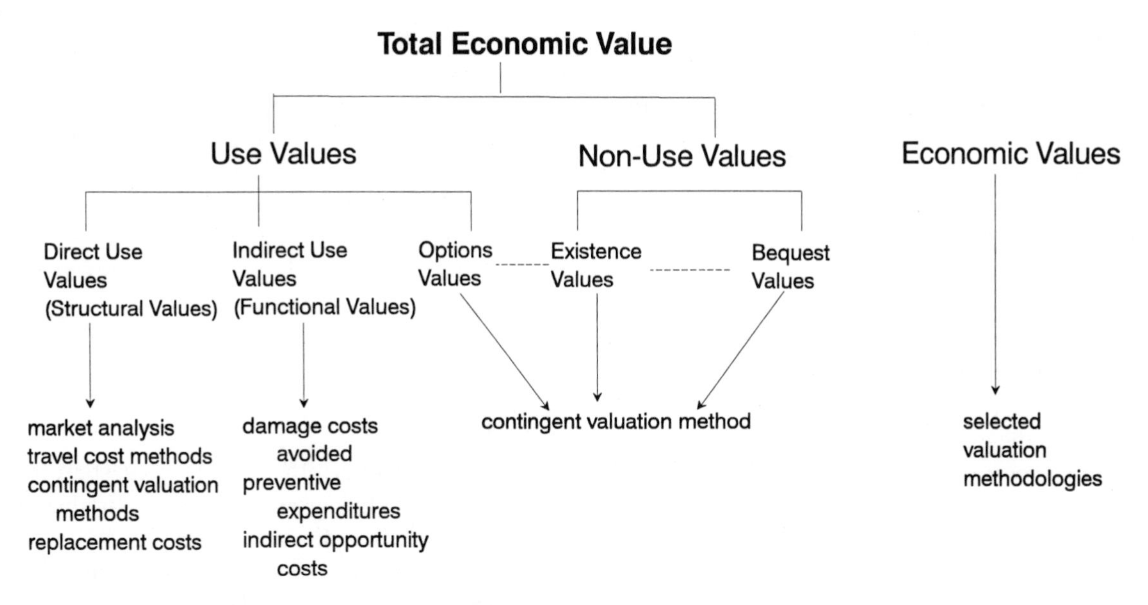

Figure 19.3 A taxonomy of economic values and valuation methods.

the cost of lost timber and 197 times greater than the cost of lost range. This outcome surprised Forest Service planners, who had traditionally emphasized range and timber production.

One proxy for describing the economic value of total forest ecosystem production is to measure values in terms of opportunity costs. Opportunity costs are the value of the next best alternative to a product foregone or traded off. For example, if a national forest adopts a new management standard for riparian zones that reduces potential sale quantities (PSQs) of timber in a watershed by 5 percent, then the opportunity cost of that decision is the reduction in PSQ times the value of timber (assuming that timber is the highest alternative use). If the reduction is 10 million board feet in the first decade and stumpage prices are $300 per thousand board feet, the opportunity costs is $3 million. In a broad sense, the value of those riparian zone management standards and guides must be worth at least $3 million to society. Land managers will need to make the costs of alternative management strategies explicit so that the public can gauge the magnitude of net benefits of those alternatives.

Another complex set of values is the real but difficult-to-quantify economic benefits that extend beyond the confines of forests and rural communities adjacent to forests. These so-called second dividends from available ecosystem services continue to attract new businesses to the Pacific Northwest (Whitelaw and Niemi 1989, Yang 1992, Niemi and Whitelaw 1994). Perceptions of environmental quality have led to economic development in future-oriented, high-tech industries. Newly arrived, urban-based economic interests generating this growth will continue to press for reduction of commodity extraction that might degrade the various amenity values of Pacific Northwest ecosystems (Niemi and Whitelaw 1994).

Economic Implications of Old Growth as a Renewable Resource

In conventional forest management, timber is harvested under one of two conditions: when the rate of increase in value of stumpage equals or is less than the chosen discount rate, or when the current (or periodic) annual increment equals or is less than the mean annual increment of tree growth (Smith 1986). However, the societal time preference becomes muddied when overlapping multiple resources and options are being managed simultaneously. One means suggested for retaining diverse structures and sustainable productivity in ecosystems is lengthening rotation times of timber harvests (see Curtis, Chapter 10). Timber with old-growth features might be viewed as a renewable resource rather than as one that is depleted to create a regulated forest of thrifty stands destined for timber production.

Important economic changes will occur if a shift is made to timber harvest rotations that approach 200 years. Harvest scheduling scenarios for the Olympic National Forest (Weigand 1994) demonstrate that traditional management liquidation of remaining old-growth acres would continue until 2010 and then drop off rapidly to virtual depletion after 2040. Under 200-year rotations, a much lower supply of old growth, both in terms of acres harvested and stumpage volume, would be available annually for harvest. The decline in supply would be 50 percent. A severe shortage of old-growth timber would occur between 2040 and 2100, but thereafter supplies would rise until 2170, when a sustainable supply of old-growth timber would equal or exceed old-growth harvest levels available in the early 1990s.

The level of sustainability would be attained sooner if a moratorium on cutting old growth in the Olympic National Forest were to begin now. The dip in old-growth timber between 2040 and 2100 would be shifted to the present, and a sustainable harvest of old growth would begin in 2050. The question of society's willingness to forego old-growth timber supplies now so that in a half-century people will have sustainable supplies is a normative ethical question that must be decided as a matter of social policy.

Managing to create a sustainable supply of old-growth-quality timber and to re-create old-growth forest ecosystems does present risks. Mensurational data are incomplete or nonexistent for even commercially important tree species such as Douglas-fir grown on long rotations (greater than 100 years) with intermediate thinnings. Existing data do not offer information about wood product quality and habitat suitability for wildlife, lichens, or understory vegetation in managed old-growth stands (Irland et al. unpublished manuscript, Curtis and Marshall 1993, Weigand and Haynes 1994). Future estimates of production rates of various ecosystem products, ranging from old-growth timber volume to the diversity of old-growth–dependent species, are extrapolations. Derived determinations of economic value for such projected future stands are highly uncertain. Yet, a considerable number of communities of interest have pressed for old-growth restoration and stand management because the risks of inaction appear to be worse than the risk of failure.

Social Forestry

In the broadest sense, the shift to ecosystem management reflects the growing interest worldwide in social forestry (see Cernea 1991). In the United States, there are two components to social forestry. The first relates to the links between land management and rural economic and social development policies. The second relates to the long-standing assertion that public forestry should consider community stability.

Strategies for Rural Development and Employment

In the 20th century, resource conservation and rural economic and social development have been largely disconnected. Indeed, much of the controversy surrounding the old-growth retention issue stems from failure to adequately consider the link between resource conservation and social and economic aspirations of rural people. As we move into the 21st century, linking these will be central to social forestry and sustainable development in particular and to ecosystem management in general. Contemporary concerns over the disparity between rural and urban economic and social opportunity already are leading to efforts to merge rural development and resource management, with the latter seen as a vehicle for partially achieving the former. Understanding the links allows resource managers to predict structural change and dynamic adjustment in local and regional economies following major changes in resource output levels.

An important aspect of the debate over old-growth

stands is that the benefits of preservation have local impacts on employment. There were 136,000 people in Washington and Oregon employed during 1989 in the primary forest products industry—logging, lumber, plywood, and pulp and paper production. This was approximately 3.3 percent of the total workforce in the two states. More important, however, is the role that the forest products industry played by providing 20 percent of manufacturing jobs in the area—35 percent in Oregon and 13 percent in Washington. During 1989, there were nine direct jobs associated with each million board feet harvested. Each direct job generated economic activity sufficient to support between one and two additional jobs in the service sector or other supporting industries. More difficult to measure is the importance of the forest products industry in communities where it provides the majority of manufacturing sector jobs.

Hence, a major policy issue centers on mitigating the concentration of negative employment impacts in regions where timber from public lands has been a critical source of jobs. There are limited opportunities to offset harvest declines—for example, through salvage cuts and mid-rotational thinnings—but reductions cannot be completely recouped. Alternative strategies for development in other rural economic sectors need to be devised. In many cases, tourism, recreation, and special forest products may provide suitable substitutes for timber jobs. However, the opportunities in terms of the number of replacement jobs, salary levels, and skill requirements are not clear. There is no guarantee that employment and wages will be constituted as previously, or that ecosystem management can sustain previous standards of living in rural areas.

Another concern is developing a workforce to manage and use forest resources in the 21st century. Ecosystem management is another manifestation of the post-modern service-oriented society (Teixeira and Swaim 1991). The importance of maintaining the ecological basis of life is now viewed as essential to the continued satisfaction of economic needs for specific goods and services (Caldwell 1990). Ecosystem management will require considerable new technologies and skills to produce the socially desired combinations of goods, services, and ecosystem states of the future. Consequently, we must reexamine the role of public agencies in promoting education and training for rural residents as well as inn promoting a rural culture of ecosystem service to our predominantly urban culture.

A second source of needed economic development will result from strategies to diversify sources of employment. Changes in access to public land resources may help expand the definition of direct forest-related employment through new jobs in recreation and nontimber forest products industries. This will require careful attention to policies to promote secondary employment through ecosystem support services, harvesting of various resources, and new ecosystem-related manufacturing. The role of new technologies in logging, in gathering nontimber forest products, and in providing recreational services will generate new kinds of jobs, both skilled and unskilled, in local rural economies. Most important, any federal attempts to promote rural employment as a tool for rural economic development will need to plan in concert with other economic forces. A favorable mix of industry for the region, the appropriate (and not necessarily high) educational level of the potential workforce, competitive labor costs, and good communications and transportation connections to urban centers (rural infrastructure)(Killian and Parker 1991) are essential to developing sustainable economies that have minimal negative impact on forest ecosystems.

Community Stability

One underlying goal in public land management has been providing for the stability of communities that are dependent on outputs from public lands. This concept has evolved along with notions of conservation and stewardship. Although an exact definition is difficult, when the word "stability" is used with the word "community," the phrase generally denotes a group of people living in a "place" and earning their livelihoods in harmony with their surroundings. Much of the underlying thought about community stability is based on the assumption that a community's economic base lies in nearby resource endowments. These concerns have motivated legislation (such as the 1944 Sustained Yield Act and the 1976 National Forest Management Act), but federal land management agencies have no specific legal mandates to provide economic stability to rural communities. Nevertheless, many agencies consider economic stability in their planning and management activities.

In reality, communities are more complex than labels such as "timber dependent" imply. Most have mixed economies, and their vitality is linked to other factors in addition to commodity production. Natural resource–dependent communities that have faced significant challenges (such as mill closures) in the past are among the most adaptable or resilient because they have successfully coped with change. In general, both communities and economies that are traditionally associated with agricultural or ranching operations are less resilient than those associated with forestry. Results also suggest that isolation from regional economies defines timber-dependent communities more than does the proportion of employment in the forest industries (Haynes and Horne 1996).

These results raise a number of potential research questions. For example, How might we sustain these isolated communities in forested regions? On what basis might we decide to save some communities while letting others decline? What is the role of earned income relative to transfer payments (some form of nonwage income)? The implication may be that an emphasis on job numbers is misplaced if a significant amount of income does not depend on wages. Also, several questions focus on the determinants of economic and social growth. In many areas of the western United States, economic and social change is being driven by population growth—a result of immigration and birth rates that are higher than the U.S. average. Economic development is tied to quality of life and encouraged by a well-trained and motivated workforce, along with state and local public and private efforts. Finally, there is the question of the efficiency of forestry as a driver of economic development. Recent work on economic well-being and experiences by financial institutions like the World Bank suggest that forestry is not an effective driver of economic development (Overdevest and Green 1995).

Conclusion

Greater reliance on or more judicious use of economic methods will not by itself quell the controversy surrounding forest management. There will always be multiple and divergent notions of the future for our forest resources. New management options, potential economic opportunities foregone, tradeoffs among values, and social resistance to economically driven management will continue to spark debate and require the analytic insight that economics provides. Economists will continually need to examine societal assumptions as they address thorny questions about the distribution of costs and benefits among people today and between the needs of successive generations and those of the present.

We wish to make four key points about the economics of ecosystem management. First, public supply and societal demand will drive future markets and future management goals for ecosystem goods, services, and states. Second, ecosystem management is new because it incorporates a conscious effort to adapt management techniques based on continual monitoring of biophysical information and evolving societal values. Failure to invest in this monitoring promises to place forest management in the same gridlocked position of the late 1980s and early 1990s. Ecosystem management places less relative emphasis on commodity outputs. In doing so, it may on occasion sacrifice the welfare of local rural communities to a more broadly defined sense of regional or national welfare where the minority concerns of rural people are subsumed. This "sacrifice" is not necessary if we include people's expectations and aspirations in ecosystem management.

Third, there is no single magic methodology to evaluate the full range of values for the goods, services, and ecosystem states that society expects from ecosystem management. Economic analysts need to explicitly link the products that society values to outputs from ecosystem management. Nonetheless, many if not most of the goods, services, and ecosystem states that society values will remain priced as opportunity costs in either direct or indirect timber values. This will persist as long as our society tolerates market failure for those ecosystem products that it values.

Finally, it is important to remember that society cannot disregard ecosystems, nor can it sustain those systems if the varied economic demands of people are not explicitly accounted for in the framework for ecosystem management.

Acknowledgments

We acknowledge the assistance of Lloyd Irland, David Iverson, David Brooks, and Amy Horne, who have helped clarify our thinking.

Literature Cited

Alexander, S., and B. J. Greber. 1991. *Environmental ramifications of various materials used in construction and manufacture in the United States.* General technical report PNW-GTR-277. Portland, OR: USDA Forest Service, Pacific Northwest Research Station.

Arrow, K. 1983. *Collected papers of Kenneth J. Arrow.* Vol. 1: *Social choice and justice.* Cambridge, MA: Harvard University Press.

Bolon, N. A. 1994. *Estimates of the values of elk in the Blue Mountains of Oregon and Washington: Evidence from the existing literature.* General technical report PNW-GTR-316. Portland, OR: USDA Forest Service, Pacific Northwest Research Station.

Bormann, B. T., M. H. Brookes, E. D. Ford, A. R. Kiester, C. D. Oliver, and J. F. Weigand. 1994a. *A framework for sustainable ecosystem management.* Vol. V of *Eastside forest ecosystem health assessment.* General technical report PNW-GTR-331. Portland, OR: USDA Forest Service, Pacific Northwest Research Station.

Bormann, B.T., P. G. Cunningham, M. H. Brookes, V. W. Manning, and M. W. Collopy. 1994b. *Adaptive ecosystem management in the Pacific Northwest.* General technical report PNW-GTR-341. Portland, OR: USDA Forest Service, Pacific Northwest Research Station.

Bradley, D. P. 1991. Ecological economics: Implications for forest management and practice: An overview. In *Ecological economics: Its implications for forest management and research,* ed. D. P. Bradley and P. O. Nilsson. Proceedings of a workshop held 2–6 April 1990 in St. Paul, MN. Research notes no. 223. Garpenberg, Sweden: The Swedish University of Agricultural Sciences, Faculty of Forestry, Department of Operational Efficiency.

Bradshaw, A. D., and M.J. Chadwick. 1980. *The restoration of land: The ecology and reclamation of derelict and degraded land.* Vol. 6 in *Studies in ecology.* Berkeley, CA: University of California Press.

Caldwell, L. K. 1990. *Between two worlds: Science, the environmental movement, and policy choice.* Cambridge, England: Cambridge University Press.

Cernea, M. M. 1991. *Putting people first.* New York: Oxford University Press.

Cliff, E. P. 1973. *Timber, the renewable material: Perspective for decision.* Washington, DC: GPO.

Curtis, R. O., and D. D. Marshall. 1993. Douglas fir rotations: Time for re-appraisal? *Western Journal of Applied Forestry* 8(3):81–85

Duerr, W. A. 1993. *Introduction to forest resource economics.* New York: McGraw-Hill.

Ervin, D. E., and R. P. Berrens. 1994. Critical economic issues in ecosystem management of a national forest. In *Ecosystem management: Principles and applications.* Vol. II of *Eastside forest ecosystem health assessment,* ed. M. E. Jensen and P. S. Bourgeron. General technical report PNW-GTR-318. Portland, OR: USDA Forest Service, Pacific Northwest Research Station.

FEMAT. 1993. *Forest ecosystem management: An ecological, economic, and social assessment.* Report of the Forest Ecosystem Management Assessment Team (FEMAT). 1993-799-071. Washington, DC: GPO.

Freeman, A. M., III. 1993. *The measurement of environmental and resource values: Theory and methods.* Washington, DC: Resources for the Future.

Haynes, R. W. 1990. *An analysis of the timber situation in the United States: 1989–2040.* General technical report RM-199. Fort Collins, CO: USDA Forest Service, Rocky Mountain Forest and Range Experiment Station.

———. 1991. Changes in timber supply in the Pacific Northwest. In *Proceedings of the 25th annual Pacific Northwest Regional Economic Conference, 2–4 May 1991, Portland, OR.* Seattle: University of Washington, Northwest Policy Center.

Haynes, R. W., and A. L. Horne. 1996. *Economic assessment of the interior Columbia Basin.* Walla Walla, WA: Interior Columbia Basin Ecosystem Management Project.

Haynes, R. W., N. A. Bolon, and D. T. Hormaechea. 1992. *The economic impact on the forest sector of critical habitat delineation for salmon in the Columbia and Snake River basin.* General technical report PNW-GTR-307. Portland, OR: USDA Forest Service, Pacific Northwest Research Station.

Haynes, R. W., D. M. Adams, and J. R. Mills. 1995. *The 1993*

RPA timber assessment update. General technical report RM 259. Fort Collins, CO: USDA Forest Service, Rocky Mountain Forest and Range Experiment Station.

Hof, J. G. 1993. *Coactive forest management.* San Diego: Academic Press, Inc.

Irland, L. C., R. S. Seymour, and D. I. Maass. Unpublished manuscript. Big tree prescriptions for growing white pine in New England and New York.

Iverson, D. C., and R. M. Alston 1994. A new role for economics in integrated environmental management. In *Implementing integrated environmental management,* ed. J. Cairns Jr., T. V. Crawford, and H. Salwasser. Blacksburg, VA: Virginia Polytechnic Institute and State University, University Center for Environmental and Hazardous Materials Studies.

Keegan, C. E., D. P. Wichman, C. E. Fiedler, and F. J. Stewart. 1993. *Logging systems and associated costs of timber harvesting on the Lolo National Forest under alternative silvicultural prescriptions.* Cooperative Agreement PNW 92-0123. Portland, OR: USDA Forest Service.

Kershaw, J., C. D. Oliver, and T. Hinckley. 1993. Effect of harvest of old-growth Douglas fir stands and subsequent management on carbon dioxide levels in the atmosphere. *Journal of Sustainable Forestry* 1:61–77.

Killian, M. S., and T. S. Parker. 1991. Education and local employment growth in a changing economy. In *Education and rural economic development: Strategies for the 1990s,* ed. R. W. Long. ERS staff report no. AGES 9153. Rockville, MD: USDA Economic Research Service, Agriculture and Rural Economy Division.

Leopold, A. 1970. *Sand County almanac.* New York: Ballantine Books.

Lee, K. N. 1993. *Compass and gyroscope.* Washington, DC: Island Press.

Montgomery, C. A. 1994. Socioeconomic risk assessment and its relation to ecosystem management. In *Ecosystem management: Principles and applications.* Vol. II in *Eastside forest ecosystem health assessment,* ed. M. E. Jensen and P. S. Bourgeron. General technical report PNW-GTR-318. Portland, OR: USDA Forest Service, Pacific Northwest Research Station.

Niemi, E., and E. Whitelaw. 1994. The potential social and economic impacts of long-rotation timber management. In *High-quality forestry workshop: The Idea of long rotations,* ed. J. F. Weigand, R. W. Haynes, and J. L. Mikowski. Proceedings of a workshop, 10–12 May 1993, Silver Falls State Park, Oregon. Seattle, WA: University of Washington, Center for International Trade in Forest Products.

O'Neill, R. V., D. L. DeAngelis, J. B. Waide, T. F. H. Allen. 1986. *A hierarchical concept of ecosystems.* Princeton, NJ: Princeton University Press.

Overbay, J. C. 1992. Ecosystem management. In *Taking an ecological approach to management.* Proceedings of a national workshop, 27–30 April 1992. WO-WSA-3. Washington, DC: USDA Forest Service, Watershed and Air Management.

Overdevest, C., and G. P. Green. 1995. Forest dependence and community well-being: A segmented market approach. *Society of Natural Resources* 8:111–131.

Pearce, D.W., and R. K. Turner. 1990. *Economics of natural resources and the environment.* Baltimore: Johns Hopkins University Press.

Richmond, B., and S. Peterson. 1994. *Stella II: An introduction to systems thinking.* Hanover, NH: High Performance Systems.

Rubin, J., G. Helfand, and J. Loomis. 1991. A benefit-cost analysis of the northern spotted owl: Results from a contingent valuation survey. *Journal of Forestry* 89(12): 25–30.

Smith, D. M. 1986. *The practice of silviculture.* 8th ed. New York: John Wiley & Sons.

Sohngen, B. L., and R. W. Haynes. 1994. *The "great" price spike of '93: An analysis of lumber and stumpage prices in the Pacific Northwest.* Research paper PNW-RP-476. Portland, OR: USDA Forest Service, Pacific Northwest Research Station.

Solow, R. M. 1991. *Sustainability: An economist's perspective.* 18th J. Seward Johnson Lecture in Marine Policy. Woods Hole, MA: Woods Hole Oceanographic Institution, Marine Policy Center.

Teixeira, R. A., and P. L. Swaim. 1991. Skill demand and supply in the new economy: Issues for rural areas. In *Education and rural economic development: Strategies for the 1990s,* ed. R. W. Long. ERS staff report no. AGES 9153. Rockville, MD: USDA Economic Research Service.

Turner, R. K. 1991. Economics and wetland management. *Ambio* 20(2):59–63.

USDA Forest Service. 1982. *An analysis of the timber situation in the United States, 1952–2030.* Forest resource report no. 23. Washington, DC.

———. 1983. *America's renewable resources: A supplement to the 1979 assessment of the forest and range land situation in the United States.* Forest Service publication no. 386. Washington, DC: USDA Forest Service.

———. 1988. *The South's fourth forest: Alternatives for the future.* Forest resource report 24. Washington, DC: USDA Forest Service.

Wack, P. 1985. Scenarios: Uncharted waters ahead. *Harvard Business Review* 63(5):73–89.

Weigand, J. F. 1994. First cuts at long-rotation forestry: Projecting timber supply and value for the Olympic National Forest. In *High-quality forestry workshop: The idea of long rotations,* ed. R. W. Long. Proceedings of a workshop, 10–12 May 1993, Silver Falls State Park, OR. Special publication 15. Seattle: University of Washington, College of Forest Resources, Center for International Trade in Forest Products.

Weigand, J. F., and R. W. Haynes. 1994. Summation: The agenda for research and management for high-quality forestry. In *High-quality forestry workshop: The idea of long rotations,* ed. R. W. Long. Proceedings of a workshop, 10–12 May 1993, Silver Falls State Park, OR. Special publication 15. Seattle: University of Washington, College of Forest Resources, Center for International Trade in Forest Products.

Whitelaw, W. E., and E. G. Niemi. 1989. The greening of the economy. *Old Oregon* 69(3):26–27.

Yang, D. J. 1992. High-tech heaven: Why the Pacific Northwest is surging with technology startups. *Business Week* (May 25): 50, 52, 54.

20 Changes in Wood Products Manufacturing

Clive G. Whittenbury

Challenges to Volume-Based Consumption 304
The Changing Timber Harvest 304
Lumber Manufacturing Today 306
Manufacturing in Grade Mills 306
Toward a New Wood Products Industry 308
Engineered Wood Products 308
One Bridge to the Future: Added Value 309
The Challenge to Industry in Forest-Dependent Communities 311
Possible Approaches to a New Grade Industry 311
The Immediate Tasks Ahead 312
The Longer View 313

The wood products industry started to undergo significant changes during the 1980s, when forest managers began to accept environmental issues as an important aspect of timber harvesting. By June 1990, when the spotted owl was featured on the cover of *Time* magazine, the Endangered Species Act had provided effective legal grounds for terminating the continued harvesting of the national forests in the Pacific Northwest at past traditional levels. Since then, the backlog of commercial timber sales has been cut, few additional sales have taken place, and many lumber mills have closed. These changes have increased the pressure on private lands to make up for the lost timber volume from the national forests.

Yet, private lands cannot, by themselves, meet the appetite of all of the forest product plants operated by the large timber landowners, and small private holdings do not have enough timber to supply those other plants that previously depended upon the national forests. Many lumber and plywood plant closings are therefore expected to be permanent, and many forest communities throughout the Pacific

Northwest that hosted these plants face economic crisis.

This situation finally has forced a serious look at the shortcomings of an industry that has served society for over 100 years using the same basic techniques for extracting wood and converting it into a myriad of wood products for the consumer. From an industrial standpoint, the wood products industry has been remarkably effective. But society has now challenged the continued use of public forests by that industry. Timber extraction has been perceived as a negative impact on other values provided by the forest, including wildlife habitat and the continuation of complex structures for the sustainability and health of natural forests.

Challenges to Volume-Based Consumption

The performance of the wood products industry has traditionally been based on the volume of material produced in various categories of products and by grades. Prices and production are stated in volumetric measure, and revenues are determined through their combination. Industry has always been geared toward the volume production of basic products that have changed little in over 100 years. Yet, within the last 10 years, that century-old foundation has been challenged by an environmental movement that has claimed alternative uses for what might have been commercial timber in the national forests. The most dramatic consequences have been in the Pacific Northwest, where commercial harvesting essentially has come to a standstill.

Although challenges to volume-based consumption of commercial timber started in the national forests, they are expected to spread to private timber and to other public sources of timber—for environmental as well as economic reasons. Economic pressure for more thrifty use of wood fiber was relatively insignificant in the past, when industry was fed an ample supply of quality commercial timber. Improvements in industry have focused on more-efficient manufacturing of traditional products rather than on diversification toward products that use less fiber or that save the valuable high-quality materials found in older trees. Yet, the emphasis now being placed on protecting complex biosystems through ecosystem management means reduced harvests for commercial purposes and different incentives for the wood products industry.

Such changes are expected to be local and to vary from place to place. Simple single solutions are not expected in an industry as socially and technically complex as the forest products industry. Overall, this industry will not change rapidly. The social and professional crisis within forest communities that depended for generations on the national forests for their economic health suggests that changes in their local industries be considered first, and with some urgency. This chapter therefore focuses upon local industries that have been related directly to the national forests and that might enjoy a renewed relationship through industrial change.

Changes in the wood products industry, which is commonly identified with forest-dependent communities, will require innovation and involve risk. Those changes will, nevertheless, point the way to solutions in the growing worldwide mismatch between available timber resources and the resources needed for traditional products. In the forest crisis of the Pacific Northwest lie the seeds of change that will ultimately benefit society in its transition from volume-based consumption of wood products to quality-based products that conserve wood volume.

The Changing Timber Harvest

The forest products industry divides along several different lines. One division exists between those companies that manufacture paper and those that manufacture solid products. Timber supplies for paper are found universally. Timber supplies for high-quality solid wood products are at the center of the Pacific Northwest crisis. Accordingly, major changes in wood products manufacturing are expected in these solid products, and are therefore featured in this chapter.

Another major division exists between those companies that have their own supply of timber and those that have been dependent on public forests. Privately owned forests will continue to supply timber either until they run out of usable trees or, if they practice sustainable harvesting, indefinitely. There is no immediate pressure for companies with access to private timber to change their manufacturing methods until they are forced by competition to accept more-efficient ways of converting their timber into wood products. With their standing investment in large manufacturing plants, many of them recently modernized, these companies will not change their plants and will continue to manufacture standard products ranging from lumber to panels. In the future, however, pioneering changes will have to be made by an industry that depends directly upon commercial wood coming from the national forests. Those changes will become economically driven and could offer future significant business opportunities to companies that today use timber from private forests.

Companies that depended upon the national forests for their timber supply have already been impacted permanently by a dramatic reduction in the amount of logs available for commercial use. Many of them have come to the end of their sales backlog, meaning that they have shut down or have resorted to piecemeal log buying from small lots and from distant forests, some of them overseas. With the revision of forest management plans, specifically from the Clinton administration's new approach to ecosystem management, a much reduced flow of logs may emerge from the national forests in the Pacific Northwest during the latter half of the 1990s. This flow may be more or less than expected, depending upon further decisions that take into account forest health in addition to habitat protection.

Simplified concepts of commercial harvesting have emphasized clearcutting at several different scales and have dominated almost all aspects of the most recent forest management debates. As a result, ecosystem damage has been equated simply with commercial harvesting in a zero-sum game—someone "loses" when someone else "wins." We can choose between owls and jobs, but we cannot have both.

Forest management, having recently adopted the basic guidelines of ecosystem management, is just at the beginning of a new phase. The goal is to develop commercial harvest practices that are compatible with ecosystem and habitat protection, thus avoiding a zero-sum game. All sides should win. But the new forest management has yet to develop clear operating principles that translate into harvesting practice. In the meantime, what should a new forest products industry be thinking about and planning for, and can the planning be part of the innovative period in which these new forest management principles are developed together with new harvesting concepts?

Finally, the wood products industry divides into segments, each dependent on different grades of wood. The primary lumber industry uses high-grade logs to supply the secondary (industrials) industry for house furnishings and fittings, and medium- to high-grade logs for construction applications. The panel industry uses medium-grade logs for plywood and low-grade logs, or waste by-products, for its other panel products. The paper industry uses low-grade logs as its feedstock.

While all of these segments operate in commodity markets, meaning that they emphasize product volume rather than individual product quality, the supply of industrial lumber grades to the secondary manufacturers is differentiated through quality. Therein lies its dependency on older trees. The primary lumber mills that supply quality industrial lumber to the secondary manufacturers are known, therefore, as grade mills. Grade mills make up the major part of the forest-dependent manufacturing industry and are the link between the secondary manufacturers and the old-growth forests. This part of the industry will therefore be the source of early change and innovation for the 21st century.

While there are exceptions, the national forests generally contain most of the remaining high-quality timber, including the old-growth stands that are a special biological class of older trees. These stands contain trees that yield clear wood for high-grade products and large boards for quality structural use. Old stands in private forests have been substantially cut. Complete protection of the old-growth component of these stands will strictly limit commercial uses, but there are still many old trees in other parts of the forest that can yield high-quality lumber.

The above factors relate grade mills to the national forests and make forest communities in which they are located dependent upon the national forests.

Grade mills therefore become another key part of the story of change in the wood products industry.

Lumber Manufacturing Today

The manufacture of lumber today depends upon the same basic principles that have been in use for over 100 years. Logs are separated into boards by grade. The lowest grades come from the inner part of the log and the best grades from the outer part where the wood is the youngest and freer from defects and knots. The largest, oldest trees have more high-grade material than smaller trees. The large trees therefore yield both high-grade lumber and low-grade lumber, while the smaller trees generally provide only low-grade lumber.

High-grade boards have beautiful, clear surfaces with few defects and straight fine grain with many rings to the inch. They are used to make wood products such as windows, doors, and furniture for which appearance and unblemished surfaces for paint are important. Low-grade boards and smaller pieces, such as two-by-fours (dimension grades), are used for construction where strength or usefulness in building (utility grades) is more important than appearance. These differences in quality are reflected in prices—clear-appearance grades are often five times the price of construction two-by-fours. This has been the basis for quite different manufacturing practices in the lumber industry.

Manufacturing construction products out of low-grade small logs simplifies the types of cuts that must be made in the log and the sorting and care required in handling the lumber. This type of manufacturing has therefore been the easiest to automate, with an emphasis on volume production in large mills. The economics of these mills are sensitive to log price and to the lumber market, with narrow margins creating a dependence on large-volume production to make any profit.

Manufacturing quality boards out of large logs means having to pay more attention to and provide more care for the higher-quality products while at the same time manufacturing low-grade lumber from the inner parts of the log. This mixed requirement complicates manufacturing procedures. The requirement limits the amount of automation that can be applied and keeps manual-processing cost levels higher than those of lower-grade construction lumber mills. This type of lumber plants (known as "grade mills") have been the most seriously impacted by the loss of commercial timber from the national forests. They are the smaller-community mills, often owned by individuals or families, and are dependent upon local skills to manage and extract grade lumber from the larger high-quality logs coming from the national forests.

In reality, manufacturing is not cleanly divided between these two types of mills—the construction lumber mills and grade lumber mills. There is some overlap. Large mills can get large logs from older trees, while small mills can get many smaller logs from harvests that are not so well stocked in large older trees. When that happens, the construction mills benefit from the log quality, while the grade mills are caught with logs undersized for their design needs. The result can be a serious economic loss, as illustrated in the following section.

Manufacturing in Grade Mills

Primary manufacturing of lumber is mainly the extraction, sorting, and finishing of flat boards from the inside of a log. Sorting and finishing make the subsequent use of the boards simple and efficient in secondary manufacturing industries such as door, window, and furniture manufacturers. The creation of flat long boards as a primary product maintains a manufacturing simplicity that has been enjoyed by the lumber industry for over 100 years. However, that simplicity has been maintained through dependency upon an unrestricted volume of feedstock. Changes in the supply now challenge that simplicity.

Wood products are defined by their structural quality, shape, and surface characteristics. These features are all developed from the simple shapes and the full initial volume of the primary lumber boards, and they become the basis for pricing the products. Structural capability traditionally comes from the quantity of natural wood in a solid piece (such as a two-by-four), rather than from more-complex shapes found in engineered structures. Decorative shapes come from the machining of simple solid pieces, often resulting in significant amounts of waste. Surface appearance comes from the high-quality material exposed from machining or the finished surface of original flat pieces of lumber.

All of these manufacturing processes show little concern for the amount of fiber in the original lumber because they start with standard-sized boards provided by the primary lumber manufacturer. The primary lumber manufacturer created standard-sized products without considering possible savings in fiber by better meeting the function of the final product. In making the wood product, the traditional manufacturing steps simplify the total process rather than conserve the original material. Simple rules for sorting materials (i.e., the grading rules) allow manual labor to both handle and visually sort lumber quickly, thereby maintaining profitable production volumes in the mill. While hydraulics, compressed air, and elementary computers have made manufacturing easier and more accurate, lumber continues to be classified by simple shapes, properties, and appearance.

In a grade mill, the main objective is the extraction of the greatest amount of high-quality wood from a log. This wood is the real source of profitability of a grade mill. The mill is optimized to cut, sort, finish, and protect this quality material even at the expense of slowing down and making less efficient the manufacture of lower-grade products from the remainder of the log. Analysis of the production costs of the different products in a grade mill shows that the common and dimension (construction) products generally lose money, while the high-grade industrial products both make up for that loss and generate the net profit for the mill. Consequently, a source of high-grade logs is essential. The loss of old trees in any timber source is a challenge to the grade mill industry and to its dependent secondary manufacturers.

A typical grade mill has the overall financial performance shown in Table 20.1 for three different quality sets of logs. The lowest quality set has more smaller logs than found in the highest quality set. Lumber volumes are measured in thousands (M) of board feet (BF). These example data were collected before the timber crisis in the Pacific Northwest, when such mills were performing well.

The importance of buying the right quality logs for a grade mill is apparent. However, the numbers become quite dramatic when financial data is separated according to the two different types of products: the construction grades and the industrial (visual-quality) grades. Table 20.2 shows that construction grades do not pay their way in a grade mill; industrial grades have to carry the profitability of the whole operation. Table 20.2 shows two selections from a grade mill operation of several years ago. Although lumber prices have increased since then, so have log prices. The net profits may be different today depending upon the mill and its location, but the figures are intended only to illustrate a general feature of the industry.

Table 20.2 indicates the real source of the gradual

Table 20.1 The overall financial performance of a typical grade mill

Averages	Low Quality $/MBF	Medium Quality $/MBF	High Quality $/MBF
Log cost	300	375	600
Manufacturing	89	93	107
Total cost	389	468	707
Product value	345	460	825
Profit	-44	-8	120

Note: MBF = thousand board feet.

Table 20.2 Comparison of the profitability of construction and industrial grades in a traditional grade mill

	Pine $/MBF	Douglas Fir $/MBF
Industrials		
Sales price	1,125	735
Cost of material	880	405
Manufacturing cost	130	130
Profit	115	200
Construction		
Sales price	210	165
Cost of material	165	90
Manufacturing cost	90	90
Profit	–45	–15

Note: MBF = thousand board feet.

demise of the traditional grade lumber industry in the Pacific Northwest. Large trees contain quality wood. They also represent major components of quality forests. Protecting the ecosystem of a quality forest means keeping many of these trees in the forest. Grade mills cannot profitably cut smaller logs that contain mostly construction grades unless the mills are converted to so-called small-log mills. That would mean very unfavorable competition with the more efficient and less costly large-capacity mills that use private timber for their feedstock.

The change in supply of commercial timber from the Pacific Northwest national forests therefore has not been simply a question of diminishing timber volume. The change has undermined the very structure of a manufacturing concept in use for generations. The many medium-sized and smaller mills that were the fabric of the northwest forest communities could not shift to smaller low-grade logs and survive financially, although many of them tried. The economic problems caused by a reduction in the size and grade of logs come from a reduction of wood product value in the market and from the high cost of manufacture relative to that product value.

The resulting economic crisis in the grade lumber industry is therefore fundamental. It is not simply a matter of replacement volume, to be provided by commercial thinning (small-diameter trees and small logs) or by the salvage of trees damaged by fire and insects (which diminishes value of the damaged outside quality wood). Continuation of the grade industry, which differentiates the skills of forest communities, requires a new search for the uses of the wood value that is still the unique product of the natural forest.

Toward a New Wood Products Industry

The simple new ground rule for a new wood products industry is to use much less tree fiber overall and, in particular, to use very much less wood from large old trees.

From the description of wood product manufacturing above, this ground rule is diametrically opposed to traditional practices in which the wood from a tree is cut into a series of solid wood products in as large a volume as possible to meet the economic principle of the lumber plant. The lumber business is described in terms of fiber volume, it is priced in terms of fiber volume, and its products are measured in terms of fiber volume.

The aircraft industry provides an instructive contrast, illustrating the fallacy inherent in manufacturing, pricing, and operating according to material quantity when material constraints are introduced. An aircraft is not valued by the amount of material in it, but rather by the cleverness through which it meets its function with the least possible material. It cannot function if it cannot fly, and it cannot fly if it has too much material in it. If furniture or houses had to fly, they would never make it today. But with economy of material, could they do so tomorrow?

The aircraft design analogy puts forward a condition that could become fundamental in wood product design—a dramatic reduction in the use of wood fiber in products while still meeting function and performance. The aircraft industry met this condition through engineered designs, a path that the construction industry is already treading.

Engineered Wood Products

Why should the structural element of a house have to be a solid two-by-four? Aircraft have thin, structural members such as I-beams, U-channels, and shells. So do wasp nests, with honeycombs of wood fiber that were copied by aircraft designers. The obvious initial answer is that it is easier and cheaper to manufacture the wood structure into a solid two-by-four. But then the next question might be, Why should it need to take 50 or even 100 years to make that structural wood element? In fact, what is the value of time for a wood product manufactured in a tree?

The value in wood products lies in the appearance of the wood and in its structural characteristics. Do these qualities have to come through the same feedstock—that is, through a tree? A piece of wood for a high-visual-quality product has attributes that need to be manufactured naturally by a tree. The clear material, with its natural color and grain, cannot be manufactured artificially without deceit. However, it can be presented in thin sheets that please the eye and hide a base material that might not look as good, but can perform as well as the original substructure—a veneer with underlying structural quality not offered by plastic or conventional wood-particle cores.

Can wood-core materials, in fact, come from sources other than older trees, but still have the quality of structure offered by those trees? This becomes the realm of engineered wood products, already found in the value-added wood industry. This branch of the industry takes basic wood materials, often low-grade, and pieces them together to manufacture high-value specialized products. Some are laminations of small solid pieces joined together to make large strong beams; others are strips of material cut out of the sheets from peeled logs and joined together in a matrix of cured resins.

These products are called engineered because they are designed and structured into something artificial, something beyond the natural state. They are the early entries for the new wood products industry; their designers have moved in the right direction by doing something innovative with natural materials. But have they begun to answer the new ground rule: to do as much, or more, with less material? This question will not be answered until the wood products industry does as the aircraft industry demanded of its own metals—meet shape, stability, and structural characteristics with the minimum amounts of material. None of today's engineered wood products achieve that objective through advanced structures or reconfigured wood fiber.

What if the new ground rule were to be translated into an engineering specification, leaving open the path to a solution? For example, what new products might be designed if we limited the amount of wood-like material to 10 percent of that previously used for the structural elements of wood products? This would require structural cores one-tenth the average density if the general shapes of products were to remain the same. Does this mean that the core materials should be like balsa wood, honeycomb, stressed-skin boxes, or webs? What about biomaterial versions of Styrofoam?

The benefits of a move to low-density engineered materials would cascade through the industry and the consumer marketplace. The weight of wood products has been accepted throughout history; the logistics of the wood product marketplace would be changed dramatically by the use of light-weight materials. Trucking and moving costs could be reduced by large amounts, as could handling equipment for manufacture. What if future wood products became one-tenth the weight for the same performed function? Tables and chairs would wobble and desks would shake—or would they? When automobiles were finally made well, they seemed to be more solid, precise, and controllable than their Jurassic ancestors. Should we expect the same from precision-engineered wood products?

If the material structure is changed from that found naturally in trees, then where will it come from? Will it start in a tree and be reengineered, or will it be built from something other than a tree?

Wood has many well-described attributes, not the least being its manufacturing efficiency using solar energy. It is one of two large-scale converters of solar energy that meet the vast material needs of society; the other is agriculture. The vast collector areas of the forests inexpensively turn water and free carbon dioxide into vast quantities of biomass via solar energy. This is why wood is so important environmentally. Substitutes for wood used in construction start at a competitive disadvantage if they are evaluated using environmental criteria.

The above questions are typical of those that will have to be answered on the way to a new forest products industry. These questions should be answered through engineering developments embedded in sound business practice. The answers must match future markets, and they must meet the needs of the wood products manufacturing communities that will use them in their economies.

One Bridge to the Future: Added Value

Added-value manufacturing, one key concept that has emerged during the last few years, may point the way toward answering some of the questions. The basic principle is simple, but its practice is not. Figure 20.1 illustrates two types of lumber plant. The traditional lumber plant is shown at the bottom right, while an added-value plant is shown at the top left. The cost and revenue lines intersect at the break-even point, where losses change to profits. The

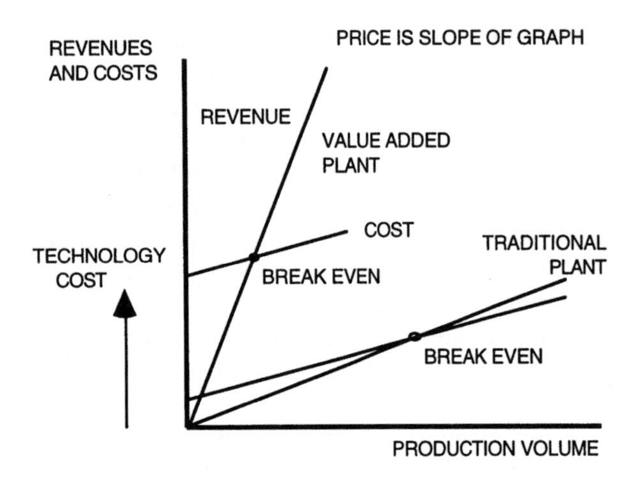

Figure 20.1 The break-even point for a traditional lumber plant versus an added-value plant.

added-value plant sells products at a higher price than the traditional plant, but requires a larger investment in fixed costs for equipment.

The added-value plant significantly changes plant economics. The larger angle of intersection means that added-value plants are less susceptible to log-cost increases or product-price decreases than traditional lumber plants. At the same time, if production cannot be maintained in an added-value plant, the losses can be large due to the fixed costs of the more expensive plant. It is easy to see the possible large excursions of the break-even point in the traditional lumber plant as lumber prices fall or logs increase in cost. It is also possible to see the importance of high production volumes to achieve profitability in the traditional lumber industry.

The role of technology in added-value manufacturing is twofold. The manufacturing process needs technology for adding value to a product—that is, through extraction, sorting, shaping, cutting, laminating, and marketing. It also requires technology to maintain production while dealing with material complexity and the many more pieces (thousands rather than hundreds) than before. Technology manages complexity through automation and computer-assisted labor. Moreover, new handling systems will be needed for the increased piece count.

A flush of added-value innovation in the 1980s started artificially shifting the balance of log yields from small amounts of natural high-quality material to larger amounts of high-quality products. This was done by building large pieces of high-quality material out of small pieces of industrial-quality material, using the part of the log between the outer high-quality wood and the inner low-quality wood. These intermediate parts have always been known for their "shop grades," which are used in remanufacturing shops to make the smaller pieces of windows and doors come from the part of the shop grade boards between the defects and knots.

Cutting clear pieces of wood from shop grade lumber and sticking them together so that the joints are barely visible is one means of creating surfaces with unblemished painted surfaces. The finger-jointed and edge-glued market grew in this way, and the extra work in making those products provided the early concept of added-value manufacturing. Although value was in fact added by the process, often it cost more in labor and in machine handling because of the now thousands rather than hundreds of pieces. Some plants have been successful in taking this approach, but others have failed as costs exceeded the added value.

Other approaches to added-value manufacturing have tackled the basic problem of converting small-tree logs into large-tree products. One of the most ingenious methods involves a combination of plywood and solid-wood lumber techniques. Small logs are peeled, and the outer highest grade material is separated from the lower-grade inside material before the sheets are pressed together into slabs. The slabs look like thick plywood, with the high-grade material at and near the two large faces of the slab. Conventional-sized lumber is then efficiently cut from these reconstituted slab logs. When the cost of feed logs is right, and the price of products is high enough, the products from this costly two-stage manufacturing process can pay their way. But the significant investment needed must be weighed against the possibility of another cheaper way of reconstituting low-grade logs being developed before the investment is recovered.

The attractive feature of added-value manufacturing is its recognition of wood value. This provides a clue to future wood product manufacturing. Added-value concepts have created the beginning of a bridge to the next century, one that promises to build an industry on dramatically reduced amounts of commercial timber.

Such a bridge requires a clear understanding of value. Lumber is discussed in terms of grades and prices, but rarely in terms of its true utility in society. If economics of the wood products industry depend upon the utility of wood, then the crisis in the lumber industry should lead to new thinking about how that utility might better be delivered through new manufacturing practices that preserve the utility while significantly reducing the amounts of high-quality tree materials used. That change is needed because the harvested volume of wood is the root cause of the commercial timber crisis.

The value to consumers of wood products lies in their appearance and structural characteristics. They look good as furniture and provide long-term stability because of their natural structure. As components of a house, they meet strength requirements and are easy to fasten together. The simplified manufacturing of the past has offered large quantities of relatively inexpensive material with these properties from the seemingly limitless forests throughout the United States. These valuable properties of wood are still needed by society. The question now is, How can they be delivered at lower total volumes?

The Challenge to Industry in Forest-Dependent Communities

The reduction in supply of commercial timber from the national forests challenges the existence of the traditional grade mill. When commercial timber again becomes available as new ecosystem management practices are applied, the grades and quantities of logs will not be clearly understood for some time. This is because harvests will be based on criteria involving forest health, habitat structure, and landscape design rather than simply on commercial timber requirements. Mixes of grades and log quantities will vary from one place to another, and the timing of harvest yields may not always be predictable. Variability and a wide range of grades are expected to characterize future harvests and introduce significant uncertainty into manufacturing planning.

At the same time, the harvests from the national forests should still provide significant amounts of quality wood. It will, however, be mixed together with a wide variety of medium- and low-grade wood. This is to be expected because forest health harvests should involve thinning of small and intermediate-sized timber and trees damaged by insects and fire. The first task in using the quality wood will be to find it and sort it. The next task will be to use it effectively in manufacturing. The last task will be to market and sell it. But before any of these tasks can be undertaken successfully, a whole new approach to manufacturing will be necessary.

Manufacturing is a bridge—made possible by the technology of the day—between society's needs and natural resources. In the case of wood products, the bridge is old. It has been improved, but fundamentally unchanged, by computer technology and machinery improvements. Knowledge of the path between the tree and the finished products lies much deeper with the people who have worked in grade mills rather than in the mill technology. Therefore, the challenge to forest-dependent communities is to apply that knowledge in a new forest products industry that will replace the traditional grade mills.

In the past, the traditional manufacture of lumber was well defined and the products well understood: There was no need for marketing, in itself. The customers knew the products, the suppliers, and their own needs—leaving just the price. That was determined through an auction-like process between the seller and buyer, almost always on the phone. In developing the new grade industry, it will be necessary to market a set of new products that are not yet understood from raw material that is yet to be defined using manufacturing techniques that have not yet been chosen.

Possible Approaches to a New Grade Industry

Grade products need grade logs, and this has traditionally required large old trees. With ecosystem management, these trees will be far less plentiful for commercial use than in the past. What exists must somehow be stretched well beyond past levels of utilization. At the same time, there is grade, suitably defined and extracted, in smaller trees. Slow growing, suppressed trees can have tight grain and quality knots, albeit in smaller pieces than those from larger trees. Will this be a source of grade products?

High-quality-appearing wood for industrials (secondary manufacturing) is traditionally manufactured into boards that are 1.25-inch thick. This serves as feedstock for secondary manufacturing. If the same wood is sliced, maintaining its surface qualities, it can be mounted on a low-cost core that provides the shape and size of the original product. The advantages are obvious in principle—with skins of one-64th of an inch, 40 boards of similar size to the original single board could now be produced. Does this mean that one old tree is now the equivalent of 40 previous old trees? Could an ecosystem management concept allow an old-tree harvest less than 3 percent of previous harvests?

Would that laminated product be successful? It has, in fact, been custom made and accepted in the market. How would it change the economics of the grade mill illustrated in the earlier table? The same plant is shown in Table 20.3 using the industrial-quality material in thin laminates on cores made from construction-grade material found in the same logs.

This table illustrates the savings in high-grade logs, with a reduction in cost of material as one moves to the high-grade column. The increased cost of manufacturing, already discussed in Figure 20.1 on added-value manufacturing, is also shown. The table does not indicate the significant reduction in timber volume that is needed for the conventional plant to reach its best financial performance, but it can be estimated from the profit figures in the high-grade column. Using one-quarter of the timber volume, the laminate plant could reach the same profit level as the conventional plant.

Table 20.3 Comparison of the economics of the grade mill for laminated and conventional products

Log grade	Low Quality $/MBF	Medium Quality $/MBF	High Quality $/MBF
Conventional			
Log cost	300	375	600
Manufacturing	89	93	107
Total cost	389	468	707
Product value	345	460	825
Profit	-44	-8	120
Laminated			
Material cost	300	313	320
Manufacturing	329	329	329
Total cost	629	642	649
Product value	960	1040	1120
Profit	331	398	471

Note: MBF = thousand board feet.

High-grade material in small trees, such as suppressed-growth Douglas-fir, cannot easily be matched to traditional manufacturing techniques. New approaches to markets and to manufacturing will be needed. Can hardwood experience provide a model? Similar changes have already taken place in the hardwood industry, where the falling harvests came from forest depletion rather than from environmental planning.

The hardwood industry has lessons and industrial practice to offer the small-scale end of the grade softwood industry. The hardwood industry long ago perfected veneers and adopted low-cost cores from low-grade materials. The principal difference between hardwood and softwood products lies in production volume. Before a transfer of manufacturing technology between the two will be successful, differences between the two should be understood and resolved. For example, hardwood manufacturing machinery is precise, strongly automated (with many similarities to that in metal machine shops), and usually takes hand-fed materials. These characteristics reflect the high value of individual hardwood products. If softwood products fall short of these values, the hardwood machinery characteristics would be unacceptably costly. However, the principles might apply in a different machine format.

The Immediate Tasks Ahead

The transition to a new high-grade forest products industry needs a base of information about future forest harvests and a new understanding of the wood products marketplace. With this underpinning, the manufacturing methods required to build the bridge between the forest and the marketplace can be determined. It will be an iterative process as understanding improves on both sides of that bridge—but the process will have to start somewhere.

The process could begin by investigating the several different components of the new industry that exist. Veneer slicers for softwoods are used at several different locations in the Northwest, and product cores can be made from many different source materials using existing manufacturing methods. Perhaps a product integration company would be an appropriate start. Such a company might emphasize market issues through custom manufacturing of different pieces brought in through separate sources.

Setting up a new manufacturing facility under the uncertain conditions of ecosystem management obviously entails risk, but so does inaction. Perhaps a compromise would be to locate potential source material in a national forest where low-risk predictions of availability can be made. Suppressed-growth timber that needs thinning would be a better commercial candidate than a few trees in an old-growth grove. The manufacturing facilities would be quite different for the two sources of wood, but the product integrator would use the same knowledge.

Alongside major changes in manufacturing lies the need for a strong link between the forest and the forest product. In the past, the manufacturing plant was designed for standard products, and the company forester sought and bid on the feedstock suited to that plant. In the future, the available commercial timber will be diverse and the plant will have to be able to respond to that diversity or deal only with a part of it. From plant design to product marketing, a thorough knowledge of the local forest and new management processes will be essential to developing and sustaining the new industry.

In the earlier discussion, sorting materials and lumber products was described as a key feature of wood product manufacturing. Sorting has been done using grading rules with a limited number of classifications; this has allowed manual labor, or machine-assisted labor, to sort the lumber as it moved rapidly through high-volume production.

Computers have been used in information industries for many years to read and sort tables of data, text, and graphics. The type of information required to sort lumber and wood products is basically similar. The lumber and secondary wood products industry could be regarded as a special type of information industry, with each piece of wood carrying its own "label" allowing the manufacturing process to sort it. Plants based upon this system would separate material handling from label sorting and would look much different than today's plants.

The Longer View

This chapter has focused on the immediate consequences of changes in timber harvests in the Pacific Northwest forests. The same issues are expected to spread through the management of other public forests and to confront the management of private timber.

Environmental issues have triggered new approaches to the management of the national forests. However, if successful resolution of these issues is based upon improved harvesting and new manufacturing solutions, future changes may well move from environmental to economic concerns. The past abundance of commercial timber meant that industrial practice could continue. With changes in timber availability, regardless of the reason, new practices will be required—and they will all come back to the diversity and lowered quality of the timber supply. Making more out of less will be the rule, whether in quantity or quality. The new approaches to be adopted in the Pacific Northwest could show the way for ecosystem management elsewhere.

Local imbalances in the supply of material volume can be solved from elsewhere. Some lumber plants will import logs into the Pacific Northwest or bring core materials for laminated structures from elsewhere. As the imbalance spreads in the future, however, the supply of wood fiber will become a universal issue. Why should the inside of wood products take decades to produce through trees? Could the material be raised in annual crops through industrial agriculture and manufactured through engineered processes into a final core product?

Answers to these types of questions will come from the basic inquiry introduced in this chapter: What are wood products? What are their basic characteristics? and Can they be provided through alternative means that still offer the environmental advantages of biomaterials but that exist in harmony with the forests of the 21st century?

Manufacturing in the 21st century will emphasize the value of wood and will deliver that value while economically and efficiently using its medium. New manufacturing technology will enable that value to be read, extracted, and delivered through wood products that use a small fraction of the fiber volumes committed from forests in the past. The technology of the wood products industry of the future will rival that found in the aerospace industry of today. In responding to the harvesting economies required by ecosystems management, wood products will have to meet similar design conditions demanded by flight: wooden structures that match a purpose with economy of quality material.

21 Special Forest Products: Integrating Social, Economic, and Biological Considerations into Ecosystem Management

Randy Molina, Nan Vance, James F. Weigand, David Pilz, and Michael P. Amaranthus

Global Perspectives and History 317
Regional Perspectives from the Pacific Northwest 318
Socioeconomic Considerations 321
Unknown Supplies 321
Changing Demands 322
Labor and Employment 323
Balancing Management Costs and Benefits 323
Adaptive Ecosystem Management Considerations for Special Forest Products 324
Understanding the Unique Biology and Ecology of Special Forest Product Species 325
Forest Community Dynamics and Landscape Considerations 327
Silviculture and Vegetation Management Approaches 328
Integrating Human Behavior 328
Conducting Necessary Inventory, Evaluation, and Research Monitoring 329
Adaptive Ecosystem Management of Commercial Mushroom Harvests 330
Acknowledgments 332
Literature Cited 332

Throughout history, forests have provided a wealth of beneficial and essential products ranging from foods and medicines to building materials. Ancient pharmacopoeias list myriad forest plants and fungi for treating various ailments. Many of these ancient remedies have evolved and continue to evolve into the important drugs of modern medicine. Use of diverse forest species remains commonplace around the world, particularly in cultures with strong rural traditions. Even in the most technologically advanced societies, traditional uses of forest products continue, often as recreational pursuits. For example, the tradition of collecting and consuming wild edible forest mushrooms by Europeans and Asians continues by their descendants in North America.

As societies modernized and depended less on the diversity of wild products from forests, many of these traditional uses diminished, some were forgotten, and others remained useful to only subsistence forest dwellers or native inhabitants. Forest management in the 20th century increasingly emphasized growing and harvesting trees for timber and fiber products as its primary objective. Despite that emphasis, a small entrepreneurial segment in forest-based communities continues to commercialize nontimber products from the forest, including foods, medicinal plants, and floral greens. Forest managers designated these as "minor" forest products, thereby reflecting an attitude that they were less important than timber in the overall scheme of forest management. But the economic impact of this industry and the quantities of products harvested can no longer be viewed as minor. In the Pacific Northwest, special forest products account for over $200 million in revenue (Schlosser et al. 1991). This amount is substantial when compared to the $2.63 billion generated from stumpage receipts to all landowners in Oregon and Washington in 1989 (data derived from Warren 1995). Thousands of tons of biological materials from dozens of species are removed annually from forest ecosystems.

The paradigm shift in forestry from timber management to ecosystem management has heightened public awareness of the importance of the special forest products industry. In addition, managers and the public have increasingly recognized that special forest product species are important components of forest ecosystems. Dramatic declines in revenue to rural forest communities from harvest of federal timber are increasing the importance of special forest products industries in rural economic recovery and development. The two largest public land agencies in the United States, the Bureau of Land Management of the U. S. Department of the Interior and the Forest Service of the U.S. Department of Agriculture, recognize this importance and are developing regional and national strategies for managing special forest products. These strategies emphasize four themes: (1) to incorporate harvesting of special forest products into an ecosystem management framework with guidelines for sustainable harvest, species conservation, and protection of ecosystem functions; (2) to involve the public, including industrial, Native American, and recreational users of these resources, in making decisions about the future of special forest products on public lands; (3) to view the management of and accessibility to special forest products as major factors in assisting rural economic diversification in formerly timber-dependent communities; and (4) to develop and implement inventory, monitoring, and research programs to ensure species protection and ecosystem health.

Blending the management of special forest products into the holistic objectives of ecosystem management will not be easy. The social structure and composition of the industry differ from the typical timber-based community. In the Pacific Northwest, large numbers of migrant harvesters from various ethnic backgrounds are major participants in the industry (Schlosser and Blatner 1995). A thorough understanding of all the groups involved in the special forest products industry is essential to developing effective communication and building common understanding about management directions. The economic structure of the industry is also poorly understood; market dynamics are difficult to track, so that trends in the industry remain clouded. If special forest products are to play a role in rural community development, these economies must be better understood so that investors can assess the risks involved in these enterprises.

The complex biology and lack of information on harvesting of special forest product species also present a significant challenge for integrative ecosystem management. Numerous federal and state laws exist to protect forest resources, including the National Forest Management Act, National Environmental Policy Act, and the Endangered Species Act. Under strong environmental regulations and in a litigious climate, resource managers require substantial data to support management decisions. Unfortunately, baseline data on the effects of harvest, on markets, and on the biology, ecology, and productivity for many special forest product species are either short-term, incomplete, or nonexistent. Also lacking is information on responses by harvesters to economic incentives and conservation measures set in place by land managers. Many of these species also play important ecosystem roles, such as providing food for wildlife and capturing and cycling nutrients. Yet, we poorly understand these complex dependencies, and the consequences of harvesting special forest products on ecosystem function and integrity are largely unknown. To incorporate sustainable harvests of special forest product species within an ecosystem management context, resource managers and researchers must develop and implement research, inventory, and monitoring protocols for these species and promote transfer of key information to affected and interested publics.

Adaptive strategies for ecosystem management provide for diverse forest stands and landscapes and for a wide range of values and products. Practices are modified as more ecosystem and societal information becomes available. Traditional forest management affects the distribution and abundance of many special forest products by shortening rotations of tree crops and growing only species and genotypes that produce the highest timber yields. Although some special forest product species can thrive in homogenized or intensively managed forests, others cannot. Ecosystem management recognizes that the dozens of genera of plants and mushrooms and assorted other materials collected in the United States as special forest products are products of natural diversity; it therefore adapts practices to help assure their sustainability. We are in the beginning of the adaptive phase of managing for special forest products. Some decisions must be made without sufficient information and adapted or modified as results from inventories, monitoring, and research become available. This chapter develops a conceptual framework for ecosystem management of special forest products and provides a model for future decision support. We draw from familiar examples in the Pacific Northwest, where the special forest products industry is expanding, but the issues, problems, and solutions likely will apply to many other forested regions.

Global Perspectives and History

Over the ages, indigenous peoples acquired an understanding of the basic ecology and usefulness of flora and fauna and explored the various foods and medicinal properties of plants and fungi. Stored in the memories of elders, healers, midwives, farmers, and fishers in the estimated 15,000 remaining indigenous cultures on Earth is broad knowledge about useful products from naturally diverse ecosystems. This knowledge has been passed on through ancient but fragile chains of oral tradition that will be broken when younger members of a society leave the traditional lifestyles. Today, much of the expertise and wisdom of these cultures has already disappeared; our understanding of special forest products gleaned from thousands of years of trial, error, and observation is consequently diminished.

In the course of cultural evolution, other socioeconomic lifestyles have emerged, especially those based on agricultural or industrial production. These agricultural and industrial societies coexist with hunter-gatherer lifestyles, but at the same time change them (Keene 1991). Today, large, single-product timber industries operate side-by-side with smaller-scale cottage industries processing diverse forest products and with individuals who gather special forest products for subsistence, cash income, or recreation. In many rural societies around the world, forested ecosystems continue to be the primary source of products used for food, fodder, fibers, housing, and medicines.

In industrial societies, advances in technology and changes in patterns of economic exchange have altered the scale and type of forest-products consumption and the relative social standing of hunter-gatherers. Over the last two centuries, industrial economies have grown by using capital gained from rapid and effective exploitation of natural resources. Industrial societies will tend to use the few products most efficiently produced from a forest ecosystem. This trend contrasts with that of subsistence societies that tend to have little capital for development and a more geographically restricted access to resources. They may find more products from a single forest, but the products are often less profitable. Almost all countries, including the United States, have substantial populations of harvesters and gatherers that lead subsistence lifestyles.

Industrialized nations such as Russia have never lost their preindustrial cultural ties to diverse forest products. Customs regarding collecting berries and mushrooms remained vital even as industrialization and urbanization proceeded rapidly in the 20th century. Research in Finland, France, India, Italy, Japan, Poland, and Russia and, more recently, in many tropical countries recognizes the importance of special forest products to those national economies. International conferences in Europe and Asia (Akerle et al. 1991, Marocke and Conesa 1981, Vänninen and

Raatikainen 1988) have called attention to the renewed economic role of special forest products in the economy of forested ecosystems.

Industrial and nonindustrial countries alike are beginning to formally recognize the disparity in wealth and material well-being among their peoples, as well as the moral obligation to practice good land stewardship. Out of this recognition, the concept of sustainable development was agreed upon by nations participating in the United Nations Conference on Environment and Development in 1992. Organizations and governments are defining and initiating operational strategies for sustainable development; this process could become a global trend in the 21st century. Special forest products will play an important role in sustainable development and are perceived as essential links to sustaining rural communities and contributing to economic diversification. At the same time, wild species that are special forest products contribute to the diversity and function of forested ecosystems. Thus, strategies must be developed that both supply and conserve these valuable species.

The focus of the biomedicinal or phytopharmaceutical industry on natural products illustrates the global scale of challenges to stewardship and sustainable development issues. Exploring biological diversity as a potential source of new, valuable phytopharmaceuticals is gaining interest and investment from some of the largest pharmaceutical companies in the world (Joyce 1994). In the synthetic drug market, about 25 percent of all prescription drugs still contain natural plant materials as active ingredients (Der Marderosian 1992). Of some 121 prescription drugs currently in use that are derived from higher plants (not including antibiotics from microorganisms), 74 percent were known in folklore (Joyce 1994); about 175 drugs used by native Americans are listed in the *U.S. Pharmacopoeia* (Robinson 1977). Thus, many plants selected for bioactive screening are sampled because ethnobotanical research indicated that the plant had a known medicinal activity or was used by an indigenous group for healing wounds or other disorders (Forlines et al. 1992, Gunther 1973). Today, large investments in drug development, technologically advanced screening capabilities, sophisticated joint ventures, cooperative agreements, and government-supported funding of research raise new issues of resource sovereignty and intellectual property rights (Joyce 1994).

Regional Perspectives from the Pacific Northwest

Pacific Northwest forests were essential to the lives of aboriginal people that migrated to North America with the receding glaciers as early as 10,000 years ago. Various forest species were important for food, tools, structures, transportation, and medicine. Large conifers were used for housing, and understory species such as Pacific yew, willows, and red alder were used for many purposes, including hunting tools, bowls, masks, and medicines (Pojar and MacKinnon 1994). Many plants and fungi were part of native traditions and mythology. Medicinal use and spiritual value were linked in medicines derived from trees, forbs, and fungi (Molina et al. 1993, Pojar and MacKinnon 1994, Smith 1983); for example, devil's club was used by Northwest tribes to cure a variety of illnesses and to ward off evil spirits.

Use of plants by indigenous people in the Pacific Northwest for food and aboriginal technology is well documented (see, e.g., Colville 1897, Gunther 1973, Reagan 1935, Turner et al. 1990). Aboriginal people not only gathered species, they also burned and selectively harvested to maintain productivity of important food plants. Use of fire promoted bracken and camas, staples of aboriginal diets, before fire was suppressed by settlers and the land agencies that followed (White 1980). The use of fire to promote berry species and forage is also well documented (Boyd 1986, Gottesfeld 1994, Norton 1979, Robbins and Wolf 1994, Turner 1991). Thus, indigenous people introduced management techniques long before the arrival of other people to the region and may have used over 100 different species (Turner et al. 1983). Indigenous people do not now rely on wildland plants as a primary food source as in the past. However, certain plant products such as berries are harvested extensively, and many traditional uses of plants as herbals, tonics, and medicines are maintained in tribal culture (Turner et al. 1983, 1990).

When settlers came to the Pacific Northwest, they too relied on wild, native species as important supplements to food, medicines, and clothing. Superimposed on wild harvesting were their own traditions and knowledge of plant species harvested or cultivated for herbs, food, and medicine. They applied knowledge from their homelands to related species in the Pacific Northwest. In addition to wild berries, they found forbs such as nettles, thistles, yarrow, and bedstraw that could be used in the same ways as their European equivalents. (Robinson 1977). Whether settlers learned from indigenous people or already knew of certain plant properties, Europeans and Native Americans often had identical uses for plants. Cascara from *Rhamnus* spp. served as a laxative for both cultures and is still commercially harvested for that purpose. Nettles provided fiber for Northwest Indians, and they were an important fiber source in Europe and Asia (Moore 1993, Robinson 1977). With development of commerce, agriculture, and urban trade centers, dependency on the local forest for its products declined; public lands began to be viewed primarily as an economic cornerstone of industrial development.

Timber, livestock, water, and minerals were the economically important products from public lands in the rapidly ascendant Euro-American economy of the Pacific Northwest. The value, if any, of other plants and fungi continued for subsistence use, largely lacking in any tradable or marketable value. The result was that such goods from public lands were essentially free nonmarket goods. Folk traditions all but disappeared as the commercial development of agriculture and timber harvesting dominated and altered the Pacific Northwest landscape. Traditional folk use of wild plants was relegated primarily to those rural people who, for economic or philosophical reasons, continued to harvest, use, and trade in native plants.

A resurgence of interest in the traditional use of native plants corresponds with increased recognition of the value of traditional healing methods (Krochmal and Krochmal 1984, Moore 1993). This interest is apparent in the growing number of publications on wildcrafting—i.e. foraging for native herbal medicinal edible and otherwise, useful wild plants. As a consequence of the public's demand for natural, unprocessed goods, products that originated in the wild are occupying an increasingly important niche in the marketplace. For example, the leaves of the sword fern, once used for medicine, insulation linings, and sleeping mats (Gunther 1973), are important in today's floral greens industry. The bark of Oregon grape provided a bright yellow dye used by native people in making baskets; its bark and berries were the source of medicinals. The plants continue to be a source of commercial pharmaceutical alkaloids, of which berberine is a primary medicinal constituent. Harvested for the floral greens industry, the plant's foliage also contributes to its current economic importance (Tilford 1993). The rush to capitalize on these markets is altering the traditional relationship between the resources and rural-dwelling harvesters, who have depended on open access to public lands. They are not only finding new markets for their products, but also experiencing increased competition for restricted resources.

Several hundred native plants, including trees, shrubs, forbs, and vascular and non-vascular plants (but excluding those used in the timber industry), are currently harvested for personal and commercial use in the Pacific Northwest (see Table 21.1). The use of these plants is diverse, falling into five general areas: (1) foods, (2) decoratives, including floral greenery and dyes, (3) herbals, (4) medicinals, and (5) specialty products such as aromatic oils and wood products. The Pacific Northwest probably leads all other regions in North America in active use of public lands as a source for diverse floral greens and botanicals (Thomas and Schumann 1993). Commercial harvesting is expanding concurrently with recreational wildcrafting. Gathering plants and plant parts for personal use as ornamentals, foods, or herbals is important to traditional rural wildcrafters, but these activities have widened to include more diverse groups of people. Renewed interest in wildcrafting appears to accompany an emerging ecological ethic. Books such as Tilford (1993) and Moore (1993) include information designed to raise environmental awareness and encourage low-impact harvest of wild plants.

Conifer boughs constitute a constant but highly seasonal market (October to December) in the Pacific Northwest. Schlosser et al. (1991) estimates that 5,000 persons were employed harvesting boughs in the region in 1989 and another 4,000 manufactured wreaths and swags during the short season. Boughs from noble fir provide the greatest volume in the region: in 1989, 9,310 tons were harvested, yielding

Table 21.1 Representative native plants and fungi from the Pacific Northwest currently important as special forest products

Use	Common Name	Scientific Name
Floral greens	Salal	*Gaultheria shallon*
	Sword-fern	*Polystichum munitum*
	Evergreen huckleberry	*Vaccinium ovatum*
	Bear grass	*Xerophyllum tenax*
	Dwarf Oregon grape	*Berberis nervosa*
	Oregon boxwood	*Pachistima myrsinites*
	Mosses	*Isothecium* spp.
		Hypnum spp.
		Neckera spp.
Christmas greens and boughs	Douglas fir	*Pseudotsuga menziesii*
	Noble fir	*Abies procera*
	Western red cedar	*Thuja plicata*
Edibles	Huckleberries	*Vaccinium* spp.
	Berries	*Rubus* spp.
	Fiddlehead	*Pteridium aquilinium*
	Mushrooms:	
	Chanterelle	*Cantharellus cibarius*
	Morel	*Morchella* spp.
	Matsutake	*Tricholoma magnivelare*
	King bolete	*Boletus edulis*
Medicinals and herbs	Pacific yew	*Taxus brevifolia*
	Cascara	*Rhamnus purshiana*
	Devil's club	*Oplopanax horridum*
	Prince's pine	*Chimaphila umbellata*
	Stinging nettle	*Urtica dioica*
Dyes	Oregon grape	*Mahonia nervosa*
	Red alder	*Alnus rubra*
	Western red cedar	*Thuja plicata*
Poles and decoratives	Lodgepole pine	*Pinus contorta*
	Noble fir	*Abies procera*
	Vine maple	*Acer circinatum*

Note: See USDA Forest Service (1993b) for a broad listing of special forest product species and their use in the United States.

$6.7 million, or roughly half the regional market value (Schlosser and Blatner 1993). Other important species commanding high prices are subalpine fir and incense cedar. The fine commercial quality of Pacific Northwest conifer boughs has not gone unnoticed elsewhere in the world. Little information has been published in North America about methods of commercial bough production in the Pacific Northwest (Murray and Crawford 1982) or elsewhere in North America (Hinesley and Snelling 1992), but Europeans have published numerous studies (e.g., Hvass 1964, Weege 1977) on determining optimal spacing, fertilization, and pruning regimes for bough production, with noble fir as an integral part of mixed-product silviculture.

The medicinal plant market in the Pacific Northwest has typically been associated with native herbs used for alternative therapies. A growing body of manufacturers in herbal and over-the-counter medicines rely on local wildcrafters, but the Pacific Northwest, with its rich biological diversity, has not been ignored by mainstream pharmaceutical industries looking for new, effective compounds. A noteworthy discovery came from the Pacific Northwest in the 1960s and 1970s when the compound taxol, extracted from Pacific yew bark, proved to be an effective agent against ovarian and breast cancer. Taxol was developed by the pharmaceutical company Bristol-Myers Squibb (BMS) under a collaborative research and development agreement with the National Cancer Institute. In 1991 and 1992, about 1.7 million pounds of yew bark were harvested from public lands in the Pacific Northwest under interim guidelines developed by the Forest Service and the Bureau of Land Management. An environmental impact statement (EIS) was issued (USDA Forest Service 1993a) that restricted yew harvest to timber-sale areas and used previously developed harvest guidelines to limit the proportion and size of yew trees taken from any harvest unit. The EIS also required replanting of harvested areas with seedlings or rooted cuttings. In 1993, BMS withdrew from federal lands in the Pacific Northwest because other yew species, harvested primarily in India, provided more cost-effective sources of taxol.

In the United States, federal laws, including the Pacific Yew Act of 1992, and environmental impact statements, such as the federal EIS on management for the northern spotted owl in the national forests, provide a legal framework for limiting harvest of Pacific yew, even though scientific knowledge of the species is lacking. Nevertheless, phytopharmaceutical industries, with their rapid and large-scale development capabilities, could quickly impact a resource when a commercially valuable product is discovered. Such discoveries are relatively rare, but they have widespread environmental and socioeconomic effects whenever they do occur. Future discoveries of phytopharmaceuticals may come from species less common than the Pacific yew. If alternative sources are neither available nor cost effective, conflicts between using native species for treating human diseases and conserving them to protect the species may be difficult to resolve.

As we approach the 21st century, the forests of the Pacific Northwest will become increasingly known for the value of their natural products. Biological diversity, one of the region's greatest ecological assets, could provide abundant opportunities for the development of new commercial products, but this could place more species at risk and increase disturbance of complex ecological relationships. Managing for biodiversity and ecosystem sustainability should provide the special forest product industry with stability and accommodate the tradition of individual collection, use, and enjoyment of native species. To do this we need to have better knowledge not only of the ecosystem but also of the special forest products industry.

Socioeconomic Considerations

Unknown Supplies

A basic gap in knowledge results from the lack of an inventory of the type and amount of existing and potential special forest products in the Pacific Northwest. Such ignorance can have negative economic effects. Without an awareness of changes in stock levels, managers cannot implement adaptive measures to adjust stocks so that a desired and sustainable level of product is available to developing markets. A reliable supply is also essential to wholesalers of special forest products (Handke 1990) and for building enduring commercial relationships that provide steady income to suppliers. For instance, loss of the German market for Washington State chanterelles was the result of a sudden and unpredicted collapse of the supply (Russell 1990). With the growth in worldwide markets for special forest products from the Pacific Northwest and elsewhere, concerns about resource depletion arise even before management of special forest products begins (Foster 1991).

Field inventories are crucial tools for (1) calculating

existing stocks of special forest products, (2) identifying existing areas of overharvest, (3) analyzing the possibility of intensified management for expanded commercial production, and (4) providing the scientific database for research on ecological and economic constraints to production (Grochowski and Ostalski 1981). For example, evidence from Europe, where mushroom picking has been intense for much longer than in North America, suggests that environmental changes and increased mushroom collection are leading to declines in mushroom populations (Arnolds 1991, Cherfas 1991, Jansen and de Vries 1988). However, few published articles document the effects of economic harvesting on special forest products (e.g., Benjamin and Anderson 1985, Geldenhuys and van der Merwe 1988, Smirnov et al. 1967). A global information system is needed to register changes in natural populations of special forest product species (Cunningham 1991). Countries such as Bulgaria, Lithuania, and Poland already incorporate national inventories of such species (Budriuniene 1988, Economic Commission for Europe 1993).

Monitoring is especially helpful for short-term prediction of harvest yields. Forecasting crop yields on the basis of weather data and phenological information can help resource managers decide on the intensity of harvest, the number of permits to issue, the prices to charge, and the allocation of harvesters and handlers in the seasonal workforce. Kujala (1988) has carried out exemplary work in phenological studies of berry and mushroom crops in Finland, but such studies are rare. Information and understanding are lacking for forecasting the productivity of North American species of economic importance, although studies address productivity of prominent commercial species such as Pacific yew (Vance et al. 1994) and chanterelles (Norvel et al. 1994). An inventorying system can provide support to managers so that better decisions can be made about allowable harvest, number of people permitted to harvest, duration of the collection period, and the cost, if any, of licenses to support ecosystem management for resource conservation.

Developing inventory and monitoring programs for special forest products will present new challenges for resource managers. Many special forest product species are irregular in occurrence on the landscape or are present for short duration (e.g., mushrooms). Other considerations include economic and uneconomic concentrations, quality characteristics of commercially valuable stock, importance of access on commercial feasibility, and how to extrapolate inventory results to information about commercial occurrence (R. Fight, Research Forester, Pacific Northwest Research Station, personal communication, 1995). Initial costs to establish new inventory and monitoring programs may be high.

Changing Demands

Tastes and preferences of people constantly change and provide impetus for innovation in special forest products industries. In the early 1960s, for example, bear grass was unknown as a floral green, and edible wild mushrooms were largely unrecognized as a commercial product (USDA Forest Service 1963). Both the evolving definition of special forest products and the reception of products by distributors and end consumers are highly subjective and difficult to know in advance. In central Europe, a market for cut flowers and floral greens endures, but markets for specific plants are subject to rapid changes in fashion (Handke 1990). The accelerating pace of information access often speeds up consumer awareness and demand for innovation. Demand for many special forest products may represent fleeting or erratic markets based on changing tastes and technology.

Cross-cultural comparisons of special forest product species provide sources of information that can lead to innovation in market development. While European countries in the north temperate zone currently use many species of native plants and fungi, closely related species in the Pacific Northwest are often underutilized or unknown for their use as special forest products. Given the receptivity of the American public to innovations in consumption and marketing, the exotic traditions of other countries might provide marketing angles for culinary, medicinal, and horticultural commodities in North America or offer new markets for products abroad. Species such as serviceberry, madrone, dogwood, hawthorn, wild rose, elderberry, mountain ash, and viburnum have relatives native to Europe and temperate Asia (Bounous and Peano 1990, Cherkasov 1988) that already are widely used. Raspberry, native to the Pacific Northwest, is commercially cultivated in forests of

Russia and Lithuania (Budriuniene 1988, Cherkasov 1988). Market niches for these species, whether for subsistence, commercial, or recreational uses, remain to be developed.

New special forest products in the Pacific Northwest might also serve as a substitute for products currently imported to the Pacific Northwest. Ruth et al. (1972) initiated studies concerning development of maple syrup production from bigleaf maple in the Pacific Northwest. Although the study was encouraging, there was no follow-up research and development. Likewise, Oregon white truffles could hold promise as a future market substitute for European sources. On the other hand, substitution for Pacific Northwest special forest products by comparable products from other regions is equally possible.

Labor and Employment

Understanding the special forest products industry and planning for its future requires that planners and managers consider the people employed in the industry. In the Pacific Northwest, the workforce in special forest products has changed rapidly. Schlosser et al. (1991) and Schlosser and Blatner (1995) surveyed processors of floral and Christmas greens and edible wild mushrooms, and Handke (1990) describes the wholesale market for Pacific Northwest floral greens in Germany. But a comprehensive picture of the participants in all sectors of the industry is not yet completed.

A particular gap in information is a thorough profile of harvesters. In the Pacific Northwest, many people consider collecting special forest products as a last resort for employment after other options have failed (McLain et al. 1994). Obtaining information about professional collectors is difficult, because many have nomadic or reclusive lifestyles and work only seasonally. The appearance of unexpected ethnic groups among commercial harvesters has required rapid cultural sensitization and response in the form of new instructional materials and outreach efforts by the Forest Service and the Bureau of Land Management. To date, however, sociologists and anthropologists have not undertaken studies for the Pacific Northwest similar in scope and intent to those in Italy (Farolfi 1990), Finland (Saastimoinen and Lohiniva 1989, Salo 1984, 1985), and Thailand (Moreno-Black and Price 1993) to characterize populations who harvest special forest products. A basic question in the context of Pacific Northwest society is whether public policy should emphasize development of special forest products within a community or promote efficiency of migratory labor.

Educational programs for harvesters can help both federal land managers and harvesters accomplish mutually beneficial goals. It is often difficult for ecosystem managers to learn about the harvester workforce in their areas. Through cooperatives, extension workshops, or community college courses, exchange of information could help to develop a sense of community and cooperation and provide a forum to address common problems and interests. Education also imparts the values and ethics of ecosystem management and can provide background training in small-business management and finance.

Balancing Management Costs and Benefits

The American public has concerns about both sustainability of special forest product resources and the equitable distribution of benefits derived from special forest products. Rapidly growing public awareness of special forest products as a source of income has been swelling the ranks of product suppliers at all levels of the market supply structure. The growth of Oregon and Washington's population to 7.7 million by 1990 and projections for its continued rapid increase preclude the practicality of open and unrestrained harvesting on public lands. The Pacific Northwest also lacks a locally evolved tradition for regulating harvests.

Private gain from collectively owned resources, such as federal forest land, should require compensation for any deterioration and for subsequent management costs for site restoration or monitoring. Likewise, on private lands, landowners deserve compensation for granting collection rights to other people. Popular perception holds that special forest products industries derived from public lands are largely unregulated, unreported, and untaxed (Molina et al. 1993). Illicit collection and loss of revenue to landowners have not been estimated; however, the rapid development of mushroom markets in the 1980s would indicate that the cost to individuals and to society is considerable. From a policy standpoint, the costs and benefits to society of an unregulated industry must be weighed against the economic costs and benefits of regulation to support sustainability and equity to landowners.

Studies of the economic costs and benefits on returns from investment in special forest products are available from only a few sources in the Pacific Northwest. Managing for a sustainable industry must take into account not just the costs of managing forest vegetation, but also the management of the labor and operations in the harvesting, transportation, grading and sorting, packaging, and distribution networks. Costs for sale of permits or contracts for long-term leases will be incurred as planning staffs prepare and supervise sale operations. Setting fair-market prices for sales is often difficult when empirical information about the value of permits or leases is unknown or highly variable. Costs also arise from the need for law enforcement, maintenance of campgrounds, and resource monitoring.

Benefits from sustainable management of special forest products should be considered at several scales simultaneously. Revenues to federal land management agencies should cover costs of planning, monitoring, and on-the-ground management. Permit holders and lessees should also benefit from the permit and leasing systems by being assured of a reliable supply. Market dynamics of supply and demand will determine the practicability of harvests by individuals as means to cover federal land management costs.

The Bureau of Land Management and the Forest Service are now exploring a uniform appraisal system for special forest products. Uniformity of legislation, regulations, and enforcement is important to develop coherent market responses from collectors and to increase returns to the land management agencies. There is the chance that regulations will not be effective if their basis and intent are not apparent. Agencies can use regulations to encourage harvest practices consistent with the goals of ecosystem management. One option might be to assign custodial harvest rights in a designated extractive reserve for a single resource for a specified term on the basis of competitive bidding (Fearnside 1989). Ecosystem management would rely on the self-interest of the permitted harvester to ensure the broader societal objectives of sustainable resource use.

There are potentially negative consequences of instituting an obligatory permit or lease system for harvest rights. Possibly the people who might most be in need of income from harvesting special forest products would lose their income source if they had to pay prohibitively high permit fees (Brown 1994). In cases where societal goals include aiding low-income people or diversifying a local rural economy, below-cost sales or sales reserved for small businesses can be effective for developing local capital and creating year-round employment. The relative social and environmental benefits and costs can be assessed as a gauge of the effectiveness of these programs to promote economic opportunities while at the same time protecting the ecosystems being managed.

International policies may provide some guidance for domestic rural development. Chambers (1983) suggests that improving the quality of life of the most economically disadvantaged is a desirable goal of rural development. If federal land management agencies adopted this goal, the economic resiliency of harvesters would be an important consideration. In addition to favorable permit systems, federal or state sources could support the establishment of special forest product cooperatives and processing businesses. Forestry has a long history of cooperatives. Generally, they stimulate entrepreneurship among people previously excluded from product development and marketing (Mater 1993).

The growing awareness of the actual and potential importance of special forest products should lead to changes in land management. Managers strive to meet dual social and ecological objectives—meeting the needs of society for special forest products and conserving the ecosystem function of special forest product species for sustainable interdependent production of all ecosystem goods, services, and conditions desired by society. We propose that adaptive ecosystem management provides the most likely context to assist land managers in meeting the dual goals.

Adaptive Ecosystem Management Considerations for Special Forest Products

Ecosystem management of special forest products on public lands requires integration of knowledge about human and ecosystem behavior. Managers and the public need to understand how factors of biological production and economic activity interact in time and space. Comprehensive knowledge of the roles of people and special forest product species in forest ecosystems will facilitate decisions that conserve, sustain, and enhance each special forest product resource and that meet human needs.

Primary considerations for management of special forest products include: (1) understanding the unique biology and ecology of special forest product species; (2) anticipating the dynamics of forest communities on a landscape level, delineating present and future areas of high production potential, and identifying areas requiring protection; (3) developing silvicultural and vegetation management approaches to sustain and enhance production; (4) integrating human behavior by monitoring and modeling people's responses to management decisions about special forest products; and (5) conducting necessary inventory, evaluation, and research monitoring.

Integrating these considerations into management decisions can appear overwhelming at first glance. Adopting the premise of a continually adapting management system clarifies the process for managers and the public. Activities in special forest products management do not proceed in linear order. Instead, information generated from one activity improves knowledge and refines direction in the other management activities. In this section, we discuss management activities derived from these considerations and how they interact. We conclude by illustrating an example of adaptive strategies for ecosystem management of commercially harvested forest fungi.

Understanding the Unique Biology and Ecology of Special Forest Product Species

The numerous native plant and fungal species of commercial value in the Pacific Northwest perform myriad critical functions in forest ecosystems. Yet, to harvesters, each species has the property of a "product unit." To resource managers each species has the property of a "resource unit." Each individual also functions as a population member, a community member, and an ecosystem member (Allen and Hoekstra 1992). All these properties and values must be considered as we develop models for managing special forest product species. Biological and ecological factors important for modeling the productivity of these species include population dynamics, regenerative ability, life cycle, genetic structure, effects of herbivory and disease, and response to site variables such as overstory canopy conditions.

Plant and fungal species of current economic importance in the Pacific Northwest are described in Molina et al. (1993), Pojar and McKinnon (1994), Schlosser et al. (1992, 1993), and Thomas and Schumann (1993). It is beyond the scope of this chapter to describe each special forest product species, its commercial value, and the biological implications of harvest. Instead we discuss three important groups of special forest product organisms—mosses, understory plants, and fungi—as examples of how the biology and ecology of these species affect their roles as ecosystem components and economic products.

Moss harvest involves the removal of entire communities of bryophytes (moss and liverworts) growing in a harvest area. The most commercially desirable moss grows on trees. A single vine maple stem may carry as many as a dozen moss and several liverwort species (N. Vance unpublished data). Traditional markets for mosses in the Pacific Northwest have increased steadily since the 1980s as demand for commercial-quality moss growing in the Coast Range has risen dramatically. The Siuslaw National Forest, for example, has issued permits for the harvest of 25,000 bushels of moss annually since 1989. Illegal harvest is believed to be at least that much. Harvesters removed an unknown quantity of biomass from public lands before harvest restrictions were imposed and before any program for research and monitoring could provide data for determining sustainable harvest levels. With the imposition of restrictions on moss harvest through permits, quantities requested are exceeding the amount harvested on Bureau of Land Management, Forest Service, and State of Oregon lands (N. Vance unpublished data).

Although research on the ecology of bryophytes in the Pacific Northwest has examined distribution and biomass (Coleman et al. 1956, McCune 1993), it has not addressed human disturbance or population depletion. Species have been well identified taxonomically and have been morphologically characterized (Pojar and MacKinnon 1994, Vitt et al. 1988). At least 30 to 40 bryophytic species are commercially harvested for packing material in the horticultural trade and for decoratives in the floral greens industry. The Pacific Northwest coastal region, with moderate temperatures and high precipitation, favors moss growth and diversity. The fog zones of the Coast Ranges support luxuriant moss growth and therefore undergo concentrated harvest. Bryophytes may take 10 or more years to reach preharvest biomass and diversity levels, depending on species, intensity of harvest, and environmental conditions.

The impact of commercial harvest on the functional role of mosses in coastal ecosystems is not well known. Moss species such as *Hylocomium splendens* can be monitored as one measure of change in forest conditions (Wiersma et al. 1987). They serve as important bioindicators of air quality because of their ability to incorporate airborne pollutants such as sulfur dioxide and nitrous oxide (Ferry et al. 1973). A large portion of commercially harvestable moss is in late-successional habitat within the range of the northern spotted owl and the marbled murrelet. One coastal forest species, *Antitrichia curtipendula,* associated with nesting sites for the marbled murrelet and red tree vole, is particularly susceptible to air pollution (USDA Forest Service 1994).

A second group of special forest product species is understory plants whose occurrence and productivity are strongly affected by forest development. For example, three important species—salal, Oregon grape, and sword fern—are common in plant associations within the western hemlock zone where Douglas-fir has been intensively managed for timber production. Foliage from these understory species has been harvested over the past 50 years for use in the floral greens industry (Schlosser et al. 1992).

Silvicultural treatments that alter densities of tree canopies may enhance growth and composition of some understory species and diminish that of others. When the overstory canopy is dense, understories are poorly developed. In that case, opening the canopy should increase understory cover and species richness. However, if understory species are established in young stands that have partial canopies, increasing openings in the canopy favors species that respond best to increased sunlight. Salal, bear grass, and Oregon grape require more light than sword fern and should benefit from moderate thinning of the overstory to prevent a dense canopy (Huffman et al. 1994, Schlosser et al. 1992).

Overstory conditions affect the product quality (growth form and appearance) of desired understory species as well as their biomass. Floral markets require that commercially desirable foliage of these species have deep green color and no blemishes. Understory plants with the highest value grow predominantly under a partial canopy (Schlosser et al. 1992). Producing plants with these desired qualities is difficult because conditions in young stands change considerably over comparatively short periods. Experienced foragers rely on dependable sources of high-quality product and often prefer to harvest in late-successional forests with more stable canopy structure.

Forest fungi form a third group of special forest product species with unique biological features and ecosystem attributes. The body of most fungi consists of one-cell-wide threads, or hyphae, (collectively know as mycelium) that grow in soil, organic matter, or host organisms, where they are hard to observe without destructive sampling. The mushrooms or truffles (collectively called fruiting bodies or sporocarps) are the reproductive portion of the fungus. Many of the commercially valuable edible fungi depend on and are important to the health of host trees because they are mycorrhizal—that is, they form distinctive fungus-root structures. The fungus transfers water and mineral nutrients to the tree, and the tree provides the fungus with carbohydrates as an energy source produced through photosynthesis. When all trees are harvested, the associated fungi die in the soil and sporocarp production ceases (Amaranthus et al. 1994). Complete removal of all host trees therefore will have immediate impact on commercial harvests of mycorrhizal fungi such as chanterelles and American matsutake.

Fungi usually fruit during a particular and a limited season, and sporocarp production varies greatly from year to year, much like cone crops on forest trees or fruit crops in domestic orchards. Within the season, fruiting often depends on local weather patterns. A given sporocarp may persist for only one to six weeks, changing in size, maturity, and commercial value, and may be eaten by wildlife or collected by humans. Frequent sampling is required to reliably characterize fruiting patterns and commercial value of a mushroom crop. Mushrooms are often clustered and unevenly distributed on all spatial scales, from local sites to drainages, landscapes, and regions. Uneven distributions require large sampling areas to derive statistically sound estimates of abundance. Relative abundance of a species's sporocarps does not necessarily reflect the importance of that fungus to host trees or the ecosystem, but fruiting-body production is the measure of human interest for managing edible fungi as special forest products.

Forest Community Dynamics and Landscape Considerations

Bormann and Likens (1979) describe the eastern U.S. hardwood forests as a shifting mosaic of irregular patches differing in composition and age and directed by processes of disturbance, growth, and decay. This way to envision natural forested landscapes is also relevant to the Pacific Northwest. Most special forest product species are adapted to disturbances that have been a normal part of their evolutionary history. Far from being a negative factor, these natural events tend to renew and diversify populations. As examples, morel mushrooms flourish after fire, and other special forest product understory species, such as salal and Oregon grape, benefit from canopy openings created by tree-root pathogens or windthrow. However, we poorly understand disturbance effects on most special forest product species. Disturbances that differ in type, frequency, or severity from historical patterns may significantly alter the abundance and quality of special forest product species. Adaptive ecosystem management of special forest products identifies the disturbance and recovery processes that sustain these species, examines impacts from timber harvest and other forest management activities, and adjusts activities in response to monitoring information and social goals.

The habitats occupied by special forest product species are characterized at the landscape scale by (1) different major forest types, (2) different successional stages within a given forest type, and (3) distinctive subcommunities such as riparian zones. Species typical of particular habitats often differ from one another in life history characteristics in ways significant to management strategies. Many early successional species are opportunistic generalists that grow rapidly and disperse widely. Compared to early successional communities, a higher proportion of late-successional species have life-cycle characteristics of more stable environments. Landscapes dominated by older forests are generally heterogeneous, or mosaics, with regard to the age classes and species that they contain. This mosaic quality provides habitat for a wide range of special forest product species. The primary objective of a landscape approach for special forest products is to create or maintain a socially desirable and ecologically sustainable mix of habitats and special forest products within an area.

In coniferous forests, most species occur in the early successional, shrub-forb-sapling stage and in the late successional, old-growth stage; the fewest species find suitable habitat in the middle, the closed canopy stages (James and Wamer 1982, Meslow 1978, Thomas et al. 1979). Intensive forest management for wood production focuses on the middle, least-diverse stage. Achieving full site occupation by commercial timber species shortens the time in early succession, although with short rotations the total amount of early successional habitat may actually increase at the landscape scale. Young stands furnish the stocks for the Christmas greens market. Dense canopies of short-rotation forests, however, can hinder development of many understory species harvested as special forest products. Harvesting trees when the mean annual increment culminates eliminates the unique ecological conditions of the old-growth stage. Species such as Pacific yew and mosses occur in greatest numbers and reach full development under the multiple canopy layers and big trees of old-growth forests.

Riparian zones are especially critical for many special forest product species. Substrate composition, soil moisture, nutrients, depth to water table, temperature, radiation, and disturbance frequency differ from upslope positions. Gravel bars, islands, and flood plains provide a habitat mosaic for a wide array of plant species and often contain distinctive associations—for example, hardwood tree species that contain mosses. Fallen trees in riparian areas contribute to structural complexity, and patches of herbs, shrubs, and deciduous and young coniferous trees produce a multilayered canopy. Openings over streams, lakes, and wetlands provide gaps and breaks in the forest canopy that promote favorable conditions for berry shrubs and ferns. Riparian plant communities also have a higher survival rate than those of nearby hillslope areas during catastrophic wildfire (Michael Amaranthus unpublished data) because higher humidities, cooler temperatures, and damper soils adjacent to streams and lakes help protect the vegetation. Special forest product species in these areas become sources of propagule dispersal for recolonization of upslope areas after fire.

An important first step in considering special forest products at the landscape scale is identifying areas with high commercial production potential or that require protection from harvesting. For example, areas with high production potential and convenient harvester access might be managed for intensive special forest product production to relieve harvesting pressure on sensitive areas. Wetlands or habitats of rare plants or animals may need harvest restrictions. Areas prone to surface erosion or mass failure might need protection. No-harvest areas may be needed to monitor effects of harvesting, and rotated harvest areas could be used to avoid unsustainable harvests and resource depletion. The inventory, monitoring, and research activities discussed below will be essential to identify specific areas for enhanced production or protection.

Silviculture and Vegetation Management Approaches

Stand-level silvicultural objectives can emphasize conditions favoring certain special forest product species. Various approaches may be used to manage for a broad range of products: posts, poles, rails, landscape transplants, shakes, cones, yew bark, boughs, mushrooms, berries, forest greens, and other special forest products. For example, leaving a cover of large trees after timber harvests (green-tree retention) allows two or more canopy layers to develop and provides shade needed by some special forest product species. The numbers of retained trees may differ depending on the needs of the special forest product species, on the ecological adaptations of the particular forest type, and on other management objectives.

Density control is an important tool for providing special forest products and diversifying forest stands (Newton and Cole 1987). Thinning of dense stands can produce posts, poles, rails, and firewood as products. Growing stands at wide spacing allows some special forest product plant species to coexist with specific timber species. Thinnings, coupled with rotation length, can be used to manage the proportion of different seral stage habitats within a landscape. For example, forests can be moved from an early successional structure to an old-growth one without ever passing through a closed-canopy stage. Careful felling and yarding practices and repeated entries are necessary, however, to minimize damage to special forest products.

Other options are available to improve production of special forest products at the stand and landscapes levels. The options might include prescribed fire for habitat rejuvenation for some species; seedbed preparation, direct seeding, and planting of special forest product species; fertilization; and avoiding introduction of exotic or unwanted species.

Integrating Human Behavior

Managers base decisions for adaptive management of special forest products on monitoring present conditions and modeling future alternatives based on best available information. Most discussions of monitoring and modeling ecosystems refer to the behavior of species and processes other than of humans and their actions. Yet, human behavior, perhaps the most complex and difficult to project and predict, plays a key role in the development of most ecosystems. Adaptive management of special forest products must expand to include detailed monitoring of the outcomes of current local human activity on the productivity and structure of ecosystems. Results of different, often innovative, management regimes involving harvesting and culturing of special forest products serve as a record for the knowledge base of the adaptive management system. Monitoring databases (continually supplemented with new data) and data analyses (continually transformed as new information is available) enables managers to better decide how to set sustainable harvest amounts and rotation lengths for special forest products.

Conversely, differing harvest levels of special forest products affect individual and community well-being. Here, explicit definition of the interest groups and communities in managerial decision making becomes vital. Management decisions at different scales (local, regional, etc.) also differently affect the well-being of specific segments of the human community. The distribution of effects on people from decisions initiated at a local geographic or ecosystem scale is usually not only local, for example. Many people employed in special forest industries, such as matsutake picking, are not residents in or near the ecosystems where the mushrooms are picked (Richards and Creasy in press). Surprisingly disjunct groups can be simultaneously affected. Managers must anticipate the effect of regulation on harvester behavior and implement regulations that result in behavior most consistent with well-thought-out objectives (R. Fight, personal communication, 1995). Monitoring total effects of management decisions on human responses in harvesting levels and benefits, both individual and collective, can easily become formidable.

Uncertainty about the soundness of management decisions regarding special forest products is great because objective measures of fairness, sustainability, and social or economic health are difficult, if not impossible, to determine. Definitions of human constraints change as society's values change over time. Given the many possibilities for harvesting across different spatial scales and the increasing number of people who participate in or are affected by harvests of special forest products, monitoring actual management practices in the existing landscape cannot cover the entire range of plausible alternatives. In many instances, particularly under conditions where adaptive management encompasses innovative management as an experiment, there may be no precedent with an attendant monitoring database to substantiate desired outcomes. The demand for innovation and new options suggests that managers should invest considerable effort in building forecasting models.

Organization of the best existing information is necessary for creating forecasting models to predict outcomes. Goals of specific scenarios for future management of special forest products should be clear. Traditional as well as new institutional arrangements for resource access and management, harvest levels, product prices, wages, and targeted interest groups are key variables for predicting future outcomes of the soundness of the management decisions. Scenarios offer insights into the distribution of benefits and costs of outcomes, but final decision making remains a singularly human choice, based on professional judgment. The continuous flow of information about current attitudes and values, and likely trends for both, can aid in making the best decision. Investing in information and information organization is costly, and benefits may not be easily linked by cause. The tradeoff is to acquire the least costly amount of information that will satisfy society and reduce dissension among interest groups involved in special forest product industries.

Conducting Necessary Inventory, Evaluation, and Research Monitoring

We are still learning how ecosystems work, and we will be for the indefinite future. Thomas Berry (1988) summed it up succinctly: "What is needed on our part is the capacity for listening to what nature is telling us." It would be a much shorter list to mention aspects of managing special forest products in a sustainable way that do not need monitoring and research rather than those that do. We do not attempt either. Most forestry research during the past decades has focused on forests managed intensively for wood, while most ecological research has focused on pristine forests. Our ability to sustain harvests and populations of special forest product species will require increased research on the ecology of managed forests. Similarly, because socioeconomic research in forestry has focused on timber market forces and timber-dependent communities, new socioeconomic research efforts are needed on specific special forest product markets and publics. Thus, ecological research will determine the role of special forest product species in an ecosystem context, and product market research will define the role of special forest products in society. The final section of this chapter provides a detailed example of integrating research and monitoring approaches for these disciplines.

Adaptive Ecosystem Management of Commercial Mushroom Harvests

Good information and the logical organization of it are the bases for human decisions on complex phenomena. Commercial mushroom harvesting exemplifies how information required for ecosystem management of a special forest product can be obtained in a stepwise, logical manner. This example presents a conceptual framework for (1) identifying concerns of managers and the public about commercial mushroom harvesting, (2) choosing appropriate studies to address those concerns, (3) designing those studies, and (4) adapting information thus obtained to the needs of management. This approach generally applies to any special forest product. The effort expended should be commensurate with the anticipated harvesting impacts.

The first step in adaptive management of a specific special forest product is estimating whether harvesting activities are sufficiently extensive (widespread) or intensive (concentrated) to be economically or ecologically significant. For example, the demand for wild mushrooms is large and the impact of harvest on ecosystems significant (Molina et al. 1993, Schlosser et al. 1991, Schlosser and Blatner 1993, Schlosser and Blatner in preparation). Many edible species are mycorrhizal, and their symbiotic association with tree roots plays a key role in forest productivity and nutrient cycling. Commercial collection of mushrooms is also socially significant because substantial competition exists among harvesting groups, such as local residents, transient harvesters, Native American tribal members, and recreational pickers (Lipske 1994). Managers must sort out the interests of these competing groups to anticipate problems, reduce conflict, and provide equitable use of the resource.

Educating managers and their constituent interest groups is equally essential. Literature searches (von Hagen et al. in preparation), published summaries of current knowledge, and counsel and opinions from experts are among the avenues for improving and supporting interim decisions when time or fiscal resources do not permit acquisition of high-quality scientific knowledge. Regrettably, political or social pressure often necessitates managerial action in the absence of adequate data. Communication of relative expectations is an important component of public education. Respect for the diversity of opinions about harvesting fungi means that no one interest group is likely to have all their expectations fulfilled.

Only a portion of knowledge about forest fungi applies to managing commercial harvests. Few mycological studies in North America have focused on marketable species. Incorporating monitoring activities into ecosystem management offers one way to ensure that good science feeds rapidly into the decision-making process. Studying harvests of commercial mushroom species informs both users and managers about the impacts of harvests on ecosystems and on economies. Several considerations apply to commercially collected fungi, especially to the predominant commercial species in the Pacific Northwest: American matsutake, morels, chanterelles, *Boletus* mushrooms, and certain truffles. The following questions confront ecosystem managers as they regulate commercial mushroom harvests.

1. *Production and distribution of mushrooms as special forest products.* How many fruiting bodies are being produced? How are they distributed across the landscape or within certain habitats? How does production differ during a season and from year to year? What is the actual or potential commercial productivity of a given area? What proportion of forest habitat is available and accessible for economically efficient harvesting? What factors determine productivity, and how might they be managed? What managerial actions can alter accessibility of the resource to meet management objectives? How does landscape design promote or impede ecological sustainability of biological production and economic harvest?

2. *Mushroom harvesting by people.* How can the sustainability of mushroom harvesting be assured? What proportion of the crop can be harvested without unacceptable impacts on the fungus itself or other resources? What techniques will mitigate those impacts? Does mushroom harvesting increase or decrease subsequent production? Is spore dispersal reduced by removal of immature mushrooms, and does it impair reproductive

success? Are fungal mycelia and subsequent mushroom production affected by search and harvest techniques such as raking, moving woody debris, or digging? Are mushrooms harmed by numerous harvesters trampling the forest floor? How important as food for wildlife are commercially valuable species, and is human competition for the resource significant? What is the demand from various markets for wild mushrooms? How will various management scenarios affect jobs, income, and revenue? How does commercial harvesting affect the relationships (for example, competition and potential conflict) between recreational and commercial harvesters?

3. *Land management decisions.* How do various timber harvesting methods (clearcutting, thinning to various densities, selection of host species) affect subsequent mushroom production over time? How does soil compaction or disturbance from logging activities affect fungal populations? How does the intensity and timing relate to subsequent mushroom production, especially for morels? How do grazing, fertilization, or pesticide application affect production? Can mushroom production be improved through habitat manipulation—for example, planting tree seedlings inoculated with specific fungi, thinning understory brush for sunlight and rainfall penetration, prescribing burns, and irrigating? Can production be increased across the landscape by managing forests to attain tree age class, structure, and composition optimal for fruiting? What types of cost-benefit analyses are needed to help managers decide about managing for special forest products within broad multiple-use objectives? Can ecologic-economic models be developed to support socially acceptable land management decisions?

4. *Biology and ecology of mushrooms as ecosystem components.* What are the important reproductive events in the life cycle of a particular species? How are new colonies or populations established and maintained? What causes them to diminish or perish? How important is spore dispersal to reproductive success, population maintenance, genetic diversity, and adaptability to unique microhabitats? How much genetic diversity exists within and among populations? Are there endemic, narrowly adapted, or unusual populations of otherwise common species? What are the growth rates of fungal colonies in soil and the degree of mycorrhizal development by specific fungi on root systems? To what degree do other mycorrhizal or saprophytic fungi compete with desired fungi for colonization sites on host roots or for space in the forest soil?

Gathering information, analyzing data, and developing adaptive ecosystem models require various investigative methods. Federal land management agencies group these methods into categories of detection, evaluation, and research monitoring.

Detection monitoring addresses the concerns in category one: production and distribution. It encompasses inventories or estimates of production that are repeated periodically to detect trends. Sampling methods may include (1) informal walk-through surveys designed to detect widely scattered populations, (2) weighing commercial collections from defined areas, or (3) systematic sampling regimes using transects or plots. Methods designed for inventories of plants or animals must be modified to meet the constraints of ephemeral, sporadic, and unevenly distributed fungal sporocarps. Some of these considerations are quite pragmatic. For instance, how do personnel frequently sample the same site without causing soil compaction or erosion? At what stage in its development should a mushroom be sampled? Should it be picked to measure weight, and if so, will that influence subsequent fruiting? If size is measured, how should the mushroom be marked for later identification? Forest mycologists are currently developing practical field procedures.

Managers also need to collect baseline social and economic data during this initial monitoring phase to assess the extent of the harvest issues within defined land bases. Variables include the community affiliations of the harvesters and the relative importance of revenue derived from special forest product harvest; the importance of commercial harvesting to local (e.g., rural) and regional economies; price signals and long-term forecasts for market demand; the importance of specific land bases for sustaining the special forest product market; the numbers of harvesters frequenting specific sites and the managerial effort needed to facilitate commercial harvests; and demands from other user groups such as recreational harvesters. Such information can be collected by managers as part of the regulation and permit system, by interviewing and surveying distributors, and by holding meetings of involved publics.

Evaluation monitoring assesses the impact of management practices and scrutinizes trends in detection monitoring; it applies to the second and third categories of concern: mushroom harvesting and land management. Examining the ecological impact of mushroom harvesting involves comparing experimental treatments. Researchers delineate recurring patches or constellations of mushrooms and then randomly assign replicated treatments, such as picking, not picking, raking duff, digging, trampling, or irrigation. The increase or decrease in sporocarp production from the treatments is compared. Similar experimental designs can be used for determining the influence of land management activities on mushroom or truffle production, but adequate replication requires expensive, stand- or landscape-level research projects. These are most cost-effective if they integrate numerous related studies.

Examining the sociological and economic impacts of management decisions requires analysis of changes in human economic behaviors, particularly how quantities harvested and accessibility to the resource affect personal, community, and market economies. Whether management objectives were achieved and considered fair by the involved publics must be considered along with the management efforts (e.g., costs) deemed commensurate with benefits to the public. Managers can develop large-scale commercial harvest experiments to evaluate effects of trial harvest contracts and techniques on ecosystem and market sustainability.

Research monitoring examines the basic biological and ecological concerns listed in the fourth category. Studies of this nature usually involve establishing secure, long-term field study sites. These sites undergo intensive scrutiny and serve as a representative sample of conditions believed to occur within the region. Investigative methods may use a wide variety of specialized equipment and techniques. Examples include trenches to observe underground mycelium; weather stations to correlate fruiting with local precipitation patterns, temperature, or humidity; vacuums or slides for spore collection; laboratory culture of fungi and mycorrhizal compatibility trials with selected host seedlings; microscopic examination and descriptions of mycorrhiza form, structure, and development; and various recently developed molecular techniques for genetic analysis.

Access to inventory and monitoring data is essential for timely and cost-effective decision support. Managers now have access to new technologies that facilitate the use of large amounts of information for analyzing ecosystem interactions and processes. Published information is available from library databases through Internet access. Mushroom distributions and abundance can be added as data layers to already sophisticated geographic information system databases. Precise field locations can be determined with global-positioning-system satellite receivers. Systems-modeling software allows managers to create contingency scenarios from matrices of alternative management options (Bormann et al. 1994). On-line expert systems can lead users to appropriate information or suggest pertinent considerations. These technologies are in various stages of development, but all are likely to become increasingly important to managers who need to consider the ecosystem ramifications of harvesting special forest products like mushrooms. The information these tools provide will help managers justify their decisions and implement adaptive modifications when new information becomes available. This thoroughness and flexibility is especially useful when managers need to balance the interests of competing or conflicting user groups. Although the task of managing special forest products in an ecosystem context is daunting, the means are becoming increasingly available. Successful integration of special forest products into ecosystem management will succeed only if managers, policy makers, and the public are committed to that goal.

Acknowledgments

We appreciate the sharing of ideas and critique of the manuscript from Keith Blatner, David Brooks, Beverly Brown, Roger Fight, Jeff Gordon, Karen Esterholdt, Richard Haynes, Leon Leigel, Les McConnell, Mike Rassbach, Mark Savage, James Trappe, Nancy Turner, and Nancy Wogen. Support from the Cascade Center for Ecosystem Management is likewise appreciated.

Literature Cited

Akerle, O., V. Heywood, and H. Synge, eds. 1991. *The conservation of medicinal plants.* Proceedings of an international consultation, 21–27 March 1988, Chiang Mai, Thailand. New York: Cambridge University Press.

Allen, T. F. H., and T. W. Hoekstra. 1992. *Toward a unified ecology.* New York: Columbia University Press.

Amaranthus, M. P., J. M. Trappe, L. Bednar, and D. Arthur. 1994. Hypogeous fungal production in mature Douglas-fir forest fragments and surrounding plantations and its relation to coarse woody debris and animal mycophagy. *Canadian Journal of Forest Research* 24:2157–2165.

Arnolds, E. 1991. Decline of ectomycorrhizal fungi in Europe. *Agriculture, Ecosystems, and Environment* 35:209–244.

Benjamin, P. K., and R. C. Anderson. 1985. Influence of collecting on the reproductive capacity of wild ginseng *(Panax quinquefolius)* in Illinois. *Bulletin of the Ecological Society of America* 66:140–141.

Berry, T. 1988. *The dream of the Earth.* San Francisco: Sierra Club Books.

Bormann, B. T., P. G. Cunningham, M. H. Brookes, V. W. Manning, and M. W. Callopy. 1994. *Adaptive ecosystem management in the Pacific Northwest.* General technical report PNW-GTR-341. Portland, OR: USDA Forest Service, Pacific Northwest Research Station.

Bormann, F. H., and G. E. Likens. 1979. *Pattern and process in a forested ecosystem.* New York: Springer-Verlag.

Bounous, G., and C. Peano. 1990. Frutti dimenticati. *Monti e Boschi* 41:23–32.

Boyd, R. 1986. Strategies of Indian burning in the Willamette Valley. *Canadian Journal of Anthropology* 5:65–86.

Brown, B. 1994. Environmental conflict, urban flight, and land tenure in the forested regions of southwest Oregon. Paper delivered at the Rural Sociological Society Annual Meeting, 12–14 August, Portland, OR.

Budriuniene, D. 1988. A model of rational non-ligneous forest resource consumption in intensive forestry in Lithuanian SSR. *Acta Botanica Fennica* 136:7–8.

Chambers. 1983. *Rural development: Putting the last first.* New York: Longman.

Cherfas, J. 1991. Disappearing mushrooms: Another mass extinction. *Science* 254:1458.

Cherkasov, A. 1988. Classification of non-timber resources in the USSR. *Acta Botanica Fennica* 136:3–5.

Coleman, B. B., W. C. Muenscher, and D. K. Charles. 1956. A distributional study of the epiphytic plants of the Olympic Peninsula, Washington. *American Midland Naturalist* 56:54–87.

Colville, F. V. 1897. Notes on the plants used by the Klamath Indians of Oregon: Contributions from the U.S. National Herbarium 5:87–108. Also reprinted in Ford, R. I., ed. 1986. *An ethnobiology source book: The uses of plants and animals by American Indians.* New York: Garland Publishing, Inc.

Cunningham, A. B. 1991. Development of a conservation policy on commercially exploited medicinal plants: A case study from southern Africa. In *The conservation of medicinal plants,* ed. O. Akerle, V. Heywood, and H. Synge. Proceedings of an international consultation, 21–27 March 1988, Chiang Mai, Thailand. New York: Cambridge University Press.

Der Marderosian, A. 1992. The status of pharmacognosy in the United States. In *Natural resources and human health: Plants of medicinal and nutritive value,* ed. S. Baba, O. Akerle, and Y. Leawaguchi. Amsterdam, The Netherlands: Elsevier.

Economic Commission for Europe. 1993. *The forest resources of the temperate zones: The UN-ECE/FAO 1990 forest resource assessment.* Vol. II, *Benefits and functions of the forest.* ECE/TIM/62 (Vol. II). New York: United Nations.

Farolfi, S. 1990. Ruolo economico dei prodotti secondari spontanei del bosco: un'indagine nel Casentino. *Monti e Boschi* 41:49–52.

Fearnside, P. 1989. Extractive reserves in Brazilian Amazonia. *BioScience* 39:387–393.

Ferry, B. W., M. S. Baddeley, and D. L. Hawksworth. 1973. *Air pollution and lichens.* London, England: The Athlone Press, University of London.

Forlines, D. R., T. Tavener, J. C. S. Malan, and J. J. Karchesy. 1992. Plants of the Olympic coastal forests: Ancient knowledge of materials and medicines and future heritages. In *Plant Polyphenols,* ed. R. W. Hemingway and P. E. Laks. New York: Plenum Press.

Foster, S. 1991. Harvesting medicinals in the wild: The need for scientific data on sustainable yields. *HerbalGram* 24:10–16.

Geldenhuys, C. J., and C. J. van der Merwe. 1988. Population structure and growth of the fern *Rumohra adiantiformis* in relation to frond harvesting in the southern Cape forests. *South African Journal of Botany* 54:351–362.

Gottesfeld, L. M. J. 1994. Aboriginal burning for vegetation management in northwest British Columbia. *Human Ecology* 22:171–188.

Grochowski, W., and R. Ostalski. 1981. Recherches sur les productions spontaneés des étages inferieurs de la foret de Pologne. In *Productions spontaneés,* ed. R. Marocka and A. Conesa. Proceedings of a seminar, 17–20 June 1980, Colmar, France. Les colloques de l'INRA 4. Paris, France: Institut Nationale des Recherches Agricoles.

Gunther, E. 1973 *Ethnobotany of western Washington: The knowledge of indigenous plants by Native Americans.* Seattle: University of Washington Press.

Handke, M. M. 1990. Market research study on the sales prospects of cut greenery foliage in Germany. Report prepared for the U.S. Agricultural Trade Office, Hamburg, Germany.

Hinesley, L. E., and L. K. Snelling. 1992. Yield of decoration greenery from Fraser fir Christmas trees. *HortScience* 27:107–108.

Huffman, D. W., J. C. Tappeiner II, and J. C. Zasada. 1994. Regeneration of salal (*Gaultheria shallon*) in the central

Coast Range forests of Oregon. *Canadian Journal of Botany* 72:39–51.

Hvass, J. 1964. Pyntegro ntproduktion: Danske ra d og erfaringer (Production of evergreen branches for decoration). *Norsk Skogbruk* 10:10–13.

James, F. C., and N. O. Wamer. 1982. Relationships between temperate forest bird communities and vegetation structure. *Ecology* 63:159–171.

Jansen, A. E., and F. de Vries. 1988. Mushrooms in decline in 10 of 18 countries. *Mushroom, the Journal* (Summer):7–10.

Joyce, C. 1994. *Earthly goods, medicine-hunting in the rainforest.* Boston: Little, Brown and Company.

Keene, A. S. 1991. Archeology and the heritage of man the hunter. *Review of Anthropology* 16:133–147.

Krochmal, A., and C. Krochmal. 1984. *A field guide to medicinal plants.* New York: Times Press.

Kujala, M. 1988. Ten years of inquiries on the berry and mushroom yields in Finland, 1977–1986. *Acta Botanica Fennica* 136:11–13.

Lipske, M. 1994. A new gold rush packs the woods in central Oregon. *Scientific American* (January):35–45.

Marocke, R., and A. Conesa, eds. 1981. *Productions spontanées.* Proceedings of a seminar, 17–20 June 1980, Colmar, France. Les colloques de l'INRA 4. Paris, France: Institut National de la Recherche Agronomique.

Mater, C. 1993. *Minnesota special forest products: A market study.* St. Paul, MN: Minnesota Department of Natural Resources, Division of Forestry.

McCune, B. 1993. Gradients in epiphytic biomass in three *Pseudotsuga-Tsuga* forests of different ages in western Oregon and Washington. *The Bryologist* 96:405–411.

McLain, R., S. Kantor, C. Robinson, and M. Shannon. 1994. Knowledge, rules, and policy development: Non-timber forest products in the Pacific Northwest. Interim draft report for U.S. Forest Service, Pacific Northwest Research Station.

Meslow, E. C. 1978. The relationship of birds to habitat structure: Plant communities and successional stages. In *Proceedings: Workshop on nongame bird management in coniferous forests of the western United States,* ed. R. M. deGraff. General technical report PNW-64. Portland, OR: USDA Forest Service, Pacific Northwest Research Station.

Molina, R., T. O'Dell, D. Luoma, M. Amaranthus, M. Castellano, and K. Russell. 1993. *Biology, ecology, and social aspects of wild edible mushrooms in the forests of the Pacific Northwest: A preface to managing commercial harvest.* General technical report PNW-GTR-309. Portland, OR: USDA Forest Service, Pacific Northwest Research Station.

Moore, M. 1993. *Medicinal plants of the Pacific Northwest.* Santa Fe: Red Crane Books.

Moreno-Black, G., and L. L. Price. 1993. The marketing of gathered food as an economic strategy of women in northeastern Thailand. *Human Organization* 52:398–404.

Murray, M. D., and P. D. Crawford. 1982. Timber and boughs: Compatible crops from a noble fir plantation. In *Proceedings of the Biology and Management of True Fir in the Pacific Northwest Symposium, 24–26 February 1981,* ed. C. D. Oliver and R. M. Kenady. Seattle: University of Washington, College of Forest Resources.

Newton, M., and E. C. Cole. 1987. A sustained yield scheme for old-growth Douglas-fir. *Western Journal of Applied Forestry* 2:22–25.

Norton, H. H. 1979. The association between anthropogenic prairies and important food plants in western Washington. *Northwest Anthropological Notes* 13:175–200.

Norvel, L., F. Kopecky, J. Lindgren, and J. Roger. 1994. The chanterelle (*Cantharellus cibarius*): A peek at productivity. In *Dancing with an elephant: Proceedings of the business and science of special forest products,* ed. C. Schnepf. Boise: University of Idaho Cooperative Extension Service.

Pojar, J., and A. MacKinnon. 1994. *Plants of the Pacific Northwest Coast.* Redmond, WA: Lone Pine Publishing.

Reagan, A. B. 1935. Plants used by the Hoh and Quileute Indians. *Transactions of the Kansas Academy of Science* 37:55–70.

Richards, R. T., and M. Creasy. In press. Cultural diversity and ecological sustainability: Wild mushroom harvesting in the Klamath Bioregion. *Society and Natural Resources.*

Robbins, W. G., and D. W. Wolf. 1994. *Landscape and the intermontane Northwest: An environmental history.* General technical report PNW-GTR-319. Portland, OR: USDA Forest Service, Pacific Northwest Research Station.

Robinson, P. 1977. *Profiles of Northwest plants.* Portland, OR: Green World Press.

Russell, K. W. 1990. Manufacturing, marketing, and regulatory considerations: Forest fungi. Talk presented at the Special Forest Products Workshop, 8–10 February 1990, Portland, OR.

Ruth, R. H., J. C. Underwood, C. E. Smith, and H. Y. Yang. 1972. Maple syrup production from big-leaf maple. Research note PNW-181. Portland, OR: USDA Forest Service, Pacific Northwest Forest and Range Experiment Station.

Saastimoinen, O., and S. Lohiniva. 1989. Picking of wild berries and edible mushrooms in the Rovaniemi region of Finnish Lapland. *Silva Fennica* 23:253–258.

Salo, K. 1984. The picking of wild berries and mushrooms by the inhabitants of Joensuu and Seinajoki in 1982 (in Finnish with English summary). *Folia Forestalia* 598:1–21.

———. 1985. Wild berry and edible mushroom picking in Suomussalmi and in some North Karelian Communes, Eastern Finland. *Folia Forestalia* 620:1–50.

Schlosser, W. E., and K. A. Blatner. 1993. Critical aspects of the production and marketing of special forest products. Unpublished manuscript prepared for the President's Forest Conference Committee, 3 May 1993, Portland, OR.

———. 1995. The wild edible mushroom industry of Washington, Oregon, and Idaho: A 1992 survey. *Journal of Forestry* 93:31–36.

———. In preparation. Special forest products: An east-side assessment.

Schlosser, W. E., K. A. Blatner, and R. C. Chapman. 1991. Economic and marketing implications of special forest products harvest in the coastal Pacific Northwest. *Western Journal of Applied Forestry* 6:67–72.

Schlosser, W. E., K. A. Blatner, and B. Zamora. 1992. Pacific Northwest forest lands' potential for floral greenery production. *Northwest Science* 66:44–54.

Schlosser, W. E., C. T. Roche, K. A. Blatner, and D. M. Baumgartner. 1993. *A guide to floral greens: Special forest products.* Pullman WA: Washington State University, Cooperative Extension Bulletin.

Smirnov, A. V., E. E. Grigoruca, and G. I. Saltymakova. 1967. Change in the abundance and yield of *Vaccinium vitis-idaea* in the forests of Siberia under the influence of anthropogenic factors (in Russian). Rasitel'ny Resursy.

Smith, G. W. 1983. Arctic pharmacognosia II: Devil's club, *Oplopanax horridus. Journal of Enthopharmacology* 7:313–320.

Thomas, J. W., R. G. Anderson, C. Maser, and E. L. Bull. 1979. Snags. In *Wildlife habitats in managed forests: The Blue Mountains of Oregon and Washington.* Agricultural handbook 553. Washington, DC: USDA Forest Service.

Thomas, M. G., and D. R. Schumann. 1993. *Income opportunities in special forest products.* Agricultural information bulletin 666. Washington, DC: USDA Forest Service.

Tilford, G. L. 1993. *The ecoherbalist's field book.* Conner, MT: Mountain Weed Publishing.

Turner, N. J. 1991. Burning mountainsides for better crops: Aboriginal landscape burning in British Columbia. *Archaeology in Montana* 32:57–73.

Turner, N. J., J. Thomas, B. F. Carlson, and R. T. Ogilvie. 1983. *Ethnobotany of the Nitinaht Indians of Vancouver Island.* Victoria, BC: British Columbia Provincial Museum.

Turner, N. J., L. C. Thompson, M. T. Thompson, and A. Z. York. 1990. *Thompson ethnobotany: Knowledge and usage of plants by the Thompson Indians of British Columbia.* Memoir 3. Victoria, BC: Royal British Columbia Museum.

USDA Forest Service. 1963. *Special forest products for profit: Self-help suggestions for rural areas development.* Agricultural information bulletin 278. Washington, DC.

———. 1992. *An interim guide to the conservation and management of pacific yew.* Pacific Northwest Region: USDA Forest Service.

———. 1993a. *Pacific yew final environmental impact statement.* Washington, DC: USDA Forest Service, USDI Bureau of Land Management, USDHHS Food and Drug Administration.

———. 1993b. *Income opportunities in special forest products: Self-help suggestions for rural entrepreneurs.* Agricultural information bulletin 666. Washington, DC.

———. 1994. *Final supplemental environmental impact statement on management of habitat for late successional and old-growth forest–related species within the range of the spotted owl.* Vol. 1. Washington, DC: USDA Forest Service, USDI Bureau of Land Management.

Vance, N. C., R. G. Kelsey, and T. T. Sabin. 1994. Seasonal and tissue variation in taxone concentrations of *Taxus brevifolia. Phytochemistry* 36:1241–1244.

Vänninen, I., and M. Raatikainen, eds. 1988. *Proceedings of the Finnish-Soviet symposium on non-timber forest resources, 25–29 August 1986, Jyväskylä, Finland.* Acta Botanica Fennica 136. Helsinki, Finland: The Finnish Botanical Publishing Board.

Vitt, D. H., J. E. Marsh, and R. B. Bovey. 1988. *Mosses, lichens, and ferns of Northwest America.* Seattle: University of Washington Press.

von Hagen, B., J. Weigand, R. McLain, R. Fight, and C. Christensen. In preparation. Annotated bibliography of literature useful for management of non-timber forest products in the Pacific Northwest.

Warren, D. D. 1995 *Production, prices, employment, and trade in Northwest forest industries: Second quarter 1994.* Resource bulletin PNW-RB-205. Portland, OR: USDA Forest Service, Pacific Northwest Research Station.

Weege, K. 1977. Bisherige Erfahrungen mit dem Anbau von

Abies procera in Nordrhein-Westfalen (Experiences to date with noble fir [*Abies procera*] silviculture in North Rhein-Westphalis). *Allgemeine Forstzeitung* 32:1155–1158.

White, Richard. 1980. *Land use, environment, and social change: The shaping of Island County, Washington.* Seattle: University of Washington Press.

Wiersma, G. B., M. E. Harmon, G. A. Baker, and S. E. Greene. 1987. Elemental composition of *Hylocomium splendens. Chemosphere* 16:2631–2645.

22 The Public Interest in Private Forests: Developing Regulations and Incentives

Frederick W. Cubbage

Public Interest Versus Private Rights 338
Policy Guidelines 340
Traditional Forestry Programs 341
Regulation 341
Financial Incentives and Cost Sharing 345
Tax Benefits 346
Technical Assistance 347
Nontraditional Forest Policy Tools 349
Landowner Cooperation and Voluntary Efforts 349
Interagency Cooperation 349
Regulatory Consolidation and Reform 350
Purchase of Development Rights and Easements 351
Incentives and Markets 352
Conclusion 353
Literature Cited 354

In the last decade, we have experienced increasing conflict over the use, management, and protection of public lands. These debates reflect deep differences about why individuals or groups value forests. These differences in values also affect the beliefs of individuals about how private forest lands should be managed and protected. Conflicts over public forest land management, however, are apt to pale compared to the problems that may be incurred if unreasonable regulation threatens the property rights and profits of private forest landowners. In fact, the issue of public interests versus private property rights has already become highly polarized. The wise use movement has advocated turning back the clock on environmental regulations; environmental groups have bemoaned the excesses of careless landowners and of property rights groups; and some public agencies are afraid to even mention, let alone scientifically analyze, the merits of regulation of private property. Moreover, ecosystem management itself has generated considerable controversy.

All forest resources are becoming scarcer as population and demands for commodity, amenity, and environmental values increase. The shift toward ecosystem management and reduced production of timber and other commodities on public lands suggests that private forests will become more important for the production of forest products and services as well as for environmental protection. Nonindustrial private forest landowners hold 59 percent of the timberland in the United States, and forest industry holds another 14 percent (Powell et al. 1993). This makes the importance of private forests apparent—but the role they should play in providing public as well as market goods is subject to wide disagreement.

This chapter is intended to provide some new insights about appropriate policy tools for encouraging ecosystem management on private lands. It progresses from the reasonably well accepted to the more speculative, as is perhaps appropriate for discussion of forestry for the 21st century. A brief review of the public interest versus private property rights begins the discussion. Criteria that can be used to choose policy also are discussed. Then, traditional forest resource regulatory and incentive policies are reviewed. Last, the prospects for developing new public policy interventions that are appropriate for ecosystem management are discussed.

Public Interest Versus Private Rights

Regulation, subsidies (incentives), or education to prompt desirable actions by private forest landowners assume that there is a discrepancy between the social benefits and costs of an individual's actions and the private benefits and costs. All of these policy tools are used to induce private forest landowners to act for the public benefit, either through moral suasion (education), financial subsidies (incentives), or coercion (regulation). One might observe that the greater the discrepancy between the private market outcomes and the perceived socially desirable outcomes, the more likelihood that stricter policy tools of coercion rather than mere exhortation will be used.

The application of such policy tools in ecosystem management is difficult because of debates about the merits of ecosystem management itself. The topic is extensive, so this chapter will simply assume that ecosystem management is one means for resource allocation and protection and will describe general policy tools that can be used for many purposes. Larger debates about ecosystem management will continue, and readers can refer to other sources for more details about the concept itself (other chapters in this book, as well as Society of American Foresters 1993, Franklin 1994, Grumbine 1994, Moote et al. 1994, Cubbage 1995b). Ecosystem management, however, does represent a change in forestry paradigms. The geographic and temporal scales implicit in ecosystem management are larger than traditional stand management. It also represents a shift from focusing on the production of commodity goods (e.g., timber, recreation visitor days) to the maintenance of forest health, biodiversity, and sustainability, with commodity output as a byproduct rather than a primary goal of management.

Any government program must demonstrate some compelling need for its existence, at least at its inception. Programs that address significant problems, that cost less, or that impose less regulatory burdens on private landowners, such as education, technical assistance, and fire control, are likely to be viewed favorably in legislative deliberations. Even where the needs for social improvements are substantial, programs that entail large administrative costs and impose large cost burdens on private landowners will be viewed less favorably. Nevertheless, many regulatory programs have been enacted, and new rules and regulations stemming from existing laws have proliferated.

The legal basis for and limits of the public interest in private lands underlie the extent to which the public (legislatures and administrative agencies) may enact and implement laws to exhort, induce, or coerce private forest landowners to act in the social good. This subject of social controls on private rights is covered in detail in other references, but it is useful to restate some of the principles involved in such debates, drawing from material by Troup (1938), Bosselman et al. (1973), Spurr (1976), Alston (1983), Cubbage and Siegel (1985), Cubbage et al. (1993), and Cubbage (1995a, 1995b). Alston (1983) undoubtedly presents the most detailed synthesis of the individual versus the public interest in forestry.

Landownership in the United States has traditionally consisted of a bundle of rights regarding the use, acquisition, and disposition of real property. As derived from our heritage of English common law, private and public landowners hold exclusive but not absolute rights. Landowners are granted title to the land on the condition that they may use, but not diminish or destroy, the value of their land; nor may they impose costs on other landowners or on society by improvident use of their land. The concept of protecting the intrinsic productivity of land infers that future generations have usufructuary rights; avoiding imposing costs on other owners refers to the concept of externalities. Private landowner rights are conditioned on the owner's wise use of the land in order to protect present and future public interests in the land and what it produces.

Many current policy debates center on how far we can go in regulating private persons in order to protect the environment, which is termed the *taking issue.* Early colonial and U.S. policy established that public regulation of private landowners was legitimate. The early laws were based on the premise that in order to protect and promote the public health, safety, morals, and general welfare, society can—through its police power—restrict the freedom with which owners may use their land (Bosselman et al. 1973). Use of police power to restrict landowners' actions stems from the common law of private nuisance or, less frequently, from the doctrine of waste (Carmichael 1975).

Police powers can be used to limit actions by private owners that might damage forest productivity, create externalities, or perhaps now, interfere with ecosystem and landscape management objectives. Taking all or most of the value of one's property without compensation, however, is proscribed by the Fifth Amendment of the U.S. Constitution and similar strictures in state laws. Landowners may challenge laws that they believe take the value of their property, but proving complete loss in the case of forestry is apt to be difficult. Alternately, regulated owners may sue for compensation for the value lost under inverse condemnation procedures in federal claims courts. In any case, the use of regulation to implement forestry goals must have clear public benefits, and regulations cannot take the entire value of private property. Balancing between some loss in private property values and compelling social need is the crux of the taking issue.

Debates on regulatory powers in forestry have been waged for decades. As early as 1919, Gifford Pinchot called for federal control of all private forests and was supported by the Society of American Foresters. About 13 cutting practices acts for state forests were enacted from 1940 to 1950. In 1947, the new 1945 Washington State Cutting Practices Act was challenged: "[It] authorizes what amounts to the taking of private property without compensation and established an unreasonable exercise of the police power . . . [and] . . . destroys private property rights" (202 P.2d 909). The Washington state supreme court upheld the right of the state to regulate private forestry actions and was upheld without comment by the U.S. Supreme Court (State v. Dexter, 32 Wash.2d 551, 202 P.2d 906, 70 S.Ct.147 (1949)).

Forestry regulations and other environmental regulations increased substantially until the 1990s. This uninterrupted trend, however, has been at least stalled and perhaps faces reversal. Private property rights groups have lobbied stridently and effectively to reduce government regulation of private land. They argue that the loss of private property rights has been excessive; that laws tend to become more inclusive, restrictive, and expensive over time; that regulators are arbitrary and capricious in enforcement; and that the public and private costs of burgeoning bureaucracies exceed the benefits of regulation. Opponents of regulation also have successfully sponsored private property rights bills in many states and in Congress. These bills try to minimize onerous regulations by requiring (1) government compensation to landowners for the loss of property values—for example, reductions in value of 30 percent to 50 percent—caused by regulation; (2) cost-benefit or risk-assessment analyses of all major new regulations; and (3) that the federal government not impose new mandates on states without providing the wherewithal to pay for them (Cubbage 1995a). The Republican congressional majority elected in 1994 passed such a federal law in 1995 and continues to consider substantial weakening of other federal regulations in water quality and endangered species laws.

Policy Guidelines

Government intervention in private forestry is usually justified on the premise that social benefits and costs differ from private benefits and costs. A spectrum of policy tools ranging from education to incentives to regulation may be used when these social and private benefits and costs differ. The tools employed will depend on the degree of the perceived problem, the public interest in the issues, the opposition to particular initiatives, and the funds available to redress problems, among other things. In general, more modest public interests and resource problems will tend to favor market approaches or education. Perceived larger problems with forest productivity or adverse environmental impacts may require more-direct and expensive intervention, such as financial incentives or regulation.

As the state of forest science and philosophy progresses, what can we say about the capacity of our laws, policies, regulations, and incentives to help achieve broader public interests while protecting private interests? Can we shift our forestry paradigms from timber production and forest management to sustainable forestry and ecosystem management? If these are deemed to be appropriate social goals, how can we develop effective policy tools to achieve them? Indeed, can the government effect meaningful changes in private forest practices to achieve public interests? The science of ecosystem management is virtually a new discipline; our ability to institutionalize the relevant public rights inherent in ecosystem management remains moot. This chapter will review several principles regarding regulations, incentives, and education, and will discuss how they may apply to an evolving view about forests for market and nonmarket goods and services.

One of the first premises of government actions is that they must be designed to achieve outcomes that are socially desirable but not currently achieved by unregulated market exchanges. This presumption has implicitly provided the foundation for most forestry laws and subsidies, including state forest practice laws in the West, federal and state forest fire and insect and disease protection programs, and the federal Forestry Incentives Program and Conservation Reserve Program. In the 20th century, the purview of appropriate and desirable government actions—to enact laws and appropriate funds—has expanded greatly.

A second criterion for judging public policies may simply be termed biological feasibility. For example, is it physically possible for government (or private) programs to achieve a social goal? Biological feasibility may be complemented by administrative practicality as a necessary condition for implementing public policies (Clawson 1975). Furthermore, biological feasibility criteria may be extended to encompass a broader set of ecosystem management concepts, including biological diversity, ecosystem stability, and sustainable development. These biocentric goals, which promise to be central in 21st-century forestry, suggest that protecting natural ecosystems per se is highly valued, or that protecting ecosystems helps preserve the health of the environment for human benefit.

A third premise is that the public benefits of government action should exceed the costs of implementing new programs or continuing and enhancing old programs. The benefits of public programs are widely touted by interest groups apt to receive their largesse, by legislators hoping to receive constituent support, and by the bureaucrats hoping to enlarge the mission and budget of their agency (these are termed iron triangle relationships). While government programs almost always have benefits that can be quantitatively or qualitatively estimated, a critical comparison of actual costs with imputed benefits must be made to measure relative benefits and costs. Additionally, given unlimited demands for government action and limited funds from taxes, fees, and other revenue sources, we must select and fund only public programs where the public benefits exceed private costs by the greatest margins possible. Additionally, the programs we implement must be efficiently and fairly administered. This set of premises may be referred to as economic efficiency criteria for evaluating government policies.

Fourth, social equity criteria underlie the selection

of government policies. These criteria may suggest that the benefits and costs of government programs are distributed fairly among individuals or groups in society. Horizontal equity criteria suggest that individuals, landowners, or interest groups in like situations be treated similarly. Vertical equity criteria suggest that poorer or less advantaged segments of society pay less to fund public programs, or that they should receive preference in public assistance. Process criteria suggest that individuals or groups affected by public policies have an adequate say in government decisions. Political criteria, of course, require that proposed policies must receive adequate legislative support to be enacted, funded, and implemented.

These general criteria should be considered when proposals to revise existing forestry programs or develop new programs are made. The social (public interest) needs for initiating government programs to achieve new forestry, ecosystem management, or sustainable forestry objectives must be compelling. Biocentric criteria, such as ecosystem diversity and stability, also underlie most of the approaches to new forestry. The benefits of any programs developed must greatly exceed their costs, and they must be implemented efficiently. Also, public policies must be equitable, ensuring that private owners of land do not bear an unreasonable burden to provide public benefits. Forcing landowners to stop pollution or damage to neighboring owners seems equitable, based on the doctrine of waste. Going so far as requiring them to provide habitat for endangered species and forego all timber income is another matter, which may fail both equity criteria and constitutional muster.

Traditional Forestry Programs

Assuming that we do settle on forest resource goals that are not currently being met by market processes, how can we implement new policies to achieve them? Common policy tools used to achieve public forestry objectives on nonindustrial and, to a lesser extent, industrial private forest lands have included government regulation, favorable tax assistance, and education. Many successful programs exist that provide assistance to private landowners or, indeed, that limit their rights to wantonly exploit their land. These programs or modifications of them will continue to provide viable mechanisms to achieve their existing objectives, and they may be expanded to broader objectives. Yet, as the pressure on government services and budgets increases, new mechanisms to achieve public goals are being examined. Achieving the broader goals of sustainable forestry and ecosystem management may inherently require broader policy tools, tailored to the unique policy objectives.

Many U.S. forest policies using regulation, incentives, or education affect the actions of private landowners. Again, a thorough discussion of each of these policies could constitute a book by itself, so readers should refer to Cubbage et al. (1993), Ellefson (1992), Dana and Fairfax (1980), or other tomes for detailed reviews of public forest policies describing forestry regulations, technical and financial assistance, and educational programs. This chapter will instead briefly recapitulate the highlights of some of these programs and their appropriateness for continuance or modification for forestry in the 21st century.

Regulation

State Forest Practice Laws

Regulation of forest practices in the United States has a long history. As mentioned, national calls for regulation of private forest lands began more than six decades ago when Gifford Pinchot chaired the Society of American Foresters' Committee for the Application of Forestry in 1919. With Pinchot's active leadership, this committee recommended federal regulation of private forest lands to control forest devastation. Several bills were introduced into Congress in the next five years to authorize federal regulation, but in the end the measures proved too strict. Instead, Congress passed the Clarke-McNary Act of 1924, which initiated federal-state cooperation to provide forest fire and insect and disease control on private lands, produce seedlings for tree planting, and provide technical assistance and education to forest landowners.

The issue of federal regulation of private forestry

did not disappear, however, until the early 1950s. A series of U.S. Forest Service chiefs publicly supported federal regulation of forestry, with varying degrees of tact and vigor. Ferdinand Silcox maintained continuous support for federal regulation throughout his tenure from 1933 to 1939. During this time, Title 10 of the National Recovery Administration Codes of 1933 authorized federal regulation of forestry, but the law was declared unconstitutional by the Supreme Court in 1935 and federal control was averted. In 1939, however, Earle H. Clapp became acting chief of the Forest Service. Unlike Silcox, who fought with a rapier for federal control, Clapp single-mindedly wielded a broadax seeking federal regulation (Greeley 1951).

In response to Clapp's aggressive pursuit of regulation, 13 states enacted "seed-tree" or state forest practice laws between 1940 and 1950 in hopes of averting threatened federal regulation. Clapp was succeeded by Lyle Watts in 1943. Watts also advocated federal control, albeit with less zeal. After Richard McArdle became chief in 1952, a national study found that forestry conditions on forest industry lands were equal to or better than those on federal lands. McArdle withdrew the long-standing Forest Service activism for federal regulation, and the issue shifted permanently to state forestry regulation (Hamilton 1965).

The forest practice laws enacted in the 1940s were generally modest seed-tree laws that were intended to require regeneration of forest lands after harvest. The western state laws in California, Oregon, and Washington were moderately detailed and enforced. The seed-tree laws in the East were seldom enforced, at least in the 1950s and 1960s. As mentioned, the constitutionality of the Washington State forest practice law to regulate timber harvests was upheld by the state courts and U.S. Supreme Court.

The seed-tree laws of the 1940s provided the framework for expansion to environmental protection in the 1970s and served as models for other state forest practice laws. Oregon, Washington, and California have revised their laws many times to incorporate more environmental protection measures. Alaska, Idaho, and Nevada have enacted similar state forest practice regulations. In 1982, Massachusetts revised its 1943 state forest practice act. In 1989 and 1991, respectively, Maine and Connecticut enacted new state forest practice acts, and a number of states in the East have enacted quasi-regulatory means of requiring best management practices.

Depending on how one classes the degree of state regulation, there are currently 10 comprehensive state forest practice acts, two seed-tree laws that are enforced with rigor, nine state laws that regulate some forest practices, and three seed-tree laws that are not enforced. Modern state forest practice laws regulate actions both to ensure regeneration for future forests and to prevent pollution or externalities from damaging other forest resources. These laws do not regulate ecosystem management per se, and given the antiregulatory environment prevalent in the mid-1990s, are not likely to be expanded to do so explicitly. Nevertheless, some of the landscape or ecosystem management concepts of limits on total area of harvests in a given time period and of watershed or "green-up" requirements before harvesting adjacent tracts have been considered in state forest practice regulations.

Local Regulation

In addition to state forest practice laws, many localities, especially in the eastern United States, have enacted their own environmental protection measures that directly regulate forest management (Cubbage and Siegel 1988). Local zoning and county laws also are becoming more important in regulating some forestry activities. As of 1995, there were more than 500 local laws in the United States that regulated forestry. More than 70 percent of these were passed in the last 10 years; 50 percent in the last 5 (Hickman et al. 1991, Greene and Haines 1994).

About three-quarters of the local ordinances found by Hickman and Martus (1991) occurred in the Northeast. Most of the other ordinances occurred in the South (about 100); only a few were found in the north-central states. In the South, almost all the local ordinances occurred at the county or parish level of government. A 1992 survey in Georgia found that 52 out of the 159 counties in the state had some type of county logging ordinance (Jackson et al. 1992). As of

1993, less than 20 local laws had been enacted in the West.

Hickman (1993) divided local ordinances into five categories based on their primary reason for enactment: (1) public property and safety-protection ordinances, (2) urban and suburban environmental protection ordinances, (3) general environmental protection ordinances, (4) special feature and habitat protection ordinances, and (5) forest land preservation ordinances.

Diverse issues have prompted consideration and enactment of local logging ordinances. Perceived problems differ somewhat between urban and rural counties. In urban counties, most ordinances that regulate tree cutting are aimed at preventing substantial soil erosion, retaining forests, maintaining property values, and protecting amenity values, particularly as new areas are developed. In large metropolitan areas, several counties have tree-protection ordinances designed to preserve trees from indiscriminate cutting or bulldozing by developers. These usually apply in areas zoned as residential or commercial. Some urban counties also address timber or pulpwood harvesting in rural land-use zones as well. These regulations are often prompted by demands from new residents who are displeased when their picturesque rural quality of life is disturbed by chain saws, skidders, and log trucks and when their neighboring forests are converted to clearcuts.

The principal issues causing enactment of ordinances in rural timber counties have been damage to county property during logging, associated problems with public safety, and the expense of repairing county roads. Loggers occasionally have damaged county ditches, roads, and bridges, requiring costly repairs at county expense. Sometimes they have gained access to a tract by filling the ditches with logs and packing them with mud or gravel, then have left the ditches blocked when finished. At other times, particularly during rainy weather, some county dirt roads have been rutted so badly during harvesting that only log trucks have been able to traverse the roads. Slick spots or frozen mud on blacktop roads, litter around logging sites, timber trespass, and damage to fences and neighboring properties have also created ill will.

Federal Environmental Laws

In addition to direct regulation, there are a host of federal laws that regulate forestry operations indirectly. Laws that affect public lands of course dictate management on those lands. As one begins to consider forests as broad ecosystems where landscape-level interactions are important, federal land laws and management actions also begin to take on importance with respect to private lands, and vice versa. These relationships open up entirely new dimensions and problems of management in a mixed public and private ownership pattern. The laws and policy tools to guide management in such areas will need to be carefully crafted so that landscape-management goals are realized and tensions among landowners are not exacerbated.

The principal federal environmental laws and amendments that affect forest management on private lands in the United States include the following:

- National Environmental Policy Act of 1969
- Clean Air Act amendments of 1970
- Federal Water Pollution Control Act amendments of 1972
- Endangered Species Act of 1973
- Coastal Zone Management Act of 1972
- Federal Environmental Pesticide Control Act of 1972

Cubbage (1993) summarized the application of these laws to private lands, as summarized below. The National Environmental Policy Act (NEPA) of 1969 mandated that the federal government, in cooperation with the states and other public and private organizations, should "use all practicable means and measures . . . to create and maintain conditions under which man and nature can exist in productive harmony, and fulfill the social, economic, and other requirements of present and future generations of Americans." In brief, NEPA is a procedural law that requires the preparation of an environmental impact statement (EIS) or environmental assessment (EA) for any federal action that may have a significant impact on the environment. As such, NEPA would seem to apply only to federal forestry or wildlife actions, such as development of national forest plans or reintroduction of rare or threatened species in national parks.

In fact, however, the use of an EIS has been tested for extension to private lands via indirect means. Most recently, an EIS was prepared by the Tennessee Valley Authority (TVA) for development of a wood-chipping and barge facility on the Tennessee River. Since the barge facility would be located on the government-owned land next to a river, it was determined that an EIS would be required. One of the alternatives considered in the EIS was that all private forest landowners in the approximately 6-million–acre region feeding the chip mill would need to obtain timber-harvesting permits for wood shipped through the loading facility. TVA eventually recommended against building the barge facilities in its EIS, thus avoiding the specter of federally triggered regulation of private forestry. The denial of the building of the barge facility, however, was challenged in a lawsuit filed by two forest industry firms and the American Forest and Paper Association. The suit, which was filed December 29, 1993, is still pending.

The amendments to the Federal Water Pollution Control Act (FWPCA) of 1972 were the first federal environmental strictures that specifically listed controlling forestry-related activities on private lands. The law was amended in 1977 and 1987 and is now commonly referred to as the Clean Water Act. Section 208 of the FWPCA mandated control of nonpoint-source pollution from development, mining, agriculture, and forestry activities. Eastern states began to develop operational methods of voluntary forestry in best management practices (BMPs) to meet the clean water goals without mandatory compliance. Most western states implemented Section 208 by incorporating water quality protection measures into state forest practice laws.

Section 404 of the 1972 FWPCA stated that a permit from the U.S. Army Corps of Engineers would be required before individuals, firms, or governments could dredge material from or deposit material in the nation's waters. The FWPCA directed the Corps to administer the Section 404 permit program, but the Environmental Protection Agency (EPA) was granted administrative oversight for the entire act, including Section 404. The definition of the nation's waters eventually expanded to include navigable rivers and streams and the nation's wetlands.

Forestry groups obtained an explicit exemption from the Section 404 permit requirements for normal silvicultural activities in the 1977 amendments to the Clean Water Act. The extent of this silvicultural exemption has been a subject of debate ever since. The final Corps and EPA regulations regarding dredge and fill activities were released in November 1986. The rules continued to exempt normal silvicultural activities from permit requirements, but stated that "[a]ctivities which bring an area into farming, silviculture, or ranching use are not part of an established operation." Thus land conversion, for example from bottomland hardwoods to soybean fields, is not exempt from permit requirements. Additionally, while normal harvesting is exempt, this "does not include the construction of farm, forest, or ranch roads." Furthermore, the exemption may be voided, or "recaptured," if forestry operations impair water quality or wetland functions and values.

The Endangered Species Act (ESA) of 1973, as amended in 1978, 1982, and 1988, mandates protection of threatened or endangered species of animals or plants. Its principal substantive requirement is that persons cannot unlawfully "take" any officially listed threatened or endangered species of animal or plant. "Take" has evolved to mean not only killing threatened or endangered species, but also destroying critical habitat. Application of the ESA to critical habitat protection means that forest management actions on private lands can be regulated to protect endangered species. The 1995 U.S. Supreme Court "Sweet Home" decision upheld the protection of critical habitat as well of the species themselves. As of 1995, almost 1,000 species had been officially listed as threatened or endangered, and over 3,000 more were under consideration. States also may list species as threatened or endangered in their comparable endangered species acts. Violation of the ESA may be prosecuted as a criminal activity.

The Clean Air Act (CAA) of 1970, as amended, is intended to reduce substances in the air that could be harmful to human, animal, or plant health by causing growth problems, sickness, mortality, or economic losses. The CAA PM10 standard will be used by EPA to develop rules governing prescribed burning and residential wood combustion. EPA was to issue PM10 guidelines for agricultural and forestry burning practices in 1992, but these have been deferred indefinitely.

The Coastal Zone Management Act of 1972 provides a means for federal involvement in and the funds for planning for coastal zone protection. The 1990 amendments expanded the definition of the coastal zone, allowing it to extend several counties inland from the ocean. In 1992, EPA developed a number of BMPs that were evaluated for their cost and effectiveness as part of a national coastal zone study. These BMPs covered protection of streamside management zones, road construction and maintenance measures, timber-harvesting practices, site preparation and reforestation, prescribed fire, and pesticide and fertilizer applications. The EPA released the management measures and economic evaluations in 1993, suggesting that they should be used to control forestry nonpoint-source pollution. These final recommendations have also been shelved indefinitely because of opposition, but could form the basis for further regulatory controls for stricter voluntary BMPs in the extended coastal zone or even in entire coastal states.

The application of pesticides and herbicides is federally controlled based on the Federal Environmental Pesticide Control Act (FEPCA) of 1972, which amended the Federal Insecticide, Fungicide, and Rodenticide Act (FIFRA) of 1947. FEPCA/FIFRA has been amended several times since. In brief, it requires licensing of pesticide operators and registration and labeling of pesticides allowed for approved uses.

Financial Incentives and Cost Sharing

Federal forestry incentive programs provide subsidies to nonindustrial private forest (NIPF) landowners to encourage desirable forest management practices. These include (1) the Agricultural Conservation Program (ACP), (2) the Forestry Incentives Program (FIP), and (3) the Conservation Reserve Program.

These programs are administered by the U.S. Department of Agriculture's Agricultural Stabilization and Conservation Service with technical assistance from the State and Private Forestry Division (now renamed the Natural Resources Conservation Service) of the U.S. Forest Service, as well as by state forestry agencies. At least 14 states also have cost sharing for tree planting or other forest practices, especially in the South.

Agricultural Conservation Program and Forestry Incentive Program

Federal financial incentives (subsidies) have provided funding under the ACP and FIP programs for decades. These programs pay part of the cost of performing a forestry practice to improve conservation (ACP) or increase timber supply (FIP). Economic evaluations of these programs have consistently found that they do prompt more timber supply and are economically efficient (Gaddis et al. 1995). The ACP promotes resource-conserving practices on farms. It began during the Depression as part of the Soil Conservation and Domestic Allotment Act of 1936 and has been maintained in various forms to the present. The ACP includes tree planting, timber stand improvement, and wildlife habitat improvement practices. The federal cost share for these activities varies, but averages about 50 percent.

The FIP was attached as a rider (Title 10) to the Agriculture and Consumer Protection Act of 1973. FIP authorizes cost-share payments to NIPFs for reforestation and timber stand improvement, site preparation for natural regeneration, and firebreak construction. State service foresters must approve plans before practices can be performed, and each county's Agricultural Stabilization and Conservation Service committee must decide which of its many applicants will receive funding. Service foresters must also approve performance of a practice before payment is made.

The federal share of costs varies, ranging up to 65 percent. Fifty percent is common in the South, which receives the most FIP funds. Individuals, groups, associations, and corporations whose stocks are not publicly traded are eligible. However, they cannot be primarily engaged in the business of manufacturing forest products or providing public utility services. After an early evaluation of the program, participation rules were adjusted to exclude owners of more than 1,000 acres and to set a minimum size of 10 acres or more for treated stands. Exceptions for ownerships of up to 5,000 acres may be granted by the Secretary of Agriculture. Land must also be classed as commercial timberland able to grow at least 50 cubic feet of wood per acre per year. No landowner may receive more than $10,000 in total cost-share funds during one FIP year (Risbrudt et al. 1983).

Conservation Reserves

The Conservation Reserve Program (CRP) of the 1950s—better known as the Soil Bank Program—helped farmers convert cropland to permanent grass or tree cover from 1955 to 1960. The program paid for tree-planting costs and made annual cash payments of up to $12 per acre per year for up to 10 years to participating landowners. Over 2.1 million acres of trees were planted in the United States. In the South, about 1.9 million acres were planted with trees—700,000 acres in Georgia alone. The program was enacted for conservation purposes and to reduce surplus farm production by reducing the land base planted in crops. It also helped make a significant contribution to timber supplies in the South. Most land planted with trees remained out of agricultural crop production, even after cash payments to farmers were discontinued (Alig et al. 1980, Kurtz et al. 1980). However, farm production did not drop perceptibly after the program.

The 1985 Farm Bill (the Food Security Act of 1985) authorized a modern CRP. Congress authorized the program for the crop years 1986 through 1990 with the goal of enrolling reserved lands of between 40 million and 45 million acres nationwide. CRP participants were paid a flat rate to perform a practice and also an annual rental payment for 10 years. Annual payments were determined through competitive bidding and averaged about $45 to $50 per acre per year for tree planting.

By January 1990, a total of 33.9 million acres of land had been enrolled in the CRP, with a total of 2.18 million acres in trees. Owners and operators in the South enrolled 2 million acres of forest land in the program, and Georgia had 608,000 acres. Other areas of the country had substantially smaller proportions of their CRP enrollments planted in trees—only 178,000 acres (about 2 percent of its total) in the North and only about 4,400 acres (0.01 percent of its total) in the Plains and Rockies states. The Pacific Coast states also had very few enrolled acres.

Although the CRP reserved substantial amounts of erodible land, the program's stated goal of planting one-eighth of the reserve with trees was not reached. In fact, only about 6 percent of the CRP land was planted in trees—not 12.5 percent as planned. These figures suggest that developing major public programs to establish tree plantations will require substantial effort and greater federal expenditures to achieve high enrollment rates.

Tax Benefits

Tax incentives also have been used to encourage timber production and forest land preservation. Rather than direct cash payments, state and federal tax incentives provide exemptions, deductions, deferments, credits, or remissions of landowners' income or property taxes. These provide indirect incentives or subsidies. Tax policy has been an important forestry concern for more than a century. The timber tax sections of the U.S. Internal Revenue Service Code are important and complex, as is our federal tax system. Capital gains treatment of timber income, investment credits, yield taxes, and modified property tax laws are the principal tax treatments that affect timber production. Timber tax exemptions and rebate laws also have been tried by some states, but with rather limited effectiveness (Skok and Gregersen 1975). Reduced property tax laws for forest lands are common in most states.

Capital Gains

In 1943 Congress passed legislation, over President Roosevelt's veto, that allowed timber income to be treated as long-term capital gains for tax purposes. Based on the 1943 legislation, for both corporate and personal income taxes, revenue from timber sales could be taxed as capital gains under Section 631 of the Internal Revenue Service Code (Siegel 1978). Before 1987, capital gains treatment allowed timber income to be taxed at more favorable rates than ordinary income, similar to the treatment of investments in the stock market.

In the Tax Reform Act of 1986, Congress eliminated the preferential taxation rate for capital gains income, including timber, by reducing the amount of income excluded from taxation to zero. It also reduced the maximum marginal tax rate for individuals from 50 percent to 28 percent (there was a special bracket of 33 percent in some cases). The landmark 1986 act was intended to eliminate many tax breaks, such as capital gains, in order to reduce tax rates for most individuals. Timber income still may be treated as a capital gain if sold according to Section 631 of the Internal Revenue Code, but from 1986 to 1990 it was taxed at the same rate as ordinary income. In 1990, Congress again revised the federal income tax laws. Higher-income taxpayers—those making approximately more than $100,000 per year—were taxed at a 31 percent marginal rate on most income. Timber income, however, still could be taxed at the 28 percent capital gains rate, providing a slight advantage. Corporations were still taxed at the new standard corporate rate of 34 percent. The highest marginal tax rates for the richest taxpayers also increased in 1993, again making the capital gains rate of 28 percent more favorable. In the 1993 budget bill, which was favored by President Clinton, the marginal tax rates were 15 percent, 28 percent, 31 percent, 36 percent, or 39.6 percent, depending on an individual's or family's income.

Reforestation Tax Incentives

In 1980, a reforestation tax incentive provision, formerly referred to as the Packwood Amendment, was enacted to allow forest landowners to receive a tax credit against their income taxes for timberland investments. Under this law, private landowners may receive both federal tax credits and deductions on their income tax for planting trees. Instead of waiting to deduct expenses at the time of harvest, the legislation allows a 10 percent investment credit plus deduction of the expenses over an eight-year schedule, up to $10,000 per year of reforestation expenses. If owners use this method, they cannot deduct the same reforestation costs from harvest sales, because they are capital expenditures and cannot be deducted twice. The method can be used by industry as well as individuals, but the $10,000 annual cap limits its application to about 50 to 100 acres per year—an amount too small for most companies to claim.

Property Tax Benefits

Property taxes are the principal funding source for county, township, local, and city governments, as well as some state governments. They provide the basis for schools, roads, police and fire protection, sanitation, parks, and other government services. Ad valorem (according to value) taxes are levied on real and personal property, including land, timber, crops, houses, vehicles, livestock, buildings, and personal belongings. Various states exempt some of these items from taxation.

As a wealth tax, rather than an income tax, property taxes may cause problems for forest landowners. The tax occurs each year, but timber income is realized only periodically. Furthermore, taxing standing timber gives older timber a higher assessed value than young timber, probably encouraging premature harvest. Similarly, taxing forest land at its highest and best use in metropolitan fringe areas may lead to tax burdens that greatly exceed timber income, encouraging landowners to convert their property to more developed uses.

About 20 states have developed yield tax laws to alleviate problems of periodic cash flows and perverse incentives to cut old stands prematurely. Hickman (1982) states that forest yield tax laws "provide for a conceptual separation of land and timber values. Land values normally remain subject to the annual property tax, although sometimes in modified form. Timber values go untaxed until the time of harvest. At this juncture, a gross income tax equal to some percentage of the stumpage value of the products cut is imposed." Almost all states have current-use property tax provisions allowing qualifying agricultural or forestland to be taxed at rates designed to prevent forced development and to preserve green space, particularly in urban areas.

Technical Assistance

State and Federal

The Clarke-McNary Act of 1924 was the forerunner of most cooperative state and federal programs designed to assist private forest landowners. It authorized federal-state cooperation to protect private forest lands from fire, to study the effects of taxes on conservation and timber production, to produce and distribute seeds and seedlings, and to educate farmers about the use and management of farm woodlots. Thirteen years later, the Cooperative Farm Forestry Act of 1937 (Norris-Doxey Act) established a program of federal funding for technical assistance to farm woodland owners that was provided by foresters employed by the states. Later, the 1950 Cooperative Forest Management Act broadened the clientele to include nonfarm private forest landowners, harvesters, and primary processors (Skok and Gregersen 1975). This was the first comprehensive technical assistance program to provide for nonindustrial private landowners. Under the program, federal funds allocated to the states must be matched by state funds. Many private companies and consultants also provide forestry assistance to nonindustrial private forest landowners.

In 1978 the Cooperative Forestry Act consolidated all previous cooperative legislation. The act authorized the Secretary of Agriculture to provide technical and financial assistance to state foresters to produce seeds and seedlings, to perform nonfederal forest planning, to protect and improve watersheds, and to provide technical and financial forestry assistance to private forest landowners, vendors, operators, wood processors, and public agencies. Funds provided by federal and state governments support state service foresters who perform the field work.

Forestry technical assistance programs have evolved and changed their emphases over the years in order to continue and grow. The early programs provided mostly timber management assistance, including aid in timber marking. Foresters in many states now provide advice on how to comply with forest practice regulations. Pine regeneration was the focus of many state forestry assistance programs in the 1980s. Eastern state foresters in the 1990s have now become much more involved in multiple-use management and the forest stewardship program—designed to promote wildlife habitat and hardwood management. Forestry agencies in other states, such as Virginia, refocused their efforts to protect water quality by aggressively helping implement BMPs. These agency redirections have occurred in response to changing public opinions and helped maintain programs in the face of budget problems.

Finally, Title 12 of the 1990 farm bill, the Forest Stewardship Act of 1990, amended the Cooperative Forestry Assistance Act of 1978. The Forest Stewardship Program (Section 1215) expanded the goals of technical assistance to include a broader array of environmental protection benefits, such as protecting, maintaining, enhancing, restoring, and preserving forest lands and multiple values; protecting wildlife and fish species; producing alternative forest crops; managing the rural-urban land interface; protecting aesthetic values of forest lands; and protecting forests from fire, insects, and disease. The broader purposes in the 1990 legislation clearly reflect an increasing recognition of environmental values, as does an effort by the State and Private Forestry (S&PF) Division of the U.S. Forest Service and state foresters to expand the number of clients and support for their programs.

Cooperative Extension

Cooperative extension programs provide another means of assistance to nonindustrial private forest landowners. Extension programs were first authorized in the Smith-Lever Act of 1914 and are now administered by state land-grant universities in cooperation with the federal government and local counties. Separate congressional authority for forestry extension services was granted under the Renewable Resources Extension Act of 1978, but only modest additional funds have been appropriated. In the 1980s, the federal government funded about 35 percent of these extension programs, local governments about 10 percent, and the states the balance. Extension programs include a substantial forestry and wildlife component in most states.

State extension foresters provide information and education for private landowners, loggers, and forest products firms through workshops, meetings, tours, demonstrations, and bulletins. In addition, extension personnel have taken a leading role in disseminating research findings to public and private foresters, informing researchers of the concerns of forestry professionals and the public, and disseminating information on timber prices in their state.

The U.S. Forest Service branch of S&PF serves an extension and administrative role at the federal level. It administers federal funds for state cooperative forestry programs and provides technical expertise to the states concerning fire, pest, and forest management programs. S&PF personnel assist in and coordinate state forest resource planning and provide advice in management of nontimber uses. In the future, extension programs could continue to play an increasing role in environmental education for forestry and in stewardship-related programs.

Nontraditional Forest Policy Tools

As the preceding review suggests, forestry has a long history of government programs designed to achieve public goals on private forests. Yet, as we move to broader forest resource management and protection goals, developing appropriate and creative new tools will be the key to achieving broad ecosystem management and landscape ecology.

Various new public and private planning and cooperative arrangements could be considered to achieve new forestry goals. These might include (1) landowner cooperation and voluntary efforts, (2) cooperative planning agreements among different levels of government (3) regulatory consolidation among governments for private lands, (4) purchase of development rights and easements for forest management or harvesting, and (5) incentives and markets. Selection among these tools must be considered carefully to ensure that they can achieve the objectives desired.

Landowner Cooperation and Voluntary Efforts

Better cooperation among landowners can help achieve landscape-level or ecosystem management goals. As zoning and regulatory environments become more complex, it may be in landowners' best interest to organize to better meet public demands and protect their private rights.

New policy tools must move beyond command-and-control regulatory approaches to private landowner cooperation and associations, perhaps patterned after cooperative western water management districts or even range management districts. Sampson (1993) believes that these approaches will be more efficient and cost less than cumbersome government programs, especially when federal and state funding is scarce. For example, voluntary cooperation among landowners to achieve landscape or ecosystem goals can help maintain wildlife corridors or prevent adjacent clearcuts.

In 1993, the President's Council on Environmental Quality (PCEQ) produced a blueprint for biodiversity protection on private lands. The report notes that conserving diversity can help private individuals or firms by generating employee or neighborhood goodwill, by engendering trust in regulatory matters, and by realizing actual operating cost savings or productivity-enhancing measures. Constraints on private efforts to conserve biodiversity include the potential to limit or halt profitable land uses, the need to meet demands for goods and services, the difficulty in measuring results and direct benefits, and real or perceived conflicts between existing regulatory approaches and proposed new ones. Fears also exist that voluntary biodiversity conservation efforts may later become codified as regulatory requirements—an effect sometimes referred to as elevating the baseline.

The PCEQ report goes on to recommended four broad goals: (1) maintain the viability of native plants and animals, (2) encourage restoration of viable plant and animal populations, (3) complement regional and global biodiversity conservation efforts, and (4) educate employees, community leaders, and the public about biodiversity conservation. Suggested strategies, illustrated by case studies, range from recognizing that no property exists in biological isolation, to focusing management on plant and animal communities, to maintaining naturally occurring structural diversity, to monitoring for impacts on biodiversity.

Interagency Cooperation

One means of achieving broader public forest resource goals on private lands might be through better cooperation among agencies that deliver existing programs to forest landowners. Cooperation among federal, state, and local governments is not easy nor costless, but it is needed. Even determining the appropriate current public values and goals for private forests requires better interagency communication.

Conflicting agency goals and interests will continue, but increased communication and cooperation are needed to improve resource management.

Improving cooperation among forestry programs, such as cooperative extension and state forestry agencies, rather than fighting turf battles over programmatic interests, would help in delivering new ecosystem management programs. The excellent and extended cooperation between the Virginia Division of Forestry and Virginia Tech Cooperative Forestry Extension to inform loggers, foresters, and landowners about nonregulatory BMPs in the state illustrates how cooperators can achieve public goals. Similarly, an agreement between the Chesapeake Bay Local Area Protection Board and the Division of Forestry provided the initial means to delegate implementation of water quality goals for forestry in the bay area to the Division of Forestry (Hawks et al. 1991). Such interagency agreements and cooperation are necessary to achieve water quality or landscape-level goals.

Consolidation of agencies may improve land management. The Clinton administration's government reinvention efforts consolidated farm programs in a new Consolidated Farm Services Agency. In 1995, proposals were made in Congress to consolidate all public land management agencies, and indeed to divest some of those lands.

Interagency cooperation also is appropriate for improvement in mixed public and private ownership situations. As we move to landscape-level management considerations, actions on private lands can affect management options on adjacent public lands, and vice versa. Far more efforts are needed to develop new cooperative policy tools in this area. American Forests is developing a new policy initiative examining the application of ecosystem management principles to mixed public and private land ownerships.

Regulatory Consolidation and Reform

Another approach to better implementing public interests on private lands is to improve on standard regulatory approaches in the future. The Forest Service has an official position of respecting private property rights; the Environmental Protection Agency publicly eschews excessive regulatory zeal; and many interest groups castigate the command-and-control regulatory approach. Yet despite the poor public image of regulation, there is little indication it will abate in the future. Certainly, one should not underestimate the desire of environmental groups to regulate forestry activities or of state and federal agencies to expand their influence, jurisdiction, and control. Thus consolidating regulations among different levels of government is a more realistic policy strategy for the future.

Regulatory reform is by no means a new idea. The 1974 Washington State Forest Practice Act stated that one of its purposes would be to consolidate regulations affecting forestry operations. President Reagan's 1988 Executive Order No. 12630 required federal agencies to submit cost-benefit analyses of proposed regulations to the Office of Management and Budget (OMB) to ensure that they did not constitute a taking of private land. Vice President Dan Quayle's and President Bush's Council on Competitiveness recommended simplifications or elimination of many federal regulations, including those governing classification of wetlands. To reinvent government, Osborne and Gaebler (1992) stress the need for mission-oriented agencies that minimize internal rules. This can be extended to external regulations as well. And the Clinton–Gore 1993 National Performance Review recommended improved interagency coordination to reduce unnecessary regulation and red tape (Gore 1993). Last, in 1995, Congress passed legislation designed to control expansion of federal regulations.

All of these efforts at regulatory reform or consolidation have had mixed success and approval. Washington State consolidated some, but not all, forestry regulations. Reagan's executive order and Quayle's council met with approbation from environmental groups and many federal agencies, but probably did reduce extreme regulatory proposals and slowed the rate of growth in new regulations. The Clinton administration slightly relaxed the interpretation of Executive Order No. 12360, so detailed cost-benefit analyses were only required for major regulatory proposals. The success of the National Performance Review in reducing regulations for federal agencies and private landowners remains to be seen.

Nevertheless, attempts at regulatory reform and consolidation must be a component of successful forest policies for the 21st century. This does not necessarily mean that there will be fewer regulations after reform. It may mean that landowners get "one-stop shopping" when they notify, seek permits from, or file plans with agencies about prospective forestry actions. Consolidation of all rules is unlikely, but consolidation of administrative authority with a forestry agency would be desirable.

If possible, elimination of redundant regulations or unreasonable rules also is desirable. The western state forest practices acts have served reasonably well to consolidate many environmental rules affecting forestry into one act and agency. They also have helped minimize the number of county or local laws by preempting their regulatory authority over forest practices. Forested states in the Northeast and South have generally adamantly opposed state forest practice acts and have been rewarded with a plethora of county, township, and local laws affecting forestry instead—over 500 by 1995. One would have to surmise that state regulatory reform that supersedes local laws benefits forest landowners and forest practice contractors in the long run.

Another promising aspect of forestry regulatory reform is shifting the focus of regulation from requiring specific actions to preventing undesirable outcomes. Most forestry regulations in particular, and other environmental regulations in general, have adopted solutions requiring the best available technology (BAT) or best management practices in order to prevent pollution. This approach stems from an old legal remedy for correcting nuisances between landowners. The advantage of BAT and BMP approaches to regulation is that they are easy to legislate and monitor; for example, it is relatively easy to see if scrubbers have been installed on coal smokestacks, or culverts placed at stream crossings during logging.

However, economists have noted (and by now even policy analysts, bureaucrats, and environmentalists often agree) that setting standards, not requiring practices, is more economically efficient. Moreover, technical advances have made the more sophisticated monitoring required by such approaches increasingly feasible. The standard-setting approach allows producers to meet standards in the cheapest way possible, and actually encourages innovation rather than freezing technology.

Although a focus on regulating outcomes rather than practices is uncommon in forestry, there are exceptions. Most states in the East have published and promoted nonregulatory BMPs to protect water quality during forestry operations. In 1993, Virginia strengthened its nonregulatory program by passing a "bad actor" law in order to prosecute loggers who violate water quality standards (WQSs). North Carolina does not require BMPs, but rather states that forestry operations in the state must meet WQSs and recommends BMPs as means to do it. Violations of the standards (not failure to use BMPs) may be prosecuted.

An emphasis on standards, not practices, makes sense if we want to achieve measurable desired outcomes. Not all foresters or forest operators support this approach, however, nor do all legislators and interest groups. Many people still feel that forest practice programs are easier to monitor and administer than the more technical approaches of measuring water quality, wildlife habitat site indexes, or other standards.

Purchase of Development Rights and Easements

The moral appeal of better landowner cooperation to meet public demands and achieve landscape-level forest resource goals is compelling. To realize such potential, however, there may need to be some creative institutionalization or transfer of private property rights. Traditional purchases of development rights essentially have removed some of the private "development sticks" in a bundle of private property rights and transferred them to the public (community or government) or to other private entities. Purchase of development rights (PDRs) has been referred to as transfer of development rights, conservation easements, and protective easements, among others. Each of these terms or variations implies that owners may hold their title to land in less than unrestricted fee absolute. Basically, such PDRs allow conservation or other agencies to use or affect management of land owned by someone else (an affirmative easement) or to compel the owner to refrain from doing certain acts on the land—such as the cutting of trees (a negative easement) (Buckland 1987).

In the case of individual or community exchanges, private forest landowners might sell or exchange some of their rights to harvest timber in a given period (or indefinitely) to protect water quality, provide fish and wildlife habitat, or retain amenity values. Buyers of the harvest rights might include other forest landowners who wanted to cut timber, local or regional governments that wanted to protect public values by not harvesting or by using special silvicultural methods, or even close neighbors or distant environmental groups who wanted to stop harvesting by purchasing the rights. For example, in order to protect water quality and fish habitat, western states have considered forest practice act revisions that would restrict the amount of area that can be harvested in a given watershed in a given time period. Such guidelines could be achieved through nonregulatory means, although they might require quasi-legal compacts among owners to ensure equitable timber harvests in each period. Otherwise, overcutting by one owner could reduce the allowable harvests of others.

PDRs and easements have advantages and disadvantages. Open-space or environmental protection through easement acquisition can help protect water quality, conserve agricultural land, provide recreation use, and control urban sprawl. Planned growth should also benefit private owners and the public. The greatest concern with PDRs is the cost of the rights—with costs averaging as much as $1,800 per acre in a 1981 study. The greater the existing development, of course, the greater the cost. High costs also may cause rather limited areas to be protected, and leapfrog development to occur. Last, landowners may oppose PDRs as inappropriate government interference in how they use their land, even if they feel growth should be controlled (Buckland 1987).

Easements and PDRs have been used in natural resource applications. Buckland (1987) summarizes 14 state and local PDR programs designed to reduce the loss of prime farmland due to urbanization. These efforts were generally able to help protect farms at great risk of development, but only a modest number of acres could be protected under the programs, costs for the development rights often equaled or exceeded the cost of the land itself, and some enrollees were dissatisfied with the rigor of the PDR controls once they entered the programs.

The purchase of easements is a more common means of obtaining some fee simple land rights. The National Park Service instigated the first major use of easements in the 1930s to purchase scenic rights along the Blue Ridge and Natchez Trace Parkways. Prices varied widely, however, and landowners often misunderstood what rights had been relinquished. Subsequent violations and adverse court decisions prompted the Park Service to discontinue easements in the 1950s (Buckland 1987).

Haapoja (1994) summarizes the application of modern conservation easements. In order to receive federal income tax benefits, the Internal Revenue Code considers four qualifying categories of easements: (1) the preservation of land for public outdoor recreation or education; (2) the protection of relatively natural habitats of fish, wildlife, or plants; (3) the preservation of open space for scenic enjoyment pursuant to an adopted government conservation policy; and (4) the preservation of historically important land or buildings. The rights the owner relinquishes are transferred to another legally qualified entity, such as The Nature Conservancy, or a government agency. The donation of a qualified conservation easement is a tax-deductible charitable gift for the amount of the decrease in the land value. Easements may be sold or donated to organizations. Once in effect, easements become binding in perpetuity for all owners, unless the development rights are repurchased in some manner.

The use of such market-based exchanges of rights among landowners might be one attractive means to achieve new forestry goals for landscape-level ecosystem management. But it is not likely to be a panacea or even widespread. Private landowners still do guard their bundle of rights quite closely—even fanatically—and will be reluctant to limit their possible management options or cloud their title without receiving substantial immediate profits. And the purchase of development rights also will have a reasonably high transfer cost in terms of legal negotiations and fees and government actions to enable and ratify such exchanges.

Incentives and Markets

Developing new, creative, and affordable strategies in the realm of large incentive programs and tax breaks for forest landowners will be difficult in the future.

While we may maintain existing programs, expansion or substantial modification seems unlikely. Even modest modification of the production-oriented FIP and ACP to incorporate multiple-use goals and additional federal strings has been resisted by landowners in some states. Broader program mandates to achieve ecosystem management, probably with more strings and controls, are not apt to be received enthusiastically. A common dictum in political science literature is that narrow, single-purpose policy instruments are more apt to succeed than broad, multiple-purpose policy instruments. As such, it seems that incentive and tax programs may be less useful for encouraging landscape-level management than cooperative or even regulatory efforts, even if we could afford them.

A last "nontraditional" means to encourage socially desirable forestry in the next century might be reliance on private markets and free enterprise. The merits of this approach depend crucially on the values that individuals or interest groups hold about forests. Those who value market goods (commodities and services) are apt to prefer market processes. Those who favor common-pool or public goods are apt to favor a number of government interventions to improve free market outcomes.

In some cases, market advocates and government advocates may be equally satisfied with laissez-faire market outcomes, albeit for different reasons. In the South, environmental groups almost universally prefer growth and management of natural hardwood stands rather than monoculture plantations of loblolly or slash pine. At the same time, NIPF landowners want to avoid the high site-preparation and planting costs of pines. Also, hardwood timber prices are increasing rapidly. Thus southern timber market processes are apt to lead to outcomes—more hardwoods and associated flora and fauna—that environmental groups favor.

Nevertheless, there appears to be a developing market nexus between environmental protection interests and commodity producers. Namely, consumers are beginning to demand that goods they purchase be produced in an environmentally acceptable manner. Thus producers of consumer goods, such as McDonalds, Time, Johnson and Johnson, and Home Depot, demand that their suppliers employ recycled products or perhaps even be certified that they manage their forests sustainably. Forest industry has become much more interested in the market demand for environmental protection, not just regulatory standards. The principal forestry trade association, American Forest and Paper Association, unveiled a sustainable forestry initiative in 1994. This initiative combines components of wildlife, water quality protection, biodiversity, and other broad objectives with traditional timber production. It is funded by donations or excises on the harvest of forest products by large firms and includes substantial technical assistance components for NIPFs. The industry hopes the effort can help them maintain a positive public image and meet consumer demands.

Markets surely will retain the merits that have made them the principal allocation mechanism for forest resources throughout U.S. history. For those who favor biological and ecological criteria for forest management, reliance on unfettered markets may be anathema. Just as economic criteria led to monocultures, short rotations, and clearcutting, markets will be equally unfavorable to green-tree retention, leaving old trees to fall down and die unused, or other green practices. Thus those who favor biocentric criteria for forest management will likely reject market allocation of forest resources.

Conclusion

The public interest in private lands in the United States has been subject to debate since the founding of our country. The Constitution authorized many enumerated and implied powers, which allowed Congress to enact laws to govern the country and to protect public health, safety, and welfare. The Bill of Rights explicitly protects many of our individual freedoms, including a prohibition of taking private property without compensation. The public interest versus individual rights has remained important ever since.

In the case of forestry, private interests and free markets remained dominant until the end of the 19th century. However, pervasive waste of forest and wildlife resources began to increase the perceived discrepancy between public and private interests, eventually leading to the setting aside of most of the national forests at the turn of the century. Since then, we have greatly diminished (but not stopped) the reservation or acquisition of more lands by government. Instead, we have developed a plethora of intermediate policy tools that encourage or require private forest landowners to act in the public interest. These tools include education, financial and tax assistance, and regulation. New tools may focus more on landowner and agency cooperation, creative modifications of property rights, and more aggressive combinations of sustainable industrial forestry initiatives prompted by market demands.

The authorization and funding of the traditional forest policy programs depends on many factors, including severity of the public need or problem with private markets, the strength of the interest groups and agencies supporting the programs, and congressional (or legislative) interest and ability to appropriate program dollars. As stated before, for traditional programs, the more severe the perceived problems, the more likely we are to move from exhortation to coercion. This principle will probably apply to the spectrum of new forestry policy tools—voluntary to regulatory—as well.

Forestry in the 21st century will need to rely on a mix of traditional and innovative policy tools to accomplish the broader strategies of landscape and ecosystem management. Traditional forest policy tools developed in the 19th century will continue to form the foundation for public programs to influence private owners, but new policies or innovative adaptations of other programs will be needed to achieve broader goals. Agency cooperation, regulatory reform, purchase of development rights, and voluntary landowner efforts can all contribute significantly to broader goals of enhancing biological diversity and managing ecosystems, as well as traditional commodity production and multiple use goals.

Choosing among appropriate forest policy tools for current and future programs must rely on explicit or implicit criteria. In the laissez-faire climate in the 19th century, criteria for resource allocation focused on individual or firm profits, abetted by massive government subsidies to settle and exploit natural resources. Setting aside the national forests was premised on equity concerns—protecting natural resources for the future and preventing wide-scale overcutting and environmental problems. Increasing timber, recreation, and game programs from the end of World War II until the 1980s rested largely on identified human (social and economic) needs and economically efficient delivery of services. New paradigms for the 21st century, at least for now, seem to focus on biocentric criteria for decision making. Preservation or promotion of biological diversity, endangered species, and old-growth stand structures are based on the importance of ecological integrity, not on anthropocentric needs. Perhaps as part of ecosystem management, biological and ecological criteria will again hold superior rank, just as they did in the early days of forestry.

However, the recent shift toward the right of the political spectrum, as evidenced by the November 8, 1994, national elections, suggests that ecosystem management goals will continue to be tempered by influential political opposition. Traditional forest policy mechanisms will continue to be employed because they are conservative and reflect only modest changes in the status quo. New cooperative efforts and adaptive market-based strategies will become more important in the future as zeal for public intervention wanes and public funds become scarce. Social and economic forces still will shape acceptable and affordable forest policies. The balance between use and preservation and between anthropocentric and biocentric goals for public and private forests will continue to represent our needs, values, and political decisions.

Literature Cited

Alig, R. J., T. J. Mills, and R. L. Shackelford. 1980. Most soil bank plantings in the South have been retained; some need follow-up treatments. *Southern Journal of Applied Forestry* 4(1):60–64.

Alston, R. M. 1983. *The Individual vs. the public interest.* Boulder: Westview Press.

Bosselman, F. P., D. Callies, and J. Banta. 1973. *The taking issue.* Washington, DC: GPO.

Buckland, J. G. 1987. The history and use of purchase of development rights in the United States. *Landscape and Urban Planning* 14:237–252.

Carmichael, D. M. 1975. Fee simple absolute as a variable research concept. *Natural Resources Journal* 15(4):749–764.

Clawson, M. 1975. *Forest for whom and for what?* Baltimore: Resources for the Future and Johns Hopkins Press.

Cubbage, F. 1993. Federal environmental laws and you. *Forest Farmer* 52(3):15–18.

———. 1995a. Regulation of private forest practices. *Journal of Forestry* 93(6):14–20.

———. 1995b. Forest resources, ecosystem management, and social science education: Promises, prospects, and problems. *Journal of Natural Resources and Life Science Education* 24(2):116–125.

Cubbage, F. W., and W. C. Siegel. 1985. The law regulating private forestry. *Journal of Forestry* 83(9):538–545.

———. 1988. State and local regulation of private forestry in the East. *Northern Journal of Applied Forestry* 5(2):103–108.

Cubbage, F. W., J. O'Laughlin, and C. C. Bullock III. 1993. *Forest resource policy.* New York: John Wiley & Sons.

Dana, S. T., and S. K. Fairfax. 1980. *Forest and range policy.* 2nd ed. New York: McGraw-Hill.

Ellefson, P. V. 1992. *Forest resource policy: Process, participants, and programs.* New York: McGraw-Hill.

Franklin, J. F. 1994. Ecological science: Basis for FEMAT. *Journal of Forestry* 92(4):21–23.

Freeman, A. D. 1975. Historical development of public restrictions on the use of private land. In *Public control of privately owned land.* Minneapolis: University of Minnesota, Center for Urban and Regional Affairs.

Gaddis, D. A., B. D. New, F. W. Cubbage, R. C. Abt, and R. J. Moulton. 1995. *Accomplishments and economic evaluations of the Forestry Incentives Program: A review.* Working paper no. 78. Research Triangle Park, NC: Southeastern Center for Forest Economics.

Gore, A. 1993. *Creating a government that works better and costs less: Report of the National Performance Review.* Washington, DC: GPO.

Greeley, W. B. 1951. *Forests and men.* Garden City, NY: Doubleday.

Greene, D., B. Jackson, and M. Baxter. 1992. *County-level logging regulation in Georgia.* Technical release 92-R-20. Washington, DC: American Pulpwood Association.

Greene, J., and T. K. Haines. 1994. Estimating the effects of state and local regulation on private timber supply. In *Proceedings: Southern Forest Economics Worker's Annual Meeting.* Research Triangle Park, NC: USDA Forest Service, Southern Station, Forest Economics.

Grumbine, R. E. 1994. What is ecosystem management? *Conservation Biology* 8(1):27–38.

Haapoja, M. 1994. Conservation easements: Are they for you? *American Forests* 100(1&2):29–31, 38.

Hamilton, L. S. 1965. The federal forest regulation issue. *Forest History* 9(1):2–11.

Hawks, L. J., F. W. Cubbage, and D. H. Newman. 1991. Regulatory versus voluntary forest water quality programs in Maryland and Virginia. In *Proceedings: 1991 Society of American Foresters Convention.* Bethesda, MD: Society of American Foresters.

Hickman, C. 1993. Local regulation of private forestry. *Forest Farmer, 29th Manual Edition* 52(3):19–21.

Hickman, C. A. 1982. Emerging patterns of forest property and yield taxes. In *Proceedings: Forest Taxation Symposium II.* Publications FWS-4-82. Blacksburg, VA: Virginia Polytechnic Institute and State University, School of Forestry and Wildlife Resources.

Hickman, C. A., and C. E. Martus. 1991. Local regulation of private forestry in the eastern United States. In *Proceedings: 1991 Southern Forest Economics Workers Meeting.* Baton Rouge: Louisiana State University, Department of Forestry, Agricultural Experiment Station.

Hickman, C. A., W. C. Siegel, and C. E. Martus. 1991. Forest practice regulatory developments in the East. In *Proceedings: 1991 Convention, Society of American Foresters.* Bethesda, MD: Society of American Foresters.

Jackson, B. D., W. D. Greene, and M. Baxter. 1992. *Field handbook of local logging and log trucking regulations in Georgia.* Athens, GA: University of Georgia.

Kurtz, W. B., R. J. Alig, and T. J. Mills. 1980. Retention and condition of Agricultural Conservation Program conifer plantings. *Journal of Forestry* 78(5):273–276.

Moote, M. A., S. Burke, H. J. Cortner, and M. G. Wallace. 1994. *Principles of ecosystem management.* Tucson: Water Resources Research Center, College of Agriculture, University of Arizona.

Osborne, D., and T. Gaebler. 1992. *Reinventing government.* New York: Plume.

Paster, J. D. 1983. Money damages for regulatory "takings." *Natural Resources Journal* 23(3):711–724.

Powell, D. S., J. L. Faulkner, D. R. Darr, Z. Zhu, and D. W. MacCleery. 1993. *Forest resources of the United States.* General technical report RM-234. Fort Collins: USDA Forest Service, Rocky Mountain Forest and Range Experiment Station.

President's Council on Environmental Quality. 1993. *Biodi-*

versity on private lands. Washington, DC: Executive Office of the President.

Risbrudt, C. D., F. H. Kaiser, and P. V. Ellefson. 1983. Cost-effectiveness of the 1979 Forestry Incentives Program. *Journal of Forestry* 81(5):298–301.

Sampson, N. 1993. Timberland investment in the environmental 90s. In *Assessing timberland investment opportunities: Prospects for the 90s and beyond,* ed. D. Newman, T. Harris, and C. DeForest. Athens, GA: University of Georgia School of Forest Resources.

Siegel, W. C. 1974. State forest practice laws today. *Journal of Forestry* 72(4):208–211.

———. 1978. Historical development of federal income tax treatments of timber. In *Proceedings: Forest Taxation Symposium.* Publication FWS 2-78. Blacksburg, VA: Virginia Polytechnic Institute and State University, School of Forestry and Wildlife Resources.

Skok, R. A., and H. M. Gregersen. 1975. Motivating private forestry: An overview. *Journal of Forestry* 73(4):202–205.

Society of American Foresters. 1993. *Task force report on sustaining long-term forest health and productivity.* Washington, DC.

Spurr, S. H. 1976. *American forest policy in development.* Seattle: University of Washington Press.

Stoebuck, W. B. 1982. Police power, taking, and due process. *Land Use and Environmental Law Review* 13: 349–392.

Troup, R. S. 1938. *Forestry and state control.* Oxford, England: Clarendon Press.

Section

V Institutions in Transition

The final seven chapters of this book focus on natural resource institutions—a very broad category that ranges from government agencies to professional and political organizations to the institution of science. The institutional questions associated with building the future of forestry are perhaps the most complex. They are questions about how we organize ourselves to a task, reconcile our differences, and assign priorities.

The natural resource institutions that we built in the 20th century were organized along traditional hierarchical lines—often they have been top-heavy and insular. If we are to achieve the kind of innovative site-specific management and landscape-level coordination described in earlier chapters, we must move in the opposite direction.

Many of the concepts discussed throughout this volume hinge on changing the focus of forest management from quantity to quality, from industrial-type production to the provision of goods and services. This paradigm shift is not unique to forestry. It is part of a much broader move from the industrial age to the information age. In the information age, natural resource institutions will need to be increasingly decentralized and integrated with the outside world. The authors in this section discuss some ideas and promising approaches that point us in that direction.

Decentralization of Authority and Responsibility

In the late 20th century, we have witnessed the collapse of the great centralized economies of Eastern Europe and Russia—something few of us would have imagined even 10 years ago. What was true for those economies is true for forestry. Highly centralized planning has serious limitations, socially and ecologically. We can no longer manage by formula.

For example, when Congress set record high timber targets in the 1980s, they did so regardless of the ecological attributes of individual sites or the idiosyncrasies of local economies. As a result, they took much of the creativity out of forest management. Not only were managers constrained in their ability to find innovative solutions to local controversies, but in many cases they found it difficult to comply with federal environmental statutes. Timber targets (expressed as allowable sale quantities or ASQs), narrowly defined budget categories, and other detailed directives characteristic of top-down decision making were incompatible and inconsistent with the concepts and practices of ecosystem management.

If we want managers to custom fit retention harvesting methods to specific sites (Franklin et al., Chapter 7), develop special forest products that help diversify local economies (Molina et al., Chapter 21), and form partnerships with other agencies and outside groups (Yaffee and Wondolleck, Chapter 24), we need to find ways to "flatten" the hierarchies of traditional bureaucracies and spread responsibility and authority out among professionals at the ground level. In Chapter 23, Meidinger describes a number of organizational models to that end.

This raises one of the most important institutional questions for forestry in the next century: How do we provide local managers with the flexibility and autonomy to be innovative while maintaining standards of quality and coordination across landscapes and regions? Creative approaches to the resolution of this question include stewardship contracts and the adoption of proposals by local partnerships such as the Quincy Library Group in California and the Applegate Partnership in Oregon.

Integration Among Managers, Scientists, and Stakeholders

Just as we need to look at biological life as a complex of interacting systems, we need to look at institutions and organizations as parts of interconnected networks. While this is not a particularly new idea, we have a long way to go toward its implementation. Making institutional networks a reality means that our long-held "us versus them" habit of formulating problems will have to change.

This can take place on many different levels. In an immediate sense, it might take the form of information sharing. In the past, government agencies have generated and controlled most of the data on public forests. They guarded that information closely and projected a professional public image of "we know best." But, as many of the recent controversies attest, interest groups are suspicious of agency conclusions and are increasingly sophisticated in generating their own information. In some cases, the data collected by interest groups are more accurate and complete than government information. For example, in the late 1980s and early 1990s, the GIS maps of old-growth forests in the Pacific Northwest developed by the Wilderness Society were both superior to and more readily available than federal agency data on old growth.

As a part of ecosystem management, we are beginning to ask questions that were previously unthinkable. What if, for example, government agencies and other public institutions opened up their databases to anyone who was interested? What if everyone had the same databases to play with? Perhaps some of the energy now focused on legal battles could be turned toward collaborative problem solving. Technologies such as computer networks and GISs make such information sharing possible and practical. The Northwest Power Planning Council provides a model for this. Data and copies of the council's computer models for fish and power production are publicly available with technical support for those who wish to use them (see Smith, Chapter 27, for a case study).

We also might foster integration by creating opportunities for outside groups to be involved in national forest management. In Chapter 24, Yaffee and Wondolleck call this bridging. They describe how new types of partnerships with outside groups can enable the Forest Service not only to tap into a broader knowledge base, but also to develop the trust it needs to carry out new management prescriptions.

Another way of integrating the work and ideas of different groups and professions is to explore alternative organizational and political models. As Meidinger points out in Chapter 23, pluralism—the dominant model of American politics—is based on interest groups competing with one another for political influence and resources. As long as competition drives the system, collaboration between traditional adversaries (for example the wood products industry and environmentalists) is extremely difficult. In contrast, a contending school of political theory, known as civic republicanism, holds that the purpose of politics is not to divide resources and influence based on political power, but rather to develop a shared understanding of the common good through public deliberation. The relevance of such ideas to ecosystem management should not be overlooked.

Securing Resources for Stewardship

Finally, a central question facing natural resource institutions is how to secure adequate and stable sources of funds to carry out the research, monitoring, and stewardship activities necessary to the practice of ecosystem management. Increasingly, natural resource management agencies are asked to perform more tasks with fewer resources. Annual appropriations will continue to be important in most countries, but given the fluctuating nature of most government spending, we also must explore alternative sources of support.

One promising approach is the establishment of trusts—that is, the creation of funds that will provide committed, dependable support for stewardship activities. The Nature Conservancy provides one of the most successful models of this. A trust fund for stewardship is created and dedicated at the same time that funds are raised for land acquisition. A trust fund approach to adaptive management was proposed by FEMAT (the Forest Ecosystem Management Assessment Team), but has been ignored thus far.

Another option for stretching existing resources is forming cooperative relationships both between government agencies and between governmental and nongovernmental organizations (Yaffee and Wondolleck, Chapter 24). Legal obstacles such as the Federal Advisory Committee Act (FACA) present a significant challenge to such strategies (see Meidinger, Chapter 23). Yet finding ways to share resources and expertise, and to eliminate bureaucratic redundancies, will be an important aspect of forestry in the 21st century.

23 Organizational and Legal Challenges for Ecosystem Management

Errol E. Meidinger

Natural Resource Institutions 362
A Law of Ecosystem Management? 363
Models of Organization 364
Models of Politics 368
Pluralism 369
Civic Republicanism 370
Legal Challenges 370
Public Land Management Laws 370
Box: Federal Laws Authorizing Ecosystem-Oriented Objectives 371
Box: Federal Laws Authorizing Cooperation with Private Land Owners and Other Governments 372
Administrative Law 372
Antitrust Law 375
Property Rights and Takings Law 376
Conclusion 377
Acknowledgments 377
Literature Cited 378

As a practical matter, ecosystem management is dependent on developing new organizational relationships for managing natural resources. While the reasons for this state of affairs cannot be fully elaborated here, many reflect the fact that our two most conventional forms of organization—markets and hierarchies—cannot achieve ecosystem management on their own.

The key inadequacies of markets are fairly obvious. First, many valuable functions and components of ecosystems are not effectively priced by markets. Thus "ecosystem services," such as storing and cycling essential nutrients, pollinating plants, maintaining hydrologic cycles and atmospheric composition, and cleansing water and air, either are unpriced by markets or are severely underpriced. There simply do not exist effective buyers whose bids reflect the full value of these ecosystem functions. Second, many beneficiaries of these ecosystem functions cannot participate in markets. These include both future generations of humans and all generations of nonhuman species. Therefore, ecosystem functions that benefit these groups are undervalued by markets.

Because of the manifest limitations of markets as generators of ecosystem management, both advocates and opponents (e.g., Fitzsimmons 1994) of the goal sometimes conclude that stronger hierarchical organization, in the form of either comprehensive regulation or unified ownership, is implied. It appears, however, that hierarchical arrangements are also seriously inadequate for ecosystem management. The reasons are more complex, but can be grasped by considering several basic requirements of ecosystem management. Among other things, successful ecosystem institutions must

- coordinate ecological and social information gathering and analysis at a variety of levels (genetic through landscape to national and global) (e.g., Grumbine 1994);
- coordinate management activities on separately owned and multiply governed lands (e.g., Sample 1992);
- spot surprises in ecological and social processes (e.g., Lee 1993, Franklin 1994);
- support ongoing social dialogue to
  - — build community understandings of eco-social issues,
  - — evaluate alternative policy possibilities, and
  - — choose courses of action (e.g., Francis 1993); and
- change course when knowledge or social values change (e.g., Walters 1986, Lee 1993).

In essence, ecosystem management requires organizational forms that can simultaneously coordinate scientific inquiry and democratic deliberation with existing institutions for managing natural resources and organizing communities. The inadequacies of traditional hierarchical institutions to this task are detailed in the next three sections.

Natural Resource Institutions

Management and control of natural resources is scattered among a variety of agents in society. At the grossest level (title holding) land is owned by many different kinds of actors—individuals, families, publicly and privately held corporations, nonprofit corporations, Indian tribes, cities, counties, states, the federal government—each operating with different incentives and under different rules. Most ecologically significant landscape units contain multiple landowners, often of different types. Less obvious given the oversimplified image of property ownership in current societal discourse, property owned by one party is also subject to interlinking property rights held by others. The most obvious of these rights are embodied in nuisance laws, which generally prohibit landowners from substantially and unreasonably interfering with the use and enjoyment of property by their neighbors or the public (e.g., Prosser and Keeton 1984). Thus, neighbors and the public have rights requiring that the property of others not be used in a way that substantially and unreasonably interferes with their interests. Although nuisance law as we know it today is largely a product of the common law courts, it has earlier roots both in community definitions of property rights (Meidinger 1993, Cappel 1995) and in formal statutes. An important general lesson is that property rights have never been absolute. Rather, they have been constructed and enforced by social institutions to include reciprocal responsibilities to other property owners and the public. As natural resource uses and social knowledge have changed over time, so have those responsibilities.

A related body of servitude law, comprised of easements, real covenants, and equitable servitudes, allows neighboring landowners to negotiate specific arrangements governing the uses of each other's property. Although servitude law is arcane and complex, with many odd limitations and hidden traps, in most states today neighbors can use servitude law to implement almost any substantive agreement about land use across ownership boundaries that they might wish to make. In addition, one of its most recent forms, the conservation easement, allows governmental and not-for-profit private organizations to negotiate voluntary arrangements with owners to limit their land uses in specified ways. This is a flexible mechanism because it permits locally variable arrangements that might allow for a certain amount of timber cutting or other natural resource exploitation according to terms specified in the easement agreement.

With the industrial revolution's creation of many new ways in which particular land uses can injure the interests of others, environmental and natural resource problems came to be seen as too complicated and changeable to be left to the care of generalist, nonexpert judges. Accordingly, successive bodies of substantive statutes and administrative regulations were promulgated to control environmental and natural resource uses. Enforcement, as well as elaboration and revision of these laws, was often left to administrative agencies, which were seen as the most sensible mechanism for handling complex, rapidly changing problems. These laws and agencies operated at a variety of levels—local, state, and federal—often in a cumulative fashion. Thus, air pollution regulation was handled entirely at the local level in the late 19th century, then also at the state level starting around World War II, and finally also at the federal level starting in the early 1970s. Today, important air pollution regulation functions continue to be carried out by all three levels of government (Krier and Ursin 1977). Much the same can be said about many other types of problems, such as water pollution and wildlife protection, although there are many local variations. Notably, the direct regulation of forest practices has not been federalized, despite efforts to do so in the middle of this century. State regulation of forest practices varies considerably. However, many of the types of regulation noted above are directly applicable to forest management, thus creating a complex and open-ended set of institutional rules governing forestry.

In sum, the legal context of forest management is multilayered and complex, beginning with a variety of types of owners (private, corporate, and governmental, all having different incentives), proceeding to include a variety of interlinking rights and duties among them, and then including many types of regulation (water, air, land use, wildlife) implemented by a variety of government bodies. Legal definitions of rights and responsibilities have always been somewhat fluid and dynamic, and may be growing increasingly so (Geisler and Kittel 1994). If ecosystem management is to be achieved, it must be done in this complex institutional context.

A Law of Ecosystem Management?

One way of attempting to institutionalize ecosystem management would be simply to add an overarching layer of law to the received system—essentially saying "thou shalt practice ecosystem management" applicable to all types of owners and governments. However, ecosystem management as a principle remains too general to translate into an enforceable law. One could seek to distill a rule from Grumbine (1994):

> Thou shalt integrate scientific knowledge of ecological relationships within a complex sociopolitical and values framework toward the general goal of protecting native ecosystem integrity over the long term.

Another could come from Moote et al. (1994):

> Thou shalt follow a management philosophy which focuses on desired states, rather than system outputs, and which recognizes the need to protect or restore critical ecological components.

Still another could come from the current chief of the U.S. Forest Service:

> Thou shalt look at things at very large scale, and across boundaries, and thereby maintain the health of the system, defined as the ability to recover from insult time after time after time. Preservation shall be the primary goal. Do not kill the goose that lays the golden eggs. Consider your individual actions in relation to the overall context.

In no case can the core principles of ecosystem management be translated into formal rules defining acceptable and unacceptable conduct. Rather, they all clearly involve considerable amounts of discretion, negotiation, deliberation, and learning. Accordingly, the challenge of ecosystem management is not in the first instance a legal one. Rather, it is an institutional one: How well is our society organized to facilitate the development of ecosystem management? Legal issues are important to the discussion, however, because institutions and organizations are partially structured by law. Because laws are often constructed as constraints on human actors and consequently on their ability to create new forms of social organization, legal requirements can act as brakes on ecosystem management. It is important to note, however, that laws also empower actors (individuals, corporations, governments) to conduct themselves in particular ways, and therefore can also be enablers for ecosystem management.

In analyzing the role of law in ecosystem management, it is not feasible simply to survey all the laws in existence and offer conclusions about whether they obstruct or foster ecosystem management. Not only are there too many laws to be analytically tractable (many having little practical effect), but their relevance is not apparent without prior information on what kinds of social organizations are useful for achieving ecosystem management. Thus, since ecosystem management is in many ways an organizational problem, it will be helpful to discuss alternative organizational structures for ecosystem management before returning to legal issues.

Models of Organization

Although organizations in this society are variable and exhibit enormous particularistic variation, they also show broad patterns of similarity (Morgan 1989). These similarities reflect and reinforce widely held normative assumptions and expectations about the appropriate structure of organizations (Dobbin 1994, Meidinger 1987). The laws structuring organizations generally reflect and reinforce the same norms. Both the laws and ecosystem management must therefore be understood in relationship to basic forms of social organization.

In the late 19th century, the model of organization that dominated the industrializing world also dominated the emerging natural resource management organizations—both private and public. The classical bureaucracy portrayed in Figure 23.1a and trenchantly analyzed by Max Weber (1978) had certain hallmarks: (1) a hierarchical system of supervisor-subordinate control, (2) departments with defined, compartmentalized jurisdictions, and (3) detailed rules governing the roles and powers of officials, operating procedures, and substantive decisions. The bureaucratic organization also (4) sought consistency across domains and (5) was staffed by appointed, salaried officials progressing through fixed career lines. (6) Policies were set by a head accountable either to commercial actors or a sovereign government. (7) Operational (lower-level) staff were responsible for configuring the materials they worked with (raw materials, information, people) into desired outputs. (8) Clear lines were drawn between the organization and its environment, and (9) staff members were required to pass frequent tests of loyalty to the organization.

As the complexity and interrelatedness of problems faced by bureaucratic organizations grew and became better understood, bureaucracies typically made several adjustments. First they created senior executive groups made up of the chief executive and directors of principal departments (Figure 23.1b) to deal with new problems that seemed to require new responses. The executive groups soon began meeting on a regular basis, as exemplified by the regular chief and staff meetings of the U.S. Forest Service. Nonmembers attending to contribute information or advice accrued an important symbol of status in the organization. Individual departments continued to exercise distinctive authority in the daily operations of the organization, however.

As it became apparent that senior executive groups could not handle all the problems raising interdepartmental or novel issues, bureaucratic organizations began establishing project teams and task forces, as shown in Figure 23.1c. Typically, however, task forces were shorter lived and weaker than departments; team members remained accountable to their own department heads rather than to their teams. Promotions were controlled by department heads, and staffers were expected to represent the views of their departments. Especially difficult or important problems were still elevated to the senior executive group.

Figures 23.1a, 23.1b, and 23.1c characterize the formal structures of most resource management organizations in North America, at least through the middle of this century, and in many cases they still do. Their shared hierarchical structure and system of accountability became a taken-for-granted feature of the modern organizational landscape, a form that people expected and relied upon at the same time that they decried its evident shortcomings. Various

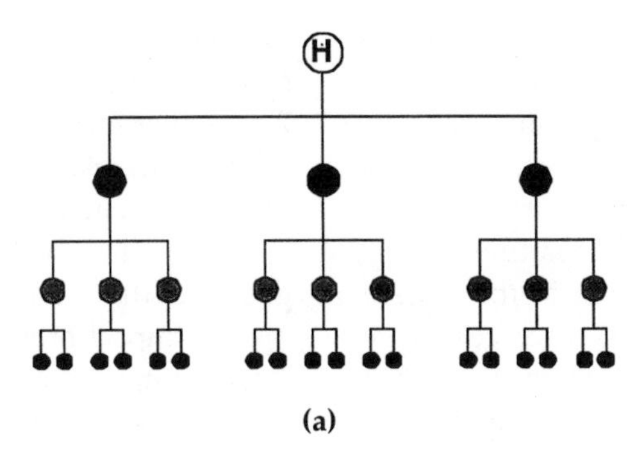

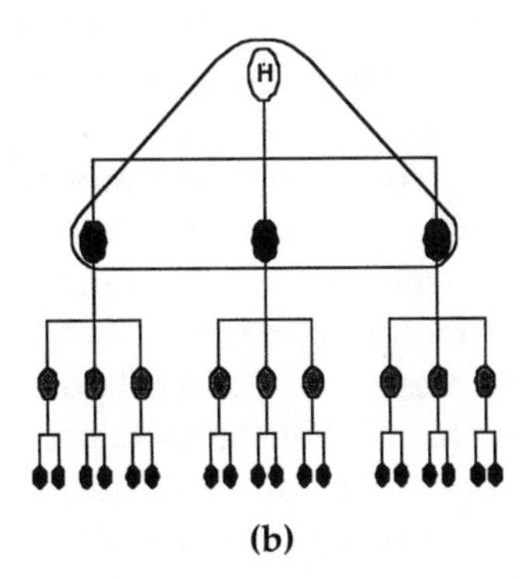

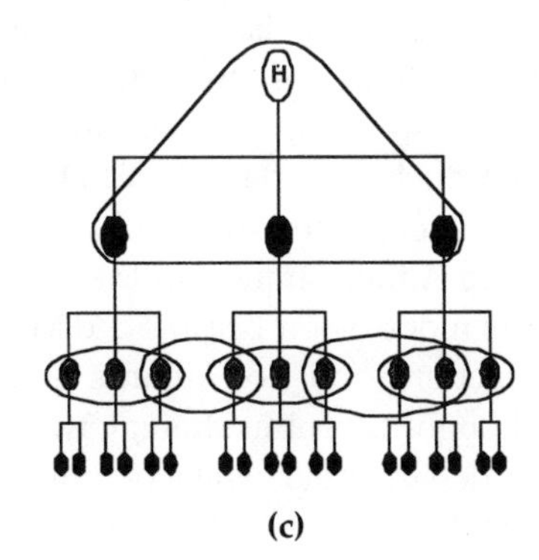

Figure 23.1 (a) The classical bureaucracy, (b) Bureaucracy with senior management team, (c) Bureaucracy with project teams and task forces.

bodies of law—administrative, corporate, proprietary, constitutional, etc.—assumed, reinforced, and gradually rigidified hierarchical bureaucratic formations.

At the same time, significant pressures for change took shape. These are often portrayed as efficiency pressures. Changes were needed, according to functional theories, because bureaucratic organizations were incapable of handling complex problems in a rapidly changing, turbulent world. While this portrayal seems accurate in many respects, it misses several important forces. First, many people found working in hierarchical bureaucracies demeaning and frustrating and began to reduce their levels of commitment, exacerbating problems of organizational productivity and adaptability. Second, the portrayal missed the complex political life of organizations, which has always involved struggles and internal divisions over the substance of organizational commitments. Third, and most important, organizations were increasingly expected to be politically responsive at both top and bottom. As the inability of chief executives and middle managers to adequately calibrate policies to changing local conditions and coalitions became increasingly apparent, bureaucracies were unable to legitimate their policies based on hierarchical directives alone. Public bureaucracies could no longer simply say "we work for Congress and the President." The fact that operational employees make policy became fairly widely understood (e.g., Lipsky 1980) and "public participation" requirements were imposed at both upper and lower organizational levels (e.g., Meidinger 1992). If participation were to mean anything, therefore, policy would have to flow up the organization as well as down.

Meanwhile, the ideal of a vertically stratified, horizontally compartmentalized organization itself began to give way. Figures 23.2 and 23.3a depict post-classical conceptions aimed at integrating organizational tasks and functions. The rather technocratic "matrix organizations" depicted in Figure 23.2 seek to give equal priority to functional departments (e.g., timber, fish, wildlife, administration, in the forestry context) in task units (Baber 1983). Each cell in the matrix thus represents a team combining different functional departments with joint responsibility for a particular task or product. Teams might be relatively permanent or very short-term. Team members are expected to maintain a dual focus on function and end-product, in the process mediating conflicting goals, processing high volumes of information, and producing intelligent responses to environmental changes, as well as producing good products. Developed by the Defense Department and the National Aeronautics and Space Administration in the 1950s and 1960s, the matrix model soon became a darling of management teachers and consultants (e.g., Chadwin 1983). Although its central strategy was to maintain functional departments while organizing work in interdepartmental task units, the matrix model was also credited with an improved capacity

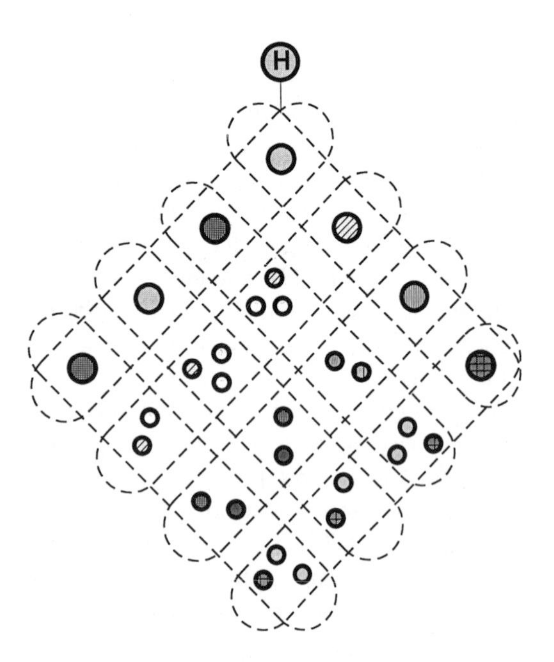

Figure 23.2 Matrix organization.

to integrate diverse perspectives from outside the organization. As is evident from the depiction in Figure 23.2, however, this was not done by formally institutionalizing relationships to external interests, but rather by relying on the diverse perspectives and commitments of interdisciplinary teams. It can be argued that the "systematic, interdisciplinary" environmental impact process required by the National Environmental Policy Act and the interdisciplinary planning processes required by several federal land management statutes in effect mandated the partial "matrixing" of several federal agencies, but it is not clear how much transformation of their traditional hierarchies has actually occurred.

Expanding on the team concept of matrix models, but loosening organizational boundaries and hierarchical accountability, the project-based organization depicted in Figure 23.3a assigns responsibility for core operational activities, much policy making, and innovation to project teams, which are regularly reconfigured when projects and environmental conditions change. Functional departments play primarily a support role, although teams are still composed with functional requirements in mind. One of the most important features of this model is that it begins to look as much like a network as like a hierarchy. Several very important implications flow from this fact. First, given the reduced importance of rules and supervisory control, Figure 23.3a-type organizations must devote more effort to creating shared understandings of organizational identity and mission. Second, the lack of a strong hierarchy means that coordination between teams must be achieved through regular and informal mutual adjustment. Since top management is lean, it cannot possibly make every judgment call that might arise in cases of disagreement. Third, as a direct consequence of the first two characteristics, good information is crucial, and relationships are designed more around information exchange than around supervision and control. Conversely, the structure of information exchange becomes a kind of control in itself, but perhaps a more fluid one than traditional hierarchy and rules.

With its linked but decentralized team organization, flexible task focus, and reliance on information flows and mutual adjustment, the model shown in Figure 23.3a might be seen as a good one for ecosys-

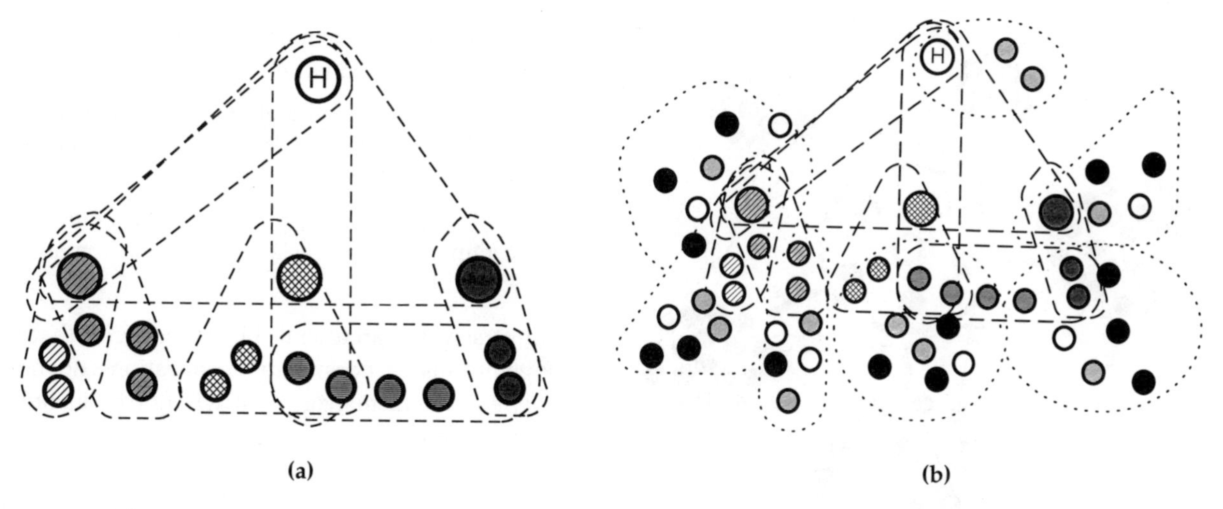

Figure 23.3 (a) Project organization. (b) Project organization with external linkages.

tem management. This could indeed turn out to be the case. But for this model to work, an important question must be addressed: What will be the significance of organizational boundaries? All the models thus far discussed treat resource management organizations as self-contained and distinct from their environments. At the same time, we know that most organizations are continually involved in complex forms of interchange and mutual dependence with actors, organizations, and resources in their environments. Indeed, part of the early success of the U.S. Forest Service was attributed to the ability of its field personnel to establish effective relationships with local community members (e.g., Kaufman 1960). However, the primary duty of the forest ranger was to adapt general forest service policies to local conditions and to gain local acquiescence to agency policies—not to empower local communities in national policy making. Primary loyalty to the agency was stressed in a variety of organizational practices, most notably, rotation of management personnel through a variety of locales.

We also know that government agencies are required to provide for public participation in all levels of politically significant decision making. We know further that government resources are likely to be inadequate to comprehend and manage most landscape-level ecosystems, thereby making cooperation with and among nongovernmental organizations essential. One plausible response is to look beyond the organizational membrane and to implement (or perhaps simply pay attention to) an extended structure like that in Figure 23.3b. This model retains some hierarchical control while allowing for variable task organization, knowledge sharing, efficient resource use, and interorganizational coordination. Recent "science assessment" efforts, such as the Interior Columbia Basin Ecosystem Management Project and the Southern Appalachian Ecosystem Assessment, can be seen as tentative efforts to develop this form of organization. Practical institutional questions confront the implementation of the model in Figure 23.3b: (1) What conditions encourage effective collaboration? and (2) Will our resource management and administrative laws facilitate this kind of organization?

A form with some of the advantages of project organization, but which allows a more fluid structure is depicted in Figure 23.4, the loosely coupled network. Best exemplified by organizations in the fashion industry, in which firms form alliances to develop product lines with the strategic leadership and support of a small core of staff, this model relies on relatively few employees and extensive subcontracting. All that really holds it together is a mutual interest in a shared enterprise—a commitment to producing a successful outcome and possibly a partial vision of how that can be done. That mutual interest can, of course, change

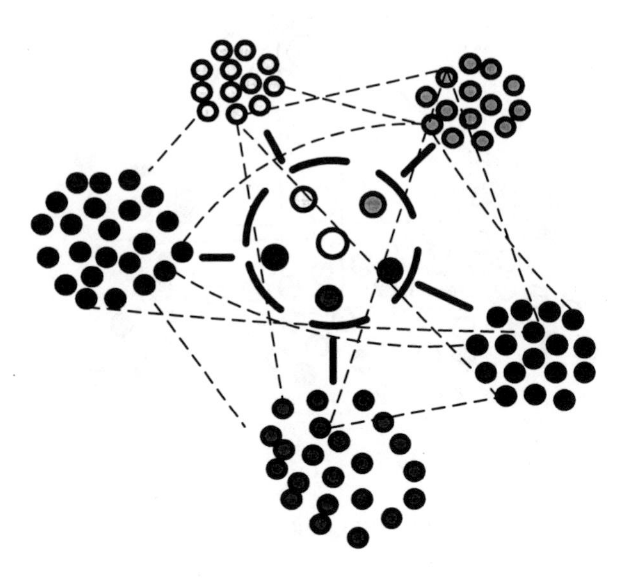

Figure 23.4 Loosely coupled organic networks.

quickly, and consequently specific organizational relations are fluid.

As an empirical matter, this model seems to bear the closest general resemblance to many landscape-level ecosystem management efforts that have been attempted to date (e.g., Lynch 1995a, 1995b; Antypas and Vanderwood 1995a, 1995b; Vanderwood 1995; see also the case examples in Wondolleck and Yaffee 1994). This may inherently be the most likely model because of the fragmented structure of ownership and governmental authority in most ecologically significant watersheds. The difficult question is whether shared interests in landscapes will be sufficient to hold such efforts together and translate them into coordinated behavior over time. In business contexts, the profit motive plays an important though not exclusive role in holding together loosely coupled networks. In the ecosystem context, incentives supporting loosely coupled networks will ordinarily be more diffuse, but will probably include a healthy dose of economic self-interest, often created in part by the perceived risk of more intrusive regulation in the absence of self-generated coordinating mechanisms.

If shared interests in place develop and deepen over time, it seems plausible that loosely coupled ecosystem management networks could become more stable and interconnected. They might thus metamorphose toward either

1. figure 23.3b extended project organizations in socially "thin" ecosystems (i.e., places organized through relatively few extant organizations), or

2. more complicated organizational forms, often labeled communities, in socially dense environments (e.g., House 1993).

Conversely, the argument thus far suggests that if stronger forms of ecosystem organizations do not evolve, ecosystem management will not be institutionally feasible.

Models of Politics

There is a suggestive parallel between the model in Figure 23.4 and what empirical researchers have been learning about the general nature of policy-making organizations in the American political system. Building on the work of Heclo (1978) and others, scholars like Sabatier and Jenkins-Smith (1993) argue that American public policy is made in "policy subsystems." Whether called by this name, or "implementation structures" (Hjern and Porter 1981), or "policy communities" (Kingdon 1984, Shannon 1991), or "regulatory communities" (Meidinger 1987), or "issue networks" (Heclo 1978, Michaels 1992), there is widespread agreement that these groupings consist of a variety of public and private actors, typically including federal and state agencies, business firms and associations, public interest groups, journalists, scholars, and others seriously interested in a given policy arena. Sabatier and Jenkins-Smith maintain that policy subsystems are ordinarily divided into two contending advocacy coalitions organized around competing core beliefs. They suggest that advocacy coalitions typify most policy arenas and can be more important in policy formation than formal authorities such as legislatures and chief executives, which often seem to serve primarily to ratify the arrangements worked out in policy subsystems.

Figure 23.5 is a simple representation of modern political organization as portrayed by Sabatier and

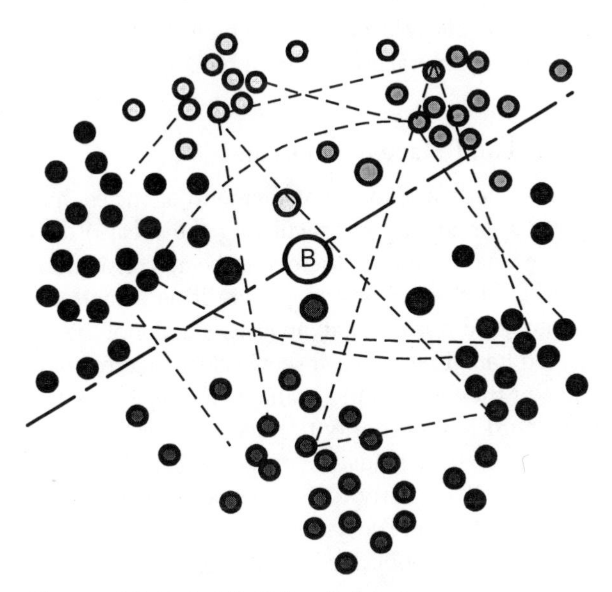

Figure 23.5 Public Policy Subsystem.

Jenkins-Smith. While retaining a weakened form of group clustering from Figure 23.4, it also allows for a variety of more diffuse relationships across subgroups. The actor labeled B represents a "policy broker," who Sabatier and Jenkins-Smith say functions to reduce intense conflict between the advocacy coalitions separated by the diagonal line. Interestingly, research by Heinz et al. (1993) indicates that in some national policy domains, policy brokers do not exist. Heinz and coauthors also argue that patterns of communication are much more complex than traditionally imagined, readily crossing interest group lines and circumventing would-be policy brokers. A graphical portrayal of their model, accordingly, would be Figure 23.5 minus the policy broker and fault line.

The discussion has moved from general to political models of organization for two reasons. First, most natural resource organizations in practice are as much political as they are purely economic or managerial. Second, politics, like markets and hierarchies, must be understood as a form of human organization, and not just as a collection of processes or rules. Because politics is often hived off from other sectors of social life in American discourse, it is important to sketch some possible connections between the alternative organizational models discussed thus far and the two dominant models of American politics, pluralism and civic republicanism.

Pluralism

By far the best-known model of American politics is often called interest group liberalism or simply pluralism (e.g., Dahl 1967, Truman 1959). Its core assumption is that interest groups with conflicting policy desires compete to obtain preferred outcomes in political forums, largely by offering or threatening to revoke valuable political resources such as votes and campaign funding for political officials. Pluralist models assume that political interests are a given (exogenous) and are not a product of political processes. They also assume that the fundamental function of politics is to distribute valuable rights, privileges, or resources among competing claimants, and not to develop shared social values or fair principles for distribution.

Pluralist models of politics are readily and often linked with bureaucratic models of organization. In these linkages, pluralist negotiation occurs at the level of the legislature and the executive—the H in Figures 23.1a through 23.1c. The role of the bureaucracy is then to implement the deal or bargain made in the political process. Many political and legal theorists focus on how to make bureaucracies more precisely implement the bargains cut in the political arena and memorialized as statutes or rules (e.g., Lowi 1969). As the above discussion of organizational models suggests, however, traditional bureaucracies may be poorly adapted to the kind of horizontal information sharing, normative dialogue, and policy adaptability implied by ecosystem management. To achieve ecosystem management, ownership organizations need to do much more than manage their own lands. They need to coordinate management with their neighbors, share information, develop shared understandings, and adjust readily to unforeseen circumstances. The post-bureaucratic models discussed above appear to emphasize these capacities. How important they will be to ecosystem management depends somewhat on the capacity of traditional political subsystems to process the detailed information and very different problems posed by highly variable landscapes. As the work of modern political researchers like Sabatier and Heinz suggests, many policy subsystems already seem to be organized in some form of loosely coupled network. What may occur with ecosystem management, accordingly, is the development of many more localized policy subsystems and a consequent shift of policymaking activity toward the local level.

Civic Republicanism

Although they are not necessarily tied to it, the post-bureaucratic models discussed above suggest parallels with America's main contending school of political theory, civic republicanism, which has enjoyed a revival in recent years (e.g., Pocock 1985, Stanley 1983, Barber 1984, Boyte 1987, Michaelman 1988). Civic republicanism is harder to characterize than the more mechanistic models of pluralism. One of its key premises is that the purpose of politics is not simply to divide spoils according to the pre-given preferences and power of interest groups, but rather to develop shared understandings of the common good through public deliberation. Because it eschews the political subservience of some citizens to others, civic republicanism resists hierarchical models of politics, implies a high degree of equality, and often has a flavor of localism. Because of its deliberative purpose, civic republicanism requires organizational structures allowing for extensive communication of knowledge and values among citizens as they solve the problems of their collective life.

While it would be premature to link the fate of ecosystem management with that of civic republicanism, the parallels should be kept in mind, since the civic republican tradition has deep roots in American politics and could turn out to be an important supporting force that will be likely ignored by bureaucratic officials seeking to achieve ecosystem management.

Legal Challenges

Although it is not possible to define a single law of ecosystem management, it is important to consider which areas of law are likely to affect its prospects most, particularly if they will inhibit it. At this stage of experience, several areas should be discussed.

Public Land Management Laws

Because federal land management agencies have played a large role in articulating the ideal of ecosystem management, there has been considerable analysis of their ability to pursue it as a matter of statutory authority. That analysis need not be rehashed here. In essence, it seems clear that the mandates of the federal land management agencies give them sufficient authority to pursue ecosystem management on their lands, but probably do not mandate it (Keiter 1994, Perry 1994; see box titled, Federal Laws Authorizing Ecosystem-Oriented Objectives).

In the widely publicized FEMAT (Forest Ecosystem Management and Assessment Team) case involving the legality of the joint BLM–Forest Service plan produced following the President's Timber Conference in the spring of 1993, Federal District Judge Dwyer concluded that, based on the confluence of the land management laws, the National Environmental Policy Act, and the Endangered Species Act and "[g]iven the current condition of the forests, there is no way the agencies could comply with the environmental laws without planning on an ecosystem basis" (*Seattle Audubon Society v. Lyons,* W.D. Wash. (1994), 871 F. Supp. 1291, 1311). This is an important dictum, but should not be read as an authoritative statement that the federal land management laws in general require ecosystem management, since it is closely tied to the factual situation in the forests of the Pacific Northwest.

The federal land management agencies also have limited authority to involve themselves with management activities on private lands (see box titled, Federal Laws Authorizing Cooperation with Private Land Owners and Other Governments). The ability of state and local land management agencies to pursue ecosystem management has not been well researched and is currently subject to some contention. Relevant laws and policies vary enormously among states. The primary questions are to what degree state agencies must concentrate on maximizing revenues as opposed to providing for other forest values and to what degree maximizing revenues would preclude ecosystem management. Based on the broad latitude generally given to public agencies, as well as the argument that ecosystem management will enhance long-term productivity, it seems implausible to read revenue-generating responsibilities generally as precluding ecologically sensitive management. An important but weak-reasoned Washington State Supreme Court decision, however, states in dictum that the state land management agency must act like a private trustee solely in the interest of the beneficiary of the land, which the court construed to be public school funds (*County of Skamania v. State,* 102 Wash.2d 127, 134, 685 P.2d 576, 580 (1984)).

Federal Laws Authorizing Ecosystem-Oriented Objectives

a. The Multiple-Use Sustained-Yield Act gives the Secretary of Agriculture broad discretion to manage the national forests for multiple use and sustained yield and defines multiple use to include "harmonious and coordinated management of the various resources, each with the other, without impairment of the productivity of the land, with consideration being given to the relative values of the various resources, and not necessarily the combination of uses that will give the greatest dollar return or greatest unit output" (16 U.S.C. sec. 531(a)).

b. The National Forest Management Act (NFMA) frames the Forest Service land-planning process that is to be used to pursue multiple-use goals. NFMA also contains authority to "provide for diversity of plant and animal communities ... in order to meet overall multiple-use objectives, and within the multiple-use objectives of a land management plan" (16 U.S.C. sec. 1604(g)(3)(B)). The forest planning process is to include comprehensive resource inventories (sec. 1604(g)(1)), "a systematic interdisciplinary approach to achieve integrated consideration of physical, biological, economic and other sciences" (16 U.S.C. sec. 1604(b)), coordination with state and local governments and other federal agencies (16 U.S.C. sec. 1604(a)), public participation (sec. 1604(d)), and continuous monitoring and evaluation (sec. 1604(g)(3)).

c. The Federal Land Policy and Management Act gives the Secretary of Interior (and by delegation the Director of the Bureau of Land Management) broad authority to manage BLM lands "on the basis of multiple use and sustained yield" (43 U.S.C. sec. 1701(a)(7)), "in a manner that will protect the quality of scientific, scenic, historical, ecological, environmental, air and atmospheric, water resource, and archaeological values," and "giv[ing] priority to the designation and protection of areas of critical environmental concern" (sec. 1712(b)(3)). The long definition of "multiple use" includes "harmonious and coordinated management of the various resources without permanent impairment of the productivity of the land and the quality of the environment with consideration being given to the relative values of the resources and not necessarily to the combination of uses that will give the greatest economic return or the greatest unit output" (43 U.S.C. sec. 1702(c)). These goals are to be accomplished through a land-use planning process involving resource inventories (sec. 1711), "a systematic interdisciplinary approach to achieve integrated consideration of physical, biological, economic, and other sciences," (sec. 1712(c)(2)), coordination with other federal agency, tribal, and state planning processes (sec. 1712(c)(9)), and public involvement (sec. 1712(f)).

d. Although applicable to all agencies, the Endangered Species Act has an especially direct bearing on federal land management agencies because many of their lands contain critical habitat for endangered or threatened species. Section 7 requires that each agency "insure that any action authorized, funded, or carried out by such agency is not likely to jeopardize the continued existence of any endangered ... or threatened species or result in the destruction or adverse modification of [critical] habitat of such species" and also requires that agencies consult with the Secretary of Interior before taking actions likely to affect such species (16 U.S.C. sec. 1536)).

e. The National Environmental Policy Act requires that, before taking any "major federal actions significantly affecting the human environment," all federal agencies produce detailed, interdisciplinary analyses (EISs) of the environmental impacts of proposed actions, alternatives to them, relationships to long-term productivity, and irreversible and irretrievable commitments of resources. They must also consult with other interested federal agencies (42 U.S.C. sec. 4332, generally known as Section 102). As agency practice and litigation have developed, agencies often find themselves producing EISs at multiple levels, from specific projects to regional plans, and have a difficult time coordinating such efforts.

Federal Laws Authorizing Cooperation with Private Land Owners and Other Governments

The Forest Service has authority to "protect trees and forests . . . [on national forest lands] and, in cooperation with others, on other lands in the United States from natural and man-made causes" (16 U.S.C. sec. 2104(a)) and to provide technical assistance to private nonindustrial landowners (sec. 2103(a)). Under the Stewardship Incentive Program, the Forest Service can use cost-sharing to "encourage long-term stewardship of non-industrial private forestlands" for goals that include forest wetland restoration, native vegetation and fish and wildlife habitat maintenance, and sustainable forest management (secs. 2103b(a), 2103(b)(4)(B)). The related Forest Legacy Program allows the Forest Service to protect nonfederal forestland from conversion to nonforest uses via conservation easements and other undefined "mechanisms" (sec. 2103(a). On BLM lands, where neighboring landowners consent or where public and private lands are intermingled, the agency has authority to develop grazing allotment management plans taking the management of private lands into account (43 U.S.C. sec. 1752(f)). The agency's mandate to coordinate with state, local, and tribal planning processes also creates capacity to link with private land management through the planning mechanisms of other governments.

The agency with the longest history of providing technical and financial assistance to private and state landowners is the Natural Resources Conservation Service (formerly known as the Soil Conservation Service), which has authority under various acts to assist with water quality improvement; water supply; wildlife habitat; soils management; erosion control; pasture, range, and crop management; flood control; and local land-use planning laws, among other things. It also can undertake natural resource inventories, wetland identification and protection, soil survey investigations, and conservation demonstration projects. Soil Conservation and Domestic Allotment Act (16 U.S.C. sec. 590g et seq.), Food Security Act (16 U.S.C. sec. 3811 et seq.), Watershed Protection and Flood Prevention Act (P.L. 83-566).

Many of the barriers to ecosystem management in federal statutes may be procedural rather than substantive. Agency planning horizons are relatively long (typically 10 to 15 years). The potential changes to plans implied by the experimental and adaptive aspects of ecosystem management will often have to go through a complicated plan amendment process, which will itself be subject to internal appeals and judicial review by affected parties relying on the plans. Moreover, since ecosystem management is just one of many allowable agency purposes, which sometimes conflict, an argument can be made that ecosystem management should be made mandatory for federal land management. However, the abstractness of the concept limits the precision with which such a mandate could be defined. In any case, larger institutional challenges to ecosystem management seem to reside in the general legal frameworks structuring federal agency interactions with the public, corporate competition, and definitions of property rights.

Administrative Law

General Provisions

In addition to the agency-specific requirements in the land management statutes, government agencies are subject to generic laws governing how they operate. After a generation of increasing agency duties to facilitate public participation, general administrative law seems to be moving toward giving agencies more autonomy. Thus, the Supreme Court has ruled in recent years that judges cannot require agency procedures beyond those clearly required in authorizing statutes (*Vermont Yankee Nuclear Power Corp. v. N.R.D.C.*, 435 U.S. 519 (1978)), that agency interpretations of the statutes they administer should receive deference unless Congress explicitly addressed the question at issue (*Chevron, U.S.A., Inc. v. Natural Resources Defense Council, Inc.*, 467 U.S. 837 (1984)), that members of the public have no standing to challenge agency decisions in court unless they directly use the land at issue, and that federal policies cannot be challenged until they are actually applied to particular pieces of land (*Lujan v. National Wildlife Federation,* 497 U.S. 871 (1990)). Since they preserve flexibility for land management agencies, these decisions could be portrayed as supportive of ecosystem management. The main question is how well the agencies will provide for public dialogue and deliberation given the flexibility allowed them by the courts. Their traditional track record is not very good. Even if the agencies seek to improve public participation, however, Congress and state legislatures may be on the verge of greatly increasing their procedural and analytic responsibilities by requiring extensive risk analysis to support any regulations significantly affecting nongovernmental interests.

The Federal Advisory Committee Act

While most administrative law focuses on either the decision-making procedures or the substantive duties of agencies, the Federal Advisory Committee Act (FACA) is virtually unique in focusing on continuing relationships between agencies and nongovernmental parties. Given the importance to ecosystem management of ongoing intergovernmental and interownership relationships, this small statute may have quite a large effect. Enacted October 6, 1972, by P.L. 92-463, 86 Stat. 770, FACA requires that whenever the federal government "establishes or utilizes" an advisory committee (broadly defined to include most types of groups) "in the interest of obtaining advice or recommendations" it must conform to a variety of requirements, including the following:

1. Chartering
   a. A formal charter by the president or agency head must exist before any duties are performed.
   b. The charter must detail committee purpose, duties, cost, usefulness, and membership.
   c. The committee expires automatically after two years unless rechartered.
2. Membership: The membership must be "fairly balanced in terms of the points of view represented and the functions to be performed." (Because this requirement is rather vague, courts have been very hesitant to override agency definitions of balance. Mainly, it requires that agencies explain why they think a committee is balanced.)
3. Open Meetings
   a. Advance notice of each meeting must appear in the *Federal Register.*
   b. Meetings must be open to the public (with narrowly limited exceptions).
   c. The public must be permitted to present views orally or in writing.
   d. There must be public review of all documents used by the committee (again subject to narrow exceptions).
   e. Detailed minutes must be kept and made available to the public.
4. Federal Control: Federal employees must
   a. call all meetings,
   b. approve all meeting agendas, and
   c. chair all meetings.

Although FACA's purposes of opening up government and providing a level playing field for all parties are understandable and laudable, the statute has several unfortunate side effects for ecosystem management:

1. *Uncertainty.* The statutory language referring to all committees "established or utilized . . . in the interest of obtaining advice or recommendations" is extremely broad and could potentially cover every group from which the government receives policy advice. Implementing regulations, promulgated by the General Services Administration in 1987, attempted to focus the statute on groups adopted "as a preferred source of advice or recommendations" or from which the government seeks "consensus advice and recommendations" (41 C.F.R. sec. 101-6.1003, 101-6.1004). Under this interpretation, most ecosystem management groups would arguably still be covered by the act. In 1989, the U.S. Supreme Court further narrowed the reach of FACA by reading "utilized" as a minor modifier of "established"; its purpose was "simply to clarify that FACA applies to advisory committees established by the Federal Government in a generous sense of that term, en-

compassing groups formed indirectly by quasi-public organizations . . . for public agencies as well as by such agencies themselves" (*Public Citizen v. Department of Justice,* 491 U.S. 440, 462 (1989)). Based on this reading, the Court found a standing committee of the American Bar Association that advised the president on the qualifications of potential federal judges to be outside the reach of FACA. However, an important element of the Court's opinion was the fact that a broader interpretation might have put FACA in conflict with the president's constitutional power to nominate federal judges (and implicitly to obtain advice in doing so from whomever he wished).

Because the constitutional issue was an element of the *Public Citizen* decision, it remains possible—although perhaps not likely—that courts could apply a more expansive definition of "utilized" to agency dealings with independently formed ecosystem management groups. Agency ecosystem management activity does not have constitutional status, in other words, courts could conceivably require FACA procedures whenever an agency deals with a free-standing group. Perhaps partly for this reason, federal land management agencies decided to stop participating in the independently formed Applegate Partnership in southwestern Oregon after losing an important but factually quite different FACA suit (Lynch 1995a). The case, *Northwest Forest Resource Council v. Espy,* (846 F. Supp. 1009 (1994)), involved the legality of the Federal Ecosystem Management Assessment Team formed as a result of President Clinton's 1993 Pacific Northwest Forest Conference. That team was clearly established by the federal government and, since it included members who were not full-time federal employees, was held subject to FACA. That decision to withdraw from the Applegate Partnership probably reflected an excess of caution, but nonetheless exemplifies one of the costs of the uncertainty of FACA's coverage. This uncertainty, of course, may also provide a handy tool for those who do not wish to see federal agencies create new ecosystem partnerships for other reasons.

In any event, FACA applies to every committee "established" by the federal government that includes individuals who are not federal, state, tribal, or local officials. If federal agencies wish to foster ecosystem management groups in places where they have not spontaneously emerged, they probably must conform to FACAs requirements. Doing so will have several problematic consequences.

2. *Federal agency control and distance.* FACA requires that the federal government define the purpose, duties, and membership of (as well as manage and largely control) any advisory committee. Although this may be understandable when an agency is acting in a national regulatory role, it creates more problems when the agency is acting as only one of many landowners and government agencies in a particular locale:

a. Ecosystem management groups will often have many more agenda issues than just advising the federal government let alone a single agency. Agency missions and authorities are likely to be inherently narrower than ecosystem management problems in mixed-ownership watersheds.

b. Federal control will alienate a number of potential participants, some of whom will simply consider federal control inappropriate and unnecessarily burdensome and others of whom will hold deeper resentments toward the federal government.

c. FACA arrangements tend to mesh poorly with existing local and regional institutions. Rather than linking up with institutions that already have organizations and norms, federal agencies under FACA must set up their own show according to their own standardized rules. This problem may be eased somewhat by a 1995 amendment to FACA exempting federal contacts with state, local, or tribal officials from FACA procedures, but will still be significant in cases where there are no appropriate preexisting state, local, or tribal organizations.

d. National partisan politics will sometimes play an inappropriate role, making the chartering process a political contest about which individuals should be members of the committee.

3. *Bureaucratic delay and cost.* In part because chartering makes an ecosystem management group an agency's responsibility, simply producing a charter takes a considerable amount of time and energy. Proposed charters inevitably cross many bureaucratic desks, creating countless opportunities for arguments about appropriate definitions of duties, funding, administrative responsibilities, and so on. Proposed membership slates can generate particular controversy, varying from the

partisan political concerns noted above to fears of "60 Minutes scandals." After a committee is chartered, other bureaucratic burdens emerge, varying from responsibility for keeping and maintaining a thorough set of records in a format readily accessible to the public to simply achieving the necessary agency administration of its meetings. It is not clear that all of these costs or concerns are necessary in the ecosystem management context, especially when participants are not receiving government compensation.

4. *Hobson's choice.* Given the situation described above, federal agencies are faced with a difficult and complex choice. First, they can accept the bureaucratic difficulties, costs, and cookie-cutter formalization of chartering a committee. Relatedly, they can use one of a variety of strategies to avoid direct application of FACA, including treating an ecosystem management group as a subcommittee of a chartered committee, contracting out the job of setting up an ecosystem management group, organizing it as an "operational committee," or essentially using the states or tribes as ecosystem management facilitators and drawing input from those governments. Each of these options, however, has many of the disadvantages of the FACA chartering process, including the federal distance and control noted above. Alternatively, the federal agencies can stand back from any ecosystem management group that might emerge on its own and try to let it function free of federal control.

Standing back has its own costs. If the agency relies on traditional public participation processes—essentially producing proposals in-house, then making them public, technically analyzing them, taking comments, and announcing decisions—it removes itself from the ongoing dialogue that can generate creative new proposals and create the conditions for efficient, consensus-based coordination with other landowners and agencies. If it attends meetings of a self-organized ecosystem management group, but leaves whenever discussion turns to matters that might involve federal policy, the agency risks exacerbating its image as arrogant and unresponsive and also undermines the capacity of the group to function at all where federal cooperation is important to the success of the overall effort.

In sum, FACA significantly constrains the organizational and coordinative options available to ecosystem management efforts. As a result, its laudable purposes of increasing public access and agency accountability may be subverted in the ecosystem management context.

Antitrust Law

As FACA regulates continuing relationships between federal agencies and others, antitrust law regulates relationships among businesses. The ideal of antitrust law is economic competition, which the law seeks to foster by challenging two types of relationships: price fixing and monopolization. Antitrust law is relevant to ecosystem management because different landowners in an ecosystem may be competing producers of a product. Therefore, the information sharing and coordination potentially involved in ecosystem management may look like a suspicious opportunity to reduce competition and raise prices. Lawyers working for timber companies have asserted that the antitrust laws pose a serious obstacle to coordinated landscape planning (Pauw et al. 1993).

Without providing definitions of the terms or examples of their meaning, antitrust laws ban "every contract, combination . . . or conspiracy, in restraint of trade or commerce" and all attempts and conspiracies to "monopolize any part of the trade or commerce" as well as successful monopolization. They may thus be even more sweeping and imprecise than FACA. Accordingly, there does appear to be some risk of exposure to antitrust suits for firms producing similar products and participating in ecosystem management groups. That risk appears to be limited, however. For the most part, a firm's antitrust liability is unlikely to be altered by ecosystem management activities and will be determined by other factors such as its competitive practices, relative market dominance, pricing policies, level of integration with firms up and down the product stream, and so on.

The only real antitrust question in most ecosystem management cases will be whether ecosystem management relationships serve as vehicles for some form of price fixing. The easiest and most unlikely case—competitors discussing and agreeing upon the prices they charge in the context of an ecosystem management proceeding—would clearly violate the antitrust laws unless immunized by other laws. There are two more plausible scenarios, however. In the first, an ecosystem management plan might lead to reduced near-term harvesting, which might in turn lead to increased prices. In the second, competitors might use ecosystem management contacts to gain new information about each other's businesses, and alter their production and pricing strategies as a result. To some extent, ecosystem processes can be designed to avoid violations in these cases. First, they should avoid setting total production levels or formally allocating production among participants. They should also minimize exchange of commercially sensitive information and rely to the degree possible on averages and statistical probabilities.

As a practical matter, several other factors reduce the risk of liability. First, markets for many forest products appear to be at least regional and often national and international in scope, meaning that local management plans will ordinarily have no discernible effect on prices. Second, few private plaintiffs will be able to show the actual harm that is required for them to bring suit. Third, public prosecutors are unlikely to bring suit where the agreements at issue promote important public policies.

Although the risks of antitrust prosecution may not be great in many cases, public officials could take action to clarify the legal situation and thereby reduce the risks of ecosystem management. Absent the most desirable solution of Congress passing legislation to limit and clarify antitrust liability for ecosystem management activities, the major potential source of protection from liability is state regulation. Under the "state action" doctrine initially enunciated by the Supreme Court in upholding California's system of price controls for raisins (*Parker v. Brown,* 317 U.S. 341 (1943)), individual states can immunize private action from antitrust liability by clearly establishing a state policy promoting that activity and actively supervising it (*California Retail Liquor Dealers Association v. Midcal Aluminum, Inc.,* 445 U.S. 97 (1980). It is not clear whether some of the existing state programs for achieving watershed planning, such as Washington State's (see Lynch 1995b), meet these requirements, but they might. Absent widespread and effective state legislation, federal agencies may have some limited capacity to protect private ecosystem management activities by using their regulatory powers to require them. This is a long shot, however, because no agency has clear authority to require (as opposed to participate in) ecosystem management and because the federal courts are hesitant to find implied immunity from the antitrust laws (*Gordon v. New York Stock Exchange,* 422 U.S. 659, 682 (1974)). On a more limited scale, the primary prosecutorial agencies, the U.S. Justice Department and the Federal Trade Commission, have authority to reduce the risk of liability by issuing guidance on the conditions under which they might bring suit, but this provides no legal immunity from private or state suits.

In sum, the fundamental preference of the antitrust laws for economic competition and for competitors working in relative ignorance of each other's plans and resources is somewhat in tension with the collaborative aspects of ecosystem management. How much of a problem this will be remains to be worked out over time, although preliminary research suggests that it will be more of an irritant than a serious barrier. Federal agencies and the states, however, have the capacity to take many actions to reduce the problem. Nonetheless, clarifying federal legislation may be in order.

Property Rights and Takings Law

As noted above, property rights are created and enforced by society and have always entailed reciprocal duties to neighbors and the public. While the security and self-interest embodied in property rights may be essential to effective ecosystem management on a wide scale, they also pose some risks, which will be discussed briefly here because they are treated at length elsewhere (e.g., Freyfogle 1993, Sax 1993, Meidinger 1993). The Supreme Court has recently suggested (while limiting its holding to a narrow and unusual set of facts) that all limitations on land use which were not explicit in pre-19th-century property law, and therefore inherent in title, may be takings (*Lucas v. South Carolina Coastal Commission,* 112 S.Ct. 2886 (1992)). In addition, state and federal legislatures have recently begun to debate and occasionally pass legislation that treats every significant reduction in property value attributable to regulation as a compensable taking.

Legislatures have the power to grant compensation in such situations, but the historical assumptions often used in the argument that they should are seriously flawed. Nuisance prohibitions have always been flexible and have adapted to changed circumstances. Property owners have never before had the right to be compensated every time the rules governing their property change in a way reducing its value. The standard has always been more complex, and has necessarily considered the reasonability of the restriction under the circumstances at the time.

In conferring such novel compensation rights on landowners, the new takings laws in effect convey a windfall to them at public expense. Moreover, and perhaps more important for ecosystem management, they impose most costs of social learning that reduce property values (such as the negative effects of destroying wetlands) on the public, while allocating most benefits of social learning that increase property value (such as new mineral uses) to landowners. This is important to ecosystem management for two reasons. First, while regulation is incapable of achieving, or perhaps even directing, ecosystem management on its own, it is an important element promoting progress through creative experimentation and negotiation among parties. Second, the contradictory incentives created by allocating the costs of new learning to the public and the benefits to private landowners are likely to increase the centrifugal forces in ecosystem management groups. The more inconsistent the economic interests of the actors are, the more difficult it will be to develop shared norms and expectations—in a word, the more difficult it will be to achieve ecosystem management.

Conclusion

The organizational issues and legal trends discussed in this chapter do not paint a rosy picture of the future of ecosystem management. True, they were selected in part to illustrate challenges and potential barriers. Effective ways of working around many of them will doubtless emerge from the creative engines of social experimentation. Exactly what organizations and legal arrangements will best facilitate ecosystem management is impossible to predict in advance and will doubtless vary by locale and scale of problem. No matter what, however, creative engines will truly be required, since so much of our amassed cultural and legal tradition favors hierarchical organization over team- and network-oriented forms, competitive relationships over collaborative ones, and centralized pluralist politics over community learning and deliberation. In addition, the sheer mass of existing laws that could potentially affect ecosystem management—the great majority of which were not discussed in the chapter—is likely to create a kind of institutional friction retarding development of the new organizational relationship discussed earlier.

Because of these conditions, ecosystem management will necessarily have an improvisational and uncomfortable feeling for some time to come. The bulk of existing laws and organizations are unlikely to give it clear form in the near term. The most that can be hoped is that they will not consistently stop it in its tracks. Accordingly, many of the most innovative efforts to achieve ecosystem management involve difficult and sometimes risky balancing acts that occur on the margins of existing legal frameworks formed by laws such as FACA and the antitrust laws. One might describe efforts like the Applegate process and some habitat conservation planning processes (Vanderwood 1995), for example, as "bargaining in the shadow of the law." While they rely heavily on existing legal mechanisms and arrangements, they also necessarily move beyond them. To some extent they succeed because the threat of returning to unproductive, progress-blocking rules is unpalatable for the parties key to success of the effort. Eventually, perhaps, the most constrictive legal structures will atrophy, but they will add friction and complexity for a long time to come.

Acknowledgments

The research in this chapter was supported by funds provided by the U.S. Department of Agriculture, Forest Service, Pacific Northwest Research Station, People and Natural Resources Program. The organizational models section was originally prepared for the Meeting on Institutional Barriers and Incentives for Ecosystem Management, Skamania, WA, October 20–22, 1994, also sponsored by the

People and Natural Resources Program. Helpful comments by the participants in the Skamania symposium, as well as by Ellen Baar, Barry Boyer, Jody Freeman, Albert Gidari, Kathleen Halvorsen, James Karr, Gary McCaleb, Michael Neff, Tom Princen, William Rodgers, John Tanaka, and the editors of this book are gratefully acknowledged. Size limitations precluded responding to all of their suggestions in this chapter, but I hope to do so in forthcoming publications.

Literature Cited

Antypas, A., and D. Vanderwood. 1995a. The central Cascades adaptive management area. Ecosystem management case study prepared for the Legal Issues in Ecosystem Management Project, University of Washington Law School, Seattle, WA.

———. 1995b. The Little River adaptive management area. Ecosystem management case study prepared for the Legal Issues in Ecosystem Management Project, University of Washington Law School, Seattle, WA.

Baber, W. F. 1983. *Organizing for the Future: Matrix models for the post-industrial polity.* University, AL: University of Alabama Press.

Barber, B. R. 1984. *Strong democracy: Participatory politics for a new age.* Berkeley and Los Angeles: University of California Press.

Boyte, H. C. 1987. *Commonwealth: A return to citizen politics.* New York: Free Press.

Brewer, R. 1988. *The science of ecology.* Philadelphia: Saunders College Publishing.

Cappel, A. J. 1995. A walk along willow: Patterns of land-use coordination in pre-zoning New Haven (1870–1926). In *Perspectives on property law,* ed. R. C. Ellickson, C. M. Rose, and B. A. Ackerman. Boston: Little, Brown.

Chadwin, M. L. 1983. Managing program headquarters units: The importance of matrixing. *Public Administration Review* 43:305–314.

Dahl, R. A. 1967. *Pluralist democracy in the United States: Conflict and consent.* Chicago: Rand-McNally.

Dobbin, F. R. 1994. Cultural models of organization: The social construction of rational organizing principles. In *The Sociology of culture: Emerging theoretical perspectives,* ed. D. Crane. Oxford, England, and Cambridge, MA: Blackwell.

Fitzsimmons, A. K. 1994. *Federal ecosystem management: A train wreck in the making.* Policy analysis no. 217. Washington, DC: The Cato Institute.

Francis, G. 1993. Ecosystem management. *Natural Resources Journal* 33:315–345.

Franklin, J. 1994. Ecosystem management: An overview. Prepared for the Conference on Ecosystem Management: Applications for Sustainable Forest and Wildlife Resources, 2–4 March, Stevens Point, WI.

Freyfogle, E. T. 1993. *Justice and the Earth: Images for our survival.* New York: The Free Press.

Geisler, C., and S. Kittel. 1994. Who owns the ecosystem? Property dimensions of ecosystem management. Prepared for the Conference on Institutional Barriers and Incentives for Ecosystem Management, 20–24 October, Skamania, WA.

Grumbine, R. E. 1994. What is ecosystem management? *Conservation Biology* 8(1):27–38.

Heclo, H. 1978. Issue networks and the executive establishment. In *The new American political system,* ed. A. S. King. Washington, DC: American Enterprise Institute.

Heinz, J. P., E. O. Laumann, R. L. Nelson, and R. H. Salisbury. 1993. *The hollow core: Private interests in national policy making.* Cambridge, MA: Harvard University Press.

Hjern, B., and D. Porter. 1981. Implementation Structures. *Organization Studies* 2:211–227.

House, F. 1993. Where inhabitory culture and institutional management converge. *Watershed Management Council Newsletter* 5(2):11–12.

Kaufman, H. 1960. *The forest ranger.* Baltimore: Johns Hopkins University Press.

Keiter, R. B. 1994. Beyond the boundary line: Constructing a law of ecosystem management. *Colorado Law Review* 65:293–333.

Kingdon, J. W. 1984. *Agendas, alternatives, and public policies.* Boston: Little, Brown.

Krier, J., and E. Ursin. 1977. *Pollution and policy: A case essay on California and federal experience with motor vehicle air pollution, 1940–1975.* Berkeley: University of California Press.

Lee, K. N. 1993. *Compass and gyroscope: Integrating science and politics for the environment.* Washington, DC, and Covelo, CA: Island Press.

Lipsky, M. 1980. *Street level bureaucracy: Dilemmas of the individual in public services.* New York: Russell Sage.

Lowi, T. J. 1969. *The end of liberalism: Ideology, policy, and the crisis of public authority.* New York: Norton.

Lynch, S. 1995a. The Applegate partnership. Ecosystem management case study prepared for the Legal Issues in Ecosystem Management Project, University of Washington Law School, Seattle, WA.

———. 1995b. Cooperative monitoring, evaluation and research, and watershed analysis. Ecosystem management case study prepared for the Legal Issues in Ecosystem Management Project, University of Washington Law School, Seattle, WA.

Meidinger, E. 1987. Regulatory culture: A theoretical outline. *Law and Policy* 9(4):355–386.

———. 1992. *Administrative regulation and democracy.* Working paper OP92.01. Buffalo: Baldy Center for Law and Social Policy, State University of New York at Buffalo.

———. 1993. The changing legal environment of northern forest policy making. In *Sustaining ecosystems, economies, and a way of life in the northern forest.* Washington, DC: The Wilderness Society.

Michaelman, F. 1988. Laws republic. *Yale Law Journal* 97:1493–1565.

Michaels. S. 1992. Issue networks and activism. *Policy Studies Review* 11(3/4):241–258.

Moote, M. A., S. Burke, H. J. Cortner, and M. G. Wallace. 1994. *Principles of ecosystem management.* Tucson: University of Arizona Water Resources Research Center.

Morgan, G. 1989. *Creative organization theory: A resource book.* Newbury Park, CA: Sage Publications.

Perry, J. P. 1994. The Clinton administration and national forest management. Presented to the American Bar Association, Section on Natural Resources, Energy and Environmental Law, San Antonio, TX.

Pauw, J., T. J. Greenan, and D. C. Ross. 1993. *Balancing endangered species regulation and antitrust law concerns.* Working paper 54. Washington, DC: Washington Legal Foundation.

Pocock, J. G. A. 1985. *Virtue, commerce, and history: Essays on political thought and history, chiefly in the eighteenth century.* New York: Cambridge University Press.

Prosser, W. L., and W. P. Keeton. 1984. *Torts.* 5th ed. St. Paul, MN: West Publishing Company.

Sabatier, P. A., and H. C. Jenkins-Smith. 1993. *Policy change and learning: An advocacy coalition approach.* Boulder: Westview Press.

Sample, V. A. 1992. Building partnerships for ecosystem management on forest and range lands of mixed ownership. In *American forestry: An evolving tradition.* Proceedings of the Society of American Foresters National Convention. Bethesda, MD: Society of American Foresters.

Sax, J. L. 1993. Property rights and the economy of nature: Understanding Lucas v. South Carolina Coastal Council. *Stanford Law Review* 45:1433–1454.

Seidenfeld, M. 1992. A civic republic justification for the bureaucratic state. *Harvard Law Review* 105:1511–1573.

Shannon, M. 1991. Resource managers as policy entrepreneurs. *Journal of Forestry* 89(6):27–30.

Stanley, M. 1983. The mystery of the commons: On the indispensability of civic rhetoric. *Social Research* 50(4): 851–879.

Truman, D. 1959. *The governmental process.* New York: Knopf.

Vanderwood, D. 1995. The Coachella Valley habitat conservation plan. Ecosystem management case study prepared for the Legal Issues in Ecosystem Management Project, University of Washington Law School, Seattle, WA.

Walters, C. 1986. *Adaptive management of renewable resources.* New York: Macmillan.

Weber, M. 1978. *Economy and society: An outline of interpretive sociology,* ed. G. Roth and C. Wittich. Berkeley: University of California Press.

Wondolleck, J. M., and S. L. Yaffee. 1994. Building bridges across agency boundaries: In search of excellence in the United States Forest Service. Report submitted to USDA Forest Service, Portland, OR.

24 Building Bridges Across Agency Boundaries

Steven L. Yaffee and Julia M. Wondolleck

Why Are Bridges Needed? 382
- Acquiring Necessary Information 383
- Creating and Implementing Effective Ecosystem-Level Management Strategies 384
- Generating Wise, Enduring Decisions 385
- Building Support for Forest Management Decisions 387
- Influencing Public Knowledge and Values 388
- Getting Work Done 388
- Developing a Workforce 389
- Focusing Forest Service Resources on Problems of Adjacent Communities 389

Examples of Successful Bridging Arrangements 390
- Negrito Ecosystem Project 390
- Alaska Recreation Plan 391
- Elk Springs Timber Sale 392
- Kiowa Grasslands 392

Building on Success: How Can Bridging Arrangements Be Fostered? 393
- Imagining the Expanded Use of Bridging Arrangements 393
- Enabling Motivated Staff to Develop and Use Bridging Arrangements 394
- Encouraging the Development of Bridging Relationships 395

Literature Cited 396

One of our favorite images of the early forestry professional is presented in a painting of a U.S. Forest Service ranger on horseback out on a western mountaintop surveying his domain. The forest is lush and unbroken, the sense of individual power and autonomy is absolute, and not another human (let alone a judge or an interest group) is in sight. It is an emotional and empowering image that may have been accurate in the early 1900s, but is inaccurate and ineffective today. Today's picture of the Forest Service ranger might place her on a mountaintop, but the forest she surveys is likely to be fragmented and cutover, the ecological and social processes that concern her are far out of range, and she is being shot at from all sides. The job of the forestry professional has changed, and it must continue to do so as we move into the next century of forest management.

As other chapters in this book suggest, no longer does forestry just deal with trees. Nor is it adequate to simply "do good science." Managing human use, responding to human demands, keeping up with and influencing human values, and dealing with the realities and vagaries of societal institutions are as much a part of a forestry professional's job as is understanding ecological processes or silvicultural techniques. Forestry in the 21st century involves creating and using information networks, facilitating effective multiparty decision making, building broad coalitions of political support, and participating in cross-jurisdictional management arrangements. Almost all conceptualizations of ecosystem management recognize these human dimensions. For example, most of the studies reviewed by Grumbine (1994) noted the importance of interagency cooperation, organizational change, human-nature interrelations, and human values to ecosystem management.

Among many challenges facing the Forest Service, building better relationships with a variety of individuals, groups, and agencies is critical to the agency's future. Through partnerships, collaborative problem-solving processes, interagency management structures, and the like, the Forest Service can rebuild the understanding and trust needed to get on with effective public resource management. Just as important, better relationships will yield better resource management decisions as managers become more informed on the full range of interests in national forest resources and the rich set of opportunities for carrying out management actions, tapping a broader knowledge base, and making choices that are more likely to be implemented down the line.

By creating an effective set of opportunities for outside groups to be involved in national forest management, agency officials also get out of the bind of having to single-handedly figure out what the public wants. Public needs and values are not absolute or ever fully defined. They emerge through discussion, commitment, and small choices made iteratively. While technical modes of surveying public values and coding public responses can provide important information for the decision-making process, they force agency officials into figuring out an answer to a question where answers are often uncertain and evolving. By creating a robust set of linkages and processes wherein various outside interests can be actively involved in numerous aspects of national forest management, Forest Service officials will promote value clarification on the part of the public and will develop a more effective understanding of public wants and values than can be achieved through other means.

This chapter focuses on activities that seek to bridge the agency–nonagency boundary. It explores the underlying reasons to build linkages with the outside world and the types of bridges that are needed, provides examples of several ongoing bridging arrangements that have been viewed by participants as highly successful, and suggests several ideas for expanding the effectiveness and use of agency-nonagency working arrangements. By arguing the importance of such linkages and providing examples of success and ways to move forward, we hope to provide ideas for those mired in traditional modes of action and empower those who are already seeking to pursue new approaches. Our recommendations should also assist those charged with mobilizing change in the Forest Service and other resource agencies, and encourage those outside the agencies who seek to help out. The future of forest management lies in much more complex networks of knowledge and relationships than have been present in the last century of conservation. Building such networks takes effort, but also unlocks the vast set of knowledge and energy present in a diverse, pluralistic society.

Why Are Bridges Needed?

There are a number of reasons for forming stronger linkages between elements of the Forest Service and the world outside the agency. Some, such as the need for educational linkages to promote diversity in the workforce, simply reflect changing social values and historical conditions. Others, such as the need to form cross-jurisdictional joint-management arrangements, respond to changing notions of effective forest management. Still others, including the formation of partnerships with nongovernmental groups, respond to the changing fiscal environment of public resource management. Some bridging arrangements, such as dispute resolution, may yield benefits almost immediately, while others, such as public education programs, may result in only long-term changes. All imply forming ties with groups outside the agency that can lead to mutually productive relationships. Table 24.1 summarizes the underlying reasons for forming linkages and the bridges that they imply.

Acquiring Necessary Information

While at one time the agency controlled much of the information it needed to manage national forests effectively, today information is ubiquitous. Relevant information is collected and stored in many places, and it can move rapidly across great distances. Ideas appropriate for effective resource management may emanate from nongovernmental groups or from other countries. Oddly enough, as information has become ubiquitous, information management—finding it, organizing it, ensuring its accuracy, and making sense of it—has become a serious task for organizational managers. Part of the challenge lies in staying up-to-date and in touch with important sources and repositories of information, and these tasks require building knowledge networks.

Forming information-seeking relationships has become particularly important for the Forest Service for several reasons. First, reporting requirements mandated by law—environmental impact statements, forest plans, etc.—have enabled those outside the agency to better understand agency intent and reasons for making decisions. In response, nongovernmental groups, state agencies, and others have developed expertise to make sense of agency information and challenge decisions. In the process of doing so, many groups have developed substantial parallel expertise such that at times the Forest Service's data or analyses may not be the best existing information on an issue.

In addition, changes in the information required to

Table 24.1 Bridge types and the underlying reasons for building them

Reasons for Bridging	Type of Bridge
1. Acquiring necessary information	• Information-sharing • Joint research and fact finding • Data negotiation
2. Creating and implementing effective ecosystem-level management strategies	• Interagency coordination and cooperation • Joint management agreements
3. Generating wise decisions, and having them endure	• Public involvement • Collaborative problem solving • Dispute resolution
4. Building support for forest management decisions	• Political linkages • Coalition formation • Media strategies
5. Influencing public knowledge and values	• Public education programs • Public outreach
6. Getting work done	• Volunteers • Resource sharing and generation • Management partnerships • Educational partnerships
7. Developing a workforce	• Educational linkages • Careers outreach
8. Focusing Forest Service resources on problems of adjacent communities	• Social services

effectively practice forest management mandate information-sharing arrangements. As concepts of ecosystem management develop within the agency, concern with larger-scale processes requires the acquisition of information across broader geographic areas. As the agency responds to the wide array of human interests and values in public forest management, agency staff need to reach out to acquire information about the nature of those values and concerns. Finally, as conservation biology and other forest management disciplines have developed, the agency's ability to craft high-quality technically based decisions requires it to do good science, and that requires the acquisition of a broader set of information at deeper levels than has ever been necessary before. Since the agency's fundamental legitimacy lies in the quality of its technical knowledge, it becomes an ineffective guardian of public resources if its information becomes outdated. And as a practical matter, an ineffective guardian will be challenged repeatedly in court.

If the agency's management ideas become outdated, it will also lose the battle for credibility and legitimacy. No longer is the Forest Service the only source of good ideas. Indeed, as the Forest Service moves toward ecosystem management, the best source of ideas may come from nongovernmental groups like The Nature Conservancy or regional planning agencies like the Adirondack Planning Agency or the Pinelands Commission. Ideas about how to employ decision-making tools, such as gap analysis, may best come from other federal agencies like the U.S. Fish and Wildlife Service or state agencies like the Idaho Game and Fish Commission, which have experimented with such techniques (Scott et al. 1991). As the Forest Service assists adjacent communities with rural economic development planning, the best sources of ideas may lie in consulting firms specializing in economic revitalization. As the agency embraces concepts of dispute resolution and collaborative problem solving, the best source of expertise may lie in third-party facilitation groups like the Keystone Center in Colorado or Resolve in Washington, DC. As it seeks to revitalize its own organizational strategies and norms, learning from firms that have built strong, decentralized, entrepreneurial structures, such as 3M, may be appropriate (Peters and Waterman 1982). Such knowledge-sharing arrangements are vitally needed to serve as conduits of information and ideas appropriate to forest management in the 1990s and beyond.

Creating and Implementing Effective Ecosystem-Level Management Strategies

A shift toward ecosystem management requires more than simply information-sharing, it necessitates a host of other activities that all require the development of relationships across the agency–nonagency boundary. By definition, ecosystem management requires cross-jurisdictional perspectives and action. Managing along ecosystem boundaries rather than administrative or political boundaries necessitates action to bridge fragmented landownership.

Bridging activities may seek a variety of ends. The simplest arrangements may provide for shared knowledge of natural systems and administrative objectives. Understanding the processes functioning on adjacent lands, and the management objectives of adjacent landowners, may help Forest Service managers craft more-effective conservation strategies on national forest lands. An intermediate arrangement may seek coordinated management, where adjacent landowners are cognizant of and responsive to the management actions of their neighbors. At the extreme, relationships that establish joint management of areas may be necessary to achieve effective ecosystem-level conservation.

All of these arrangements require the development of networks of relationships between many landowners. There are literally millions of decision makers influencing the long-term direction and viability of land management in the United States, including some 80,000 units of government and millions of individuals and corporations. The Forest Service has the advantage of controlling some unusually large chunks of land, but effective management of the national forests requires interaction with governmental and nongovernmental groups who control inholdings, rights to resources within public lands, and adjacent properties. Management dilemmas and controversies resulting from fragmented management in areas like the Yellowstone ecosystem (which exhibits relatively concentrated ownership) bespeak the need to implement regional management strategies.

Even after ecosystem management fades as a buzzword, and other terms take its place, effective resource management will require management across fragmented landscapes. Problems like global climate change and the displacement of native species by exotics mandate problem-solving approaches that are regional in scale. Increased understanding of the wide-ranging nature of certain sensitive species, like the northern spotted owl, Pacific salmon, and red-cockaded woodpecker, force managers to look beyond their own units. The Forest Ecosystem Management Assessment Team (FEMAT 1993, see also Johnson, Chapter 25) might have been an extraordinary event in the history of federal resource management, but its approach will become more the norm in the next century. Effective resource management will necessitate more-effective landscape-level working arrangements, and such approaches require a network of human relationships to make them work.

Generating Wise, Enduring Decisions

For much of the Forest Service's history, the range of interests in national forests was relatively small, and the agency's statutory objectives narrow, such that decisions could be made by agency officials using technical tools and simple objective functions. Today, the range of interests in public forestry and the scope of statutory objectives for national forest management are huge, and decisions cannot be made by applying simple decision rules. The diverse set of individuals and groups interested in and affected by national forest management includes traditional commodity groups, nontraditional commodity groups (such as manufacturers of engineered wood products), environmental groups, recreation groups, professional associations, educators, local governments, local economic development interests, state governments, other federal agencies, and international trade concerns, among others. Even within these categories lies a great deal of diversity of interest and style, including considerable differences between a group like EarthFirst! and the National Wildlife Federation.

A wise decision is one in which the decision maker carefully considered the problems to be solved and objectives to be sought, collected and reviewed as much relevant information as possible, and considered the likely future so that decisions would be relevant to changing times. In today's forest management environment, wise decision making requires managers to understand the full spectrum of interests in national forest resources held by numerous public and private groups, and the complete set of capabilities existing inside and outside the agency for implementing any decision. For example, being aware of the concerns and cultures of adjacent human communities may suggest directions different from those traditionally considered by agency staff or may suggest management strategies that are more effective over the long term. To develop this understanding requires decision makers to form relationships with groups outside agency walls.

With many legitimate objectives for national forest management and a fixed land base, wise decision making recognizes that, while there may be better and worse directions, there is no single right answer to the question of how to manage a national forest landscape. Rather, there are various directions that benefit different interests in different ways, and this decision-making reality runs head-on into the traditional view of the Forest Service as a technical decision-making body. Forest managers will not find the answers to their management dilemmas in FORPLAN outputs (a computerized model that the Forest Service used in its planning efforts during the 1980s) or in closed-door interdisciplinary team meetings. Active and ongoing involvement with stakeholders in decision-making processes is needed to fully expose the interests and opportunities associated with a diverse society.

A technical decision-making paradigm must yield to a more pluralistic and consensus-building one, where groups influence direction based on the legitimacy of their interests and the information they muster. The measure of success of such processes is not having everyone equally angry at the Forest Service or having more people happy than angry, as has sometimes been the case in recent Forest Service attempts at dispute resolution. Rather, the measure of success is whether they produce good decisions: those that are consistent with national policy objectives, that stakeholders can live with, and that agency staff assess as technically sound.

One test for the wisdom of such decisions is whether they endure over time—whether they solve the problem they aimed to solve, whether they can be adapted appropriately as new knowledge develops, whether they can withstand challenge, and whether they are supported by those affected by them and can be implemented. The agency scorecard over the past decade on administrative and judicial appeals of agency decisions has not been good. Too many decisions have been challenged, and the agency has won too few of those challenges. Simply to preserve the sanity of Forest Service employees, this record must change.

More important, a different decision-making approach is needed so that the agency can more consistently generate wise decisions. In today's environment, wise decisions are more likely to be achieved through decision-making processes that are collaborative and that involve stakeholders in forums designed to expose concerns, build relationships, establish trust, and encourage creative problem solving. Such processes are informed by technical information, but technical inputs are just one of several sets of needed information.

Such processes can also promote ownership of resulting decisions. Effective implementation depends on the support and contributions of involved parties, and this support is more likely if the decision-making process itself is more collaborative and the settlement more cognizant of legitimate interests and concerns. Since forest management decisions are implemented over many years, and conditions and needs change, many of these processes create ongoing involvement for stakeholders through periodic meetings or hands-on implementation. The understanding and trust that can develop through the creation of long-term relationships helps decisions endure and provides a seedbed for future decision making.

This style of decision making is also consistent with changes in decision-making and organizational management theory. As we have developed a more pluralistic and educated society and as the industrial age workplace has given way to the information age, it is argued that styles of decision making that are top-down, authoritative, and designed for military and industrial production must give way to more-decentralized and consensual approaches targeted to quality control and the provision of services. In *Reinventing Government,* David Osborne and Ted Gaebler note that "[f]ifty years ago centralized institutions were indispensable. Information technologies were primitive, communication between different locations was slow, and the public work force was relatively uneducated. . . . But today information is virtually limitless, communication between remote locations is instantaneous, many public employees are well educated, and conditions change with blinding speed. There is no time for information to go up the chain of command and decisions to come down" (Osborne and Gaebler 1992).

Their prescription for the changing management condition was decentralization of authority and responsibility. According to Osborne and Gaebler, "Entrepreneurial leaders instinctively reach for the decentralized approach. They move many decisions to the 'periphery' . . . into the hands of customers, communities, and nongovernmental organizations. They push others 'down below' by flattening their hierarchies and giving authority to their employees." The result is the creation of organizations that are more flexible and innovative, with better employee morale and productivity. Most deploy employees in problem-oriented teams and reduce their insulation from the "real world."

Others have argued that the Forest Service in particular should change its decision-making paradigm—from viewing itself as a technical expert with the right answers to seeing itself as the facilitator of a collaborative decision-making process in which many stakeholders, including the agency itself, need to participate in a meaningful way (Wondolleck 1988). Agency officials have a unique function in such a process by having interests, information, and statutory bottom-lines that other groups do not have. But still, they are one among many parties with interests, values, and capabilities in national forest management. Jeff Sirmon (1993) called these networks of groups "communities of interest" and highlighted the importance of shared power in decision making. In fact, the Forest Service has been experimenting with collaborative decision-making processes for a number of years and has used collaborative negotiations to resolve a number of forest plan appeals. One of the principal objectives of such processes is to build relationships among the community of interests so that problem solving and subsequent implementation can occur.

Building Support for Forest Management Decisions

Relationships built among the community of interests in national forest management can do more than generate wise enduring decisions, they can also help generate support for the Forest Service in political arenas. Whether agency officials like to believe it or not, they operate in a political environment. Forest management decisions influence the outcome of important resource allocation decisions, and the ability of the Forest Service to implement directions on the ground is influenced significantly by individuals and institutions outside the agency who have political stakes: Congress, the president and executive officers, the Secretary of Agriculture, state governors and legislatures, and others.

Political influences on national forest management can be viewed in several ways. Since the Forest Service is not self-sustaining and manages publicly owned resources, it is appropriate that elected representatives of the broader populace review and guide direction. Further, if political processes are viewed as one means of expression of human values, then such influences are good. They help provide social value input into technical decisions. Indeed, the Forest Service's current efforts at organizational redirection and renewal are at least partly the result of changes in national politics.

Regardless of whether one views politics as good or bad, the reality is that public forest managers function in a political environment where there is competition for resources and direction. That reality means that agency leaders at least must build concurrence, if not active support, for desired direction. If they do not work to build support for agency direction in the political environment, they will not succeed in sustaining desired courses of action. Agency direction will be dictated by political leaders, and the resources for carrying out important agency priorities will not always be available. To the extent that ecosystem management requires more emphasis on resource protection and less on job creation in adjacent communities, the task of building political support for desired direction becomes even more important, since political arenas place considerable value on tangible economic benefits.

While politics have always been a part of public resource management—starting with Gifford Pinchot's relationship with President Theodore Roosevelt and battles with the Interior Department—the politics of national forest management have changed demonstrably in recent years. Declining slack in the natural resource base at a time of rising demands has generated a competition between interests of a magnitude unseen in earlier years. While in past years forest managers could give everyone much of what they demanded, they are unable to do so today.

At the same time, the mantle of expertise long worn by agency officials has worn thin, prompting challenges that would never have taken place 30 years ago. Nongovernmental groups have unprecedented power to challenge administration decisions, and the weakening of traditional agency constituents and the rise of new political forces have destabilized the politics underlying national forest management, opening up the possibility of directions once thought inconceivable. That old-growth preservation forces could secure a dramatically reduced emphasis on timber production in the Pacific Northwest (represented by the Option 9 decision; see FEMAT [1993] and Johnson, Chapter 25) was unthinkable 10 years ago.

The changing political context of national forest management creates unprecedented opportunities for agency leaders to craft new coalitions of support for desired direction. The opportunity exists for agency officials to use the diversity of interests in national forest management as a positive force, rather than being stymied by it. They can do so by working hard to build coalitions of interests supportive of agency direction while keeping them in flux so that they do not become stagnant and binding. And because political relationships are interconnected across many levels of government and social organization, building support means pushing for desired directions at the local, state, and national levels, and working with the media that influence opinion at all of these levels.

Ultimately, building political support involves developing relationships that lead to a two-way flow of information. By understanding the hopes and dreams of the diverse set of groups in American society, appropriate direction can be crafted. By increasing people's understanding through explanation and involvement, their hopes and dreams can be developed in a knowledgable manner. Both require the careful construction of bridging relationships.

Influencing Public Knowledge and Values

Building relationships that help inform the public has short- and long-term benefits. In the short term, it can lead to political concurrence on agency direction. In the long term, such relationships can contribute to the knowledge base of the public and influence the values they hold for public natural resources. There is no doubt that as the American population has urbanized, it has become less in touch with natural systems and the infrastructure that supports their lifestyles. Paradoxically, Americans have become more supportive of environmental protection at the same time their demand for the consumption of natural resources has risen. To moderate these demands and promote conservation of natural resources requires closing this loop, so that American consumers are more aware of and responsible for the impacts of their actions on public resources.

Building relationships across agency boundaries that lead to public education can also assist Forest Service staff in explaining the realities and constraints they face and their objectives for forest management. Observing a public hearing, it is difficult not to sympathize at times with agency officials who have to contend with a lack of knowledge and understanding on the part of some of the public. Agency officials have legitimate perspectives and ideas for management direction, but these are not always obvious to people outside the agency, and often they are not communicated very effectively in traditional public meetings or hearings. Explaining ideas helps, but building relationships that foster understanding over time is more effective.

Such relationships also form the basis for an important public service. To a country concerned about science education and knowledge of life skills such as conflict management, the national forests provide an ideal training ground. Abundant public lands provide a host of opportunities for the public to observe natural processes in action and participate in experiments in managing those processes. They also provide marvelous opportunities for discussing the societal implications of various decisions and to experiment with various decision-making processes. Even more significant, involving the public more fully in national forests and forest management has important value-forming functions: helping to shape the values perceived in natural resources, reinforcing the need for democratic institutions to make choices about collective goods, and fostering a sense of responsibility for common property resources. Building bridges that inform and involve many different types of publics can assist in developing this knowledge and in the value-formation process.

Getting Work Done

National forest management today involves carrying out more tasks with fewer resources. This creates a dilemma for the Forest Service: Either less gets done, what gets done is done poorly, agency workers get much more productive, or they find alternative ways to acquire needed capabilities and resources. It is unlikely that expanded resources will come from a federal government dealing with serious budget constraints, including deficits and high fixed costs. While operating efficiencies need to be pursued, perhaps the best opportunity for dealing with resource constraints lies in forming relationships with groups outside the agency that lever needed capabilities and resources.

Whether in the form of partnerships, volunteer work crews, interagency management groups, or resource-sharing arrangements, such relationships have a number of benefits. Since they involve few fixed costs, they are flexible. Short-term work projects can be initiated, carried out, and terminated without incurring the administrative costs of regular civil service workers. Bridging relationships can provide access to expertise and resources not currently available inside the agency. By involving users and individuals affected by forest management, partnership arrangements can create a sense of ownership in public resources that may help to protect public lands. Since cooperative working arrangements require a commitment of time and energy on the part of outside groups, such arrangements also provide a market test for services requested from public lands. It requires users to demonstrate commitment, not just make demands.

Clearly, there are drawbacks to cooperative working arrangements (Yaffee 1983). They can result in work being done of varying quality. A workforce that is not paid may be less stable. For example, volunteers may not show up on a rainy day even if work is scheduled. Agency employees needed to supervise and train cooperators may have to do so during evenings and weekends. The agency will have less control over the direction and methods used on a project, and while work may be accomplished, it may not take a form that agency officials would like. And since there is competition for people's time, the tasks most needed and able to be done by less-skilled workers, such as routine maintenance work, may not motivate the involvement of partners.

In spite of these drawbacks, the benefits of cooperative working arrangements far outweigh their liabilities. Besides, given federal reductions in force and budget limitations, the agency may simply have no other choice for getting needed work done on the national forests. In a time when understanding of the agency and trust in its leadership are limited, the subsidiary benefits of partnership arrangements make them well worth the outreach activities needed to implement them. Such arrangements are not new. The Forest Service has used the cost-sharing program and other partnership arrangements for some time. But the role for such arrangements has increased due to the changing context of public forest management.

Developing a Workforce

One of the tasks facing the Forest Service as it downsizes and reinvents itself is to build a workforce appropriate to the challenges of the next century, and building bridges to outside institutions and cultures can assist this organizational change process. To broaden employees' understanding of the range of values and attitudes about public resources, outreach activities should aim for accessing the diversity of cultures present in modern-day America.

Developing the knowledge and capabilities of existing workers requires access to a host of educational resources, including those in universities and other agencies. New forest science, expanded social science, and evolving ideas of organizational management and decision making suggest the need for current employees to access short courses, certificate programs, and the like. Short-term rotations to other organization work sites (including state departments of natural resources, other federal agencies, and nongovernmental groups like the Nature Conservancy) can expose Forest Service employees to a range of experiences, perspectives, and skills. In addition, rather than conducting training totally in-house, as has become the norm, sending employees to training opportunities that include participants from other agencies, organizations, and corporations has the great side benefit of building contacts and understanding across organizational lines.

Sooner or later, new employees will be needed as attrition takes place. Mechanisms for recruiting educated workers are necessary, and these will involve relationships outside the agency. Recruiting women and members of minority groups requires understanding and access not currently present in the agency. Identifying and motivating urban Americans to seek employment in public forestry calls for new marketing and recruitment strategies. Using outside organizations as selection, training, and feeder mechanisms may be one of the most effective ways to build cultural diversity and expand the knowledge base in the agency's workforce.

Focusing Forest Service Resources on Problems of Adjacent Communities

A final reason for expanding linkages between the Forest Service and its neighbors lies in the quality of the resources and expertise within the agency, and the critical needs of communities adjacent to agency lands. The Forest Service possesses an extraordinary wealth of opportunities, skills, and ideas not duplicated in many areas. Opportunities exist to use national forest lands as the base for a variety of socially desirable activities. Hosting school class field trips, employing prison work groups in Civilian Conservation Corps–like encampments, serving as a base for national community-service work groups, and providing test sites for college researchers are all ways to expand the impact of national forest resources on other societal objectives.

In addition, Forest Service employees are a resource that can be used to contribute more to the communities adjacent to national forests. State and Private Forestry have viewed outreach as their primary mission for some time, but the sizable workforce in the national forest system represents an additional set of resources of value to adjacent communities. Agency staff may be some of the most highly educated individuals in an area, with substantial knowledge of science, planning, and economic development. Visibly contributing these skills to a region can also generate positive role models for children. For example, seeing women or racial minorities in professional and organizational leadership roles may be empowering and eye-opening for children in many small rural communities. Building the bridging relationships needed to support these arrangements are worth the effort for many reasons. For the agency, however, such relationships will help build understanding and support, both sorely needed at this point in the agency's history.

Examples of Successful Bridging Arrangements

There are numerous examples of ongoing bridging relationships in national forest management. Many show innovation, commitment, and productivity, and are perceived as successful by both Forest Service and nonagency participants. Four are described below to provide illustrations of the varying kinds of arrangements that are possible and to explain the reasons for their success. The quotations in the following four examples are taken from Wondolleck and Yaffee (1994).

Negrito Ecosystem Project

In rural southwestern New Mexico, the Reserve Ranger District of the Gila National Forest is no stranger to controversy. It is located in Catron County, a "national leader in the county rights movement." Fed up with federal government regulation, this county is in the process of bringing lawsuits against the U.S. Fish and Wildlife Service regarding the Mexican spotted owl and the U.S. Forest Service regarding its sensitive species list. It has also passed ordinances that threaten the arrest of Forest Service personnel who make decisions contrary to local "custom and culture." The county has written its own land-use plan for the Gila National Forest and claims the Forest Service must comply.

Despite this troubled context, the Reserve Ranger District is also the location of an innovative and collaborative ecosystem management initiative. District timber and recreation specialist Gary McCaleb noted that the district was approaching a crisis situation that demanded an innovative approach: "The planning wasn't working. Projects weren't being implemented. We were walking into a meat grinder. [The district ranger] thought a different approach might work. It sounded like ecosystem management would be a good thing to do."

The Negrito Ecosystem Project was initially proposed by a small conservation group in the area and adopted by the district. A diverse but balanced coordinating group began meeting in 1991 to develop and redefine the proposal. This core group includes Forest Service staff, the president of Friends of the Gila, a timber company representative, a county extension agent, and a grazing permittee. The group follows a consensus approach to decision making and is committed to grappling with the huge task of managing this watershed in a holistic manner consistent with overall ecosystem health and human needs. The group met about 25 times in its first two years and has done most of the legwork and documentation for the project, including writing National Environmental Policy Act decision memos for small projects and drafting budgets and project proposals. In addition to this core group, an informal working group of about 24 people from a variety of backgrounds and interests has participated in four or five all-day meetings. They have helped develop the project's philosophy, participated in identifying issues and concerns, and joined in team building. Group participants serve as information conduits between the project and community organizations and people.

Those involved in the Negrito initiative recognized at the outset that, while the parties were diverse in backgrounds and interests, they shared a major underlying objective of seeing a healthy social and biological ecosystem in their community. When given the opportunity by the Forest Service to work toward this common goal, they recognized the benefit of setting aside their previously adversarial stances and began cooperating. The responsiveness of Forest Service employees to this idea and the commitment of the district ranger and forest supervisor began laying the foundation from which trust and communication among these groups could be built. One district staff person commented that one of the major successes of the project so far is the "interest shown by all participants and the relationships that have grown as a result of this."

District Ranger Gardner explains his approach: "I told them that the only bounds on the group is that their decisions can't adversely affect people who work on the district and they can't conflict with laws and resource directives from Congress. That kicked the door open and people came in." Gardner admits that he had to overcome some barriers within himself regarding the unconventional consensus process used with the project. "It's not a traditional Forest Service process, I guarantee that. I had a lot of concerns at first. I've been in the Forest Service for 18 years and I've always had a real structured process for projects." But he soon realized, "If groups are willing to sit down and work things out between themselves, the Forest Service should be flexible enough to respond to them."

Alaska Recreation Plan

The Petersburg Ranger District on Alaska's Tongass National Forest has long been plagued by controversy over timber planning and distrust that has been exacerbated by the diversity of the region's residents. In 1989, District Ranger Pete Tennis decided to confront this situation head-on by using an uncontroversial project to help build positive relationships within area communities and pave the way for better interactions in future initiatives. Tennis decided a five-year recreation plan would meet this criteria and directed his staff to figure out the best way to accomplish a cooperative plan.

The first step of the process was to canvas the area communities for their ideas on desired recreation projects. The team embarked on an aggressive and highly visible public involvement campaign. In addition to a series of public meetings, they used a variety of mechanisms to inform the public about the opportunity to share their ideas. According to District Forester Maria Durazo-Means:

> We found out that traditional scoping methods such as public meetings were not very effective, so we took a different approach. We tried nontraditional methods such as talking with people informally at post offices and grocery stores and visiting schools. This enabled us to reach a segment of the public that normally doesn't attend public meetings. We kept the recreation plan in the public eye through the use of various media (television, radio, CB radio to reach fishermen, information brochures to post office box holders, and posting flyers up around town). Eventually, it paid off and people started talking to us.

One innovation of this project was in tailoring public involvement techniques to meet the specific needs and customs of different cultures. "For example, in a native village, public meetings have historically not been a good forum, so other methods were used such as setting up tables and booths at grocery stores, at the post office, and talking one-on-one to people to get their input" (District Forester Maria Durazo-Means). An archaeologist who had worked with the residents of one small town suggested that the agency try to reach adult residents through their children. So, the district put on Smokey Bear and Woodsy Owl skits about recreation for the native school children and sent them home with information for their parents about the recreation plan. Schoolwide assemblies were held in middle and high schools to collect project suggestions. According to the local mayor, the Forest Service's coming out one-on-one to talk to the villagers created much more enthusiasm for the plan and helped generate consensus. According to Durazo-Means, the "community opened up to us . . . they were excited about something we were doing together."

This planning effort received high marks from the involved communities. "I've seen more recreation help from the Forest Service in one year than has been done in the last 20," commented the local mayor. "For the first time it gave outlying communities an opportunity for input on the type of recreation projects we would be interested in. . . . It becomes the people's recreation area, not just the Forest Service's."

Success in this case was facilitated by a particularly creative and innovative ranger as well as the district's willingness and desire to try something different. As the plan itself describes:

> In the past, recreation project planning has been a bit like the job of the lonely Maytag repairman: a lone person or two trying to think of new ways to serve that big public out there. This plan represents our effort to break loose from our Maytag mode.

Elk Springs Timber Sale

The Targhee National Forest is a place where controversy and antagonism have often been the norm. "We're at war with the Targhee on most issues. We've been on the brink of litigation several times," commented a representative of the Greater Yellowstone Coalition. Facing severe criticism from many sides, the Targhee National Forest's Island Park District Assistant Ranger John Councilman decided to initiate an unconventional timber sale planning process.

To accomplish his goals, Councilman collaborated with the Idaho Department of Fish and Game, small timber operators, and environmental groups, including the Greater Yellowstone Coalition. He met with individuals from each of these groups several times, in the field and in his office, in order to get their opinions on what the sale should look like. He also allowed them to comment on a draft sale plan, and changed the draft according to their suggestions. In addition, he went out to the site several times with the district's wildlife biologist to discuss wildlife needs in the sale area. "We thought about migration patterns, forage areas, nesting sites," said Councilman. There are several threatened and endangered species in the area, including the gray wolf, grizzly bear, bald eagle, and peregrine falcon. The resulting timber sale plan was completely unlike the forest's previous clearcut-only plans. It was also the only one on the forest that has not been appealed.

Councilman attributes the success of the effort to the participants' willingness to try something new by working together. Moreover, he noted that simply "forming lines of communication" between groups that had seldom talked to one another "brought people together quite a bit." The process was not trouble-free, and Councilman noted, "There were a lot of psychological barriers on my part. I've been in the practice of silviculture for 10 years. A lot of this goes against my training." By promoting face-to-face communication and allowing other agencies and groups to share information and discuss their concerns and how they might be met, it was possible to develop an acceptable plan.

Kiowa Grasslands

On the Kiowa National Grasslands in New Mexico, an experimental collaboration between the Forest Service, Soil Conservation Service, and a handful of local ranchers has greatly improved the quality of this area's rangeland. Under this jointly administered program, rather than having different management schemes for public and private lands and a mix of advice from Soil Conservation Service and Forest Service staff, a rancher sits down with the two agencies together to develop a long-range plan for the entire area being utilized. By managing all of one rancher's land as a single operating unit, needs of wildlife, cattle, and environmental restoration are being addressed as a whole.

A typical plan has multiple stages, but almost always begins with the development of a comprehensive water program. Instead of the usual handful of larger watering holes for cattle, water under this effort is piped in to create multiple small watering areas. By having many smaller watering areas spread out over the area being grazed, the rancher can then employ time-controlled grazing, a system that rotates the herd through a series of small parcels for short periods of time. Unlike many traditional ranching systems that apply lower but constant pressure on a large parcel, this system of short but intense grazing followed by long periods of rest is believed to better facilitate natural revegetation; it was designed to better mimic the natural grazing pattern of pre-domesticated large browsers.

Improvements in environmental quality have been staggering. Some native grasses that were thought to have been locally extinct have reappeared on previously degraded parcels, and cottonwoods and willow seedlings are sprouting in the riparian areas. Wildlife habitat has been improved so much that over 50 species of birds were recently recorded where previously there were only a handful. An old creek bed that had been dry since the 1950s is once again running with water. Both conception and birth rates for the cattle have improved as has the carrying capacity of the parcels.

According to Kiowa Grasslands District Ranger Alton Bryant, a positive attitude and a proactive approach to addressing mutual problems are what made this program work: "You always hear about how harmful grazing is on the land and how agencies can't seem to coordinate their activities, but we know it doesn't have to be that way." Both the ranchers and the agency representatives agree that the main ingredient in the success of this program has been the human factor. Their frequent contact and, moreover, genuine interest and concern about the economic and ecological viability of local ranches, has enabled both Forest Service and Soil Conservation Service staff to develop a relationship of trust and mutual problem solving with the permittees. According to one rancher, there is a feeling now, more than ever, "that we are in this together. . . . Alton and Mike [Delano of the SCS]'s doors are always open and I feel as if they really care."

Building on Success: How Can Bridging Arrangements Be Fostered?

The successes evident in the above examples can be replicated elsewhere by fostering an understanding of how bridging occurs, assisting motivated staff members to build agency–nonagency linkages, and working to change attitudes that defeat bridging arrangements. Based on a study of some 230 bridging arrangements between the Forest Service and outside agencies and groups (Wondolleck and Yaffee 1994), we offer three sets of ideas to those seeking to promote bridging.

Imagining the Expanded Use of Bridging Arrangements

One way to foster bridging is to document ongoing agency–nonagency linkages and describe these linkages to motivated staff members as well as potentially interested groups outside the land management agencies. Just as role models provide guiding images for our personal and professional development, images of successful programs provide guidance to individuals who are seeking effective solutions to ongoing problems. Ideas precede action; it can be empowering to show staff how and why someone in a comparable situation was successful at building bridges.

Documenting and disseminating "stories" of effective bridging can be highly effective as a means of technology transfer. Quite simply, people learn from stories: They conform to the way humans think (Sarbin 1986), they engage us emotionally, and they are highly memorable (Hidi and Baird 1986). The use of stories is more likely to motivate ground-level staff, who may find policy prescriptions vague and unconnected to reality. Having a collection of images and ideas in their heads may be more compelling than any set of equations or prescriptions. Humans have a centuries-old history of conveying ideas and norms via storytelling, but only a decades-old experience with more rational modes of analysis. Indeed, one study suggests that officials at the highest levels of government tend to make decisions more from generalizations based on their own personal experiences and situations than on rational grounds. They rely on stories more than objective, analytic decision-making processes (Neustadt and May 1986).

There are many ways to convey positive imagery to interested agency and nonagency personnel. Videotapes, newsletters, guidebooks, and computer-based files of success stories are cheap and easy mechanisms to disseminate ideas. Using "bridging pioneers" to travel to other areas for short periods and explain their actions can help ground the ideas of bridging in the realities of personalities and comparable experiences. It can also help build networks of individuals involved in linkages—relationships that can be informative and supportive. Since bridging is largely a way to achieve better substantive ends, the topic of bridging should be presented in conferences or workshops as an integral component of other, more "normal" aspects of forest management (such as timber sale planning or sensitive species management), rather than having it viewed as a separate activity. Finally, communicating the possibilities inherent in a variety of bridging relationships to nonagency individuals and groups will mobilize a constituency for such relationships. Having agency staff go to town meetings, attend meetings of community and interest groups, and prepare materials for local and regional media can trigger ideas, interest, and support in the world around the agency.

Enabling Motivated Staff to Develop and Use Bridging Arrangements

Even if agency staff have a clear image of the types of relationships they would like to build with the nonagency world, significant barriers stand in their way. In our study of more than 200 situations where individuals had engaged in bridging activities, participants highlighted resource and administrative constraints as two of three key barriers to their effectiveness. In particular, lack of time, money, and personnel were seen as limiting the potential for building linkages. In a time of reductions in force, when agencies tend to retrench into core activities, building external relationships was seen as secondary to necessary work. In addition, lack of flexibility in administrative procedures and budgetary categories and excessive red tape were viewed as problematic to activities that tended to crosscut traditional program areas and deviate somewhat from standard operating procedures.

Some resource constraints can be improved by small changes in the allocation of funds. Creating a discretionary fund for small expenses connected with bridging activities and a start-up fund for new bridging arrangements can help. Many bridging entrepreneurs in the Forest Service complained of having to cover the costs of simple hosting expenses, such as coffee and donuts, out of their own pocket; agency funds were not available or not worth the paperwork involved in securing them. Others complained of the "hurry up, wait, and hurry up" phenomenon involved in initiating new activities—that is, they are pressured to get an activity going, then have to wait a long time for approvals and funding, then have little time to actually carry out the work. For groups outside the agency, delays in receiving funding can be particularly damaging since they often have limited operating funds to pay expenses. In other situations, adequate funding may be present, but narrow budget categories preclude crosscutting activities, and bridging just does not fit comfortably in a budget based on narrow programmatic or line items. Finding ways to bridge budget categories could help provide needed flexibility in funding.

Other ways to provide flexibility in operating at the local level would help foster bridging arrangements. There are various ways to open up the potential for innovation without sacrificing accountability and performance, including providing more discretion at the local level so that staff can run with some ideas without having to wait for lengthy clearances, facilitating the use of flexible working hours for employees so that they can participate in bridging activities conducted in the evenings or on weekends, and cutting the red tape that many potential nonagency partners dread. Other federal administrative procedures, such as those prescribed by the Federal Advisory Committee Act, unduly constrain interactions between federal land management agencies and outside groups and need revision now that "agency capture," corruption, and other historical problems with advisory committees are of less concern (Wondolleck and Yaffee 1994).

There are other ways to expand the capability of agency workforces to build and maintain effective relations with nonagency individuals and groups. Recognizing that bridging is an aspect of many employees' jobs, not just the province of public affairs officers or land management planners, has important implications for personnel development. Developing a broad set of training opportunities in communications, interpersonal dynamics, meeting facilitation, negotiation, and conflict management could help existing staff members, particularly if the training is implemented with entire management teams utilizing examples from forest management. Appropriate topics include both people skills, such as how to interact with different individuals or groups under varying situations, and process skills, such as how to structure and facilitate collaborative problem-solving groups.

The importance of bridging to the future effectiveness of resource management agencies suggests other personnel policy changes. New employees must come with a better understanding of people and process skills, and this reality has relevance to how forestry professionals are trained in academic institutions. In addition, designated "linkers," including facilitators and mediators, partnership coordinators, and rural development specialists, need to be hired for their skills and their ability to serve as tutors and advisors to others in the agency. Finally, policies that encourage cross-agency transfers of personnel need revision, since such transfers break up relationships that are critical to agency effectiveness.

Encouraging the Development of Bridging Relationships

The other primary barrier to the development of additional agency–nonagency relationships is a set of attitudes, fears, and norms on the part of agencies and outside groups that limit participation in linkages. These include excessive reliance on a "we know best" paradigm among agency managers, resistance to new modes of action, and a lack of trust by either side. Often these attitudes result in new initiatives by field personnel that are then stifled by a lack of support by supervisors. Such interactions tend to backfire and result in more mistrust and hard feelings on the part of external groups.

As a new generation of forestry professionals and leaders emerge, it is important that these attitudes change: they diminish the rich set of opportunities inherent in bridging and doom efforts at ecosystem management. A clearer understanding needs to be developed about how agency–nonagency linkages fit into updated notions of resource management. This understanding would include an awareness of the broad set of legitimate values in forest management, an expanded view of the role of the public and nonagency groups in decision making, and a more humble image of the Forest Service as one member of a large community of interests in forest management. If bridging activities are viewed as marginal to achieving agency missions and objectives, they will be marginalized. Agency leaders need to demonstrate commitment to building these relationships, not just for the sake of building relationships, but for the better decisions and enhanced effectiveness that such relationships can yield.

These attitudes are reinforced by a set of organizational incentives, and these can be changed to encourage the development of bridging arrangements. Making bridging a part of many employees' jobs, recognizing the time involved as a legitimate use of agency time, insuring supervisory support of relationship-building, and broadening personnel evaluation procedures to assess the effectiveness of bridging activities—all would help remove some of the disincentives that currently exist. Recognizing and rewarding dedicated groups and individuals inside and outside the agency and including bridging in unit-level planning and evaluations would help create positive incentives. In addition, as the normative base for forest management changes to emphasize a diverse set of goals, including maintenance of ecosystem integrity, rather than a narrow set of commodity production goals, the larger spatial and temporal scale implied by these goals will of themselves encourage the development of a network of relationships.

To individuals seeking to build expanded relationships with groups outside the agency, our advice sounds much like a Nike "Just Do It" advertisement:

1. Numerous opportunities currently exist to build an expanded network of agency–nonagency relationships. We suggest you start by capitalizing on prevailing opportunities. At the same time, initiating contacts with outside groups, not just responding to their concerns, is important. Sometimes even small things can yield considerable benefits and can lead to additional success. As one Forest Service public affairs officer commented, "good relationships seem to beget good relationships."

2. There are many different types of bridging arrangements, and there is no one right way to build bridges. We suggest an adaptive management approach: Try something and learn from it.

3. Attempts at bridging are not always going to work. Don't take it personally. Due to diverse interests and perspectives, conflict will be inevitable in such interactions. What we suggest is to work through the conflict by listening, being open-minded, and focusing on how to creatively pursue forest management objectives, not on carrying out a certain way to achieve those ends.

4. Individuals who have been involved in successful bridging arrangements are not superhuman. Rather, they are individuals who tried doing something, stuck with it, and were willing to put a fair amount of energy into the endeavor. While some were frustrated by their

efforts, most ultimately found the activities very rewarding. They were proud of their efforts and convinced that they had accomplished on-the-ground action, often in spite of the agency bureaucracy.

It is clear that the forestry of the future will work through networks of relationships between public and private individuals and groups to generate needed knowledge, to achieve public support, to lever appropriate resources, and to achieve on-the-ground management. Evolving knowledge in the area of ecosystem science points to the need to incorporate larger geographic scales and longer temporal scales into decision making. Current political realities demand increased efforts to build political coalitions to allow management to proceed. Public finance considerations make creative financing and staffing arrangements inevitable. While this image of forestry is more complex, making some managers long for days gone by, it is also rich in possibilities. It is possible to build a public forestry that is ecosystem oriented, science based, multiple-value oriented, and consensus seeking. Building better relationships between agencies and the world around them is a step in the right direction.

Literature Cited

FEMAT. 1993. *Forest ecosystem management: An ecological, economic, and social assessment.* Report of the Forest Ecosystem Management Assessment Team (FEMAT). 1993-793-071. Washington, DC: GPO.

Grumbine, R. E. 1994. What is ecosystem management? *Conservation Biology* 8:27–38.

Hidi, S., and W. Baird. 1986. Interestingness: A neglected variable in discourse processing. *Cognitive Science* 10(2): 179–194.

Neustadt, R. E., and E. R. May. 1986. *Thinking in time: The uses of history for decision makers.* New York: Free Press.

Osborne, D., and T. Gaebler. 1992. *Reinventing government: How the entrepreneurial spirit is transforming the public sector.* New York: Plume/Penguin.

Peters, T. J., and R. H. Waterman Jr. 1982. *In search of excellence: Lessons from America's best-run companies.* New York: Warner Books.

Sarbin, T. R. 1986. *Narrative psychology: The storied nature of human conduct.* New York: Praeger.

Scott, J. M., B. Csuti, K. Smith, J. E. Estes, and S. Caicco. 1991. Gap analysis of species richness and vegetation cover: An integrated biodiversity conservation strategy. In *Balancing on the brink of extinction: The Endangered Species Act and lessons for the future,* ed. K Kohm. Washington, DC: Island Press.

Sirmon, J. M. 1993. National leadership. In *Environmental leadership: Developing effective skills and styles,* ed. J. K. Berry and J. C. Gordon. Washington, DC: Island Press.

Wondolleck, J. M. 1988. *Public lands conflict and resolution.* New York: Plenum.

Wondolleck, J. M., and S. L. Yaffee. 1994. *Building bridges across agency boundaries: In search of excellence in the U.S. Forest Service.* A research report to the USDA Forest Service, Pacific Northwest Research Station. Ann Arbor, MI: School of Natural Resources and Environment, University of Michigan.

Yaffee, S. L. 1983. Using nonprofit organizations to manage public lands. *Transactions: North American Wildlife and Natural Resources Conference* 48:413–422.

25 Science-Based Assessments of the Forests of the Pacific Northwest

K. Norman Johnson

The 1960s and 1970s: Sustainability of Commercial Timber Outputs 398
Duerr Report 398
Douglas-fir Supply Study 398
Beuter Report 399
The 1980s: Regional and Forest Planning by the Forest Service 399
The 1990s: Sustainability of Fish and Wildlife Resources and Ecosystem Processes 400
Thomas Report 400
Gang of Four Report 400
SAT Report 401
FEMAT 402
Comparison of Bio-Regional Assessments 402
Measures of Sustainability of Forests and Ecosystems 402
Modeling Vegetative Dynamics, Growth, and Yield 403
Comprehensiveness 403
Ownerships Considered 403
Ecological, Economic, and Social Context 404
Linkage to the Outside World 404
Ecological, Economic, and Social Effects 404
Resource Analysis 405
Delineation of Alternatives 405
Efficiency Analysis 405
Role of the Public, Policy Makers, and Managers 406
The FEMAT Case 406
The Influence of FACA 407
From Technocratic to Democratic Approaches 407
The Practice of Science in Science-Based Assessments 408
Conclusions 408
Literature Cited 409

Regional assessments of the forests of the Pacific Northwest by scientists have been conducted for many decades. Early studies focusing on the sustainability of commercial timber outputs have given way to studies focusing on the sustainability of fish and wildlife resources. Of late, these studies have taken center stage in the debate over Northwest forests—due in large part to court challenges of the adequacy of fish and wildlife protection in federal forest plans. These studies culminated recently in FEMAT, a series of options for the management of federal forests in the Pacific Northwest written directly for the president of the United States (FEMAT 1993).

This chapter focuses on studies of Pacific Northwest forests that were done by groups of scientists at the bequest of policy makers or to address particular policy problems. Gordon (1993) identifies this type of study as a "science-based assessment"—which he defines as "attempts to use science-derived information and techniques to answer, or to help answer, questions formulated by politicians and other policy makers." Science-based assessments have become increasingly popular in the 1990s, ever since a group of owl biologists came up with a plan for Northwest forests that shook federal forest management in the region to its roots (Thomas et al. 1990). But such an approach has been used off and on for decades. For example, the Beuter report (Beuter et al. 1976) in the 1970s gained almost as much publicity in Oregon as did the Thomas report in the early 1990s.

This chapter begins with several earlier studies to provide a perspective on how people at different times have approached science-based assessments of the sustainability of Northwest forests and ecosystems. In total, seven studies are examined in detail, with six more mentioned in passing. In addition, Forest Service efforts at forest planning in the 1980s are reviewed. While not inclusive of all bio-regional studies in the Pacific Northwest over the last 30 years, these cases illustrate the evolution of science-based assessments in the region.

The search for sustainability dominates all of these studies, as it dominates policy making for Northwest forests. While measures of sustainability have shifted from commercial timber outputs to fish and wildlife resources, the search for policies that sustain forests goes on unabated.

The 1960s and 1970s: Sustainability of Commercial Timber Outputs

In the two decades following World War II, public forests in the Northwest increased their harvests until they were cutting their full allowable cuts. Similarly, private industrial forests appeared to be harvesting at levels beyond what they could maintain indefinitely. By the 1960s, questions began to be raised about the sustainability of regional wood supplies. Over the next two decades, several studies were commissioned to address the matter.

Duerr Report

Questions about the sustainability of regional timber supplies led to a study in the early 1960s by William Duerr, a professor of forestry from Syracuse, in cooperation with the U.S. Forest Service's Pacific Northwest Research Station. Duerr chose the Douglas-fir region as his study area—roughly, western Oregon and Washington and the Klamath Mountains of northern California. The study projected the coming decline in private timber harvest for the Douglas-fir region along with the timber harvest levels that would result from a federal even-flow policy. Adding these two flows together, Duerr asserted that regional timber supplies would decrease until second-growth timber on industry land matured.

Duerr also looked at an alternative policy in which federal timber filled the gap while private second growth matured, thus smoothing regional timber supply. In short, he found that the national forests would harvest for a time above the level they could maintain indefinitely. While consideration of this alternative may seem mundane now, it caused enormous consternation within the Forest Service when the report was distributed for internal review. As a result, the chapter on a temporary departure from even flow was deleted from the final report, and Duerr removed his name from the report.

The resulting report was entitled *Timber Trends for Western Oregon and Western Washington* (USDA Forest Service 1963). It drew immediate fire from Congress in the person of Wayne Morse of Oregon, who held hearings in which the "suppressed" chapter was presented and discussed. Through the work of Duerr and the outcry of Morse, the legitimacy of examining alternative scenarios for the timber harvest on the national forests of the Northwest was firmly established.

Douglas-fir Supply Study

The Pacific Northwest Experiment Station and Region 6 of the Forest Service soon embarked on a study of the potential for maintaining and increasing timber offerings in the Douglas-fir region, partly in response to the outcry over the Duerr report (USDA Forest Service 1969). The study had two important conclusions: (1) the allowable cut levels, under which the Forest Service was operating, could not be sustained into the second rotation under the extensive management practices then in place and (2) a commitment to intensify management in future stands could allow continuation of these allowable cut levels indefinitely, especially when allowable cuts were measured in cubic feet. Based on this study and subsequent work, the Forest Service mandated a "non-declining yield" of timber through a number of rotations as the guide for managing timber resources on the national forests. In addition, they began calculating allowable cuts in cubic feet and committed to a policy of intensive management sufficient to maintain harvest levels near historic amounts. A new definition of sustainability of commercial timber crops was born—nondeclining yield of timber volume over several rotations—that continues to this day.

Beuter Report

As the Forest Service solidified its estimate of sustainable timber harvest levels, questions again arose about the sustainability of overall regional timber supplies. The controversy was especially heated in Oregon, where members of the State Board of Forestry questioned the ability of the Bureau of Land Management (BLM) and private landowners to maintain their timber offerings, especially in southern parts of the state. The board asked the College of Forestry at Oregon State University to undertake a study of the state's timber supplies. That study, conducted under the leadership of John Beuter, broke the state into 10 "timbersheds" and examined timber resources and likely timber availability for five owner groups—the Forest Service, the BLM, state and other public agencies, the forest industry, and nonindustrial private landowners (Beuter et al. 1976). To do these projections, scientists built a nonspatial, deterministic simulation model that used growth and yield coefficients to update forest inventories managed through even-aged regeneration harvest methods. This model (TREES) and its descendants are still in use today.

The Beuter report confirmed that private timber harvests in western Oregon could not be maintained at their current levels. Short of an increase in harvest levels by the Forest Service, the report suggested that regional timber harvests would decline until a resurgence of industry second growth occurred after the year 2000. It also established the usefulness of subregional projections.

In the late 1980s, Sessions et al. (1990) updated the Beuter report with similar methods and obtained similar results. Also, an assessment of timber supply was done for western Washington (Adams et al. 1992). By this time, though, sustainability had begun to be measured through fish and wildlife resources and ecosystem properties more than commercial timber harvest levels. While these studies improved knowledge about likely timber availability from private land under the polices then in effect, they contributed only modestly to the debate over forest management in the Northwest.

The 1980s: Regional and Forest Planning by the Forest Service

By the 1980s, resources for analyses of forest ecosystems were largely consumed by the regional and forest planning efforts of the Forest Service. Regional plans were designed to carry out the regulations implementing the National Forest Management Act—including conservation plans for species whose protection had regional dimensions. Appeals of the regional plans triggered much of the spotted owl litigation that, in turn, led to the science-based assessments of the 1990s.

Forest plans became the major battleground over management of Northwest forests in the 1980s. Enormous energy went into these plans as the Forest Service divided each national forest into land-use zones and developed standards and guidelines for the management of each zone. Much of the battle focused on where timber harvesting would be permitted. Harvest levels were scheduled to fall about 20 percent as a result of the daunting and time-consuming exercise.

For this effort, the Forest Service used FORPLAN, a model that focused on the sustainability of commercial timber supplies. In retrospect, this choice may seem surprising. However, much of the debate over the National Forest Management Act and its implementing regulations turned on the sustained yield of timber products and compliance or departure from nondeclining yield. FORPLAN, though, lacked the ability to simulate the ecosystem dynamics and spatial relationships that are of interest in forecasting the viability of species and the continuance of critical ecosystem processes (Johnson 1992).

During this time, the Forest Service became increasingly aware of the need to address issues pertaining to the sustainability of fish and wildlife resources. Indeed, the Forest Service was directly mandated to do so in regulations implementing the National Forest Management Act and the Endangered Species Act, and indirectly mandated to do so in a number of other laws. In these planning efforts, the Forest Service continued to determine which models were to be used in estimating species viability, with only occasional advice from the scientific community. As part of the process, managers selected among different protection schemes for fish and wildlife species, with a general focus on finding the "minimum management requirements" necessary to maintain habitats. Acceptable levels of risk were rarely explicitly defined.

The 1990s: Sustainability of Fish and Wildlife Resources and Ecosystem Processes

As the forest plans neared completion, attention shifted to management of the northern spotted owl. In 1989, as part of an appropriation rider that set timber harvest levels on federal lands in the Northwest for two years, Congress called for federal agencies to reexamine their plan for the northern spotted owl. In response, an Interagency Scientific Committee (ISC) was established to develop a scientifically credible conservation strategy for the northern spotted owl (see Yaffee 1994, Caldwell et al. 1994).

Thomas Report

The ISC was chaired by Jack Ward Thomas. The Thomas report turned federal forest management in the Northwest on its ear. It defined an ecological planning area for analysis—the range of the northern spotted owl—and turned the focus from sustainability of timber supplies to habitat for a wildlife species. The ISC's strategy called for large reserves with dispersal areas between them. Timber harvest levels from federal lands, although not estimated in the report, obviously had to be reduced. By changing the focus from timber supplies to species viability, the Thomas report fundamentally altered the focus of federal forest management in the region. From that point on, providing a stable level of timber sales from federal lands would be difficult indeed.

Gang of Four Report

The Thomas report, as powerful as it was, explicitly considered only one species, albeit a wide-ranging and demanding species. Congress, in its desire to get the biodiversity issue in the Northwest behind it once and for all, soon commissioned another study of federal land in the region of the northern spotted owl. This new study was to suggest forest management strategies to protect old-growth forests, the species associated with these forests, and threatened fish stocks. The so-called Gang of Four report (Johnson et al. 1991) portrayed three dozen alternative combinations of reserves and matrix management, along with their implications for five measures of biodiversity, timber harvest levels, and timber-related employment.

Seven of those alternatives are portrayed in Figure 25.1. The study demonstrated that a strategy concentrating on one species (i.e., the northern spotted owl) would not adequately protect biodiversity in the region (alternative 4a in Figure 25.1). In addition, it suggested that providing at least a 50 percent probability of retention for 100 years (alternatives 8a and above) would require a significant reduction in timber harvest beyond that associated with the owl strategy.

The Forest Service's immediate reaction, under the Bush administration, was to ignore the Gang of Four report. Repeated attempts by various congressmen to craft bills based on its findings went nowhere.

Instead, the Forest Service proceeded with implementation of the Thomas report. The final environmental impact statement for the northern spotted owl, largely based on the Thomas report, was successfully challenged in court. The challenge succeeded in part because the environmental impact statement acknowledged, but did not refute, the Gang of Four report's assertion that protecting the

Timber Harvest

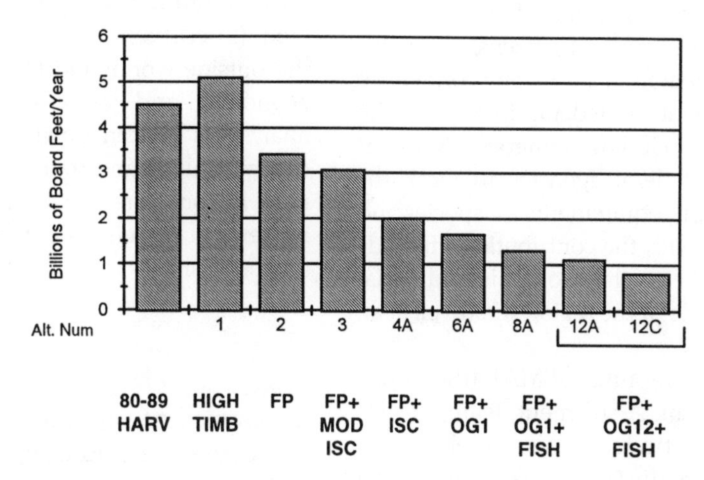

Risk Analysis

Category	Probability of Retention for a Century or Longer								
Function LS/OG Network	--	VL	VL	L	ML	M	M	MH	VH
Viab Spotted Owl Population	--	VL	L	ML	H	H	H	VH	VH
Marb Murrelet Nest Habitat	--	VL	L	L	ML	M	M	MH	H
Other LS/OG Species Habitats	--	VL	VL	L	ML	M	MH	H	VH
Sensitive Fish Habitat	--	VL	VL/L	VL/L	L	ML	MH	MH	VH

Figure 25.1 Implications for timber harvest and ecosystem protection of some of the choices for management of late-successional and old-growth (LS/OG) forests of the Pacific Northwest. Note: FP = Forest Plans; ISC = Interagency Scientific Committee; OG1 = highly significant LS/OG; OG12 = highly significant and significant LS/OG; FISH = fish habitat protection plan; VL = very low; L = low; ML = medium low; M = medium; MH = medium high; H = high; VH = very high (from Johnson et al. 1991).

northern spotted owl would not protect major elements of biodiversity (Caldwell et al. 1994).

SAT Report

The Forest Service next convened a scientific analysis team (SAT), again under the leadership of Jack Ward Thomas (Thomas et al. 1993). Its report addressed the court's criticisms of the Thomas report relative to protecting biodiversity on federal forests. The SAT report added an extensive riparian protection plan, along with other standards for particular groups of species, to the Thomas report's owl protection plan. It also increased the number of measures of diversity (species examined) to more than 100.

FEMAT

Soon after release of the SAT report, President Clinton convened a diverse group of citizens, scientists, and experts from the region in a day-long conference to discuss management of federal forests in the Northwest. In his concluding remarks, President Clinton called upon federal agencies to work together to reconcile legal requirements for species and ecosystem protection with the contribution that federal forests could make to the economic and social well-being of the people of the region (Caldwell et al. 1994).

Over the next three months, FEMAT (the Forest Ecosystem Management Assessment Team) developed nine management alternatives, largely drawn from the Thomas report, the Gang of Four report, and the SAT report. Each alternative was analyzed from biological, economic, and social perspectives (FEMAT 1993). In all, more than 500 measures of biodiversity were evaluated. In addition, one last alternative (known as option 9) was added near the end of the effort, under the leadership of Dr. Jerry Franklin. Option 9 integrated many of the elements of the other alternatives, as well as ideas of the three previous reports (Thomas, Gang of Four, SAT). After some modification, option 9 was chosen by the Clinton administration as its plan for federal forests of the Northwest within the range of the northern spotted owl. Under it, timber harvest levels from federal forests were expected to fall more than 75 percent from the 1980s (see the *Journal of Forestry*, April 1994, for a presentation and critique of FEMAT).

Comparison of Bio-Regional Assessments

The seven studies highlighted above reflect the evolving scope and purpose of science-based assessments of Northwest forests. They are compared and evaluated below according to a number of properties. First, they are compared relative to properties specific to assessments of forests and natural ecosystems: (1) measures of sustainability and (2) methods of projecting forest dynamics. Four broader properties of assessments are (1) comprehensiveness, (2) use of the principles of resource analysis, (3) openness, and (4) use of scientific methods. Comprehensiveness is divided into four categories: (1) ownerships considered, (2) economic and social context, (3) linkage to the outside world, and (4) description of ecological, economic, and social effects. Similarly, resource analysis is broken down into two categories: efficiency analysis and trade-off analysis. Finally, openness is considered in terms of inclusion of policy makers, managers, and the public in deliberations and analysis.

Measures of Sustainability of Forests and Ecosystems

Bio-regional studies in the 1960s and 1970s used commercial timber harvest as their measure of sustainability, while those of the 1990s have switched to species and ecosystem attributes. In the Northwest, each recent study has used more measures of sustainability than the last: Thomas report (1), Gang of Four (5), SAT (100-plus), FEMAT (500-plus).

In the process, the notion that a few species can represent the rest has disintegrated. Despite the call for general ecosystem properties to guide landscape design and management, efforts to consolidate species measures into a few higher-order ones have met with little success. This is not surprising given that the Endangered Species Act and the National Forest Management Act emphasize species—nevertheless, it is a disturbing trend.

The more criteria that must be evaluated before developing a plan, the more difficult it is to innovate and move away from the default, generic prescriptions often outlined in the studies. In addition, many of the evaluations of species in the FEMAT report, especially the lesser-known ones, were highly subjective and difficult for managers to understand. Thus, the explosion in the number of measures of sustainability, combined with the lack of objective information on how actions influence these measures, often leaves the managers with little choice but to follow the generic prescriptions.

The challenge for the future is twofold: First, we need to aggregate the multitude of measures of ecosystem performance into a few broad-based measures, and second, we must increase the objectivity of descriptions of these measures in terms of the effects of actions upon them.

Modeling Vegetative Dynamics, Growth, and Yield

All of the studies, except the Thomas and SAT reports, modeled timber harvest and forest dynamics using traditional growth and yield models and traditional harvest-scheduling models. These systems, such as FORPLAN and TREES, are based on control of the forest to produce commercial timber crops. Spatial relationships between stands are relatively unimportant, as are projections of natural processes such as disturbance and succession.

As long as timber yields are determined by the relationship between forest inventory and growth in stable and slowly changing ecosystems, and timber harvests are regularly scheduled, such an approach may be appropriate. Recently, however, these "tree-farming approaches" have lost their usefulness for a number of reasons. First, in recent plans, large areas of forest are left to develop under natural processes without much human intervention. Thus, projecting natural succession at the stand and landscape levels becomes increasingly important. Second, traditional approaches do not deal effectively with disturbance processes, such as fire, that move across the landscape. Thus, their effectiveness is sharply reduced in places that have a relatively high fire frequency, such as the Sierra Nevada or Blue Mountains. Finally, the lack of spatial detail in traditional models makes it difficult to deal with the relationships between cover and forage, with fragmentation, with patch size, and with other spatial issues related to the effectiveness of habitat for different species.

Effective portrayal of vegetative dynamics in forested ecosystems of the Pacific Northwest awaits the development of a new class of stand and landscape models that can portray succession, disturbance across the landscape, and spatial relationships among patches of different characteristics. Promising work is being done to develop these models in the Sierra Nevada Ecosystem Project (SNEP 1994), but it will be a few years before its success can be determined.

Comprehensiveness

Ownerships Considered

The older timber-supply studies generally considered all owners in their analyses. Of the three examined here, only the Douglas-fir supply study (USDA Forest Service 1969) focused on a subset of owners. Recent bio-regional studies in the Northwest, on the other hand, have concentrated on federal land. This focus is understandable because federal environmental laws apply most directly to federal land and because those lands are most directly under the command of the groups that requested the studies. Some of these studies, however, have been directed to examine nonfederal lands. In his charge to FEMAT, the president said, "[Y]ou should note specific nonfederal contributions that are essential to or could significantly help accomplish the conservation and timber supply objectives of your assessment" (FEMAT 1993).

Yet, even with such a broad charter, FEMAT was quite timid in its analysis and prescriptions for nonfederal land. A number of reasons might explain this: Less publicly available data generally exists for private lands as compared to federal lands. The rules and regulations setting limits on private land actions are often quite vague and indirect, making it difficult to predict their likely effects. And the FEMAT scientists exhausted themselves on the federal portion of their assignment. Also, a political reluctance to have the private portions of the ecosystems studied may have been a contributing factor.

Nevertheless, many of the challenges to ecosystem management in the Northwest lie on private forests. These include issues over management of forests adjacent to streams with anadromous fish stocks in coastal Oregon and Washington. In addition, many challenges lie on lands outside of commercial forests, such as agricultural lands or human settlement areas.

The Botkin report (Botkin et al. 1995) is one of the

first of the ecological studies in the Northwest to look at all ownerships within an ecological planning area. In its analysis of the status and future of salmon in western Oregon and northern California, the assessment team identified agriculture, forestry, and urban areas as leading causes of anadromous fish declines. The report recommended that riparian rules, similar to those already in place for forests, should be written for streams running through agricultural and urban areas. This came as no surprise to students of the issue. It may, however, have surprised the study's clients, including the Oregon State Legislature.

Ecological, Economic, and Social Context

Forest management in the Northwest is played out in the context of a much larger set of ecological, economic, and social forces, including climate change, population growth, change in the ethnic- and age-structure of the population, economic growth, and economic diversification. All of these forces can alter the significance of changes in forest management. For example, the conclusion in the FEMAT report, which stated that the economies of the Pacific Northwest would continue to grow with or without federal timber supplies, changed the federal timber supply issue from a regional economic concern to a local one—a profound change in political terms.

By and large, the studies before FEMAT did not describe the ecological, economic, and social forces within which changes in forest management would be played out. The FEMAT report made an important contribution by outlining some aspects of economic context. However, it provided little information on population and other social changes expected in the Northwest, or on the implications of global warming for the different policy options. A more recent effort, the Sierra Nevada Ecosystem Project, has attempted to project population growth in California and the Sierra Nevada Range as a variable in the demand for goods and services and the intensity and pattern of human use in the area. These projections and analyses will provide an important context for changes considered in the study.

Certainly a challenge for future science-based, bioregional assessments will be to consider more fully the ecological, economic, and social forces that form the context within which forest polices will be made and implemented. These considerations will both complicate and enrich future assessments.

Linkage to the Outside World

Forest management alternatives considered in bioregional studies can have significant effects outside the region. Change in timber availability can affect lumber and stumpage prices in other regions. It also can affect harvest levels in other regions and thus impact environmental quality and biodiversity there. Likewise, changes in air and water quality associated with changes in policies can affect other areas.

None of the reports reviewed here dealt comprehensively with effects outside the study region. The FEMAT report did, however, estimate lumber price effects on consumers nationally. The finding that these price effects would be small contributed important information to the policy debate. In addition, FEMAT scientists found that two-thirds of the region's historic timber availability was on nonfederal land, and that harvesting on those lands would increase in the short-term to somewhat offset the large federal decrease projected in most options. This also had important implications for policy making.

Ecological, Economic, and Social Effects

Generally, the timber supply studies of the 1960s and 1970s focused on economic effects, while later bioregional studies concentrated on ecological effects. The Gang of Four report and the FEMAT report attempted to provide integrated looks at ecological and economic effects. The FEMAT report was the first of these studies to directly address social effects.

Recent studies have measured ecological effects in terms of the impact of various management scenarios on the habitat of particular species; vertebrates, invertebrates, plants, mosses, and lichens have been included as these analyses have broadened. While both scientists and policy makers have called for more-integrated, transcendental measures of ecosystem health, progress has been slow. The old-growth ecosystem measures of Gang of Four and FEMAT reports are among the few such measures to surface. The focus on species has occurred for a number of reasons: Laws tend to emphasize protection of species rather than ecosystems, and scientists have found it easier to measure elemental parts of ecosystems rather than more-integrated components.

Discussion of economic effects has largely been confined to timber harvesting and processing. The general inability to estimate effects of management options on recreation, tourism, and general quality of life has often made for one-sided analyses. General social effects have been largely ignored or assumed to be measured by economic effects. The FEMAT report broke new ground by attempting to measure community capacity to absorb job loss as well as overall community risk of significant deleterious social impacts from the policy options.

The Gang of Four and FEMAT reports have firmly established that a comprehensive set of measures for ecological, economic, and social effects of alternative policies has a place in science-based assessments. The search for the best set of measures, however, will continue.

Resource Analysis

According to the principles of resource analysis (Davis and Johnson 1987), forest policy analysis should delineate a wide range of choices and then systematically vary the key elements to display the marginal benefits and costs of different combinations. In addition, analysts should find the least-cost approach to achieving the goals of each alternative.

Future science-based assessments could learn much from the principles of resource analysis and recent attempts to implement them, such as the Gang of Four report and FEMAT. Principles guiding delineation of alternatives and an analysis of their efficiency should be part of the guiding principles of future assessments.

Delineation of Alternatives

Some of the studies, such as the Beuter report, focused on the implications of current policy, with little prescriptive analysis of other alternatives. Others, such as the Thomas and SAT reports, presented a single plan for dealing with the problems they uncovered. Still others, such as the Douglas-fir supply study, the FEMAT report, and the Gang of Four report, considered a range of options. On further inspection, however, FEMAT's alternatives were only slight variations on a theme. Only the Gang of Four report and the Douglas-fir supply study presented a wide range of choices.

Only the Gang of Four study systematically varied one feature at a time, thus allowing policy makers to understand what they were gaining and losing among the different choices. Three major elements were varied in the analysis: (1) reserve design (five options), (2) fish-protection plan (two options), and (3) management of the matrix (three options). In total, over 30 options were analyzed so that policy makers could see the marginal effects of varying one element at a time.

Figure 25.1 shows some of the information presented in the Gang of Four report. The number in the name of each alternative represents a particular combination of reserve and fish-protection systems, while the letters represent variations in matrix management (i.e., area between reserves). Thus, 8A reserves highly significant old forest, mandates a more protective fish habitat strategy, and uses forest plan standards for the matrix. Alternative 12A reserves highly significant old forests and significant old forests and uses forest plan standards for the matrix; 12C reserves highly significant old forests and significant old forests and uses long rotations in the matrix. In addition, a set of overlays that went with the report allowed the decision maker to construct a map of any particular alternative—much like one would select the ingredients in a pizza.

Through this sort of systematic variation of elements, the Gang of Four team attempted to minimize the role of scientists in selecting alternatives. It is still true, however, that the scientists defined the range of choices that were examined.

Efficiency Analysis

Many of the recent ecological studies have suggested new allocations and rules for federal forests. By and large, these parameters were superimposed on existing forest plans, without reexamining and updating those plans. Thus, there may be restrictions on commodity production in the forest plans that are no longer needed with new allocations and rules. Usually, revisiting forest plans is left to some future effort at the forest level. Unfortunately, this reoptimization rarely occurs. Analysts on the forests generally lack the scientific credibility to justify the increased efficiencies that might occur.

In option 9 of the FEMAT report there was a modest attempt at improving efficiency. First, areas designated for protection of late-successional species and those designated for special fish protection were overlapped. Second, there was an attempt to evaluate the need for forest plan measures to protect threatened species given the large reserves associated with option 9.

Role of the Public, Policy Makers, and Managers

At their best, scientists can be very effective in describing problems in large ecosystems, the reasons for those problems, and the consequences of different options for resolving them. They also can, on occasion, describe creative options for resolving the problems that they and others uncover. Policy makers, managers, and the general public, however, all have a role, at least equal to scientists, in suggesting options for sustaining ecosystems.

Generally, the science teams reviewed here took their overall instructions from policy makers and allowed suggestions from the public and managers at the beginning of their studies. Then, the scientists went behind closed doors until the study was finished—although the ISC did allow the public into its meetings when there was discussion of threats to the northern spotted owl, data on its survival, and suggestions for its recovery. None allowed an interactive relationship between the scientists and the public where scientists would try out alternatives, get reactions, revise their work, get more reactions, and so on.

The FEMAT Case

At the beginning of the FEMAT process, the Clinton administration made a conscious decision not to allow time for public suggestions beyond the discussion in the president's summit on Northwest forests. Administration officials were concerned that public involvement was a spigot that could not be turned off. In addition, they felt that another step should not be added to a process that already was stretched to deliver a product in the time allowed. Thus, the public had little opportunity (beyond previous public comment) to offer its own detailed proposals for Northwest forests. As a result, the scientists relied chiefly on their own ideas and those from previous efforts.

The public's chance came during the comment period on the draft environmental impact statement written to implement option 9. While the alternatives were largely unmodified by this review, strong challenges to the legal adequacy of the species protection provisions in option 9 helped insure that these provisions would be strengthened in the Record of Decision that followed.

It can be argued that the FEMAT process was weakened by its methods for dealing with policy makers. The process allowed partial access to policy makers who commissioned the study—that is, a representative of the Clinton administration was generally present at the internal meetings of scientists. The administration's presence at these meetings gave the appearance of political interference and provided the potential for it to happen in reality. It might have been much more effective for the science team to make periodic presentations to the Clinton administration.

Managers were largely excluded from the FEMAT process, except for occasional, informal contacts between individual scientists and managers, and a brief review of option 9 near the conclusion of the effort. This occurred for a number of reasons: First, managers, at least in theory, led the design of the forest plans, which had been judged inadequate in the protection of species (Thomas et al. 1990, Johnson et al. 1991, Thomas et al. 1993). FEMAT was designed in part to develop a more protective plan for these species. Second, many scientists were concerned that managers would be unable to objectively look at the options that might eliminate or alter timber sales that they had already designed.

Exclusion of the managers, however, had two major costs. First, managers who would be in charge of implementing the president's plan had little initial ownership in it. The difficulty encountered by the Clinton administration in getting its plan off the ground reflects this. Second, option 9 set up rules for forest management that were comforting to the scientists, but very difficult to implement. This difficulty was compounded by the addition of more species-protection rules in the Record of Decision. Involving managers from the beginning might have improved the implementation of these rules.

The Influence of FACA

Recent rulings that the FEMAT process violated the Federal Advisory Committee Act (FACA) will have an impact on how studies are done that provide advice to federal agencies. Among other things, FACA requires that meetings in which a mixture of federal and nonfederal scientists are involved follow specific open-meeting procedures. The FEMAT process was determined to have violated this aspect of the law. Such rulings will limit the involvement of nonfederal scientists in future scientific analyses, and it will make any public involvement much more formal.

From Technocratic to Democratic Approaches

On the one hand, setting scientists up as philosopher kings is inconsistent with our democratic processes. Scientists in these studies are at best hired hands and should not usurp the roles of decision maker, manager, or individual citizen in weighing public values. On the other hand, insulating the analysis from political interference is also important. Without the assuredness of independent work, these projects lack credibility.

Jasanoff (1990) suggests that scientific assessments can be classified as technocratic or democratic processes. Technocratic scientific assessments convey enormous responsibility on scientists to formulate and evaluate an approach or alternative approaches to solving some problem for policy makers. This work is done far removed from the rest of the world. It is based on the notion that the organization of knowledge is a key element in solving policy problems, that facts and values can be separated in the policy-making process, and that scientists should be best left alone to organize these facts. Only after the scientists are done, can values be applied in the policy-making process. Democratic scientific assessments, on the other hand, emphasize incorporation of a full range of values in the development of options for policy making. Moreover, they recognize the difficulty of separating facts and values, the need for a broad and balanced membership in technical advisory groups, and the need for open review and decision making.

Most of the efforts described here, especially the recent ecologically based ones, have been more technocratic than democratic. While this partially can be explained by the perceived need to insulate scientists from political pressure, the potential costs of this approach in terms of developing effective, implementable policies must be acknowledged.

Currently, one federal and one nonfederal ecosystem assessment effort have tried to open their processes up to policy makers, managers, and the public. The Interior Columbia Basin Ecosystem Management Project (ICBEMP) has regularly scheduled meetings with the public at which they present their analysis techniques and answer questions. In other meetings, they ask for ideas for alternatives. The Sierra Nevada Ecosystem Project (SNEP) has regular meetings with a group of selected "key contacts" and also puts out a general call for scenarios.

When the public is asked to change from critic to problem solver and suggest policies that will improve ecosystem management, problems rarely arise due to an inundation of ideas, as the Clinton administration apparently feared in FEMAT. SNEP's call to the public for suggestions on the management of the Sierra Nevada, for example, yielded about 40 responses. Still, these responses were an important contribution to the SNEP effort. Policies were suggested that the scientists had not conceived, and the comment process as a whole allowed the scientists to assess how well they were meeting their charge to provide a wide range of alternatives.

The best procedure for working with policy makers, managers, and citizens remains to be developed. It should not be overlooked, however, that the science-based studies that have recently resulted in major changes in forest management in the Northwest—the Thomas report, SAT report, Gang of Four report, FEMAT report, and the East-Side Screens—

have been done with almost no public interaction during the critical stages of developing and evaluating options. Other processes with extensive public involvement, such as the national forest plans, have been mired in the status quo and have not proven to be effective vehicles for solving ecosystem problems.

The Practice of Science in Science-Based Assessments

Classic research science is organized around developing and testing hypotheses and then submitting the results of this experimentation to a blind (anonymous) peer review process. A referee, without a vested interest in the outcome, decides whether the research paper survives peer review.

The policy-driven science of the studies described in this chapter has been based on a different paradigm. These studies have attempted to bring the knowledge and informed opinion of scientists to bear upon policy questions. Little formal hypothesis testing occurs—although the Thomas report did attempt to compare its proposals to available information about spotted owls.

The peer review process also was greatly truncated. Peer review, to the degree that it occurred, often was not blind—particular individuals were sought out for their opinion on the work. Both the Thomas report and FEMAT, however, did have elements of a blind peer review, albeit on a very abbreviated schedule.

Independent referees were rarely employed. Rather, the agency commissioning the work, or the scientists themselves, decided how to react to review comments. By the time any review had occurred, policy makers were usually waiting anxiously for results. Outright rejection of these results as technically inadequate was not a possibility; at most, the scientists were able to quickly patch up the weakest parts and push on.

Such cavalier treatment of traditional approaches to science in many of the studies discussed here has led to renewed calls for peer review and related processes. The Sierra Nevada Ecosystem Project has taken these calls to heart. A three-stage sequential review process has been used for review papers prepared by the scientists in the project: (1) review of papers by other team members, (2) review by key contacts (i.e., a group of interested citizens and local experts), and (3) blind peer review. The scientists who did the work decided how to react to the first two reviews; a steering committee acted as a referee for the third. These reviews took three to five months to complete.

While SNEP's process dealt with some of the criticisms leveled at past science-based assessments, it did so at some cost. Science-based assessments call for scientists to move beyond their individual disciplines and blend their knowledge with others to address policy issues. The emphasis on review of the efforts by individual scientists and small groups of scientists leaves little energy for the integrative work that is needed.

Protocols for how science should be practiced in science-based assessments of forest ecosystems are still evolving. Past assessments often strayed quite far from traditional methods; future assessments will most probably be called upon to use a process closer to classic peer review science.

Conclusions

As forest management issues evolve in the Northwest, scientists will continue to be called upon by policy makers for science-based assessments. Despite occasional statements by politicians that we have had enough scientific studies, calls for them will continue for a number of reasons: First, policy makers often do want an objective assessment of options and tradeoffs because they realize that policy based on unsound premises can come back to haunt them. Scientific assessment teams can be a logical source of this information. Second, sending knotty problems out for study is a time-honored political tradition. These days, the committee to which these problems are sent is often a committee of scientists. Third, scientists often encourage these requests from friendly politicians as a way to interject themselves into the policy debate.

From the many science-based assessments of

northwest forests conducted over the last 30 years, we have learned much about how to structure processes to produce useful policy information. Some of those lessons are discussed in this chapter. It is also true, however, that many key issues about how to do science-based assessments remain unanswered; thus we can expect that they will continue to evolve well into the next century.

Literature Cited

Adams, D. M., R. J. Alig, D. J. Anderson, J. A. Stevens, and J. T. Elik. 1992. *Future prospects for western Washington's timber supply.* Institute of Forest Resources contribution no. 74. Seattle: University of Washington, College of Forest Resources.

Beuter, J. K., N. Johnson, and L. Scheurman. 1976. *Timber for Oregon's tomorrow.* Corvallis, OR: Forest Research Lab, College of Forestry, Oregon State University.

Botkin, D. B., K. Cummins, T. Dunne, H. Regier, M. Sobel, L. Talbot, and L. Simpson. 1995. Status and future of salmon of western Oregon and northern California: An overview of findings and options. *Research report 951002.* Santa Barbara, Calif: Center for the Study of the Environment.

Caldwell, L. K., C. F. Wilkinson, and M. A. Shannon. 1994. Making ecosystem policy: Three decades of change. *Journal of Forestry* 94(4):7–10.

Davis, L., and K. N. Johnson. 1987. *Forest management.* 3rd edition. New York: McGraw-Hill.

Franklin, J. F. 1995. Scientists in wonderland. In *Science and biodiversity policy: A supplement to BioScience* S74–S78.

FEMAT. 1993. *Forest ecosystem management: An ecological, economic, and social assessment.* Report of the Forest Ecosystem Management Assessment Team (FEMAT). 1993-793-071. Washington, DC: GPO.

Gordon, J. 1993. Assessments I have known: Toward science-based forest policy. In *Starker lectures: Communications, natural resources, and policy.* Corvallis, OR: College of Forestry, Oregon State University.

IFMAT (Indian Forest Management Assessment Team). 1993. *An assessment of Indian forests and forest management in the United States.* Portland, OR: Intertribal Timber Council.

Jasanoff, S. 1990. *The fifth branch: Science advisors as policy makers.* Cambridge, MA: Harvard University Press.

Johnson, K. N. 1992. Consideration of watersheds in long-term planning: The case of FORPLAN and its use on the national forests. In *New Perspectives for Watershed Management,* ed. R. Naiman. Seattle: University of Washington Press.

Johnson, K. N., J. F. Franklin, J. W. Thomas, and J. Gordon. 1991. *Alternatives for management of late successional forests of the Pacific Northwest.* A report to the Agricultural Committee and the Merchant Marine Committee of the U.S. House of Representatives. Corvallis, OR: College of Forestry, Oregon State University.

Sessions, J., K. N. Johnson, J. Beuter, G. Lettman, B. Greber. 1990. *Timber for Oregon's tomorrow: The 1989 update.* Corvallis, OR: Forest Research Lab, College of Forestry, Oregon State University.

Sierra Nevada Ecosystem Project. 1994. Progress Report. Davis: Center for Water and Wildland Research, University of California.

Thomas, J. W., E. D. Forsman, J. B. Lint, E. C. Meslow, B. R. Noon, and J. Verner. 1990. *A conservation strategy for the northern spotted owl: A report of the Interagency Scientific Committee to address the conservation of the northern spotted owl.* Portland, OR: USDA Forest Service, USDI Bureau of Land Management,USDI Fish and Wildlife Service, USDI National Park Service.

Thomas, J. W., M. Raphael, R. Antony, E. Forsman, G. Funderson, R. Holthausen, B. Marcot, G. Reeves, J. Sedell, and D. Solis. 1993. *Viability assessments and management considerations for species associated with late successional and old-growth forests of the Pacific Northwest.* Washington, DC: USDA Forest Service.

U.S. Department of Interior. 1992. *Recovery plan for the northern spotted owl: Final draft.* Portland, OR: USDI Fish and Wildlife Service.

USDA Forest Service. 1963. *Timber trends in western Oregon* Research paper PNW-5. Washington, DC.

USDA Forest Service. 1969. *Douglas-fir supply study: Alternative programs for increasing timber supplies from national forest lands.* Portland, OR: USDA Forest Service, Regions 5 and 6 and Pacific Northwest Forest and Range Experiment Station.

Yaffee, S. L. 1994. *The wisdom of the spotted owl: Policy lessons for a new century.* Washington, DC: Island Press.

26 Scarcity, Simplicity, Separatism, Science—and Systems

R. W. Behan

Scarcity 412
Simplicity 412
Separatism 413
Science 414
Systems 415
Conclusion 417
Literature Cited 417

Linking environmental policy to forest planning is already a matter of law. The National Forest Management Act of 1976 explicitly refers to and accommodates the National Environmental Policy Act (NEPA). In a law-based—some would say law-obsessed—society, there can be no stronger directive: The linkage is a statutory mandate.

NEPA directed the federal agencies to seek and maintain a state of harmony between humanity and its biophysical environment. Since the forest-planning enterprise was to accommodate NEPA, we should have today (had the laws been faithfully executed) healthy, vigorous forests and healthy, happy, and prosperous communities, all coexisting in a state of harmony.

Eighteen years after the National Forest Management Act was passed, no observer (even one of moderate sensitivity) would assert that the intent of NEPA has been achieved by the forest-planning process. A state of harmony between the national forests and the people who use and appreciate them—particularly in the Pacific Northwest—is, for now, nowhere to be observed.

On a net basis, the forest-planning adventure has been disastrous. Achievements have been grossly outweighed by the environmental, social, managerial, and political damages and costs. Indicting the Forest Service for this travesty of professional management and public administration is indeed inescapable—but it is also insufficient. Also at fault is the obsolete paradigm of professional forestry based on producing a maximum sustained yield of timber. This is the objective the agency ostensibly pursued. Some will claim that multiple use has been equally important as a driver of Forest Service action. I cannot disagree, but as I have argued elsewhere, multiple use has been far more a policy of land use than a technique of land management.

Maximum sustained yield of timber might well be called the forestry of the 20th century—and it differs little from that of the 19th or 18th. It is freighted, conditioned, influenced, defined, and described by the first four words in the title above: scarcity, simplicity, separatism, and science. The task of a forestry for the 21st century will be to unload the freight.

Scarcity

Twentieth century forestry has sought preeminently to project an assured supply of timber, as large as possible, into an infinite future. In short, it is a scarcity-avoidance strategy.

The classic escapes from scarcity—discover more, develop more, import more, or substitute—were all denied for various reasons to the political economies of western Europe in the 18th century. Maximum sustained yield was developed as an alternative, and only in the last two decades in the United States has it been seriously scrutinized.

The escapes, which are fully accessible, are richly rewarded in the global economy developing today. A scarcity-driven strategy is not only anachronistic but, as we have seen in the Pacific Northwest, absolutely dysfunctional and counterproductive. Far too many contemporary values were sacrificed to make the practice of providing a maximum future supply of timber economically justifiable and politically acceptable.

Perhaps the greatest challenge to forestry for the 21st century will be abandoning the presumption of scarcity and the corollary prescription of maximum sustained yield. Rescuing one's society from the certain misfortune of a timber famine has always been a source of great self-esteem for the profession of forestry, but we must learn to aspire far less to heroism. Instead, foresters in the next century must assure the continuity of the forests and human institutions that appreciate and depend upon them.

Simplicity

Twentieth century forestry is elegant in the simplicity of its thinking. Foresters chronically and uncritically have defined timber as a renewable resource in simplistic, purely biophysical terms. Because a new stand of merchantable trees would grow up to replace the ones cut down after a sufficient period of time had passed, renewability was axiomatic. Renewability has always been the preeminent rationale for cutting trees down. With the exception of forest economists, who understood the sobering power of compound interest, little or no thought was given to the importance of time. Foresters have been quite at ease talking about rotations of a century or more.

By this simple calculus, petroleum is also a renewable resource. The rotation age is measured in millions of years, not centuries, but the rejuvenation of petroleum is a biophysical certainty, if only we can be patient. We are not sufficiently patient to wait for a new crop of petroleum, however, and I suspect our society is not sufficiently patient enough to wait for a new crop of old-growth trees. The political evidence to support this claim is massive.

Society's patience—or impatience—should weigh heavily in the determination of renewability. We have a fairly decent measure of social impatience. Economists like to call it "social time preference rate." Others call it the rate of interest. Which interest rate to use in problem analyses always makes for a spirited debate. Suppose we take the rate currently paid on long-term public debt as a reliable indicator of social impatience; 30-year treasuries have been hovering around 8 percent for several years now. If we use this datum to define renewability, a forest growing at a rate equal to or greater than 8 percent per year could be called renewable. Otherwise we will call it a stock resource.

Another example of simplicity is the historic notion that a group of old, tall, straight Douglas fir trees constitutes a sawtimber resource. Applying single-cause, single-effect thinking, we clearcut Douglas-fir because it is shade intolerant. Yet, foresters are coming to understand that a group of old, tall, straight Douglas-fir trees is a complex of coexistent and simultaneous resources. We are coming to understand as well that a simplistic clearcutting of the group of trees provokes a complicated set of consequences we are only beginning to comprehend—not only in the forest itself, but in society at large.

This trend in thinking must continue. In the 21st century, foresters must deal rigorously and consciously with the probably chaotic complexity of forest and institutional systems.

Separatism

The father of categorization and classification was Linnaeus, the great Swedish naturalist. Or perhaps he was just a major contributor to a profound separatism that has dominated Western thought for centuries, perhaps millennia. It goes far beyond the differentiation of kingdoms, phyla, classes, orders, families, genera, and species.

We speak of timber, water, recreation, wildlife, and forage, for example, as if they were separate and independent resources and uses. We have created university curricula, textbooks, professional societies, and scientific journals for each of them. Land management agencies create organizational niches, budget line items, and career ladders for the various separate "functions." Moreover, the agencies devote considerable energy to interfunctional boundary disputes, budget competition, and warfare. In addition, we can identify resource separatism among the forestry profession's clientele groups—timber associations, wilderness societies, water users associations, wildlife organizations, livestock groups—all dedicated advocates of single and separate resources.

In the wildland landscape, there are designated recreation areas, wildlife refuges, and protected metropolitan watersheds. Separate, single-use designations are popular and ubiquitous. Foresters have invested great amounts of capital in the conversion of wild forests into timber plantations, and range conservationists have made pastures out of wild grasslands with considerable success.

In much of the western United States, forests, both private and public, have been transformed into a mosaic of separated, single uses—clearcut plantations abut wilderness areas. On a routine flight into Portland, Oregon, or Seattle, Washington, one can see a patchwork quilt of separatism on the landscape that is a result of 20th-century forestry. In the 21st century, the surrealism of separatism will become more and more evident. We will need to redefine forestry accordingly—and we will need to do a lot of restoration work on the patchwork quilt.

The most basic, profound, and pernicious separation in Western cultures is the distinction between mankind and the biophysical environment. The words "artificial" and "natural" display and clarify the separation—the one is preeminently a product of humanity, the other specifies its absence. This man/nature dichotomy has so confused and confounded the politics, policy, and practice of professional forest management that little can be done beyond the margin until it is confronted and dissolved.

Consider the words "natural resource," a term of separation *par excellence.* Substances and services occurring spontaneously in the biophysical environment are deemed natural resources—for example, tall, straight Douglas-fir trees are a timber resource; undeveloped, unoccupied wild land (where man is a visitor who does not remain) is a wilderness resource; and white-tailed deer is a wildlife resource. The resource is equivalent to the spontaneously occurring biophysical substance or service; there is no necessary human factor involved at all. That would be artificial. This is a separatist notion, and it also carries on the tradition of simplicity discussed above.

A brief reflection reveals that not every tall, straight tree is a natural resource. Twenty or thirty years ago, tall, straight western red cedar trees in Alaska were cut and then left in place to rot. They were not natural resources. There was no commercial use in Alaska for cedar—no market, no value. Yet, cedar today is a prime timber resource. It is an excellent technical substitute for redwood decking, and split-cedar roofing has become a prestigious, high-status means of staying dry. What happened in the intervening decades? The value of cedar from Alaska became apparent.

Spontaneously occurring substances become resources when they take on value, and they are no longer resources when they lose it. The value is created in the minds of people and aggregated in the institutions of society. It is the social perceptions of utility, and the techniques of using the substances, that distinguish between resources and what was elegantly termed by geographer Erich Zimmerman "neutral stuff." In this view of resources, the man/nature dichotomy utterly dissolves; resources are seen not as natural, but as biosocial constructs .

Forestry for the 21st century must adopt such a view. We must see that separatism is true artifice, and create management strategies and practices accordingly. A decent beginning is visible in the ecosystem concept, but we are only halfway home.

Science

Foresters have been consistently proud of and comforted in the assertion that forestry is a scientific enterprise. It is a claim that has been used historically, at least since the days of Bernhard Fernow's experience in the Adirondacks, as a first line of defense in times of conflict and controversy.

Mr. Fernow was the dean of the Forestry School at Cornell University in the first decade of the 20th century. He was demonstrating good forestry around Saranac Lake by clearcutting hardwoods to favor and nurture the commercially superior spruce; he installed a cooperage factory to utilize the low-grade hardwood. The local citizens who appreciated the beauty of the mountains objected vigorously. In their complaints, they used such terms as "forest devastation." Today we would call such malcontents environmentalists, I suspect. Mr. Fernow demurred, suggesting that the lay citizenry simply did not understand the science of silviculture; he was privately outraged that his professional, scientific expertise was so heavily subject to discount, indeed override. For overridden it was. The governor of New York, notwithstanding some historical evidence of inebriation at the critical political moment, set Mr. Fernow's subsequent annual budget at zero, oversaw the closure of the Forestry School, and witnessed the dean's departure for Canada.

A number of contours of the Adirondack controversy are visible today in the forest turmoil of the Pacific Northwest. The lesson not yet learned from Mr. Fernow's time until the present might be formulated several ways:

1. Science is necessary to the practice of forestry, but not sufficient.

2. Forestry is as much a political enterprise as it is scientific.

3. The stubborn reliance on orthodox science is a major cause of the contemporary difficulties of professional forestry.

Professional forestry has long been dominated by biological, especially ecological, science. More recently, management science has become prominent in both the teaching and practice of forest management. Both display and adhere to the scientific standards of objectivity, rationality, and quantification, and both rely on the scientific traditions of reductionism and specialization.

Reductionism and specialization are a species of separatism. As a strategy of inquiry, it has proven to be a good means of learning how the separate pieces *work*. Almost by definition, however, dividing the focus of our inquiry into ever smaller parts precludes our learning how things *relate*. Perhaps we know enough in the forestry community about how things work—at least for the near future. We know far too little about how things relate. Foresters did not know how spotted owls related to clearcutting, or how their ignorance might be related to political processes. The failure to understand relationships has been costly.

Exploring how things relate is the means by which the nature of systems can be understood. But the tradition of science is in running in absolutely the wrong direction to facilitate doing so. We need to redirect our science to systematically to build blocks of knowledge about ever larger, not ever smaller, elements of forest systems. We need to recombine, to reorder, and to integrate the knowledge base into larger and larger aggregations and to study rigorously how things relate—that is, the behavior of entire systems. This will require the pooling and interpersonal sharing of intellectual and inquisitive talent, not just an affiliation of specialists.

Both ecological and managerial science insist—as the canons of science dictate they should—on objectivity, rationality, and the quantitative. This is altogether left-brain stuff, and the dominance by science of the teaching and practice of professional forestry is accordingly conditioned. Scarcely any attention in teaching or practice is paid to the right-brain attributes and the products of subjectivity, intuition, and the qualitative. Nor do we distinguish the unknown from the unknowable, or dwell on the awful uncertainties and complexities that characterize reality. Where, in the curricula or applications of forestry, is spirituality, creativity, metaphysics, magic, or history?

Professional forestry rather comprehensively ignores both the intrapersonal origin of values and the social processes that aggregate them. We do not see why and how resources are identified among the universe of neutral stuff. Nor do we understand as professionals how society accommodates and resolves the conflicts over differing resource identifications. To put the matter in more general terms, we see very little of *people.* Clearly, a forestry for the 21st century must correct this failure.

Closely related to this shortcoming is a monumental oversight in the study of ecology. Ecologists often are guilty of the more profound and pernicious separatism mentioned above—the separation of man from nature. Indeed, ecological succession takes place explicitly and exclusively in the absence of human intervention. The mosses replace the lichens, and Douglas-fir overtops the red alder—only if the contriving mind of man is utterly ignored. Ecology is the study of the biosphere, but it systematically excludes the species with the greatest potential for and the greatest history of defining its configuration.

This systematic oversight makes the study of ecology far more convenient. But it has a sinister and unfortunate consequence: It confounds and confuses the politics, policy, and practice of forest management.

The convenient datum of a man-exclusive system soon becomes a normative datum—the ecosystem is best which has been the least altered by human activity. Or we read about indicators of ecosystem health and find the greatest pathogen is human. When ecologists voice these arguments, they have violated the canon of objectivity. There is no such thing as a scientific rationale for a normative standard. Science can tell us what is right, not what is good. Subtly, but persuasively, ecosystem management becomes the restoration and maintenance of natural systems with artificial impacts eliminated or mitigated.

In clearcuts as well as wilderness areas, humans have perceived values. We have applied appropriate technologies and institutions to acquire and distribute those values, and we have been coequal determinants of landscapes, along with biophysical entities and processes. Humanity is inescapably influential in every landscape, managed or natural. The science of ecology needs systematically to take this into account.

Systems

A systems view of forest management has been a long time arriving. Over forty years ago Aldo Leopold published his landmark work, *A Sand County Almanac* (Leopold 1949). In it, he spoke about land as a commodity and land as a community, and argued strongly his preference for the latter. The view of land as a community, including humans as responsible members of the community, was the basis for Leopold's articulation of a land ethic. Clearly, he was remarkably prescient.

Leopold also described two sorts of forestry: Type A forestry saw land as commodity, grew trees like cabbages, and viewed them as cellulose. Type B forestry treated land as community, caring for all the elements of the forest and the interacting relationships between the parts. Type B forestry, Leopold avowed, had the stirrings of an ecological conscience.

Forty-three years later, in June 1992, the U.S. Forest Service announced the adoption of something akin to type B forestry; it was to be called ecosystem management. It was decades past due, the tardiness displaying a shameful case of bureaucratic intransigence, political cowardice, and professional inertia.

But only 20 years after *A Sand County Almanac,* a

vivid controversy arose over cabbage patch forestry on the Bitterroot National Forest in Montana. It was replicated in West Virginia on the Monongahela National Forest shortly thereafter. Several of the subsequent attendant analyses explicitly rejected cabbage patch forestry and suggested a more comprehensive approach (see Bolle 1970).

The forestry profession took notice. In 1970, at the suggestion of Dean Arnold Bolle from the University of Montana's School of Forestry, the Society of American Foresters sponsored a Forestry Curriculum Development Project. The result was a textbook explicitly adopting a systems view of forest management (Duerr et al. 1979). In addition, a number of forestry schools began experimenting beyond the cabbage patch.

Congress also took notice of cabbage patch forestry—particularly after the Monongahela judge invited it do so. In his introduction to the bill that became the National Forest Management Act, Senator Humphrey noted that we could no longer view the forest only as a source of commercial timber. Senator Humphrey described type B forestry precisely.

Certainly the National Forest Management Act specifies a system view, and the standards and guidelines in the forest plans constituted a laudable beginning. But timber targets were set, agency budgets were tilted nearly exclusively to road building and timber sales, and forest plans were appealed and litigated as soon as they appeared. The wholesale public rejection of the forest plans was a continuing reaction to cabbage patch forestry. First heard in the Bitterroot controversy, such criticism now had a legal outlet, and it was pursued with increasing skill and effectiveness.

Finally, with forest management in the Pacific Northwest virtually paralyzed with plan appeals, lawsuits, and restraining orders, an imminent presidential election, and the old-growth of the Pacific Northwest essentially liquidated (industry having achieved its goal), the Forest Service announced its intention to adopt ecosystem management and reduce clearcutting dramatically. Perhaps, over 40 years after Aldo Leopold articulated the notion, type B forestry has arrived—just in time for the 21st century.

Ecosystem management describes a quantum jump in professional thinking. Yet, it is critical for us to see and accommodate a wider system than current manifestations of ecosystem management encompass. Ecosystem management is explicitly and appropriately based on the science of ecology. But as long as ecology maintains, even implicitly, the separation of man and nature, system forestry will not fare much better than cabbage patch forestry.

System forestry in the 21st century will be able to predict biophysical responses to management inputs prior to their application. We will understand before we do something what the response of the entire forest system will be. To avoid the social turmoil that cabbage patch forestry has generated, however, we also will need to be able to predict social responses to management inputs. We will need to include humanity, people, social elements, and institutions in our characterization of an ecosystem. In short, we will need to correct the strategic oversight of ecology.

Ecosystem management as we define it today is certainly a complicated proposition. But, it deals with just one of the coexisting, coordinate subsystems. In a forestry for the 21st century, we need to incorporate the biophysical subsystem, which shapes and is shaped by the social subsystem—which, in turn, shapes and is shaped by the biophysical. This is the biosocial perspective (Bonnicksen 1991). Mutual adjustment is the rule, which foresters will need first to understand and then encourage, guide, and predict.

But the level of complexity expands several orders of magnitude as a result. How can we handle the information loading? In the near term, the answer appears to be computer simulation models. We have many such models already, of course. Virtually all of the simulation models to date are science-based, quantitative, linear-programming optimizing applications. This represents a terrible constraint—one that biosocial forestry cannot tolerate. It is trite to observe that huge quantities of important things cannot be quantified, or that science cannot provide all the answers, let alone identify the important questions.

In the past several years, one promising alternative to quantitative linear programming applications has emerged. A group of systems analysts at the University of British Columbia developed a truly ingenious modeling technique called K-SIM (Kane et al. 1973). K-SIM produces a novel and fundamentally different sort of model of a kind that is indispensable to the practice of biosocial system forestry. Rather than the science-based, data-hungry, measurement-dependent cardinal models typical of linear programming applications, K-SIM models are judgment-based and ordinal. That means they can characterize systems of virtually any complexity and include variables from virtually any source—the rational, objective, value-free output of the left-brain, articulated through science, or the intuitive, subjective, value-laden output of the right-brain, articulated socially through economic and political institutions. K-SIM can handle what might happen to Canadians' sense of national sovereignty if too much Columbia River water is allowed to the United States by international treaty, can determine the value to U.S. society of maintaining the subculture of the West Coast logger, or indicate the value of maintaining viable populations of the northern spotted owl.

A software package known as EZ-IMPACT, developed at the University of Wisconsin, incorporates K-SIM into a structured model-building context (see Behan 1994). EZ-IMPACT guides the user through the steps of identifying relevant variables and specifying their interactions; it forces the user, in short, to characterize the system at hand. Then, various alterations can be specified in any number of the variables to see what will happen to the entire system as a consequence. EZ-IMPACT is an example of the next generation of modeling software, a type that will be critical to the practice of professional forestry in the 21st century.

Conclusion

There is not much doubt in the forestry community today that ecosystem management is on the right track. The suggestion that plantation forestry needs only some minor tweaking is no longer tenable or even fashionable.

As we come more and more to understand the nature of the biosocial forest system with which we work, and become more skilled at building and using simulation models to anticipate the outcomes of proposed actions, there will be fewer and fewer surprises. If we have the sense to include our constituent groups in both the construction and operation of the models—that is to say, if we grant them participant status in decision processes—the political turmoil in which we practice forestry today can be greatly reduced.

Moreover, decision processes must be centered in the forest at the local level. We must reject the practice of managing forests by federal statute, a practice on which we have relied heavily in the closing decades of the 20th century. The self-serving, mutually reinforcing triad of bureaucrats, congressional committee staffs, and professional career lobbyists must be confronted.

It really is just one system. When a forestry for the 21st century explicitly acknowledges and exploits that reality, and when forest planning is built on that sort of forestry, environmental policy can be executed with fidelity and good will.

Literature Cited

Behan, R. W. 1994. Multiresource management and planning with EZ-IMPACT: From linear optimization to ordinal judgment. *Journal of Forestry* 92:2.

Bolle, A. 1970. *A university view of the Forest Service.* Reprinted as Senate document no. 91-115. Washington, DC: GPO.

Bonnicksen, T. M. 1991. Managing biosocial systems. *Journal of Forestry* 89:10–15.

Duerr, W. A., D. E. Teeguarden, N. B. Christiansen, and S. Guttenberg. 1979. *Forest resource management: Decision-making principles and cases.* Philadelphia, London, and Toronto: W. B. Saunders Company.

Kane, J., I. Vertinsky, and W. Thompson. 1973. K-SIM: A methodology for interactive policy simulation. *Water Resources Research* 9(1):65–79.

Leopold, A. 1949. *A Sand County almanac and sketches here and there.* London, Oxford, and New York: Oxford University Press.

27 Making Decisions in a Complex and Dynamic World

Gordon R. Smith

Where Have We Been? Rational Comprehensive Planning 420
What We Have Learned 421
Complex Systems Are Unpredictable 421
People Satisfice 422
Preferences Are Socially Constructed Through Iteration 422
Sources of Conflicts 423
Separation of Costs and Benefits 424
Where Do We Go from Here? Learning from Experience and Developing Goals 424
Probing Problems 424
Interaction Processes for Decision Making 425
Changing Interaction Processes and Restructuring Conceptions of Problems 429
Managing Resources by Learning and Constructing Goals 430
Northwest Power Planning Council 430
Timber, Fish, and Wildlife Interdisciplinary Teams 431
Conclusion 433
Literature Cited 434

During most of the 20th century, forestry has been viewed largely as a technical problem to be solved by gathering data and using linear, rational analyses to choose and implement management strategies. In some ways this approach has worked well—notably in producing large volumes of wood fiber, especially from industrial forest lands. However, in other important ways, it has fallen short of expectations. For example, we have found that we cannot predict some resource conditions that we care about, that goals change as people contest them, and that we need to learn as we manage resources and to quickly adapt our actions in light of unanticipated changes in resource conditions.

This chapter addresses ways of making forest resource management decisions in situations that are constantly changing, complex, and without clear goals. The chapter is composed of three parts: The first part sketches the approach of rational comprehensive planning. The second part outlines some things we have learned by struggling with the limits and failures of rational planning. The third part explores evolving approaches to dealing with resource management problems that are poorly addressed by rational planning approaches. Two situations are briefly described where participants used interaction and learning to manage natural resources. Hopefully these lessons can inform and improve our forest management in the 21st century.

Where Have We Been? Rational Comprehensive Planning

During the 19th century, Americans developed principles of managing work by listing the specific tasks necessary to do a job, counting the time and materials required, and centralizing command of workers. In 1910, Frederick W. Taylor applied the name "scientific management" to this reductionistic approach to understanding and controlling activity (Shafritz and Ott 1992). Its major premise was that the manager can know all the important constituent dynamics of a system, and that by designing and overseeing each of these parts, the manager can both control what happens and achieve desired outcomes. Simplification of production systems to ease control was central to this strategy—whether it involved creating interchangeable parts and assembly lines or establishing even-aged, monocultural tree stands.

From scientific management developed rational comprehensive planning. Rational planning was the dominant approach to natural resource management from the 1950s arguably through the early 1990s. Three key assumptions underlie rational planning: First, as in scientific management, rational planning assumed that the world is both predictable and knowable in a mechanistic and deterministic way. If one can just get enough information about the system being managed, one can accurately predict the consequences of management actions. Second, rational planning assumes that the goal of management is clear. And third, rational planning assumes that not only is there one best "right answer," but that it can be discovered through an objective technical process.

Typically, these processes involve collecting data on resource conditions, modeling resource dynamics, and calculating costs and benefits. Using the ideal form, the planner lists all objectives, identifies the full range of alternative actions, comprehensively evaluates the consequences of each alternative, and then chooses the alternative that maximizes net benefit (Simon 1976). Even the most enthusiastic advocates of comprehensive planning recognized that an infinite number of options exist and that it is not possible to evaluate all options. The assumption was, however, that planners could identify the most significant options and the consequences of pursuing those options.

The 1970s were the heyday of rational comprehensive planning, epitomized in forest resource management by the enactment of the federal forest planning laws: the Forest and Rangeland Renewable Resources Planning Act (RPA) of 1974 and the National Forest Management Act (NFMA) of 1976. RPA directed the Forest Service to prepare a renewable resource program and update the program every five years, including inventorying resources and investment needs and specifying output benefits expected from each investment. NFMA modified RPA and added language that recognized agency discretion in implementation of congressional direction, but added guidelines intended to protect nontimber resources and limit logging to sustainable levels. Legislation also required the Forest Service to comply with the National Environmental Policy Act (NEPA) of 1969 when carrying out RPA and NFMA. Forest plans were to contain detailed analysis of management alternatives prepared by interdisciplinary teams.

To comply with these requirements to perform complex calculations of the costs and benefits of multiple alternatives, the Forest Service adopted quantitative models, most notably the computerized linear programming model FORPLAN. Early on, it became clear that the cost of acquiring the data to prepare comprehensive analyses was prohibitive, and that the planning models were based on value-laden assumptions (Cortner and Schweitzer 1983).

In 1976, Congress mandated the development of forest plans by 1985. However, by 1986, of the anticipated 123 plans, only 50 had been issued in final form and 46 had been issued in draft form, with more than 200 appeals filed against the final plans (Hunt

1986). Plans intended to last for a period of 10 years sometimes took 15 years to produce. Overall, the planning process has been estimated to cost more than $200 million per year (Cubbage et al. 1993). Each final plan typically consisted of a stack of volumes three to six inches high. Despite this compilation of data and analysis, the court cases continued to pile up.

Notwithstanding the lack of resolution of management problems, our resource management schools continued to instill a belief that technical analyses provided a comprehensive and objective means to achieving resource management goals. Often when plans were adopted, Congress did not appropriate enough money to implement them (Sample 1990). Moreover, when resource managers attempted to implement some plans, they failed because of unanticipated spatial interactions. For example, stands slated for harvest sometimes were in basins that had been heavily cut in recent years. To do additional cutting prior to regrowth on nearby cut stands would likely result in violation of water quality standards or other environmental standards. The fact that the world does not conform to the assumptions of rational comprehensive planning came back to bite the heels of resource planners in the form of conflicts over management goals, lawsuits, and court injunctions.

What We Have Learned

Over the past decades, we have learned many powerful lessons about how the world works that show rational comprehensive planning to be of limited value. These lessons show us that quantitative analysis alone cannot resolve resource management conflicts. This chapter discusses complexity and effects of context on choice. Empirically, we have seen that the rest of society often does not accept the choices that scientific planning teams produce, despite volumes of analysis. One lesson we have learned is that systems can be surprisingly simple and still be too complex for us to predict exact conditions only moderately far into the future. Instead of evaluating outcomes of all possible actions, people choose the first apparently acceptable option or what appears to be the best of a small number of options. A second lesson we have learned is that people's value judgments depend a lot on the context in which the judgment is made—rather than being given beforehand and external to the decision-making process.

Complex Systems Are Unpredictable

The rational-planning, reductionist view of the world holds that if one just knows initial conditions well enough, one can predict exact future conditions of a system. But we have learned that small differences in initial conditions can result in large differences in subsequent conditions. Instead of small differences canceling each other out, or not having a significant effect on larger patterns, we sometimes observe small differences having spreading and increasing effects. Sometimes these effects are nonlinear, where small causes do not uniformly cause small effects or where the magnitude of responses is not in linear proportion to the magnitude of causes. Nonlinearity causes great difficulties in forest planning when planners attempt to specify the outputs and resource states resulting from alternative management choices.

For example, we have made great strides in predicting fire behavior (see Agee, Chapter 12). However, despite these advances in predictive capabilities, our predictions are accurate only at certain scales and for relatively short timeframes. Small causes sometimes produce large effects. Forest fires typically start from a small ignition source, such as a lightning strike, a campfire, a tossed cigarette, or sparks or heat produced by an engine. Most forest fires go out or are put out by people before they burn a large area. However, we only have to look at the news in late summer to know that sometimes fires get very large. Above certain temperatures, wind speeds, and fuel levels, and below certain humidity levels, fires burn hot and spread. With low winds, high moisture, and breaks in fuel, fires usually go out quickly. But the response is not linear. At some small increment of change in conditions, a fire will switch from creeping along or going out and become a raging conflagration. Nonlinearity causes particular challenges to resource management in large, rare events such as major fires, disease outbreaks, windthrow, appearance of new diseases, and spread of aggressive and invasive species.

The fire example demonstrates that local interactions are important. Particular details of ignition and heat, fuel supply and contiguity, and moisture and air all may be significantly different at points only a few feet apart. These variations in small, local interactions can make the difference between a fire burning out and building to an intensity where dynamics at larger scales are important.

Local interactions of different variables have been important as the Forest Service attempted to meet timber harvest levels projected in forest plans and developed using the linear programming model FORPLAN. The model does not consider what stands are adjacent to other stands, and when foresters tried to locate cuts while limiting the cumulative size of adjacent clearcuts, maintaining desired distributions, and meeting other standards, they found that mature timber that was scheduled by FORPLAN was not necessarily available for harvest. Newer computer programs, including various geographical information system (GIS) programs and the harvest scheduling program SNAP II (see Sessions et al., Chapter 18), which are spatially explicit, do better at addressing the spatial interactions of multiple variables. However, problems still remain with needing more detailed data than is crunchable and with the unpredictability of nonlinear interactions.

People Satisfice

Rational planning theory holds that one can predict even nonlinear effects just by getting more detailed information about local conditions and interactions. However, we have learned that even if it were possible to know interactions and initial conditions in enough detail to predict outcomes, it would not be rational to do so. Gathering detailed information and calculating outcomes requires expenditures of time and effort. The more information gathered and processed, the more time and effort required. In practice, people do not consider all possible options or all marginally relevant information. The cost of considering everything is too high. Instead, people engage in a process that is often called "satisficing" (Simon 1976). People choose a relatively small subset of information, which they believe to be most relevant and available. Similarly, they consider a relatively small set of alternative actions from the infinite set possible. Using a limited set of data and evaluating a limited set of options, people choose the most acceptable option, not what is optimal. Forest plans typically include detailed analysis of about 10 management alternatives. If none are acceptable, people consider more options or lower their standard of what is acceptable.

The combined effects of nonlinearity of interactions and limited knowledge of specific interactions illustrate how quickly uncertainty and errors compound to the point that our predictions are meaningless. The faster the system in question changes and the more specific the thing we want to predict, the shorter the time for which we will be able to make reliable predictions. We typically cannot collect enough data to make much more than commonsensical predictions.

Preferences Are Socially Constructed Through Iteration

Often conflicts over resource management are framed as issues of uncertainty about the biological effects of management. However, the primary conflict is often about social and moral values. We cannot look in a book and find a set of equations that describes the interactions of society's marginal benefit functions for each resource. Neither can we study the spending choices of every citizen, sum their individual marginal benefit functions, and come up with an equation or set of equations specifying society's valuation of various stocks and flows of resources. This is because people, while having core world views developed through their unique personal history, develop their preferences in the process of interacting with others to address a situation. People's preferences are dramatically subject to the context in which they are acting.

Public choices and preferences apply not just to public lands, but also to private lands. Actions on private lands have effects that go beyond the boundaries of those lands, ranging from contributing to aggregate log supply to affecting whether or not there are floods downstream or whether species are sustained. Because effects spill across ownership boundaries, forest practices are regulated to keep conditions within socially acceptable limits.

Wildavsky has examined where people's preferences come from. He concludes that the explanation that best fits the evidence is that "preferences must come from inside, not outside, our ways of life" (Wildavsky 1987). He argues that we make a fundamental choice when we choose who we associate with:

> [P]references are endogenous—internal to organizations—so that they emerge from social interaction in defending or opposing different ways of life. When individuals make important decisions, these choices are simultaneously choices of culture—shared values legitimating different patterns of social practice . . . [T]here are no disembodied values apart from the social relations they rationalize, and there are no social relations in which people do not give reasons for or otherwise attempt to justify their behavior . . . [P]eople discover their preferences by evaluating how their past choices have strengthened or weakened (and their future choices might strengthen or weaken) their way of life. . . .They construct their culture in the process of decision making. Their continuing reinforcement, modification, and rejection of existing power relationships teaches them what to prefer.

Put more concretely, Bachrach comments:

> Persons who in their everyday life—in their clubs, professional organizations, and social activity—have the opportunity of formulating and honing their opinions are in a position to determine where their interests lie. . . . And it is only through self-awareness that they can identify their political interests. In short, political participation plays a dual role: it not only catalyzes opinion but also creates it. (Bachrach 1975)

Many researchers have documented what we commonly experience—that people tend to support decisions that they have been involved in making more than those that they have not been involved in making (Crowfoot and Wondolleck 1990, Fisher et al. 1991, Benveniste 1989, Shafritz and Ott 1992). Formation of preferences through interaction is a partial explanation of this phenomenon.

Social construction of preferences does not imply that any state is possible if only we perceive the world correctly. Some resource flows or conditions are not biologically or physically possible, even if people prefer them.

Sources of Conflicts

For several reasons, people who interact will not necessarily form compatible preferences. The structure of the conflict may support conflict and inhibit agreement. People may have material interests that are so strong and so conflicting with the material interests of others that no solution is mutually acceptable. Competition for goods is the fundamental premise of economic analyses of conflict.

Not all preferences are likely to be changed through interactions with others. People also may have core values that conflict and hence form preferences in opposition to those with whom they interact. Preferences less central to people's beliefs are more subject to change than those that pertain to core world views (Sabatier and Jenkins-Smith 1993). Fundamentally different views of the resilience of the world and efficacy of human action are another source of deep conflict (Schwarz and Thompson 1990).

As a case in point, consider the conflict over the northern spotted owl. Arguments in this conflict reflect widely divergent views on how the world works and on what is important. Some say the owl is not endangered or that protecting individual owls is enough and protecting large areas of habitat is overkill. Typically, the implied argument for added owl protection is that owls are important because they are part of a fragile system. If we extirpate the owl, we do not know what worse things will follow. A less mainstream but still fairly frequently held view is the deep ecology perspective in which other species have as much right to exist as humans; therefore, people have no right to cause other species to go extinct. In Schwarz and Thompson's typology of world views (1990), these are both preservationist views holding that nature is fragile and ephemeral and that if people are not careful, they will cause irreparable harm. Most who argue for less protection of the owl assert that nature will do just fine with human activities (whether or not the owl goes extinct), and it would be foolish to impose significant owl preservation costs on ourselves. This world view holds that nature is benign and the world very resilient.

The conflict over the owl has gained strength from this conflict of world views. The Forest Service and other agencies have painfully learned that the most thorough technical analysis does not cause the combatants to change their world views, agree that the agency has made the best possible choice, or support agency decisions. In fact, agency efforts to use technical planning processes to address the conflict seem to have exacerbated the polarization and intensity of the fight (Yaffee 1994).

Separation of Costs and Benefits

Some conflicts occur when those who reap the benefits of a situation or action seek to impose the costs on others. For example, when logging and hydropower dams provide benefits to timber companies and electric utilities and their consumers, they may decimate fish stocks, thus imposing losses on those who fish and consume fish. Separation of costs and benefits also may take place when individuals or groups appropriate public goods through political processes without compensating the public. For example, timber interests have gotten Congress to require the U.S. Forest Service to spend millions of dollars each year subsidizing the sale of timber in the Tongass National Forest. The timber industry captures value by selling the timber, while taxpayers bear the cost of providing that timber. These conflicts are especially likely when benefits occur in the form of unmarketed environmental services—such as the hydrologic control of forests in providing clean, even flows of water.

Empirical study has shown that biological resources are much more likely to be managed sustainably if those who benefit from management have to live with those who bear the costs (Ostrom 1990). If the recipient of benefits is the same as the bearer of costs, then that actor is unlikely to take an action unless the benefits are greater than the costs.

Where Do We Go from Here? Learning from Experience and Developing Goals

Robust strategies for resource management must deal with unpredictability and the social construction of preferences. Moreover, they must accommodate limits on time, expertise, and money. Toward these ends, a three-part approach to resource management is proposed here: (1) probing the problem and possible solutions, (2) analyzing the nature of the problem to determine if it is being addressed in a suitable forum, and (3) if the forum is not suitable, attempting to switch interaction processes or reformulate the problem.

Probing Problems

Often scientists assert that scientific experimentation or formal processes of adaptive management are needed to successfully manage complex resources (Lee 1993, Naiman 1992, Noss and Cooperrider 1994). While these approaches are extremely valuable, for many resource management situations, they are not appropriate. Formal experimentation takes a long time, is expensive, and risks important variables being held constant or not measured. Adaptive management measures the effects of management, and thus is more likely than experimentation to detect important but unanticipated interactions between variables. However, formal adaptive management still requires expensive data collection and scarce quantitative analysis skills.

Charles Lindblom (1990) has examined how individuals and societies learn, persist, and change. He concludes that most social problem solving is done through social interactions where choices and agreements are formulated and evaluated through informal processes of observation, hypothesizing, and data comparison—what he calls "probing." Lindblom asserts that competition among ideas can facilitate probing, and that probing by the public as well as employees of public and private organizations is central to grappling with social problems.

People widely engage in probing, bringing all available knowledge to bear on a problem. Specialized skills and knowledge help, but effort and collaborative questioning go a long way. The probing that Lindblom describes is a widely applicable strategy for resource management. Because it is open to all available information, probing sometimes can avoid oversights caused by excluding information from outside a strictly defined scientific discipline.

The iterative aspect of adaptive management provides another strategy for increasing the robustness of decision making. Because the outcomes of complex systems are unpredictable, ongoing monitoring allows adjustment of activities to mitigate unplanned and undesirable outcomes. Empirically, this iterative approach appears to be important to maintaining the productivity of resources (Ostrom 1990).

Iterative management also allows for adaptation when goals change, which they will do because they are socially constructed. Also, relative abundances of resources and technologies for using resources change—raising or lowering the value of individual resources. Changes in the value of resources—including defining as resources things that were not considered such before—changes management goals. Hence, in subsequent iterations, different outcomes would be sought.

Interaction Processes for Decision Making

Probing of social choices and management problems requires forums in which to gather information, question proposed explanations and choices, and make decisions. People have created diverse forums for decision making. Synthesizing from the experience of hundreds of efforts to manage common pool resources around the world, Ostrom (1990) concluded that interaction processes that successfully manage natural resources have the following attributes:

- Defined bounds of the resource and access to the resource
- Resource use rules congruent with the local situation
- Affected people participating in making resource use rules
- Monitoring of access
- Graduated sanctions
- Conflict-resolution mechanisms
- Legitimization by external authorities
- Integration with organizations performing larger- and smaller-scale functions

These elements may be viewed as variations on three dimensions of forums or interaction processes: (1) strength of bounds, (2) internalization of effects, and (3) adaptability through time. Bounds of systems include definition and knowledge of the biological and physical parameters of resources being managed. Bounds also describe limits on access to resources and actors' participation in making or enforcing rules for resource use. Internalization of effects addresses whether those who take actions must account for or bear the consequences of actions. Lack of internalization may result in people taking action even when the costs are much greater than the benefits (Samuelson and Nordhaus 1985). Adaptability through time refers to the capacity to observe changes in resource conditions, dynamics, or values and make appropriate changes. The experiences of hundreds of cases examined by Ostrom and others have shown that the productivity of resources is more likely to be sustained if the situation has a moderately high degree of boundedness, a fairly high degree of internalization, and a moderate degree of adaptability. However, not all resource management problems have these attributes. There may be a way of addressing a problem even though it does not have optimal attributes.

These three dimensions—boundedness, internalization, and adaptation—can be used to evaluate the suitability of an interaction process for addressing a specific resource management issue (Table 27.1). Not considering adaptation through time, interaction processes can be mapped by their relative suitability for addressing problems with varying degrees of boundedness and internalization of effects (Figure 27.1). This can help both in choosing a forum appropriate to an issue and in addressing issues that have been brought to a given forum.

Table 27.1 Attributes of different interaction processes for resource decision making

Interaction Process	Boundedness	Internalization	Adaptation
Unitary Ownership	Strictly limited to owners, as enforced by the state	Strong, if property geographically encompasses effects	Depends on skills and resources of owner
Authoritarian Rule	Limited to ruler	Limited; ruler can impose costs on subjects	Low, because of centralized decisions
Consensus	Limited to parties and issues agreed to by participants	Strong for issues affecting participants	Can be fast, but may be slow if requiring major change in agreements
Mediation	Typically defined by mediator, who selects actors with perceived power	Strong for issues affecting participants	Typically has limited duration, thus limited
Courts	Limited by scope of law	Variable, depending on who has standing	Strongly limited by law and precedent
Scientific Peer Review	Limited to experts; favors actors with data	Strong when quantifiable effects are observable	Moderately slow and not necessarily linking knowledge and action
Public Relations	Open to all with skill or resources or funds to purchase services	Variable because of varied access to exposure and emotional power of stories	May occur quickly after catalytic event
Legislative Processes	Open to citizens, but weighted to those with money and evocative stories	Variable, but may be high	Often slow because of durability of coalitions
War	Open to all	Low, with focus on winning and imposing costs on others	Low, with respect to maintaining natural resources

Boundedness

There are multiple dimensions by which boundedness may be assessed. Problems are unbounded when goals are not agreed upon, when it is not clearly defined who does and does not have a say in setting goals, and when the effects of management actions are unclear. Lack of agreed-upon goals is easy to observe and is often evidenced by arguments about what problem should be addressed. In the case of the spotted owl, some people argue that the problem is how to preserve ancient forests; some argue that it is how to provide a large supply of cheap logs to mills; and still others argue that participation is about how to provide steady, high-wage jobs in rural areas. Although it is often not difficult to assess who has a say in setting goals, it sometimes changes in unexpected ways. In clearly defined administrative processes, such as court proceedings, written rules specify who can participate and in what capacity. In issues concerning clearly defined property rights,

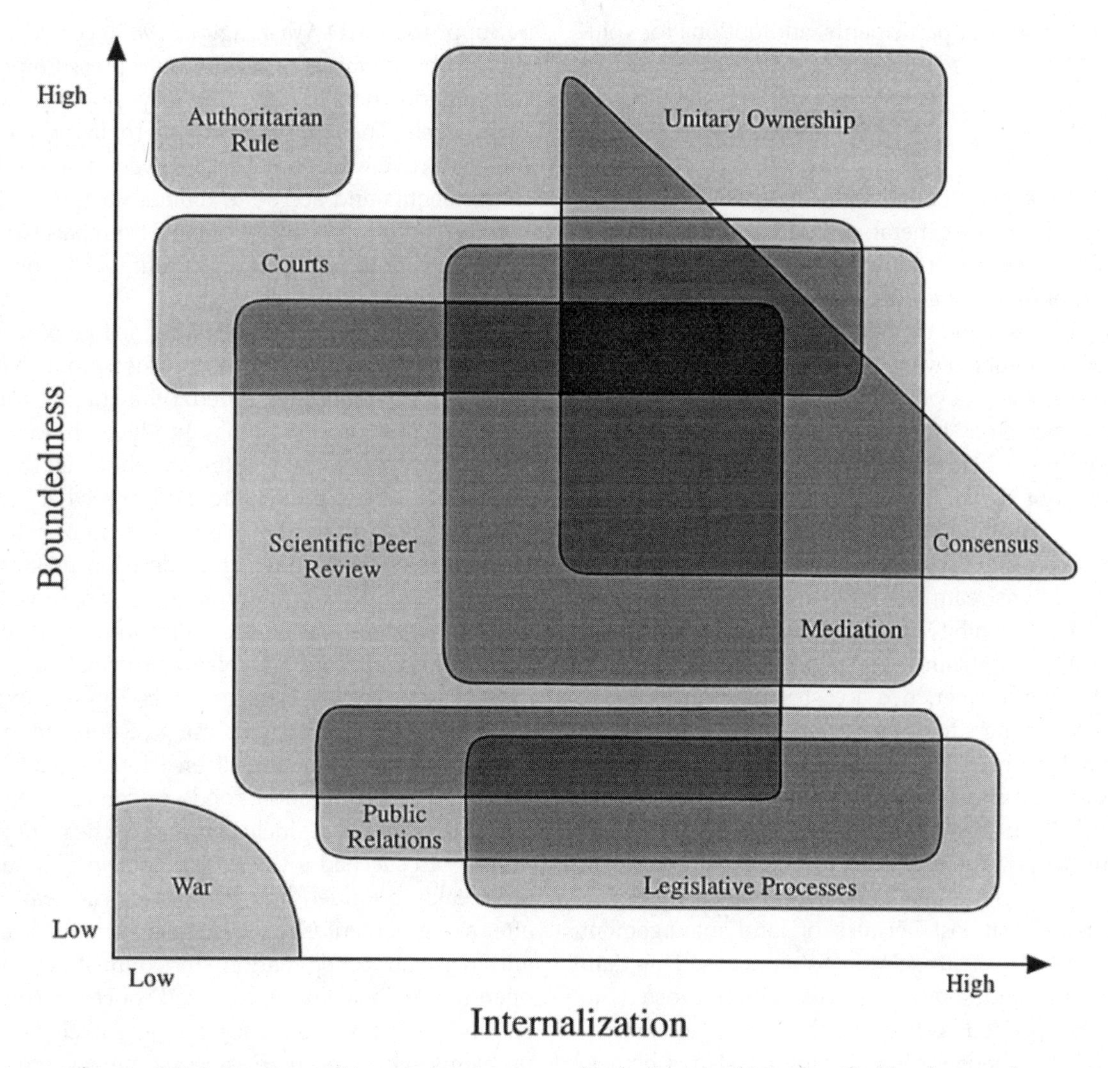

Figure 27.1 Typology of interaction processes for decision making classified by degree of boundedness of the forum and internalization of effects of actions.

participation is defined by law or court precedent. When anyone can come and participate, such as in some hearings processes, media events, and public protest events, participation is weakly bounded.

Where there is agreement about who participates in goal setting, scientific peer review or similar technical processes often can settle a problem. If many players can and do attempt to be involved in decision making, contests are likely to play out in forums with relatively open access, such as legislative bodies and public relations campaigns. When the effects of management occur in the distant future or are clouded by many other causes, there may be little alternative to addressing the problem in a political arena.

If problems and goals are clearly defined and agreed to, then standardized bureaucratic processes usually can work. If a standardized process works, it typically works faster and with less cost than an open process. If the nature of the problem is not agreed to, or if participants have conflicting goals, then open processes (with ground rules of civil behavior) have a higher likelihood of creating agreement. Typically, open processes have relatively little bounding and involve investigating the nature and extent of the problem, the values of participants, and options for solutions.

Internalization

Internalization is not as directly observable as boundedness. The most general approach to assessing internalization is determining who gets what benefits from a condition or action, and who bears what costs. However, costs and benefits are often hidden. More-obvious benefits may include rights to use or remove resources. Less-obvious benefits might be indirect subsidies, such as holding the cost of access to natural resources below market rates. Depending on allocations of rights, avoided harms may be a cost to the actor who must avoid causing harm or a benefit to those having a right to be free from harm. For example, downstream landowners may benefit from rules that prohibit upstream owners from doing things that might increase floods, but downstream people would experience a cost if they have to pay upstream owners to not engage in flood-causing activities. Like benefits, costs can be direct or indirect, marketed or unmarketed. Direct, marketed costs are readily observable and include things such as users having to pay for access to resources. Unmarketed costs may include health damage caused by pollution, recreation lost because of land management choices, or losses to salmon fishers caused by dam operations and forest practices. Often those who bear unmarketed costs make their existence known through the media or by lobbying legislative bodies. But some victims may not even know the source of costs, such as people who suffer cancer years after exposure to radioactive fallout.

Typically discussed as "tenure" issues, some people argue that individual ownership is necessary to internalize costs and thus sustain natural resources. More important than individual ownership, however, is that both rights to access and benefits be clear, bounded, and stable (Ostrom 1990). In addition, affected parties must have secure tenure in whatever is affected by the management actions of others (Dasgupta 1982, Romm 1993). For example, when Indian tribes in the Pacific Northwest obtained a court decision stating that the tribes were entitled to have fish available for harvest (*United States v. Washington,* 506 F. Supp. 187 (W.D. Washington, 1980), cert. denied), those tribes obtained a legal tool to pressure private forest landowners to harvest in ways not likely to extirpate fish. The approach of internalizing costs is most effective when costs are in a very different form from benefits and accrue to different actors (such as workers suffering health problems from participating in a production process that is profitable to company owners).

Relatively highly bounded interaction processes, such as in the courts, may sometimes produce decisions that can be implemented. While highly bounded interaction processes may produce choices with institutional or social legitimacy, they tightly prescribe who may participate and on what issues. When a situation is highly bounded, some who are affected may be excluded from decision making. By avoiding exclusion, unboundedness can serve as a guard against failures of internalization. Those seeing that they are affected are able to participate.

Mediation and consensus processes are currently popular approaches to addressing resource management problems involving diverse interests and unmarketed environmental goods or services. Despite the appeal of these inclusive approaches, they are unlikely to succeed if the problems being addressed are weakly bounded and the parties can wait until after an agreement is reached, then enter the process and demand renegotiation. In contrast, by being open to previously unrepresented concerns, consensus and legislative processes are more likely to settle problems which occur because some outcome has not been internalized.

Increasing the internalization of costs may reduce boundedness. Often, some parties are able to externalize the costs of their actions by bounding the situation so that those who bear the costs have no say in the process. The Indian fishing example illustrates this point. Before the court decision granting tribes protection of fish habitat and populations, forest practice decisions excluded Indians who were concerned that logging would reduce fish populations. The logging operations were forced to internalize costs by changing the bounds of the debate; the courts gave tribes who fished downstream from areas proposed for logging standing to participate in decisions about forest practices. Common ways of bounding problems include legal recognition of the authority of actors to use resources in certain ways and the establishment of legal definitions limiting who has standing to seek action.

Boundedness, however, may slow adaptation. Highly bounded processes often can hinder the use of new information or can fail to take advantage of new opportunities to solve problems. An adaptive organization adjusts goals and procedures in response to changes in resource conditions or demands. Even adaptation is not an absolute good. Very high rates of adaptation may hinder organizational effectiveness by escalating costs more than necessary and breaking down institutional memory or working relationships between individuals in those organizations. Also, high rates of adaptation may impede completion of long-term management activities.

While extremely high rates of internalization, boundedness, and adaptation can have negative consequences, the consequences are typically worse in cases with extremely low rates. War is the ultimate unbounded system, being without internalization of effects on others.

Changing Interaction Processes and Restructuring Conceptions of Problems

Not infrequently, people find themselves in interaction processes ill-suited to addressing the resource management problems they face. In practice, efforts to change the kind of interaction process used to address a problem often makes the problem more or less bounded or internalized. Switching interaction processes, or restructuring conceptions of a problem, often takes substantial time or money or requires the extraordinary skill of an individual who can opportunistically use other events as catalysts for change.

Usually the switching of interaction processes is more a matter of getting an issue addressed in a new forum than of stopping an old process. An actor with time, energy, and skills can often transfer an issue to a new interaction process by framing it in terms relevant to that new process. Less-bounded interaction processes are, by definition, easier to enter. For example, even if a problem does not have large economic consequences, it may get on the agenda of a legislative body through use of an emotionally or symbolically powerful story. In contrast, courts are highly bounded. Before they will address a case, a plaintiff must claim that a specific law has been violated and harm done to a specific legal entity.

Using an interaction process that does not fit with existing social processes can be problematic. For example, just as it is difficult to get litigants to accept mediation, it would be difficult in a democratic nation to establish an authoritarian ruler to address a natural resource management problem. If property rights encompass the problem, one could establish unitary ownership, but natural resource issues often involve conflicting rights and few contestants have the wealth required to buy out all others. This strategy is sometimes used, however, and it is a major one for the Nature Conservancy, a land trust that uses donations to purchase land parcels important for maintaining biological diversity.

An alternative to switching interaction processes is restructuring the definition of the problem. A number of researchers have focused on interaction processes in which people and organizations interact and choose actions (Stanley 1990, Bachrach and Baratz 1970, John 1994, Benveniste 1989, Majone and Wildavsky 1978, Smith et al. 1995). Restructuring may be accomplished by changing how participants perceive the situation or perceive other participants. Any issue can be viewed from multiple perspectives, and restructuring an issue persuades people to see it another way (Stone 1988). For example, environmentalists have used public relations, political efforts, and education to change broadly held perceptions of old-growth forests. They largely have been successful at redefining the issue from replacing decadent old stands to maintaining ancient forest as a national heritage and a reservoir of biodiversity and environmental services.

Consensus groups also are touted as means for redefining problems. Through interaction in consensus groups, participants often change their views on the other actors and on the problems they collectively face. Much has been written about how to structure interactions to facilitate the development and adoption of social choices that serve participants well (Fisher et al. 1991, Susskind and Cruikshank 1987, Cormick 1989, Smith et al. 1995). Typically, consensus groups are formed only if adversaries can find common goals, such as sustaining the productivity of a resource or sustaining the persistence of the actors themselves. Perception is (at least in part) shifted from viewing others as competitors to seeing them as fellow members of a group with important shared goals.

Trust among actors and symbolic reasoning are very important to structuring and restructuring people's conceptions of problems. Most investigators of interaction processes for social decision making emphasize the importance of repeated interaction between actors for developing trust and commitment to actions (Firey 1977, Moote et al. 1993, Korten 1981, Landy and Plotkin 1982). Putting it succinctly, natural resources are more likely to be sustained when it is clear who gets the benefits from resources and when those who get the benefits have to live with those who bear the costs.

Managing Resources by Learning and Constructing Goals

There are many examples of efforts to manage natural resources that facilitate learning about the resources and using social interactions to choose management goals. While many of these efforts can claim some success, none appear to be entirely without criticism. Brief descriptions of two of these efforts are presented here, with a focus on developing resource management goals and dealing with uncertainty through an explicit learning process.

Northwest Power Planning Council

The central purpose of the Northwest Power Planning Council is to balance fisheries concerns with hydropower generation on portions of the Columbia River basin in Washington, Oregon, Idaho, and Montana—an area covering more than 160 million acres. Major uses of the river and its riparian areas include hydropower generation, irrigation, navigation, flood control, fish production, wildlife habitat, and recreation.

The council was created by the federal Northwest Electric Power Planning and Conservation Act of 1980 and was given the mission to "develop and adopt . . . a program to protect, mitigate, and enhance fish and wildlife, including related spawning grounds and habitat." The organic legislation also directed the council to treat the river as an ecosystem and consult with interested groups, including tribes. Congress took this action in response to pressure to reverse the decline in salmon runs in the Columbia and its tributaries. Runs have declined from an estimated 16 million fish per year before white settlement to about 2.5 million fish when the act was passed to less than 1.5 million fish in the early 1990s (Northwest Power Planning Council 1992). Congress perceived that dam operators were focused on producing high revenues from power generation with little concern for the resulting loss of fish. Creating the council reduced the bounding of the interaction process by giving a stronger role to tribal and environmental interests, but it also bounded the dispute to technical and scientific discourse and consultation between bureaucratized organizations. The council represents an attempt to move a conflict from a national legislative process to a regional scientific and collaborative process.

The council has assets and authority that give it influence in the Columbia River Basin. The Northwest Power Planning Act directs federal agencies to take the council's plans "into account at each relevant stage of decision-making processes to the fullest extent possible." This gives the council leverage to get other agencies to implement the plans it develops. The council receives federal funding to pay for technical staff, data gathering, and purchase of land and services for fish and wildlife habitat protection. The council also has authority over the amount of water to be added to river flows to benefit fish—which decreases revenue from power generation. Recent costs of council programs, including forgone power revenues, have been about $160 million per year. The council has developed sophisticated models of water flows, power demand, and economic effects of flow and power generation regimes. In addition, it conducts or contracts for research on related topics, including assessments of the effectiveness of various fish enhancement practices.

In correspondence with what we have learned about resource management in recent decades, the council uses multiple approaches for learning about the environment, ranging from experimental research to probing of problems. Council staff or contractors conduct research on water flow and hatchery effects of wild fish stocks using experimental and quasi-experimental research designs. The council addresses the unpredictability of the complex Columbia River by managing adaptively, measuring the effects of barging fish around dams, of lowering reservoir levels to speed fish travel to the ocean, and of screening dam intakes to divert fish from passing through turbines. It monitors fish populations, water flows, and changes in demand for electricity. It probes for previously uninvestigated ways of increasing wild fish populations while continuing to generate power, irrigate farmlands, and maintain water levels appropriate for barge traffic.

The council also uses information sharing and iterative interaction with other actors in an effort to construct social agreements about the nature of problems and desirable actions. It circulates draft reports and projections for public comment. Moreover, data and copies of the council's computer models for fish and power production are publicly available, along with technical support for those who wish to use them. The council is required by law to adopt comments by other agencies into its plans or to explain why it does not. Council staff expend significant effort involving state and local governments (including fish, wildlife, and land management agencies) and tribes in its planning activities. There also appears to be significant informal interaction between council members and staff and individuals from other organizations. All of the larger groups of private users of the river are organized and have representatives who interact with council staff on a regular basis.

The council has attempted to internalize the environmental costs of decreasing fish runs, but has met strong resistance, particularly from large, industrial power users (Direct Service Industries, Inc. 1992). The power users object to losing the economic benefits of engaging in practices that reduce fish runs. As a result, the council has had very limited success in securing tribal tenure rights to fish. However, the council does appear to have contributed to a perception that no group of river users should be entirely driven out by other users.

The problem of declines in fish stocks is clear, but there is less agreement about how to solve the problem. The council uses structured planning and research processes to measure causes of fish declines and the effectiveness of efforts to increase fish runs. The council uses a combination of scientific and consultative processes to address apparently irreconcilable conflicts between river users. The political sensitivity of council members contributes to the council's capacity to use these processes along with informal interaction between staff and other actors.

The council also appears to be able to adapt over time, within the focus of its statutory mandate. Members and to some degree staff may add new issues to the agenda, and previously uninvolved actors may participate through formal comment on draft products and informal interaction with council staff.

Credibility has been built through use of explicit processes for gathering and analyzing data and through ongoing interaction with interested parties. This credibility appears to have contributed to the agencies' implementing plans developed by the council, and plan implementation appears to have increased since the council was created. However, despite the relatively substantial resources of the organization and its strong commitment to gathering data to develop resource management knowledge, there is still significant resistance to paying the additional management costs involved in gathering the data necessary to manage adaptively (Lee 1993). If formal adaptive management is difficult in this organization, it is likely to be even more difficult in other, less well endowed organizations. While fish runs may not have declined as much as they would have if the council had not acted, most runs are not recovering. Some environmentalists oppose experimentation because they believe it results in fish loss that could be avoided if stringent, but little-tested, fish-protection activities were undertaken. Finally, the council appears to have done little to involve actors who are not organized to be active in public, bureaucratic processes.

Timber, Fish, and Wildlife Interdisciplinary Teams

The Timber, Fish, and Wildlife Agreement of Washington State (TFW) was established in 1987 after six months of mediated negotiation between state agencies, Native American tribes, environmental organizations, and the forest products industry. TFW is an informal, unsigned agreement to make specific changes in state forest practices laws and rules, and sets forth processes for resolving disputes over forest practice activities on state and private lands. TFW negotiations were undertaken to address an intensely litigated controversy about what forest practices would be acceptable and how to limit harm to anadromous fish caused by forest management activities. The underlying idea was to shift the dispute from the courts to consensus committees and scientific processes. A central driver in the controversy was tribal rights to continue to have fish in the rivers available for harvest. Other contributing factors included efforts of the state Department of Natural Resources (DNR) to enforce uniform forest practice regulations across the state, and efforts by environmentalists to change forest practice regulations to provide more protection to fish and wildlife. These efforts decreased the boundedness of the prior conflict and set the stage for the TFW agreement. The costs and delays of continuing litigation, plus the possibility of losing in court, were great enough to bring the parties to the negotiating table.

Participants in the TFW negotiation agreed to cooperatively seek processes and forest practice regulations that would provide more protection to fish and wildlife while allowing landowners to profitably grow and cut timber. The agreement includes proposed changes to allowed forest practices that focus on protecting riparian areas and avoiding the negative cumulative effects of harvesting. It also specifies a new process for reviewing, conditioning, and approving applications to conduct forest practices made by private forest owners to the DNR. The agreement further specifies future cooperative activities aimed at developing new practices that are compatible with fish and wildlife conservation. Finally, provisions were included to review and adapt implementation of the agreement (TFW 1987). The agreement transformed the actors from antagonists to collaborators. The state legislature and DNR endorsed the agreement, and TFW cooperators were charged with implementing it.

An important new element of the forest practice application review process was the use of interdisciplinary teams, often referred to as ID teams. The DNR assembles these ID teams to conduct on-site review of proposed practices that "have a potential for a substantial impact on the environment" (Washington Forest Practices Board 1993). ID teams include professionals in such fields as fisheries, wildlife biology, geomorphology, hydrology, soils, and forest engineering. In addition to DNR staff, other participants on ID teams include habitat biologists from the state Department of Fisheries and Game, and staff from the Department of Ecology who deal with forest practices. Participants in the TFW agreement can have representatives on ID teams, and participation of technical experts from industry, tribes, and other organizations is welcomed (TFW 1987).

Tribal and environmental specialists are frequently invited to participate on ID teams, and tribes have obtained significant permanent funding for participation. This involvement significantly expands participation in decisions about sensitive forest practice applications and greatly increases site-specific knowledge and conditioning of permits through site visits. ID teams offer a scientific and technical process for addressing disputes on a site-specific basis. The presumption is that experts representing multiple values can agree to management actions that internalize environmental costs rather than simply maximize timber revenue. ID teams also bound the conflict, effectively limiting participation to technical experts representing the TFW cooperators. They moved disputes from legal, legislative, and public relations efforts to an avowedly technical process.

On the suggestion of the ID team, DNR may attach recommendations to forest practice permits beyond those legally required of the landowner. Typically, these recommendations are negotiated with the landowner, but compliance is not legally enforced and there are no provisions for monitoring (Halbert and Lee 1990).

ID teams embody the important element of constructing social agreement through ongoing interaction. Team interactions in the woods both provide an informal setting as well as opportunities to gather otherwise unavailable data in the form of on-site observations by technical experts. The informality of interactions and the cooperative background of the TFW agreement should contribute to flexible, creative thinking about solving resource management problems.

Only tangentially may ID teams contribute to defining secure tenure for a full range of economic and environmental values. The TFW agreement itself was an attempt by all participants to reduce uncertainty about future forest practice regulations. ID teams may reduce some of the uncertainties about forest practices that landowners had faced prior to the TFW agreement. However, with the TFW agreement already adopted, each ID team site visit becomes a forum to negotiate what activities will occur on that site within the broad latitude of regulations. Depending on the inclination and social skills of participants (especially the DNR forester who calls the team meeting), the interaction may be more or less competitive or cooperative. It does not, however, necessarily internalize environmental costs in forest management decisions, nor does it guarantee satisfaction of participants' needs and goals. While the DNR forester handling the forest practice application may invite technical specialists to participate on an ID team and while TFW cooperators are granted a place in the site-specific negotiation, others are excluded without consent of the landowner. Still, the structure of the ID team process is clear and allows flexibility to address previously unanticipated biological and physical concerns.

The ability of actors who were not original TFW cooperators to join the process is highly constrained. The ability of cooperators to address issues not included in the original TFW agreement is limited by the unlikelihood of reconstructing the intensity and commitment felt by original TFW participants, but this could be supported as the group's history of cooperative action grows. Participation on ID teams, however, strains the resources of environmental organizations, which have difficulty funding professional staff positions to participate in dozens of reviews across the state. Also, environmentalists frequently feel that their concerns are not equally addressed in forest practice permits (Wondolleck and Yaffee 1994).

As positions on ID teams are filled with individuals new to the TFW process and without the same commitment of early members and as new conflicts arise that are not addressed through TFW processes, bases for cooperation on ID teams will erode. The exclusiveness of TFW processes opens it to criticism that it delegates protection of public resources to a select group of participants who use processes that are not open and publicly accountable. Still, despite these criticisms, TFW cooperators have much more information about and involvement in forest practice decisions than similar actors in other states.

As with the Northwest Power Planning Council, many of the attributes of TFW ID teams correspond with what we have learned in recent decades about robust strategies for natural resource management. ID teams are one of several approaches used by TFW cooperators to address unpredictability and the need for local interactions; other approaches include cooperative research and joint monitoring programs. ID teams are a loosely constrained approach to integrating locally occurring interactions between biological and physical elements of ecosystems. Ideally, ID teams bring together expertise from a range of disciplines to probe the dynamics of a particular site, make professional judgments about effects of various resource management activities, and recommend actions tailored to local conditions. However, there appears to be no regular mechanism for monitoring results and learning from experience. In theory, ID teams are very adaptable. But as the political context changes, and different actors seek to participate in addressing cross-jurisdictional resource management issues, it is unclear whether ID teams will be able to adapt and whether participants will continue to view them as the best tool for addressing their problems.

Conclusion

In natural resource management, we have come a long way from rational comprehensive planning, with its strong assumptions of predictability and the social power of technical processes. In addition to socially constructing goals and facilitating fast learning about the conditions of complex systems, robust strategies for operating under uncertainty include preserving options to increase the likelihood of having desirable alternatives when unforeseen events occur. While the processes of interaction and adaptation may look messy to believers in rational planning, empirical evidence suggests that these methods are more likely to succeed in dealing with the complexity of forest ecosystems and the diversity of views and values of the people who are concerned with and depend on forests. In many places throughout the world, people are using these processes to address natural resource conflicts: the Great Lakes Water Quality Authority's attempts to deal with nonpoint and transnational pollution; the Chesapeake Bay Program's work with homeowners across several states to improve the water quality of the bay; the work of interdisciplinary teams in Washington State to examine site-specific issues in forest management; and the Northwest Power Planning Council's willingness to make its computer models available to all contestants in the controversy over fish, power, irrigation, and other resource issues in the Columbia River Basin. For all their difficulties, these approaches appear to address uncertain situations and conflicts over resources more effectively than rational planning approaches.

Literature Cited

Bachrach, P. 1975. Interest, participation, and democratic theory. In *Participation in politics,* ed J. R. Pennock and J. W. Chapman. New York: Lieber-Atherton.

Bachrach, P., and M. S. Baratz. 1970. Power and poverty: *Theory and practice.* New York: Oxford University Press.

Benveniste, G. 1989. *Mastering the politics of planning: Crafting credible plans and policies that make a difference.* San Francisco: Jossey-Bass Publishers.

Cormick, G. 1989. Strategic issues in structuring multiparty public policy negotiations. *Negotiation Journal* 5(2):125–132.

Cortner, H. J., and D. L. Schweitzer. 1983. Institutional limits and legal implications of quantitative models in forest planning. *Environmental Law* 13:493–516.

Crowfoot, J. E., and J. M. Wondolleck. 1990. *Environmental disputes: Community involvement in conflict resolution.* Washington, DC: Island Press.

Cubbage, F. W., J. O'Laughlin, and C. S. Bullock III. 1993. *Forest resource policy.* New York: John Wiley & Sons.

Dasgupta, P. 1982. *The control of resources.* Cambridge, MA: Harvard University Press.

Direct Service Industries, Inc. 1992. *Northwest Salmon: Recovery in the balance.* Folder of position papers and reprints. Portland, OR.

Firey, W. 1977. *Man, mind, and land: A theory of resource use.* Westport, CT: Greenwood Press.

Fisher, R., W. Ury, and B. Patton. 1991. *Getting to yes: Negotiating agreement without giving in.* New York: Penguin Books.

Halbert, C. L., and K. N. Lee. 1990. The Timber, Fish, and Wildlife Agreement: Implementing alternative dispute resolution in Washington state. *The Northwest Environmental Journal* 6:139–175.

Hunt, F. A. 1986. We asked for it: Or did we? *American Forests* 92(10):19–21, 60–63.

John, D. 1994. *Civic environmentalism: Alternatives to regulation in states and communities.* Washington, DC: CQ Press.

Korten, D. C. 1981. The management of social transformation. *Public Administration Review* 19(Nov./Dec.):609–618.

Landy, M. K., and H. A. Plotkin. 1982. Limits of the market metaphor. *Society* 19(4):8–17.

Lee, K. N. 1993. *Compass and gyroscope: Integrating science and politics for the environment.* Washington, DC: Island Press.

Lindblom, C. E. 1990. *Inquiry and change: The troubled attempt to understand and shape society.* New Haven, CT: Yale University Press.

Majone, G., and A. Wildavsky. 1978. Implementation as evolution. *Policy Studies Review Annual* 2:103–117.

May, P. J. 1992. Policy learning and failure. *Journal of Public Policy* 12(4):321–354.

Moote, A., S. Burke, H. J. Cortner, and M. G. Wallace. 1993. *Principles of ecosystem management.* Tucson: Water Resources Research Center, University of Arizona.

Naiman, R. J., ed. 1992. *Watershed management: Balancing sustainability and environmental change.* New York: Springer-Verlag.

Northwest Power Planning Council. 1992. *Columbia River Basin Fish and Wildlife Program: Strategy for salmon.* Vol. I. Portland, OR: Northwest Power Planning Council.

Noss, R. F., and A. Y. Cooperrider. 1994. *Saving nature's legacy: Protecting and restoring biodiversity.* Washington, DC: Island Press.

Ostrom, E. 1990. *Governing the commons: The evolution of institutions for collective action.* New York: Cambridge University Press.

Romm, J. 1993. Sustainable forestry, an adaptive social process. In *Defining sustainable forestry,* ed. G. H. Aplet, N. Johnson, J. T. Olson, and V. A. Sample. Washington DC: Island Press.

Sabatier, P. A., and H. C. Jenkins-Smith. 1993. *Policy change and learning: An advocacy coalition approach.* Boulder: Westview Press.

Sample, V. A. 1990. *The impact of the federal budget process on national forest planning.* New York: Greenwood Press.

Samuelson, P. A., and W. D. Nordhaus. 1985. *Economics.* 12th ed. New York: McGraw-Hill.

Schwarz, M., and M. Thompson. 1990. *Divided we stand: Redefining politics, technology, and social choice.* Philadelphia: University of Pennsylvania Press.

Shafritz, J. M., and J. S. Ott. 1992. *Classics of organization theory.* Belmont, CA: Wadsworth Publishing Company.

Simon, H. A. 1976. *Administrative behavior: A study of decision-making processes in administrative organization.* 3rd ed. New York: Free Press.

Smith, G., C. Robinson, and M. Shannon. 1995. *Institutional strategies for managing resources across jurisdictions and ownerships: A theoretical assessment of ten cases.* Seattle: Institute for Resources in Society, University of Washington.

Stanley, M. 1990. The rhetoric of the commons: Forum discourse in politics and society. In *The Rhetorical Turn,* ed. H. W. Simon. Chicago: University of Chicago Press.

Stone, D. A. 1988. *Policy paradox and political reason.* Glenview, IL: Scott, Foresman and Company.

Susskind, L., and J. Cruikshank. 1987. *Breaking the impasse: Consensual approaches to resolving public disputes.* New York: Basic Books.

TFW. 1987. *Timber/Fish/Wildlife Agreement: A better future in our woods and streams.* Final report. Olympia, WA: State Department of Natural Resources.

Washington Forest Practices Board. 1993. *Washington forest practices: Rules, board manual, and act.* Olympia, WA: Department of Natural Resources, Forest Practices Division.

Wildavsky, A. 1987. Choosing preferences by constructing institutions: A cultural theory of preference formation. *American Political Science Review* 81(1):3–21.

Wondolleck, J. M., and S. L. Yaffee. 1994. *Building bridges across agency boundaries: In search of excellence in the United States Forest Service.* Ann Arbor, MI: School of Natural Resources and Environment, University of Michigan.

Yaffee, S. L. 1994. *The Wisdom of the spotted owl: Policy lessons for a new century.* Washington, DC: Island Press.

28 Open Institutions: Uncertainty and Ambiguity in 21st-Century Forestry

Margaret A. Shannon and Alexios R. Antypas

The quest for certainty is a quest for peace which is assured, an object which is unqualified by risk and the shadow of fear which action casts.

—*John Dewey ([1929] 1960)*

The Dilemma of Uncertainty and Ambiguity in Modern Society 437
Building Social Institutions in the Image of Science 439
How Managers and Citizens Became More Like Scientists: Learning from Elders or Experiments? 440
Managing: From Routine or Through Experiments? 441
Creating the Experimenting Society 442
Civic Science and Adaptive Management 443
Literature Cited 444

Uncertainty is the resource supporting and creating the modern economic state. When economic theorists argued two centuries ago that land and natural resources should be valued beyond the confining limits of biological productivity, they loosened land and natural resources from the certainty of predetermined allocations linked to fixed patterns of social relationships.

Ambiguity is the resource supporting and creating the modern pluralistic political state. When political theorists argued two centuries ago that political preferences and interests could shift regardless of the bounds of material needs or bonds of kinship, they loosened the political order from the constraints of rigid class boundaries and fixed relationships of superiority and subordination.

The Dilemma of Uncertainty and Ambiguity in Modern Society

We just want some certainty as to what the requirements are so that we can follow them.

—Private forest landowner, Washington State

> I thought multiple use was difficult to implement, but ecosystem management is impossible—what am I supposed to do?
>
> —Federal forest manager, Washington State

Landowners, resource managers, foresters, policy makers, and citizens experience uncertainty and ambiguity every day. Most of them have frustrations that are focused on a perceived loss of clear direction. For many foresters, the purpose of the forests had seemed clear: produce a continuous supply of timber and support local economies in doing so. It had seemed that the priorities among resource uses was clear: produce timber and provide for other consistent uses. But a shift in perspective—for example to wildlife management—muddied this clarity with other values and objectives. Moreover, society has continued to add objectives for using, conserving, and protecting forest resources.

Recognition of this dilemma of many objectives led to the idea of multiple use and to efforts to develop management models capable of analytically solving multiobjective problems. However, this approach did not address the maintenance of the whole system; it simply allocated the parts to their highest and best uses, often measured by their contribution to economic measures of public benefit. In recent years, concerns for the whole ecosystem, as expressed in the concepts of biodiversity and sustainability, have become increasingly popular.

One interesting story of the last 30 years is the emergence of new institutions rising from concerns for the maintenance of the whole ecosystem and a rejection of the technicism created by replacing political debate with analytical models. Technicism is defined as "a state of mind that rests on an act of conceptual misuse, reflected in myriad linguistic ways, of scientific and technological modes of reasoning. This misuse results in the illegitimate extension of scientific and technical reasoning to the point of imperial dominance over all other interpretations of human existence" (Stanley 1981). We must be cautious, however; in forestry, the gloss of technicism is only now beginning to wane. For some, the newest definition of the problem—ecosystem management—can still be a technocrat's dream: knowledge of the entire forest system can lead to methods for assuring controlled conditions, both biophysical and social, which generate predictable outputs.

There is, however, another possibility inherent in the concept of ecosystem management when the lessons of ecology are truly taken to heart. Ecology teaches not a technology of control, but rather a more humble, tentative, and interpretive approach to understanding relationships and the human role in ecosystems. Moreover, since every ecosystem is unique, ecology's focus is on the concrete and the particular, raising diversity to the level of a virtue rather than treating it as an unwelcome complication.

This chapter takes up the twin problems of ambiguity and uncertainty in the context of forest ecosystem management in the 21st century. Ambiguity in modern society is rapidly increasing as specialized organizations find that they must reconfigure themselves to effectively address the new problems of sustaining ecosystems and conserving biodiversity. Likewise, uncertainty is growing because predictability of outputs is no longer the focus in these new problems; rather, maintenance of relationships is now of central concern to forest managers.

The good news is that these shifts in how problems are defined can enhance the humanistic qualities of forestry. Today there are many indications that a civic science may be invigorated as questions in forestry are increasingly addressed within a larger context and with specific places in mind. Thus, it may be that ambiguity will result in increasing opportunities for open civic inquiry, and uncertainty will prompt the founding of new institutions that will be responsible for maintaining certainty through the maintenance of relationships. This is the essence of the meaning of sustainability as it is currently used throughout the world today.

The founding ideas of the 20th century must be our starting point if the patterns of change leading to their transformation are to make any sense. Thus, this chapter begins with an exploration of the 20th-century revolution—that is, using the institution of science as a model for social and political institutions. Next, a book published in 1968, which marked the turning point to the new era, is briefly discussed to illustrate that what may seem to be sudden developments have actually been taking shape for several decades. The final section of the chapter looks directly at the problem of ecosystem management from the perspective of ambiguity and uncertainty. The argument put forward is that only by fully embracing the twin resources of ambiguity and uncertainty within a deliberative civic dialogue can ecosystem management or adaptive management lead the way toward a 21st-century forestry. Overall, this chapter is an attempt to understand the underlying dynamics of change, which are creating new forestry institutions capable of a 21st-century forestry.

Building Social Institutions in the Image of Science

> The institution of science is the only institution based on and geared for change. It is built not only to adapt to change, but to overthrow and create change.
>
> —Warren G. Bennis and Philip E . Slater (1968)

Given that creating knowledge through science is by definition a process of overthrowing old ideas, it is not surprising that the institution of science became the model for other 20th-century institutions. Characteristic of early 20th-century American institutions in politics, economics, and social life are the dynamics of change and the embracing of new ideas, new opportunities, new alignments, new resources, new policies, new organizations, new roles for government, new technologies, new forms of organization, and new ways of life. However, there are several quite different ways of thinking about what science is and about methods for carrying out scientific investigations. These various mind-sets can inform and sustain different organizational forms under the broad umbrella of the institution of science.

The initial enterprise of the physical and biological sciences, which grew from Newtonian physics, was a conception of science based on linear, deterministic models of causality. If reality was a mechanical universe, then the role of scientists was to discover its causal mechanisms. Once discovered, these mechanisms could be reliable instruments for technical control. Studies of different factors affecting tree growth are a good example of this approach in forestry (see generally, Regier 1992, Maruyama 1980, Caley and Sawada 1994).

Once the causes were clear, the work of managers was also clear. For this reason, bureaucratic organizations embraced "scientific management." The task of management was to get the conditions right so that the predicted outcome would result. Bureaucracy was an ideal organizational form for this in that it could institutionalize procedures to meet preset goals and objectives (Kaufman 1960). Bureaucracy could also work because science was viewed as a means to discover what was already a causally predetermined reality. Hence the knowledge created by science was viewed as fairly permanent—it could be added to, but only rarely profoundly changed, as in moments of scientific revolution. Thus, the project of science—the accumulation of a systematic body of knowledge based upon empirical testing and observation of the world—could provide a reliable and stable basis for making decisions.

By the middle of the 20th century, another conception of the reality and related scientific mind-set began to emerge. In many fields, scientists began to replace linear patterns of causality with nonlinear, stochastic relationships. Reality seemed much more chance driven. New methods of science were developed based upon probabilistic relationships. This model of science supported the creation of the idea of multiple use as the approach to multipurpose scientific management. Which "use" was preferred depended upon many conditions in the locality. Thus, many different solutions could solve the problem of multiple use.

Forestry organizations adapted to this new mind-set by incorporating a multiplicity of disciplines in their organizational structures. To develop a plan for integrated multiple-use management, the disciplinary specialists would be "locked in a room, told to defend 'their' resource, but to optimize across all resources and work out a management plan" (Shannon 1987, 1989). Computer simulations were used to develop and analyze a variety of management options. Such tools were attempts to mimic the stochastic nature of reality, thereby allowing some managerial control of the conditions in order to raise the probability of desired results actually occurring.

In the last few decades another conception of science has emerged that is based on holistic models of emergent systems. Thinking of reality as an emergent system means that it evolves when elements are brought into relationships with one another and interact. The exact nature of the outcome is unpredictable because the workings of the system as a whole cannot be predicted by analysis of its parts. Moves in forest ecology toward nonequilibrium theories are part of this change in mind-set.

To gain a glimpse of the future, we can lay side by side two lists of commonly used concepts, one describing ecosystems and one describing societies (Table 28.1). These lists are neither definitive nor comprehensive; rather, they are instructive in that they are built by analyzing the current literature in the natural and social sciences. Also instructive are the emerging principles of sustainability, which are emanating from many disciplines and many places around the world. These principles were developed by analyzing a broad range of current international forestry literature (Table 28.2).

Do these concepts hold the seeds of future organizations and institutions? The kinds of new organizational structures and institutional frameworks emerging in the last few years are frequently loose networks of groups coalescing around specific problems. A 21st-century forestry will be the emergent outcome of the relationships among numerous institutions and organizations, all associated with different problems in different contexts. In this way, a 21st-century forestry will be characterized by the same kinds of uncertainty and ambiguity as other emergent systems. This is a profound shift from the more mechanical expectations of predictable supplies of forest resources and from production-oriented institutions of the recent past.

Table 28.1 Commonly used concepts describing ecosystems and societies

Ecosystems	Institutions
Non-linear	Knowledge
Complex	Integration
Reciprocal	Community
Probabilistic	Inclusion
Contingent	Difference
Interactive	Anticipation
Synergistic	Translation
Holistic	Deliberation

Table 28.2 Emerging principles of sustainability

- Maintain ecological functions, conditions, and/or biodiversity
- Evaluate and adapt social processes and governance structures
- Adapt to change
- Integrate ecological, cultural, and economic systems
- Ensure intergenerational equity
- Accept ambiguity of the concept of sustainability

Foresters, resource managers, landowners, and citizens all express anxiety over the nature and pace of change today. Yet, 20th-century institutions were built to incorporate change and to continuously adapt. Is there something new in the voices of anxiety today? The next section focuses on changes in learning and education, on ideas of management and the role of managers, and on politics and the nature of citizenship. The purpose of this brief discussion is to provide some insights into the dynamics of societal change and to lead into our discussion of adaptive management as a central part of a 21st-century forestry.

How Managers and Citizens Became More Like Scientists: Learning from Elders or Experiments?

It is a distinctive feature of the 20th century that formal education replaced experience as the primary method for gaining the practical knowledge needed to live and work in society. Prior to this century, ordinary people learned the skills of living and working through practical experience. Knowledge gained through experience was passed down through practice. Such knowledge tended to be limited to the skills necessary in everyday life or to the maintenance of the spiritual connections between people and their environs. From a societal perspective, this means that elders generally knew more.

It is useful to recall in this context that the knowledge of forest and wildlife management originates from experience and practice. Apprenticeship was (and in many places remains) the route to acquiring it. Indeed, the arguments at the beginning of the 20th century about how to teach forestry grew from the efforts to supplant the apprentice approach with the modern methods of experimental science. Most famous was the argument at the turn of the century between Carl Schenck and Gifford Pinchot. Both were European-educated foresters who had had a combination of formal education and apprenticeship with observation. But under Schenck's direction, the forestry school at the Vanderbilt estate continued the German model—apprenticeship combined with formal learning carried out in the forest. Pinchot sought to place forestry among the practical professions of 20th-century America. With a family endowment at Yale University, he created a program in forestry based upon a university education in a classroom setting. This learning was augmented by field experience and training, beginning the tradition of "summer field camp" as part of most American forestry programs.

The rise of forestry education within universities around the country led to the supplanting of apprenticeship with formal classroom education. This paralleled broader 20th-century trends of using formal methods of experimental science to build a foundation of rational knowledge beyond the limitations of tradition and individual idiosyncrasies. In many spheres, including the agricultural and forest sciences, reliance on technical methods of managing the environment increased the likelihood of guaranteed and predictable outcomes (e.g., predetermined quantities of wood). The very reason for producing knowledge about how trees grow, what conditions enhance their growth, and what factors improve their quality was to develop knowledge that could produce systematic and predictable outcomes. Formal management practices could then be designed to translate this knowledge into management routines that produced continuous amounts of wood for society.

Managing: From Routine or Through Experiment?

Because modern science rests on an empirical basis, managers, like scientists, must come to terms with the practical consequences of hypothesized outcomes. What happens when experiments lead to knowledge that is contrary to standard operating procedures? Clearly, management must be released from the confines of routine and allowed to embrace new information and work toward new management approaches. Acceptance of change as routine is the core of the idea of being "modern." Thus, modern management, modern science, and modern society all embrace the uncertainty of outcomes necessary for testing hypotheses. This means that managers, scientists, and citizens have to become comfortable with the patterns and processes of a variable and changing world.

John Dewey, an American philosopher writing at the turn of the 20th century, revealed these inherent contradictions in 20th-century American society. The mainstay of his philosophy was that the emergence of experimental learning formed the basis of the polity as well as of management (Dewey [1929] 1960). One of his central tenets was that the experimental method necessitated defining, understanding, and valuing experience differently. From the viewpoint of a traditional society, the test for accepting new knowledge is its conformance to a given and immutable set of ideas and premises. Dewey led 20th-century philosophers against this approach and toward a method of reason, experiment, and change. Doing so required distinguishing two forms of experience: the kind of experience gained through simple acquisition of a skill through practice and the kind based on experimental approaches to empirical testing and inquiry. The distinction is subtle; it turns on the conception of control.

Is the world to be conceived as a hostile environment of uncontrolled change wherein humans can supplicate or seek to influence, but never control? Or is it a system of knowable relationships that, given adequate knowledge, humans can control? Modern science and management are predicated on the latter conception. Its method is to induce change and observe the results. Whether the change is induced by the methods of measurement or by physical intervention does not matter; the net result is the same. Knowledge is based on the observed differences in state. Thus, experience is a function of time and observation over many different settings. In this way, modern scientists have much in common with the apprentice of old—they both learn through experience. Yet, the kind of experience is radically different. For the apprentice, it is to learn to replicate the methods of the master and slowly acquire skills. For the scientist, it is to test hypotheses over many different settings and isolate that which is similar from that which differs. Seniority remains respected in this setting. When seniority translates into power over what is tested and over standards of validity, seniority can inhibit change. Nonetheless, when seeing what is new and different is the purpose of inquiry, the inexperienced often have the advantage.

The larger point in this discussion is that gaining knowledge through experimental approaches relies on the suspension of expectation and anticipated outcomes. The values of stability and surety are replaced by the values of change and movement. Anticipate surprise! Expect the unexpected! These are the mantras of an experimenting society (Campbell 1969).

Creating the Experimenting Society

> Democracy becomes a functional necessity whenever a social system is competing for survival under conditions of chronic change. . . . The institution of science is the only institution based on and geared for change. It is built not only to adapt to change, but to overthrow and create change. . . . In order for the spirit of inquiry, the foundation of science, to grow and flourish, a democratic environment is a necessity. Science encourages a political view that is egalitarian, pluralistic, liberal. It accentuates freedom of opinion and dissent. It is against all forms of totalitarianism, dogma, mechanization, and blind obedience.
>
> — Warren G. Bennis and Philip E. Slater (1968)

In 1968, Bobby Kennedy and Martin Luther King Jr. were assassinated, riots burned parts of Los Angeles and Washington, DC, police battled antiwar protesters in the streets of Chicago, citizen protests against clearcutting on the Bitterroot National Forest reached Congress, and Senator Henry M. Jackson introduced the National Environmental Policy Act. In this swirling milieu, two leading social thinkers, Warren G. Bennis and Philip E. Slater, captured the essence of 20th-century American society in *The Temporary Society.* By this title, they did not mean that America was on the verge of collapse—quite the opposite. They argued that as a political form, democracy was based on the models of inquiry derived from science. What were threatened were other political, administrative, and organizational forms based upon the bureaucratic models of hierarchy and command and control.

In retrospect, it is not surprising that their lead essay, "Democracy Is Inevitable," was originally published in the *Harvard Business Review* in 1964. More rapid change in economic, social, and political arrangements occurred in the 1960s than any other decade. The term "the 60s generation" denotes the coming of age of those whose formative experiences occurred in this rapidly changing world. As a generation, their strength still lies in their ability to cope with change, but their weakness is patience when change is slow or when history might be a good teacher. This point is not just a quick aside: it is this generation who now heads companies and agencies, holds high government offices, leads scientific societies and universities, and guides thinking in many disciplines. What might be viewed as an individual characteristic is better understood as one of the dominate social forces affecting the 21st century.

In a chapter entitled "Beyond Bureaucracy," Bennis and Slater hypothesized that several societal forces were necessitating organizational changes: (1) the increasing pace of social change, (2) the expanding size of organizations, (3) the increasing workforce diversity, and (4) new theories of management based upon participation, collaboration, and reason. In response to these social forces, they listed several problems that require new nonbureaucratic forms of organization:

Integration—integrating individual needs and organizational goals

Social influence—distributing power and sources of power and authority

Collaboration—producing mechanisms for the control of conflict

Adaptation—responding appropriately to changes induced by the environment

Identity—achieving clarity, consensus, and commitment to organizational goals

Revitalization—dealing with growth and decay

Bennis and Slater ended this chapter with the recognition that the future would not be "happy" in some utopian sense, but rather would reverberate with the tensions and anxieties produced by major social change:

> Teaching how to live with ambiguity, to identify with the adaptive process, to make a virtue out of contingency, and to be self-directing—these will be the tasks of education, the goals of maturity, and the achievement of the successful individual.... In these new organizations of the future, participants will be called upon to use their minds more than at any other time in history. Fantasy, imagination, and creativity will be legitimate in ways that today seem strange. Social structures . . . will increasingly promote play and freedom on the behalf of curiosity and thought.

Clearly, in their view, we must live with ambiguity, identify with the adaptive process, make a virtue of contingency, and be self-directing. The challenge is to create institutions and organizations that are designed to operate in the shifting context of ambiguity and uncertainty. Thus, 21st-century institutions are likely to be democratic, decentralized, localized, problem-specific, and place-oriented. From these changes arise the ideas of adaptive management and an enriched civic science.

Civic Science and Adaptive Management

> The essential need is the improvement of the methods and conditions of debate, discussion, and persuasion. That is the problem of the public.
>
> —John Dewey ([1927] 1946)

In *Compass and Gyroscope: Integrating Science and Politics for the Environment,* Kai Lee (1993) develops a concept of civic science embedded within democratic deliberation. While Dewey focused upon the role of the public, Lee is interested in the role of the expert. He argues that experts need to provide the scientific analysis necessary for informed public debate. He goes on to suggest that experts should think of themselves as teachers and interpreters who can help decipher the technological world for citizens and enable them to make sensible political judgments.

The concept of civic science might be extended to include citizens as "lay social scientists," a term first suggested by Mary Stanley (1988) in her seminal analysis of American citizenship. In this way, an essential role for citizens is to clearly, objectively, and honestly assess themselves, their communities, their environment, and the institutions affecting their lives. In addition to revitalizing democracy by making citizens more aware of and understanding of each other and their world, civic science is a catalyst for adaptive management (Shannon and Antypas 1996, Shannon 1992).

The idea of adaptive management emerged in strong form in the 1970s (McLain and Lee in press). Through it, the managerial task became more a scientific and political task than simply a bureaucratic one. In other words, rather than simply aligning resources and capacity in the service of known objectives, planning required choosing among objectives. Similarly, rather than simply holding management against bureaucratic rules and procedures, monitoring necessitated evaluating the actual consequences of different management activities.

The choice of objectives is a political choice and thus the work of democratic institutions. Forestry is becoming more democratic because the ambiguity inherent in choosing among different objectives requires the engagement of citizens in moral deliberation. This new way of doing business is reflected in the concept of ecosystem management:

"In ecosystem management we are talking about decision space that is at least three-dimensional. A good decision has to be economically feasible, socially acceptable, and ecologically sustainable" (Elizabeth Estill, personal communication, April 25, 1996).

Similarly, when management activities are treated as experiments subject to failure as well as success, managers must learn to think like scientists. This shift in orientation is reflected in commonly cited definitions of adaptive management. For example, the Forest Ecosystems Management Assessment Team, gathered by President Clinton to develop alternatives for management of Pacific Northwest forests, defined adaptive management as "the process of implementing policy decisions as scientifically driven management experiments that test predictions and assumptions in management plans, and use the resulting information to improve the plans" (FEMAT 1993).

Taken together, ecosystem management and adaptive management transform forestry so that it is no longer an isolated, technical activity. A 21st-century forestry will crisscross institutions, disciplines, organizations, specializations, roles, and identities. Foresters will enter the debates on when, where, why, and how forests should be managed, including for what purposes and by whom. No longer will foresters be specially authorized to determine how forests are managed. Rather, they will join in creating the open and deliberative civic science necessary for sustaining forests.

Rather than forestry being a predictable set of techniques designed to efficiently maintain productivity, it is now becoming an open institution oriented toward sustainability. Generally, we think of institutions as stable patterns of relationships over time. But as Bennis and Slater suggest, today's institutions are built around ambiguity, uncertainty, and surprise. In this context, forestry must embrace the ambiguity of multiple objectives, the uncertainty of diverse perspectives, and the possibility of surprising outcomes. Working within open institutions, 21st century foresters will constantly be charting new paths, forging new relationships, and adapting to a rapidly shifting social environment.

With this we conclude by revisiting the concepts of uncertainty and ambiguity as they become the twin resources supporting and creating the 21st century. *Uncertainty* is the resource supporting the institutional flexibility and adaptability that leads to sustained social learning. Social learning depends upon and respects many different forms of knowledge and replaces the ideal of technical control with stewardship. *Ambiguity* is the resource that encourages a diversity of political viewpoints while facilitating the kind of civic deliberation that leads to a community of interests and responsibility for the commonwealth.

Literature Cited

Bennis, W. G., and P. E. Slater. 1968. *The temporary society.* New York: Harper and Row.

Caley, M. T., and D. Sawada. 1994. *Mindscapes: The epistemology of Magoroh Maruyama.* London, England: Gordon and Breach.

Campbell, D. T. 1969. Reforms as experiments. *American Psychologist* 24:409–429.

Dewey, J. [1927] 1946. *The public and its problems: An essay in political inquiry.* Based on the Larwill Foundation lectures, Kenyon College, January 1926. Reprint, Chicago: Gateway Books.

———. [1929] 1960. *The quest for certainty: A study of the relation of knowledge and action.* Gifford lectures 1929. Reprint, New York: Capricorn Books, G. P. Putnam and Sons.

FEMAT. 1993. *Forest ecosystem management: An ecological, economic, and social assessment.* Report of the Forest Ecosystem Management Assessment Team (FEMAT). 1993-793-071. Washington, DC: GPO.

Kaufman, H. 1960. *The forest ranger: A study in administrative behavior.* Baltimore: Johns Hopkins University Press.

Lee, K. N. 1993. *Compass and gyroscope: Integrating science and politics for the environment.* Washington, DC: Island Press.

Maruyama, M. 1980. Mindscapes and science theories. *Current Anthropology* 21:589–608.

McLain, R. J., and R. G. Lee. In press Adaptive management: Pitfalls and promises. *Environmental Management.*

Regier, H. A. 1992. Ecosystem integrity in the Great Lakes Basin: An historical sketch of ideas and action. *Journal of Aquatic Ecosystem Health* 1:25–37.

———. 1995. The Great Lakes–St. Lawrence River Basin case study. Paper presented at symposium, At the Crossroads of Science, Management, and Policy: A Review of Bioregional Assessment, 6–9 November, Portland, OR.

Shannon, M. A. 1987. Forest planning: Learning with people. In *Social science in natural resource management sys-*

tems, ed. M. L. Miller, R. P. Gale, and P. Brown. Boulder, CO: Westview Press.

———. 1989. *Managing public resources: Public deliberation as organizational learning.* Doctoral dissertation, University of California, Berkeley.

———. 1992. Foresters as strategic thinkers, facilitators, and citizens. *Journal of Forestry* 90(10):24–27.

Shannon, M. A., and A. R. Antypas. 1996. Civic science is democracy in action. *Northwest Science* 70(1):66–69.

Social Science Research Group. 1994. *Principles of sustainability: A review of the literature.* Working paper. Seattle: Institute for Resources in Society, College of Forest Resources, University of Washington.

Stanley, M. 1981. *The technological conscience: Survival and dignity in an age of expertise.* Chicago, IL: University of Chicago Press.

———. 1988. Six types of citizenship. *The Civic Arts Review* (Fall):12–15.

29 The Emerging Role of Science and Scientists in Ecosystem Management

John C. Gordon and James Lyons

Congress and Science-Based Policies 448
Scientists as Agency or Congressional Advisors 449
Scientists and Debate over the Fate of the Northern Spotted Owl 449
Rationale for Convening the Gang of Four 449
A Product of Scientists, Not Science 450
Lessons Learned 450
The Function of Scientists in a Political Environment 451
Benefits to Policy Users 451
What Do Scientists Do Next? 452
Conclusions 452
Literature Cited 452

Questions about the state of scientific knowledge and about scientists themselves have played an increasingly prominent role in the debate over the fate of forests and forest resources in the Pacific Northwest and elsewhere (NRC 1990, 1994). Interestingly, the concept of ecosystem management, although written about by scholars for at least three decades (Van Dyne 1969), has emerged in response to legal and societal pressures rather than scientific ones (Healy and Ascher 1995). These pressures have created a demand for science-based answers to questions formulated by citizens and lawyers. At the same time, the forestry profession itself is undergoing profound changes, including a greater reliance on science for the simultaneous management and conservation of a great array of forest resources (NRC 1990, Gordon 1994). There is little doubt that these trends will continue into the 21st century.

However, the linkage between science, scientific opinion, and public policy is as yet poorly understood and developed. A central task facing those who urge the implementation of ecosystem management is to forge better linkages between science, management, and policy making (Gordon 1994). In this chapter, we examine the efforts of the Scientific Panel on Late Successional Forest Ecosystems, created at the request of two congressional committees, which provides one example of how scientists and Congress have worked together to achieve greater mutual understanding and progress in ecosystem management. We then draw some conclusions about how the relationship between science, management, and policy might be improved.

Congress and Science-Based Policies

Congress's record on developing science-based policies for environmental issues is mixed, and procedures and processes through which scientists provide effective advice on specific environmental legislation often prove inadequate (Lawler 1995). The National Academy of Sciences, through its action arm, the National Research Council, provides formal advice to the federal government on a wide array of science-related questions, many of them environmental. Usually advisories are issued as committee reports, which tend to be highly credible and often get considerable visibility (although that varies). However, environmental and natural resources scientists are not numerically prominent in the academy—and academy attempts to target those areas of science are not always successful. Moreover, National Research Council reports rarely are produced rapidly enough to respond to questions that arise during the genesis of specific pieces of legislation (Lawler 1995).

The hearing process, although it produces copious quantities of testimony, seems poorly suited to marshaling strong and continuing scientific credibility. Presentations are brief, and scientific expertise is captured ephemerally.

The Congressional Research Service (CRS) has provided highly focused, short-term answers to questions related to specific pieces of legislation. However, CRS does not have the breadth of expertise required to address complex ecosystem management issues. In recognition of this, it convened an Ecosystem Management Symposium in 1994. The symposium assembled information from a broad array of federal agencies and experts. But, consistent with its mission, it was a one-time event with no continuing attempt to monitor and interpret the rapidly moving field of ecosystem management. The Office of Technology Assessment (OTA) was capable of longer-term studies of greater technical and scientific depth, including some focused on forests and even ecosystem management (Gorte and Alston 1993). However, OTA no longer exists.

Thus when confronted with contentious, science-based environmental issues, Congress has often sought, and probably will continue to seek, other mechanisms for scientific input. Perhaps the most striking example is the establishment of the National Acid Precipitation Assessment Program (NAPAP), whose findings were used in developing the Clean Air Act reauthorization of 1990. In 1980, public concern for the effects of so-called acid rain on forests and lakes helped promote the enactment of the Acid Precipitation Assessment Act. This legislation provided for the establishment of an interagency, interdisciplinary group to fully research the effects of acidic deposition on a range of resources and materials, including bodies of water and forests. Working under the auspices of the President's Council on Environmental Quality, NAPAP developed a framework for a broad-based, multiyear research program to investigate how acid precipitation might be impacting human health and the environment. The program involved leading scientists from the United States as well as visiting scholars from abroad.

Research related to the effects of acid precipitation on forests was coordinated by the U.S. Forest Service. While the initial focus of this research was on the relationship between acid precipitation and forest health, the program soon expanded to include an assessment of a broad range of air pollutants. NAPAP studies indicated that the relationship between acid rain and damage to lakes and forests was at best unclear. However, research on the effects of other air pollutants on forests confirmed the previously hypothesized link between high ozone levels and forest damage.

The NAPAP research program yielded information at a relatively slow pace by congressional standards. However, by the end of a 10-year period, NAPAP offered rather convincing scientific evidence (at least to some) that acid precipitation and forest damage are not strongly linked. Other air pollutants, particularly ozone, are of much greater concern.

The NAPAP research effort was nearly completed

as congressional debate over reauthorization of the Clean Air Act began in earnest. Despite the lack of certainty regarding the impact of acid rain on forests and other natural resources, the study was used as one rationale for strong measures to curb the emissions of acid rain precursors, which eventually became a cornerstone of the reauthorized Clean Air Act (ORB 1991, Russell 1992).

Scientists as Agency or Congressional Advisors

Scientists often have served as advisors to Congress and administrative agencies, helping to set agendas, establish program priorities, and clarify information. The Environmental Protection Agency, for example, regularly convenes a scientific advisory panel to help set agency priorities and evaluate program efforts. During hearings before the Senate Environment and Public Works Committee in 1991, EPA Administrator William Reilly reported that the priorities recommended by his scientific advisory committee were somewhat at odds with the congressional priorities.

Scientists also are called upon to serve as expert witnesses to aid Congress in assessing new information and determining priorities. For example, a great deal of the credit for bringing the issue of global warming to the forefront can be credited to a NASA scientist whose testimony before a Senate committee generated headlines and sparked congressional interest in the issue.

Scientists and Debate over the Fate of the Northern Spotted Owl

The decision to list the northern spotted owl as a threatened species was by law based solely on the biological status of the owl. This thrust biological and particularly ecological scientists directly into the middle of this controversy.

Interest in the protection of old-growth forests had been evident for some time. However, it was not until a petition to list the northern spotted owl was accepted that a legal foothold existed to challenge the scientific basis for forest management in the Pacific Northwest—and concurrently its effect on the survival of the owl. Ironically, however, the injunctions against harvesting national forest old-growth timber in the region were not the result of legal challenges brought under the Endangered Species Act. Rather, they were based on the failure of the Forest Service to comply with its own regulations under the National Forest Management Act. Those regulations required the Forest Service to maintain the owl's viability.

It was questions about the status and viability of the northern spotted owl, the relationship between the owl and old-growth forests, and later the relationship between old-growth ecosystems and the survival of other animal and plant species (particularly salmon) that brought science and scientists directly into the debate. In particular, Congress sought the expertise of the Scientific Panel on Late Successional Forest Ecosystems (known as the Gang of Four) to assist in resolving this difficult issue (Scientific Panel on Late-Successional Forest Ecosystems 1991).

Rationale for Convening the Gang of Four

There were several reasons for requesting the assistance of a group of scientists to aid in resolving the old-growth and spotted owl controversy. The most obvious was the complexity of this issue. While some people attempted to narrowly focus the dispute on owls versus jobs, the full scope of the debate included the effects of past and proposed logging not only on the viability of the owl, but also on other old-growth-related species. Members of Congress recognized the danger of formulating a solution to the spotted owl dilemma that failed to adequately address the needs of other fish and wildlife species. As one representative from Washington State noted, a solution that failed to address the potential for the "endangered species a month" problem, would not be a viable solution. With this recognition came the realization that a multispecies protection strategy could only succeed if it were founded on sound, ecosystem management principles that addressed whole systems instead of single species.

At its earliest stages, the debate focused on how to define old growth and on efforts to determine how much remained. Despite clear signals that a complete inventory of remaining old growth on federal lands was needed, the Forest Service provided little information. Moreover, the value of the information they did provide was limited since it was based on timber-stand inventories. Only after a major mapping effort was initiated by a national environmental organization did the Forest Service issue a contract to attempt to conduct a special old-growth inventory.

Based on its inventory of remaining old-growth forests, The Wilderness Society produced individual maps for all national forests in the region. Each map, based on a review of aerial survey information and subsequent ground inspections, depicted the location of remaining old-growth forest stands 200 acres or more in size. The accuracy of these maps was later proven to be quite high. However, given that the project was undertaken by an environmental organization that had taken a clear position in support of protection for most of the remaining old growth, the utility of The Wilderness Society maps was brought into question.

Lacking a sound Forest Service inventory and distrusting information produced by an organization with a clear stake in the old-growth debate, members of Congress requested an independent group of scientists to review the status of old-growth forests in the Pacific Northwest. In this way, they reasoned, the biases of both the environmental community and the Forest Service could be held in check. In addition, since the review was to be prepared by a panel of scientists, each with expertise related to the old-growth issues, the credibility of the report would be high.

What the Gang of Four generated, however, went beyond a simple, independent inventory of old-growth forests in the region. In response to guiding questions posed by the congressional committees, the panel report provided the framework for a unique, new approach to the management of federal forest resources in the region—and it implied a strategy for the conservation of biological diversity on both public and private lands.

A Product of Scientists, Not Science

It is important to keep the product of this effort in perspective. The report and associated alternatives developed by the Gang of Four panel was not itself the product of primary science—at least not in the traditional sense. There were no hypotheses to be tested, nor was there an experimental design, a control, field data gathering, or analysis. Instead, the panel's report was the product of the scientists' attempt to answer congressional questions based upon existing information, their own research experience, and the expertise and experience of hundreds of agency managers, scientists, and technicians who participated in the time-bounded effort.

In some ways one might argue that the science upon which they based their judgment and the resulting report was in part the product of the experiment that has come to be known as forestry in the Pacific Northwest. To date, forest management in the region has been based on research, traditional principles of forestry, and professional judgment. However, given the complexities of forest ecosystems and the difficulties involved in assessing the effects of forest management on plants, animals, and ecological processes, the outcome of this huge experiment has been difficult to quantify. We can only draw conclusions from the smaller designed experiments conducted in different parts of the landscape. This realization has led to calls for adaptive management—that is, the structuring of management decisions and their implementation as deliberate learning experiences.

Critics have argued that the Gang of Four report simply reflects the opinions of the scientists involved, and that the process failed to provide an adequate opportunity for participation of those with alternative views—including scientists.

With regard to the charge that the product was largely opinion, the critics are correct. However, it was opinion formed by a massive, short-term effort. The panel called on federal agency staff to use the information they already had to answer very specific questions. The primary science that would have been needed to test and verify the observations and hypotheses of the panel report would have taken several decades to conduct and a commitment of resources that neither Congress nor the agencies have shown any willingness to make.

Lessons Learned

We must recognize that the management of the national forests in the Pacific Northwest and elsewhere now is based only in small measure on what might be called traditional science. The principles of forest management that are applied are, in large measure, based on professional judgment and empirical observation. Their effect on specific ecosystems and the functions and interactions within those ecosystems is as yet poorly understood.

What drives forest management is the wisdom and judgment of individuals who share a common training and expertise in natural resource management. Often, these forest managers are given a set of goals and told to achieve them. Sometimes these goals are within the realm of possibility given the known limits of the resources at hand. But at times these goals can only be achieved by stretching other resources, which may or may not be able to cope with the strain. To date, little scientific effort has been organized to answer questions of ecological risk, or even to provide a common methodology for approaching them.

Science can be used to forecast how different management approaches impact individual resources or species. Increasingly, however, questions apply to the fate of whole systems, within which one species may prosper as another declines or one resource may be enhanced while another is degraded. In addition, resource decisions are not solely, or even principally, driven by science; rather, they are driven by a vector of political interests (Herrick 1992). Currently, no interest has exclusive or very effective use of science and scientists. The challenge for 21st-century science will be to provide a better, more integrated information base accessible to all, without losing the measure of objectivity and credibility it now has.

The Function of Scientists in a Political Environment

A "no free lunch" solution emerged from the Gang of Four assessment of existing information. Any solution that seemed to maintain the long-term sustainability of the late-successional old-growth forest ecosystem and associated animal and plant species seemed also to result in a significant reduction in timber harvests. This in turn was predicted to have severe economic and social consequences for affected individuals and communities.

A perfect political solution, one that responded to the needs of members of Congress and the administration, would have had little or no negative impact on people. Obviously, such a solution did not exist. Although people are accustomed to "miracles" from science (drugs, weapons, even trees), 21st-century ecology and assessments based on it are not likely to produce any. Rather, ecological science is more likely to produce somber estimates of alternative consequences in which some values are lost or poorly preserved.

In the short term, the most valuable contribution of the panel may have been in demonstrating to those involved in the political process that a biologically sound solution would require a change in traditional forest management practices. In the longer view, the value of the panel's report may be its demonstration of scarcity and the need to deal with scarcity through ecosystem management.

The alternatives developed by the panel—each of which is regionwide, ecosystem-based, and risk-driven—can serve as a basis for a new science-based forest management paradigm. This approach breaks from the traditional strategy of limiting or eliminating management from specific areas of the forest and permitting intensive management in others. It also transforms the political debate from that of acres reserved versus board feet available to preventing multiple species management problems by integrating management across the landscape. The hope is that landscape-level management, which is more sensitive to the positive and negative impacts of harvesting on multiple species and resources, might reduce conflicts and provide greater certainty, both for timber and for nontimber resources.

Benefits to Policy Users

The panel report aided Congress in a number of ways. First, the panel's findings served as a wake-up call for some in Congress who had been led to believe that the status quo was politically and biologically the best that could be done. Second, for those members of Congress who accepted the need for a change in management direction, the panel's efforts helped them reassess their expectations for a solution. This enabled legislators to "ground truth" what they believe are reasonable alternatives and to determine for themselves what "reasonable" means. Finally, the report diminished the desire among some to continue to search for "free lunch" solutions, which, by all accounts, do not exist.

What Do Scientists Do Next?

Scientists have served as expert witnesses in the courts and in administrative procedures for many years. Now Congress seems willing to embrace scientific opinion as one basis for making policy judgments (Carnegie Commission 1991). In fact, it may have been the courts, acting to halt timber sales on public lands in the West, that hastened this acceptance.

The testimony of expert, scientific witnesses has played a major role in the outcome of each of the lawsuits pending in the PNW that affect spotted owls and old-growth forests. For example, in *Northern Spotted Owl v. Hodel* (716 F. Supp. 479, 481 (W. D. Wash. 1988)), Judge Thomas S. Zilly stated,

> [T]he Service disregarded all the expert opinion on population viability, including that of its own expert, that the owl is facing extinction, and instead merely asserted its expertise in support of its conclusions.

Judge Zilly then ordered the U.S. Fish and Wildlife Service to reconsider its decision not to list the northern spotted owl as threatened or endangered.

Science and scientific opinion are but one component of the political processes that must work to resolve issues. Politicians, like scientists doing assessments, must make informed judgments regarding what may or may not work—and hope their judgments are correct. However, members of Congress and the administration have an advantage over scientists. They need not wait for years of data collection and analysis to ascertain if their judgments are correct. Elections, on two-, four-, or six-year cycles, provide a ready means for evaluating the "accuracy" of the political judgments that these individuals must make if the issue is to be resolved.

Conclusions

Forestry in the next century will be science-based and policy will be science-driven to a much greater degree than in the past, but science will be different (Gordon 1992). Science-on-demand will be a much greater component of forest and natural resource science, and basic science will have to justify its place in the budget without much in the way of initial advantage. Only if forest science can cope with these pressures effectively will it be able to serve society adequately. The old model of securing research support—unkindly summarized as "give us the money, shut up, and wait for the miracles"—will certainly not work in the next century (Brown 1993). But just as certain, the demand for science-based information for managing ecosystems will grow rapidly and require new thinking by those engaged in forestry and natural resources research (NRC 1990).

We need a new model for linking science, management, and policy—that is, a structure that synthesizes science and management and that can inform public policy. Implementation of the new model will require a changed and reinvigorated leadership (Berry and Gordon 1993). Science must be more directed to the needs of managers, and managers must accept that science implies experimentation and hence uncertainty. Policy makers, in particular, need to realize the costs and limits of science as a policy tool. Universities should play a major role in creating the institutional base for the new model (Gordon 1994, Danforth 1995). To do this, they will have to be more interconnected among themselves and more connected to the society around them. This should be a natural path for ecosystem scientists.

Literature Cited

Berry, J., and J. Gordon, eds. 1993. *Environmental leadership: Developing effective skills and styles.* Washington, DC: Island Press.

Brown, G. 1993. Can scientists make change their friend? *Scientific American* (June):152.

Carnegie Commission. 1991. *Science, technology, and Con-*

gress: Expert advice and the decision-making process. Report of the Carnegie Commission on Science, Technology, and Government. New York.

Cunningham, P., J. Gordon, and B. Bormann. *Designing adaptive management experiments.* Corvallis, OR: USDA Forest Service.

Danforth, W. H. 1995. Universities are our responsibility. *Science* 269:1651.

Gordon, J. 1992. The role of science in resolving key natural resource issues. In *Proceedings: Seeking Common Ground, a Forum on Pacific Northwest Natural Resources, Feb. 24–25, 1992.* Portland, OR: USDA Forest Service, Region 6.

———. 1994. From vision to policy: A role for foresters. *Journal of Forestry* 92:16–19.

Gorte, R. W., and R. M. Alston. 1993. The role of information, science, and technology in resolving the challenges of ecosystem management. In *Proceedings: SAF National Convention.* Bethesda, MD: Society of American Foresters.

Healy, R. G., and W. Ascher. 1995. Knowledge in the policy process: Incorporating new environmental information in natural resources policy making. *Policy Sciences* 28:1–19.

Herrick, C. N. 1992. Science and climate policy: A history lesson. *Issues in Science and Technology* 8(2):56–59.

Lawler, A. 1995. NRC pledges faster delivery on reports to government. *Science* 270:22–23.

National Research Council. 1990. *Forestry research: A mandate for change.* Washington, DC: National Academy Press.

———. 1994. *Rangeland health: New methods to classify, inventory, and monitor rangelands.* Washington, DC: National Academy Press.

ORB. 1991. *The experience and legacy of NAPAP.* Report of the Oversight Review Board of the National Acid Precipitation Assessment Program presented to the Joint Chairs Council of the Interagency Task Force on Acidic Deposition.

Russell, M. 1992. Lessons from NAPAP. *Ecological Applications* 2(2):107–111.

Scientific Panel on Late-Successional Foresty Ecosystems. 1991. Alternatives for management of late-successional forests of the Pacific Northwest: A report to the U.S. House of Representatives, Committee on Agriculture, Subcommittee on Forests, Family Farms, and Energy, Committee on Merchant Marine and Fisheries, Subcommittee on Fisheries and Wildlife, Conservation, and the Environement.

USDA Forest Service. 1995. Navigating into the future. Rensselaerville Roundtable: Integrating science and policy making. Washington, DC.

Van Dyne, G., ed. 1969. *The ecosystem concept in natural resource management.* New York: Academic Press.

Index

Abies procera, 320
Abies spp., 92
Acacia, 119
Acer circinatum, 320
Acer macrophyllum, 92, 158
ACP (Agricultural Conservation Program), 345, 353
Adaptation, in decision making, 425, 426, 431
Adaptive management, 5, 324–332, 424–425, 443–444
Added-value manufacturing, 309–312
Adirondack Planning Agency, 384
Advisory committees, science, 373–375, 407–408, 448–449
Age
  culminations, 167–168
  susceptibility to disease and insects with, 180, 181
Age class, 132–133
Agencies
  autonomy of, 372
  building bridges across boundaries, 381–396
  cooperation between, 251, 349–350
  enforcement by, 363
  modeling organization of, 364–368
  organizational and legal challenges for, 361–377
  regional, 384
  staff, 394–396
Aggregate retention, 123
  federal guidelines for, 124
  size and shape of aggregate, 128, 134–135
  vs. dispersed retention, 12, 122–123, 128, 131, 134, 222
Agricultural Conservation Program (ACP), 345, 353
Agriculture
  edge effect from, 63
  shifting cultivation, 33, 46
Alaska, 119
  alluvial valley bottoms, 213
  riparian management regulations, 73, 74
  Tongass National Forest, 391–392
Alaska Forest Resources and Practices Act of 1978, 74
Alaska Recreation Plan, 391–392
Albedo, 38
Alder, 158, 159, 255, 258, 320
Algae, nitrogen fixation by, 177–178
Allelopathy, 46
Allowable sale quantities (ASQ), 282, 357
Alnus rubra, 20, 155, 158, 159, 320
Ambiguity, 437, 438, 443, 444
American Indians
  culture and salmon, 95
  fishing rights, 428
  Menominee Indian Reservation, 125
  philosophy, 101
Amphibians, 95
  in Pacific Northwest late-successional forests, 89, 96
  in riparian areas, 98, 99
Anthropogenic factors, 415. *See also* Clearcuts; Fragmentation
  cultures and institutions, 241–242
  watersheds and, 239, 240
Antitrichia curtipendula, 91–92, 326
Antitrust law, 375–376
Ants. *See* Arthropods
Appalachian Mountains, old-growth stands in, 42
Applegate Partnership, 358, 374, 377
Arbutus menzeisii, 92
Arcto-Tertiary Geoflora, 92
Armillaria. See Root rot
Arthropods, 93–94. *See also* Insects
  in canopy, age of stand and, 181
  in Douglas-fir stands, 174
  in Pacific Northwest late-successional forests, 89, 93
  in retention harvest, 130
Ash yellows disease, 39

ASQ (allowable sale quantities), 282, 357
Atlantic conveyor belt, 34, 36
ATLAS modeling program, 263
Attrition, 62, 63–64, 65
Augusta Landscape Project, 236
Australia, 118, 119, 135

Balance of nature, 33
Basins, river, 78
Bats, 95–96
Bear grass, 320. *See also* Floral greens; Understory
Beetles. *See* Arthropods
Belowground. *See* Subsurface area
Bennis, Warren G., 439, 442–443
Berberis nervosa, 320
Berries, 45, 46, 320
Best available technology (BAT), 351
Beuter, John, 398, 399
Big-leaf maple, 92, 159
Biodiversity, x, 7–8
 artificial *vs.* natural, 100
 disturbance and, 33
 ecosystem function and, 40–42
 eight key lessons of, 96–100
 during establishment phase, 14
 extended rotation and, 169
 management and, 49, 100–101
 of old forests of the West, 87–109
 retention harvesting and, 116, 133, 134, 156–157
 stability and, 48–49
 during thinning phase, 15
Biological control of pests, 179
Biological integrity, 39
Biological landscape, 57–66
Biological legacies, 8, 43–46, 112, 176
Biomass
 ecosystem function and, 22–23, 24
 metric tons of bryophytes/hectare, 91
Bio-regional assessments, 402–403
Bioregulation, 47
Birds, 95
 in control of insects, 95
 creation of cavities for, 147
 monitoring of, Washington, 121
 in Pacific Northwest late-successional forests, 89, 96
 with retention harvesting, 133–134
 in riparian areas, 98, 99
 use of stands by, 22
Black stain root disease, 180
Boletus edulis, 320
Bolle, Arnold, 416
Botkin report, 403–404
Bottom-up controls, 42
Boundaries (ecotones), 70, 217–218
Boundedness, in decision making, 425, 426–428
Boxwood, 320
British Columbia
 alluvial valley bottoms, 213
 landscape design, case study, 261–269
 retention harvest in, 118, 124, 135
Bryophytes, 22, 91–92, 320
 biology and ecology, 325–326
 in Pacific Northwest late-successional forests, 89, 91
 as special forest products, 325–326
Budworms, 38
Buffer strips, 72, 221
Bureau of Forestry, 82
Bureau of Land Management. *See also* Policies and regulations, federal
 appraisal of special forest products, 316, 324
 forest lands of
  evaluation of sustainability by Reuter report, 399
  Northwest Forest Plan alteration of, 233
 legal dictates to, for forest planning, 72, 370
Butt rot disease, 174

"Cabbage patch" forestry, 415–416
Cable logging, 121
CAI (current annual increment), 166
California
 low-retention harvest in, 126
 riparian management regulations, 73, 74
 species richness in late-successional forests, 98
California Forest Practices Act of 1973, 74
California spotted owl, 126. *See also* Spotted owl
California Spotted Owl Technical Group, 127
Canopy
 composition and qualities of, 22
 during establishment phase, 14
 multiple layers, 8, 119, 123
 niches associated with, 8
 openings in the, 8, 156–158, 174, 175, 327
 during shifting mosaic phase, 19
 during thinning phase, 15

topping, to reduce windthrow susceptibility, 128
during transition phase, 21
Cantharellus cibarius, 320
Capital gain, timber income as, 346–347
Carbon, storage of, 22, 23–24, 167
Cascade Center for Ecosystem Management, 127
Cascade Mountains, 127, 179–180, 199, 200. *See also* Pacific Northwest
Cascara, 320
Catastrophes, 63. *See also* Disturbances
Cattails, 60
Cavities, creation of, 147
Cedar Camp, OR, 46–47
Cedar River Watershed, WA, 135
CFI (Continuous forest inventory), 125
Channel structure, 80, 81, 159
Chaos, 20, 34
Chequamegon National Forest, WI, 218–219
Chestnut trees, 43, 48
Chiloscyphus polyanthos, 92
Chimaphila umbellata, 320
Christmas greens and boughs, 320
Civic republicanism, 370
Civic science, 443–444
Clams. *See* Mollusks
Clapp, Earle H., 342
Clarke-McNary Act of 1924, 341, 347
Clean Air Act amendments of 1970, 343, 344–345
reauthorization of 1990, 448–449
Clean Water Act of 1972, 72, 74, 229
Clearcuts
clustered, 64
costs, 292
effects of, 3
on bryophytes, 91
flooding, 114
loss of soil community, 116
historical background, 152, 166
regeneration and, 35, 46, 47, 111
with reserves, 112
in Sweden, 112
visual impact of, and public relations, 166
Climate
Atlantic conveyor belt and the, 36
as disturbance factor, 36–37, 60–61
ecosystems and, 35, 36, 142
El Niño oscillation, 36
fire management and, 199
geographic range of organisms, 60, 62
global changes in, 36, 199
during origin of old-growth forests, 142
Climax, use of term, 19
Clinton, William. *See* Forest Ecosystem Management Assessment Team
Coarse woody debris (CWD). *See also* Dead wood
as habitat, 119, 180
small mammal migration through, 134
structural retention and, 112, 118, 119, 123
in watershed spatial planning, 274
Coastal Zone Management Act of 1972, 343, 345
Colonist species, watershed as source of, 75
Colonization surfaces, 241
Columbia River/East-Side Ecosystem Project, 233
Community
definition, 57
forest service resources and the, 389–390
role in new manufacturing processes, 311
stability of, 297–298
Compass and Gyroscope: Integrating Science and Politics for the Environment (Lee), 443
Competition
during establishment phase, 14
patch arrangement and, 220
during recovery phase, 45–46
seedling mortality and, 154
Complex systems, 2–3, 34–35
Composition, 22. *See also* Development
disturbance and, 33
effects of fire and, 195
time as factor in, 24
Conflicts
resolution of, 425
source of, 423–424
Connectivity
enhancing, 115, 117–118, 124, 135
patch arrangement and, 9, 220
uncertainty and, 241
Consensus decision making, 385, 426, 429–430
Conservation
landscape alteration and, 59, 62, 64–66
management and, 49–50
Conservation biology, 13
Conservation easement, 362–363
Conservation Reserve Program (CRP), 346
Constraints and opportunities, in landscape design, 261, 265

Consumers
  per capita consumption of wood, 1970-1988, 82
  timber prices for, 293, 404
Continuous forest inventory (CFI), 125
Cooperative extension programs, 348–349
Cooperative Farm Forestry Act of 1937, 348
Coral reefs, 35
Core populations, 61
Corridors, 65
  definition, 217
  early design of, 117–118
  as keystone structures, 42
  patch arrangement and, 9, 123
Cost. *See* Economics
Cost sharing, 345–346
Cottonwood, black, 155
Cougar Ramp harvest unit, 121, 135
Councilman, John, 392
Coweeta Experimental Forest, NC, 7
Coweeta Hydrologic Laboratory, NC, 216
Credibility, 431
Crowding, effects of, 183
CRP (Conservation Reserve Program), 346
Cull trees, 80, 81
Cultivation, shifting, 33, 46
Culverts, 71
Current annual increment (CAI), 166
Cutting Practices Act of 1945, 339
Cyathea australis, 119

Data
  collection, 422
  management, 4
  sharing, 358, 383, 431, 450
Dead wood. *See also* Coarse woody debris; Snags
  as a biological legacy, 44
  during establishment phase, 14–15, 16
  as keystone structure, 43
  management of, 146–147
  role of, 8
  role of fungi in, 90
  during thinning phase, 17
  during transition phase, 18
Decentralization
  of authority and responsibility, 357–358
  institutional, 4, 386
Decision making, 419–434
  bridges across boundaries for, 383, 384, 385–388
  building support for, 386–387
  delineation of alternatives, 405
  efficiency analysis, 405–406
  historical background, 420–424
  interaction processes for, 425–429, 431
  modeling for, 3, 4, 272, 287
  multi-party, 4
  in watershed management, 242, 251, 272
Decomposition
  by arthropods, 94, 178
  by fungi, 90, 178
  by invertebrates, 92, 178
Defoliator problems, 192
Deforestation, of rain forests, 38
DEMO (Demonstration of Ecosystem Management Options), 135
Democracy, 442, 443
Demographic stochasticity, 63
Demonstration of Ecosystem Management Options (DEMO), 135
Density control, 141, 143, 168, 328
Desertification, 35
Design. *See also* Landscape analysis and design
  of corridors, 117–118, 123
  elements in variable retention harvest systems, 118–124
  in watershed areas, 271–283
"Desired future dynamics," 99–100
Development
  general models of, 14–19
  implications for forest management, 25–27
  stand, shaping through silviculture practices, 141–147
  variability of, in space and time, 19–21, 24
Development rights, purchase of, 351–352
Devil's club, 320
Dewey, John, 441, 443
Dicksonia antartica, 119
Digital terrain model (DTM), 261–263
Discount rates, 286
Disease, 80, 145. *See also* Root rot
  resistance to, 32, 155, 172
  retention harvest and, 130
  susceptibility to, 177
Disease control, pesticide practices for, 8, 131, 179
Dispersal, 59. *See also* Seed dispersal
  among subpopulations and habitat patches, 60
  fragmentation and, 62–64, 65–66
  of fungi, 90

matrix conditions and, 117–118
pollen, 204–205
Dispersed retention, *vs.* aggregate retention, 12, 122–123
Dispute resolution, 382, 384
Dissection
of landscape, 62–63
maintaining landscape movement and, 59–60, 64
Disturbance regimes, 234–235, 236
Disturbances, 8–9, 31–32. *See also* Disease; Fire
climate and, 36–37, 60, 62
definition, 31
floods, 7, 80, 81, 114, 119
historic, 31
initial conditions of, 39
insects and pathogens as, 80, 174–176
in landscape analysis, 259, 261–263
new concepts of, 35–39
nonequilibrium and, 32–34
restoration and, 80–81
stand development and, 21
Diversity. *See* Biodiversity
Dominant species
in establishment phase, 14
out of equilibrium with environment, 39
role of, 32
Douglas-fir
as floral greens, 320
grazing on seedlings of, 46
in mixed-species plantations, 158
root system and windthrow, 128
soil and establishment of, 45, 47
sustainable supply study, 398–399
Douglas-fir stands. *See also* Douglas-fir–western hemlock stands; Stands
canopy arthropods of, 181
in Cascade Mountains, 179–180
mature forest, 18, 76, 117
pollination in, 205
regeneration and density of, 158–159
stem exclusion phase, 16
uneven-age management of, 152
Douglas-fir–western hemlock stands. *See also* Spotted owl
composition, 22
ecology, 174
growth with retention harvesting, 132
nutrient retention in, 24
Drought, 36–37, 44, 63
DTM (digital terrain model), 261–263
Duerr, William, 398
Durazo-Means, Maria, 391
Dwarf mistletoe, 130
Dyes, 320
Dynamic mosaic. *See* Shifting mosaic
Dynamic view of nature, 32–34
biodiversity conservation and, 99–100
in management, 49–50, 234

Easements, 362
conservation, 362–363
purchase of, 352
Eastern hemlock
loss of seed source, 48
transition phase, 19
Ecological assessments
regional, 233
science-based, 367, 397–402, 408
Ecological fitness, mycorrhizae, 176–177
Ecology
landscape, 216–217
as new view of nature, 32–34
process elements of biodiversity, 88
processes and principles, 7–9
role of hidden species, 96
role of insects, pathogens and mycorrhizae, 171–186
of special forest products species, 325–327
Economics. *See also* Incentives
added-value manufacturing, break-even point, 310
appraisal of special forest products harvest, 177, 323–324
in bio-regional assessment, 404, 405
context in the 21st century, 285–298
cost of retention harvest practices, 131
discounted present net worth, 114
logging costs, 292–293
of long rotations, 113, 114
performance of a grade mill, 307–308
period of rotation, 168
reinventing the wood products industry, 281–282
value of edible fungi, 177
value of residual trees, 132
Economics analysis
costs and benefits, 424, 428
of ecosystem management, 286–287, 305
scale and, 287–290

Ecosystem management
adaptive, 424–425, 443–444
of commercial mushroom harvests, 330–332
of special forest products, 324–330, 424
economics analysis of, 286–287, 305
emerging role of science and scientists in, 447–452
fire management integration with, 193–200
forest genetics for, 203–210
forest stands and, 12–13
fundamental goals, 240
historical background, x–xi, 229–230, 305
individual species *vs.* community, 97–98
at larger spatial scales, 255–256
a law of?, 363–364
organizational and legal challenges for, 361–377
physical environment as basis for, 229–237
during regeneration, 112
separatism in, 413–414
systems view of, 415–417
in time and space, 215–226
Ecosystems
complexity of, 2–3, 34–35, 48
function of, 7–8, 12–13, 22–24
in riparian management, 76–78
stability and, 40–42
global, regional and landscape processes, 35–37
structure of, 7–8
vs. landscapes, 57–58
Ecotones, 70, 217–218
Edge effects, 9, 63, 217, 221
Edges, 9
effects of fragmentation, 63, 220
softening of, 221
temporal changes in managed landscape model, 223, 224, 225
Education and training
approach to problems, 424–425, 431
cooperative extension programs, 348–349
creating the experimenting society, 442–443
for loggers, 131, 231, 297
for managers, 440–442
for the public, 250–251, 382, 388
for special forest products harvesters, 323, 330
vs. experience, 440–441
Efficiency analysis, 405–406
EIS (environmental impact statement), 343–344, 371
Elk Springs timber sale, 392
El Niño oscillation, 36
Emmigration, 60. *See also* Migration
Empetrum hermaphroditum, 46
Employees
training of, 131, 231, 297, 323, 330, 440–442
workforce issues, 388–389
Employment, 296–297, 323
Endangered spaces, 218–219
Endangered species. *See also* Endangered Species Act of 1973
extended rotation and, 169
management of, 100, 209, 230
Endangered Species Act of 1973, ix, 72, 74, 229, 343
overview, 344, 371
shift in focus to ecosystems, 100, 370
Endemism, patterns of, 88
Energy flow, 47
Engineered/recycled wood products, 291, 308–309
Engineer species, 42–43
Enrichment. *See* Structural enrichment
Entries per rotation, 132–133
Environmental impact statement (EIS), 343–344, 371
Environmental stochasticity, 63
Epiphytes, 22
dispersal rate, 24
water-collection in, 43
Equilibrium view of nature, 32–34
Equitable servitude, 362
Equity criteria, 340–341
ERDAS GIS software, 222
Erosion, 22
dead wood and, 44
in riparian management, 70, 71
Establishment phase, 14–15, 16
rotation period and, 26
"safe sites" for, 21, 47
Eucalyptus, 33, 129
Even-age management systems, 152
Evolution, concepts of, 204, 208
Evolutionary sustainability, 204
Exotics, 32, 182
biodiversity conservation and, 100–101
grasses, 35, 38, 47
trees, 183
Expert witnesses, 452
Extant species, 101
Extinction, 61, 63, 101
Extinction debt, 35
EZ-IMPACT, 417

Farm Bill of 1985, 346
Federal Advisory Committee Act of 1972, 359, 373–375, 394, 407
Federal Environmental Pesticide Control Act of 1972, 343, 345
Federal Land Policy and Management Act of 1976, 72, 74, 371
Federal policies and regulations, 72–74, 129, 343–345, 370–372. *See also individual legislation;* Policies and regulations
Federal Water Pollution Control Act, amendments of 1972, 343, 344
Felling techniques, 71
FEMAT. *See* Forest Ecosystem Management Assessment Team
Fencerows, 220
Fertilization (nutrient), 141, 145–146, 184
Fertilization (reproductive), 205–206
Fiddlehead fern, 320
FIP (Forestry Incentives Programs), 345–346, 353
Fir, 92, 205, 320. *See also* Douglas-fir; Douglas-fir stands
Fire
  acreage of, 36, 192
  biodiversity and, 133
  biological legacy of, 3, 112, 180
  climate and, 36, 37
  composition and, 33, 48
  effects of suppression, 33, 80, 183, 197
  establishment phase following, 14
  let-burn, 192
  Mann-Guleb fire, 194
  outlier event of, 199–200
  prediction of effects, 194–195
  regeneration after, 196
  retention harvest and, 129–130
  Storm King fire, 194
  threshold transition and, 48
  Tillamook fire, 153, 154
  topography and, 37
  Wenatchee fires, 200
  Yellowstone fires, 193
Fire behavior, 193–195, 421–422
Fire breaks, 78
Fire management, 191–200, 196–198
Fish, 94–95
  habitat for, 92
  hydropower generation and, 430–431
  metapopulations of, 206–207
  in Pacific Northwest late-successional forests, 89, 94
  sustainability of, reports, 400–402
Fitness, 174, 204, 208–209
Floodplains
  processes and protection, 70, 71, 77
  restoration, 80
Floods
  rain-on-snow, 7, 114, 119
  restoration and, 80, 81
Floral greens, 319, 320, 322, 326
Flows, in landscape analysis, 257–258
Fluminicola, 93
Food resources, 43, 70, 318–321
  fungi, 90, 177, 178, 320, 321, 337
  lichens, 90
  salmon, 95
Food Security Act of 1985, 346, 372
Forest and Rangeland Renewable Resources Planning Act of 1974, 420
Forest development. *See* Development
Forest Ecosystem Management Assessment Team (FEMAT), x, 74, 78, 233
  definition of adaptive management in, 434–444
  effects of, 403–404
  legal case, 370
  report, 4, 75–76, 402, 403–404, 406–407, 408
  riparian area classes, 73
  watershed management, 271, 272
Forest floor, 118, 119
Forest interior, model of changes in time, 223, 224, 225
Forest Legacy Program, 372
Forest Plan, 72, 76, 78, 124, 134
Forest Practices Acts, state, 74, 271, 350
Forest products
  changing nature of, 290–296
  manufacture, changes in, 303–314
  producers of, 293
  reinventing the wood products industry, 281–282
  special. *See* Special forest products
  timber supply, 169, 281–282, 291–294, 304–306, 412
Forestry Curriculum Development Project, 416
Forestry Incentives Program (FIP), 345–346, 353
Forest stands. *See* Douglas-fir stands; Stands
Forest Stewardship Act of 1990, 348
FORPLAN model, 213, 399, 403, 420, 422

Fragmentation, 62, 63–64, 221
  arthropod diversity and, 93
  effects of, 62–64
    on fungi, 184–185
  maintaining landscape movement and, 64–65, 220
Frost Meadows harvest unit, 121
Fuelwood consumption, 291
Funding
  for bridging arrangements in agencies, 394
  definition of local extinction and, 101
Fungi, 3, 8, 22, 89–90. *See also* Mycorrhizae
  biology and ecology, 172–173, 325, 326–327, 331
  edible, value of, 44, 177
  effect of nutrient fertilization, 184
  guilds and, 45
  introduced, 182
  medicinal potential, 90, 318, 320
  in Pacific Northwest late-successional forests, 89, 90
  role in recovery, 32
  in soil, 44, 45, 116
  as special forest products, 44, 177, 325, 326–327, 330–332
  stability and, 49
Future of forest management, 2, 76–82. *See also* Research needs

Gang of Four
  rationale for convening, 449–450
  report, 4, 400–401, 402, 404, 405, 408, 450
Gaultheria shallon, 320
Genetics
  artificial regeneration and applied research, 155
  of forest populations, 203–210
  genetic composition of residual trees, 131–132
  genetic diversity, tradeoffs among and within populations, 207–208
  genetic drift, 60
  genetic interchange, 60
  genetic traits, 209–210
  homozygous *vs.* heterozygous individuals, selection for, 206
  mutations, 172
  risk of genetic homogeneity, 32, 182–183
  stochasticity, 63
Geographic information systems (GIS), 195–196, 214, 222, 263, 275
Geographic positioning satellites (GPS), 214
Geographic range, 59, 60, 62, 64, 66
Georgia, 221
Gila National Forest, 390–391
GIS. *See* Geographic information systems
Global climate change, 36, 199
Goals
  construction of, 430–433
  linking to activities, 273–275
GPS (geographic positioning satellites), 214
Grade mills, 305, 306–308
Grades of wood, 305, 306, 307, 308
  possible approaches to, 311–312
  "shop grades," 310
Grazing animals, 46
"Great Green Wall," China, 38
Great Lakes Water Quality Authority, 434
Greenhouse effect, 36
"Green" marketing, 82
Ground, below. *See* Subsurface area
Group selection, 154, 156
  costs, 292
  historical discontinuation of, 152–153
  with low retention, 126–127
Growth, of residual trees, 132, 145
Guidelines. *See* Policies and regulations
Guilds, 45, 49, 77

Habitat, 70
  coarse woody debris, 119, 180
  conflicting objectives to maintain, 101
  creation of, 81, 92
  extended rotation and, 169
  loss and degradation of, 39, 94
  microhabitats as links, 98–99
  monitoring of, 250
  multiple canopy layers as, 119
  of special forest products species, 327
  thinning and, 169
  trees as, 118
  understory plants as, 119
  watershed planning and, 274
Hardwoods, 155, 158, 183, 312
Harvest
  low retention, 124, 126–127
  wood quality of long rotation stands, 113
Harvest cutting. *See also* Clearcuts
  board feet, 1940–1992, 70, 72
  effects on riparian areas, 69, 70
  global aspects of, 11–12

retention strategies, 4
variable retention systems, 111–136
HARVEST model, 222–225
Harvest units, fragmentation and, 220, 222, 225–226
Hemlock
eastern, 19, 48
western, 22, 24, 119, 128, 132, 174
Hemphillia, 93
Herbicides
application of, 131
effect of, 47
long rotations and less need for, 113, 167
Herbivory, 22, 23, 46
Historical background, ix–xi, 1–2
decision making by rational comprehensive planning, 420–424
early logging practices, 152–153, 165–166
ecosystem management, x–xi, 229–230
ecosystem science, 7
fire management, 191–192
landscape patterns, 232–233
public interest *vs.* private rights, 339, 341–342
riparian areas in forest management, 69–70, 72, 74, 75, 78
science-based ecological assessment, 398–402
special forest products, 317–318
spotted owl debate, ix–x, xi, 75, 303
H.J. Andrews Experimental Forest, OR, 7, 128, 216
Home range movements, 59, 63, 64, 65
Hornworts. *See* Bryophytes
Host-specific species, 98
Hubbard Brook Experimental Forest, NH, 7, 47, 216
Huckleberry, 320. *See also* Floral greens; Food resources; Understory
Human activity, effects of. *See* Anthropogenic factors
Hurricanes, 63, 112
Hydrologic cycle, 34
Hydrology, 48
Hylocomium splendens, 326
Hypnum spp., 320. *See also* Bryophytes

Idaho
protection of species in parks, 100
riparian management regulations, 73, 74
Targhee National Forest, 392
Idaho Forest Practices Act of 1974, 74
Idiosyncratic-response hypothesis, 41
Immigration, 60. *See also* Migration
Inbred embryos, 206
Incentives, 282–283, 352–353
financial, and cost sharing, 345–346
public interest *vs.*, 338–340
for stewardship, 372
tax, 346–347
Indicators. *See* Monitoring, biological indicators
Influence of Forestry Upon the Lumber Industry, The (Price), 82
Inocula, 116, 130, 134
Insects. *See also* Arthropods
biological control by birds, 95
biology, 172–173
budworms, 38
defoliating, 33
ecological role of, 171–177, 179, 180–186
effect of fertilization, 184
introduced, 182
outbreaks, 80
pollination by, 178–179
susceptibility to
diversity and, 42
topography and, 37
variable retention and, 130–131
Institutions, 362–363
building bridges across boundaries, 381–396
cooperation between, 251, 349–350
creating networks within, 358
enforcement by, 363
open, 437–452
organization
of modeling, 364–368
and watershed management, 241–242, 247
organizational and legal challenges for, 361–377
in transition, 357–359
Integrity, biological, 39
Interagency Scientific Committee (ISC), 400
Interior Columbia Basin Ecosystem Management Project (ICBEMP), 407
Internalization of effects, in decision making, 425, 426, 427, 428–429, 431
Invasiveness, of ecological attributes, 240–241
Invertebrates, 3, 8, 92–94
Douglas-fir seedling survival and, 47
stability and, 49
use of stands, 22
Island biogeographical theory, 118
Island effect, 45
Isle Royale National Park, MI, 42
Isothecium spp., 320. *See also* Bryophytes

Jobs, 296–297, 323
Journal of Forestry, 8
Juga, 93

Keystone Center, 384
Keystone concept, 42–43
Kiowa National Grasslands, NM, 392–393
K-SIM model, 416–417

LAD. *See* Landscape analysis and design
Lakes. *See* Riparian areas
Land. *See* Property ownership; Property rights
Land exchange, 79
LANDSAT Thematic Mapper satellite imagery, 223
Landscape analysis and design (LAD), 218, 255–269
  analysis phase, 256–259
  case study, 261–269
  creating "intentional" landscapes, 256
  design phase, 257, 258–261
Landscape matrix, 218–219
Landscape patterns, 4, 9, 13
  biological, 57–66, 230
  boundaries between, 217–218
  in fire management, 195, 198–200
  historical background, 232–233
  influence on disturbance, 37–39, 234, 235
  as keystone factor, 42
  management and, 26, 49, 76, 78–79, 215–226, 233–237
  pattern-process relations, 231–232
  in structural retention design, 124
  temporal dynamics and spatial patterning in, 222–224
Landscape structure, 217–220
Land trusts, 359
Land-use practices, 64. *See also* Property rights; Social factors
  legacy of, 221, 232–233
  private *vs.* public land, 82
  regulations for, 72
  separatism in defining, 413–414
  valuation of, 294–295
La Niñas, 36
Lanx, 93
Late-successional forests, 18, 76, 88. *See also* Old-growth stands
  ecosystem functions of, 89
  species richness in, 96, 98
Lawsuits, 4, 416
Legacies. *See* Biological legacies
Legal challenges, 370–377
Leopold, Aldo, 200, 204, 415
Let-burn fires, 192
Lichens, 22, 90–91
  nitrogen fixation by, 134, 178
  in Pacific Northwest late-successional forests, 89, 90
  propagation of, 27
  retention harvesting and, 134
Lifeboat species and processes, 115–116, 122, 133, 135
Lindblom, Charles, 424
Linnaeus, Carolus, 413
Litter, 70
  arthropods of the, 93
  role of, in retention harvest system, 118, 119
  as a structural requirement, 116
Live oak, 92
Liverworts. *See* Bryophytes
Lodgepole pine, 320
Loggers
  training for, 131, 231, 297
  worker input into harvest design, 131
Long rotations
  biodiversity and, 169
  economics, 113, 114
  endangered species and, 169
  habitat and, 169
  road systems in, 114
  snags, 114
  vs. retention harvest methods, 113–115
  in watershed areas, 114
  wood quality, 113
Long-Term Ecological Research Program, 7
Lumber mills, 305, 306–308

Madro-Tertiary Geoflora, 92
MAI (mean annual increment), 113, 166, 167, 168
Mammals, 133
  in late-successional forests, 89, 95–96
  percent in riparian areas, 98, 99
  use of salmon by, 95
  use of stands by, 22
Management. *See also* Ecosystem management; Landscape patterns, management and; Watershed management
  adaptive, 5, 324–332, 424–425, 443–444
  of biodiversity, 100–101

cross-jurisdictional, 4
dynamic view of nature and, 49–50, 77
even-age, 152
implications of stand development, 25–27, 77–78
integration among managers, scientists, and stakeholders, 358
interdisciplinary teams, 431–433
multiple-use, 72, 74, 294, 371, 438
narrative objectives for, 260–261, 266, 269
retention harvest and, 131–133
riparian, 69–82
role and limitations of science in, 414–415
routine *vs.* experiment, 441–442
separatism in, 413–414
systems view, 415–417
uneven-age, 152–153
Managers
education of, 440–442
integration with scientists and stakeholders, 358
role of, in policy making, 406–408
scientists as, 450, 451
Mann Gulch fire, 194
Manufacturing of wood products, changes in, 303–314
Maples, 46
bigleaf, 92, 158
vine, 320
Marbled murrelets, x, 91
Markets, 353, 376
MASS (Montane Alternative Silvicultural Systems), 135
Matrix lands, 117–118
in landscape analysis and design, 257–259
management of, 400–401
Northwest Forest Plan guidelines, 124
Matrix organizations, 366
Maturation subphase, 18
Mature forests, U.S. acreage of, 88. *See also* Late-successional forests
McArdle, Richard, 342
McCaleb, Gary, 390
Mean annual increment (MAI), 113, 166, 167, 168
Mediation, 426, 428
Medicinals, 91, 92, 282, 318–321
Megomphix, 92–93
Menominee Indian Reservation, 125
Metapopulations, 60–61, 206–207
Metastability, 34
Michigan, 42, 222
Microbial organisms, 8, 39, 92
effect of fertilization, 184
nitrogen fixation by, 177–178
in soil beneath clearcut, 45
stability and, 49
Microclimate
at edges, 63
resources and, 21–22
structural retention and, 115, 116, 133
Microsites, for seedlings, 153
Migration, 59, 63, 65, 134, 392
Mills, 305, 306–308
Mistletoe, dwarf, 130
Mites. *See* Arthropods
Mixed-species planting, 156, 157, 158
Modeling. *See also* Geographic information systems
disturbance, 234–235, 403
dynamics, 77, 403
in economics analysis, 286–287
fire behavior, 195, 403
harvest planning, 213
institutional organization, 364–368
integration of science and social variables, 244, 245–246, 416–417
managed landscapes in time and space, 222–224
politics, 368–370
river basin management, 78
vegetative dynamics, 403
Moisture
capture of, by lichens, 90
soil, 21–22, 91, 155
Mollusks, 89, 92–93
Monadenia, 92
Monitoring, 5, 77, 185
access to resources, 425
biological indicators used in, 94, 101, 185
bryophytes, 92, 326
fungi, 90
lichens, 91
public stewardship and, 249
special forest product supply, 322, 325, 329–330, 331–332
sustainable forestry, Monominee Indian Reservation, 125
Monocultures, 25–26, 182–183
Montane Alternative Silvicultural Systems (MASS), 135
Moquah Barrens, WI, 218, 219
Morchella spp., 320

Mortality, from roadkill, 63
Mosaic. *See* Shifting mosaic
Mosses. *See* Bryophytes
Mountain ash forest, 118, 119
Mount Hood National Forest, OR, 38
Mount St. Helens, species recovery from, 43
Movements of organisms, 59–60, 62. *See also* Dispersal; Migration
Multiple Use-Sustained Yield Act of 1960, 72, 74, 371
Mushrooms. *See* Fungi; Mycorrhizae
Mutualism, 220
Mycorrhizae
ecological fitness and, 176
ecological role of, 116, 171–178, 179–186
productivity and, 177

National Academy of Sciences, 448
National Acid Precipitation Assessment Program (NAPAP), 448–449
National Council of the Paper Industry for Air and Stream Improvement (NCASI), 272
National Environmental Policy Act of 1969, ix, 72, 74, 229
ecosystem-oriented management objectives, 370, 371
overview, 343–344
rational planning in the, 411
National Forest Management Act of 1976 (NFMA), ix, x, 72, 74, 229
ecosystem-oriented management objectives, 371
failure of forest planning with, 411
management of exotics, 100
rational planning in the, 420
in regional planning, 399
National Hierarchy of Ecological Units, 218
National Park Service, 352
National Performance Review, 350
National Recovery Administration Codes of 1933, 342
National Research Council, 448
National Science Foundation, 7
National Wilderness Act of 1964, 72, 74
Natural Resources Conservation Service, 372
Nature, new view of. *See* Dynamic view of nature
Nature Conservancy, The, 352, 359, 384
Neckera spp., 320. *See also* Bryophytes
Negrito Ecosystem Project, NM, 390–391
NEPA. *See* National Environmental Policy Act of 1969
Nest boxes, 147
Net primary productivity (NPP), 22, 24
Nettles, 320
New Mexico
Kiowa National Grasslands, 392–393
Negrito Ecosystem Project, 390–391
NFMA. *See* National Forest Management Act of 1976
Nicolet National Forest, WI, 218–219
Nitrogen fixation, 43, 177–178
by alder, 158
by fungi, 89
introduction of plants for, 145–146
by lichens, 91
by vascular plants, 92
Noble fir, 320
Noise, in ecosystem theory, 34
Nonfederal land, role in biodiversity conservation, 99
Norris-Doxey Act, 348
North Carolina
Coweeṭa Experimental Forest, 7
Coweeta Hydrologic Laboratory, 216
Northeastern United States, 11–12
Northern spotted owl. *See* Forest Ecosystem Management Assessment
Team; Spotted owl
Northwest Electric Power Planning and Conservation Act of 1980, 430
Northwest Forest Plan for Management of Habitat for Late-Successional and Old-Growth Forest Related Species Within Range of the Northern Spotted Owl, 72, 76, 78, 124, 134
Northwest Power Planning Council, 430–431, 434
NPP (net primary productivity), 22, 24
Null hypothesis, 41
Nurse plants, 46
Nutrient cycles, 48. *See also* Nitrogen fixation
arthropods and, 94
bryophytes and, 91–92
ecosystem function and, 22, 23, 24
fish and, 94–95
mycorrhizae and, 173, 178
role of fungi in availability of, 89–90
Nutrient management, 145–146

Oak, 92
Octea whiteri, 37
Office of Management and Budget (OMB), 350

Office of Technology Assessment (OTA), 448
Old-growth stands. *See also* Spotted owl
  age of, 117
  biodiversity within, 87–109
  definitions, 25
  economic implications of, 295–296, 305
  origin of, 142
  perspectives on, 24–25
  status, 12
  top-down controls in, 42
  "true old-growth stage," 17
Old-growth subphase, 18
Olympic Mountains, 199–200
Oncorhynchus kisutch, 95
Oncorhynchus nerka, 95
Operability, in watershed planning, 272–273
Operations inventory (OI), 125
Oplopanax horridum, 320
Oregon
  bird diversity, 134
  Douglas-fir mortality from root rot, 174, 175
  low-retention harvest in, 126
  percent of population using public land resources, 82
  riparian management regulations, 73, 74, 75, 76
  Siuslaw National Forest, 325
  species richness in late-successional forests, 98
  stumpage prices, 291–292
  Willamette National Forest, 236
Oregon boxwood, 320. *See also* Floral greens; Understory
Oregon Forest Practices Act of 1972, 74
Oregon grape, 320. *See also* Floral greens; Understory
Oregon State Forest Practices Act of 1941, 154–155
Organizational systems
  complexity of, 2–3
  decentralization of, 4, 386
  responsibility within, 357–358
Ovules, survival of, 206

PACFISH (Pacific Anadromous Fish Habitat Management Strategy of 1994), 73, 74
Pachistima myrsinites, 320
Pacific Anadromous Fish Habitat Management Strategy of 1994 (PACFISH), 73, 74
Pacific madrone, 92
Pacific Northwest. *See also* Douglas-fir–western hemlock stands; Old-growth stands; Spotted owl
  biodiversity of old forests, 87–109
  landscape analysis and design, 255–269
  special forest products, 318–321
    revenue from, 316
  timber harvest, 70, 72, 303, 304
Pacific yew, 320, 321
PAI (Periodic annual increment), 166, 168
Paleoecological trends, 88
Parasitism, 220
Partial cutting, discontinuation of, 152–153
Patches, 9. *See also* Fragmentation
  aggregate retention of, 113, 117, 118, 120–122
  coalescence of, 66
  definition, 217
  dispersal among, 60
  environmental resources, 217
  landscape context of, 213–214, 220, 257–259
  origin, 217
  size and shape of, 123, 220, 222
  staggered cuttings as, 153
Patchiness, 8
  biological legacies and, 44
  during establishment phase, 14
  metapopulations and, 60, 220
  during shifting mosaic phase, 19
Pathogens, 8, 80, 130
  climate and disturbance by, 37
  ecological role of, 171–172, 174–177, 181, 182, 183–186
  humans as, 415
  root. *See* Root rot
  succession after disturbance by, 174
  windthrow *vs.*, in storm falls, 174
PDR (purchase of development rights), 351–352
Perforation
  of landscape, 62, 63
  maintaining landscape movement and, 64
Periodic annual increment (PAI), 166, 168
Persistence, of ecological attributes, 240–241
Pesticides
  application of, 131
  biodiversity and, 8
  biological control, 179
Pharmaceuticals. *See* Medicinals
Phelinus, 155
Photographs, for monitoring, 249
Photosynthesis, 48
"Piggy-backing," 65
Pinchot, Gifford, 339, 341, 441

Pine barrens, 218, 219
Pinelands Commission, 384
Pines, 14, 19, 218, 219, 320
Pinus contorta, 320
Pinus sylvestris, 14
Plantations
hardwood, 155
management of, 25, 26, 142–147
young, 155, 158, 166
Plant communities, 80. *See also* Bryophytes; Understory; Vascular plants
Planting density, 141, 143, 168, 328
Plum Creek Timber Company, aggregate retention by the, 120, 121, 135
Pluralism, 369
Plywood, in added-value manufacturing, 310, 312
Policies and regulations, ix, 242, 247, 258–259, 263, 316. *See also individual legislation*
administrative law, 372–373
antitrust law, 375–376
coalition development, 4, 383, 392
Congress and science-based, 448–449
consolidation and reform of, 350–351
cost of, 293, 294–295
cutting practices acts, 74, 339
development of, 337–354
modeling, 368–370
federal, 72–74, 124, 343–345, 370–372
guidelines for evaluation, 340–341
international, 324, 434
local, 342–343
nontraditional tools for, 349–353
organizational and legal challenges for, 361–377
role of policy makers and managers, 406–408
social forestry, 297
state, 73–75, 341–342, 376
Politics
building support for decision making, 386–387
modeling, 368–370
network development, 383
scientists in a political environment, 451
Pollination, 92, 178–179, 204–205
Pollution, 35. *See also* Clean Air Act amendments of 1970; Clean Water Act of 1972
Polystichum munitum, 320
Poplar, 155
Population dynamics
coping with uncertainty, 208–210
dispersal and, 60
genetic diversity and, 207–208
Population regulation, 48
Potential sale quantities (PSQ), 295
Prairie communities, 41, 48
Predation, 220
at edges, 63
herbivore control and tree growth, 42
movements of predators, 59
role in recovery, 32, 46
structural retention facilitation of, 118
Prescribed natural fire, 193, 195
President's Council on Environmental Quality (PCEQ), 349, 448
Primary forest, percent of preindustrial amount, 12
Primary production, 22, 23, 24
Prince's pine, 320
Private forests, 72, 74
effect of checkerboard ownership, 78
floodplain management, avoidance of, 77
riparian management requirements, *vs.* public land, 82
Productivity
disturbance of the soil and, 24
mycorrhizae and insects and, 177
net primary, 22, 24
Products. *See* Forest products
Property ownership, 362
considerations in bio-regional assessments, 403–404
cooperation and voluntary efforts, 349
easements and development rights, 351–352
land trusts, 359
tax benefits, 347
technical assistance, 347–349
timberland, percent private *vs.* forest industry, 338
Property rights
public interest and, 337–354, 362, 370–372
"takings" law and, 376–377
Pruning, 141, 146
Pseudotsuga menziesii, 20. *See also* Douglas-fir
PSQ (potential sale quantities), 295
Pteridium aquilinium, 320
Public acquisition of land, 79
Public interest. *See* Social factors
Public outreach, 250–251
Public perception, 388, 429–430
Public relations and communication, 121, 166, 168, 263

Public stewardship, 248, 249–251, 285
Purchase of development rights (PDR), 351–352

Quantitative measures, 244, 245
Quercus chrysolepis, 92
Quincy Library Group, 358

Radiation, 21
Rain forests, 33, 38, 46
Rain-on-snow floods, 7, 114, 119
Range
 of natural conditions, 100
 of organisms, 59, 63, 64, 65
Rarity, patterns of, 88
Rational comprehensive planning, 420–421
Real covenants, 362
Recolonization rate, 27
Recovery, 3, 8–9, 32. *See also* Succession
 biological legacies and, 43, 112
 inhibition during, 45
 mechanisms of, 39, 40
 from Mount St. Helens eruption, 43
Recreation, 82. *See also* Social factors
Recreation plan, case history, 391–392
Recycled products, 291, 308–309
Red alder, 155, 158, 159, 320
Red pine savanna, 219
Red ring rot, 180
Redundancy, in key processes, 48–49
Redundant-species hypothesis, 41
Redwood, 127
Reforestation, 217, 218, 347
Refugia, 26, 40
 dominant tree species as, 120
 retention harvest and, 114, 115–116, 130, 134
Regeneration, 151–160
 advanced, effectiveness of, 153–154
 after fire, 196
 artificial conifer, 154–156
 competition and, 46
 methods for, 111
 natural, studies of, 153–154
 soil fungi and, 45
Regional ecological assessments, 233, 399–400, 402
Regional patterns, influence on disturbance, 37–39
Regional planning agencies, 384
Regulations. *See* Policies and regulations
Renewability, 412
Renewable Resources Extension Act of 1978, 348
Reproductive biology, trees, 205–206
Reptiles, 89, 95
Research needs, 134–135, 160. *See also* Future of forest management
Reserves, 12
 conservation, 346
 creation of, in 1993, x
 large, does not ensure conservation, 97
 old-forest elements between, 97
 timber harvest within, x–xi
Residual Trees as Biological Legacies (Hunter), 127
Resilience, 40, 48
Resistance to disease, mechanisms of, 39, 40, 172
Resolve, 384
Resource analysis, 79, 405
Resources, of the stand, 21–22
Restoration, 12, 76, 79–81
 cost-benefit analysis of, 286
 definition, 79, 101
 of endangered areas, 218–219
 fundamental principles of, 79–80
 need for assessment prior to, 80
 use of translocation in, 65
Restoration zone, 221
Retention harvest. *See* Variable retention harvest system
Rhamnus purshiana, 320
Richness, 98
Riparian areas, 8, 42
 cross-section of, 71
 effects of disturbances on, 236
 monitoring surveys, 250
 role in biodiversity, 98
 special forest products in, 327–328
Riparian reserves, 75
Riparian zone management
 current policies and practices, 70–76
 definition, 70, 71
 examples, 236
 functional bases of delineation, 76–78
 future directions, 76–82
 historical background, 69–70, 72, 74, 75, 78
 producing mixed-species stands in, 157
 width of, 71–72, 75
Risk assessment
 in decision making, 242
 of property rights regulations, 339
 systems modeling for, 287
 wildlife resources, 400–401

River basin patterns, 78
Rivers. *See* Riparian areas
Rivet hypothesis, 41, 42
Roadkills, 63
Road systems, 71, 78
  effects of, on landscape patterns, 231–232
  in landscape design, 260–261
  reduction of, in long rotations, 114
Robustness, 33–34
Root rot, 33
  Armillaria, 39, 184
  creation of canopy gaps by, 145
  host preferences of, 173
  immunity of alder to, 155
  percent of land affected by, 175
  spread of, 184
  susceptibility to, 39
Rotations, 27
  definition, 166
  effects of short, 168–169
  with retention harvesting, 132–133
  role of extended, 165–170
  timber supply problem and, 169
Rubus spectabilis, 20
Rubus spp., 320

"Safe sites," 21, 47
Safety, 131
Salal, 320. *See also* Floral greens; Understory
Salmon, 69, 94–95
Salmonberry, 45, 46
Sand County Almanac, A (Leopold), 415
Satellite imagery, 223
Satellite populations, 61
Satellites, geographic positioning, 214
Satisficing, 422
SAT report, 401, 402, 408
Scale
  economics analysis and, 287–290
  ecosystem management and, 255–256
Scaling criterion, 75
Scarcity, 412
Scheduling and Network Analysis Program (SNAP), 275
Schenck, Carl, 441
Science
  advisory committees, 373–375, 407–408, 448–449
  building social institutions in the image of, 439–440
  civic, 443–444
  ecological assessment based on, 397–399, 408
  emerging role in ecosystem management, 447–452
  policy based on, 448–449
  role and limitations of, in forest management, 414–415
Scientific Agency Team (SAT), 271
Scientific analysis team (SAT), 401, 402, 408
Scientific Panel for Sustainable Forest Practices, 124, 126, 131, 134
Scientists
  emerging role in ecosystem management, 447–452
  as expert witnesses, 452
Seasonal factors
  animal ranges, 59
  wildlife use of riparian areas, 77
Seed dispersal, 33, 204–205. *See also* Dispersal
  by bats, 96
  by birds, 45, 205
  by invertebrates, 92
  by mammals, 205
Seedlings
  density of, 159
  producing planting stock, 155
  survival of, 46, 154
Seed orchards, 207
Seeds, 128, 154. *See also* Biological legacies
Seed source, loss of, 47–48
Seed-tree regeneration, 111, 152
  in Douglas-fir forests, 128
  state regulations, 328
Self-organizing systems, 34, 35, 47
Separatism, 413–414
Sequoia, 37
Shade
  seed survival and, 154
  stream, 70, 71, 74, 80
Shasta fir, 205
Shelterwood, 111, 152, 154
  costs, 292
  in Douglas-fir forests, 128
  in production of mixed-species stands, 156, 157
  in retention harvest system, 115
Shifting cultivation, 33, 46
Shifting gap phase, 19
Shifting mosaic concept, 32
Shifting mosaic phase, 13, 19
  biomass during, 24
  resources during, 22

"Shop grades," 310
Shrubs. *See* Understory
Sierra Nevada Ecosystem Project, 407, 408
Sierra Nevada forests, 126, 127
Silcox, Ferdinand, 342
Silen, Roy R., 128
Silviculture
  common objectives, 124
  definition, 111
  regeneration methods and, 151–160
  second-growth retention prescription, 127
  of special forest products, 328
Silviculture Systems Project, 135
Simplicity, 412–413
Simulations. *See* Modeling
Single large or several small (SLOSS) debate, 134
Single-tree selection, 154, 224
Sink populations, 60, 61
Site-potential tree height, 75, 76, 77
Site-specific knowledge, development of, 3–4, 234, 235
Sitka spruce–western hemlock stands percent selective harvest per stand, 128
  role of understory in, 119
Siuslaw National Forest, OR, 325
Size. *See* Tree size
Slash burns, 113, 129, 193, 198–199
Slater, Philip E., 439, 442–443
Slugs. *See* Mollusks
Smith-Lever Act of 1914, 348
Smokey the Bear, 192, 195
Snags, 80. *See also* Coarse woody debris; Dead wood
  animal territory and, 114
  creation of, 146–147, 175
  fire management and, 198
  in long rotations, 114
  structural retention and, 112, 117, 118, 119, 123
Snails. *See* Mollusks
SNAP (Scheduling and Network Analysis Program), 275
Snap Creek Project Area, 275–279
SNAP II model, 3, 4, 275–279
Social factors. *See also* Land-use practices
  in bio-regional assessment, 404, 405
  creating the experimenting society, 442–443
  in economics analysis, 289–290
  equity criteria of policies, 340–341
  institutions, organization of, 241–242, 247
  integration with biological information, 4–5
  per capita consumption of wood (1970–1988), 82
  public expectations, 141, 142–143
  public interest in private rights, 337–354
  public perception, 388, 429–430
  renewability and society's impatience, 412–413
  restructuring conceptions of problems, 429–430
  for riparian resources, 76, 81–82
  of salmon fishing, 95
  source of conflicts, 423–424
  source of public preferences, 422–423
  in special forest products harvest, 328–329
Social forestry, 296–298
Society of American Foresters, 339, 341
Socioeconomics
  monitoring of conditions, 250
  of special forest product harvesting, 321–324
Socioenvironmental factors
  indices of, 244, 246–247
  integrated models, 244, 245–246, 416–417
  shared visions, 247–248
Soil
  arthropods of the, 93
  compaction of, and disease, 180
  Douglas-fir seedling survival and, 46
  ecosystem of the, 40, 118–119
  erosion, 22, 44, 70, 71
  management and, 49
  moisture in, 21–22, 91, 155
  organic matter in, 44
  productivity and disturbance of the, 24
  role as substrate, 116
Soil aggregates, 44
Soil Bank Program, 346
Soil chemistry, species diversity and, 41
Soil Conservation and Domestic Allotment Act, 372
Soil ecosystem, 118–119
Soil expectation value, 167, 168
Source populations, 60, 61
Spacial heterogeneity. *See* Patchiness
Spacing (density control), 141, 143, 168, 328
Spatial analysis and design
  in economics analysis of ecosystem management, 287–289
  of landscapes, 259–261
  in watersheds, 271–283
Spatial patterns
  in managed landscapes, 220–222
  projection of, by modeling, 222–224
  in watersheds, 240, 241, 273

Special forest products, 315–332
changing demand for, 322–323
estimation of supply, 321–322
revenue from, 282
Species composition. *See* Composition
Species diversity. *See* Biodiversity
Species richness, 98
Species-specific management, 77
Spiders. *See* Arthropods
Spotted owl, 133
area required for, 26
fire management and the, 196–198
historical background, ix–x, xi, 75, 303
low-retention harvest and, 126
number of species closely associated with, 89
Plum Creek Company habitat retention for, 121
rationale for convening Gang of Four, 449–450
scientists and the debate, 449
Spruce, 119, 128
Squirrels, 133, 147
Stability, 39
process, 40
role of resistance and resilience, 40
species, 39, 48–49
structural, 40
Stands, 12. *See also* Development; Douglas-fir stands; Structural enrichment; Structural retention; Succession
age of, 115
enrichment of, 115
environmental and ecosystem characteristics of, 21–24
homogeneous *vs.* heterogeneous, 25–26, 182–183
production of, 14, 156–159
shaping development of, 141–147
structure, composition and function, 11–27
State policies and regulations, 73–75, 341–342, 376. *See also individual legislation;* Policies and regulations
Steady state phase, 19
Stem exclusion phase, 15, 16, 142, 166
Stewardship, 248, 249–251, 348, 358, 359
Stewardship Incentive Program, 372
Stinging nettle, 320
Stochasticity, 63, 100
Stocks, of native species, 79, 89, 94
Storage
of carbon, 23–24, 167
of conifer seeds, 153
Storm King fire, 194
Strange attractors, 20
Streams. *See also* Riparian areas
channel structure of, 80, 81, 159
ephemeral, 77
sediment in, 274
Stream shade, 70, 71, 74, 80
Streptomyces, 45, 46, 47
Strix. See Spotted owl
Structural enrichment, 115, 116–117, 122, 133, 135
Structural retention, 115–124
examples, 121, 124–127
guidelines, 124
how much to retain, 119–120
lack of quantitative data, 119–120
during long rotation, 114
spatial pattern for, 120, 122–123
what to retain, 118–119
Stumpage owners, 293
Stumpage prices, 291–292
Subpopulations, 60, 61
Subsidies. *See* Incentives
Substrate-specific species, 98
Subsurface area, 8
site-specific species, 98, 99
during transition phase, 21
Subtropical forests, arthropods of, 93
Succession, 32
after pathogen disturbance, 174
biological legacies and, 44, 45
modern concepts of, 13–14, 21
mycorrhizae and, 175–176
natural, 159–160
rates of, implications for management, 78
site-specific approach to, 3
Sustainability, 318
evolutionary, 204
of fish and wildlife resources, 400–402
measures of, 402–403
principles of, 440
Sustainable forestry, x, 8, 125, 398–399
"Sweet Home" decision, 344
Sword-fern, 320. *See also* Floral greens; Understory
Symbiosis, of mycorrhizae, 173, 178
Systems view, 3, 415–417

"Takings" law, 339, 376–377
Tanoak, 158

Targhee National Forest, ID, 392
Tax incentives, 346–347
Tax Reform Act of 1986, 346–347
Taxus brevifolia, 320, 321
Taylor, Frederick W., 420
Technical assistance, 347–349
Technicism, 438
Temperature, 21
Temporal factors. *See also* Succession
  changes in forest interior, 223, 224, 225
  dynamics of time and space in managed landscapes, 220–222
  dynamics of time and space in watersheds, 240, 241
  in economics analysis, 289
  ecosystem management and, 215–226
  in forest development, 19–21, 24
  of genetic changes, 209
Tennessee Valley Authority, 344
Tennis, Pete, 391
Thermodynamic equilibrium, 34
Thinning
  in extended rotations, 114, 167
  precommercial and commercial, 25–26, 81, 132, 143–145, 157
  regeneration and, 158–159
  special forest products and, 328
  susceptibility to disease with, 183
  vs. harvest, 166
Thinning phase, 15–17
Thomas, Jack Ward, 400, 408
Threatened species, ix, x, 100. *See also* Endangered species
Threshold transitions, 35, 46–48, 209
Thuja plicata, 128, 158, 320
Timber. *See* Forest products
Timber, Fish, and Wildlife Agreement of Washington State, 75, 272, 431–433
Timber income, as capital gain, 346–347
Timber Trends for Western Oregon and Western Washington (USDA Forest Service), 398
Time. *See* Temporal factors
Tolt River watershed, WA, 243, 244, 245
Tongass National Forest, AK, 391–392
Top-down controls, 42
Topography, 37, 128, 260–261, 264–265
Training. *See* Education and training
Transition phase, 17–19, 21, 25
  biomass during, 23
  nutrients during, 24
  threshold transitions, 35, 46–48, 209
Transpiration, 38
Tree size, 40, 118
  in group harvest, 126–127
  site-potential tree height, 75, 76, 77
  structural retention guidelines and, 126
  value per unit volume, rotation and, 168
TREES model, 399, 403
Triad model, 64
Tricholoma magnivelare, 320
Trilobopsis, 92
Tropical forests, 33, 38, 43, 46
"True old-growth stage," 17, 19
Trust funds, 359
Tsuga heterophylla, 20
Turnover, 61

Ulota, 92
Uncertainty, 5
  of connectivity, 241
  in predictability of complex systems, 421–422
  social learning and, 444
  in 21st-century forestry, 437–452
Understory
  during establishment phase, 15
  microclimates of, 21
  mycorrhizae of plants in the, 178
  in natural succession, 160
  retention of, 118, 119
  shrub and hardwood management, 158, 159
  special forest products as, 326
  thinning and development of, 15–17, 143–145
  during transition phase, 17, 18
Uneven-age management systems, 152–153
University of Washington, 121
Uphill cable logging, 121
Upslope forests
  assumptions derived from, 77
  in landscape perspective, 78–79
Urtica dioica, 320
U.S. Constitution, Fifth Amendment, 339, 353
U.S. Department of Agriculture, 316
U.S. Department of the Interior, 316
U.S. Environmental Protection Agency, science advisory committees, 449
U.S. Fish and Wildlife Service. *See* Spotted owl

U.S. Forest Service, 233, 316. *See also* Forest Ecosystem Management Assessment Team; Policies and regulations, federal
  appraisal of special forest products, 324
  challenges facing the, 381–382
  changing role of, in decision making, 386, 449–450
  State and Private Forestry Division, 348
U.S. International Biological Program, 7
U.S. Office of Management and Budget, 350
U.S. Office of Technology Assessment, 448
U.S. Pharmacopoeia, 318
User fee systems, 82

Vaccinium ovatum, 320
Valuation methods, 286, 294–295
Variable retention harvest system, 111–113, 115–136
  examples of, 124–127
  fire management, 198–199
  management issues, 127–133
  mortalities of residual trees, 129
  scientific issues, 133–135
Vascular plants, 92. *See also* Understory
  mycorrhizal association and, 116, 172–173
  in Pacific Northwest late-successional forests, 89, 92
  as special forest products, 318–321
Veneer slicers, 313
Vespericola, 93
Vigor, 40
Vine maple, 320
Virginia, cooperation between agencies, 350
Viruses, 92
Visual impact, 166, 167, 275
Volume-based consumption of wood products, 304
Voluntary programs, 82

Wagon Wheel Gap experiments, 7
Washington
  Cedar River Watershed, 135
  percent of population using public land resources, 82
  Plum Creek Timber Company retention harvest in, 121
  riparian management regulations, 73, 74, 75
  species richness in late-successional forests, 98
  stumpage prices, 291–292
  Tolt River watershed, 243, 244, 245
  Wenatchee fires, 200
Washington Department of Ecology, 245
Washington Department of Natural Resources, 120, 243, 245, 432–433
Washington Forest Practice Act of 1974, 74, 271
Washington State Timber, Fish, and Wildlife Agreement, 75, 272, 431–433
Water quality, 70, 351
Water samples, for monitoring, 249–250
Watershed analysis units (WAU), 243, 244
Watershed management, 75, 230, 239–251
  case history of consensus process, 390–391
  generation of alternatives, 273–275
  implementing spatial planning, 271–283
  quantitative approaches for implementation, 242–247
Watershed Protection and Flood Prevention Act, 372
Watersheds
  assessment of, 272
  landscape ecology and, 216
  long rotations and, 114
Watts, Lyle, 342
WAU (watershed analysis units), 243, 244
Wenatchee fires, WA, 200
West Arm Demonstration Forest, 261–269
Western hemlock–red cedar stands, 128
Western red cedar, 128, 158, 320
Wetlands, 60, 77, 98. *See also* Riparian areas
Weyerhaeuser Company, 245
White pine, 19
Wild and Scenic Rivers Act of 1968, 72, 74
Wildcrafting, 319
Wilderness Society, The, 450
Wildfire. *See also* Fire; Birds; Fish; Insects; Mammals
  extended rotation and, 169
  as game species, xi
  riparian functions for, 77
  sustainability of, reports, 400–402
Willamette National Forest, OR, 236
Willapa Alliance, 246
Windthrow, 33, 80
  retention harvest and, 127–129
  vs. pathogens, in storm falls, 174
Wisconsin, 222
  Chequamegon National Forest, 218–219
  logging and slash fires in, 47–48
  Menominee Indian Reservation, 125
  Moquah Barrens, 218, 219
  Nicolet National Forest, 218–219
Wolves, 42, 46

Woody debris. *See* Coarse woody debris
Wood yield, 167
  by age, 168
  definition, 166
  with retention harvesting, 132, 151
  thinning and, 159
  timber supply, 169, 281–282, 291–294, 304–306, 412
Worker safety, 131
Wyethia mollis, 46

Xerophyllum tenax, 320

Yarding corridors, 71
Yellowstone fires, 193
Yew, 320, 321

Zoning regulations, 342–343